COMPUTATIONAL METHODS
FOR TRANSIENT ANALYSIS

MECHANICS AND MATHEMATICAL METHODS

A SERIES OF HANDBOOKS

General Editor

J. D. ACHENBACH

Northwestern University, Evanston, Illinois, USA

First Series

COMPUTATIONAL METHODS IN MECHANICS

Editors

T. BELYTSCHKO

Northwestern University, Evanston, Illinois, USA

K. J. BATHE

Massachusetts Institute of Technology, Cambridge, Massachusetts, USA

NORTH-HOLLAND
AMSTERDAM · NEW YORK · OXFORD · TOKYO

COMPUTATIONAL METHODS FOR TRANSIENT ANALYSIS

Volume 1 in
Computational Methods in Mechanics

Edited by

Ted BELYTSCHKO

Northwestern University
Evanston, Illinois, USA

Thomas J.R. HUGHES

Stanford University
Stanford, California, USA

NORTH-HOLLAND
AMSTERDAM · NEW YORK · OXFORD · TOKYO

First edition: 1983
Second printing: 1986

ISBN: 0 444 86479 2

Published by:
ELSEVIER SCIENCE PUBLISHERS B.V.
P.O. Box 1991
1000 BZ Amsterdam
The Netherlands

Sole distributors for the U.S.A. and Canada:
ELSEVIER SCIENCE PUBLISHING COMPANY, INC.
52 Vanderbilt Avenue
New York, N.Y. 10017
U.S.A.

Library of Congress Cataloging in Publication Data
Main entry under title:

Computational methods in mechanics.

(Mechanical and mathematical methods)
Bibliography: v. 1., p.
Includes index (v. 1)
Contents: v. 1. Computational methods for transient analysis -- .
1. Mechanics, Applied--Data processing--Addresses, essays, lectures. I. Belytschko, Ted, 1943- II. Hughes, Thomas J. R. III. Series.

TA350.3.C65 1983   620.1'0724   83-4072
ISBN 0-444-86479-2

PRINTED IN THE NETHERLANDS

Preface

Recent decades have witnessed the emergence and proliferation of an entirely new analytical tool, computer simulation. Applications abound in fields ranging from the social sciences through the fields which are traditionally quantitatively oriented, such as engineering mechanics.

In mechanics, an important area of computer simulation is the transient response analysis of structures and continua. Applications encompass areas such as the ground motion and behavior of structures during earthquakes, safety studies of nuclear reactors in hypothetical accidents, behavior of structures subjected to conventional and nuclear weapons effects, automotive and aircraft crash phenomena, and many other areas. Simulations are increasingly used to supplant tests in situations where they are either very expensive or impossible. Although it would be injudicious to rely entirely on computer simulations, the current direction of engineering practice is one in which computer simulations are used to make design decisions, while experiments are used to gain better insight into the basic physical phenomena and to validate the simulation computer programs.

The modeling of mechanical phenomena often requires direct time integration of the governing equations and the treatment of diverse nonlinearities. Even with today's decline in the cost of computational resources, costs are often exorbitant and thus the usefulness in the design and decision process has not reached fruition. Therefore, considerable impetus has been given to development of more efficient time integration procedures. Furthermore, it has been found in many problems, such as in the interactions of a structure with an infinite domain, that specialized techniques can lead to substantial savings in computer time and are in fact essential if these problems are to be solved economically. In addition, computer simulation has brought about intensified interest in nonlinear analysis, because even nonlinearities which are so severe as to be completely intractable by classical analysis techniques can often be handled with ease.

In spite of the rapid growth of this field, little has been published in book form to guide the practitioner. Many texts limit their treatment to an introduction of methods for linear transient analysis, with at most a perfunctory discussion of few time integration operators. Texts in applied mathematics, on the other hand, tend to focus on the finite difference method, even though finite element methods are used more widely in engineering mechanics because they are able to deal more

Computational Methods for Transient Analysis
Edited by T. Belytschko and T.J.R. Hughes

effectively with complex geometries and boundary conditions. Furthermore, concise algorithmic structures have evolved for finite elements which are ideally suited to effective software development. Thus, treatments of computer simulation which focus exclusively on regular meshes discretized by finite difference methods, while of benefit to gaining an abstract knowledge of numerical methods, fail to deal with many of the important and challenging aspects of engineering software.

The purpose of this book has been to provide comprehensive treatments of several topics pertinent to the simulation of transient response of structures and solids. Emphasis has been placed on areas for which a reasonable state of development has been attained. Among the topics treated are time integration procedures, partitioned methods, in which different field equations, parts of mesh or operators are integrated with different methods, the development of methods for problems involving infinite domains, and the treatment of material and geometric nonlinearities. Because structural mechanics often involve contiguous fluid domains, fluid-structure interaction analysis has also been included.

This book is primarily intended for readers with some background in solid mechanics and finite elements. Efforts have been made to consolidate and simplify material which has heretofore appeared only in diverse journal articles, and to provide the reader with a perspective of the state of the art. Numerous references are included to original sources and to additional topics not treated in depth herein.

Ted Belytschko
Thomas J.R. Hughes

Contents

Computational Methods for Transient Analysis
Edited by T. Belytschko and T.J.R. Hughes

CHAPTER 1

An Overview of Semidiscretization and Time Integration Procedures

Ted BELYTSCHKO

Department of Civil Engineering
Northwestern University
Evanston, IL 60201, USA

Computational Methods for Transient Analysis
Edited by T. Belytschko and T.J.R. Hughes

Introduction

The rapid evolution of computer software has brought about significant changes in the numerical methods for the solution of transient problems. On the one hand, the computer software which implement these methods have evolved into systems of great size which represent substantial investments. To justify these costs, it is necessary for these packages to be able to treat large classes of problems involving a wide variety of materials, geometries and modes of response. Furthermore, in order to meet the escalating requirements of engineering analysis, these computer programs have incorporated capabilities for treating phenomena of great complexity, such as highly nonlinear materials and multi-field interactions. To provide these capabilities in a usable form, software developers have developed many techniques which differ substantially from the classical techniques for numerical treatment of partial differential equations. Although these developments were often initially of an ad hoc character, the theoretical basis for these methods has gradually evolved, providing a fertile area of research and at the same time strengthening the methods.

In this book, the theoretical basis and implementation of computer methods for the transient analysis of solids and structures are described. In addition, certain classes of fluid-structure problems, in which the response of the structure is of primary interest and the behavior of the fluid can be simplified extensively, are considered. The latter include situations such as acoustic models of the fluid or situations where the flow velocity and effects of viscosity are small.

With minor exceptions, the emphasis will be on methods which will be called semidiscretization methods. In these methods, the partial differential equations in space and time are first discretized in space, yielding a system of ordinary differential equations in time. The semidiscretization in space can be accomplished through finite element or finite difference methods. In some cases, the equations resulting from finite difference and finite element semidiscretizations will be identical, while in other cases they will have distinctive properties. These will be examined in Section 1 of this chapter.

Section 2 treats the semidiscretization of problems of solid and structural mechanics. The formulation of methods for large displacements and large strains is emphasized, since these aspects are crucial in nonlinear problems. Fluid-structure interaction is also discussed to provide a perspective for Chapters 3, 4 and 10, which deal with topics largely motivated by this class of problems.

Constitutive equations are not discussed for the theoretical basis of many equations commonly used is given in Chapter 8.

Section 3 outlines explicit, implicit and mixed forms of time integration. The relationship between the spectral characteristics of a mesh and an appropriate integration method is emphasized, and methods of obtaining useful bounds for a variety of elements are described. This is also studied in Chapter 6, which examines the dispersive properties of various semidiscretizations, that is how the spectral character of a solution as represented by the phase velocity of waves is affected by a semidiscretization. Stability of various time integration procedures is primarily discussed in Chapter 2, and some of these results are used in Section 3. The relative merits of explicit and implicit methods are then discussed in Section 3 which leads naturally into the type of operator-splitting and partitioning methods; the latter are extensively described in Chapter 3.

The popularity of explicit methods has led to iterative solution methods for static solutions which rely almost totally on the software of a typical explicit code, these methods are known as dynamic relaxation and are described in Chapter 5.

An aspect of semidiscretization methods which makes the modeling of large wave propagation problems very burdensome is the expense of treating large meshes and the relatively large loss in accuracy which develops when a wave passes through many elements. For this reason, methods which can confine the modeling to the structure or portion of the medium which is of interest are of great importance. Two methods are examined in this book: boundary element methods in Chapter 4 and 'silent' boundaries in Chapter 7. The former treatment is restricted to acoustic media, while the latter treats elastic media, which can be reduced to acoustics. Chapter 4 also developes various interesting and useful approximations to the complete boundary element method.

In Chapter 8, 9 and 10, three classes of solution algorithms are described: explicit, Lagrangian finite difference; implicit finite element (Lagrangian) and explicit Arbitrary Eulerian Lagrangian finite element. These three classes of algorithms are widely used in nonlinear transient analysis and they exemplify the applications of the methods and theory described in this book.

1. Semidiscretizations of the diffusion equation

1.1. Nomenclature and governing equation

The simplest form of time-dependent partial differential equations are those in which the spatial operator is the Laplacian, which collectively are called diffusion equations. Since the purpose of this section is to illustrate the salient features of finite element and finite difference semidiscretizations, the diffusion equation and its spatial component, the Laplace operator provide an ideal setting. The Laplace operator appears in many technologically important problems, such as heat conduction, flow through porous media, and acoustics; however, we will use the nomenclature of heat conduction.

In the first part of this section, we will give the governing equations for heat conduction. By simply altering the meaning of various variables, these equations can also be applied to other physical problems. The finite element discretization is then developed. Although finite element semidiscretizations are now quite standard, we include this material so that some important consequences of the finite element semidiscretization on transient algorithms can be explained.

The finite difference method is then outlined and compared with the finite element method. Two forms of the finite difference method are considered, one developed through Taylor series, the second through contour integrals. The second is shown to be identical to a reduced quadrature finite element and to possess the same inherent rank deficiency, which results in the well known hourglass modes. The control of these modes is briefly discussed, for they are of importance in both finite element and finite difference methods. Although their appearance in the Laplace operator is not widely recognized, it provides a particularly simple setting to precisely identify its nature and control.

The notation used throughout this chapter employs lower case subscripts for Cartesian components and upper case indices for node numbers. Repeated indices are summed over their range. Commas followed by a lower case subscript designate spatial derivatives with respect to that spatial variable, and time derivatives are indicated by superposed dots. Where convenient, column matrices and vectors will be designated by boldface lower case symbols, such as $\boldsymbol{a}$, and matrices and second order tensors by boldface upper case symbols, such as $\boldsymbol{K}$. When a quantity pertains to a particular element, this will be indicated by the letter e, which will be enclosed in parenthesis except when it it attached to a scalar.

We consider a body Ω enclosed by a surface Γ with unit normal $\boldsymbol{n}$. The following nomenclature will be used:

θ = temperature,
$g_i = \theta_{,i}$ the gradient of the temperature field,
s = source per unit volume,
t = time,
q_i = heat flux,
x_i = space coordinates,
ρ = density,
α = specific heat,
$\bar{\rho} = \rho\alpha$.

The governing equation is the equation of energy conservation

$$-q_{i,i} + s = \bar{\rho}\dot{\theta} \quad \text{in } \Omega . \tag{1.1}$$

Initial conditions provide

$$\theta(0, \boldsymbol{x}) = \theta_0 \quad \text{in } \Omega . \tag{1.2}$$

A measure of the driving force for the flux is provided by the gradient of the temperature field

$$g_i = \theta_{,i} . \tag{1.3}$$

The heat law relates the temperature gradient to the flux by

$$q_i = q_i(g_j) . \tag{1.4}$$

A linear heat law can be written as

$$q_i = -k_{ij} g_j \tag{1.5}$$

where k_{ij} are the heat transfer coefficients.

For purpose of defining the boundary conditions, the surface Γ is subdivided into Γ_θ and Γ_q and

$$\theta = \theta^* \quad \text{on } \Gamma_\theta , \tag{1.6a}$$

$$q_i n_i = q^*(\theta) \quad \text{on } \Gamma_q , \tag{1.6b}$$

where the asterisks designate prescribed values and the possible dependence of q^* on θ is specifically indicated because this is true for convection and radiation boundaries.

Equations (1.1), (1.3) and (1.5) may easily be combined for linear heat conduction to obtain

$$(k_{ij}\theta_{,j})_{,i} + s = \bar{\rho}\dot{\theta} \tag{1.7a}$$

or for an isotropic law as

$$k\theta_{,ii} + s = \bar{\rho}\dot{\theta} \tag{1.7b}$$

where the first term on the left hand side is the Laplacian. This is a parabolic partial differential equation and its mathematical properties have been studied extensively. A key consequence of its parabolic character is that its solutions are smooth, i.e. discontinuities in θ are not expected.

1.2. Finite element semidiscretization

Two essential ingredients are required for obtaining finite element semidiscretizations of this equation: a variational or weak form of Eq. (1.1) and a construction of the approximate solutions. In a semidiscretization, we obtain the weak form of Eq. (1.1) at each instant in time, and in effect convert the partial differential equation (PDE) to a system of ordinary differential equations (ODE) with time as the independent variable.

Before going through this process in detail, we will outline its major features. The domain Ω is first subdivided into elements Ω_e interconnected by nodes. The approximate solution is then described within each element as product of the shape functions $N_I(\boldsymbol{x})$ and the nodal temperatures $\theta_I^e(t)$ of element e by

$$\theta(\boldsymbol{x}, t) = \theta_I^{(e)}(t) N_I(\boldsymbol{x}) \tag{1.8}$$

where the repeated subscripts are summed over the nodes of the element. The shape functions must be chosen so that the function $\theta(\boldsymbol{x}, t)$ is continuous across element boundaries, but its derivatives need not be continuous; the approximating function is therefore called a C^0 function.

On Γ_θ boundaries, θ_I are specified by the boundary condition and are not unknowns. Note that the shape functions are independent of time and the temporal dependence is entirely ascribed to the nodal temperatures. This in fact constitutes a local (element-by-element) separation of variables, which becomes significant for systems considered subsequently where discontinuities in space-time are expected.

The weak form of Eq. (1.1) is

$$\mathscr{m}(\theta, v) + \mathscr{f}(\theta, v) = \mathscr{s}(q^*, s, v)\,, \tag{1.9a}$$

$$\theta = \theta(\boldsymbol{x}, t), \qquad v = v(\boldsymbol{x})\,, \tag{1.9b}$$

where each of the above terms represents an integration over Ω and the second line is included to stress that θ is a function of x and t but v only a function of x. This weak form is required to govern the weak solution at all times, so the integration over Ω eliminates all spatial dependence and yields a system of ordinary differential equations.

To do this in detail, we define

$$\mathscr{m}(\theta, v) = \int_\Omega \bar{\rho}\dot{\theta} v \, \mathrm{d}\Omega\,, \tag{1.10a}$$

$$\mathscr{f}(\theta, v) = -\int_\Omega v_{,i} q_i \, \mathrm{d}\Omega\,, \tag{1.10b}$$

$$\mathscr{s}(q, s, v) = -\int_{\Gamma_q} q^* v \, \mathrm{d}\Gamma + \int_\Omega s v \, \mathrm{d}\Omega\,. \tag{1.10c}$$

The test function is defined in each element by

$$v(\boldsymbol{x}) = v_I^{(e)} N_I(\boldsymbol{x}) \tag{1.11}$$

and

$$v = 0 \quad \text{on } \Gamma_\theta\,. \tag{1.12}$$

Since v is defined by the same shape functions as θ, it is also C^0.

The element nodal variables are related to the system nodal variables by a matrix $\boldsymbol{L}^{(e)}$, so that

$$\boldsymbol{v}^{(e)} = \boldsymbol{L}^{(e)}\boldsymbol{v}, \qquad \boldsymbol{\theta}^{(e)} = \boldsymbol{L}^{(e)}\boldsymbol{\theta}. \tag{1.13}$$

The elements of the matrix $\boldsymbol{L}^{(e)}$ are strictly 0 or 1, so the operation of Eq. (1.15) consists simply of the extraction of appropriate elements from the larger array.

Substituting Eqs. (1.8), (1.10) and (1.11) into Eq. (1.9a) we then obtain the following system of ordinary differential equations

$$\boldsymbol{M}\dot{\boldsymbol{\theta}} + \boldsymbol{f} = \boldsymbol{s} \tag{1.14}$$

where $\boldsymbol{M}$, $\boldsymbol{f}$ and $\boldsymbol{s}$ are defined in terms of the following element matrices:

$$\boldsymbol{M}^{(e)} = [M_{IJ}]^{(e)} = \int_{\Omega^e} \bar{\rho} N_I N_J \, \mathrm{d}\Omega, \tag{1.15}$$

$$\boldsymbol{f}^{(e)} = [f_I]^{(e)} = -\int_{\Omega^e} N_{I,i} q_i \, \mathrm{d}\Omega, \tag{1.16}$$

$$\boldsymbol{s}^{(e)} = [s_I]^{(e)} = \int_{\Omega^e} N_I s \, \mathrm{d}\Omega - \int_{\Gamma_q^e} N_I q^* \, \mathrm{d}\Gamma. \tag{1.17}$$

The vectors (column matrices) $\boldsymbol{s}^{(e)}$ and $\boldsymbol{f}^{(e)}$ of the elements are related to the system matrices $\boldsymbol{s}$ and $\boldsymbol{f}$ by vector assembly

$$\boldsymbol{f} = \sum_{e=1}^{\mathrm{NELE}} \boldsymbol{L}^{(e)\mathrm{T}} \boldsymbol{f}^{(e)}, \tag{1.18a}$$

$$\boldsymbol{s} = \sum_{e=1}^{\mathrm{NELE}} \boldsymbol{L}^{(e)\mathrm{T}} \boldsymbol{s}^{(e)}, \tag{1.18b}$$

and the matrix $\boldsymbol{M}$ by matrix assembly

$$\boldsymbol{M} = \sum_{e=1}^{\mathrm{NELE}} \boldsymbol{L}^{(e)\mathrm{T}} \boldsymbol{M}^{(e)} \boldsymbol{L}^{(e)}, \tag{1.19}$$

These can also be written as

$$\boldsymbol{f} = \mathop{\mathsf{A}}_{e} \boldsymbol{f}^{(e)}, \tag{1.20a}$$

$$\boldsymbol{s} = \mathop{\mathsf{A}}_{e} \boldsymbol{s}^{(e)}, \tag{1.20b}$$

$$\boldsymbol{M} = \mathop{\mathsf{A}}_{e} \boldsymbol{M}^{(e)}, \tag{1.21}$$

where A is the assembly operator. These operations can be performed strictly as additions, rather than matrix multiplications, because the matrices $\boldsymbol{L}^{(e)}$ are

Boolean. Readers not familiar with this can find excellent accounts in [3] and [46]. The development of Eqs. (1.14) to (1.19) requires the replacement of the integral over the entire domain by the sum of the integrals over the element domains. To make this possible, the first derivatives and q_i in Eqs. (1.10) cannot be singular, which is why C^0 is required in the trial and test functions.

For the purposes of computer implementation it is convenient to define the discrete form of the gradient operator and using Eq. (1.8) write Eq. (1.3) in the matrix form

$$\boldsymbol{g} = \boldsymbol{B}\boldsymbol{\theta}^{(e)} \tag{1.22}$$

where

$$\boldsymbol{g} = \begin{Bmatrix} g_1 \\ g_2 \\ g_3 \end{Bmatrix} = \begin{Bmatrix} g_x \\ g_y \\ g_z \end{Bmatrix}, \tag{1.23}$$

$$B_{iI} = \frac{\partial N_I}{\partial x_i}. \tag{1.24}$$

A flux matrix $\boldsymbol{q}$ is then defined by

$$\boldsymbol{q} = \begin{Bmatrix} q_1 \\ q_2 \\ q_3 \end{Bmatrix} = \begin{Bmatrix} q_x \\ q_y \\ q_z \end{Bmatrix}. \tag{1.25}$$

Eq. (1.16) can then be written in the form

$$\boldsymbol{f}^{(e)} = -\int_{\Omega_e} \boldsymbol{B}^{\mathrm{T}}\boldsymbol{q}\,\mathrm{d}\Omega. \tag{1.26}$$

The linear heat law, Eq. (1.5), can be written in matrix form as

$$\boldsymbol{q} = -\boldsymbol{D}\boldsymbol{g}, \tag{1.27a}$$

$$D_{ij} = k_{ij} \quad \text{or} \quad k\delta_{ij} \text{ (isotropy)}. \tag{1.27b}$$

Combining Eqs. (1.22), (1.26) and (1.27), we obtain

$$\boldsymbol{f}^{(e)} = \boldsymbol{K}^{(e)}\boldsymbol{\theta}^{(e)} \tag{1.28a}$$

where

$$\boldsymbol{K}^{(e)} = \int_{\Omega_e} \boldsymbol{B}^{\mathrm{T}}\boldsymbol{D}\boldsymbol{B}\,\mathrm{d}\Omega \tag{1.28b}$$

with $\boldsymbol{D}$ given by (1.27b) and $\boldsymbol{B}$ by (1.24). This matrix is assembled exactly like $\boldsymbol{M}$. The linear system is governed by

$$\boldsymbol{M}\dot{\boldsymbol{\theta}} + \boldsymbol{K}\boldsymbol{\theta} = \boldsymbol{s}. \tag{1.29}$$

This can easily be verified by combining Eqs. (1.14), (1.13), (1.19) and (1.28). A schematic of the finite element semidiscretization process is given in Table 1. Here the correspondence between the terms of the original PDE and the ODE is brought out by keeping terms in the same order and indicating the independent variables in each term.

In nonlinear implicit methods, a predictor of the updated column vector $\boldsymbol{f}$ is also often needed. This is provided by

$$\dot{\boldsymbol{f}}^{(e)} = \mathring{\boldsymbol{K}}^{(e)}\dot{\boldsymbol{\theta}}^{(e)} \tag{1.30}$$

where $\mathring{\boldsymbol{K}}^{(e)}$ is given by

$$\mathring{\boldsymbol{K}}^{(e)} = \int \boldsymbol{B}^{\mathrm{T}}\mathring{\boldsymbol{D}}\boldsymbol{B}\,\mathrm{d}\Omega \tag{1.31}$$

where $\mathring{\boldsymbol{D}}$ relates the rates of $\boldsymbol{q}$ and $\boldsymbol{g}$ through

$$\dot{\boldsymbol{q}} = \mathring{\boldsymbol{D}}\dot{\boldsymbol{g}}\,. \tag{1.32}$$

Note that $\mathring{\boldsymbol{D}}$ and $\mathring{\boldsymbol{K}}^{(e)}$ *are not* rates; the superscript indicates that they relate rate quantities. The matrix $\mathring{\boldsymbol{K}}^{(e)}$ is known as the *element tangential conductance* matrix. It is assembled like any other element matrix, e.g. Eq. (1.19), to obtain the system matrix. Note that the form of the linear conductance $\boldsymbol{K}$, Eq. (1.28) and $\mathring{\boldsymbol{K}}$ are identical except that $\boldsymbol{D}$ is replaced by $\mathring{\boldsymbol{D}}$.

A large variety of finite elements in one, two or three dimensions can be developed from the equations given in the preceding development [60]. Any interpolation function which is a C^0 interpolant is a valid candidate for the shape functions N_I. Since a treatment of these aspects of the finite element methods is not the goal of this chapter we give in Tables 2 and 3 two forms of the finite element equation: the one dimensional form with linear interpolants N_I and the two dimensional form with the bilinear isoparametric shape functions.

In addition to the integral form of the element matrix $\boldsymbol{K}^{(e)}$ for this element, we have also given an approximation developed in [37] in Table 4. This form is exact for a rectangle with $\varepsilon = 1.0$ and can be used for quadrilateral elements because it

Table 1
Finite element semidiscretization

PDE: $\dfrac{\partial q_i}{\partial x_i} + \bar{\rho}\dot{\theta}(\boldsymbol{x}, t) - s(\boldsymbol{x}, t) = 0$

Finite element semidiscretization: $\theta(\boldsymbol{x}, t) = \boldsymbol{N}(\boldsymbol{x})\boldsymbol{L}^{(e)}\boldsymbol{\theta}(t)$, $v(\boldsymbol{x}) = \boldsymbol{N}(\boldsymbol{x})\boldsymbol{L}^{(e)}\boldsymbol{v}$

$$\int_\Omega v(\boldsymbol{x})\left[\underbrace{\frac{\partial q_i}{\partial x_i}} + \underbrace{\bar{\rho}\dot{\theta}(\boldsymbol{x}, t)} - \underbrace{s(\boldsymbol{x}, t)\right]\mathrm{d}\Omega + (\text{natural boundary terms})} = 0$$

ODE: $\boldsymbol{f}(t) + \quad \boldsymbol{M}\dot{\boldsymbol{\theta}}(t) \qquad\qquad - \boldsymbol{s}(t) = 0$

If the system is linear

ODE: $\boldsymbol{K}\boldsymbol{\theta}(t) + \boldsymbol{M}\dot{\boldsymbol{\theta}}(t) \qquad\qquad - \boldsymbol{s}(t) = 0$

Table 2
One dimensional finite element equations with linear shape functions

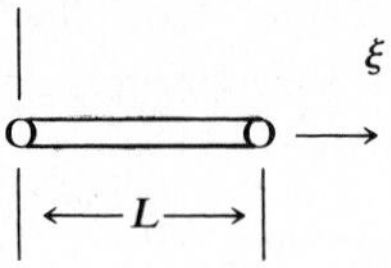

$$N_1 = 1 - \xi, \quad N_2 = \xi, \quad \xi = \frac{x}{L} \tag{T2.1}$$

$$\boldsymbol{B} = \frac{1}{L}[-1 \;\; +1] \tag{T2.2}$$

$$\boldsymbol{q} = [q_x] \quad \boldsymbol{g} = [g_x] \tag{T2.3}$$

$$\text{Eq. (1.26)} \Rightarrow \boldsymbol{f}^{(e)} = \begin{Bmatrix} q_x \\ -q_x \end{Bmatrix} \tag{T2.4}$$

$$\text{Eq. (1.28)} \Rightarrow \quad \boldsymbol{K}^{(e)} = \frac{k}{L}\begin{bmatrix} +1 & -1 \\ -1 & +1 \end{bmatrix} \tag{T2.5}$$

$$\text{Eq. (1.17)} \Rightarrow \quad \boldsymbol{M}^{(e)} = \frac{\bar{\rho}L}{6}\begin{bmatrix} 2 & 1 \\ 1 & 2 \end{bmatrix} \tag{T2.6}$$

$$\text{diagonal } \boldsymbol{M}^{(e)} = \frac{\bar{\rho}L}{2}\begin{bmatrix} 1 & 0 \\ 0 & 1 \end{bmatrix} \tag{T2.7}$$

meets the patch test. This form of the element relation was developed specifically for efficiency in nonlinear calculations. Its effectiveness results from the fact that it only requires a 1-point quadrature of Eqs. (1.26) and (1.28). However, as will be shown later, it requires an additional term, a stabilization matrix, which is also given in Table 4, to avoid rank deficiency. Complete quadrature of the element matrices for this element is usually performed by a 2×2 Gauss quadrature [3], which requires approximately 4 times as many computations per element. Table 4 will also be used to compare finite element and finite difference equations.

1.3. Taylor series finite difference method

In the finite difference method, the spatial discretization is usually performed by use of a Taylor series. There is a large variety of difference formulas for the heat conduction equation, but the most widely used are the 5-point and 9-point formulas. Only the linear form of the equation will be considered to facilitate comparison. Using a Taylor series gives

$$u_{,xxI/J} = \frac{1}{\Delta x^2}(u_{I+1/J} - 2u_{I/J} + u_{I-1/J}) + \mathrm{O}(\Delta x^2)\,, \tag{1.33a}$$

Table 3
Quadrilateral finite element for diffusion equation

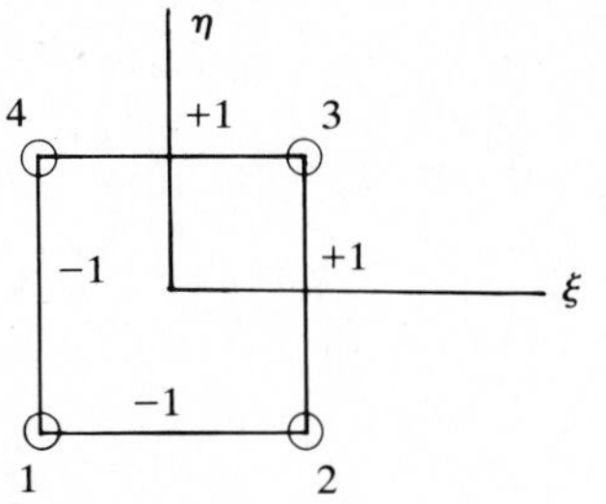

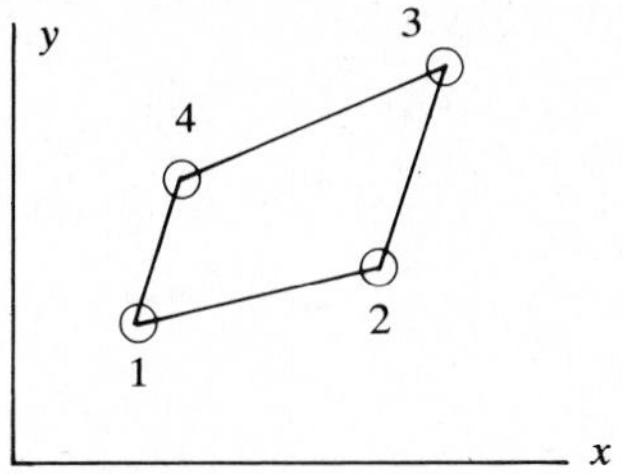

Reference plane $N_I = \frac{1}{4}(1 + \xi_I\xi)(1 + \eta_I\eta)$ no sum on I ξ_I, η_I coordinates of node I	Physical plane $x = x_I N_I(\xi, \eta)$ $y = y_I N_I(\xi, \eta)$	(T3.1)
Gradient $\boldsymbol{g} = \boldsymbol{B\theta}^{(e)} \quad \boldsymbol{B} = [B_{iI}] = \begin{bmatrix} N_{1,x} & N_{2,x} & N_{3,x} & N_{4,x} \\ N_{1,y} & N_{2,y} & N_{3,y} & N_{4,y} \end{bmatrix}$		(T3.2)
Flux $\boldsymbol{q} = -\boldsymbol{Dg} \quad \boldsymbol{D} = \begin{bmatrix} k_{11} & k_{12} \\ k_{21} & k_{22} \end{bmatrix}$		(T3.3)
Nodal flux $\boldsymbol{f}^{(e)} = \int_{\Omega^{(e)}} \boldsymbol{B}^{\mathrm{T}}\boldsymbol{q}\, \mathrm{d}\Omega$		(T3.4)
Conductance matrix (discretization of Laplacian) $\boldsymbol{K}^{(e)} = \int_{\Omega^{(e)}} \boldsymbol{B}^{\mathrm{T}}\boldsymbol{DB}\, \mathrm{d}\Omega$		(T3.5)

$$u_{yy\,I/J} = \frac{1}{\Delta y^2}(u_{I/J+1} - 2u_{I/J} + u_{I/J-1}) + \mathrm{O}(\Delta y^2)\,, \tag{1.33b}$$

where O indicates the order to truncation error. As usual in this chapter, upper case indices designate node numbers and two indices are used because a regular array of nodes in the x and y directions is assumed. The first upper case index refers to x, the second to y. For a regular square mesh with spacing h in the x and y directions, we obtain

$$u_{,ii\,I/J} = \frac{1}{h^2}(u_{I+1/J} + u_{I-1/J} + u_{I/J+1} + u_{I/J-1} - 4u_{I/J})\,. \tag{1.34}$$

This expression is usually described by the computational molecule shown in Table 5. The truncation error is of the order of the mesh spacing squared, h^2.

A 9-point difference formula which is of order h^6 is given in Table 5. According

Table 4
2D four-node quadrilateral with reduced (one-point) quadrature and stabilization

Semidiscretized gradient operator

$$\left.\begin{array}{ll} \boldsymbol{B} = \dfrac{1}{A}\begin{bmatrix} \boldsymbol{b}_1^{\mathrm{T}} \\ \boldsymbol{b}_2^{\mathrm{T}} \end{bmatrix} & A = \tfrac{1}{2}(x_{31}y_{42} + x_{24}y_{31}) \\ & x_{IJ} \equiv x_I - x_J \\ & y_{IJ} \equiv y_I - y_J \end{array}\right\} \tag{T4.1}$$

Basis vectors

$$\left.\begin{array}{ll} \boldsymbol{b}_1^{\mathrm{T}} = \tfrac{1}{2}[y_{24}, y_{31}, y_{42}, y_{13}] & \boldsymbol{x} \equiv \boldsymbol{x}_1^{\mathrm{T}} = [x_1, x_2, x_3, x_4] \\ \boldsymbol{b}_2^{\mathrm{T}} = \tfrac{1}{2}[x_{42}, x_{13}, x_{24}, x_{31}] & \boldsymbol{y} \equiv \boldsymbol{x}_2^{\mathrm{T}} = [y_1, y_2, y_3, y_4] \\ \boldsymbol{s}_1^{\mathrm{T}} = [1, 1, 1, 1] & \\ \boldsymbol{s}_2^{\mathrm{T}} \equiv \boldsymbol{h}^{\mathrm{T}} = [1, -1, 1, -1] & \end{array}\right\} \tag{T4.2}$$

Nodal flux matrix and conductance matrix from Eqs. (T3.4) and (T3.5)

$$\boldsymbol{f}^{(e)} = \underbrace{-q_x\boldsymbol{b}_1 - q_y\boldsymbol{b}_2}_{\text{one-point quadrature}} - \underbrace{\tilde{q}\boldsymbol{\gamma}}_{\text{stabilization matrix, see Eq. (1.51)}} \tag{T4.3}$$

$$\boldsymbol{K}^{(e)} = \frac{1}{A}(\overbrace{\boldsymbol{b}_i k_{ij} \boldsymbol{b}_j^{\mathrm{T}}} + \overbrace{\bar{\varepsilon}\boldsymbol{\gamma}\boldsymbol{\gamma}^{\mathrm{T}}}) \tag{T4.4}$$

where

$$\boldsymbol{\gamma} = A\boldsymbol{h} - (\boldsymbol{h}^{\mathrm{T}}\boldsymbol{x})b_1 - (\boldsymbol{h}^{\mathrm{T}}\boldsymbol{y})\boldsymbol{b}_2 \tag{T4.5}$$

Specialization to rectangle with $k_{ij} = k\delta_{ij}$

$$\boldsymbol{K}^{(e)} = \frac{k}{A}\boldsymbol{b}_i\boldsymbol{b}_j^{\mathrm{T}} + \bar{\varepsilon}A\boldsymbol{h}\boldsymbol{h}^{\mathrm{T}} \quad \text{with } \bar{\varepsilon} = \frac{k\boldsymbol{b}_i^{\mathrm{T}}\boldsymbol{b}_i}{12}\varepsilon \tag{T4.6}$$

to [26] the advantage of this finite difference operator over the 5 point does not persist when used in the Poisson equation, but numerical experiments suggest that it often does possess better accuracy.

These finite difference operators can also be assembled from element equations. For example, the element matrix obtained from Table 4 when $\varepsilon = 3$ for a square element is

$$\boldsymbol{K}^{(e)} = \frac{k}{2}\begin{bmatrix} 1 & -\frac{1}{2} & 0 & -\frac{1}{2} \\ & 1 & -\frac{1}{2} & 0 \\ & & 1 & -\frac{1}{2} \\ \text{symmetric} & & & 1 \end{bmatrix}. \tag{1.35}$$

This is equivalent to the 5-point molecule; an element stiffness equivalent to the 9-point molecule is obtained by setting $\varepsilon = 2$ in Table 4. It is worth noting, however, that neither of these element conductance matrices can be developed through Eq. (1.28) by standard shape functions.

Finite difference forms for nonuniform mesh spacing in both the x and y direction are also available; see for example [19]. If the boundaries are irregular

Table 5

(a) 5-point finite difference molecules for Laplacian

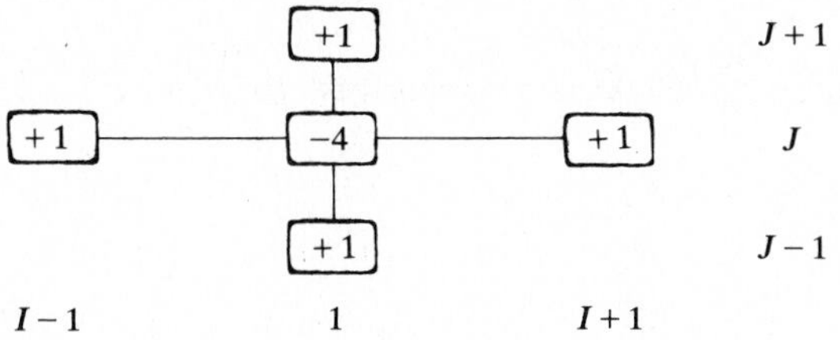

(b) 9-point molecule

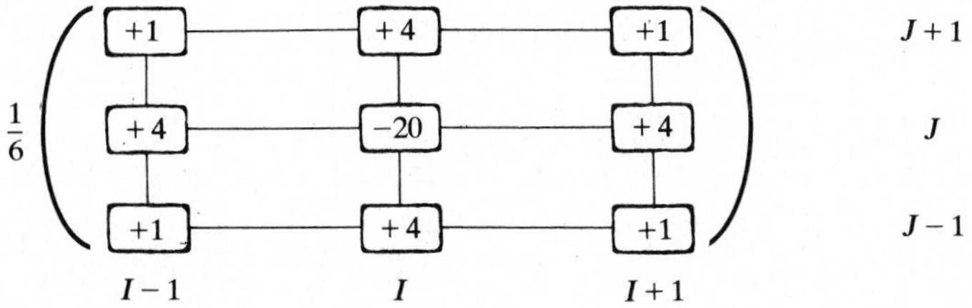

and curved, even nonuniform mesh spacing leads to difficulties in satisfying boundary conditions on derivatives. An alternative is to map the physical domain into a rectangular domain and treat an equation with variable coefficients [17]. This provides a powerful technique when many problems of a similar type are to be solved but is quite unwieldy in general purpose software.

1.4. Contour integral finite difference method

Since mapping methods are rather awkward in general purpose software, finite difference methods in a large number of Lagrangian continuum codes are developed by means of contour integral formulas. The method will here be illustrated for Eqs. (1.1) and (1.3). The basic idea in these methods is to use Gauss' theorem to express the derivative in a zone in terms of a contour integral [57]. Thus for a zone $A_{(e)}$ enclosed by the contour $C_{(e)}$ indicated by solid lines in Fig. 1

$$\int_{A_{(e)}} \theta_{,i} \, \mathrm{d}A = \int_{C_{(e)}} \theta n_i \, \mathrm{d}S \,. \tag{1.36}$$

If the gradient $\theta_{,i}$ is assumed constant in the zone (or element) A_e, then

$$\theta_{,i} = \frac{1}{A_{(e)}} \int_{C_{(e)}} \theta n_i \, \mathrm{d}S \,. \tag{1.37}$$

The right hand side of Eq. (1.37) is evaluated by assuming θ to be linear along each edge of the zone. Taking a typical zone as shown in Fig. 1, we obtain

$$\theta_{,x} = \tfrac{1}{2} \sum_{\substack{I=1 \\ J=I+1 \text{ when } I<4 \\ J=1 \quad \text{when } I=4}}^{4} (\theta_I + \theta_J)(y_I - y_J) \tag{1.38}$$

where the local node numbers of the zone are used. Rearrangement of terms enables us to write

$$\theta_{,x} = \frac{1}{A} \boldsymbol{b}_1^{\mathrm{T}} \boldsymbol{\theta}, \tag{1.39a}$$

$$\theta_{,y} = \frac{1}{A} \boldsymbol{b}_2^{\mathrm{T}} \boldsymbol{\theta}, \tag{1.39b}$$

$$\boldsymbol{\theta}^{\mathrm{T}} = [\theta_1 \ \ \theta_2 \ \ \theta_3 \ \ \theta_4],$$

where $\boldsymbol{b}_1$ and $\boldsymbol{b}_2$ are given in Table 4.

For the divergence term, the dashed path in Fig. 1 is used to establish the discrete form. Thus the counterpart of Eq. (1.36) for any scalar q is

$$\int_{A^*} q_{,i} \, \mathrm{d}A = \int_{C^*} q n_i \, \mathrm{d}s \tag{1.40}$$

where A^* is the area enclosed by C^* and q is considered constant in each zone. Because q is constant in each zone, this equation is somewhat ambiguous since the divergence will be zero within each zone and infinite between the zones. Nevertheless Eq. (1.40) can be written

$$\begin{aligned} A^* q_{,x} = {} & q^{(1)}(y_{\underline{2}} - y_{\underline{1}}) + q^{(2)}(y_{\underline{3}} - y_{\underline{4}}) \\ & + q^{(3)}(y_{\underline{4}} - y_{\underline{3}}) + q^{(4)}(y_{\underline{1}} - y_{\underline{2}}) \end{aligned} \tag{1.41}$$

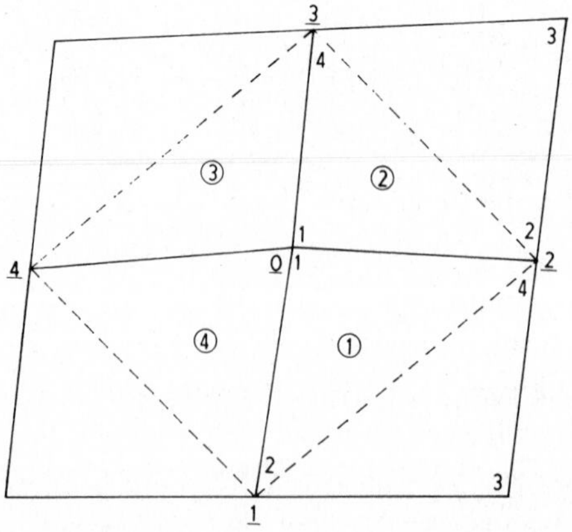

Fig. 1. Typical mesh for development of contour integral finite difference method. Global node numbers are underlined, local node numbers not underlined. Zone (element) numbers are circled.

where the underlined numbers are global node numbers in Fig. 1. A similar expression can be written for $q_{,y}$ which when combined yield

$$A^* q_{i,i} = \sum_{i=1}^{4} (q_x^{(i)} B_{11}^{(i)} + q_y^{(i)} B_{21}^{(i)}) \tag{1.42}$$

where the local node number 1 of each element is assumed to be the center node. This expression is almost identical to the finite element stencil given in Table 4 for the bilinear element with one point quadrature; the only difference is the weighting A^*. This is irrelevant in static problems, but it may be different in finite element equations depending on how the mass is lumped. A similar correspondence has been established for the equations of continuum mechanics [11].

1.5. Diagonal forms of capacitance (mass) matrices

The capacitance matrix $\boldsymbol{M}$ has the same bandwidth and sparsity characteristics as $\boldsymbol{K}$. In explicit time integration procedures, this is a severe disadvantage, since as will be seen in Section 3, when a diagonal form of the matrix $\boldsymbol{M}$ is used, no equations need be solved. On the other hand, nondiagonal forms of $\boldsymbol{M}$ require the solution of equations even with explicit time integration. Since diagonal forms of $\boldsymbol{M}$ have been used with considerable success in finite difference methods, ad hoc procedures for lumping $\boldsymbol{M}$ have also been adopted in finite element methods.

The simplest procedure for diagonalizing $\boldsymbol{M}$ is to simply replace Eq. (1.15) by

$$[M_{IJ}]^{(e)} = \frac{\delta_{IJ}}{\text{NODELE}} \int_{\Omega^e} \bar{\rho}\, \mathrm{d}\Omega \tag{1.43}$$

where NODELE is a number of nodes in the element. In continuum mechanics, this procedure corresponds to assigning an equal fraction of the total mass of each element to each node. Therefore a diagonal mass matrix is often called a lumped mass, while the nondiagonal form given by Eq. (1.15) is called a consistent mass matrix because it emerges from a consistent application of the weak form both to the temporal term and spatial terms of the governing equation.

The relative accuracy of the consistent and lumped mass matrices for the wave equation is studied in Chapter 6. In particular, comparing Figs. 8 and 9 of Chapter 6, we see that for the bilinear 4 node element, the consistent mass is only slightly more accurate. This lack of improvement is, however, not always the case. For example, in the advection-diffusion equation, whenever transport is dominant, diagonal mass matrices may introduce substantial errors; Donea and Giuliani [15] has examined this problem and proposed a rather novel and effective procedure for overcoming this drawback.

For higher order elements, simple lumping schemes such as given by Eq. (1.43) are quite ineffective. Several alternative procedures are available. Zienkiewicz [60] for example, recommends that each diagonal element of the lumped mass be computed by summing the corresponding row of the matrix, cf.

$$[M_{II}]^e = \sum_{J=1}^{n} M^e_{IJ} = \sum_{I=1}^{n} M^e_{IJ} \tag{1.44}$$

where n is the order of the $\boldsymbol{M}^e$ matrix. In an element where no derivatives are used as nodal variables, or in other words, where all nodal unknowns are the function itself, this gives

$$[M_{II}]^e = \int_{\Omega^e} \bar{\rho} N_I \sum_{J=1}^{n} N_J \, d\Omega \tag{1.45a}$$

$$= \int_{\Omega^e} \bar{\rho} N_I \, d\Omega \tag{1.45b}$$

where the last term follows because the sum of the shape functions must be equal to unity. The same fact can also be used to show that

$$\sum_{I=1}^{n} M^e_{II} = \bar{\rho}\Omega^e \tag{1.45c}$$

so this mass lumping scheme conserves $\bar{\rho}$.

Another scheme for mass lumping is to use integration schemes in which the quadrature points coincide with the nodes. For example, in the reference plane, (ξ, η) see Table 3, the quadrature points for 2×2 trapezoidal integration coincide with the 4 nodes of the element. Since the shape function N_I vanishes at all integration points other than I, it follows from Eq. (1.15) that

$$M_{II} = \bar{\rho}(\xi_I, \eta_I) J(\xi_I, \eta_I) \,. \tag{1.46}$$

This scheme does not conserve the total mass but appears to be more accurate for irregularly shaped elements than Eq. (1.43).

Nether of these schemes is effective for high order elements, such as the 8-node 2D serendipity or the 9-node 2D Lagrange interpolation. However, these elements are not particularly well suited to problems where explicit schemes are suitable.

An alternative scheme for mass diagonalization is to choose the mass terms so that the eigenvalues of the semidiscretization best match the eigenvalues of the continuous system. That technique was first proposed in [35] and is described further in Chapter 6.

1.6. Hourglass modes and their control

An important property of both the finite element method with reduced quadrature for the quadrilateral and the contour integral finite difference method is that the discrete form of the gradient will predict a zero gradient even for fields which are not constant, and in fact, are far from being constant. As a consequence, these fields do not generate any flux and hence they may grow arbitrarily large. In continuum mechanics, a similar phenomenon occurs because certain displacement fields which are not rigid body motions are associated with zero strains. Hence, they cause no stresses or nodal forces, so that these motions are unresisted by the discrete model. The form of these unresisted motions changes a rectangular Lagrangian mesh to a pattern of hourglass shapes, so this phenomenon is often called 'hourglassing'. Since the absence of stresses or strains implies that these modes have no energy, they are also called spurious zero-energy modes or kinematic modes.

We will here examine the nature of these modes in the context of the Laplace operator, which provides the simplest setting to examine their characteristics and control. To begin, we need the following preliminaries. Using the vectors given in Table 4, we note that

$$\boldsymbol{b}_i^{\mathrm{T}} \boldsymbol{s}_j = 0\,, \tag{1.47a}$$

$$\boldsymbol{b}_i^{\mathrm{T}} \boldsymbol{x}_j = A\delta_{ij}\,. \tag{1.47b}$$

From the linear independence of the vectors $\boldsymbol{b}_i$, $\boldsymbol{s}_i$ and $\boldsymbol{x}_i$, $\boldsymbol{s}_i$, which is easily established, it follows that both groups of 4 vectors span the 4-dimensional vector space R^4.

The orthogonality properties given above enable us to easily demonstrate the pathology of these formulations. If we let the nodal values of $\boldsymbol{\theta}^{(e)}$ for an element to be given by $\boldsymbol{s}_1$, then using (1.22) and (1.47) we obtain

$$\boldsymbol{g} = \boldsymbol{B}\boldsymbol{\theta}^{(e)} = \begin{bmatrix} \boldsymbol{b}_1^{\mathrm{T}} \\ \boldsymbol{b}_2^{\mathrm{T}} \end{bmatrix} \boldsymbol{\theta}^{(e)} = \boldsymbol{0}\,. \tag{1.48}$$

This result is expected and in fact necessary since it indicates that for a constant field, the gradient vanishes. The one dimensional space of the vector $\boldsymbol{s}_1$ will be called the proper null-space of $\boldsymbol{B}$.

If we let $\boldsymbol{\theta}^{(e)} = \boldsymbol{h} = \boldsymbol{s}_2$, then Eq. (1.47a) again implies that the gradient $\boldsymbol{g}$ vanishes. This result is quite unexpected since the field associated with these nodal values is not constant. The contradictory nature of this result can be appreciated further by noting that for a square element coincident with the reference plane element in Table 4, these nodal displacements correspond to $\theta(x, y) = xy$, which has nonzero gradient throughout the element. This one dimensional space $\boldsymbol{h}$ will be called the improper null-space of $\boldsymbol{B}$.

Similarly it can be shown from Eqs. (T4.4) and (1.47a) that for the quadrilateral with one-point quadrature

$$\boldsymbol{K}^{(e)}\boldsymbol{s} = \boldsymbol{K}^{(e)}\boldsymbol{h} = 0\,. \tag{1.49}$$

Thus, the matrix $\boldsymbol{K}^{(e)}$ is of rank 2, whereas the fact that only constant $\boldsymbol{\theta}$ fields should give zero flux (and hence zero nodal fluxes $\boldsymbol{f}^{(e)}$) suggests that $\boldsymbol{K}^{(e)}$ should be of rank 3. For this reason, this pathology is often called a rank deficiency of the element.

These shortcomings can be rectified by augmenting the gradient vector $\boldsymbol{g}$ by a generalized gradient $\tilde{g}$ and the flux matrix by an additional flux term $\tilde{q}$ which are given by

$$\tilde{g} = \boldsymbol{\gamma}^{\mathrm{T}}\boldsymbol{\theta}^{(e)}, \tag{1.50a}$$

$$\tilde{q} = -\bar{\varepsilon}\tilde{g}, \tag{1.50b}$$

where $\boldsymbol{\gamma}$ and $\bar{\varepsilon}$ remain to be defined. The discrete gradient operator is then

$$\boldsymbol{B}^* = \frac{1}{A}\begin{bmatrix} \boldsymbol{b}_1^{\mathrm{T}} \\ \boldsymbol{b}_2^{\mathrm{T}} \\ \boldsymbol{\gamma} \end{bmatrix} \tag{1.51}$$

The construction of $\boldsymbol{\gamma}$ should take into account that the discrete gradient operator $\boldsymbol{B}$ gives the correct result for linear fields, which can easily be seen by letting $\boldsymbol{\theta}^{(e)} = c_1\boldsymbol{x} + c_2\boldsymbol{y}$ and using (1.47b).

Therefore, $\boldsymbol{\gamma}$ will be chosen so that

(i) for any nodal values associated with linear θ fields, $\tilde{g} = 0$.

(ii) if $\boldsymbol{\theta}^{(e)}$ is in the complement of the proper null-space of $\boldsymbol{B}$, then $\tilde{g} = 0$.

Since $\boldsymbol{\gamma}$ is also in R^4, its most general form is

$$\boldsymbol{\gamma} = a_1\boldsymbol{b}_1 + a_2\boldsymbol{b}_2 + \boldsymbol{a}_3\boldsymbol{s} + \boldsymbol{a}_4\boldsymbol{h} \tag{1.52}$$

and the form that meets conditions (i) and (ii) can be deduced by determining the constants a_i appropriately. For this purpose, we note that for any linear θ field, the nodal values can be written as

$$\boldsymbol{\theta}^{(e)} = c_1\boldsymbol{x} + c_2\boldsymbol{y} + c_3\boldsymbol{s}\,. \tag{1.53}$$

Using condition (i) and Eqs. (1.47), (1.50), (1.52) and (1.53), we obtain

$$\boldsymbol{\gamma} = A\boldsymbol{h} - (\boldsymbol{h}^{\mathrm{T}}\boldsymbol{x})\boldsymbol{b}_1 - (\boldsymbol{h}^{\mathrm{T}}\boldsymbol{y})\boldsymbol{b}_2\,. \tag{1.54}$$

Since $\boldsymbol{\gamma}$ as given above is linearly independent of $\boldsymbol{b}_i$, $\boldsymbol{B}^*$ must be of rank 3 and hence span the complement of the proper null-space, thus satisfying condition (ii).

This form of the hourglass control operator was first developed in [20] by subtracting out the bilinear part of the field. Formulas for both two and three dimensional continuum mechanics were given there. The method of derivation given here is based on [37]. The resulting element stiffness matrix is given in Table 4. The parameters $\bar{\varepsilon}$ and ε are related so that for $\varepsilon = 1.0$, the $\boldsymbol{K}^{(e)}$ matrix corresponds exactly to that of the fully integrated $\boldsymbol{K}^{(e)}$ matrix for the rectangle.

The parameter ε also enables some interesting comparisons to be made with various finite difference molecules for the Laplacian. When $\varepsilon = 2.0$, the Eq. (T4.6) corresponds to the 9-point finite difference molecule, whereas $\varepsilon = 3.0$ corresponds to the 5-point molecule. These correspondences hold only for uniform rectangular meshes. Since the discrete gradient operators and the associated element stiffness matrices are rank deficient only when $\varepsilon = 0$, these spurious singular modes do not occur in the standard finite difference formulas; they are found only in underintegrated finite elements and contour-integral finite difference formulas.

The form of the $\boldsymbol{K}^{(e)}$ matrix with this hourglass control is given in Table 4. The additional contribution which results from the hourglass control is sometimes called a stabilization matrix [13]. Recommended values of ε vary from 0.05 to 1.0; see [37] and [20], depending on the type of problem which is semidiscretized.

Rank deficiency does not occur in triangular finite elements or when 2×2 quadrature is used for the quadrilateral. Neither of these alternatives is very appealing: triangles tend to be too stiff while 2×2 quadrature is too expensive. This is particularly true in three dimensions, where the counterparts of these alternatives are tetrahedra and $2 \times 2 \times 2$ quadrature.

2. Solid mechanics semidiscretization

2.1. Small deformation equations

We will first consider the analysis of small deformations of solid continua. Only small deformations are considered so that specific definitions of the stress and strain tensors which are needed to treat large deformations are avoided. These will be dealt with later.

For this class of problems, the governing equations are those of momentum conservation, which are often called the equations of motion. These can be written in tensor form as

$$T_{ij,j} + b_i = \rho \ddot{u}_i \tag{2.1}$$

where T_{ij} is the stress tensor, b_i the body force per unit volume, ρ the density, and u_i the displacements; superposed dots denote time derivatives.

In addition, a set of strain-displacement equations

$$\varepsilon_{ij} = \tfrac{1}{2}(u_{i,j} + u_{j,i}) \tag{2.2}$$

and a constitutive law, which relates the stress-rates to the strain-rates,

$$\dot{T}_{ij} = C_{ijkl}\dot{\varepsilon}_{kl} \tag{2.3a}$$

is needed. The matrix $\boldsymbol{C}$ usually has the symmetries

$$C_{ijkl} = C_{jikl} = C_{ijlk} = C_{klij} \,. \tag{2.3b}$$

The first two equalities are called minor symmetries, the last equality the major symmetry. In plasticity, nonassociated flow laws lack the major symmetry. If the material is linear, we may write

$$T_{ij} = C_{ijkl}\varepsilon_{kl} \,. \tag{2.3c}$$

In small displacement problems, the conservation of matter (continuity equation) plays no role. Conservation of energy can be written as

$$\rho\dot{e} = T_{ij}\dot{u}_{i,j} + s - q_{k,k} \tag{2.4}$$

where e is the internal energy and s and q_k the heat source and flux; see Section 1. Since the time scale of heat conduction usually differs substantially from that of the mechanical response, we can usually uncouple thermal and mechanical phenomena. Therefore conservation of energy can be used to monitor how well the evolution of the mechanical system is treated and Eq. (2.4) simply serves as a definition of internal energy.

2.2. Finite element semidiscretization

The finite element semidiscretization procedure is performed as follows. The solid volume Ω is subdivided into elements Ω_e and the displacement field in each element is approximated by shape functions $\boldsymbol{N}_I(\boldsymbol{x})$ which are functions of space alone and nodal displacements $u_{iI}(t)$ which incorporate the time dependence

$$u_i(\boldsymbol{x}, t) = \sum u_{iI}(t) N_I(\boldsymbol{x}) \,. \tag{2.5}$$

The discrete form of the gradient operator in the strain–displacement equations, Eqs. (2.2), can be written as

$$u_{i,j} = B_{jI} u_{iI} \tag{2.6a}$$

where

$$B_{jI} = \frac{\partial N_I}{\partial x_j} \,. \tag{2.6b}$$

The discrete form of the strain–displacement equation is

$$\varepsilon_{ij} = \tfrac{1}{2}(B_{iI} u_{jI} + B_{jI} u_{iI}) \,. \tag{2.7}$$

Let the solid continuum be Ω with a boundary Γ consisting of Γ_u and Γ_τ where

$$u_i = u_i^* \quad \text{on } \Gamma_u \,, \tag{2.8a}$$

$$T_{ij} n_j = \tau_i^* \quad \text{on } \Gamma_\tau \,, \tag{2.8b}$$

and let $v = 0$ on Γ_u. Then the weak form (variational form) of Eq. (2.1) is

$$\mathcal{W}^{\text{int}}(\boldsymbol{u}, \boldsymbol{v}) + \mathcal{M}(\boldsymbol{u}, \boldsymbol{v}) = \mathcal{W}^{\text{ext}}(\boldsymbol{u}, v) \tag{2.9}$$

for all v that vanish on Γ_u where

$$\mathcal{W}^{\text{int}}(\boldsymbol{u}, \boldsymbol{v}) = \int_\Omega v_{i,j} T_{ij}(\boldsymbol{u}) \, \mathrm{d}\Omega \,, \tag{2.10}$$

$$\mathcal{M}(\boldsymbol{u}, \boldsymbol{v}) = \int_\Omega \rho v_i \ddot{u}_i \, \mathrm{d}\Omega \,, \tag{2.11}$$

$$\mathcal{W}^{\text{ext}}(\boldsymbol{u}, \boldsymbol{v}) = \int_\Omega v_i b_i \, \mathrm{d}\Omega + \int_{\Gamma_\tau} v_i \tau_i^* \, \mathrm{d}\Gamma \,. \tag{2.12}$$

This form is identical with the principle of virtual work: $\mathcal{W}^{\text{int}}$ is the increment of internal work associated with $\boldsymbol{v}$, $\mathcal{M}$ is the d'Alembert inertial force, and $\mathcal{W}^{\text{ext}}$ is the increment of external work. The test function $\boldsymbol{v}$ is often called the variation; the trial function is $\boldsymbol{u}$. The continuum application of the weak form to a finite element requires C^0 continuity in the test functions. The natural boundary conditions, Eq. (2.8b), are also enforced in a weak sense by the variational form so that they need not be imposed on the trial functions.

If we approximate the test functions by the same shape functions as $\boldsymbol{u}$, Eqs. (2.5), we obtain that

$$\boldsymbol{v}^{\mathrm{T}}(\boldsymbol{f}^{\text{int}} + \boldsymbol{M}\ddot{\boldsymbol{u}} - \boldsymbol{f}^{\text{ext}}) = 0 \tag{2.13}$$

where $\boldsymbol{M}$, $\boldsymbol{f}^{\text{int}}$, and $\boldsymbol{f}^{\text{ext}}$ are defined in Table 6. Because this equation must hold for arbitrary $\boldsymbol{v}$, it can be deduced that

$$\boldsymbol{M}\ddot{\boldsymbol{u}} + \boldsymbol{f}^{\text{int}} = \boldsymbol{f}^{\text{ext}} . \tag{2.14}$$

The above is a system of N ordinary differential equations (ODE) of second order in time.

To aid in the intuitive interpretation of these equations, note that they can be

Table 6
Finite element semidiscretization of continua

Equation of motion

$$\boldsymbol{M}\ddot{\boldsymbol{u}} + \boldsymbol{f}^{\text{int}} = \boldsymbol{f}^{\text{ext}} \tag{T6.1}$$

Element mass matrix (consistent)

$$\boldsymbol{M}^{(e)} = [M_{iIjJ}]^{(e)} = \int_{\Omega_{(e)}} \rho N_I N_J \delta_{ij} \, \mathrm{d}\Omega \tag{T6.2}$$

Element nodal force matrix

$$\boldsymbol{f}^{\text{int}}_{(e)} = [f_{iI}]^{(e)} = \int_{\Omega_{(e)}} B_{jI} T_{ij} \, \mathrm{d}\Omega \tag{T6.3}$$

$$B_{jI} = \frac{\partial N_I}{\partial x_j} \tag{T6.4}$$

External nodal force matrix

$$\boldsymbol{f}^{\text{ext}}_{(e)} = [f_{iI}]^{\text{ext}}_{(e)} = \int_{\Omega_{(e)}} N_I b_i \, \mathrm{d}\Omega + \int_{\Gamma_{\gamma(e)}} N_I \tau_i^* \, \mathrm{d}\Gamma \tag{T6.5}$$

Element stiffness matrix

$$\boldsymbol{K}^{(e)} = [K_{ikIK}]^{(e)} = \int_{\Omega_{(e)}} B_{jI} C_{ijkl} B_{lK} \, \mathrm{d}\Omega \tag{T6.6}$$

written in a form which is identical to Newton's Second Law

$$\boldsymbol{M}\ddot{\boldsymbol{u}} = \boldsymbol{f}, \tag{2.15a}$$

$$\boldsymbol{f} = \boldsymbol{f}^{\text{ext}} - \boldsymbol{f}^{\text{int}}. \tag{2.15b}$$

The negative sign on the internal forces arises from the convention used for its definition. The forces $\boldsymbol{f}$ are net forces on the nodes; similarly external nodal forces $\boldsymbol{f}^{\text{ext}}$ are considered to act on the nodes, but the internal nodal forces $\boldsymbol{f}^{\text{int}}$ are defined to act on the elements, so the resulting forces on the nodes are equal and opposite.

For a linear material, the above becomes

$$\boldsymbol{M}\ddot{\boldsymbol{u}} + \boldsymbol{K}\boldsymbol{u} = \boldsymbol{f}^{\text{ext}} \tag{2.16a}$$

where

$$\boldsymbol{K} = \underset{e}{\mathsf{A}}\boldsymbol{K}^{(e)}, \tag{2.16b}$$

where $\boldsymbol{K}^{(e)}$ is given in Table 6. If $\boldsymbol{C}$ has the symmetries of (2.3b), $\boldsymbol{K}$ will be symmetric.

2.3. Large displacement – Large strain analysis of continua

In the previous section, the question of large deformations was deliberately bypassed because it introduces a large variety of choices which are more clearly treated separately, and which furthermore have no direct bearing on the weak form of the equations of motion. However, the treatment of large deformations does entail careful treatment of certain aspects of the computational procedure. Many of the questions that arise concern both nonlinear continuum mechanics and finite element methodologies. The continuum mechanics which is summarized here is based on [38] and [45]; other discussions of these topics may be found in [3], [6], [44] and [50].

Three basic choices need to be made in the development of a large deformation semidiscretization scheme:

(i) the type of mesh (mesh description)

(ii) the kinematic description, i.e. how the deformation is measured

(iii) the kinetic description, i.e. how the stresses are measured (in most semidiscretization schemes, choice (ii) determines (iii) because the kinetic and kinematic measures should be conjugate as described later).

To describe these options further, we introduce three sets of coordinates:

(i) spatial (Eulerian) coordinates $\boldsymbol{x}$

(ii) material (Lagrangian) coordinates $\boldsymbol{X}$

(iii) referential coordinates $\boldsymbol{\chi}$

The deformation of a continuum can be described by a mapping $\boldsymbol{\phi}$ which gives the current coordinates in terms of the material coordinates; the latter are assumed to be coincident with the material coordinates in the undeformed configuration. Hence

$$\boldsymbol{x} = \boldsymbol{\phi}(\boldsymbol{X}, t)\,. \tag{2.17}$$

The displacement of the continuum is given by

$$\boldsymbol{u} = \boldsymbol{\phi}(\boldsymbol{X}, t) - \boldsymbol{X} \tag{2.18}$$

which can also be written (at the cost of some confusion) as

$$\boldsymbol{u} = \boldsymbol{x} - \boldsymbol{X}\,. \tag{2.19}$$

2.4. Mesh descriptions

In a computational procedure, the referential coordinates $\boldsymbol{\chi}$ can be used to define the nodes and elements (or zones) of the mesh. Thus, if we let

$$\boldsymbol{\chi}_I = \boldsymbol{X}_I \tag{2.20}$$

each node will remain coincident with the same material particle and each element will contain the same domain of material throughout the deformation. This type of mesh is called *Lagrangian* and the salient features of its behavior are shown in Fig. 2, where a one dimensional rod is shown in its undeformed and deformed configuration along with a Lagrangian mesh. Although each element contains an infinite number of material points, only one point is indicated, for this suffices to show that material points within a particular element remain within that element throughout the deformation process and that *node points* and *material points* remain *coincident* in a *Lagrangian mesh.* It can also be observed from this figure that a node point always remains on the boundary of the domain, so that the motion of the boundary will not present any difficulties. However, it is evident even in this one dimensional example that as the mesh becomes deformed, large changes in element dimensions may take place, which is a disadvantage.

If the referential coordinate system is chosen to coincide with the spatial coordinates

$$\boldsymbol{\chi}_I = \boldsymbol{x}_I\,, \tag{2.21}$$

then the mesh is called *Eulerian,* and as shown in Fig. 2, each node point will remain fixed in space. Therefore, material points may migrate from one element to another and the boundary of the domain does not remain coincident with a node point but instead moves through the element. This makes the imposition of boundary conditions more complicated. Thus, while an Eulerian mesh handles large interior deformation easily, it is less suited to large motions of boundaries or interfaces.

In a Lagrangian or Eulerian mesh, there is no need to explicitly introduce the referential coordinate system, since the variables $\boldsymbol{X}$ and $\boldsymbol{x}$ already play the role of referential coordinates. However, as can be seen from the preceding two examples, both Lagrangian and Eulerian meshes have disadvantages when large

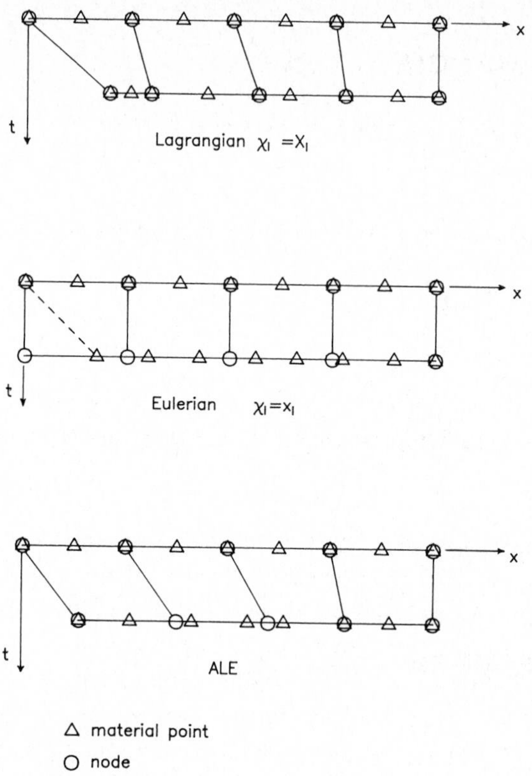

Fig. 2. One dimensional illustration of Lagrangian, Eulerian and arbitrary Lagrangian–Eulerian mesh.

displacements take place. For example, in this one dimensional problem it is possible to choose a referential coordinate system so that boundary node points remain coincident with the material points and hence the boundaries, while interior node points remain equispaced, so that the distortion of the mesh is spread more uniformly. This is illustrated in Fig. 2. This type of mesh is known by various appellations: Arbitrary Lagrangian–Eulerian (ALE) in the finite difference literature [27], quasi-Eulerian methods in [32] (because of the similarity of the structure of its equation of motion to that for an Eulerian mesh), and mixed Lagrangian–Eulerian [29], particularly in applications where one coordinate direction is Eulerian, the other Lagrangian. The advantages of these methods are most apparent in large deformation fluid-structure problems, such as described in Chapter 10, where a Lagrangian mesh calculation cannot be completed without 'rezoning' the mesh to remove excessively distorted elements. This is indicated in Fig. 3, which compares a simulation of an expanding bubble in a fluid. As can be seen, the fluid elements adjacent to the bubble in a Lagrangian mesh become very narrow and the computation fails. In the ALE mesh, the elements have been programmed to move so that their relative dimensions do not change with the deformation, and mesh failure is avoided. Nodes on the free surface

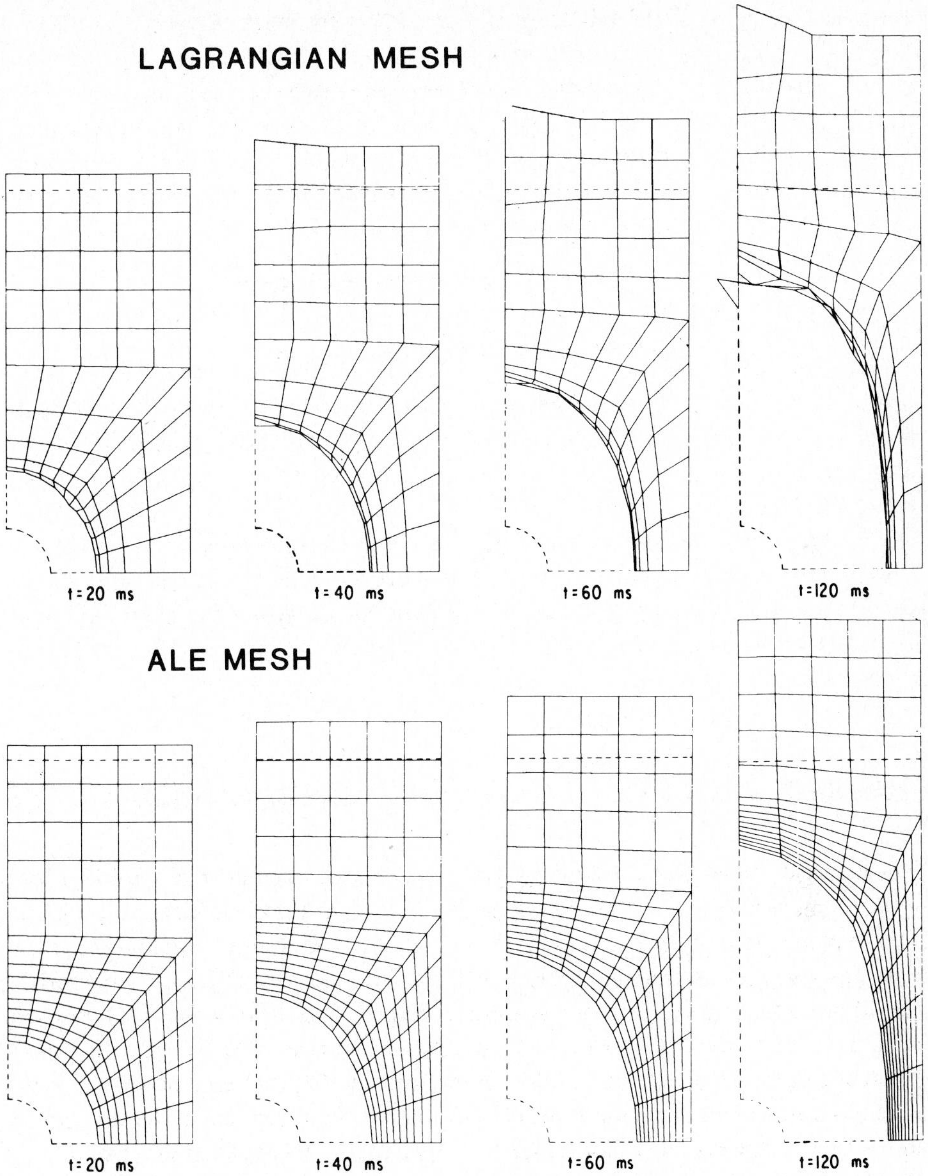

Fig. 3. Deformed meshes for a large deformation problem (bubble expanding in a fluid) showing failure of Lagrangian mesh and a successful ALE computation, from [32].

have been treated as Lagrangian, so the difficulties of Eulerian treatments of free surfaces are avoided.

A byproduct of the choice of mesh description is that it establishes the independent variables. Thus, for a Lagrangian mesh, the independent variable is $\boldsymbol{X}$. At a quadrature point which is used in evaluating the internal forces, cf. Eq. (T9.4), the coordinate $\boldsymbol{X}$ remains invariant regardless of the deformation of the structure, therefore the stress has to be defined as a function of the material coordinate $\boldsymbol{X}$; this is natural in a solid since the stress in a path-dependent material will depend on the 'history' observed by a material point. On the other hand, for an Eulerian mesh, the stress will be treated as a function of $\boldsymbol{x}$, which means that the history of the point will need to be 'convected' throughout the computation. This leads to additional numerical diffusion. In a linear-viscous, compressible fluid, on the other hand, the history of a material particle has no effect on the stresses.

2.5. Kinematic descriptions

The term kinematics is here used to described how the deformation of the continuum is measured. The most well known measure of strain in engineering practice is the 'linear' strain tensor, which has been defined by Eq. (2.2). The linear strain tensor can only be used in problems where the rigid body rotations are small.

When rigid body rotations are large, the linear strain tensor becomes nonzero in the absence of deformation, which therefore obviously makes it useless as a measure of deformation. This can be seen by simply considering the displacements associated with a rigid body rotation. The displacements in any rigid body rotation can be written as

$$u_i = (R_{ij} - \delta_{ij})x_j \tag{2.22}$$

where R_{ij} is the orthogonal transformation matrix consisting of the direction cosines. In two dimensions, for example

$$\boldsymbol{R} = \begin{bmatrix} \cos\theta & -\sin\theta \\ \sin\theta & \cos\theta \end{bmatrix} \tag{2.23}$$

where θ is the rotation of the body.

Substituting Eq. (2.22) into (2.23), gives

$$\varepsilon_{ij} = \tfrac{1}{2}(R_{ij} + R_{ji} - 2\delta_{ij}) \tag{2.24}$$

or in two dimensions

$$\boldsymbol{\varepsilon} = \begin{bmatrix} \cos\theta - 1 & 0 \\ 0 & \cos\theta - 1 \end{bmatrix}.$$

Neither of the above vanish if θ is nonzero, which means that if the linear strain tensor is used for large rotation problems a nonzero strain, and hence a stress, will be predicted when the solid is simply rotated as a rigid body.

If we use a Taylor expansion of $\cos\theta$ we can write

$$\boldsymbol{\varepsilon} = \begin{bmatrix} \tfrac{1}{2}\theta^2 + \mathrm{O}(\theta^4) & 0 \\ 0 & \tfrac{1}{2}\theta^2 + \mathrm{O}(\theta^4) \end{bmatrix} \tag{2.25}$$

so that we can see that the spurious strain arising from rigid body rotation is of order θ^2. Thus, if the rotations are of the same order as the strains, they can be neglected when strains are of the order 10^{-2} or less, since this will only lead to 1% errors due to rigid body rotation. It is worth noting that linear strain can also be used for moderately large strains of the order of 0.1 provided that the *rotations are small*, for the linear strain's error in the measure of deformation arises entirely out if its failure to disregard rigid body rotation. On the other hand, for slender structures which are quite inextensible, such as thin metallic shells, nonlinear kinematics must be used even when the rotations are order 10^{-2} because we are often interested in strains of order 10^{-4} to 10^{-3} so the error due to the rotations would be larger than the strains.

Large deformation problems, those in which the rigid body rotations and deformations are large, are called *geometrically nonlinear*. For geometrically nonlinear problems an infinite number of measures of deformation are available but most theoretical work and computer software employs the following three measures:

(i) the velocity strain (stretching or rate-of-deformation)

$$D_{ij} = \tfrac{1}{2}\left(\frac{\partial \dot{u}_i}{\partial x_j} + \frac{\partial \dot{u}_j}{\partial x_i}\right); \tag{2.26}$$

(ii) the Green strain tensor (also called the Lagrangian strain tensor)

$$E_{ij} = \tfrac{1}{2}\left(\frac{\partial u_i}{\partial X_j} + \frac{\partial u_j}{\partial X_i} + \frac{\partial u_k}{\partial X_i}\frac{\partial u_k}{\partial X_j}\right); \tag{2.27}$$

(iii) the Almansi strain tensor (also called the Eulerian strain tensor)

$$E^A_{ij} = \tfrac{1}{2}\left(\frac{\partial u_i}{\partial x_j} + \frac{\partial u_j}{\partial x_i} - \frac{\partial u_k}{\partial x_i}\frac{\partial u_k}{\partial x_j}\right). \tag{2.28}$$

Any of these strain tensors can be used with any type of mesh. The relationships between these measures of deformation are given in Table 7.

These measures of deformation have some important distinctions. The velocity-strain measures the current rate of deformation, but it gives no information about the total deformation of the continuum. Furthermore, its integral in time for a material point does not yield a well-defined, path-independent tensor so that information about phenomena such as total stretching (or maximum strain) are not available in an algorithm that employs only the velocity-strain. Therefore, to obtain a measure of total deformation, the velocity strain has to be transformed to some other strain rate, i.e., $\dot{\boldsymbol{E}}$ by formulas such as given in Table 7; this rate can then be integrated to yield a measure of the total deformation.

The second two measures, the Green strain and the Almansi strain, do measure the total deformation. However, as can be seen from the above equations, the relationship between these strains and the displacements is nonlinear, which adds to the complexity of the numerical algorithm. Furthermore, as will be seen later, the rate of $\boldsymbol{E}$ is not conjugate to the physical stress.

Table 7
Measures of deformation and their transformations and conjugate stresses

Strain Measure of deformation	Definition	Conjugate stress	Rate of work	Relationship to $\boldsymbol{E}$	Relationship to $\boldsymbol{E}^A$	Relationship to $\boldsymbol{D}$
$\boldsymbol{E}$ Green strain	$2\,\mathrm{d}\boldsymbol{X}^{\mathrm{T}}\boldsymbol{E}\,\mathrm{d}\boldsymbol{X} = \mathrm{d}\boldsymbol{x}^{\mathrm{T}}\,\mathrm{d}\boldsymbol{x} - \mathrm{d}\boldsymbol{X}^{\mathrm{T}}\,\mathrm{d}\boldsymbol{X}$	$\boldsymbol{S}$ (2nd P.K.)	$S_{ij}\dot{E}_{ij}$	*	$\boldsymbol{E} = \boldsymbol{F}^{\mathrm{T}}\boldsymbol{E}^A\boldsymbol{F}$	$\dot{\boldsymbol{E}} = \boldsymbol{F}^{\mathrm{T}}\boldsymbol{D}\boldsymbol{F}$
$\boldsymbol{E}^A$ Almansi strain	$2\,\mathrm{d}\boldsymbol{x}^{\mathrm{T}}\boldsymbol{E}^A\,\mathrm{d}\boldsymbol{x} = \mathrm{d}\boldsymbol{x}^{\mathrm{T}}\,\mathrm{d}\boldsymbol{x} - \mathrm{d}\boldsymbol{X}^{\mathrm{T}}\,\mathrm{d}\boldsymbol{X}$	$\boldsymbol{T}$ (Cauchy)	$T_{ij}\overset{\nabla}{E}_{A_{ij}}$ [a]	$\boldsymbol{E}^A = (\boldsymbol{F}^{-1})^{\mathrm{T}}\boldsymbol{E}\boldsymbol{F}^{-1}$	*	$\dot{\boldsymbol{E}}^A = \boldsymbol{D} - [(\boldsymbol{E}^A)^{\mathrm{T}}\boldsymbol{L} + \boldsymbol{L}^{\mathrm{T}}\boldsymbol{E}^A]$
$\boldsymbol{D}$ Velocity strain	$2\,\mathrm{d}\boldsymbol{x}^{\mathrm{T}}\boldsymbol{D}\,\mathrm{d}\boldsymbol{x} = \dfrac{\mathrm{d}}{\mathrm{d}t}(\mathrm{d}\boldsymbol{x}^{\mathrm{T}}\,\mathrm{d}\boldsymbol{x})$	$\boldsymbol{T}$ (Cauchy)	$T_{ij}D_{ij}$	$\boldsymbol{D} = (\boldsymbol{F}^{-1})^{\mathrm{T}}\dot{\boldsymbol{E}}\boldsymbol{F}^{-1}$	$\boldsymbol{D} = \overset{\nabla}{\boldsymbol{E}}_A = \dot{\boldsymbol{E}}^A + [(\boldsymbol{E}^A)^{\mathrm{T}}\boldsymbol{L} + \boldsymbol{L}^{\mathrm{T}}\boldsymbol{E}^A]$	*

$F_{ij} = \dfrac{\partial x_i}{\partial X_j}$ $\quad L_{ij} = \dfrac{\partial \dot{u}_i}{\partial x_j}$ $\quad J = \det(\boldsymbol{F})$

[a] $\overset{\nabla}{\boldsymbol{E}}_A$ is the Rivlin–Ericksen rate of the Almansi strain, which is defined in column 6 of the last row

2.6. Stress descriptions and constitutive equations

The stresses conjugate to the strain measures are given in Table 7, and the transformation laws between these stress measures in Table 8. A stress is called conjugate to the strain if its scalar product with the strain gives work.

An important aspect of the measure of stress is the frame indifference (also called objectivity or frame-invariance) of its rates. A frame-indifferent rate must transform according to the law for a second-order tensor when the spatial reference frame is rotated. This is important because constitutive equations are often expressed in rate form, so if the rates are not frame indifferent, the material behavior will depend on the choice of the spatial coordinate system. Such behavior is obviously physically unrealistic and thus must be avoided. For example, the time derivative of the Cauchy stress is not frame invariant, so a linear relationship between $\dot{\boldsymbol{T}}$ and $\boldsymbol{D}$ is unrealistic. This may easily be seen by taking a solid which is in a state of stress $\boldsymbol{T}$ and rotating the solid without deforming it. The stress tensor transforms according to

$$T^*_{ij} = R_{ik}R_{jl}T_{kl} \tag{2.29}$$

when R_{ik} gives the transformation by

$$x^*_i = R_{ik}x_k\,. \tag{2.30}$$

When $\boldsymbol{x}^*$ and $\boldsymbol{x}$ are coincident but rotating relative to each other

$$\dot{R}_{ik} = W_{ik} = \tfrac{1}{2}\left(\frac{\partial \dot{u}_i}{\partial x_k} - \frac{\partial \dot{u}_k}{\partial x_i}\right) \tag{2.31}$$

and

$$R_{jk} = \delta_{jk} \tag{2.32}$$

so

$$\dot{T}_{ij} = W_{ik}T_{kj} + W_{jl}T_{il} \tag{2.33}$$

Table 8
Transformation between stresses

	$\boldsymbol{T}$ (Cauchy stress)	$\hat{\boldsymbol{T}}$ (corotational Cauchy stress)	$\boldsymbol{S}$ (2nd. P.K.)
$\boldsymbol{T}=$		$\boldsymbol{R}\hat{\boldsymbol{T}}\boldsymbol{R}^{\mathrm{T}}$	$J^{-1}\boldsymbol{FSF}^{\mathrm{T}}$
$\hat{\boldsymbol{T}}=$	$\boldsymbol{R}^{\mathrm{T}}\boldsymbol{TR}$		$J^{-1}\boldsymbol{USU}^{\mathrm{T}}$
$\boldsymbol{S}=$	$J\boldsymbol{F}^{-1}\boldsymbol{T}(\boldsymbol{F}^{-1})^{\mathrm{T}}$	$J\boldsymbol{U}^{-1}\boldsymbol{S}(\boldsymbol{U}^{-1})^{\mathrm{T}}$	
$F=\dfrac{\partial x_i}{\partial X_j}$	$J=\det \boldsymbol{F}$	$\boldsymbol{F}=\boldsymbol{RU}$ by polar decomposition [38]	

which is not zero if there is an initial state of stress $\boldsymbol{T}$. A frame indifferent rate can be obtained by subtracting precisely the quantities given in the above equation. It is called the Jaumann rate, which will be denoted by $\boldsymbol{T}^{\nabla}$ and is given by

$$T_{ij}^{\nabla} = \dot{T}_{ij} - W_{ik}T_{kj} - W_{jk}T_{ki}\,. \tag{2.34}$$

Many other frame indifferent rates can be constructed. This diversity in the choice of rates is analogous to the multiplicity of choices that is found in measuring strain and is one of the major reasons for the confusion that has bedeviled large-deformation computational mechanics; for more detailed clarification of the rates, see [44] and [51].

One important frame invariant rate is $\dot{\boldsymbol{S}}$. Its frame invariance can be deduced from the relationship between $\boldsymbol{S}$ and the Cauchy stress $\boldsymbol{T}$ given in Table 8; when the material is simply rotated $\boldsymbol{F} = \boldsymbol{R}$ and $\dot{\boldsymbol{S}} = 0$. To develop this stress rate in terms of $\boldsymbol{T}$, $\dot{\boldsymbol{S}}$ is compared to $\dot{\boldsymbol{T}}$ as given in [44]. Taking time derivatives of the $\boldsymbol{S}-\boldsymbol{T}$ relationship in Table 8 yields

$$\begin{aligned}\dot{S}_{ij} &= J\frac{\partial X_i}{\partial x_a}\frac{\partial X_j}{\partial x_b}\dot{T}_{ab} + J\frac{\partial \dot{u}_b}{\partial x_b}\frac{\partial X_i}{\partial x_c}\frac{\partial X_j}{\partial x_a}T_{ca}\\ &\quad - J\frac{\partial X_i}{\partial x_c}\frac{\partial \dot{u}_c}{\partial x_a}\frac{\partial X_j}{\partial x_b}T_{ab} - J\frac{\partial X_i}{\partial x_a}\frac{\partial \dot{u}_c}{\partial x_b}\frac{\partial X_j}{\partial x_c}T_{ab}\,.\end{aligned} \tag{2.35}$$

If the coordinates $\boldsymbol{X}$ are treated as a referential system and are considered to be instantaneously coincident with $\boldsymbol{x}$, it follows that $\partial X_i/\partial x_j = \delta_{ij}$ and $J = 1$ so we obtain another frame-invariant rate called the Truesdell rate [50]

$$T_{ij}^{\bar{\nabla}} = \dot{T}_{ij} + \frac{\partial \dot{u}_b}{\partial x_b}T_{ij} - \frac{\partial \dot{u}_i}{\partial x_a}T_{aj} - \frac{\partial \dot{u}_j}{\partial x_b}T_{ib}\,. \tag{2.36}$$

The difference between the Truesdell rate and the Jaumann rate can be shown through Eqs. (2.34) and (2.36) to be given by

$$T_{ij}^{\bar{\nabla}} - T_{ij}^{\nabla} = T_{ij}D_{bb} - D_{ia}T_{aj} - D_{jb}T_{bi}\,. \tag{2.37}$$

Thus while both the Truesdell rate and Jaumann rate are frame invariant, constitutive relations expressed in terms of the two rates will be of a different form.

Any frame invariant stress and strain measures may be used in the construction of a constitutive equation. The most popular rate form is to express the Jaumann rate in terms of the velocity strain

$$T_{ij}^{\nabla} = C_{ijkl}D_{kl}\,. \tag{2.38}$$

The above form can be used for nonlinear materials in that $\boldsymbol{C}$ can be a function of frame invariant measures of stress or strain, i.e. $\boldsymbol{T}$, $\boldsymbol{T}^{\nabla}$, $\boldsymbol{D}$ or other state variables. Note that it cannot be a function of the material time derivative of $\boldsymbol{D}$ because that quantity *is not* frame-indifferent [38].

A second popular class of material laws express the second Piola–Kirchhoff stress in terms of the Green strain, i.e.

$$S_{ij} = S(E_{ij})\,. \tag{2.39}$$

If the Truesdell rate is used instead of the Jaumann rate, the constitutive equation is of the form

$$T_{ij}^{\bar{\nabla}} = \bar{C}_{ijkl} D_{kl}\,. \tag{2.40}$$

Bars appear on both the $\boldsymbol{C}$ and the rate designation to indicate that this is a constitutive equation expressed in terms of the Truesdell rate.

The $\bar{\boldsymbol{C}}$ and $\boldsymbol{C}$ matrices can from Eqs. (2.37), (2.38) and (2.39) be seen to be related by

$$\bar{C}_{ijkl} = C_{ijkl} + T_{ij}\delta_{kl} - \tfrac{1}{2}(T_{ik}\delta_{jl} + T_{il}\delta_{jk} + T_{jk}\delta_{il} + T_{jl}\delta_{ik}) \tag{2.41}$$

where the minor symmetry in the k, l indices, which always holds because of the symmetry of $\boldsymbol{D}$, has been invoked. In elastic–plastic problems, $\boldsymbol{C}$ and $\boldsymbol{T}$ may be of the same order of magnitude, so the two material matrices may differ substantially and Eq. (2.41) should be used to transform between the two. Note that if $\boldsymbol{C}$ is assumed to have the major symmetry, $\bar{\boldsymbol{C}}$ will not have the major symmetry, and vice versa.

The corotational components of the Cauchy stress and the velocity strain are related by

$$\hat{\dot{T}}_{ij} = \hat{C}_{ijkl}\hat{D}_{kl}\,. \tag{2.42}$$

In structures, where certain components of stress are required to vanish (for example, the through-the-thickness stresses in a shell in plane stress), a corotational formulation provides conceptual simplifications and facilitates software implementation.

When a material is anisotropic, the $\boldsymbol{C}$ matrix changes with time as the material rotates. Thus

$$\begin{aligned} C_{ijkl}^{\nabla} = \dot{C}_{ijkl} &- W_{ia}C_{ajkl} - W_{jb}C_{ibkl} \\ &- W_{kc}C_{ijcl} - W_{ld}C_{ijkd}\,. \end{aligned} \tag{2.43}$$

Therefore, in addition to the change in $\boldsymbol{C}$ that results from the change in the constitutive response of the material and is reflected in $\dot{\boldsymbol{C}}$, it is necessary to account for the effect of the rigid body rotation on the components of $\boldsymbol{C}$ when it is expressed in a fixed coordinate system. This procedure is analogous to the Jaumann rate on the stress tensor. If the constitutive law is employed in a corotational framework, as in Eq. (2.42), this is not necessary.

The appropriate choice of kinematic and kinetic descriptions for computer software depends on the class of materials to be treated and to some extent, on the type of time integration to be used. If the original state of the material can be characterized by a few state variables and the response depends only on the current state of stress, a Cauchy stress/velocity-strain formulation seems most suitable. Examples of such materials are simple fluids, where the stress rate depends on $\boldsymbol{D}$ or isotropically-hardening, elastic–plastic materials, where the stress-rate depends on $\boldsymbol{T}$, $\boldsymbol{D}$ and the yield stress, a scalar state variable. The original shape of the body usually plays a minor role on the response; witness the effect of the original shape of the raw ingot on a piece of sheet metal. On the other hand, for rubber, the original shape is quite important.

2.7. *Equations of motion for Lagrangian mesh – Cauchy stress formulation*

In this formulation, the Cauchy stress is employed. Note that all dependent variables are functions of $\boldsymbol{X}$, however no derivatives with respect to $\boldsymbol{X}$ need to be taken.

The weak form of the momentum equation is identical to that for small deformation (linear) theory (compare (2.44) with Eq. (2.9) to (2.12))

$$\int_{\Omega} \frac{\partial v_i}{\partial x_j} T_{ij}\, \mathrm{d}\Omega - \int_{\Omega} \rho v_i \ddot{u}_i\, \mathrm{d}\Omega = \int_{\Omega} \rho v_i b_i\, \mathrm{d}\Omega + \int_{\Gamma_\tau} v_i \tau_i^*\, \mathrm{d}\Gamma . \tag{2.44}$$

Note that the first integrand is equivalent to $D_{ij}(\boldsymbol{v})T_{ij}$, the rate of work of the test function $\boldsymbol{v}$. The following form is obtained for the internal nodal forces

$$f_{iI}^{(e)\,\mathrm{int}} = \int_{\Omega_e} B_{jI} T_{ij}\, \mathrm{d}\Omega . \tag{2.45}$$

The equations for this formulation with a Jaumann stress rate are summarized in Table 9, along with the second Piola–Kirchhoff stress formulation for purposes of comparison. Equations for a uniform stress element are given in Table 10.

2.8. *Lagrangian mesh – $\boldsymbol{S}$ formulation*

The equations of motion expressed in terms of the material coordinates and second Piola–Kirchhoff stress in PDE form are

$$\frac{\partial}{\partial X_j}\left(\frac{\partial x_i}{\partial X_k} S_{jk}\right) + \rho_0 b_i = \rho_0 \ddot{u}_i \tag{2.46}$$

and the traction boundary conditions are

$$S_{ij} n_j^0 = \frac{\partial X_i}{\partial x_k} \tau_k^* \quad \text{on } \Gamma_\tau \tag{2.47}$$

where the superscript 0 on the normal vector $\boldsymbol{n}$ is used to designate the fact that this is the normal to the original surface of the body. Any alternative set of equations of motion can be deduced from the variational principle of Eq. (2.44) by simply replacing the first term by a conjugate strain rate–stress pair. Thus this

Table 9
Summary of Lagrangian mesh finite element equations

General equations Cauchy stress formulation	2nd P.K. stress ($\boldsymbol{S}$) formulation
$B_{iI} = \dfrac{\partial N_I}{\partial x_i}$	$\bar{B}_{iI} = \dfrac{\partial N_J}{\partial X_i}$
$D_{ij} = \frac{1}{2}(B_{iI}\dot{u}_{jI} + B_{jI}\dot{u}_{iI})$	$E_{ij} = \frac{1}{2}(\bar{B}_{jI}u_{iI} + \bar{B}_{iI}u_{jI} + \bar{B}_{iI}\bar{B}_{jJ}u_{kI}u_{kJ})$
$\boldsymbol{T}^{\nabla} = \boldsymbol{C}\boldsymbol{D}$	
$\dot{T}_{ij} = T^{\nabla}_{ij} + W_{ik}T_{kj} + W_{jk}T_{ki}$	$\boldsymbol{S} = \boldsymbol{S}(\boldsymbol{E})$
$f^{(e)\text{int}}_{iI} = \int_{\Omega^{(e)}} B_{jI}T_{ij}\, \mathrm{d}\Omega$	$f^{(e)\text{int}}_{iI} = \int_{\Omega^{(e)}_0} \bar{B}_{jI}x_{iJ}\bar{B}_{kJ}S_{kj}\, \mathrm{d}\Omega_0$

$\boldsymbol{f}^{\text{int}} = \underset{e}{\mathbf{A}}\, \boldsymbol{f}^{(e)\text{int}}$

$\boldsymbol{M}\ddot{\boldsymbol{u}} + \boldsymbol{f}^{\text{int}} = \boldsymbol{f}^{\text{ext}}$

equation of motion and traction boundary condition can be deduced from the variational form, Eq. (2.44) with the first term replaced by

$$\int_{\Omega_0} E_{ij}(\boldsymbol{u}, \boldsymbol{v})S_{ij}\, \mathrm{d}\Omega_0 = \int_{\Omega_0} \frac{\partial v_k}{\partial X_j}\frac{\partial x_k}{\partial X_i} S_{ij}\, \mathrm{d}\Omega_0\,. \tag{2.48}$$

The finite element form of the equation of motion can then be developed. For this formulation the shape functions are expressed in terms of the Lagrangian coordinates $\boldsymbol{X}$, so

Table 10
Specialization of Lagrangian mesh equations to uniform strain elements

Tensor notation procedure	(for $\boldsymbol{S}$ formulation, replace x, y in $\boldsymbol{b}_i$ by X, Y to obtain $\bar{B}_{iI}$)
$B_{iI} = \dfrac{1}{A}\begin{bmatrix} \boldsymbol{b}^{\mathrm{T}}_1 \\ \boldsymbol{b}^{\mathrm{T}}_2 \end{bmatrix} \quad \begin{matrix} i=1 \\ i=2 \end{matrix}$	
$\boldsymbol{b}_i$ given in Table 4	
$f^{(e)\text{int}}_{iI} = AB_{iI}T_{ij}$	$f^{(e)\text{int}}_{iI} = A\bar{B}_{jI}x_{iJ}\bar{B}_{kJ}S_{kj}$
Matrix procedure	
$\boldsymbol{B} = \dfrac{1}{A}\begin{bmatrix} \boldsymbol{b}^{\mathrm{T}}_1 & \boldsymbol{0} \\ \boldsymbol{0} & \boldsymbol{b}^{\mathrm{T}}_2 \\ \boldsymbol{b}^{\mathrm{T}}_2 & \boldsymbol{b}^{\mathrm{T}}_1 \end{bmatrix}$ $\boldsymbol{\sigma} = \begin{Bmatrix} T_x \\ T_y \\ T_{xy} \end{Bmatrix}$	
$\boldsymbol{f}^{(e)\text{int}} = A\boldsymbol{B}^{\mathrm{T}}\boldsymbol{\sigma}$	

$$u_i = N_I(\boldsymbol{X})u_{iI}, \tag{2.49a}$$

$$v_i = N_I(\boldsymbol{X})v_{iI}, \tag{2.49b}$$

and note that for isoparametric elements

$$x_i = N_I(\boldsymbol{X})x_{iI}. \tag{2.50}$$

It is now convenient to define a $\bar{\boldsymbol{B}}$ matrix that gives the derivatives with respect to the Lagrangian coordinates $\boldsymbol{X}$ by

$$\bar{B}_{iI} = \frac{\partial N_I}{\partial X_i}. \tag{2.51}$$

Substituting Eqs. (2.49–50) into (2.48), and using Eq. (2.51) and the arbitrariness of $v_{iI}^{(e)}$, we obtain

$$f_{iI}^{(e)\,\mathrm{int}} = \int_{\Omega_0^e} \bar{B}_{jI} x_{iJ} \bar{B}_{kJ} S_{kj} \, \mathrm{d}\Omega_0 . \tag{2.52}$$

As indicated by the subscripts on Ω, this integral must be evaluated over the domain of the undeformed element. It is quite obvious that the integral in Eq. (2.52) is far more involved than that in Eq. (2.45), but nevertheless is identical! This is employed with Lagrangian meshes. The equations of motion are identical to the previous formulation, (2.15).

2.9. Tangential stiffnesses for Lagrangian – T formulation

For implicit time integration, it is necessary to construct a predictor for the nodal forces. An idiosyncracy of continuum mechanics, as compared to nonlinear heat conduction, is that the relation between rates of nodal forces and nodal velocities depends on the state of stress. Therefore, it cannot be simply obtained by replacing the total stress–strain relation, such as Eq. (2.3c), by an incremental relation such as Eq. (2.3a).

The predictor is usually obtained by linearization and expressed as a tangential stiffness matrix which relates the rate of nodal forces to the nodal velocities

$$\dot{f}^{\mathrm{int}} = \mathring{\boldsymbol{K}}\dot{\boldsymbol{u}} \tag{2.53}$$

which can then be used to predict $\boldsymbol{f}_{\mathrm{int}}^{N+1}$. It has been shown that the Truesdell rate gives a symmetric stiffness matrix [31], so we shall first use that rate here. To obtain the matrix $\mathring{\boldsymbol{K}}$, we take the time derivatives of both sides of (2.45)

$$\dot{f}_{iI}^{(e)} = \int_{\Omega_e} (B_{jI}\dot{T}_{ij} + \dot{B}_{jI}T_{ij} + B_{jI}T_{ij}\dot{J}) \, \mathrm{d}\Omega \tag{2.54a}$$

$$= \int_{\Omega_e} (B_{jI}\dot{T}_{ij} - B_{lI}\dot{u}_{l,j}T_{ij} + B_{jI}T_{ij}\dot{u}_{l,l}) \, \mathrm{d}\Omega . \tag{2.54b}$$

Combining Eqs. (2.37) and (2.40) gives

$$\dot{T}_{ij} = \bar{C}_{ijkl} D_{kl} - T_{ij} D_{pp} + T_{ip} \dot{u}_{j,p} + T_{jp} \dot{u}_{i,p} \tag{2.55}$$

where commas denote derivatives with respect to spatial coordinates. Substituting the above into (2.54) and simplifying yields

$$\dot{\boldsymbol{f}}^{(e)} = (\mathring{\boldsymbol{K}}^{(e)\text{inc}} + \mathring{\boldsymbol{K}}^{(e)\text{str}}) \dot{\boldsymbol{u}}^{(e)} \tag{2.56}$$

where

$$\mathring{\boldsymbol{K}}^{(e)\text{inc}}_{ikIK} = \int_{\Omega_e} B_{jI} \bar{C}_{ijkl} B_{lK} \, \mathrm{d}\Omega\,, \tag{2.57}$$

$$\mathring{\boldsymbol{K}}^{(e)\text{str}}_{ikIK} = \int_{\Omega_e} B_{jI} T_{jl} B_{lK} \delta_{ik} \, \mathrm{d}\Omega\,. \tag{2.58}$$

The first is known as the incremental stiffness matrix, the second as the initial stress matrix. For computational convenience, the two can be combined in the form

$$\mathring{K}_{ikIK} = \int_{\Omega_{(e)}} B_{jI} D_{ijkl} B_{lK} \, \mathrm{d}\Omega \tag{2.59}$$

with

$$D_{ijkl} = \bar{C}_{ijkl} + T_{jl} \delta_{ik}\,. \tag{2.60}$$

This is an elegant form first given in [28]; it is easily implemented in computer software and its structure is almost identical to that of a linear program. Note that the major symmetry properties of $\boldsymbol{D}$ are identical to that of $\bar{\boldsymbol{C}}$, so if $\bar{\boldsymbol{C}}$ has the symmetries of (2.3b), $\mathring{\boldsymbol{K}}$ is symmetric.

The tangential stiffness matrix for the Jaumann rate can be obtained by the same procedure. Using Eqs. (2.41) and (2.45), we obtain that the stiffness for the Jaumann rate is given by (2.59) with

$$D_{ijkl} = C_{ijkl} + T_{ij} \delta_{kl} + \tfrac{1}{2}(T_{jl} \delta_{ik} - T_{jk} \delta_{il} - T_{ik} \delta_{jl} - T_{il} \delta_{jk})\,. \tag{2.61}$$

Note that this tangential stiffness matrix involves the hypoelastic terms $\boldsymbol{C}$ rather than $\bar{\boldsymbol{C}}$ and that it lacks the symmetries in the ij, kl indices, i.e.

$$D_{ijkl} \neq D_{klij}\,. \tag{2.62}$$

As a consequence, if $\boldsymbol{C}$ possesses major symmetry, the tangential stiffness matrix associated with the Jaumann rate is not symmetric. However, if $\bar{\boldsymbol{C}}$ possesses major symmetry and $\boldsymbol{C}$ is expressed in terms of $\bar{\boldsymbol{C}}$ by Eq. (2.37), then $\boldsymbol{D}$ recovers the major symmetries. In fact $\boldsymbol{D}$ becomes identical to Eq. (2.60). Thus the use of a Truesdell rate appears to be more natural than a Jaumann rate in finite element implementations of nonlinear mechanics.

2.10. Tangential stiffness for Lagrangian – $\boldsymbol{S}$ formulation

Formulations which employ the Green strain tensor and its conjugate stress, the second Piola–Kirchhoff stress, are often called Lagrangian because these stress and strain measures are Lagrangian in character. Thus when the Green strain and second Piola–Kirchhoff stress are used with a Lagrangian mesh, the formulation is often called total Lagrangian to emphasize this fact. It is stressed that the other formulations given here have also been employed with a Lagrangian mesh.

The tangential stiffness matrices have been derived in [3], [45], [25] and [51]. We will give a slightly more straightforward derivation based on rates. Taking the rates of both sides of Eq. (2.52), we obtain

$$\dot{\boldsymbol{f}}_{iI}^{(e)} = \int_{\Omega_e} \bar{B}_{jI}\bar{B}_{kJ}(x_{ij}\dot{S}_{kj} + \dot{x}_{iJ}S_{kj})\, \mathrm{d}\Omega_0 . \tag{2.63}$$

Note that the $\dot{\boldsymbol{B}}$ and $\dot{J}$ terms which occur in (2.54a) are absent here because the shape functions are expressed in terms of $\boldsymbol{X}$ and the integral is over the original volume. We then use a linearized form of the constitutive equation (2.39),

$$\dot{S}_{ij} = \tilde{C}_{ijkl}\dot{E}_{kl} = \tilde{C}_{ijkl}\frac{\partial x_p}{\partial X_k}\frac{\partial \dot{u}_p}{\partial X_l} . \tag{2.64}$$

We note that since

$$\frac{\partial x_i}{\partial X_k} = \delta_{ik} + \frac{\partial u_i}{\partial X_k} , \tag{2.65}$$

we have

$$x_{iI}\bar{B}_{kI} = \delta_{ik} + u_{iI}\bar{B}_{kI} . \tag{2.66}$$

Substituting Eqs. (2.64)–(2.66) into (2.63) yields

$$\dot{\boldsymbol{f}}^{(e)} = (\mathring{\boldsymbol{K}}^{(e)\text{inc}} + \mathring{\boldsymbol{K}}^{(e)\text{str}} + \mathring{\boldsymbol{K}}^{(e)\text{rot}})\dot{\boldsymbol{u}}^{(e)} \tag{2.67}$$

where

$$\mathring{K}_{ijIJ}^{(e)\text{inc}} = \int_{\Omega_0^{(e)}} \bar{B}_{kI}\tilde{C}_{ikjl}\bar{B}_{lJ}\, \mathrm{d}\Omega_0 , \tag{2.68}$$

$$\mathring{K}_{ijIJ}^{(e)\text{str}} = \int_{\Omega_0^{(e)}} \bar{B}_{kI}S_{kl}\bar{B}_{lJ}\delta_{ij}\, \mathrm{d}\Omega_0 , \tag{2.69}$$

$$\mathring{K}_{ijIJ}^{(e)\text{rot}} = \int_{\Omega_0^{(e)}} \bar{B}_{kI}\left(\frac{\partial u_i}{\partial X_l}\tilde{C}_{kljs}\bar{B}_{sJ} + \frac{\partial u_j}{\partial X_r}\tilde{C}_{kjls}\bar{B}_{sJ}\frac{\partial x_i}{\partial X_l}\right) \mathrm{d}\Omega_0 . \tag{2.70}$$

As indicated by the superscript notation, these three ingredients constitute the incremental stiffness matrix, the initial stress matrix, and the initial rotation matrix. The incremental stiffness matrix gives the change in nodal forces due to the incremental deformation of the material, while the initial stress matrix gives the change in nodal forces due to the rotation of initially stressed element. The initial rotation matrix gives the effect on the change in nodal forces which arise from the computation of $\boldsymbol{K}^{(e)\mathrm{inc}}$ and $\boldsymbol{K}^{(e)\mathrm{rot}}$ with respect to the undeformed domain of the solid. A fourth tangential stiffness which accounts for the change of $\boldsymbol{f}^{\mathrm{ext}}$ due to the rotation or change in area of any surface loaded by a pressure or follower force is also useful, see [45] for this matrix.

2.11. Updated Lagrangian formulation

A widely used formulation is the updated Lagrangian formulation of [59] and [4]. In this formulation, an approach similar to the total Lagrangian formulation is used but the Lagrangian coordinates are considered as referential coordinates which are instantaneously coincident with $\boldsymbol{x}$. This is analogous to the procedure employed to derive the Truesdell rate from the rate of the second Piola–Kirchhoff stress tensor in Eqs. (2.35) and (2.36). This leads to the following simplifications:

$$\frac{\partial x_i}{\partial X_j} = \frac{\partial X_j}{\partial x_i} = \delta_{ij}, \qquad \mathrm{d}\Omega = \mathrm{d}\Omega_0, \qquad \bar{B}_{iI} = B_{iI}, \tag{2.71}$$

so $\boldsymbol{T} = \boldsymbol{S}$ and the expression for the internal forces, Eq. (2.52), becomes identical to (2.45).

Similarly, since $\boldsymbol{u} = \boldsymbol{x} - \boldsymbol{X} = 0$ in that case, so all derivatives of $\boldsymbol{u}$ vanish and the initial rotation matrix vanishes; the incremental stiffness matrix, Eq. (2.68), and the initial stress matrix, Eq. (2.69), reduce to Eqs. (2.57) and (2.58), respectively. Hence the updated Lagrangian formulation is identical to the Lagrangian mesh Cauchy stress formulation. The tangential stiffness matrix is symmetric and tacitly implies a constitutive law of the form of Eq. (2.40), namely, a relation between the Truesdell rate and the velocity strain. This is of course immediately apparent from the fact that when the Lagrangian and Eulerian coordinates instantaneously coincide, the rate of $\boldsymbol{S}$ coincides with the Truesdell rate, cf. Eqs. (2.35) and (2.36). Hence, even for small-strain, large-rotation problems, the $\bar{\boldsymbol{C}}$ matrix should be used in an updated Lagrangian formulation.

2.12. Corotational formulation

In the semidiscretization of structures, expressions for the nodal forces in terms of the global components of stresses are awkward because stresses normal to the plane of the structure are required to vanish. Since Eq. (2.44) holds in any coordinate system it is easy to show that

$$f_{kI}^{(e)} = \int_{\Omega_e} R_{ki}\hat{B}_{jI}\hat{T}_{ij}\,\mathrm{d}\Omega = R_{ki}\int_{\Omega_e} \hat{B}_{jI}\hat{T}_{ij}\,\mathrm{d}\Omega \tag{2.72a}$$

where

$$\hat{B}_{kI} = \frac{\partial N_I}{\partial \hat{x}_k}\,. \tag{2.72b}$$

The third term in Eq. (2.72) represents an approximation which can be made whenever the rigid body rotation is constant or almost constant in an element.

Alternative corotational formulations can be developed in terms of the corotational components of the stretch tensor; see [1], [2] and [9]. These have the advantage that the measure of deformation is defined in a total sense, in contrast to $\boldsymbol{D}$, which is not integrable, yet yield simple nodal force relations of a type similar to (2.72).

An important property of these formulations is that all of the tangential stiffnesses are mechanically identical and differences which arise are due to differences in the mechanical properties which are reflected in Eqs. (2.38), (2.40) and (2.64). By appropriate transformations, such as Eq. (2.41), the same tangential stiffness can be obtained.

2.13. *Fluid-structure problems*

Many engineering problems are concerned with structures which enclose a fluid or are embedded within a fluid, where, in many cases, the load originates in the fluid. The presence of the fluid can significantly alter the behavior of the structure, and at the same time, the deformation of the structure changes any loads transmitted from the fluid. These phenomena are collectively called fluid-structure interaction.

Two classes of problems are of wide interest in engineering:

(i) small-displacement response of inviscid fluids,

(ii) large-displacement response of fluids.

Most methods developed to date have dealt with problems where the Mach number is small and viscosity can be neglected; the latter assumption is usually valid because the effects of viscous stresses on structures are small and because whenever fluid loads are large enough to affect the structural response, the hydrostatic stresses must be dominant.

Both of the above classes of problems can be treated by velocity formulations, where the velocity (or displacement) of the fluid is an independent variable. For the velocity formulation, the semidiscretization for a fluid-structure model which employs a Lagrangian mesh is identical to that described in the preceding sections. Since $\boldsymbol{T}$–$\boldsymbol{D}$ constitutive relations are usually used in fluids, equations of motion of the form in Table 10 are preferred.

In the modeling of fluids which undergo even moderately large motions, mesh distortion becomes quite troublesome. Therefore the use of Eulerian or arbitrary Eulerian–Lagrangian meshes for fluids is recommended in such circumstances. In this class of problems, structures are almost always treated by Lagrangian meshes, so special interface procedures must be developed when an Eulerian mesh is used for the fluid; see for example [57]. A coupled ALE fluid-structure method is described in Chapter 10. The forms given in Chapter 10 may readily be simplified to those for an Eulerian mesh. The significant difference in the equation of motion is the appearance of a transport term, which is always unsymmetric. Furthermore, for any formulation which does not employ a Lagrangian mesh, the mass conservation equation must be enforced.

When the displacements of the fluid are small, a pressure formulation is generally advantageous because it involves only a single dependent variable. If the equation of state for the fluid is linear, this governing equation is the acoustic equation

$$p_{,ii} = \frac{1}{c^2}\ddot{p} \quad \text{in } \Omega_F, \tag{2.73a}$$

where p is the pressure (positive in compression) and c the wave speed, given by

$$c^2 = \frac{\beta}{\rho} \tag{2.73b}$$

where β is the bulk modulus. The boundary conditions are

$$p = p^* \quad \text{on } \Gamma_p, \tag{2.74a}$$

$$p_{,n} = p^*_{,n} \quad \text{on } \Gamma_{pn}, \tag{2.74b}$$

$$p_{,n} = p_{,i}n_i^{\mathrm{F}} = -\rho\ddot{u}_n^{\mathrm{F}} \quad \text{and} \quad T_{ij}n_j^{\mathrm{S}} = \rho n_i^{\mathrm{F}} \quad \text{on } \Gamma_I. \tag{2.74c}$$

Here n_i^{F} is the outward normal to the fluid domain and asterisks designate variables which are prescribed along the boundary. The surface Γ_I is the interface between the fluid and structure and the continuity of normal displacements between the fluid and structure is obtained by combining (2.74c) with the condition $\ddot{u}_i^{\mathrm{F}} = \ddot{u}_i^{\mathrm{S}}$

$$p_{,n} = -\rho\ddot{u}_i^{\mathrm{F}}n_i^{\mathrm{F}} = \rho\ddot{u}_i^{\mathrm{S}}n_i^{\mathrm{S}}. \tag{2.74d}$$

The variational or weak form of the combined field equations for the fluid and structure is

$$\int_{\Omega_{\mathrm{S}}} \left(\delta u_i\rho\ddot{u}_i + \frac{\partial\delta u_i}{\partial x_j}T_{ij} - \delta u_i b_i\right)\mathrm{d}\Omega - \int_{\Gamma_\tau} \delta u_i\tau_i^*\,\mathrm{d}\Gamma + \int_{\Gamma_I} (\delta p\rho\beta\ddot{u}_i^{\mathrm{F}}n_i^{\mathrm{F}} - \delta u_i p n_i^{\mathrm{F}})\,\mathrm{d}\Gamma$$

$$+ \int_{\Omega_{\mathrm{F}}} \left(\delta p\rho\ddot{p} + \frac{\partial\delta p}{\partial x_i}\beta p_{,i}\right)\mathrm{d}\Omega - \int_{\Gamma_{pn}} \delta p\beta p^*_{,n}\,\mathrm{d}\Gamma = 0. \tag{2.75}$$

Here Ω_{S} and Ω_{F} are the structural and fluid domains respectively, and the prefix δ designates the test function (or variation).

If we let

$$u_i = u_{iI}N_I^u(\boldsymbol{x}), \tag{2.76a}$$

$$p = p_I N_I^p(\boldsymbol{x}), \tag{2.76b}$$

we obtain the following semidiscretized equations

$$M_S\ddot{u} + f_S^{int} + R^T p = f_S^{ext}, \qquad (2.77a)$$

$$M_F\ddot{p} + K_F p - \beta\rho R\ddot{u} = f_F^{ext}, \qquad (2.77b)$$

where

$$R_{iIJ}^{(e)} = \int_{\Gamma_I^{(e)}} N_I^p N_J^u n_i^S \,d\Gamma \qquad (2.77c)$$

and K_S, M_S and f_S are given in Table 6; M_F and K_F are given by Eqs. (1.15), (1.28), respectively. We have written the equations for a nonlinear structure interacting with an acoustic fluid to emphasize that material and geometric nonlinearities can occur in a structure even when the acoustic approximation for the fluid is valid.

The linear form of Eqs. (2.77) can be written in the form given in [61]

$$\begin{bmatrix} M_S & 0 \\ -\rho k R & M_F \end{bmatrix} \begin{Bmatrix} \ddot{u} \\ \ddot{p} \end{Bmatrix} + \begin{bmatrix} C_S & 0 \\ 0 & C_F \end{bmatrix} \begin{Bmatrix} \dot{u} \\ \dot{p} \end{Bmatrix}$$

$$+ \begin{bmatrix} K_S & R^T \\ 0 & K_F \end{bmatrix} \begin{Bmatrix} u \\ p \end{Bmatrix} = \begin{Bmatrix} f_S^{ext} \\ f_F^{ext} \end{Bmatrix}. \qquad (2.78)$$

Here damping has been added to the fluid and structural subdomains. Similar linearized incremental equations can be written for the interaction of a nonlinear structure with an acoustic fluid.

An important feature of the above equations is their lack of symmetry. This poses some difficulties in extracting eigenvalues and in implicit time integration of the systems. This is one of the motivations for the development of staggered and partitioned time integration procedures which are described in Chapter 3.

Symmetric forms of the coupled structural acoustic equations have been developed in [41] and [18]. In the former, the problem was symmetrized by adding the potential function for the fluid displacements; this matrix can then be eliminated to yield a symmetric system. In the latter, symmetrization is accomplished by replacing p in Eq. (2.78) by a vector q given by $p = \dot{q}$. This vector is, except for a multiplicative constant, the velocity potential.

If the acoustic frequencies are all greater than the maximum structural frequency of interest, then the fluid can be considered incompressible. This corresponds to letting the bulk modulus β be infinite in Eq. (2.73b), so the governing equation (2.73a), becomes the Laplacian. If the second equation in (2.78) is divided by β, the M_F matrix then also vanishes since it contains the term ρ/β (see Eq. (1.15)).

Hence, the second of Eq. (2.78) becomes

$$-\rho\beta R\ddot{u} + K_F p = f_F^{ext} \qquad (2.79)$$

which can be combined with the first of Eqs. (2.78) to obtain

$$(\boldsymbol{M}_S + \underbrace{\rho \boldsymbol{R}^T \boldsymbol{K}_F^{-1} \boldsymbol{R}}_{\text{added mass}})\ddot{\boldsymbol{u}}_S + \boldsymbol{C}_S\dot{\boldsymbol{u}}_S + \boldsymbol{K}_S\boldsymbol{u}_S = \boldsymbol{f}_S^{ext} - \boldsymbol{R}^T\boldsymbol{K}_F^{-1}\boldsymbol{f}_F^{ext} . \tag{2.80}$$

As can be seen from the above, the fluid structure system can then be described completely in terms of only structural nodal displacements. The terms which are designated 'added mass' reflect the total effect of the incompressible fluid on the response of the structure. This is often also called the 'hydrodynamic' mass. It is interesting to observe that the assumptions of incompressible, small deformation behavior (the latter assumption was made for the acoustic approximation) of the fluid eliminates the need for a coupled solution. The presence of the fluid is entirely reflected by the added mass. While at first glance this may appear to offer tremendous savings, this is not altogether true. Whereas the matrices $\boldsymbol{M}_S$, $\boldsymbol{C}_S$ and $\boldsymbol{K}_S$ are usually sparse and banded, the added mass matrix is full over all wetted nodes of the structure.

This added mass matrix is used to model the low end of the response in the doubly-asymptotic expansion method in Chapter 4. The added mass is obtained by the boundary element method, which is necessary when the fluid domain is infinite. The boundary element method may also be used to obtain the added mass for enclosed fluids; it will have the same lack of sparsity.

3. Time integration

3.1. Explicit Euler time integration

The simplest integration formula for this first order system is the Euler integration formula. If let $\boldsymbol{\theta}(N\Delta t) = \boldsymbol{\theta}^N$, then the forward Euler formula gives

$$\boldsymbol{\theta}^{M+1} = \boldsymbol{\theta}^M + \Delta t \dot{\boldsymbol{\theta}}^M \tag{3.1}$$

and combining with Eq. (1.14), we obtain

$$\boldsymbol{\theta}^{M+1} = \Delta t \boldsymbol{M}^{-1}(\boldsymbol{s}^M - \boldsymbol{f}^M) + \boldsymbol{\theta}^M . \tag{3.2}$$

The algorithm for this procedure is shown in Table 11. As can be seen, the complexity of the algorithm is independent of whether the system is linear or nonlinear. If $\boldsymbol{M}$ is diagonal (lumped), the procedure involves no solution of any equations, but simply the evaluation of the discrete approximation to the gradient of the temperature field $\boldsymbol{g}$, the computation of the flux $\boldsymbol{q}$, and $\boldsymbol{f}$ in each element. The importance of using a diagonal $\boldsymbol{M}$ matrix in this procedure can easily be appreciated.

In return for this simplicity, the method is only *conditionally stable*, and as shown in Chapter 2, numerical stability limits the time step in linear systems to

$$\Delta t \leqslant \frac{2}{\lambda_{max}} = \Delta t_{stab} \tag{3.3}$$

Table 11
Euler explicit integration of diffusion equation

1. Set initial conditions: $\boldsymbol{\theta}^0 = \theta(0)$; $N = 0$
2. Compute $\boldsymbol{f}^N$ by looping over all elements for each element e
   (a) compute gradient $\boldsymbol{g}^N = \boldsymbol{B\theta}^{(e)N}$; Eq. (1.22)
   (b) compute flux $\boldsymbol{q}^N$ by Eq. (1.4)
   (c) compute $\boldsymbol{f}^{(e)N} = -\int_{\Omega^{(e)}} \boldsymbol{B}^{\mathrm{T}}\boldsymbol{q}^N \, \mathrm{d}\Omega$, Eq. (1.26)
   (d) assemble $\boldsymbol{f}^{(e)N}$ into $\boldsymbol{f}^N$
3. Compute nodal sources $\boldsymbol{s}^N$
4. $\boldsymbol{\theta}^{N+1} = \Delta t\boldsymbol{M}^{-1}(\boldsymbol{s}^N - \boldsymbol{f}^N) + \boldsymbol{\theta}^N$
5. $N \leftarrow N + 1$, go to step 2

where $\lambda_{\max}$ is the maximum eigenvalue of the system

$$\boldsymbol{Kd} = \lambda\boldsymbol{Md}\,. \tag{3.4}$$

The maximum stable time step, as given by Eq. (3.3) will be called Δt_{stab}.

3.2. *Central-difference explicit integration*

The most popular explicit method for the second-order ordinary differential equations which result from a semidiscretization of the momentum equation is based on the central difference formulas. If the response of the continuum is strongly nonlinear, the central difference method should be used with a varying time increment, so we will give the equations in that form. Let Δt_N be the time increment between t_{N+1} and t_N, and let $\boldsymbol{u}^N = \boldsymbol{u}(t_N)$. In addition midstep values for velocities are defined by

$$t_{N+1/2} = \tfrac{1}{2}(t_N + t_{N+1}) \tag{3.5a}$$

and

$$\dot{\boldsymbol{u}}^{N+1/2} = \dot{\boldsymbol{u}}(t_{N+1/2}) \tag{3.5b}$$

The central difference formula for the velocity is then

$$\dot{\boldsymbol{u}}^{N+1/2} = \frac{1}{\Delta t_N}(\boldsymbol{u}^{N+1} - \boldsymbol{u}^N) \tag{3.6}$$

and the acceleration is given by

$$\ddot{\boldsymbol{u}}^N = \frac{1}{\Delta t_{N+1/2}}(\dot{\boldsymbol{u}}^{N+1/2} - \dot{\boldsymbol{u}}^{N-1/2}) \tag{3.7}$$

where

$$\Delta t_{N+1/2} = \tfrac{1}{2}(\Delta t_N + \Delta t_{N+1})\,. \tag{3.8}$$

Combining Eqs. (3.6) and (3.7), the acceleration can be written

$$\ddot{\boldsymbol{u}}^N = \frac{1}{\Delta t_{N-1}\Delta t_{N-1/2}\Delta t_N}[\Delta t_{N-1}(\boldsymbol{u}^{N+1} - \boldsymbol{u}^N) - \Delta t_N(\boldsymbol{u}^N - \boldsymbol{u}^{N-1})]\,. \tag{3.9}$$

If the time step is constant, the above reduces to the well known expression

$$\ddot{\boldsymbol{u}}^N = \frac{1}{\Delta t^2}(\boldsymbol{u}^{N+1} - 2\boldsymbol{u}^N + \boldsymbol{u}^{N-1})\,. \tag{3.10}$$

If we consider an undamped system governed by Eq. (2.15), then Eq. (3.10) yields

$$\boldsymbol{u}^{N+1} = \Delta t^2 \boldsymbol{M}^{-1}\boldsymbol{f}^N + 2\boldsymbol{u}^N - \boldsymbol{u}^{N-1}\,. \tag{3.11}$$

Since the external and internal forces which determine $\boldsymbol{f}^N$ depend only on the stresses and loads at time step N, all of the terms on the right hand side are historical in that they are known at time step N. Therefore the new displacements can be determined directly without solving any equations provided that $\boldsymbol{M}$ is diagonal. The use of a diagonal mass matrix is particularly critical in structural problems when different shape functions may be used for different components of the displacements. For example, the consistent mass for the beam element in Table 14 changes in the global frame as the element rotates.

A flowchart for an explicit algorithm for the nonlinear analysis of continua is given in Table 12. This algorithm is based on a Lagrangian mesh – Cauchy stress formulation.

In step 3(b) of Table 12, the constitutive law is evaluated using the state at N; this is indicated by the functional dependence of $\boldsymbol{C}$ on $\boldsymbol{T}^N$ and $\boldsymbol{E}^N$. Furthermore, a forward Euler integration is partially used in step 3(b)–(c). Note that the time lag and

Table 12
Flowchart for explicit integration of equations of motion

1. Set initial conditions: $\boldsymbol{u}^0 = \boldsymbol{u}(0)$, $\dot{\boldsymbol{u}}^{1/2} = \dot{\boldsymbol{u}}(0)$, $\boldsymbol{T}^0 = \boldsymbol{T}(0)$; $N = 0$
2. Update displacements: $\boldsymbol{u}^{N+1} = \boldsymbol{u}^N + \Delta t_N \dot{\boldsymbol{u}}^{N+1/2}$
3. Compute internal force matrix by looping over all elements for each element
   (a) $V_{ij} = B_{jI}\dot{u}_{iI}^{(e)N+1/2}$, $D_{ij}^{N+1/2} = \frac{1}{2}(V_{ij} + V_{ji})$, $W_{ij}^{N+1/2} = \frac{1}{2}(V_{ij} - V_{ji})$
   (b) constitutive law: $\boldsymbol{T}^{\nabla N+1/2} = \boldsymbol{C}(\boldsymbol{T}^N, \boldsymbol{E}^N, \boldsymbol{D}^{N+1/2})\boldsymbol{D}^{N+1/2}$
   (c) $\dot{\boldsymbol{T}}^{N+1/2} = \boldsymbol{T}^{\nabla N+1/2} + \boldsymbol{W}^{N+1/2}\boldsymbol{T}^N + \boldsymbol{T}^N(\boldsymbol{W}^T)^{N+1/2}$
   (d) update stress $\boldsymbol{T}^{N+1} = \boldsymbol{T}^N + \Delta t_N \dot{\boldsymbol{T}}^{N+1/2}$
   (e) compute $\omega_{\max}^{(e)}$ for setting next time step, Δt^{N+1}
   (f) $\boldsymbol{f}_{(e)\text{int}}^{N+1} = \int \boldsymbol{B}^{\mathrm{T}}\boldsymbol{T}\,\mathrm{d}\Omega$
   (g) assemble $\boldsymbol{f}_{(e)\text{int}}^{N+1}$ into $\boldsymbol{f}_{\text{int}}^{N+1}$
4. Compute external nodal forces $\boldsymbol{f}_{\text{ext}}^{N+1}$
5. Update accelerations: $\ddot{\boldsymbol{u}}^{N+1} = \boldsymbol{M}^{-1}(\boldsymbol{f}_{\text{ext}}^{N+1} - \boldsymbol{f}_{\text{int}}^{N+1})$
6. $N \leftarrow N + 1$
7. Update velocities $\dot{\boldsymbol{u}}^{N+1/2} = \dot{\boldsymbol{u}}^{N-1/2} + \Delta t_{N+1/2}\ddot{\boldsymbol{u}}^N$ impose essential boundaries conditions on velocities
8. Go to step 2

accuracy of the integration can be improved by subcycling, i.e. using a time step $\Delta t/m$ to perform steps 3(b) to 3(d), since $\boldsymbol{E}^{N+1}$ can be determined at step 3(b). However, if the stress depends directly on $\boldsymbol{D}$ (linear damping), the value of $\boldsymbol{D}$ lags by a half time step, which cannot be ameliorated by subcycling since $\boldsymbol{D}$ can only be extrapolated to $N+1$.

The central difference procedure also includes an asynchronization on the initial conditions on the velocity (step 1). The procedure given here is one of several ways of handling this problem. In this procedure, step 3 is unnecessary for $N=0$ if initial velocities are all zero.

The stable time step for the central difference method is given by

$$\Delta t \leqslant \min \frac{2}{\omega_n}\left(\sqrt{1+\xi_n^2}-\xi_n\right)=\Delta t_{\text{stab}} \tag{3.12}$$

where ω_n are the natural frequencies of the mesh and ξ_n is the fraction of critical damping in the nth mode. This corresponds to Eq. (22), p. 128 with $\beta=0$, $\gamma=\frac{1}{2}$. Since structural systems are usually quite underdamped, the minimum of (3.12) is usually given by the maximum frequency, so:

$$\Delta t \leqslant \frac{2}{\omega_{\max}}\left(\sqrt{1+\xi^2}-\xi\right) \tag{3.13}$$

where ξ is the fraction of critical damping in the maximum frequency. The maximum frequency is obtained from Eq. (3.4) with

$$\omega=\sqrt{\lambda}\,. \tag{3.14}$$

The decrease in the stability limit caused by damping is due to its explicit treatment; note from Table 12 that the damping forces are based on velocities which lag by half a time step. The term in the parenthesis of Eq. (3.13) is absent when the damping terms are treated implicitly.

3.3. *Eigenvalue bounds for semidiscretizations*

It is generally quite inconvenient to determine a stable time step by actually finding the maximum eigenvalue of Eq. (3.4). This is particularly true for nonlinear problems, where the eigenvalues change as the solution evolves. In finite element methods, the maximum eigenvalue of the system may be bounded with sufficient accuracy by the maximum eigenvalue of any element, that is

$$\lambda_{\max} \leqslant \max_{\text{for all } e} \lambda_{\max}^{(e)} \tag{3.15}$$

where $\lambda_{\max}^{(e)}$ is the eigenvalue of

$$\boldsymbol{K}^{(e)}\boldsymbol{d}^{(e)}=\lambda_{\max}^{(e)}\boldsymbol{M}^{(e)}\boldsymbol{d}^{(e)}\,. \tag{3.16}$$

Proofs for this inequality have been given in [22], [30] and [21]. We will sketch the proof as given in [21]. The proof begins with the expression of the Rayleigh quotient for the maximum eigenvalue, which is associated with the eigenvector $\boldsymbol{d}_m$

$$\lambda_{\max} = \frac{\boldsymbol{d}_m^{\mathrm{T}} \boldsymbol{K} \boldsymbol{d}_m}{\boldsymbol{d}_m^{\mathrm{T}} \boldsymbol{M} \boldsymbol{d}_m} . \tag{3.17}$$

We note that the nodal displacements of any element which correspond to the eigenvector $\boldsymbol{d}_m$ can always be expressed in terms of the element eigenvectors $\boldsymbol{d}_I^{(e)}$ which satisfy Eq. (3.16) in the form

$$\boldsymbol{L}^{(e)} \boldsymbol{d}_m = \sum_{I=1}^{\mathrm{NDOF}} c_{eI} \boldsymbol{d}_I^{(e)} \tag{3.18}$$

where NDOF is the element's number of degrees of freedom. If we expand $\boldsymbol{K}$ and $\boldsymbol{M}$ in terms of their element contributions, Eqs. (1.19), Eq. (3.18) becomes

$$\lambda_{\max} = \frac{\sum_e (\boldsymbol{L}^{(e)} \boldsymbol{d}_m)^{\mathrm{T}} \boldsymbol{K}^{(e)} (\boldsymbol{L}^{(e)} \boldsymbol{d}_m)}{\sum_e (\boldsymbol{L}^{(e)} \boldsymbol{d}_m)^{\mathrm{T}} \boldsymbol{M}^{(e)} (\boldsymbol{L}^{(e)} \boldsymbol{d}_m)} \tag{3.19}$$

and using Eq. (3.18), we can write the above as

$$\omega_{\max} = \frac{\sum_e \sum_I \lambda_I^{(e)} c_{eI}^2 (\boldsymbol{d}_I^{(e)})^{\mathrm{T}} \boldsymbol{K}^{(e)} \boldsymbol{d}_I^{(e)}}{\sum_e \sum_I c_{eI}^2 (\boldsymbol{d}_I^{(e)})^{\mathrm{T}} \boldsymbol{M}^{(e)} \boldsymbol{d}_I^{(e)}} . \tag{3.20}$$

The eigenvectors of both the element and assembled matrices are orthogonal and the eigenvectors $\lambda_I^e \geqslant 0$ (see Chapter 2, Section B.I.4). Arranging the element eigenvalues in ascending order $0 \leqslant \lambda_1 \leqslant \cdots \leqslant \lambda_m$, the orthogonality enables the above to be simplified to

$$\lambda_{\max} = \frac{\sum_e \sum_I c_{eI}^2 \lambda_I^e}{\sum_e \sum_I c_{eI}^2} = \lambda_m \frac{1 + (\lambda_{m-1}/\lambda_m)(c_{m-1}/c_m)^2 + \cdots}{1 + (c_{m-1}/c_m)^2 + \cdots} . \tag{3.21}$$

From the above it can immediately be concluded that

$$\lambda_{\max} \leqslant \lambda_m \max_{\text{for all } e} (\lambda_{\max}^{(e)}) . \tag{3.22}$$

The above inequality is in fact a form of Rayleigh's bounding theorem which relates the eigenvalues of any two systems which are identical except for linear constraints. For example, let the eigenvalues of a matrix $\boldsymbol{A}$ of order N be given by $\lambda_1 \leqslant \lambda_2 \leqslant \cdots \leqslant \lambda_N$, so that

$$\boldsymbol{A} \boldsymbol{d}_I = \lambda_I \boldsymbol{d}_I . \tag{3.23}$$

If the vector $\boldsymbol{d}$ is required to satisfy a linear constraint

$$\boldsymbol{h}^{\mathrm{T}} \boldsymbol{d}_I = \boldsymbol{c} , \tag{3.24}$$

then the $N-1$ eigenvalues of the eigenvalues problem with 1 constraint, $\tilde{\lambda}^{[1]}$,

$$\boldsymbol{A}\boldsymbol{d} = \tilde{\lambda}^{[1]}\boldsymbol{d}\,, \tag{3.25a}$$

$$\boldsymbol{h}^{\mathrm{T}}\boldsymbol{d} = \boldsymbol{c}\,, \tag{3.25b}$$

are nested between the eigenvalues of the original system:

$$\lambda_1 \leqslant \lambda_1^{[1]} \leqslant \lambda_2 \leqslant \cdots \leqslant \lambda_{N-1}^{[1]} \leqslant \lambda_N\,. \tag{3.26}$$

If M constraints are imposed, these patterns of inequalities are repeated M times, and it follows that the eigenvalue of the system with M constraints, $\tilde{\lambda}_N^{[M]}$ will satisfy

$$\lambda_{\min} \leqslant \lambda_{\min}^{[M]} \leqslant \lambda_{\max}^{[M]} \leqslant \lambda_{\max}\,. \tag{3.27}$$

Consider a set of unconnected elements; the assembly operation consists of imposing the linear constraints that the nodal values $\boldsymbol{d}^{(e)}$ of all elements connected to a common node by equal. Thus, the inequality (3.22) is simply a corollary of Rayleigh's theorem. However, this finite element form of Rayleigh's theorem is important. It applies to *any linear finite element* model governed by *symmetric and positive semi-definite matrices*, and is applicable to a variety of matrices found in finite element numerics, such as iteration matrices in iterative methods. Furthermore, if a nonlinear system is studied or treated by local linearization, it can be used to find the eigenvalues of the linearized system. It can also be applied to finite difference equations, as shown later.

The stable time step can be determined within the time integration loop. This is indicated in Table 12, where the maximum frequency $\omega_{\max}^{(e)}$ is computed within the element loop. Since the element frequency is an upper bound, that is, always greater than the maximum frequency of the system, this procedure provides an estimate on the time increment which is conservative in the sense that it must be stable for a linear system, for if the timestep for a second order system is set by

$$\Delta t = \frac{2}{\omega_{\max}^{(e)}} \tag{3.28}$$

(we consider an undamped system for simplicity) it follows that

$$\Delta t \leqslant \frac{2}{\omega_{\max}} = \Delta t_{\text{stab}}\,. \tag{3.29}$$

Similar arguments can also be made for first order systems in terms of $\lambda_{\max}$.

Although it may be feared that this may sometimes lead to a time step which is much smaller than the stable time step, this fear is usually not justified. As will be shown shortly, for a uniform mesh, this method often leads to the exact maximum eigenvalue, and even for irregular meshes, it provides a very good estimate. In a nonlinear problem, it would certainly take much more effort to recompute this eigenvalue at each time step than could possibly be saved by the slight increase in the time increment that could be obtained by using the actual eigenvalue of the system. In a linear problem, the savings may justify the eigenvalue computation. For example, if the maximum eigenvalue is obtained by 100 iterations in the power method, this is equivalent in cost to advancing the computation 100 time steps. Thus, if Eq. (3.29) underestimates the time step by 20%, in any computation with more than 500 time steps the use of a time increment based on an exact eigenvalue will be more efficient. However, the advantage is probably not sufficient to justify the added cost and software which is required, and furthermore, most explicit programs are designed for both linear and nonlinear problems; in the latter, an accurate maximum eigenvalue provides almost no benefit because even Eq. (3.28) does not guarantee stability.

It is in fact usually not practical to compute even the maximum eigenvalues of the element directly, for this would undoubtedly increase the cost of the computation considerably. Instead, formulas which give the maximum element eigenvalues or upper bounds on the maximum element eigenvalues $\lambda_{\max}^{(e)}$ are used. An upper bound for $\lambda_{\max}^{(e)}$ always gives a stable time step; the only drawback lies in the possibility that the upper bound may be much larger than the eigenvalue, so that the eigenvalues are drastically overestimated, and hence the time increment is much smaller than it need be.

Formulas for the maximum element eigenvalues (or upper bounds) for a family of one, two and three dimensional elements are given in Table 13 for the diffusion equation, wave equation, and the equations of motion. These formulas are limited to one-point quadrature and a lumped (diagonal) mass matrix of the form given in Eq. (1.43). Exact formulas are also given for the 4 node (reduced quadrature) element in two dimensions.

It should be noted that these bounds are applicable regardless of the boundary conditions provided that the boundary conditions do not involve a relationship between the dependent variable and its derivatives. Thus it applies to boundary conditions (1.6a) and (1.6b) as long as q^* is *not a function of* θ; if so, this relationship must be incorporated in the element stiffness. This generality is in marked contrast to Fourier methods used in finite difference methods, which apply only to uniform meshes under particular boundary conditions.

These formulas are obtained by noting that the range of $\boldsymbol{K}^{(e)}$ for these elements is spanned by $\boldsymbol{b}_i$; see Eqs. (T4.3), (T4.4). Expanding the eigenvectors in $\boldsymbol{b}_i$ then yields the results [37].

As can be seen from Table 13 by examining the results for rectangular elements, when the equations are parabolic (diffusion equation), the stable time step varies with the square of the element dimension, while for hyperbolic problems (equations of motion for continua and wave equation) it varies directly with the element dimension.

Estimates of this type have not been developed for higher order elements or consistent mass matrices. As noted in Chapter 2, the eigenvalues associated with consistent masses are always greater, so the stable time step is smaller.

Table 13
Eigenvalue bounds and eigenvalues for uniform gradient elements

Diffusion and wave equation (Eqs. (1.7) and (2.73))
For $\boldsymbol{K}^{(e)}$ given by Eq. (1.28b) with 1-point quadrature and $\boldsymbol{M}^{(e)}$ by Eq. (1.43)

$$\lambda_{\max}^{(e)} \le \frac{n_N}{\rho} k_{ij} B_{iI} B_{iI} \begin{cases} \text{sum } i = 1 \text{ to number of dimensions } n_D \\ \text{sum } I = 1 \text{ to number of nodes } n_N \end{cases} \tag{T13.1}$$

Special cases of Eq. (T13.1)
1D element from Table 2, $n_N = 2$, $c^2 = k/\bar{\rho}$ or β/ρ

$$\lambda_{\max}^{(e)} \le 4c^2/L^2 \quad \text{(equality holds; see Chapter 2)} \tag{T13.2}$$

2D quadrilateral from Table 4, isotropic: $k_{ij} = k\delta_{ij}$, $n_N = 4$

$$\lambda_{\max}^{(e)} \le \frac{2c^2}{A^2}(l_{24}^2 + l_{31}^2), \quad l_{24}^2 = x_{42}^2 + y_{42}^2, \quad l_{31}^2 = x_{31}^2 + y_{31}^2 \tag{T13.3}$$

For $a \times b$ rectangle (T13.3) yields

$$\lambda_{\max}^{(e)} \le \frac{4c^2(a^2 + b^2)}{a^2 b^2} \tag{T13.4}$$

Exact values for $\lambda_{\max}$ for $a \times b$ rectangle with stabilization [37]

$$\lambda_{\max}^{(e)} = \max[4c^2/\min(a^2, b^2), 4\varepsilon c^2(a^2 + b^2)/3A^2] \tag{T13.5}$$

Stable time steps for rectangle based on (T13.5)

Diffusion equation

$$\Delta t \le \frac{1}{2c^2}\min[a^2, b^2, 3a^2b^2/\varepsilon(a^2 + b^2)] \tag{T13.6}$$

Wave equation

$$\Delta t_{\text{stab}} = \frac{1}{c}\min[a, b, ab/\sqrt{\tfrac{1}{3}\varepsilon(a^2 + b^2)}] \tag{T13.7}$$

Continuum element (uniform strain)

$$\lambda_{\max}^{(e)} \le \frac{n_N}{\rho}(\lambda + 2\mu)B_{iI}B_{iI} \quad \lambda, \mu \text{ Lamé constants} \tag{T13.8}$$

Rectangular element (no stabilization)

$$\lambda_{\max}^{(e)} \le 4c^2(a^2 + b^2)/A^2, \qquad c^2 = (\lambda + 2\mu)/\rho \tag{T13.9}$$

Table 14 gives eigenvalues for a structural beam element [7]. As can be seen from those results, one frequency (the axial mode) varies directly varies with the length, a second (the bending mode) with the square of the length. As a mesh is refined, the bending mode will eventually govern the time step. This is a consequence of the parabolic nature of the equations which govern the bending of structures, such as Euler–Bernoulli beams and Kirchhoff plates. When a structure is modeled within a soft medium, this characteristic of structural theories leads to

Table 14
Maximum frequencies for a beam element

Stiffness matrix (flexural)

$$\boldsymbol{K}^{(e)} = \frac{EI}{L^3}\begin{bmatrix} 12 & 6L & -12 & 6L \\ & 4L^2 & -6L & 2L^2 \\ & & 12 & -6L \\ \text{symmetric} & & & 4L^2 \end{bmatrix} \qquad \text{(T14.1)}$$

Diagonal of lumped mass matrix

$$M_{II}^{(e)} = \frac{\rho AL}{24}[12 \quad L^2 \quad 12 \quad L^2] \qquad \text{(T14.2)}$$

Axial stiffness and mass: Table 2 with $k = \rho A$

$$\lambda_{\max}^{(e)} = \max\left(\frac{4c^2}{L^2}, \frac{192c^2 I}{L^4 A}\right) \qquad \text{(T14.3)}$$

$$\Delta t_{\text{stab}} = \min\left(\frac{L}{c}, \frac{\sqrt{3}L^2}{12cr_g}\right) \qquad \text{(T14.4)}$$

A = cross-sectional area of beam
E = Young's modulus
I = momentum of inertia of cross-section
L = length
$c^2 = E/\rho \qquad r_g^2 = I/A$

rather large disparities in the eigenvalues, although the ratio is seldom large enough to call for stiffly-stable integrators.

The Rayleigh method can also be used to determine the maximum eigenvalues of finite difference semidiscretizations. If a finite difference molecule can be generated from an element stiffness, then the bound of Eq. (3.15) applies. Since we have shown in Section 1.3 that $\varepsilon = 2$ and 3 give the 9 and 5-point molecules for square meshes, it follows immediately that in this case the finite difference methods are governed by the ε condition and have larger maximum eigenvalues than the finite element; their stability limits are smaller. These results also apply to rectangular meshes for the 5-point molecule.

The stability limit for the time step given in Table 13 with $\varepsilon = 3$ agrees exactly with the result of a Neumann analysis for the 5-point difference formula given in [52]. The contour integral method has the same structure as the finite element with $\varepsilon = 0$, so it has the same maximum eigenvalue.

This technique is quite useful for determining maximum eigenvalues for nonuniform finite difference meshes. As long as the equations at a point can be generated by an FEM type assembly, the maximum eigenvalue of the molecule is bounded by the maximum eigenvalues of the 'elements' used to generate this mesh. Thus since 5-point formulas can be generated from 1D elements, Eq. (T13.2) can easily be used to obtain bounds for irregular meshes.

3.4. Some practical considerations in explicit integration

An interesting aspect of linear explicit computations is that even for large systems, the computation will be stable with a time increment computed by Eq.

(3.12) with the *limited precision of a typical computer* (6 to 16 significant digits) but will become wildly unstable even if 1.0001 Δt_{stab} is used. As can be seen from the stability proof in Chapter 2, the highest mode for an unstable computation will grow exponentially in a linear system, so once it is initiated it will trigger 'overflow traps' after 40 to 100 steps. The unstable nature of a calculation in a linear system should be obvious to any user after examining the results.

In nonlinear calculations, a numerical instability is not always as dramatic or obstreperous and in fact may avoid detection. The reasons for this are that (i) the eigenvalues vary with time and may decrease after an instability so that the computation regains stability and (ii) nonlinear processes often are capable of dissipating a large amount of energy, so the instability is limited to a small part of the mesh. This process has been called an 'arrested' instability [5].

An example of a physical model in which an arrested instability may occur is the elastic-plastic response of a structure with large displacements. During the elastic part of the response, the effect of the geometric nonlinearity can trigger an instability which will result in growth of the highest mode in a region, and hence the strains. The growth in the strains causes plastic material response, and hence a decrease in the stiffness and the maximum eigenvalues, so the calculation recovers stability if the time increment remains constant. The spurious energy generated by the instability will be absorbed as excessive plastic work.

An arrested instability will not be apparent from a perusal of the final results since the solution will not be unreasonably large. In fact, it is possible for the error to be of the order of 10% to 50%, so the solution may appear quite reasonable.

An arrested instability is associated with the generation of spurious energy (which is dissipated as plastic work) and leads to a loss of energy balance. Loss of energy balance is in fact associated with any unstable calculation of a mechanical system, so this provides an excellent way for checking the stability of an explicit calculation both during its process and after its completion.

Energy balance can be computed by the following procedure. Let W_{int} designate the internal energy, the global counterpart of e in Eq. (2.4), W_{ext} the work performed by the external nodal forces and W_{kin} the kinetic energy. The internal and external work should then be integrated by

$$W_{int}^{M+1} = W_{int}^{M} + \frac{\Delta t^M}{2} \sum_e \int_{\Omega_e} D_{ij}^{M+1/2}(T_{ij}^{M} + T_{ij}^{M+1})\, d\Omega\,, \tag{3.30}$$

$$W_{ext}^{M+1} = W_{ext}^{M} + \frac{\Delta t^M}{2} (\dot{\boldsymbol{u}}^{M+1/2})^{T}(\boldsymbol{f}_{ext}^{M} + \boldsymbol{f}_{ext}^{M+1})\,. \tag{3.31}$$

The kinetic energy can only be computed at half time steps

$$W_{kin}^{M+1/2} = \tfrac{1}{2}(\dot{\boldsymbol{u}}^{M+1/2})^{T}\boldsymbol{M}\dot{\boldsymbol{u}}^{M+1/2}\,, \tag{3.32}$$

so we define W_{kin}^{M} by

$$W_{\text{kin}}^{M} = \tfrac{1}{2}(W_{\text{kin}}^{M-1/2} + W_{\text{kin}}^{M+1/2})\,. \tag{3.33}$$

Energy balance then requires that

$$|W_{\text{int}}^{M} + W_{\text{kin}}^{M} - W_{\text{ext}}^{M}| \leq \delta \|W\| \tag{3.34}$$

where δ is a specified tolerance and for convenience $\|W\|$ is some norm of the energy so that δ can be problem independent. A useful form for $\|W\|$ is

$$\|W\| = |W_{\text{ext}}| + |W_{\text{int}}| + |W_{\text{kin}}|\,, \tag{3.35}$$

since this quantity is indicative of the total energy in the system. The absolute value sign on the left hand side is not essential since an instability usually manifests itself in the growth of internal and kinetic energies, so the left hand sign is usually positive in cases of trouble. Simply using $|W_{\text{int}}|$ or $|W_{\text{kin}}|$ leads to problems when the response is oscillatory, because these quantities are then oscillatory and may become very small or vanish; using $|W_{\text{ext}}|$ is inconvenient for problems where the loading is impulsive and prescribed as an initial velocity.

The internal energy computation is given here in terms of the velocity strain and Cauchy stress. Any of the other conjugate measures of strain rate and stress given in Table 7 could also be used.

For the form of energy balance given above, δ should be smaller than 10^{-2} if the computation is stable even for a 7 digit computer. In hundreds of computations carried out by the author and coworkers, this level of accuracy was maintained for as many as 50 000 time steps on meshes with 200 to 1000 elements. Even for δ on the order of 0.05, instability should be suspected, and particularly for large meshes, the results should be examined with care since a local arrested instability may lead to severe errors in parts of the mesh. If the energy balance is performed using the internal and external energies at time step M and the kinetic energy at $M + \frac{1}{2}$ rather than an average such as given in Eq. (3.32), the energy does not balance as well.

For the Euler integration of the heat conduction equation, alternative norms must be used to check stability. Energy balance also provides a criterion for heat conduction, so defining

$$W_{\text{int}}^{N+1} = W_{\text{int}}^{N} + \boldsymbol{M}(\boldsymbol{\theta}^{N+1} - \boldsymbol{\theta}^{N})\,, \tag{3.36}$$

$$W_{\text{ext}}^{N+1} = W_{\text{ext}}^{N} + \frac{\Delta t}{2}(\boldsymbol{s}^{N} + \boldsymbol{s}^{N+1})\,, \tag{3.37}$$

energy balance requires that

$$|W_{\text{int}}^{N} - W_{\text{ext}}^{N}| \leq \delta \|W\| \tag{3.38}$$

where $\|W\|$ is defined analogously to Eq. (3.35) except that the kinetic energy is omitted.

The conditional stability of explicit methods is a nagging drawback and numerous efforts have been made to circumvent it. However, Krieg [34] has shown that within the class of linear multistep methods, none has a larger stable time step than the central difference method.

Unconditionally stable explicit methods which are not linear multistep have been described by Wamberg [56]. These integration procedures are called rational Runge–Kutta methods. For example, the simplest of these methods when applied to Eq. (1.29) çan be written as follows:

$$\boldsymbol{M}\boldsymbol{v}_1 + \boldsymbol{K}\boldsymbol{\theta}^N = \boldsymbol{s}^{N+1}\,, \tag{3.39a}$$

$$\boldsymbol{M}\boldsymbol{v}_2 + \boldsymbol{K}(\boldsymbol{\theta}^N + \tfrac{1}{2}\Delta t\boldsymbol{v}_1) = \boldsymbol{s}^{N+1}\,, \tag{3.39b}$$

$$\boldsymbol{b} = 2\boldsymbol{v}_1 - \boldsymbol{v}_2\,, \tag{3.39c}$$

$$\boldsymbol{e} = 2(\boldsymbol{v}_1^{\mathrm{T}}\boldsymbol{b})\boldsymbol{v}_1 + (\boldsymbol{v}_1^{\mathrm{T}}\boldsymbol{v}_1)\boldsymbol{b}\,, \tag{3.39d}$$

$$\boldsymbol{\theta}^{N+1} = \boldsymbol{\theta}^N + \Delta t\boldsymbol{e}/(\boldsymbol{b}^{\mathrm{T}}\boldsymbol{b})\,. \tag{3.39e}$$

Thus if $\boldsymbol{M}$ is diagonal, the method is completely explicit and it is furthermore unconditionally stable. However, the performance of these methods beyond the stability limit of the Euler method tends to be disappointing for the heat conduction equation because the accuracy deteriorates significantly whenever $\Delta t \geqslant 2\Delta t_{\mathrm{stab}}$. This deterioration in accuracy for large time steps seems to be characteristic of unconditionally stable explicit methods. Since this method involves twice as many evaluations of $\boldsymbol{f} = \boldsymbol{K\theta}$ per time step as Euler integration, it therefore offers little advantage. One area in which this method is promising is in adaptive meshing techniques where rather stiff components are sometimes introduced to control mesh overlap.

3.5. *Implicit integration*

The implicit Euler formula gives

$$\boldsymbol{\theta}^{M+1} = \boldsymbol{\theta}^M + \Delta t\dot{\boldsymbol{\theta}}^{M+1} \tag{3.40}$$

and combining with Eq. (1.14), we obtain

$$\boldsymbol{\theta}^{M+1} = \boldsymbol{\theta}^M + \Delta t\boldsymbol{M}^{-1}(\boldsymbol{s}^{M+1} - \boldsymbol{f}^{M+1})\,. \tag{3.41}$$

The above is a nonlinear set of algebraic equations for a nonlinear problem, since $\boldsymbol{f}$ is then a nonlinear function of $\boldsymbol{\theta}$.

For a linear system, the above equations can be rewritten by using $\boldsymbol{f} = \boldsymbol{K\theta}$ (cf. Eq. (1.29)) in the form

$$(\boldsymbol{M} + \Delta t\boldsymbol{K})\boldsymbol{\theta}^{M+1} = \Delta t\boldsymbol{s}^{M+1} + \boldsymbol{M}\boldsymbol{\theta}^M\,, \tag{3.42}$$

where the right hand side consists strictly of historical terms (which are known at time step $M+1$) and the forcing term at $M+1$.

This integrator is *unconditionally stable* for linear systems, which means that it can be used with Δt chosen entirely by considerations of accuracy. For a one-step method, more accurate integrators can be chosen; see for example Eq. (5) in Chapter 2. This one will be used in the present discussion because it minimizes the algebra. Implicit integrators for the equations of motion are discussed in Chapters 2 and 9.

One of the most important aspects of implicit methods is the solution of the linear and nonlinear algebraic semidiscretized equations, (3.42) and (3.41), respectively.

There are basically two means for treating these equations:

(i) Newton–Raphson methods, which for linear systems become direct elimination (triangulation and back-substitution),

(ii) iterative methods.

Elimination methods are most commonly used in finite element programs, whereas in finite difference programs iterative methods are more prevalent. Although this is to some extent motivated by historical developments, there are also some good reasons for this.

An important feature of direct elimination is that the triangulation, which requires the most computation, need only be performed once for linear systems. The back substitution, which is required in every time step, requires far fewer computations. However, this advantage can be exploited only when Δt is constant. Whenever Δt changes during the simulation, the system matrix must be retriangulated; one method for avoiding this is to use scaling techniques such as proposed in [47]. For nonlinear systems, the matrix often needs to be retriangulated if reasonable convergence is to be achieved.

Alternatively, these equations may be solved by iterative methods. The simplest of these methods are the Jacobi and Gauss–Seidel iterative methods. Let

$$\boldsymbol{A} = \boldsymbol{M} + \Delta t\boldsymbol{K}, \tag{3.43a}$$

$$\bar{\boldsymbol{f}} = \Delta t \boldsymbol{s}^{M+1} + \boldsymbol{M}\boldsymbol{\theta}^{M}, \tag{3.43b}$$

so that Eq. (3.42) becomes

$$\boldsymbol{A}\boldsymbol{\theta} = \bar{\boldsymbol{f}}. \tag{3.43c}$$

Let the matrix $\boldsymbol{A}$ be split by

$$\boldsymbol{A} = \boldsymbol{N} + \boldsymbol{S}. \tag{3.44}$$

Then an iterative procedure can be formed by

$$\boldsymbol{N}\theta_{[n+1]} = -\gamma \boldsymbol{S}\boldsymbol{\theta}_{[n]} + \bar{\boldsymbol{f}} \tag{3.45}$$

where bracketed indices denote iteration numbers, and γ is a relaxation factor; it constitutes an extrapolation which can be used to accelerate the convergence in many situations. The two methods are given by

$$\boldsymbol{N} = \begin{cases} \boldsymbol{A}_D & \text{Jacobi}, \\ \boldsymbol{A}_D + \boldsymbol{A}_U & \text{Gauss–Seidel}, \end{cases} \tag{3.46}$$

where subscripts D and U denote the diagonal and strictly upper triangular parts of $\boldsymbol{A}$. Since the solution must satisfy (add $\boldsymbol{N\theta}$ to both sides of the governing equation (3.43c))

$$\boldsymbol{N\theta} = \boldsymbol{N\theta} - \boldsymbol{A\theta} + \bar{\boldsymbol{f}} \tag{3.47a}$$

subtracting (3.45) from (3.47a) and premultiplying by $\boldsymbol{N}^{-1}$ gives that the error $\boldsymbol{e}$ is governed by

$$\boldsymbol{e}_{[n+1]} = (\boldsymbol{I} - \gamma \boldsymbol{N}^{-1}\boldsymbol{A})\boldsymbol{e}_{[n]} = \boldsymbol{P}\boldsymbol{e}_{[n]} . \tag{3.47b}$$

The requirement for convergence is similar to that for stability as given in Chapter 2; namely, the spectral radius of $\boldsymbol{P}$ must be less than 1. Therefore, if we denote the eigenvalues of $\boldsymbol{P}$ by λ, it follows that

$$|\lambda| = |1 - \gamma\mu| \leq 1 \tag{3.48}$$

where μ is the eigenvalue of

$$\boldsymbol{N}\boldsymbol{d} = \mu \boldsymbol{A}\boldsymbol{d} . \tag{3.49}$$

For the Jordan methods, the maximum eigenvalues, which govern convergence, may be estimated by procedures such as given in Section 3.4. Similar equations may be written for solids with integration by the trapezoidal rule (see Chapter 2) which gives $\boldsymbol{A} = \boldsymbol{M} + \frac{1}{4}\Delta t^2 \boldsymbol{K}$. The maximum eigenvalues for a beam with both flexural and axial stiffness in this case are given in Fig. 4 [21]; the stiffness and mass matrices are based on Table 14. Note that the axial eigenvalue is within the convergence limit for all values of the Courant number q, so that acceleration can be used. On the other hand the flexural eigenvalues are outside the convergence limit for even moderate values of q_f so that γ must be less than one. Thus it can be seen that for elements of this type, it is difficult to achieve good convergence, since γ must be chosen to bring the flexural eigenvalue to the convergence domain, which slows the convergence of the axial response. Similar behavior is

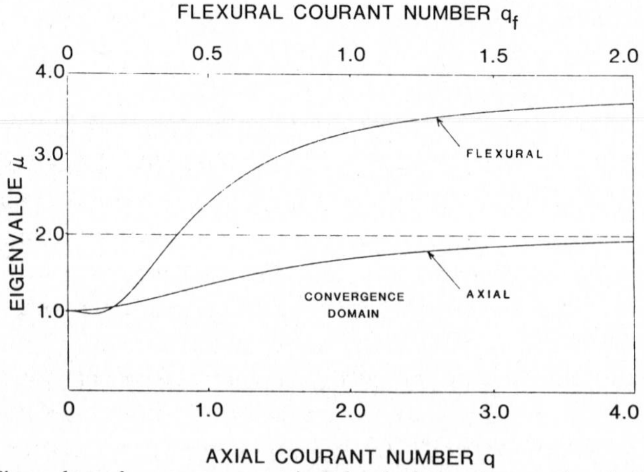

Fig. 4. Eigenvalues of convergence matrix $\boldsymbol{P}$ for the beam element given in Table 14.

found in plates and shells and in higher order elements, which introduce an 'optical' branch of the spectrum; see Chapter 6 or [7].

The eigenvalues of the iteration matrix $\boldsymbol{P}$ for the Gauss–Seidel method are the squares of those for the Jacobi method. Thus for uniform meshes, the Gauss–Seidel method is far superior. Even greater gains can be achieved if the extrapolation is made at each step of the matrix multiplication, rather than only after a complete sweep as indicated by γ in Eq. (3.45). This is the basis of the SOR (sequential overrelaxation) method; see [19] for a description in the finite difference context.

However, the adequacy of all of these methods depends on the absence of very high eigenvalues and the availability of good estimates on μ so that γ can be optimally selected. Furthermore, they perform well only when $\mu_{max} < 2$. Thus their effectiveness for general purpose finite element codes is limited. On the other hand, the ratio of eigenvalues only affects the condition number in elimination methods, so large ratios are not as deleterious.

Iterative methods which do not require eigenvalue estimates appear to be more promising. For example, the conjugate gradient acceleration method [33] seems to perform quite well in finite element meshes: a numerical study is given in [21]. The quasi-Newton methods described in Chapter 9 are also quite promising.

3.6. Comparison explicit and implicit time integration

Let us begin first by listing the advantages of explicit time integration. They are as follows:

1. Few computations are required per time step.
2. The algorithm is simple in logic and structure, so it easily handles complex nonlinearities.
3. Compared to direct elimination procedures (Newton–Raphson) it requires little core storage.
4. It is ideal for testing out new ideas because it requires less coding.
5. It is very reliable as to accuracy and completion of the computation.

The explicit method has only one notable disadvantage, but this drawback can be devastating in certain situations: it is only conditionally stable so that a very large number of time steps may be required.

The advantages and disadvantages of implicit methods are almost the converse of those of the explicit method. The main advantage is that because of its enhanced stability, the time step may be much larger. The disadvantages are:

1. greater complexity and size in the software, particularly if Newton–Raphson techniques are used,
2. less reliability,
3. greater core storage requirements if direct elimination (Newton–Raphson methods) are used.

By reliability we here refer to the ability of an algorithm to complete a computation without an abort and with the requisite accuracy. Accuracy is an important factor because, particularly in problems which are predominantly linear, where most practical implicit methods are unconditionally stable, the time steps that are used may be sufficiently large so that time integration errors become dominant. By contrast, in explicit methods, spatial discretization errors are always dominant, because the truncation errors due to time integration in the stable domain are usually far less than those associated with the spatial discretization; see Chapter 6.

We will now discuss the suitability of explicit and implicit methods for various classes of problems, bearing in mind the advantages and disadvantages of the explicit and implicit methods which have been summarized. In drawing guidelines from this discussion, it is important to keep in mind that certain classes of problems require small time steps in order to achieve suitable accuracy, whereas others do not. It is also worthwhile to bear in mind that exploratory software developments in which new concepts of material or phenomenological modeling are tried and in which the software is experimental dictates different time integration procedures than software which is developed for a production basis.

Returning to the relationship between time integration methods and the phenomenology being studied, it is useful to categorize problems governed by the equations of motion into

1. Wave propagation problems and
2. Inertial (structural dynamics) problems.

Wave propagation problems are those in which the behavior at the wave front and an accurate replication of the wave front is of engineering importance. Problems that fall into this category are the shock response from impact and explosions and problems in which wave effects such as focusing, reflection and diffraction are important. The decision as to whether wave propagation is important requires considerable engineering judgement. For example, in studying waterhammer effects in piping systems, if one is simply interested in the response of the piping, then the high frequency components associated with the wave front are usually unimportant; on the other hand, if one is interested in tracing a waterhammer through a long loop in order to examine whether a diaphram will burst, then an accurate resolution of the wave front may be important. The same dilemma is found in analyzing the response of many structures: the overall deformations are not governed by the shape of the wave front but depend primarily on the total impulse and the lower frequency portions of a loading, whereas brittle fracture may depend on the magnitude of the peak stress and hence an accurate resolution of the wave front.

The second class of problems are here called inertial problems. This class of problems is often called structural dynamics problems, to indicate that low frequencies dominate the response, but we have chosen the nomenclature inertial to indicate that the name also applies to fluid response. Problems that are typically of the inertial type are seismic response and the elastic–plastic, large deformation response of thin structures. The rise time and duration of the load relative to the time required for a wave to traverse the structure, though not a decisive factor, can be used as a preliminary criterion for categorizing the problem: if the rise time and duration of the load exceed several traversal times of the wave across the body of interest, the problem is usually of the inertial type.

As a rough guideline, wave propagation problems are best solved by explicit techniques, whereas implicit techniques are more appropriate for intertial problems. The reasons for this may be appreciated from a perusal of Fig. 5, which summarizes some results given in [40]. This figure shows the dispersion, and hence the spectral error, in a uniform finite element mesh of linear displacement elements for the wave equation. Ideally the curve should coincide with that labeled 'no dispersion'. The curve labeled 'exact' corresponds to an exact integration of the finite element semidiscretization, and it can be seen that for larger wave numbers (frequencies) substantial error develops due to the finite element approximation in space. The dispersion for an explicit integration of the finite element equations with Courant number $q = 0.5$ (half the stable time step), in fact improves on the exactly integrated equations. The result for $q = 1.0$ is not shown because it exhibits no dispersion; this is an anomaly which occurs only for uniform meshes in one dimensional linear problems, so it is of no practical significance. Note however, that the implicit integration of the finite element equations with $q = 2$ already introduces large additional errors. Therefore if high frequency components of the solution are relevant, then time steps of the order required for stability in explicit methods are required simply to capture these high frequencies with sufficient accuracy. On the other hand, when the high frequency components are unimportant, then considerable benefits can be gained by exploiting the greater stability of implicit methods by using larger time steps.

In addition to the spectral content of interest in the problem, the size of the

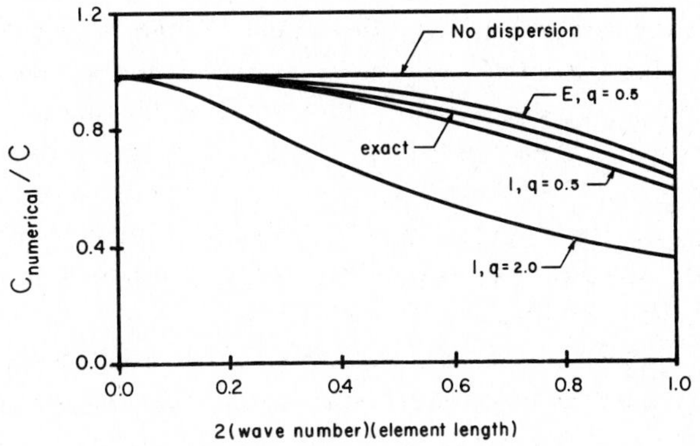

Fig. 5. Comparison of numerical dispersion in a one-dimensional linear u mesh (uniform) for explicit (E) and implicit (I) time integration for various Courant numbers $q = \Delta t l/c$.

mesh and the severity of the nonlinearities is of relevance in deciding which approach to choose. This is seen by comparing the computational requirements of the two methods for the four examples given in Table 15. The first example is an axisymmetric, cylindrical shell modeled by a 100 element mesh. For this example, the computational requirements of the implicit and explicit methods for a time step are comparable because the band width is very small. The highest frequency of a shell mesh is quite high because of its parabolic character, so the stability limits on the time step for explicit integration are severe. Therefore, problems of this type are usually best solved by implicit methods with a Newton–Raphson solver. On the other hand, for a three dimensional problem, implicit integration with elimination would require far more computations per time step, and in addition would probably require out-of-core storage, which reduces the speed of an implicit method even more. Hence the advantage of implicit method for three dimensional problems becomes marginal even if the higher frequencies are not of interest.

The advantage of implicit methods is also degraded when severe nonlinearities are important in the problem. For example, as shown in Chapter 2, the introduction of step functions, which characterize phenomena such as contact-impact, can cause a sawtooth type of response in implicit methods. Similar difficulties are found in characterizing the early-time response of elastic–plastic structures to severe loads. In these situations, an accurate simulation of the response generally requires a time step of order of the stability limit, so little advantage is gained by an implicit method.

In parabolic problems, the impact of the conditional stability of explicit integrators is more severe, for as can be seen from Table 13, the stable time step

Table 15
Estimates of computations (multiplications) per time step in explicit and implicit integration of nonlinear problems; from [7]

	Formula	Example 1	Example 2	Example 3
Explicit	$(2m_B + m_M)eM_I$	2×10^4	1×10^5	5×10^5
Implicit	$nb^2 + 2nb +$ stiffness comp $+ (2m_B + m_M)eM_I$	3.5×10^4	1.5×10^7	2×10^9

Nomenclature for table: e is the number of elements, m_B the number of multiplications in strain-displacement relations, M_I the number of integration points per element, m_M the number of multiplications in constitutive equation, n the degrees of freedom in mesh, and b the semi-bandwidth of mesh.

Example 1: 100 node, cylindrical, axisymmetric shell; $M_I = 10$, $m_B = 12$, $m_M = 8$.

Example 2: 31×62 node two-dimensional, plane mesh; constant strain elements; $M_I = 1$, $m_B = 12$, $m_M = 8$.

Example 3: $11\times 11\times 31$ node three-dimensional mesh; constant strain element; $M_I = 1$, $m_B = 724$, $m_M = 36$.

decreases with the square of the mesh size. Therefore, the trend today is definitely toward implicit or semi-implicit integrators; some of the latter are briefly discussed in the next section. Nevertheless, explicit methods are still quite useful, particularly for 'temporary' software that is developed to explore a problem.

It should be noted that for wave propagation problems, the use of semidiscretizations with explicit time integration can still be extremely expensive. As a rule of thumb, it is generally advised that three to five constant-strain elements span the shortest wave of interest. Although this is feasible in one dimensional problems, it is not in most three dimensional problems. Therefore, alternative methods are often employed in wave propagation problems. The methods of characteristics can easily achieve excellent resolution of the wavefront but is not as versatile. For example, if any softening occurs in the stress–strain relation, that is, if the modulus becomes negative, the hyperbolic character of the equations is lost and characteristics no longer exist, so the method of characteristics is not applicable. Furthermore, the programming of general characteristics codes in two and three dimensions is quite formidable.

Another approach to obtaining better resolution is to include the discontinuity within the elements as the shockfront passes through; this has been pursued by Wellford and Oden [58]. Solutions have been given for one dimensional and simple two dimensional cases, but as in the method of characteristics, development of general algorithms for two and three dimensions seems to be difficult.

A class of methods which have substantial promise with severe gradients are adaptive grid techniques where the motion of the nodes is chosen so that better accuracy is achieved in each time step. This procedure is a feasible option, for instance, in the ALE techniques described in Chapter 10. Recently, very general methods of this type have been proposed by [40] where the nodal coordinates are included among the unknowns in the weak form. One dimensional solutions have been reported in [23].

3.7. Operator splitting and partitioning methods

From the discussion of the merits of explicit and implicit time integrators, it is apparent that the treatment of many problems could be facilitated through the development of methods which combine their attractive features. The ideal method is one which can deliver the desired accuracy without requiring complex software or an excessive amount of computer time or core storage.

In divising such a method, the following features of practical problems must be considered.

1. The spectral content of elements within a mesh may differ substantially; for example, if a hard structure is embedded in a soft medium, the elements which model the structure will generally have much higher frequencies than the medium.

2. Strong nonlinearities and discontinuous response may appear in different parts of the governing equations; for example, if impact and separation are

possible in parts of the mesh, the equations of motion for these parts will at times have discontinuities. Similarly, elastic–plastic response causes loss of smoothness in the constitutive equations whenever transitions between elastic and plastic behavior occur.

3. Many engineering problems involve the interaction of different physical processes, such as for example, the interaction of chemical process and mechanical response. The two fields may be parabolic and hyperbolic, and the characteristic time scales associated with these processes may be substantially different, but they are nevertheless solved simultaneously in order to minimize data transfer problems and to capture interactions which may occur.

These properties naturally suggest partitioned methods, where different time integrators are used on different parts of the mesh or for the different field equations. Partitioned methods are discussed extensively in Chapter 3.

We will here introduce these methods in conjunction with another feature, subcycling, in which different time steps are used in different parts of the mesh [14], [8]. For this purpose, we write the linear one-step integration equation (3.1) in the form

$$\boldsymbol{\theta}^{N+1} = \boldsymbol{\theta}^{N} + \Delta t(1-\alpha)\dot{\boldsymbol{\theta}}^{N} + \Delta t\alpha\dot{\boldsymbol{\theta}}^{N+1} \tag{3.50}$$

where α is a parameter which gives the explicit integrator for $\alpha = 0$ and the implicit integrator for $\alpha = 1$. In fact, as discussed in Chapter 2, the above form gives an entire family of one-step methods but we will limit our attention to these two.

The matrix θ is now partitioned into two parts as

$$\boldsymbol{\theta} = \begin{Bmatrix} \boldsymbol{\theta}_A \\ \boldsymbol{\theta}_B \end{Bmatrix} \tag{3.51}$$

where nodes in A and B are integrated with time steps $m\Delta t$ and Δt, respectively. For this purpose, Eq. (3.50) is written as

$$\boldsymbol{\theta}^{N+1} = \boldsymbol{\theta}^{N} + \boldsymbol{W}_1^N\dot{\boldsymbol{\theta}}^{N} + \boldsymbol{W}_2^N\dot{\boldsymbol{\theta}}^{N+1} \tag{3.52}$$

where for $N \neq ml$

$$\boldsymbol{W}_1^N = \Delta t\begin{bmatrix} \boldsymbol{0} & \boldsymbol{0} \\ \boldsymbol{0} & (1-\alpha_B)\boldsymbol{I} \end{bmatrix}, \qquad \boldsymbol{W}_2^N = \Delta t\begin{bmatrix} \boldsymbol{0} & \boldsymbol{0} \\ \boldsymbol{0} & \alpha_B\boldsymbol{I} \end{bmatrix} \tag{3.53}$$

and for $N = ml$

$$\boldsymbol{W}_1^N = \Delta t\begin{bmatrix} m(1-\alpha_A)\boldsymbol{I} & \boldsymbol{0} \\ \boldsymbol{0} & (1-\alpha_B)\boldsymbol{I} \end{bmatrix}, \quad \boldsymbol{W}_2^N = \Delta t\begin{bmatrix} m\alpha_A\boldsymbol{I} & \boldsymbol{0} \\ \boldsymbol{0} & \alpha_B\boldsymbol{I} \end{bmatrix} \tag{3.54}$$

where the $\boldsymbol{W}$ matrices are partitioned like $\boldsymbol{\theta}$. Equations (3.52) to (3.54), when applied to (1.14), describe an algorithm where the solution is advanced $m-1$ times with a time step Δt on nodes in B, and in every mth cycle, the nodes in A and B are advanced with the time step $m\Delta t$ and Δt, respectively. Whether the nodes are integrated implicitly or explicitly is determined by the choice of the parameters α_A and α_B. Thus the integration procedures shown in Table 16 result from the above.

These explicit–implicit and implicit–implicit nodal partitions given here are only special cases of these classes of partitions; for example element partitions can also be developed as shown in Chapters 2 and 3.

The equations for an explicit partition with explicit subcycling are as follows for $N \neq ml$

$$\boldsymbol{M}_B\boldsymbol{\theta}_B^{N+1} = \boldsymbol{M}_B\boldsymbol{\theta}_B^N - \Delta t(\boldsymbol{f}_B^N - \boldsymbol{s}_B^N) \tag{3.55}$$

and for $N = ml$

$$\boldsymbol{M}\boldsymbol{\theta}^{N+1} = \boldsymbol{M}\boldsymbol{\theta}^N - \Delta t\begin{Bmatrix} m\boldsymbol{f}_A \\ \boldsymbol{f}_B \end{Bmatrix}^N - \begin{Bmatrix} m\boldsymbol{s}_A \\ \boldsymbol{s}_B \end{Bmatrix}^N . \tag{3.56}$$

Thus it is only necessary to update the nodes in A for the $m-1$ cycles of the small time step and to update both sets of nodes every mth cycle by using (3.56). An interesting feature of this procedure is that stability is assured in the linear system if

$$m\Delta t \leqslant \frac{2}{\lambda_A}, \quad \lambda_A = \max_{\text{for all } e \in S_A} \lambda^{(e)}, \tag{3.57a}$$

$$\Delta t \leqslant \frac{2}{\lambda_B}, \quad \lambda_B = \max_{\text{for all } e \in S_B} \lambda^{(e)} \tag{3.57b}$$

where S_A is the set of all elements with all nodes in A and S_B is the set of all elements with at least one node in B. Although this development is given in terms of two partitions for convenience, in applications it is common to use anywhere from 2 to 20 partitions.

An alternative to partitioning methods is operator splitting methods. In these methods an implicit integrator is generally chosen as a starting point and the operator, namely $\boldsymbol{A}$ in Eq. (3.43c), is split so that the solution of equations is avoided or the system to be solved is reduced in size. The motivation for this procedure is to retain the stability of the implicit integrator and achieve the computational efficiency of an explicit integrator.

Table 16

m	α_A	α_B	integration procedure
1	0	1	explicit–implicit
>1	0	0	explicit with subcycling
1	1	1	implicit–implicit
>1	1	0	implicit with explicit subcycling
>1	1	1	implicit with implicit subcycling

Split operator methods have been developed extensively in finite difference methods. In many cases, these developments have exploited the data structure of finite difference methods, that is, the fact that the node number locates the node within the mesh.

As an example of this class of methods, consider the method of Saul'yev [53] applied to the one-dimensional diffusion equation (cf. Eq. (1.7b))

$$\bar{\rho}\dot{u} = ku_{,xx} . \tag{3.58}$$

In applying this method, the implicit formula is applied to the left hand side and the right hand side is split into two parts, one part applied at $N+1$, the other at N so that

$$\bar{\rho}\,\frac{u_J^{N+1} - u_J^N}{\Delta t} = \frac{k}{\Delta x^2}\,(\underbrace{u_{J+1}^N - u_J^N}_{u_{,xJ+1/2}} - \underbrace{u_J^{N+1} + u_{J-1}^{N+1}}_{u_{,xJ-1/2}}) . \tag{3.59}$$

Solving for u_J^{N+1} we obtain

$$u_J^{N+1}(1+r) = u_J^N + r(u_{J+1}^N - u_J^N + u_{J-1}^{N+1}) , \tag{3.60a}$$

$$r = \frac{k\Delta t}{\bar{\rho}\Delta x^2} . \tag{3.60b}$$

Thus if the nodes are numbered in sequence from left to right and the computation proceeds from left to right, the right hand side of the above equation is always known, and no equations need be solved. In a similar manner an equation can be developed which is explicit in going from right to left.

A finite element form of this procedure has been given by Trujillo [55]. The conductance matrix $\boldsymbol{K}$ is split into upper and lower matrices

$$\boldsymbol{K} = \boldsymbol{K}_L + \boldsymbol{K}_U, \quad \boldsymbol{K}_L = \boldsymbol{K}_U^{\mathrm{T}} \tag{3.61}$$

and the vector $\boldsymbol{\theta}$ is updated by

$$(\boldsymbol{M} + \Delta t/2\boldsymbol{K}_L)\boldsymbol{\theta}^{J+1/2} = (\boldsymbol{M} - \Delta t/2\boldsymbol{K}_U)\boldsymbol{\theta}^J + \frac{\Delta t}{2}\,\boldsymbol{s}^J , \tag{3.62}$$

$$(\boldsymbol{M} + \Delta t/2\boldsymbol{K}_U)\boldsymbol{\theta}^{J+1} = (\boldsymbol{M} - \Delta t/2\boldsymbol{K}_L)\boldsymbol{\theta}^{J+1/2} + \frac{\Delta t}{2}\,\boldsymbol{s}^J . \tag{3.63}$$

The sweep in Eq. (3.62) is made from the top to the bottom of the $\boldsymbol{\theta}$ matrix, and in (3.63) it is reversed. Thus, no equations need be solved. For a one-dimensional mesh of linear $\boldsymbol{\theta}$ elements with node numbers numbered in sequence, (3.61) is identical to (3.60). However, an interesting feature of this method is that it retains its stability regardless of the numbering of the nodes. It is also interesting to observe that the Trujillo method is similar to Gauss–Seidel iteration, consisting essentially of 2 iterations. Using the concepts of Chapter 3, this method can also be cast as a matrix partition method. Thus the distinction between these two classes of methods is quite fuzzy.

Trujillo has also reported a form of his method applicable to structural dynamics problems [54]. In this case, the method can also be made unconditionally stable, but its accuracy is quite poor. For example, in [43] it is shown that regardless of the time increment Δt, the wave apparently advances only one spatial node per time step, so that the dispersive errors in higher frequencies are extremely severe. Park [48] has made numerous modifications and extensions of these methods to improve accuracy.

The finite difference literature abounds with other split operator methods such as the method of lines and alternating direction methods [49, 16]. Some of these have been cast in a finite element format, for example [24], gives a finite element alternating direction method, but the research in this aspect is quite embryonic. Another promising avenue in finite elements may be to devise splitting methods which take advantage of the finite element data structure rather than to adapt methods devised in a different context.

References

[1] J.H. Argyris, S. Kelsey and H. Kamel, Matrix Methods of Structural Analysis: A Precis of Recent Developments, in: *Matrix Methods of Structural Analysis*, B.F. deVeubeke, ed., AGARDograph 72 (Pergamon Press, New York, 1964) pp. 1–164.

[2] J.H. Aryris et al., Finite Element Method – The Natural Approach, *Computer Methods in Applied Mechanics and Engineering*, **17/18**, 1–106 (1979).

[3] K.J. Bathe, *Finite Element Procedures in Engineering Analysis* (Prentice-Hall, Englewood Cliffs, NJ, 1982).

[4] K.J. Bathe, E. Ramm and E.L. Wilson, Finite Element Formulations for Large Deformation Dynamic Analysis, *Int. J. Numerical Methods in Engineering* **9**, 353–386 (1975).

[5] T. Belytschko, Transient Analysis, in: *Structural Mechanics Computer Programs*, W. Pilkey et al., eds. (Univ. Press of Virginia, 1974) pp. 255–276.

[6] T. Belytschko, Nonlinear Analysis – Description and Stability, in: *Computer Programs in Shock and Vibration*, W. Pilkey and B. Pilkey, eds. (Shock and Vibration Information Center, Washington, DC, 1975) pp. 537–562.

[7] T. Belytschko, A Survey of Numerical Methods and Computer Programs for Dynamic Structural Analysis, *Nuclear Engineering and Design* **37**, 23–24 (1976).

[8] T. Belytschko, Partitioned and Adaptive Algorithms for Explicit Time Integration, in: *Nonlinear Finite Element Analysis in Structural Mechanics*, by W. Wunderlich et al., eds. (Springer, Berlin, 1981) pp. 572–584.

[9] T. Belytschko and B.J. Hsieh, Nonlinear Transient Finite Element Analysis with Convected Coordinates, *Int. J. Num. Methods in Engr.* **7**, 255–271 (1973).

[10] T. Belytschko and J.M. Kennedy, Computer Models for Subassembly Simulation, *Nuclear Engineering and Design* **49**, 17–38 (1978).

[11] T. Belytschko, J.M. Kennedy and D.F. Schoeberle, On Finite Element and Difference Formulations of Transient Fluid Structure Problems, *Proc. Comp. Methods in Nuclear Engineering*, American Nuclear Society, Savannah, GA, Vol. 2 (April 1975) pp. IV 39–IV 53.

[12] T. Belytschko, R. Mullen and N. Holmes, Explicit Integration, Stability, Solution Properties and Cost, in: *Finite Element Analysis of Transient Nonlinear Structural Behavior* (ASME, 1975) pp. 1–21.

[13] T. Belytschko, C.S. Tsay and W.K. Liu, A Stabilization Matrix for the Bilinear Mindlin Plate Element, *Computer Methods in Applied Mechanics and Engineering* **29**, 313–327 (1981).

[14] T. Belytschko, H.J. Yen and R. Mullen, Mixed Methods for Time Integration, *Computer Methods in Applied Mechanics and Engineering* **17/18**, 259–275 (1979).

[15] J. Donea and S. Giuliani, A Simple Method to Generate High-order Accurate Convection

Operators for Explicit Schemes Based on Linear Finite Elements, *International Journal for Numerical Methods in Fluids* **1**, 63–79 (1981).

[16] J. Douglas and H.H. Rachford, On The Numerical Solution of Heat Conduction Problems and Two and Three Space Variables, *Trans. Amer. Math. Soc.* **82**, 421 (1956).

[17] P.R. Eiseman, A Multi-surface Method for Coordinate Generation, *Journal of Computational Physics* **33**, 118–150 (1979).

[18] G.C. Everstine, A Symmetric Potential Formulation for Fluid-Structure Interaction, *Journal of Sound and Vibration* **79**(1), 157–160 (1981).

[19] J.H. Ferziger, *Numerical Methods for Engineering Applications* (John Wiley, New York, 1982).

[20] D.P. Flanagan and T. Belytschko, A Uniform Strain Hexahedron and Quadrilateral with Orthogonal Hourglass Control, *Int. Journal for Numerical Methods in Engineering* **17**, 679–706 (1981).

[21] D.P. Flanagan and T. Belytschko, Simulations Relaxation in Structural Dynamics, *Journal of the Engineering Mechanics Division, ASCE* **107**, 1039–1055 (1981).

[22] I. Fried, Discretization and Round-off Errors in the Finite Element Analysis of Elliptic Boundary Value Problems and Eigenvalue Problems, Ph.D. Thesis, MIT (1971).

[23] R.J. Gelinas, S.K. Doss and K. Miller, The Moving Finite Element Method: Applications to General PDE's with Multiple Large Gradients, *Journal of Computational Physics* **40**, 202–249 (1981).

[24] L. Hayes, Implementation of Finite Element Alternating-Direction Methods in Nonrectangular Regions, *International Journal of Numerical Methods in Engineering* **16** 35–50 (1980).

[25] H.D. Hibbitt, P.V. Marcal and J.R. Rice, A finite element formulation for problems of large strain and large displacement, *Int. Journal of Solids and Structures* **6**, 1069–1086 (1970).

[26] F.B. Hildebrand, *Finite Difference Equations and Simulations* (Prentice-Hall, Englewood Cliffs, NJ, 1968).

[27] C.W. Hirt, A.A. Amsden and J.L. Cook, An Arbitrary Lagrangian–Eulerian Computing Method for all Flow Speeds, *Journal of Computational Physics* **14**, 227–253 (1974).

[28] T.J.R. Hughes and W.K. Liu, Nonlinear Finite Element Analysis of Shells. Part I, Three Dimensional Shells, *Computer Methods in Applied Mechanics and Engineering* **26**, 331–362 (1981).

[29] T.J.R. Hughes, W.K. Liu and T.K. Zimmerman, Lagrangian–Eulerian Finite Element Formulation for Incompressible Viscous Flows, *Computer Methods in Applied Mechanics and Engineering* **29**, 329–349 (1981).

[30] T.J.R. Hughes, K.S. Pister and R.L. Taylor, Implicit–Explicit Finite Elements in Nonlinear Transient Analysis, *Computer Methods in Applied Mechanics and Engineering*, **17/18**, 159–182 (1979).

[31] T.J.R. Hughes and J. Winget, Finite Rotation Effects in Numerical Integration of Rate Constitutive Equations arising in Large-Deformation Analysis, *Int. Journal for Numerical Methods in Eng.* **29**, 1862–1867 (1980).

[32] J.M. Kennedy and T.B. Belytschko, Theory and Application of a Finite Element Method for Arbitrary Lagrangian–Eulerian Fluids and Structures, *Nuclear Engineering and Design* **68**, 129–146 (1981).

[33] D.R. Kincaid and D.M. Young, Survey of Iterative Methods, Center for Numerical Analysis, The University of Texas at Austin, NA-135 (April 1978).

[34] R.D. Krieg, Unconditional Stability in Time Integration Methods, *J. Appl. Mechanics* **40**, 417–421 (1973).

[35] R.D. Krieg and S.W. Key, Transient Shell Response by Numerical Time Integration, *Int. J. Num. Methods in Eng.* **17**, 273–286 (1973).

[36] W.K. Liu, Development of Finite Element Procedures for Fluid-Structure Interaction, Ph.D. Thesis, Cal. Tech. (1981).

[37] W.K. Liu and T. Belytschko, Efficient Linear and Nonlinear Heat Conduction with a Quadrilateral Element, *Int. Journal for Numerical Methods in Engineering*, to appear.

[38] L.E. Malvern, *Introduction to the Mechanics of a Continuous Medium* (Prentice-Hall, Englewood Cliffs, NJ, 1969).

[39] R.M. McMeeking and J.R. Rice, Finite Element Formulations for Problems of Large Elastic–Plastic Deformations, *International Journal of Solids and Structures* **11**, 601–616 (1975).

[40] K. Miller and R. Miller, Moving Finite Elements, *SIAM J. Numerical Analysis* **18**, 1019–1032 (1981).

[41] H. Morand and R. Ohayon, Substructure Variation Analysis of the Vibrations of Coupled Fluid-structure Systems – Finite Element Results, *Int. Journal for Numerical Methods in Engineering*, **14**(5), 741–755 (1979).

[42] R.L. Mullen and T. Belytschko, Dispersion Analysis of Finite Element Semidiscretizations of the Two Dimensional Wave Equation, *Int. Journal for Numerical Methods in Engineering*, **18**, 11–29 (1982).

[43] R.L. Mullen and T. Belytschko, An Analysis of an Unconditionally Stable Explicit Method, *Computers and Structures* **16**, 691–696 (1983).

[44] S. Nemat-Nasser, Continuum Bases for Consistent Numerical Formulations of Finite Strains in Elastic and Inelastic Structures. in: *Finite Element Analysis of Transient Nonlinear Structural Behavior*, by T. Belytschko et al. eds., AMD Vol. 14 (ASME, New York, 1975) pp. 85–98.

[45] J.T. Oden, *Finite Elements of Nonlinear Continua* (McGraw-Hill, New York, 1972).

[46] J.T. Oden and J.N. Reddi, *An Introduction to The Mathematical Theory of Finite Elements* (Wiley–Interscience, New York, 1976).

[47] K.C. Park, The Solution of Variable Step Implicit Difference Equations for Dynamic Systems Analysis, in: Innovative Numerical Analysis for the Engineering Sciences, R. Shaw et al. eds. (University Press of Virginia, 1980).

[48] K.C. Park, An Improved Semi-implicit Method for Structural Dynamic Analysis, *Journal of Applied Mechanics* **49**(3), 589–593 (1982).

[49] D.W. Peaceman and H.H. Rachford, The Numerical Solution of Parabolic and Elliptic Differential Equations, *SIAM Journal* **3**, 28–41 (1955).

[50] W. Prager, *Introduction to the Mechanics of Continua* (Ginn, Boston, 1961) and (Dover, New York, 1973).

[51] J.R. Rice, R. McMeeking, D.M. Parks and P. Sorensen, Recent Finite Element Studies in Plasticity and Fracture Mechanics, *Computer Methods in Applied Mechanics and Engineering*, **17/18**, 411–442 (1979).

[52] R. Richtmyer and K.W. Morton, Difference Methods for Initial Value Problems (Interscience, New York, 1967).

[53] V.K. Saul'yev, *Integration of Equations of the Parabolic Type by the Method of Nets* (Pergamon Press, New York, 1964).

[54] D.H. Trujillo, An Unconditionally Stable Explicit Algorithm for Structural Dynamics, *Int. Journal for Numerical Methods in Engineering* **11**, 1579–1592 (1972).

[55] D.M. Trujillo, An Unconditionally Stable Explicit Algorithm for Finite Element Heat Conduction Analysis, *Nuclear Engineering and Design* **32**, 110–120 (1975).

[56] A. Wamberg, Rational Runge–Kutta Methods for Solving Systems of Ordinary Differential Equations, *Computing* **20**, 259–275 (1978).

[57] C.Y. Wang, Analysis of Nonlinear Fluid Structure Interaction Transient in Fast Reactors, Report ANL-78-103, Argonne National Laboratory, Argonne, IL (1975).

[58] L.C. Wellford and J.T. Oden, Discontinuous Finite-Element Approximations for the Analysis of Shades in Nonlinearly Elastic Materials, *Journal of Computational Physics* **19**(2), 179–210 (1975).

[59] S. Yaghmai and E.P. Popov, Incremental Analysis of Large Deflections of Shells of Revolution, *Int. Journal of Solids and Structures*, **7**, 1375–1393 (1971).

[60] O.C. Zienkiewicz, *The Finite Element Method*, 3rd Edition (McGraw-Hill, New York, 1977).

[61] O.C. Zienkiewicz and P. Bettess, Fluid-structure Dynamic Interaction and Wave Forces. An Introduction to Numerical Treatment, *Int. Journal Numerical Methods in Engineering* **13**, 1–16 (1978).

CHAPTER 2

Analysis of Transient Algorithms with Particular Reference to Stability Behavior

Thomas J.R. HUGHES*

Division of Applied Mechanics
Durand Building
Stanford University
Stanford, CA 94305, USA

*Professor of Mechanical Engineering.

Computational Methods for Transient Analysis
Edited by T. Belytschko and T.J.R. Hughes

A. Introduction

This chapter treats the analysis of transient algorithms. The emphasis, as with the rest of the book, is on problems of structural dynamics, although some other physical models are dealt with to a lesser extent. For example, heat conduction is discussed since it serves as a vehicle for introducing many aspects of the general theory simply, and because it is of interest in its own right. Additionally, a few results are presented for fluids, the motivation being the current interest in fluid-structure interaction problems.

The aims of this chapter are to be both tutorial and a state-of-the-art compilation of techniques and results. Thus sufficient detail is presented on basic points so that a newcomer to the subject can use the chapter for self-study, whereas the advanced researcher will find a compendium of information, brought together for the first time, and a number of areas identified in which further work needs to be done. Emphasis has been put on techniques which are basic to understanding the subject, and results included are felt to be of essentially permanent value.

Analytical techniques and results are presented for algorithms which are, by and large, in the mainstream from the points of view of past and present use and contemporary research interest. Unavoidably, certain topics had to be omitted to keep the present chapter within reasonable length. These included 'higher-order implicit methods', such as those introduced by Argyris et al. [1, 2], Geradin [15] and Hilber [20], for example. Also omitted are so-called alternating-direction, fractional-step and splitting methods. Although widely used in finite differences, these techniques have not yet significantly impacted structural dynamics which is dominated by finite element procedures. Already some valuable finite element research has been performed and the importance of these topics in structural dynamics is anticipated to increase in the ensuing years.

The preponderant philosophy in structural dynamics discretizations has been one in which spatial and temporal discretizations are done *independently*. One usually thinks of the spatial discretization as having been performed first, resulting in a system of *ordinary* differential equations to be solved. Although clearly the intent is to solve the original system of *partial* differential equations, the thrust of analytical interest in structural dynamics has been with the properties of the temporal discretization operator, given certain specific features of the spatially discrete system. This is the point of view taken in the present chapter. Thus, for the most part, the concern is with the solution of particular types of ordinary differential equation systems. One still occasionally hears the assertion that the mathematical theory of discrete methods for ordinary differential equation is complete and that nothing new needs to be said about particular systems such as those arising in structural dynamics. It is, however, now generally agreed that this assertion is false and that one will find little, if anything, in the classical ordinary differential equation literature of any value in modern structural dynamics computations.

Particular attention is focused herein on the concept of numerical stability. Stability plays several roles in the analysis and performance of temporal algorithms. Some notion of stability (generally a weak one, such as 'continuous dependence') is necessary to prove convergence. Somewhat stronger notions are generally desirable in order to assure qualitatively reasonable response consistent with the physical behavior of the governing partial differential equations. For example, in heat conduction solutions should be required to 'decay'. In linear analysis there seems to be little dispute as to what the correct definition of stability is. In nonlinear analysis there has been some controversy. In fact, algorithms which are stable according to one definition may be found unstable according to another. Throughout, every attempt has been made to be precise about the definitions of stability employed and the relevance of these concepts to the practical performance of algorithms.

An outline of the remainder of this chapter follows.

The chapter is divided into four major sections, labelled A, B, C and D. Section B, the longest, is divided into four major subsections, labelled I, II, III and IV. The sections and subsections are further divided into various topics. A decimal system is used for referring to the sections (e.g., B.III.4, D.2, etc.). Equation numbers are enclosed within parentheses. If in a section an equation is referenced which is outside the section, then the section is also given (e.g., "(35) of Section B.I", or as "(35) of Section I" if the referencing is done within Section B).

Figures and tables are numbered consecutively within the chapter, and references are ordered alphabetically according to the last name of the first author. Reference numbers appear in square brackets.

In Section B, linear systems are treated. In Section B.I, first-order symmetric systems, which arise in heat conduction, are considered. Standard one-step algorithms are applied to the spatially discrete ('semi-discrete') heat equation. The fundamental role of the single-degree-of-freedom (SDOF) model problem is illustrated, appropriate notions of stability, consistency and accuracy are introduced, and convergence is proved. This section is quite brief and serves to quickly introduce basic concepts to the reader in the simplest setting.

In Section B.II, second-order systems arising in linear structural dynamics are discussed. To illustrate important ideas, the Newmark family of methods [55] is introduced to time-discretize the semi-discrete equation of motion. Two members of the Newmark family are currently perhaps the most widely used algorithms in structural dynamics, namely, the average acceleration method (trapezoidal rule) and central difference method. In addition, several other members of the Newmark family are among the most widely discussed in the classical structural dynamics literature (e.g., the linear acceleration and Fox-Goodwin methods).

Mathematical concepts necessary for a precise analysis of the second-order case are introduced in Section B.II (e.g., natural matrix norms, spectral stability, etc.) and convergence is proved. Again, the fundamental importance of the associated SDOF model problem is stressed. A detailed discussion of stability conditions for the Newmark family is presented, and related topics, such as oscillatory response, high-frequency behavior and viscous damping effects, are treated. Appropriate measures of accuracy are introduced and discussed.

It is shown in both Sections B.I and B.II that conditional stability of finite element systems reduces to estimates of element eigenvalues. A brief summary of some known results for simple elements is presented.

Perhaps the most widely studied methods in the mathematics literature are the linear multi-step (LMS) methods, which are discussed in Section B.III. Standard LMS methods for first-order systems are presented and it is shown how to recast the semi-discrete heat equation and equation of motion within the first-order formalism. Stability analysis of LMS methods is performed by reduction to SDOF model equations. Different notions of stability, prominent in the mathematics literature, are described and compared with those commonly adopted in structural dynamics. The so-called stiffly-stable methods [14] are presented and their realm of applicability is delineated. Park's stiffly A-stable method [57], which has gained popularity in structural dynamics in recent years, is also presented.

A class of LMS methods for second-order systems is introduced which contains many structural dynamics algorithms as special cases. As examples, the following algorithms may be mentioned: Houbolt's method [26]: collocation methods [21] (e.g., Wilson-θ [62]); α-method [22] and the Newmark methods. A comparison of the various schemes is made based upon accuracy measures.

In Section B.IV, a brief discussion of some algorithms based upon operator and mesh partitions is presented. Attention is focused upon predictor-corrector [35] and predictor/multi-corrector explicit methods [40], and implicit-explicit finite element mesh partitions [35, 36]. These methods do not fall within the LMS category, and thus require different analytical techniques. The energy method is found useful for deducing stability behavior and a procedure for proving convergence is presented. (In another chapter of this book, Park and Felippa present a general theory of partitioned analysis procedures.)

In Section C nonlinear symmetric systems are presented. A simple class of nonlinear structural dynamics models is introduced and a practical approach to stability is described based upon consistent linearization. Convergence is established, accounting for the possibility of implicit-explicit mesh partitions. A stronger notion of stability is proposed – energy conservation – and an algorithm which possesses this property is presented.

The stability behavior of one-step methods in nonlinear heat conduction is also treated in Section C. A global energy criterion is used to illustrate the improper use of model-equation arguments in nonlinear analysis.

The chapter closes in Section D with a discussion of nonsymmetric operators, which arise in the modelling of fluid subdomains. The theory for nonsymmetric systems is in a much more primitive state than the theory for symmetric systems treated in previous sections. It is shown however, that typical implicit methods which are unconditionally stable for symmetric operators, retain this property in the nonsymmetric case, even in the presence of convective nonlinearity. However, matrix analyses, which work for symmetric operators, do not seem to be able to yield practical, stable time-step information for conditionally stable algorithms. As a result, classical finite difference procedures, such as von Neumann's method, are presently the only viable alternatives. There are many shortcomings to von Neumann's method (e.g., it is applicable only to constant coefficient cases on regular rectangular meshes, ignores boundary effects, etc.), nevertheless, it is capable of yielding practically useful information. Some recently derived results for the linear convection-diffusion equation are presented [18, 23, 51] and their use in practical computing is discussed.

B. Linear systems

B.I. First-order symmetric systems

B.I.1. Semi-discrete heat equation

The semi-discrete heat equation is

$$\boldsymbol{M}\dot{\boldsymbol{d}} + \boldsymbol{K}\boldsymbol{d} = \boldsymbol{F} \tag{1}$$

where $\boldsymbol{M}$ is the capacity matrix, $\boldsymbol{K}$ is the conductivity matrix, $\boldsymbol{F}$ is the heat supply vector, $\boldsymbol{d}$ is the temperature vector, and $\dot{\boldsymbol{d}}$ is the time (t) derivative of $\boldsymbol{d}$. The matrices $\boldsymbol{M}$ and $\boldsymbol{K}$ are assumed symmetric, $\boldsymbol{M}$ is positive-definite, and $\boldsymbol{K}$ is positive-semidefinite. The heat supply is a prescribed function of t. We write $\boldsymbol{F} = \boldsymbol{F}(t)$ for $t \in [0, T]$. Equation (1) is to be thought of as a coupled system of n_{eq} ordinary differential equations.

The initial-value problem consists of finding a function $\boldsymbol{d} = \boldsymbol{d}(t)$ satisfying (1) and the initial condition

$$\boldsymbol{d}(0) = \boldsymbol{D}_0 \tag{2}$$

where $\boldsymbol{D}_0$ is given.

B.I.2. One-step algorithms for the semi-discrete heat equation: generalized trapezoidal method

Perhaps the most well-known and commonly used algorithms for solving (1) are members of the so-called *generalized trapezoidal family of methods* which consists of the following equations:

$$\boldsymbol{M}\boldsymbol{v}_{n+1} + \boldsymbol{K}\boldsymbol{d}_{n+1} = \boldsymbol{F}_{n+1}\,, \tag{3}$$

$$\boldsymbol{d}_{n+1} = \boldsymbol{d}_n + \Delta t \boldsymbol{v}_{n+\alpha}\,, \tag{4}$$

$$\boldsymbol{v}_{n+\alpha} = (1-\alpha)\boldsymbol{v}_n + \alpha \boldsymbol{v}_{n+1}\,, \tag{5}$$

where $\boldsymbol{d}_n$ and $\boldsymbol{v}_n$ are the approximations to $\boldsymbol{d}(t_n)$ and $\boldsymbol{v}(t_n)$, respectively; $\boldsymbol{F}_{n+1} = \boldsymbol{F}(t_{n+1})$; Δt is the time step, assumed constant in the present circumstances; and α is a parameter, taken to be in the interval $[0, 1]$.

Some well-known members of the generalized trapezoidal family are identified in Table 1.

Table 1

α	Method
0	Forward differences; forward Euler
$\frac{1}{2}$	Trapezoidal rule; midpoint rule; Crank–Nicolson
$\frac{2}{3}$	Galerkin
1	Backward difference; backward Euler

In the case of $\alpha = 0$, the method is said to be *explicit*. The value of explicit methods may be seen from (3)–(5) if $\boldsymbol{M}$ is assumed 'lumped' (i.e., diagonal). In this case the solution may be advanced without the necessity of equation solving.

If $\alpha \neq 0$, the method is said to be *implicit*. In these cases a system of equations, with coefficient matrix $(\boldsymbol{M} + \alpha \Delta t \boldsymbol{K})$, needs to be solved at each step since $\boldsymbol{K}$ will not be diagonal in practice.

B.I.3. Convergence: definition

The primary requirement of the above algorithms is that they converge. We shall call an algorithm convergent if for t_n fixed and $\Delta t = t_n/n$, $\boldsymbol{d}_n \to \boldsymbol{d}(t_n)$ as $\Delta t \to 0$.

To establish the convergence of an algorithm, two additional notions must be considered: *stability* and *consistency*.We shall show later on that once stability and consistency are verified convergence is automatic.

In addition, we shall be concerned with the accuracy of an algorithm, especially the rate of convergence, and allied topics such as the behavior of the (spurious) higher modes and the locations of optimal points in time.

There are several techniques which can be employed to study the characteristics of an algorithm. In the present context the most revealing approach appears to be the modal approach (sometimes called spectral, or Fourier, analysis).

B.I.4. Reduction to a single degree-of-freedom (SDOF) model equation

In this section we decompose the problem under consideration into n_{eq} uncoupled scalar equations. The complete analysis of an algorithm thus reduces to the analysis of the simple single degree-of-freedom case, and concomitantly the behavior of the individual modes which comprise the solution is at once understood.

We recall some properties of the related eigenvalue problem which we will need in the subsequent developments.

Let $\boldsymbol{M}$ be symmetric positive definite and let $\boldsymbol{K}$ be symmetric positive-semidefinite (usually $\boldsymbol{K}$ is positive-definite too). Then

$$(\boldsymbol{K} - \lambda_l \boldsymbol{M})\boldsymbol{\psi}_l = \boldsymbol{0}, \quad l \in \{1, 2, \ldots, n_{eq}\} \tag{6}$$

where

$$0 \leq \lambda_1 \leq \lambda_2 \leq \cdots \leq \lambda_{n_{eq}} \tag{7}$$

and

$$\boldsymbol{\psi}_l^{\mathrm{T}} \boldsymbol{M} \boldsymbol{\psi}_m = \delta_{lm} \qquad \text{(orthornormality)} \tag{8}$$

where δ_{lm} is the Kronecker delta.

Furthermore, the eigenvectors $\{\boldsymbol{\psi}_l\}_1^{n_{eq}}$ constitute a *basis* for $\mathbb{R}^{n_{eq}}$, meaning that any element in $\mathbb{R}^{n_{eq}}$ can be written as a linear combination of the $\boldsymbol{\psi}_l$'s.

From the orthonormality property, it immediately follows that

$$\boldsymbol{\psi}_l^{\mathrm{T}} \boldsymbol{K} \boldsymbol{\psi}_m = \lambda_l \delta_{lm} \quad \text{(no sum)}. \tag{9}$$

The plan is to use the above properties of the eigenvalues and eigenvectors to decompose, first, the system of ordinary differential equations (i.e., (1)) then, second, the generalized trapezoidal algorithm (i.e., (3)–(5)).

Let

$$\boldsymbol{d}(t) = \sum_{m=1}^{n_{eq}} d_{(m)}(t) \boldsymbol{\psi}_m \tag{10}$$

from which it follows that

$$\dot{\boldsymbol{d}}(t) = \sum_{m=1}^{n_{eq}} \dot{d}_{(m)}(t) \boldsymbol{\psi}_m . \tag{11}$$

The scalar-valued functions $d_{(m)}(t)$ (Fourier coefficients) are obtained by premultiplying (10) by $\boldsymbol{\psi}_l^{\mathrm{T}} \boldsymbol{M}$ and invoking the orthonormality property. The result is

$$d_{(l)}(t) = \boldsymbol{\psi}_l^{\mathrm{T}} \boldsymbol{M} \boldsymbol{d}(t) . \tag{12}$$

(We have included the subscript in parentheses to avoid notational confusion.) The coefficients in (11) are obtained by differentiating (12):

$$\dot{d}_{(l)}(t) = \boldsymbol{\psi}_l^{\mathrm{T}} \boldsymbol{M} \dot{\boldsymbol{d}}(t)\,. \tag{13}$$

Employing these expressions in (1) and premultiplying by $\boldsymbol{\psi}_l^{\mathrm{T}}$ yields

$$\sum_{m=1}^{n_{\mathrm{eq}}} (\dot{d}_{(m)} \boldsymbol{\psi}_l^{\mathrm{T}} \boldsymbol{M} \boldsymbol{\psi}_m + d_{(m)} \boldsymbol{\psi}_l^{\mathrm{T}} \boldsymbol{K} \boldsymbol{\psi}_m) = \boldsymbol{\psi}_l^{\mathrm{T}} \boldsymbol{F} \tag{14}$$

from which it follows that

$$\dot{d}_{(l)} + \lambda_l d_{(l)} = F_{(l)}\,, \tag{15}$$

the lth modal equation, where $F_{(l)}(t) = \boldsymbol{\psi}_l^{\mathrm{T}} \boldsymbol{F}(t)$.

The initial condition for (15) is obtained by premuliplying (2) by $\boldsymbol{\psi}_l^{\mathrm{T}} \boldsymbol{M}$:

$$d_{(l)}(0) = \boldsymbol{\psi}_l^{\mathrm{T}} \boldsymbol{M} \boldsymbol{d}(0) = \boldsymbol{\psi}_l^{\mathrm{T}} \boldsymbol{M} \boldsymbol{D}_0 =: D_{0(l)}\,. \tag{16}$$

Solution of (15) for each mode allows us to construct the solution $\boldsymbol{d}$ of the original problem via (10).

To further simplify the subsequent writing we shall omit the lth modal subscript in (15) and (16). Thus our typical modal initial-value problem consists of the following equations

$$\dot{d} + \lambda d = F, \quad t \in [0, T]\,, \tag{17}$$

$$d(0) = D_0 \tag{18}$$

(This is sometimes called the single degree-of-freedom (SDOF) model problem.)

The steps for the decomposition of the generalized trapezoidal method are similar. The pertinent results are summarized below:

$$\boldsymbol{d}_n = \sum_{m=1}^{n_{\mathrm{eq}}} d_{n(m)} \boldsymbol{\psi}_m\,, \tag{19}$$

$$\boldsymbol{d}_{n+1} = \sum_{m=1}^{n_{\mathrm{eq}}} d_{n+1(m)} \boldsymbol{\psi}_m\,, \tag{20}$$

$$d_{n(l)} = \boldsymbol{\psi}_l^{\mathrm{T}} \boldsymbol{M} \boldsymbol{d}_n\,, \tag{21}$$

$$d_{n+1(l)} = \boldsymbol{\psi}_l^{\mathrm{T}} \boldsymbol{M} \boldsymbol{d}_{n+1}\,, \tag{22}$$

$$\sum_{m=1}^{n_{\mathrm{eq}}} [d_{n+1(m)} \boldsymbol{\psi}_l^{\mathrm{T}} (\boldsymbol{M} + \alpha \Delta t \boldsymbol{K}) \boldsymbol{\psi}_m - d_{n(m)} \boldsymbol{\psi}_l^{\mathrm{T}} (\boldsymbol{M} - (1-\alpha) \Delta t \boldsymbol{K}) \boldsymbol{\psi}_m] = \Delta t \boldsymbol{\psi}_l^{\mathrm{T}} \boldsymbol{F}_{n+\alpha}\,, \tag{23}$$

$$(1 + \alpha \Delta t \lambda_l) d_{n+1(l)} = (1 - (1-\alpha) \Delta t \lambda_l) d_{n(l)} + \Delta t F_{n+\alpha(l)}\,, \tag{24}$$

$$d_{0(l)} = D_{0(l)}\,. \tag{25}$$

As before, we will omit the lth modal subscript in (24) and (25). Thus we may write

$$(1+\alpha\Delta t\lambda)d_{n+1}=(1-(1-\alpha)\Delta t\lambda)d_n+\Delta tF_{n+\alpha}\,, \tag{26}$$

$$d_0=D_0\,. \tag{27}$$

(This is sometimes called the temporally discretized SDOF model problem.)

Remark 1. Note that directly applying the generalized trapezoidal method to (17) also results in (24). This fact is depicted in the commutative Diagram 1. That is, either path produces the same result.

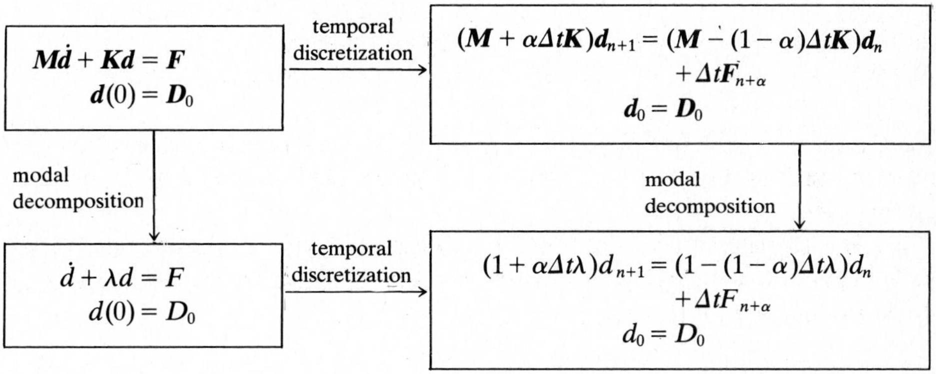

Diagram 1.

Remark 2. The convergence of $\boldsymbol{d}_n$ to $\boldsymbol{d}(t_n)$ can be established by showing the Fourier coefficients converge (i.e., $d_{n(l)}$ converges to $d_{(l)}(t_n)$ for each $l\in\{1,2,\ldots,n_{\text{eq}}\}$). The proof of this goes as follows. Let $\boldsymbol{e}(t_n)=\boldsymbol{d}_n-\boldsymbol{d}(t_n)$, the error in $\boldsymbol{d}_n$, and let $e_{(l)}(t_n)=d_{n(l)}-d_{(l)}(t_n)$ denote the lth Fourier component of $\boldsymbol{e}(t_n)$. Then

$$\begin{aligned}\boldsymbol{e}(t_n)^{\mathrm{T}}\boldsymbol{M}\boldsymbol{e}(t_n)&=\sum_{l,m=1}^{n_{\text{eq}}}(e_{(l)}(t_n)\boldsymbol{\psi}_l)^{\mathrm{T}}\boldsymbol{M}(e_{(m)}(t_n)\boldsymbol{\psi}_m)\\&=\sum_{l,m=1}^{n_{\text{eq}}}e_{(l)}(t_n)e_{(m)}(t_n)\boldsymbol{\psi}_l^{\mathrm{T}}\boldsymbol{M}\boldsymbol{\psi}_m\\&=\sum_{l,m=1}^{n_{\text{eq}}}e_{(l)}(t_n)e_{(m)}(t_n)\delta_{lm}\quad\text{(orthonormality)}\\&=\sum_{l=1}^{n_{\text{eq}}}(e_{(l)}(t_n))^2\end{aligned} \tag{28}$$

and so $\boldsymbol{e}(t_n)^{\mathrm{T}}\boldsymbol{M}\boldsymbol{e}(t_n)\rightarrow 0$ if and only if $e_{(l)}(t_n)\rightarrow 0$ for each $l\in\{1,2,\ldots,n_{\text{eq}}\}$. Since $\boldsymbol{M}$ is positive definite, $\boldsymbol{e}(t_n)^{\mathrm{T}}\boldsymbol{M}\boldsymbol{e}(t_n)\rightarrow 0$ if and only if $\boldsymbol{e}(t_n)\rightarrow\boldsymbol{0}$.

B.I.5. Stability

To motivate the appropriate notion of stability for the case under consideration, we shall investigate the behavior of the homogeneous model equation

$$\dot{d} + \lambda d = 0\,. \tag{29}$$

This is a first-order ordinary differential equation which can be easily solved. The solution at time t_{n+1} for initial value $d(t_n)$, $t_{n+1} > t_n$, is

$$d(t_{n+1}) = \exp(-\lambda(t_{n+1} - t_n))d(t_n) \tag{30}$$

from which it follows that

$$\begin{aligned} |d(t_{n+1})| &< |d(t_n)|\,, \quad \lambda > 0\,,\\ d(t_{n+1}) &= d(t_n)\,, \quad \lambda = 0\,. \end{aligned} \tag{31}$$

The conditions, (31), are what we wish to mimic in the temporally discrete case. The homogeneous temporally discrete model equation is

$$(1 + \alpha\Delta t\lambda)d_{n+1} = (1 - (1-\alpha)\Delta t\lambda)d_n\,. \tag{32}$$

Noting that $(1 + \alpha\Delta t\lambda) > 0$ for all allowable values of the parameters, we can write (32) as

$$d_{n+1} = Ad_n \tag{33}$$

where $A = (1 - (1-\alpha)\Delta t\lambda)/(1 + \alpha\Delta t\lambda)$ is the *amplification factor.*

Our stability requirements will be that

$$\begin{aligned} |d_{n+1}| &< |d_n|\,, \quad \lambda > 0\,,\\ d_{n+1} &= d_n\,, \quad \lambda = 0\,. \end{aligned} \tag{34}$$

From the definition of A, the second of (34) is automatic. The first condition is equivalent to insisting that

$$|A| < 1 \tag{35}$$

for $\lambda > 0$.

To determine the restrictions imposed upon α, Δt, and λ, it is convenient to rewrite (35) as $-1 < A < 1$, viz.

$$-1 < \frac{(1 - (1-\alpha)\Delta t\lambda)}{(1 + \alpha\Delta t\lambda)} < 1\,. \tag{36}$$

The right-hand inequality is satisfied for all allowable values of the parameters, and the left-hand inequality is satisfied whenever $\alpha \geq \frac{1}{2}$. However, if $\alpha < \frac{1}{2}$, the left-hand inequality requires $\lambda\Delta t < 2/(1 - 2\alpha)$. For a given λ this imposes an upper bound on the size of the allowable time step. The greater the λ, the smaller the time step required.

Remark. An algorithm for which stability imposes a time step restriction is called *conditionally stable.* An algorithm for which there is no time step restriction imposed by stability is called *unconditionally stable.*

The significance of the stability concept introduced may be seen from the following example:

Example 1. Note that the solution of (33) may be written

$$d_n = A^n d_0 . \tag{37}$$

If d_n is to behave like the solution of (29) (i.e., decay), A^n must converge (i.e., as $n \to \infty$, $A^n \to 0$). If $|A| < 1$, clearly this will be the case. On the other hand, if $|A| > 1$, growth will occur. Even if the value of $|A|$ is only slightly larger than 1, disastrous growth can occur.

To see what can happen, consider the numerical data in Table 2.

Table 2
A^n for various values of A and n

A	$n = 100$	$n = 1000$
0.99	0.37	4.32×10^{-5}
1.01	2.70	2.09×10^{4}
0.9	2.66×10^{-5}	1.75×10^{-46}
1.1	1.39×10^{4}	2.47×10^{41}

The necessity of keeping $|A| < 1$ should be clearly apparent, or else virtually unbounded errors will enter a computation.

Remark 1. In the conditionally stable case, the stability condition $\Delta t < 2/[(1-2\alpha)\lambda]$ must hold for all modes (i.e., all λ_l, $l \in \{1, 2, \ldots, n_{eq}\}$) in the system. The greatest λ_l, namely $\lambda_{n_{eq}}$ imposes the most stringent restriction upon the time step (i.e., $\Delta t < 2/[(1-2\alpha)\lambda_{n_{eq}}]$). In fact, for the heat conduction problem it can be shown that $\lambda_{n_{eq}} = O(h^{-2})$ where h is the mesh parameter and thus the critical time step amust satisfy $\Delta t <$ const. h^2. In a large system of equations (i.e., $n_{eq} \gg 1$), this condition is a severe constraint, thus unconditionally stable algorithms are generally preferred.

Remark 2. If (1) emanates from a finite element discretization, then $\lambda_{n_{eq}}$ may be conservatively bounded by the maximum eigenvalue of the individual elements [40, 43].

Remark 3. In Figure 1 the behavior of A as a function of $\lambda\Delta t$ is depicted for several values of α. The value of $\lambda\Delta t$ for which $A = 0$ is called the *oscillation limit* because for greater values, the sign of A^n changes from step to step. For the unconditionally stable algorithms (i.e., ones for which $\alpha \geq \frac{1}{2}$), the asymptotic value of the amplification factor, A_∞, satisfies $|A_\infty| \leq 1$. Thus for all fixed $\lambda\Delta t$, $|A| < 1$ and thus all spurious high modal components decay. However, for $\alpha = \frac{1}{2}$ (or very near

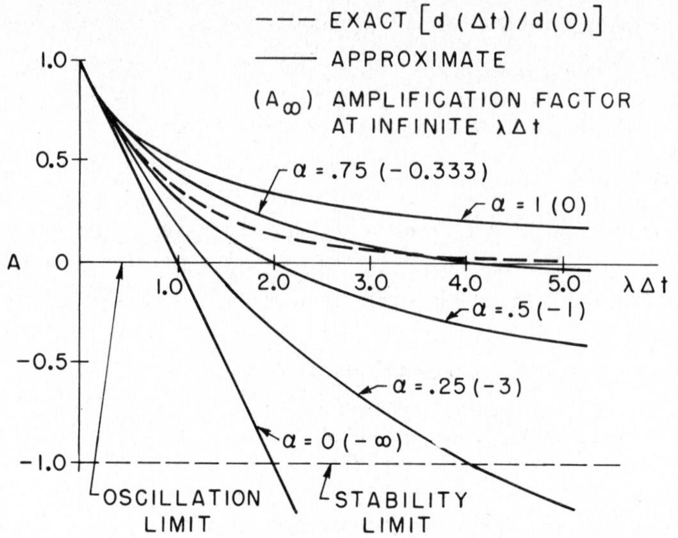

Fig. 1. Amplification factor for typical one-step methods.

$\frac{1}{2}$) and $\lambda\Delta t \gg 1$, $A \simeq -1$, and thus high modal components will behave like $(-1)^n$. This 'saw-tooth' pattern in time manifests itself frequently in computations.

In reporting data for a calculation in which this is the case, these spurious higher modes may be filtered out by reporting the step-to-step averages $(\boldsymbol{d}_{n+1} + \boldsymbol{d}_n)/2$, since $(-1)^{n+1} + (-1)^n = 0$. Other filtering possibilities (via time averaging the solutions and then using these averages in subsequent calculations) have been investigated by Wood and Lewis [63] and Lindberg [52].

Remark 4. Note that if $\alpha = 1$ and $\Delta t \to \infty$, the equilibrium, or steady state solution (i.e., one for which $\dot{\boldsymbol{d}} = \boldsymbol{0}$) is obtained, viz.

$$\boldsymbol{K}\boldsymbol{d}_{n+1} \to \boldsymbol{F}_{n+1}\,. \tag{38}$$

This fact can be exploited in a transient analysis computer program if the equilibrium solution is desired. Just select a value at Δt large enough so that (38) holds to machine precision.

We summarize the preceding in Table 3.

Table 3
Summary – Stability for the generalized trapezoidal methods

amplification factor:	$A = \dfrac{(1-(1-\alpha)\Delta t\lambda)}{(1+\alpha\Delta t\lambda)}$
stability requirement:	$\lvert A\rvert < 1$ for $\lambda = \lambda_{n_{eq}}$ (maximum eigenvalue)
unconditional stability:	$\alpha \geq \frac{1}{2}$
conditional stability:	$\alpha < \frac{1}{2}$, $\Delta t < 2/[(1-2\alpha)\lambda_{n_{eq}}]$

B.I.6. Consistency and convergence

The temporally discrete model problem may be written in the form

$$d_{n+1} - Ad_n - L_n = 0 \tag{39}$$

where the *load* $L_n = \Delta t F_{n+\alpha}/(1 + \alpha\Delta t\lambda)$. If we replaced d_n and d_{n+1} in the left-hand side of (39) by the corresponding exact values, we obtain an expression of the form

$$d(t_{n+1}) - Ad(t_n) - L_n = \tau(t_n) \tag{40}$$

where $\tau(t_n)$ is called the *local truncation* error. If $|\tau(t)| \leq c\Delta t^{k+1}$, for all $t \in [0, T]$, where c is a constant independent of Δt, and $k > 0$, the algorithm defined by (39) is called *consistent*; k is called the *order of accuracy* or *rate of convergence.*

Proposition 1. *The generalized trapezoidal methods are consistent, and furthermore* $k = 1$ *for all* $\alpha \in [0, 1]$, *except* $\alpha = \frac{1}{2}$, *in which case* $k = 2$.

The proof is a simple exercise. (Sketch: Expand $d(t_{n+1})$ and $d(t_n)$ about $t_{n+\alpha}$ in finite Taylor expansions and use the model equation of eliminate t-derivatives of $d(t_{n+\alpha})$.)

Remark 1. Thus the trapezoidal rule ($\alpha = \frac{1}{2}$) is the only member of the family of methods which is second-order accurate.

Theorem. *Consider equations* (39) *and* (40). *Let* t_n *be fixed* (n, *and consequently* Δt, *are allowed to vary*), *and assume the following conditions hold*:

(i) $|A| \leq 1$ (*stability*),
(ii) $|\tau(t)| \leq c\Delta t^{k+1}$, $t \in [0, T]$, $k > 0$ (*consistency*).

Then $e(t_n) \to 0$ *as* $\Delta t \to 0$.

Proof. Subtract (40) from (39):

$$e(t_{n+1}) = Ae(t_n) - \tau(t_n) \quad \text{('error equation')}. \tag{41}$$

Replace $e(t_n)$ on the right-hand side by the error equation for the previous step, i.e.,

$$e(t_n) = Ae(t_{n-1}) - \tau(t_{n-1}) \tag{42}$$

to obtain

$$e(t_{n+1}) = A^2e(t_{n-1}) - A\tau(t_{n-1}) - \tau(t_n). \tag{43}$$

Now repeat this procedure to eliminate $e(t_{n-1})$ from the right-hand side of (43), viz.

$$e(t_{n+1}) = A^3e(t_{n-2}) - A^2\tau(t_{n-2}) - A\tau(t_{n-1}) - \tau(t_n) \tag{44}$$

and so on. The final result is

$$e(t_{n+1}) = A^{n+1}e(0) - \sum_{i=0}^{n} A^i\tau(t_{n-i}) . \tag{45}$$

The first term on the right vanishes since $e(0) = d_0 - d(0) = 0$. We take absolute values of (45) evaluated at t_n instead of t_{n+1} and use some elementary facts as follows:

$$\begin{aligned} |e(t_n)| &= \left|\sum_{i=0}^{n-1} A^i\tau(t_{n-1-i})\right| \\ &\leq \sum_{i=0}^{n-1} |A|^i|\tau(t_{n-1-i})| \\ &\leq \sum_{i=0}^{n-1} |\tau(t_{n-1-i})| \qquad \text{(stability)} \\ &\leq n \max|\tau(t)|, \quad t \in [0, T] \\ &\leq nc\Delta t^{k+1} \qquad \text{(consistency)} \\ &= t_n c\Delta t^k . \end{aligned} \tag{46}$$

Therefore $e(t_n) \to 0$ as $\Delta t \to 0$, and furthermore the rate of convergence is k (i.e., $e(t_n) = O(\Delta t^k)$).

Remark 2. This result establishes the convergence of the generalized trapezoidal methods. The optimal rate of convergence is 2 and is attained by the trapezoidal rule.

Remark 3. This theorem is a particular example of perhaps the most celebrated theorem in numerical analysis, the Lax equivalence theorem, which may be stated as "consistency plus stability is necessary and sufficient for convergence." The Lax equivalence theorem and appropriate generalizations have significant consequences in other areas of mathematics (see, for example, Chorin et al. [8]).

B.II. Second order symmetric systems

B.II.1. Semi-discrete equation of motion

The semi-discrete equation of motion is

$$\boldsymbol{M}\ddot{\boldsymbol{d}} + \boldsymbol{C}\dot{\boldsymbol{d}} + \boldsymbol{K}\boldsymbol{d} = \boldsymbol{F} \tag{1}$$

where $\boldsymbol{M}$ is the mass matrix, $\boldsymbol{C}$ is the viscous damping matrix, $\boldsymbol{K}$ is the stiffness matrix, $\boldsymbol{F}$ is the vector of applied forces, and $\boldsymbol{d}$, $\dot{\boldsymbol{d}}$ and $\ddot{\boldsymbol{d}}$ are the displacement, velocity and acceleration vectors, respectively. We take $\boldsymbol{M}$, $\boldsymbol{C}$, and $\boldsymbol{K}$ to be symmetric, $\boldsymbol{M}$ is postive-definite, and $\boldsymbol{C}$ and $\boldsymbol{K}$ are positive-semidefinite.

The initial-value problem for (1) consists of finding a displacement, $\boldsymbol{d} = \boldsymbol{d}(t)$, satisfying (1) and the given initial data:

$$\boldsymbol{d}(0) = \boldsymbol{D}_0 , \tag{2}$$

$$\boldsymbol{v}(0) = \boldsymbol{V}_0 . \tag{3}$$

B.II.2. One-step algorithms for the semi-discrete equation of motion: *Newmark method*

Perhaps the most widely used family of direct methods for solving (1)–(3) is the Newmark family [55], which consists of the following equations:

$$\boldsymbol{M}\boldsymbol{a}_{n+1} + \boldsymbol{C}\boldsymbol{v}_{n+1} + \boldsymbol{K}\boldsymbol{d}_{n+1} = \boldsymbol{F}_{n+1} , \tag{4}$$

$$\boldsymbol{d}_{n+1} = \boldsymbol{d}_n + \Delta t \boldsymbol{v}_n + \frac{\Delta t^2}{2}\{(1-2\beta)\boldsymbol{a}_n + 2\beta \boldsymbol{a}_{n+1}\} , \tag{5}$$

$$\boldsymbol{v}_{n+1} = \boldsymbol{v}_n + \Delta t\{(1-\gamma)\boldsymbol{a}_n + \gamma \boldsymbol{a}_{n+1}\} , \tag{6}$$

where $\boldsymbol{d}_n$, $\boldsymbol{v}_n$, and $\boldsymbol{a}_n$ are the approximations of $\boldsymbol{d}(t_n)$, $\dot{\boldsymbol{d}}(t_n)$, and $\ddot{\boldsymbol{d}}(t_n)$, respectively. Eq. (4) is simply the equation of motion in terms of the approximate solution, and (5) and (6) are finite difference formulas[1] describing the evolution of the approximate solution. The parameters β and γ determine the stability and accuracy characteristics of the algorithm under condieration. Eqs. (4)–(6) may be thought of as three equations for determining the three unknowns $\boldsymbol{d}_{n+1}$, $\boldsymbol{v}_{n+1}$, and $\boldsymbol{a}_{n+1}$, it being assumed that $\boldsymbol{d}_n$, $\boldsymbol{v}_n$, and $\boldsymbol{a}_n$ are known from the previous step's calculations.

The Newmark family contains as special cases many well-known and widely used methods. Some of these classical methods are discussed in the following examples:

Example 1. (*Constant*) *average acceleration method.* The average acceleration method corresponds to parameter values $\beta = \frac{1}{4}$ and $\gamma = \frac{1}{2}$.[2] The method is implicit, but unconditionally stable, and is one of the most effective and popular techniques for structural dynamics problems. The average acceleration method may also be derived by applying the trapezoidal rule (see Section I.3) to the first-order form of the equations of motion, as follows. Let

$$y = \begin{Bmatrix} \boldsymbol{d} \\ \dot{\boldsymbol{d}} \end{Bmatrix} \tag{7}$$

and

$$f(y, t) = \begin{Bmatrix} \dot{\boldsymbol{d}} \\ \hline \boldsymbol{M}^{-1}(\boldsymbol{F}(t) - \boldsymbol{C}\dot{\boldsymbol{d}} - \boldsymbol{K}\boldsymbol{d}) \end{Bmatrix} . \tag{8}$$

[1]Zienkiewicz [64] has shown how to derive the Newmark methods from finite element concepts in the time domain.

[2]The Newmark methods are second-order accurate if and only if $\gamma = \frac{1}{2}$. Notions of accuracy and stability are made precise in subsequent sections.

Then

$$\dot{y} = f(y, t) \tag{9}$$

defines a first-order system of $2n_{\mathrm{eq}}$ ordinary differential equations equivalent to (1). The trapezoidal-rule approximation of (9) may be written in terms of the following two equations:

$$z_{n+1} = f(y_{n+1}, t_{n+1}), \tag{10}$$

$$y_{n+1} = y_n + \frac{\Delta t}{2}(z_n + z_{n+1}). \tag{11}$$

Here y_n and z_n are the approximations of $y(t_n)$ and $\dot{y}(t_n)$. It may be easily verified that (10) and (11) are equivalent to (4)–(6) with $\beta = \frac{1}{4}$ and $\gamma = \frac{1}{2}$.

Example 2. *Linear acceleration method.* This method corresponds to parameter values $\beta = \frac{1}{6}$ and $\gamma = \frac{1}{2}$. It is implicit and conditionally stable.

Example 3. *Fox–Goodwin method.* The parameter values for this scheme are $\beta = \frac{1}{12}$ and $\gamma = \frac{1}{2}$. It is implicit and conditionally stable. If the viscous damping term is absent (i.e., $\boldsymbol{C} = \boldsymbol{0}$), the Fox–Goodwin method is fourth-order accurate. Lambert [49] calls this technique the 'royal road method'.

Remark 1. By virtue of the fact that the methods of Examples 2 and 3 are implicit and *conditionally* stable, they are not felt to be economically competitive for large-scale systems when compared to implicit and *unconditionally* stable techniques such as the average acceleration method.

Example 4. *Central difference method.* The central differences method corresponds to parameter values $\beta = 0$ and $\gamma = \frac{1}{2}$. The name of the method derives from the fact that (5) and (6) may be combined to yield

$$\boldsymbol{v}_n = (\boldsymbol{d}_{n+1} - \boldsymbol{d}_{n-1})/(2\Delta t), \tag{12}$$

$$\boldsymbol{a}_n = (\boldsymbol{d}_{n+1} - 2\boldsymbol{d}_n + \boldsymbol{d}_{n-1})/\Delta t^2, \tag{13}$$

the classical first and second central difference approximations, respectively.

The central difference method is conditionally stable. However, when $\boldsymbol{M}$ and $\boldsymbol{C}$ are diagonal, it is explicit. When the time step restriction is not too severe, such as is often the case in elastic wave-propagation problems, the central difference method is generally the most economical direct integration procedure and is thus widely used.

Remark 2. The methods discussed in the previous examples are numerically compared in Goudreau and Taylor [16].

B.II.3. Convergence analysis of the Newmark methods

The convergence analysis of the Newmark family of methods is similar to the convergence analysis given in Section I: (i) reduction to an SDOF model problem; (ii) a suitable notion of stability is defined and shown to hold under certain circumstances; (iii) the local truncation error is determined, from which the order of accuracy is obtained; and (iv) the latter two conditions are used to prove convergence for the SDOF problem.

B.II.3.a. Reduction to an SDOF problem

The reduction to an SDOF problem is facilitated by way of the so-called 'undamped eigenproblem', given as follows:

$$(\boldsymbol{K} - \lambda \boldsymbol{M})\boldsymbol{\psi} = \boldsymbol{0}\,. \tag{14}$$

The eigenpairs $\{\lambda_l, \boldsymbol{\psi}_l\}_1^{n_{eq}}$ are used in the reduction of (1)–(6) to SDOF form. The only additional complication when compared with the first-order case (see Section I) concerns the viscous damping matrix which we shall assume is a linear combination of the mass and stiffness matrices, viz.

$$\boldsymbol{C} = a\boldsymbol{M} + b\boldsymbol{K} \qquad (\text{'Rayleigh damping'}) \tag{15}$$

where a and b are non-negative constants. Since

$$\boldsymbol{\psi}_l^{\mathrm{T}}\boldsymbol{M}\boldsymbol{\psi}_m = \delta_{lm}\,, \tag{16}$$

$$\boldsymbol{\psi}_l^{\mathrm{T}}\boldsymbol{K}\boldsymbol{\psi}_m = \lambda_l\delta_{lm} \quad (\text{no sum})\,, \tag{17}$$

it follows that

$$\boldsymbol{\psi}_l^{\mathrm{T}}\boldsymbol{C}\boldsymbol{\psi}_m = (a + b\lambda_l)\delta_{lm} \quad (\text{no sum})\,. \tag{18}$$

In keeping with the standard terminologies of the problem at hand, we define $\omega_l = (\lambda_l)^{1/2}$, the lth *undamped frequency of vibration*, and $\xi_l = (a/\omega_l + b\omega_l)/2$, the lth *modal damping ratio*.

Applying the modal decomposition procedure to (1)–(6) yields the corresponding SDOF equations which, if we suppress the modal index l, appear as

$$\ddot{d} + 2\xi\omega\dot{d} + \omega^2 d = F\,, \tag{19}$$

$$d(0) = D_0\,, \tag{20}$$

$$v(0) = V_0\,, \tag{21}$$

$$a_{n+1} + 2\xi\omega v_{n+1} + \omega^2 d_{n+1} = F_{n+1}\,, \tag{22}$$

$$d_{n+1} = d_n + \Delta t v_n + \frac{\Delta t^2}{2}\{(1 - 2\beta)a_n + 2\beta a_{n+1}\}\,, \tag{23}$$

$$v_{n+1} = v_n + \Delta t\{(1 - \gamma)a_n + \gamma a_{n+1}\}\,, \tag{24}$$

$$d_0 = D_0\,, \tag{25}$$

$$v_0 = V_0\,. \tag{26}$$

In the present case, we shall be concerned with the convergence of displacements and velocities. Following similar lines of reasoning to those used in Section I.5, it can be easily verified that

$$\boldsymbol{d}_n \to \boldsymbol{d}(t_n) \Leftrightarrow d_n \to d(t_n)\,, \tag{27}$$

$$\boldsymbol{v}_n \to \dot{\boldsymbol{d}}(t_n) \Leftrightarrow v_n \to \dot{d}(t_n)\,. \tag{28}$$

That is, proof of convergence for the generic SDOF case implies convergence for the corresponding matrix problem. The situation is summarized in the commutative Diagram 2.

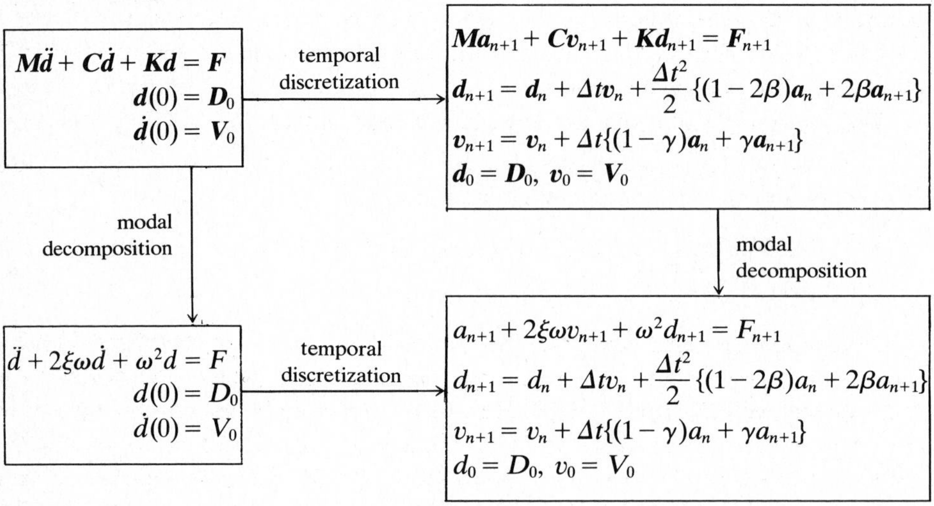

Diagram 2.

The next step in the analysis is to recast (22)–(24) in a first-order form analogous to equation (39) of Section I.7:

$$y_{n+1} = \boldsymbol{A}y_n + \boldsymbol{L}_n \tag{29}$$

where $\boldsymbol{A}$ is the *amplification matrix*, $\boldsymbol{L}_n$ is the *load vector*, and

$$y_n = \begin{Bmatrix} d_n \\ v_n \end{Bmatrix}. \tag{30}$$

by eliminating a_n and a_{n+1} from (23) and (24) through repeated use of (22), we arrive at the following expression for $\boldsymbol{A}$ and $\boldsymbol{L}_n$:

$$\boldsymbol{A} = \boldsymbol{A}_1^{-1}\boldsymbol{A}_2\,, \tag{31}$$

$$\boldsymbol{L}_n = \boldsymbol{A}_1^{-1}\begin{Bmatrix} \frac{\Delta t^2}{2}[(1-2\beta)F_n + 2\beta F_{n+1}] \\ \Delta t\ [(1-\gamma)F_n + \gamma F_{n+1}] \end{Bmatrix} \tag{32}$$

where

$$A_1 = \begin{bmatrix} 1+\Delta t^2\beta\omega^2 & 2\Delta t^2\beta\xi\omega \\ \Delta t\gamma\omega^2 & 1+2\Delta t\gamma\xi\omega \end{bmatrix}, \tag{33}$$

$$A_2 = \begin{bmatrix} 1-\frac{\Delta t^2}{2}(1-2\beta)\omega^2 & \Delta t(1-\Delta t(1-2\beta)\xi\omega) \\ -\Delta t(1-\gamma)\omega^2 & 1-2\Delta t(1-\gamma)\xi\omega \end{bmatrix}. \tag{34}$$

Remark 1. A thorough and incisive analysis of the Newmark method is contained in Hilber [20]. Much of the analysis in this and subsequent sections taps heavily upon [20].

Remark 2. Difference equations such as (29) are often referred to as 'one-step, multivalue methods' [14], the number of values being equal to the dimension of the vector $\boldsymbol{y}$, which in our case is 2.

The vector of local truncation errors, $\boldsymbol{\tau}$ is defined by

$$\boldsymbol{y}(t_{n+1}) = \boldsymbol{A}\boldsymbol{y}(t_n) + \boldsymbol{L}_n + \boldsymbol{\tau}(t_n) \tag{35}$$

where

$$\boldsymbol{y}(t_n) = \begin{Bmatrix} d(t_n) \\ \dot{d}(t_n) \end{Bmatrix}. \tag{36}$$

Obtaining an explicit expression for $\boldsymbol{\tau}$ is tedious, due to the complicated nature of $\boldsymbol{A}$ and $\boldsymbol{L}_n$. However, one can use an analysis similar to the one described in Section I to infer the correct exponent of Δt in the entries of $\boldsymbol{\tau}$. It may be argued that $\boldsymbol{\tau}(t) = O(\Delta t^{k+1})$ for all $t \in [0, T]$, where $k = 2$ if $\gamma = \frac{1}{2}$ and $k = 1$ otherwise.

The stability of the Newmark methods is determined by properties of the amplification matrix. The convergence proof follows along the lines of the theorem presented in Section I.6. Here, the equation

$$\boldsymbol{e}(t_n) = \boldsymbol{A}^n\boldsymbol{e}(0) - \sum_{i=0}^{n-1} \boldsymbol{A}^i\boldsymbol{\tau}(t_{n-1-i}) \tag{37}$$

determined from (29) and (35), above, is used in place of (41), Section I.6.

To make these ideas more precise requires familiarity with some preliminary results which are described as follows.

B.II.3.b. Natural matrix norms

The present discussion is based upon Noble [56], pp. 425–431.

Let $\boldsymbol{y} \in \mathbb{R}^m$ denote an m-dimensional column vector, i.e.,

$$\boldsymbol{y} = \begin{Bmatrix} y_1 \\ y_2 \\ \vdots \\ y_m \end{Bmatrix} \tag{38}$$

where each y_i is a real number. Under the usual rules of vector addition and scalar multiplication, $\mathbb{R}^m$ is a linear space. Various norms, $\|\cdot\|$, may be introduced. For instance, we have the following well-known examples:

1-norm:

$$\|y\|_{(1)} = |y_1| + |y_2| + \cdots + |y_m| . \tag{39}$$

2-norm or Euclidean norm:

$$\|y\|_{(2)} = (y_1^2 + y_2^2 + \cdots + y_m^2)^{1/2} . \tag{40}$$

∞-norm or max norm:

$$\|y\|_{(\infty)} = \max_i |y_i| . \tag{41}$$

These are all special cases of the so-called *p-norm*:

$$\|y\|_{(p)} = (|y_1|^p + |y_2|^p + \cdots + |y_m|^p)^{1/p} . \tag{42}$$

It is a well-known result that all norms on finite-dimensional vector spaces are equivalent. This means convergence in one norm implies convergence in any other, and so we may employ a generic norm $\|\cdot\|$ in our subsequent analyses.

Let $\mathbb{R}^{m^2}$ denote the set of all $m \times m$ matrices.

Definition 1. A *matrix norm* is a map $\|\cdot\|: \mathbb{R}^{m^2} \to \mathbb{R}$ such that for all $\boldsymbol{A}$ and $\boldsymbol{B} \in \mathbb{R}^{m^2}$ and $\alpha \in \mathbb{R}$ the following conditions hold:

(i) $\|\boldsymbol{A}\| \geq 0$ and $\|\boldsymbol{A}\| = 0 \Leftrightarrow \boldsymbol{A} = \boldsymbol{0}$ (positive definiteness),

(ii) $\|\alpha \boldsymbol{A}\| = |\alpha| \, \|\boldsymbol{A}\|$,

(iii) $\|\boldsymbol{A} + \boldsymbol{B}\| \leq \|\boldsymbol{A}\| + \|\boldsymbol{B}\|$ (triangle inequality),

(iv) $\|\boldsymbol{AB}\| \leq \|\boldsymbol{A}\| \, \|\boldsymbol{B}\|$.

Let $\|\boldsymbol{A}\|$ denote any matrix norm of $\boldsymbol{A}$ and let $\|\boldsymbol{y}\|$ denote any vector norm of $\boldsymbol{y}$. A matrix norm and vector norm are said to be *compatible* if

$$\underbrace{\|\boldsymbol{Ay}\|}_{\substack{\text{vector}\\ \text{norm}}} \leq \underbrace{\|\boldsymbol{A}\|}_{\substack{\text{matrix}\\ \text{norm}}} \underbrace{\|\boldsymbol{y}\|}_{\substack{\text{vector}\\ \text{norm}}} . \tag{43}$$

The question arises: Are there any matrix norms compatible with a given vector norm? The answer is affirmative and a compatible matrix norm may be defined by the following relation:

$$\underbrace{\|\boldsymbol{A}\|}_{\substack{\text{matrix}\\ \text{norm}}} = \max(\underbrace{\|\boldsymbol{Ay}\|}_{\substack{\text{vector}\\ \text{norm}}} / \underbrace{\|\boldsymbol{y}\|}_{\substack{\text{vector}\\ \text{norm}}}) . \tag{44}$$

The right-hand side is well defined (see Noble [56] for additional details). The matrix norm obtained from (44) is called the *natural matrix norm* compatible with the given vector norm. Examples of natural matrix norms corresponding to (39)–(41) are (resp.)

$$\|A\|_{(1)} = \max_j \left(\sum_{i=1}^{m} |A_{ij}| \right) \quad \text{(maximum absolute column sum)}, \tag{45}$$

$$\|\boldsymbol{A}\|_{(2)} = (\text{maximum eigenvalue of } \boldsymbol{A}^{\mathrm{T}}\boldsymbol{A})^{1/2}, \tag{46}$$

[3]

$$\|\boldsymbol{A}\|_{(\infty)} = \max_i \left(\sum_{j=1}^{m} |A_{ij}| \right) \quad \text{(maximum absolute row sum)}. \tag{47}$$

(Proofs may be found in Noble [56].)

Let $\boldsymbol{A}$ be a given $m \times m$ matrix with eigenvalues $\{\lambda_i\}_1^m$. In general, the λ_i's will be complex numbers. The modulus of λ_i, denoted $|\lambda_i|$, is defined to be $(\lambda_i\bar{\lambda}_i)^{1/2}$, where the bar denotes complex conjugate. The *spectral radius* of $\boldsymbol{A}$, denoted $\rho(\boldsymbol{A})$, is defined by

$$\rho(\boldsymbol{A}) = \max_i |\lambda_i| . \tag{48}$$

If $\|\cdot\|$ denotes *any* natural matrix norm, then it can be shown that

$$\rho(\boldsymbol{A}) \leq \|\boldsymbol{A}\| . \tag{49}$$

The proof of this fact is contained in Noble [56]. The following example illustrates that the spectral radius of a matrix may be arbitrarily small (in fact, zero) while its (natural) norm may be arbitrarily large.

Example 1. Consider the matrix

$$A = \begin{bmatrix} 0 & k \\ 0 & 0 \end{bmatrix} \tag{50}$$

where $k \gg 1$. The eigenvalues of $\boldsymbol{A}$ are zero, consequently $\rho(\boldsymbol{A}) = 0$. On the other hand,

$$\|\boldsymbol{A}\|_{(1)} = \|A\|_{(2)} = \|A\|_{(\infty)} = k . \tag{51}$$

Henceforth we shall employ natural matrix norms, unless otherwise specified. By the remarks made following (42), this choice represents no loss of generality.

Definition 2. A *Jordan block*, $\boldsymbol{J}$, is a square matrix which takes the form:

$$\boldsymbol{J} = \begin{bmatrix} \lambda & 1 & & & \\ & \lambda & 1 & & \\ & & \ddots & \ddots & \\ & & & \ddots & 1 \\ & & & & \lambda \end{bmatrix} \tag{52}$$

where λ is a complex number and all omitted terms are zero.

[3] The matrix $\boldsymbol{A}^{\mathrm{T}}\boldsymbol{A}$ is symmetric and positive semi-definite, so its eigenvalues are real and non-negative. Thus the right-hand side of (46) is well defined.

Definition 3. Given an arbitrary square matrix $\boldsymbol{A}$, there exists a non-singular matrix $\boldsymbol{Q}$ such that

$$\boldsymbol{Q}^{-1}\boldsymbol{A}\boldsymbol{Q} = \boldsymbol{J} = \begin{bmatrix} \boldsymbol{J}_1 & & & \\ & \boldsymbol{J}_2 & & \\ & & \ddots & \\ & & & \boldsymbol{J}_k \end{bmatrix} \tag{53}$$

where the $\boldsymbol{J}_i$'s are Jordan blocks and all omitted terms are zero; $\boldsymbol{J}$, in (53), is called the *Jordan form* of $\boldsymbol{A}$. Furthermore, the diagonal elements of the $\boldsymbol{J}_i$'s are the eigenvalues of $\boldsymbol{A}$.

Remark 2. An important special case of the Jordan form arises when $\boldsymbol{A}$ has linearly independent eigenvectors. Let $\boldsymbol{P}$ denote the matrix of eigenvectors of $\boldsymbol{A}$. Then

$$\boldsymbol{P}^{-1}\boldsymbol{A}\boldsymbol{P} = \boldsymbol{\Lambda} = \operatorname{diag}(\lambda_1, \lambda_2, \ldots, \lambda_m) \tag{54}$$

where the λ_i's are the eigenvalues of $\boldsymbol{A}$ and $\operatorname{diag}(\lambda_1, \lambda_2, \ldots, \lambda_m)$ is the diagonal matrix of eigenvalues. A sufficient condition for $\boldsymbol{A}$ to have linearly independent eigenvectors is that it have distinct eigenvalues. However, this condition is not necessary. For example, the identity matrix

$$\begin{bmatrix} 1 & 0 \\ 0 & 1 \end{bmatrix} \tag{55}$$

has multiple eigenvalues (i.e., $\lambda_1 = \lambda_2 = 1$), but its eigenvectors

$$\begin{Bmatrix} 1 \\ 0 \end{Bmatrix}, \quad \begin{Bmatrix} 0 \\ 1 \end{Bmatrix} \tag{56}$$

are linearly independent.

B.II.3.c. Spectral stability

Stability in the present situation is concerned with the rate of growth, or decay, of powers of the amplification matrix. To prove a convergence theorem we will need that

$$\|\boldsymbol{A}^n\| \le \text{const.}, \quad \forall n. \tag{57}$$

It turns out that this requirement is guaranteed if the following conditions are satisfied:

(i) $\rho(\boldsymbol{A}) \le 1$.

(ii) Eigenvalues of $\boldsymbol{A}$ of multiplicity greater than one, are strictly less than one in modulus.

A matrix $\boldsymbol{A}$ satisfying (i) and (ii) is said to be (*spectrally*) *stable.*

To see that (i) and (ii) imply (57) for the case of a 2×2 amplification matrix, we will employ the Jordan form of $\boldsymbol{A}$ and any natural matrix norm. The demonstration will be divided into two parts corresponding to (53) and (54). We begin with the latter case.

Case 1. ***A*** *has linearly independent eigenvectors.* For this case we have

$$\begin{aligned}\boldsymbol{A}^n &= (\boldsymbol{P}\boldsymbol{\Lambda}\boldsymbol{P}^{-1})^n \qquad \text{(by eq. 54)}\\ &= \boldsymbol{P}\boldsymbol{\Lambda}^n\boldsymbol{P}^{-1}. \end{aligned} \tag{58}$$

Taking the norm of (58) results in

$$\begin{aligned}\|\boldsymbol{A}^n\| &= \|\boldsymbol{P}\boldsymbol{\Lambda}^n\boldsymbol{P}^{-1}\|\\ &\le \|\boldsymbol{P}\|\,\|\boldsymbol{\Lambda}^n\|\,\|\boldsymbol{P}^{-1}\| \qquad \text{(by (iv), Def. 1)}. \end{aligned} \tag{59}$$

From (59) we see that (57) will hold if

$$\|\boldsymbol{\Lambda}^n\| \le \text{const.}, \quad \forall n. \tag{60}$$

Since

$$\boldsymbol{\Lambda}^n = \begin{bmatrix} \lambda_1^n & 0 \\ 0 & \lambda_2^n \end{bmatrix} \tag{61}$$

and (i) is assumed to hold, the eigenvalues satisfy $|\lambda_i^n| \le 1$, $i = 1, 2$. Thus we conclude that (57) holds.

It is also clear from (61) that if either $|\lambda_1|$ or $|\lambda_2|$ is greater than 1, then $\|\boldsymbol{\Lambda}^n\|$ would be unbounded.

Case 2. ***A*** *has linearly dependent eigenvectors.* Analogous to the derivation of (59), starting with (53), we obtain:

$$\|\boldsymbol{A}^n\| \le \|\boldsymbol{Q}\|\,\|\boldsymbol{J}^n\|\,\|\boldsymbol{Q}^{-1}\|. \tag{62}$$

Thus we need that

$$\|\boldsymbol{J}^n\| \le \text{const.}, \quad \forall n, \tag{63}$$

to insure that (57) holds. A simple calculation reveals that

$$\boldsymbol{J}^n = \begin{bmatrix} \lambda^n & (n-1)\lambda^{n-1} \\ 0 & \lambda^n \end{bmatrix}, \quad n > 1. \tag{64}$$

Invoking (ii) guarantees that $|\lambda|^n < 1$ and $|(n-1)\lambda^{n-1}| < \text{const.}$ In fact, as $n \to \infty$, $\boldsymbol{J}^n \to \boldsymbol{0}$. Thus we see that (57) holds.

The necessity of (ii) can be seen from the term $(n-1)\lambda^{n-1}$ in (64). If $|\lambda| = 1$, then $(n-1)|\lambda|^{n-1} = n-1$ and thus $\|\boldsymbol{J}^n\|$ would become unbounded as $n \to \infty$.

Remark 3. The preceding arguments generalize to an $m \times m$ matrix $\boldsymbol{A}$. See Gear [14], p. 177 for the general case.

B.II.3.d. Convergence

Now we are in a position to prove a convergence theorem.

Theorem. *Consider* (37) *at a fixed* t_n *and assume that* $\boldsymbol{A}$ *is spectrally stable. Furthermore, assume that the local truncation error satisfies the following consistency condition*:

$$\|\boldsymbol{\tau}(t)\| \leq c \cdot \Delta t^{k+1}, \quad t \in [0, T]. \tag{65}$$

Then $\boldsymbol{e}(t_n) \rightarrow \boldsymbol{0}$ *as* $\Delta t \rightarrow 0$.

Proof. We take the norm of (37):

$$\begin{aligned}
\|\boldsymbol{e}(t_n)\| &= \left\| \boldsymbol{A}^n \boldsymbol{e}(0) - \sum_{i=0}^{n-1} \boldsymbol{A}^i \boldsymbol{\tau}(t_{n-1-i}) \right\| \\
&\leq \sum_{i=0}^{n-1} \|\boldsymbol{A}^i\| \, \|\boldsymbol{\tau}(t_{n-1-i})\| \quad \text{(by compatibility, i.e., eq. (43))} \\
&\leq \text{const.} \sum_{i=0}^{n-1} \|\boldsymbol{\tau}(t_{n-1-i})\| \quad \text{(stability)} \\
&\leq \text{const. } n \max \|\boldsymbol{\tau}(t)\|, \quad t \in [0, T], \\
&\leq \text{const. } t_n \cdot \Delta t^k \qquad \text{(consistency)},
\end{aligned} \tag{66}$$

from which the assertion follows.

Remark 4. Observe that the theorem holds for an arbitrary one-step multivalue method of the form (29). Later on, when we study multistep methods, we will show that they can be recast in the form of one-step multivalue methods, thus one may invoke the preceding theorem to deduce convergence.

B.II.4. Stability conditions for the Newmark method

We will now determine the stability characteristics of the Newmark method by examining the spectral properties of $\boldsymbol{A}$. By virtue of the fact that $\boldsymbol{A}$ is a 2×2 matrix in the present case, a particularly simple and elegant analysis may be given [20]. The eigenvalues of $\boldsymbol{A}$ are determined by the characteristic equation

$$0 = \det(\boldsymbol{A} - \lambda \boldsymbol{I}) = \lambda^2 - 2A_1\lambda + A_2 \tag{67}$$

where

$$A_1 = \tfrac{1}{2} \operatorname{trace} \boldsymbol{A} = \tfrac{1}{2}(A_{11} + A_{22}), \tag{68}$$

$$A_2 = \det \boldsymbol{A} = A_{11}A_{22} - A_{12}A_{21}. \tag{69}$$

The roots of (67) are given by

$$\lambda_{1,2} = A_1 \pm \sqrt{A_1^2 - A_2}. \tag{70}$$

From (70) we see that if $A_1^2 < A_2$ the roots are complex conjugate; if $A_1^2 = A_2$ the roots are real and identical, and if $A_1^2 > A_2$ the roots are real and distinct.

Since the eigenvalues of $\boldsymbol{A}$ are functions of A_1 and A_2 only, the requirements of spectral stability may be viewed as restricting the allowable values of A_1 and A_2 to a 'stability region' in A_1,A_2-space. The boundary of the stability region consists of all points at which $\rho(\boldsymbol{A}) = 1$ and it may be determined by setting $\lambda = \lambda(\alpha) = e^{-i\alpha}$ in (67), where $i = \sqrt{-1}$ and $\alpha = \in [0, 2\pi]$; see Figure 2. (This accounts for all eigenvalues of unit modulus.) In the subsequent development we need to employ a few well-known trigonometric identities which are summarized as follows:

$$e^{i\alpha} = \cos\alpha + i\sin\alpha\,, \tag{71}$$

$$\cos 2\alpha = 2\cos^2\alpha - 1\,, \tag{72}$$

$$\sin 2\alpha = 2\sin\alpha\cos\alpha\,. \tag{73}$$

Proceeding as indicated above, we obtain

$$\begin{aligned} 0 &= e^{i2\alpha} - 2A_1 e^{i\alpha} + A_2 \\ &= (\cos 2\alpha - 2A_1\cos\alpha + A_2) + i(\sin 2\alpha - 2A_1\sin\alpha) \\ &= [2\cos\alpha(\cos\alpha - A_1) + A_2 - 1] + i[2\sin\alpha(\cos\alpha - A_1)]\,. \end{aligned} \tag{74}$$

The real and imaginary parts of (74) must each vanish identically. Therefore

$$2\cos\alpha(\cos\alpha - A_1) + A_2 - 1 = 0 \tag{75}$$

and

$$\sin\alpha(\cos\alpha - A_1) = 0\,. \tag{76}$$

The implications of (75) and (76) may be ascertained by considering the following three cases:

(i) $\alpha = 0$ ($\lambda = +1$). For this case (75) become

$$1 - 2A_1 + A_2 = 0 \tag{77}$$

whereas (76) is identically satisfied.

(ii) $\alpha = \pi$ ($\lambda = -1$). This time (75) becomes

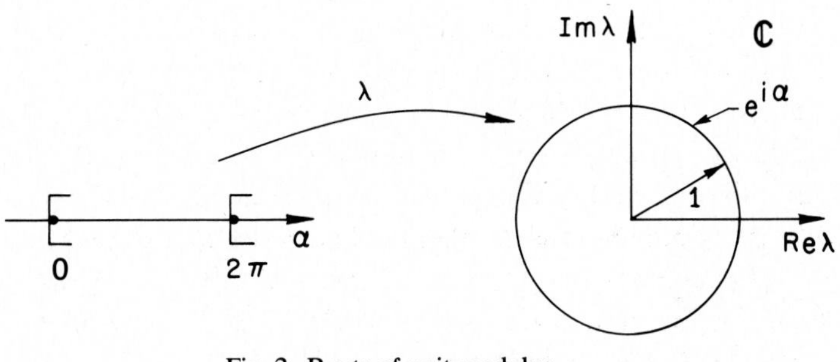

Fig. 2. Roots of unit modulus.

$$1+2A_1+A_2=0 \tag{78}$$

and, again, (76) is identically satisfied.

(iii) $0<\alpha<\pi$ or $\pi<\alpha<2\pi$ (λ is complex). For these values of α, $\sin\alpha\neq 0$. Hence (76) implies $A_1=\cos\alpha$. Using this result in (75) yields

$$A_2=1\,. \tag{79}$$

Observe that (77)–(79) define three straight lines in A_1, A_2-space; see Figure 3. The preceding arguments reveal that $\rho(\boldsymbol{A})=1$ on the boundary of the triangle in Figure 3, but at no point interior to it. A simple topological argument indicates that $\rho(\boldsymbol{A})<1$ on the interior of the triangle: At the origin $A_1=A_2=0$, consequently $\rho(\boldsymbol{A})=0$. Furthermore, $\rho(\boldsymbol{A})$ is a continuous function of A_1 and A_2, nowhere equal to 1 on the interior of the triangle. Therefore $\rho(\boldsymbol{A})$ must be less than 1 throughout the interior.

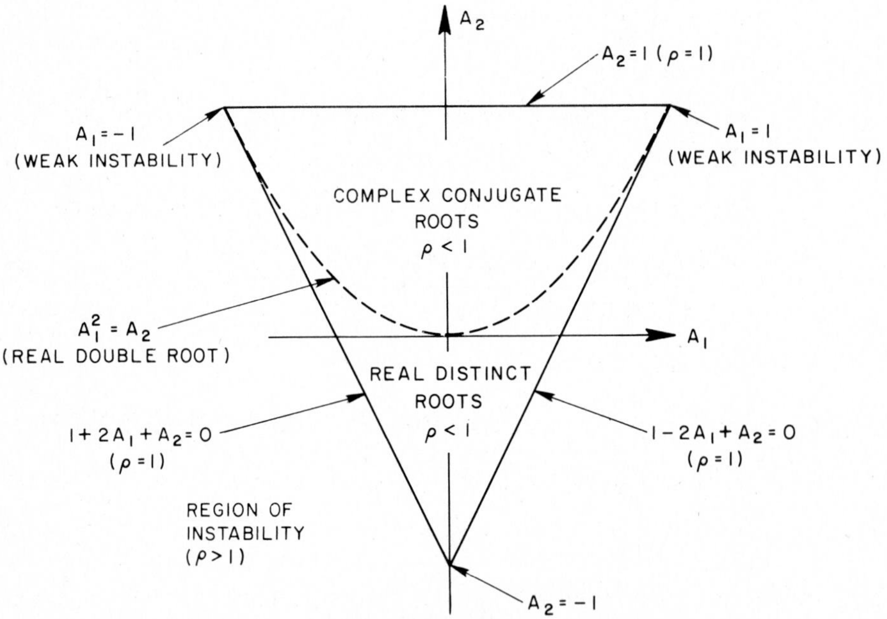

Fig. 3. Region of stability in the A_1, A_2-plane [20].

To precisely define the stability region we must exclude from the triangle all points which give rise to double roots of modulus unity. By (70), these may only occur along the parabola $A_2=A_1^2$. This curve intersects the boundary of the triangle at two points, namely $(A_1, A_2)=(\pm 1, 1)$, and double roots of modulus unity occur at each of these points, possibly giving rise to the situation described at the end of Case 2 of Section 3.c above. Instabilities of this kind are often referred to as *weak instabilities*, and can occur in practice. Hence the stability region is the closed triangle minus the two points $(A_1, A_2)=(\pm 1, 1)$. (It is a simple exercise to verify that points exterior to the triangle result in $\rho(\boldsymbol{A})>1$.)

Perhaps the most useful analytical description of the stability region is given in terms of the following two sets of conditions:

$$-(A_2+1)/2 \le A_1 \le (A_2+1)/2\,, \qquad -1 \le A_2 < 1\,, \tag{80}$$

and

$$-1 < A_1 < 1\,, \qquad A_2 = 1\,. \tag{81}$$

Condition (80) pertains to the interior and side boundaries of the triangle (see Figure 4), whereas (81) pertains to the upper boundary (see Figure 5).

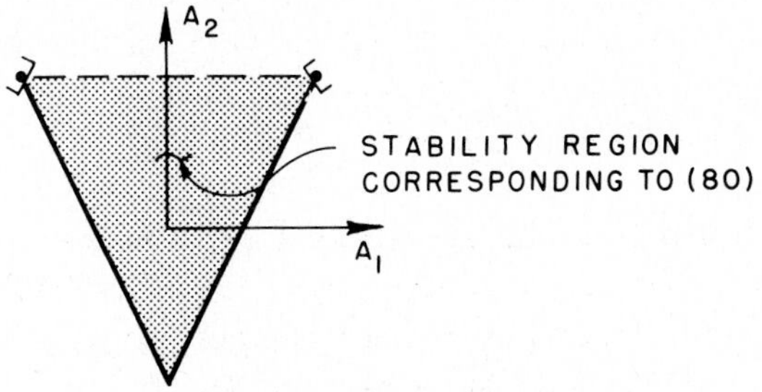

Fig. 4.

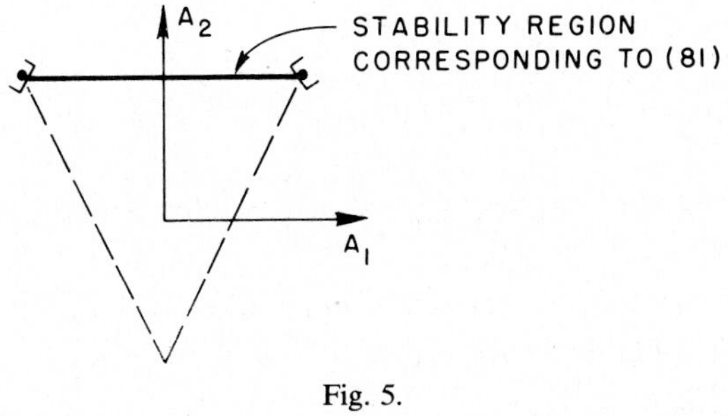

Fig. 5.

Now we wish to apply (80) and (81) to the Newmark methods to determine the stability conditions. We need the quantities A_1 and A_2 which may be determined from (31), (33) and (34), namely,

$$A_1 = 1 - [\xi\Omega + \Omega^2(\gamma + \tfrac{1}{2})/2]/D\,, \tag{82}$$

$$A_2 = 1 - [2\xi\Omega + \Omega^2(\gamma - \tfrac{1}{2})]/D \tag{83}$$

where

$$D = 1 + 2\gamma\xi\Omega + \beta\Omega^2\,, \tag{84}$$

$$\Omega = \omega\Delta t \qquad (\text{'sampling frequency'})\,. \tag{85}$$

(Recall that ξ and $\omega \ge 0$.) The results are summarized as follows:

Summary of stability conditions for the Newmark method

Unconditional:

$$2\beta \geq \gamma \geq \tfrac{1}{2}. \tag{86}$$

Conditional:

$$\gamma \geq \tfrac{1}{2}, \tag{87}$$

$$\beta < \tfrac{1}{2}\gamma, \tag{88}$$

$$\Omega \leq \Omega_{\text{crit}}, \tag{89}$$

where

$$\Omega_{\text{crit}} = \frac{\xi(\gamma - \tfrac{1}{2}) + [\tfrac{1}{2}\gamma - \beta + \xi^2(\gamma - \tfrac{1}{2})^2]^{1/2}}{(\tfrac{1}{2}\gamma - \beta)}. \tag{90}$$

The stability conditions must be satisfied for each mode in the system. Consequently, the highest natural frequency, $\omega_{n_{\text{eq}}}$, is critical and therefore must satisfy (89). If (1) arises from a finite element discretization, $\omega_{n_{\text{eq}}}$ may be bounded by the maximum frequency of the individual elements [40, 43].

Remark 1. It is often convenient to express conditional stability in terms of the *period of vibration*, $T = 2\pi/\omega$, rather than the frequency, in which case (89) becomes

$$\frac{\Delta t}{T} \leq \frac{\Omega_{\text{crit}}}{2\pi}. \tag{91}$$

Eq. (91) must hold for the smallest period of vibration in the system.

Remark 2. To get a feeling for the stability conditions of the Newmark methods, let us consider the undamped case (i.e., $\xi = 0$) in which (90) simplifies to

$$\Omega_{\text{crit}} = (\gamma/2 - \beta)^{-1/2}. \tag{92}$$

The situation for the classical examples discussed in Section 2 is summarized in Table 4. Observe that the critical time step restrictions for the linear acceleration

Table 4
Stability of well-known members of the Newmark family of methods

Method	β	γ	Stability condition
Average acceleration (trapezoidal rule)	$\frac{1}{4}$	$\frac{1}{2}$	unconditional
Linear acceleration	$\frac{1}{6}$	$\frac{1}{2}$	$\Omega_{\text{crit}} = 2\sqrt{3} \simeq 3.464$
Fox–Goodwin (royal road)	$\frac{1}{12}$	$\frac{1}{2}$	$\Omega_{\text{crit}} = \sqrt{6} \simeq 2.449$
Central difference	0	$\frac{1}{2}$	$\Omega_{\text{crit}} = 2$

and Fox–Goodwin methods are not substantially greater than for the central difference method. Under the circumstances which render the central difference method explicit, it generally results in considerable savings in computer cost when compared with the linear acceleration and Fox–Goodwin methods.

Remark 3. Note that if $\gamma = \frac{1}{2}$ viscous damping has *no* effect on stability. Furthermore, when $\gamma > \frac{1}{2}$, the effect of viscous damping is to *increase* the critical time step of conditionally stable Newmark methods. Thus the undamped critical sampling frequency (92) serves as a conservative value when an estimate of the modal damping coefficient is not available. (Recall that, in the case of Rayleigh damping, ξ is determined by ω; see eq. (15).)

B.II.4.a. Oscillatory response

Although a stable computation is guaranteed by (86)–(90), it is generally advisable in practice to satisfy slightly more stringent conditions. (The reasons for this will be made apparent shortly.) These conditions emanate from restricting the eigenvalues of the amplification matrix to be complex conjugate. Recall that this is the case if

$$A_1^2 < A_2 \tag{93}$$

which corresponds to the subregion of the stability triangle indicated in Figure 6.

The Newmark methods satisfy (93) and the spectral stability requirement when the following conditions hold:

Unconditional:

$$0 \le \xi < 1, \qquad \gamma \ge \tfrac{1}{2}, \qquad \beta \ge (\gamma + \tfrac{1}{2})^2/4. \tag{94}$$

Conditional:

$$0 \le \xi < 1, \qquad \gamma \ge \tfrac{1}{2}, \qquad \Omega < \Omega_{\text{bif}} \tag{95}$$

where

$$\Omega_{\text{bif}} = \frac{\frac{1}{2}\xi(\gamma - \frac{1}{2}) + [\frac{1}{4}(\gamma + \frac{1}{2})^2 - \beta + \xi^2(\beta - \frac{1}{2}\gamma)]^{1/2}}{\frac{1}{4}(\gamma + \frac{1}{2})^2 - \beta}. \tag{96}$$

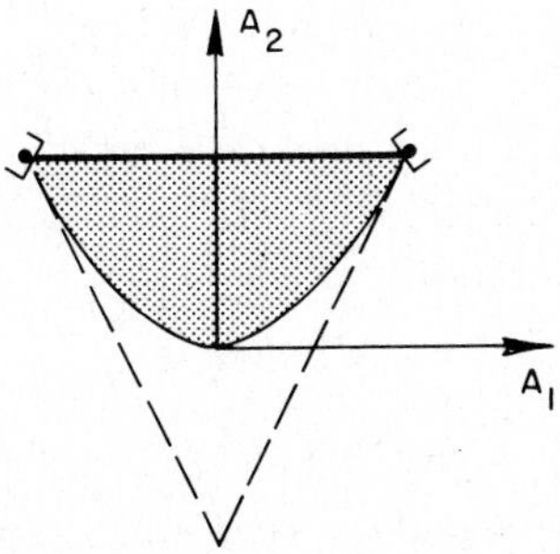

Fig. 6. Stability region for which roots of the amplification matrix are complex conjugate.

In the undamped case, (96) becomes simply

$$\Omega_{\text{bif}} = [\tfrac{1}{4}(\gamma + \tfrac{1}{2})^2 - \beta]^{-1/2}\,. \tag{97}$$

Ω_{bif} is the value of Ω at which complex conjugate eigenvalues bifurcate into real, distinct eigenvalues.

B.II.4.b. High-frequency behavior

As is well known, the high frequencies and mode shapes of the spatially discretized equations generally do not accurately represent the behavior of the original problem. As in the case of first-order systems, (see Section I), it is often desirable to remove (i.e., 'filter') the response of these modes in transient analysis. To study this phenomenon it is appropriate to consider the spectral radius.

As may be gleaned from (61) and (64), if $\rho(\boldsymbol{A}) < 1$, $\boldsymbol{A}^n$ decays like $\rho(\boldsymbol{A})^n$. The closer $\rho(\boldsymbol{A})$ is to one, the slower is the decay. Thus, if it is desired that the high modes be 'damped out', $\rho(\boldsymbol{A})$ should be strictly less than one for these modes. That is, we view $\rho(\boldsymbol{A})$ as a function of Ω (or equivalently $\Delta t/\mathrm{T}$) and investigate its behavior for large values of Ω (i.e., the high modes).

Consider the spectral radius plots shown in Figure 7. In each case $\gamma = 0.9$ and $\xi = 0$. Let

$$\rho_\infty = \lim_{\Delta t/\mathrm{T}\to\infty} \rho(\boldsymbol{A})\,. \tag{98}$$

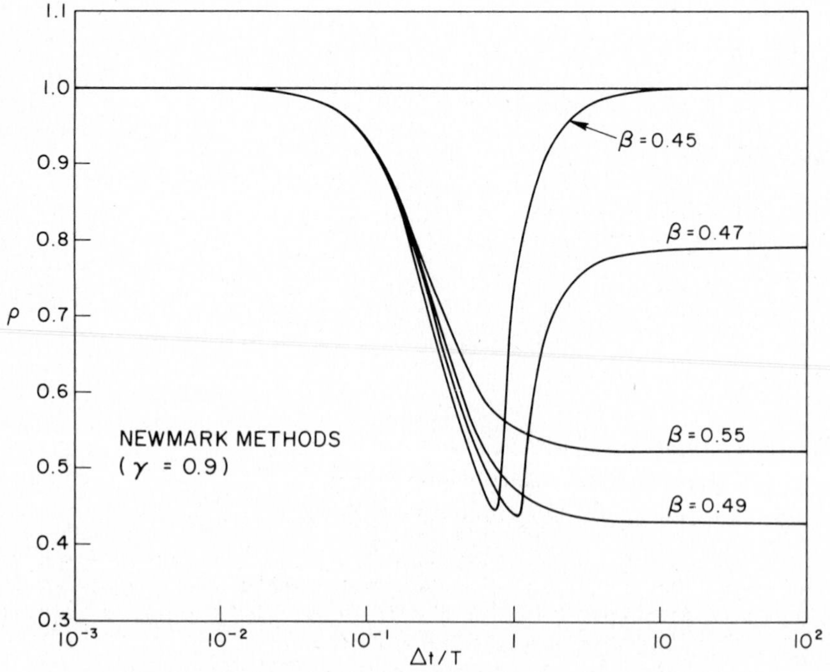

Fig. 7. Spectral radii for Newmark methods for varying β [20].

If β is selected according to (94), with equality holding, $\rho_\infty = 0.43$ and thus there is strong damping of the high modes.[4] Increasing, or decreasing, β reduces ρ_∞. The minimum value of β which insures unconditional stability is $\beta = \frac{1}{2}\gamma = 0.45$. However, for this value $\rho_\infty = 1$ and the high modes are virtually undamped. The minimum points of the $\beta = 0.45$ and $\beta = 0.47$ curves represent the values of $\Delta t/T$ at which bifurcation of the complex conjugate roots occurs. (These values may be computed from (82) and (83) with $\Omega \to \infty$.) By virtue of the fact that the $\beta = 0.49$ and $\beta = 0.55$ cases satisfy (94), no bifurcation occurs. The behavior illustrated in Figure 7 is, in fact, representative of the general case and for this reason it is usually recommended to select β in accordance with (94) in preference to (86). If (94) is not satisfied, then Δt should be selected according to (95) rather than (89), for similar reasons.

B.II.4.c. Viscous damping

It is natural to consider the addition of viscous damping for purposes of filtering the high modes. Unfortunately this strategy is often ill advised. To see this, consider Figure 8 in which the value $\xi = \frac{1}{2}$ (i.e., 50% of 'critical') is employed

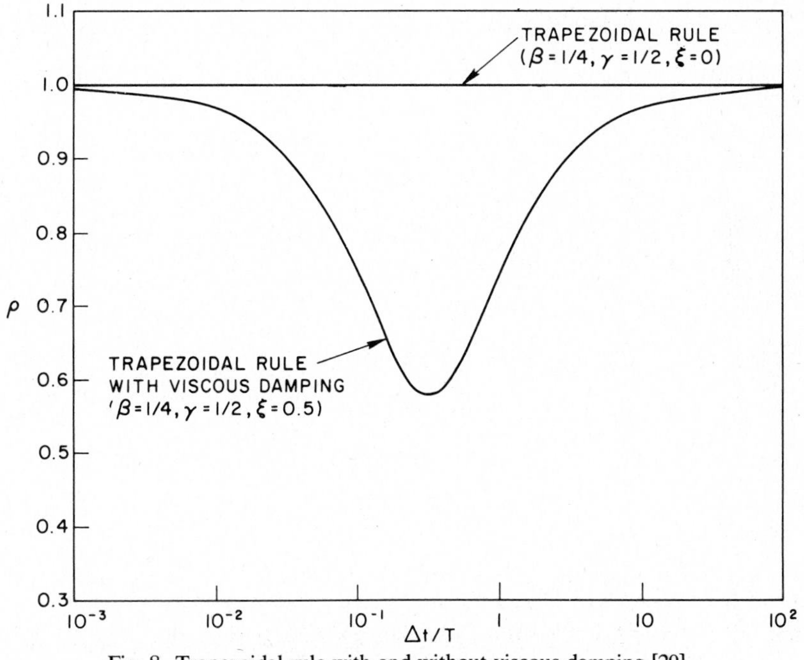

Fig. 8. Trapezoidal rule with and without viscous damping [20].

[4] It may be easily shown that $\beta = \frac{1}{4}(\gamma + \frac{1}{2})^2$ implies $\rho_\infty = |1 - 2/(\gamma + \frac{1}{2})|$. It is an immediate consequence that $\beta = 1$, $\gamma = \frac{3}{2}$ entails $\rho_\infty = 0$, maximal damping of the high frequencies. Unfortunately, this algorithm achieves very poor accuracy in the low modes and thus it is not useful for transient analysis. However, the strong damping characteristics may be exploited in quickly bringing a transient solution to equilibrium.

in conjunction with the average acceleration method. As can be seen, only a middle band of frequencies is substantially affected. The higher frequencies, in fact, are virtually unaffected since $\rho_\infty = 1$. The reader may wonder if it is realizable in practice that $\Omega \gg 1$ for the higher modes. For unconditionally stable algorithms, such as the average acceleration method, it is often the case in large multi-degree-of-freedom applications that the Δt is chosen small enough only to accurately represent the lower modes and, as a result, Δt is very large compared with the periods of the higher modes, that is $\Delta t/T \gg 1$ for these modes.

Stiffness proportional damping, which results in ξ being proportional to ω, also yields $\rho_\infty = 1$ and thus is ineffective as a mechanism for removing high-frequency response.

B.II.5. *Measures of accuracy: numerical dissipation and dispersion*

As a prelude to studying the accuracy properties of the Newmark method, we review the exact solution of the homogeneous SDOF model equation

$$\ddot{d} + 2\xi\omega\dot{d} + \omega^2 d = 0 \tag{99}$$

subjected to initial conditions $d(0) = D_0$ and $\dot{d}(0) = V_0$. There are three cases:

case (i). $0 \leq \xi < 1$ '*underdamped*':

$$d(t) = \mathrm{e}^{-\xi\omega t}\left(D_0 \cos \omega_d t + \frac{(V_0 + \xi\omega D_0)}{\omega_d} \sin \omega_d t\right) \tag{100}$$

where

$$\omega_d = (1 - \xi^2)^{1/2}\omega \qquad \text{(damped natural frequency)}. \tag{101}$$

case (ii). $\xi = 1$, '*critically damped*':

$$d(t) = \mathrm{e}^{-\omega t}(D_0 + (V_0 + \omega D_0)t)\,. \tag{102}$$

case (iii). $\xi > 1$, '*overdamped*':

$$d(t) = \mathrm{e}^{-\xi\omega t}\left(D_0 \cosh \hat{\omega} t + \frac{(V_0 + \xi\omega D_0)}{\hat{\omega}} \sinh \hat{\omega} t\right) \tag{103}$$

where

$$\hat{\omega} = (\xi^2 - 1)^{1/2}\omega\,. \tag{104}$$

Each of these cases has a discrete analog, summarized as in Table 5.

To see this, we will write the algorithm in so-called displacement difference equation form. To this end, consider the expanded, homogeneous analog of (29):

$$d_{n+1} = A_{11}d_n + A_{12}v_n\,, \tag{105}$$

$$v_{n+1} = A_{21}d_n + A_{22}v_n\,. \tag{106}$$

Replacing n by $n-1$ in (105) and (106) yields

Table 5

continuous	discrete
underdamped	complex conjugate roots (λ_1, λ_2)
critically damped	real and identical roots
overdamped	real and distinct roots

$$d_n = A_{11}d_{n-1} + A_{12}v_{n-1}\,, \tag{107}$$

$$v_n = A_{21}d_{n-1} + A_{22}v_{n-1}\,. \tag{108}$$

The displacement difference-equation form is obtained by solving (107) for v_{n-1}, substituting in (108), and using the result to eliminate v_n from (105), i.e.,

$$d_{n+1} - 2A_1d_n + A_2d_{n-1} = 0\,. \tag{109}$$

(The velocity can be shown to satisfy an identical relation by solving (108) for d_{n-1}, substituting into (107), and using the result to eliminate d_n from (106).)

If the roots of (67) are distinct (i.e., if $\lambda_1 \neq \lambda_2$), then the general solution of (109) is

$$d_n = c_1\lambda_1^n + c_2\lambda_2^n\,. \tag{110}$$

(Throughout this section, c_1 and c_2 play the role of generic constants whose values may change from line to line.)

On the other hand, if the roots of (67) are identical (i.e., $\lambda_1 = \lambda_2 = \lambda$), then the general solution of (109) takes the form

$$d_n = (c_1 + c_2n)\lambda^n\,. \tag{111}$$

Both (110) and (111) may be verified by substitution into (109) and by using (67). It can be seen from (111), that if $|\lambda| = 1$, the possibility of a 'weak instability' arises. In this case, the growth of d_n would be linear in n. In general, the terminology 'weak instability' applies to polynomial growth in n of arbitrary order (e.g., n^2 or n^3). This type of growth is considerably weaker than that of the instabilities characterized by $|\lambda|^n$, where $|\lambda| > 1$, hence the name.

The most typical, and therefore most important, case in structural dynamics is *underdamping*. We shall develop the discrete analog of (99) and in so doing establish appropriate measures which fully characterize the accuracy of the time integration algorithm. To carry out this program we shall begin with some considerations of the exact solution and then follow with a parallel development for the discrete case.

Let us write

$$\boldsymbol{y}(t_{n+1}) = \boldsymbol{E}\boldsymbol{y}(t_n) \tag{112}$$

where $\boldsymbol{E}$, the *evolution operator*, may be thought of as the exact amplification matrix. The definition of $\boldsymbol{E}$ may be obtained with the aid of (99). The eigenvalues of $\boldsymbol{E}$ take on the following form:

$$\lambda_{1,2}(\boldsymbol{E}) = \mathrm{e}^{-\xi\Omega \pm \mathrm{i}\Omega_d} \tag{113}$$

where $\Omega = \omega\Delta t$ and $\Omega_d = \omega_d\Delta t$. The invariants of $\boldsymbol{E}$ are thus

$$E_1 = \tfrac{1}{2}\,\mathrm{trace}\,\boldsymbol{E} = \mathrm{e}^{-\xi\Omega}\cos\Omega_d\,, \tag{114}$$

$$E_2 = \det\boldsymbol{E} = \mathrm{e}^{-2\xi\Omega}\,, \tag{115}$$

and the spectral radius is

$$\rho(\boldsymbol{E}) = E_2^{1/2} = \mathrm{e}^{-\xi\Omega}\,. \tag{116}$$

It is easily verified that the exact analog of (110), namely

$$d(t_n) = c_1\lambda_1(\boldsymbol{E})^n + c_2\lambda_2(\boldsymbol{E})^n\,, \tag{117}$$

leads back to (99) upon substitution of (113) and use of the initial conditions.

In the discrete case, the eigenvalues of $\boldsymbol{A}$ can be shown to take on a form similar to (113), viz.

$$\lambda_{1,2}(\boldsymbol{A}) = \mathrm{e}^{-\bar{\xi}\bar{\Omega} \pm \mathrm{i}\bar{\Omega}_d} \tag{118}$$

where $\bar{\Omega} = \bar{\omega}\Delta t$ and $\bar{\Omega}_d = (1-\xi^2)^{1/2}\bar{\Omega}$, and

$$\bar{\Omega}_d = \arctan\left(\frac{A_2}{A_1^2} - 1\right)^{1/2}, \tag{119}$$

$$\bar{\xi} = \frac{-1}{2\bar{\Omega}}\ln A_2\,. \tag{120}$$

Likewise the invariants and spectral radius can be written in analogous fashion to (114)–(116):

$$A_1 = \mathrm{e}^{-\bar{\xi}\bar{\Omega}}\cos\bar{\Omega}_d\,, \tag{121}$$

$$A_2 = \mathrm{e}^{-2\bar{\xi}\bar{\Omega}}\,, \tag{122}$$

$$\rho(\boldsymbol{A}) = \mathrm{e}^{-\bar{\xi}\bar{\Omega}}\,. \tag{123}$$

Employing (118) in (110), and evaluating the result at $n = 0$ and $n = 1$, leads to

$$d_n = \mathrm{e}^{-\bar{\xi}\bar{\omega}t_n}(D_0\cos\bar{\omega}_d t_n + \bar{c}\sin\bar{\omega}_d t_n) \tag{124}$$

where

$$\bar{c} = \frac{d_1 - A_1 D_0}{(A_2 - A_1^2)^{1/2}} = \frac{\frac{1}{2}(A_{11} - A_{22})D_0 + A_{12}V_0}{(A_2 - A_1^2)^{1/2}}\,. \tag{125}$$

As is self-evident, $\bar{\xi}$ and $\bar{\omega}$ are the algorithmic counterparts of ξ and ω, respectively. The values of $\bar{\xi}$ and $\bar{\omega}$ may be defined by substituting explicit definitions of A_1 and A_2 into (119) and (120). For the Newmark method, these are given by (82) and (83).

As measures of the numerical dissipation and dispersion, we take $\bar{\xi}$, the *algorithmic damping ratio,* and $(\bar{T}-T)/T$ the *relative period error,* respectively, where $\bar{T}=2\pi/\bar{\omega}$ and $T=2\pi/\omega$. Generally, analytical expressions for $\bar{\xi}$ and $(\bar{T}-T)/T$ are difficult to obtain. Consequently, one must resort to computer evaluation. Later on, after having introduced other commonly used algorithms in structural dynamics, we shall include a comparison of their dissipative and dispersive properties with those of the Newmark method.

In passing, we note that the following partial analytical results may be obtained for the Newmark method:

$$\bar{\xi}=\xi+\Omega(\gamma-\tfrac{1}{2})/2+O(\Omega^2)\,, \tag{126}$$

$$\frac{\bar{T}-T}{T}=O(\Omega^2)\,. \tag{127}$$

From (126) and (127), it is interesting to observe that first-order errors resulting from $\gamma\neq\frac{1}{2}$ manifest themselves only in the form of excess numerical dissipation, and not in period discrepancies.

The following results for Newmark's method may be easily verified:

(1) If $\xi=0$ (i.e., no physical damping) and $2\beta\geq\gamma=\frac{1}{2}$, then $\bar{\xi}=0$(i.e., no numerical damping).

(2) If $\xi=0$ and $2\beta\geq\gamma>\frac{1}{2}$, then $\bar{\xi}>0$ (i.e., valves of $\gamma>\frac{1}{2}$ may be associated with numerical damping).

Other commonly used measures of algorithmic dissipation are the *logarithmic decrement* $\bar{\delta}=\ln[d(t_n)/d(t_n+\bar{T})]$ and *amplitude decay function* $\mathrm{AD}=1-d(t_n+\bar{T})/d(t_n)$. Either of these measures determines the other as $\mathrm{AD}=1-\exp(-\bar{\delta})$. As is clear from their definitions, AD and $\bar{\delta}$ can only be determined from the discrete solution of an initial-value problem. This entails post-processing involving approximate interpolation to ascertain consecutive peak values. Since $\bar{\xi}$ is defined in terms of the eigenvalues of the amplification matrix, it seems to be the preferable measure of dissipation. For small time steps all three measures are equivalent for practical purposes. This can be seen as follows: First of all, as $\Delta t/T\to0$, $\bar{\delta}\to0$; therefore for sufficiently small $\Delta t/T$, the definition of AD implies that $\mathrm{AD}\approx\bar{\delta}$. Furthermore, for small $\Delta t/T$, (124) yields $\bar{\delta}\approx2\pi\bar{\xi}$.

Another point, which may be verified for the Newmark method, is that as $\Delta t\to0$, $\bar{c}$, which is defined in (124) and used in (125), converges to the corresponding exact quantity, namely $(V_0+\xi\omega D_0)/\omega_d$; see (100).

Similar procedures to the ones above may be used to develop the discrete analogs of the critically damped and overdamped cases, but since these are not germane to the present topics, we will dispense with them.

B.II.6. *Summary of time-step estimates for some simple finite elements*

As remarked upon previously, sufficient conditions for stability in finite element analysis may be obtained from estimates of the maximum eigenvalues of individual elements. We give some examples below:

B.II.6.a. *Two-node linear rod element*

If we assume a *lumped* (i.e., diagonal) mass matrix, then

$$\omega_{\max} = 2c/h \tag{128}$$

where h is the element length and $c = \sqrt{E/\rho}$ is the so-called bar-wave velocity, in which E is Young's modulus and ρ is density. The critical time step for the Newmark method with $\beta = 0$, $\gamma = \frac{1}{2}$ (central difference method) is

$$\Delta t \leq 2/\omega_{\max} = h/c \tag{129}$$

which is the time required for a bar wave to traverse one element.

If we assume a *consistent* mass matrix, then

$$\omega_{\max} = 2\sqrt{3}\, c/h\,, \tag{130}$$

resulting in a reduced critical time step, viz.

$$\Delta t \leq h/(\sqrt{3}\, c)\,. \tag{131}$$

This result is typical: *Consistent mass matrices tend to yield smaller critical time steps than lumped mass matrices.*

B.II.6.b. *Three-node quadratic rod element*

For this case, if we assume a lumped mass matrix based on a Simpson's rule weighting [13] (i.e., the ratio of the middle node mass to the end node masses is 4), we get

$$\omega_{\max} = 2\sqrt{6}\, c/h\,, \tag{132}$$

$$\Delta t \leq h/(\sqrt{6}\, c)\,. \tag{133}$$

Comparison of (133) with (129) reveals that the allowable time step in about 0.4082 that for linear elements with lumped mass. This is based upon equal element lengths. Perhaps a more equitable comparison is one based upon equal nodal spacing. In this case the ratio doubles to 0.8164, but still the advantage is with linear elements.

Remark. Results for one-dimensional heat conduction may be deduced from the preceding cases by employing

$$\Delta t \leq 2/\lambda_{\max} \tag{134}$$

where $\lambda_{max} = (\omega_{max})^2$ and c^2 is set equal to k, the diffusivity coefficient. For example, corresponding to (129) we have the classical result

$$\Delta t \leq h^2/(2k)\,. \tag{135}$$

B.II.6.c. The linear beam element

This element has been described in [42] and [34]. Transverse displacement, w, and face rotation, θ, are assumed to vary linearly over the element.

One-point Gauss quadrature exactly integrates the bending stiffness and appropriately underintegrates the shear stiffness to avoid 'locking' [42]. We assume the trapezoidal rule is used to develop the lumped mass matrix. The matrices for a typical element describing bending in the plane are:

$$\boldsymbol{k}_b = \frac{EI}{h}\begin{bmatrix} 0 & 0 & 0 & 0 \\ 0 & 1 & 0 & -1 \\ 0 & 0 & 0 & 0 \\ 0 & -1 & 0 & 1 \end{bmatrix} \quad \text{(bending stiffness)}\,, \tag{136}$$

$$\boldsymbol{k}_s = \frac{\mu \hat{A}_s}{h}\begin{bmatrix} 1 & \frac{1}{2}h & -1 & \frac{1}{2}h \\ \frac{1}{2}h & \frac{1}{4}h^2 & -\frac{1}{2}h & \frac{1}{4}h^2 \\ -1 & -\frac{1}{2}h & 1 & -\frac{1}{2}h \\ \frac{1}{2}h & \frac{1}{4}h^2 & -\frac{1}{2}h & \frac{1}{4}h^2 \end{bmatrix} \quad \text{(shear stiffness)}\,, \tag{137}$$

$$\boldsymbol{m} = \tfrac{1}{2}\rho\hat{A}h\begin{bmatrix} 1 & 0 & 0 & 0 \\ 0 & I/\hat{A} & 0 & 0 \\ 0 & 0 & 1 & 0 \\ 0 & 0 & 0 & I/\hat{A} \end{bmatrix} \quad \text{(mass)}\,, \tag{138}$$

where $\hat{A}$ is the cross-section area, $\hat{A}_s$ is the shear area, I is the moment of inertia, and μ is the shear modulus. The degree-of-freedom ordering is w_1, θ_1, w_2, θ_2, where the subscript is the node number.

Solution of the eigenvalue problem results in

$$\omega_{max} = \max\left\{2c/h,\ (2c_s/h)\left(1 + \frac{\hat{A}}{I}(\tfrac{1}{2}h)^2\right)^{1/2}\right\} \tag{139}$$

where c is the bar wave velocity and $c_s^2 = \mu\hat{A}_s/(\rho A)$, the beam shear-wave velocity. Thus the critical time step for the central difference operator is given by

$$\Delta t \leq \min\left\{h/c,\ (h/c_s)\left(1 + \frac{\hat{A}}{I}(\tfrac{1}{2}h)^2\right)^{-1/2}\right\}. \tag{140}$$

To get a feeling for these quantities, we shall take a typical situation. Assume the cross-section is rectangular with depth t and width 1. This results in $\hat{A} = t$ and $I = t^3/12$. We assume the ratio of wave speeds $c/c_s = \sqrt{3}$. This corresponds to a Poisson ratio of $\frac{1}{4}$ and shear correction factor $\kappa = \hat{A}_s/\hat{A} = \frac{5}{6}$, so it is a reasonable approximation for most metals. The time step incurred by the bending mode will be critical when

$$h/t \leq \sqrt{2/3} \simeq 0.8165\,. \tag{141}$$

This would only be the case for a very deep beam or an extremely fine mesh and is thus unlikely in practice. The more typical situation in structural analysis is when $t \ll h$ (i.e., very thin beams or coarse meshing). In this case the critical time step is slightly less than t/c, the time for a bar wave to traverse the thickness. As this is an extremely small time step, the cost of explicit integration becomes prohibitive.

A more favorable condition can be derived by adopting different values for the rotational lumped mass coefficients. Specifically, take

$$\boldsymbol{m} = \tfrac{1}{2}\rho\hat{A}h \begin{bmatrix} 1 & 0 & 0 & 0 \\ 0 & \alpha & 0 & 0 \\ 0 & 0 & 1 & 0 \\ 0 & 0 & 0 & \alpha \end{bmatrix} \tag{142}$$

and select α to achieve a more desirable critical time step, without upsetting convergence. Beams mass matrices of the above type were introduced by Key and Beisinger [46]. This time, solution of the eigenvalue problem yields

$$\omega_{\max} = \max\left\{(2c/h)(I/\alpha\hat{A})^{1/2},\ (2c_s/h)\left(1 + \frac{1}{\alpha}\left(\tfrac{1}{2}h\right)^2\right)^{1/2}\right\} \tag{143}$$

and thus

$$\Delta t \leq \min\left\{(h/c)(\alpha\hat{A}/I)^{1/2},\ (h/c_s)\left(1 + \frac{1}{\alpha}\left(\tfrac{1}{2}h\right)^2\right)^{-1/2}\right\}. \tag{144}$$

Taking the value $\alpha = \frac{1}{8}h^2$ in (144), and again assuming $\hat{A}/I = 12/t^2$ and $c/c_s = \sqrt{3}$, results in

$$\Delta t \leq \min\left\{(h/c)\left(\sqrt{\frac{3}{2}\frac{h}{t}}\right),\ h/c\right\}. \tag{145}$$

As long as

$$h/t > \sqrt{2/3} \cong 0.8165\,, \tag{146}$$

bar-wave transit time or better is achieved. Condition (146) will almost certainly be the case in practice.

Analogous procedures may be applied to plate and shell elements (see, e.g., Hughes, Liu and Levit [38]).

The original paper of Key and Beisinger [46] may be consulted for a treatment of the cubic, Bernoulli–Euler beam element, and analogous shell element considerations.

B.II.6.d. Quadrilateral and hexahedral elements

Flanagan and Belytschko [12] have performed a valuable analysis of the uniform strain (4-node) quadrilateral and (8-node) hexahedron, applicable to arbitrary geometric configurations of the elements. They obtain the following estimate of maximum element frequency:

$$\omega_{\max} \leq c_{\mathrm{D}} g^{1/2} \tag{147}$$

where $c_{\mathrm{D}}^2 = (\lambda + 2\mu)/\rho$, dilatational wave velocity; λ and μ are the Lamé parameters; and g is a geometric parameter. For example, in the case of the quadrilateral, g is defined as follows:

$$g = \frac{4}{A^2} \sum_{i=1}^{2} \sum_{a=1}^{4} B_{ia} B_{ia} , \tag{148}$$

$$[B_{ia}] = \frac{1}{2} \begin{bmatrix} (y_2 - y_4) & (y_3 - y_1) & (y_4 - y_2) & (y_1 - y_3) \\ (x_4 - x_2) & (x_1 - x_3) & (x_2 - x_4) & (x_3 - x_1) \end{bmatrix} \tag{149}$$

where x_a and y_a are the coordinates of node a. (For the definition of g in the case of the hexahedron, see [12].) The estimate, (147), leads to a *sufficient* condition for stability. For the central difference method (147) results in

$$\Delta t \leq 2/(c_{\mathrm{D}} g^{1/2}) . \tag{150}$$

As an example of the restriction imposed by (150), consider a rectangular element with side lengths h_1 and h_2. In this case (150) becomes

$$\Delta t \leq 1 \bigg/ \left(c_{\mathrm{D}} \left(\frac{1}{h_1^2} + \frac{1}{h_2^2} \right)^{1/2} \right) . \tag{151}$$

For higher-order elements, there appears that little of a precise nature has been done. Most results are of the form (150) where the geometric factor is approximated by trial-and-error. One would hope that, with the aid of automatic symbolic manipulators, improved time step estimates will become available in the ensuing years for more complex elements, material properties, etc.

B.III. Linear multi-step (LMS) methods

B.III.1. LMS methods for first-order equations

Consider a system of first-order, linear, ordinary differential equations

$$\dot{\boldsymbol{y}} = \boldsymbol{f}(\boldsymbol{y}, t) = \boldsymbol{G}\boldsymbol{y} + \boldsymbol{H}(t) \tag{1}$$

where $\boldsymbol{G}$ is a given constant matrix and $\boldsymbol{H}$ is a given vector-valued function of t. A *k-step linear multi-step method for* (1) is defined by the following expression:

$$\sum_{i=0}^{k} \{\alpha_i \boldsymbol{y}_{n+1-i} + \Delta t \beta_i \boldsymbol{f}(\boldsymbol{y}_{n+1-i}, t_{n+1-i})\} = \boldsymbol{0} . \tag{2}$$

The α_i's and β_i's are parameters which define the method. Note that the word 'linear' in 'linear multi-step method' has nothing to do with the linearity of (1).

Indeed, (2) is perfectly well defined for a nonlinear function $f(y, t)$. Rather, the linearity pertains to the form of (2).

An excellent reference for LMS methods of this type is Gear [14].

An LMS method is called *explicit* if $\beta_0 = 0$, otherwise it is called *implicit.* It is called a backward-difference method if $\beta_i = 0$ for all $i \geq 1$.

An example of a 1-step LMS method of first-order type is the generalized trapezoidal family discussed in Section I.2, for which $\alpha_0 = -\alpha_1 = 1$, $\beta_0 = -\alpha$, and $\beta_1 = \alpha - 1$. Recall that $\alpha = 0$ defines an explicit method, and $\alpha = 1$, a backward difference method.

The greater the number of steps involved in the definition of an LMS method, the greater the 'historical data pool' which must be stored in computing – a practical disadvantage.

B.III.2. Semi-discrete heat equation and equation of motion in first-order form

In order to study LMS methods applied to the cases considered thus far, we need to first put these equations in the form of (1), and then ascertain the spectral properties of the resulting $\boldsymbol{G}$-matrices.

B.III.2.a. Semi-discrete heat equation

In this case,

$$\boldsymbol{y} = \boldsymbol{d}, \tag{3}$$

$$\boldsymbol{G} = -\boldsymbol{M}^{-1}\boldsymbol{K}, \tag{4}$$

$$\boldsymbol{H}(t) = \boldsymbol{M}^{-1}\boldsymbol{F}(t). \tag{5}$$

The spectrum of $\boldsymbol{G}$ is determined from the eigenproblem

$$\boldsymbol{0} = (\boldsymbol{G} - \lambda(\boldsymbol{G})\boldsymbol{I})\boldsymbol{\psi}(\boldsymbol{G}) \tag{6}$$

which becomes upon substitution of (4) and multiplication by $\boldsymbol{M}$

$$\boldsymbol{0} = (\boldsymbol{K} + \lambda(\boldsymbol{G})\boldsymbol{M})\boldsymbol{\psi}(\boldsymbol{G}). \tag{7}$$

By comparison with (6) of Section I,

$$\lambda(\boldsymbol{G}) = -\lambda \leq 0, \tag{8}$$

$$\boldsymbol{\psi}(\boldsymbol{G}) = \boldsymbol{\psi}. \tag{9}$$

Thus in this case the spectrum of $\boldsymbol{G}$ *resides upon the negative real axis in the complex plane,* $\mathbb{C}$.

B.III.2.b. Semi-discrete equation of motion

In this case,

$$y = \begin{Bmatrix} d \\ \dot{d} \end{Bmatrix}, \tag{10}$$

$$G = \begin{bmatrix} 0 & I \\ -M^{-1}K & -M^{-1}C \end{bmatrix}, \tag{11}$$

$$H(t) = \begin{Bmatrix} 0 \\ M^{-1}F(t) \end{Bmatrix}. \tag{12}$$

If we let

$$\psi(G) = \begin{Bmatrix} \psi_1(G) \\ \psi_2(G) \end{Bmatrix} \tag{13}$$

then the eigenproblem can be expressed as

$$\begin{Bmatrix} 0 \\ 0 \end{Bmatrix} = \left(\begin{bmatrix} 0 & I \\ -M^{-1}K & -M^{-1}C \end{bmatrix} - \begin{bmatrix} \lambda(G)I & 0 \\ 0 & \lambda(G)I \end{bmatrix} \right) \begin{Bmatrix} \psi_1(G) \\ \psi_2(G) \end{Bmatrix} \tag{14}$$

or, equivalently,

$$0 = \psi_2(G) - \lambda(G)\psi_1(G), \tag{15}$$

$$0 = -M^{-1}K\psi_1(G) - (M^{-1}C + \lambda(G)I)\psi_2(G). \tag{16}$$

Combining (15) and (16), and multiplying by M, yields a quadratic eigenproblem:

$$0 = (K + \lambda(G)C + \lambda(G)^2 M)\psi_1(G). \tag{17}$$

Using the results and notations of Section II for the undamped eigenproblem (see (14)–(18) of Section II and discussion thereabout), (17) can be reduced to

$$\lambda(G)^2 + 2\xi\omega\lambda(G) + \omega^2 = 0 \tag{18}$$

by premultiplying by ψ^{T} and expanding $\psi_1(G)$ in terms of the ψ's.

The solutions of (18) take on the following forms:

(i) $0 \le \xi < 1$, *underdamped*:

$$\lambda_{1,2}(G) = -\xi\omega + \mathrm{i}\omega\sqrt{1-\xi^2}. \tag{19}$$

(ii) $\xi = 1$, *critically damped*:

$$\lambda_{1,2}(G) = -\omega \quad \text{(double root)}. \tag{20}$$

(iii) $\xi > 1$, *overdamped*:

$$\lambda_{1,2}(G) = -\xi\omega \pm \omega\sqrt{\xi^2 - 1}. \tag{21}$$

Note that in the practically important case of no damping (i.e., $\xi = 0$), $\lambda_{1,2}(G) = \pm\mathrm{i}\omega$, conjugate imaginary roots. *In all cases, since ξ and $\omega \ge 0$, the eigenvalues of G are confined to the negative half-plane of $\mathbb{C}$ including the imaginary axis.*

In considering LMS methods, it is important to insure that the stability region in the complex plane encloses the spectrum of the matrix G for the case in question.

B.III.3. *Stability of LMS methods*

B.III.3.a. *Reduction to an SDOF problem*

As in the cases treated in Sections I and II, reduction to a model problem plays an essential role. For these cases, $\boldsymbol{G}$ possesses linearly independent eigenvectors, which we denote by the matrix $\boldsymbol{P} = \boldsymbol{P}(\boldsymbol{G})$. Employing the notations $\lambda = \lambda(\boldsymbol{G})$ and $\boldsymbol{\Lambda} = \boldsymbol{\Lambda}(\boldsymbol{G})$, for simplicity, we have that

$$\boldsymbol{P}^{-1}\boldsymbol{G}\boldsymbol{P} = \boldsymbol{\Lambda} = \operatorname{diag}(\lambda_1, \lambda_2, \ldots, \lambda_N) \tag{22}$$

where N is the dimension of $\boldsymbol{G}$. Defining

$$\boldsymbol{z} = \boldsymbol{P}^{-1}\boldsymbol{y} \tag{23}$$

enables us to obtain from the homogeneous form of (1), the uncoupled system

$$\dot{\boldsymbol{z}} = \boldsymbol{\Lambda}\boldsymbol{z} \tag{24}$$

which is characterized by the generic scalar equation

$$\dot{z} = \lambda z\,. \tag{25}$$

Likewise, the homogeneous form of the algorithm, (2), can be uncoupled, resulting in

$$\sum_{i=0}^{k} (\alpha_i \boldsymbol{z}_{n+1-i} + \Delta t \beta_i \boldsymbol{\Lambda} \boldsymbol{z}_{n+1-i}) = \boldsymbol{0}\,, \tag{26}$$

$$\sum_{i=0}^{k} (\alpha_i + \Delta t \lambda \beta_i) z_{n+1-i} = 0\,. \tag{27}$$

Stability is phrased in terms of roots of the polynomial associated to (27), viz.

$$\sum_{i=0}^{k} (\alpha_i + \Delta t \lambda \beta_i) \zeta^{n+1-i} = 0\,. \tag{28}$$

B.III.3.b. *Concepts of stability*

Definition 1. An LMS method of the form (2) is said to be *absolutely stable*, at a fixed $\lambda \Delta t$, if all ζ satisfying (28) are such that $|\zeta| \leq 1$.

Definition 2. The *region of absolute stability* of an LMS method is the set of $\lambda \Delta t \in \mathbb{C}$ at which it is absolutely stable.

Remark. The definition of absolute stability is very close to that of spectral stability, given earlier, except no special account is given to multiple roots of unit modulus, which potentially give rise to weak instabilities. That the roots of (28), and the eigenvalues of an associated amplification matrix, are the same, can be realized by constructing a one-step, multivalue method equivalent to the given LMS method, viz.

$$\underbrace{\begin{Bmatrix} z_{n+1} \\ z_n \\ \vdots \\ z_{n+2-k} \end{Bmatrix}}_{k\times 1} = \underset{k\times k}{\underbrace{\boldsymbol{A}}} \underbrace{\begin{Bmatrix} z_n \\ z_{n-1} \\ \vdots \\ z_{n+1-k} \end{Bmatrix}}_{k\times 1} \tag{29}$$

where

$$\boldsymbol{A} = \begin{bmatrix} \dfrac{-(\alpha_1+\lambda\Delta t\beta_1)}{(\alpha_0+\lambda\Delta t\beta_0)} & \dfrac{-(\alpha_2+\lambda\Delta t\beta_2)}{(\alpha_0+\lambda\Delta t\beta_0)} & \cdots & \cdots & \dfrac{-(\alpha_k+\lambda\Delta t\beta_k)}{(\alpha_0+\lambda\Delta t\beta_0)} \\ 1 & 0 & \cdots & 0 & 0 \\ 0 & 1 & \cdots & 0 & 0 \\ \vdots & \vdots & \ddots & \vdots & \vdots \\ 0 & 0 & \cdots & 1 & 0 \end{bmatrix}. \tag{30}$$

Another important notion of stability is given in the following definition.

Definition 3. An LMS method is *A-stable* if solutions of (27)$\rightarrow 0$ as $n\rightarrow\infty$ when $\operatorname{Re}\lambda<0$.

Physically speaking, an A-stable algorithm is one which produces solutions which decay to zero, whenever the corresponding exact solutions of (25) decay to zero. Clearly, if an algorithm is A-stable then the region of absolute stability contains the left half-plane of $\mathbb{C}$.

The condition of A-stability places no limitation on the size of Δt, consequently it is closely related to what we termed unconditional (spectral) stability in Sections I and II. The only difference is that spectral stability allows eigenvalues of unit modulus which potentially could conserve a solution rather than contract it to zero.

It should be apparent that A-stable algorithms are important for many physical problem classes. However, within the class of LMS methods, the subclass of A-stable methods is severely delimited. This is the content of a celibrated theorem due to Dahlquist [9], which states:

1. *An explicit A-stable LMS method does not exist.*
2. *A third-order accurate A-stable LMS method does not exist.*
3. *The second-order accurate A-stable LMS method with the smallest error constant*[5] *is the trapezoidal rule.*

[5]The error constant is the coefficient of Δt^{k+1} in the local truncation error; see Section I.6.

The upshot of this theorem is that if we seek an A-stable LMS method which possesses some special feature (such as high-frequency numerical dissipation), we necessarily entail some loss of accuracy with respect to the trapezoidal rule. Since the spectrum of the $\boldsymbol{G}$-matrix for the equation of motion falls within the left-half complex plane and, in the undamped case, falls on the imaginary axis, it would appear that the trapezoidal rule is the canonical A-stable LMS method for structural dynamics. The widespread use and popularity of the trapezoidal rule in this context appears to confirm this observation. However, there are other useful A-stable LMS methods for structural dynamics. For example, we may mention Park's method which will be described shortly. Many structural dynamics algorithms do not fall within the class of first-order LMS methods considered so far. The Newmark family of methods may be mentioned in this regard. Another class of LMS methods, for second-order equations, will be described later on which contains the Newmark family among others.

Since A-stable LMS methods are at most second-order accurate, the question arises "Is there some weakened notion of A-stability which pertains to a physically relevant class of problems and allows for the development of higher-order-accurate methods?" An affirmative answer is provided in the next section.

B.III.3.c. Stiffly-stable methods

The stiffly stable methods to be described were proposed by Gear. His text [14] may be consulted for further details.

Definition 4. A method is *stiffly stable* if it is absolutely stable in the region of the $\lambda\Delta t$-plane defined by $\mathrm{Re}(\lambda\Delta t) < -\delta$, where δ is a positive constant.

Gear has developed a family of k-step, k^{th}-order accurate LMS methods which are stiffly stable. The coefficients which define these methods are given in Table 6 ($\alpha_0 = -1$ in each case).

Table 6
Coefficients of stiffly-stable methods [14]

k	2	3	4	5	6
β_0	$\frac{2}{3}$	$\frac{6}{11}$	$\frac{12}{25}$	$\frac{60}{137}$	$\frac{60}{147}$
α_1	$\frac{4}{3}$	$\frac{18}{11}$	$\frac{48}{25}$	$\frac{300}{137}$	$\frac{360}{147}$
α_2	$-\frac{1}{3}$	$-\frac{9}{11}$	$-\frac{36}{25}$	$-\frac{300}{137}$	$-\frac{450}{147}$
α_3		$\frac{2}{11}$	$\frac{16}{25}$	$\frac{200}{137}$	$\frac{400}{147}$
α_4			$-\frac{3}{25}$	$-\frac{75}{137}$	$-\frac{225}{147}$
α_5				$\frac{12}{137}$	$\frac{72}{147}$
α_6					$-\frac{10}{147}$

Note that each method is a backward-difference LMS method. In fact, $k = 1$ corresponds to the classical backward-difference 1-step algorithm; see Section I.

The stability regions are illustrated in Figure 9 and 10. As may be concluded from Figure 9, for $k = 1$ and $k = 2$, the methods are in fact A-stable. However, both of these methods are significantly less accurate than the trapezoidal rule. For $k \geq 3$, the regions of instability include portions of the imaginary axis. Consequently, these stiffly stable algorithms also appear inappropriate for typical structural dynamics applications. Jensen [44, 45] has investigated their use in this context and proposed appropriate modifications.

Since the regions of stability of each of the methods, $k = 1, 2, \ldots, 6$, includes the negative real axis, one would anticipate good behavior in application to problems of heat conduction.

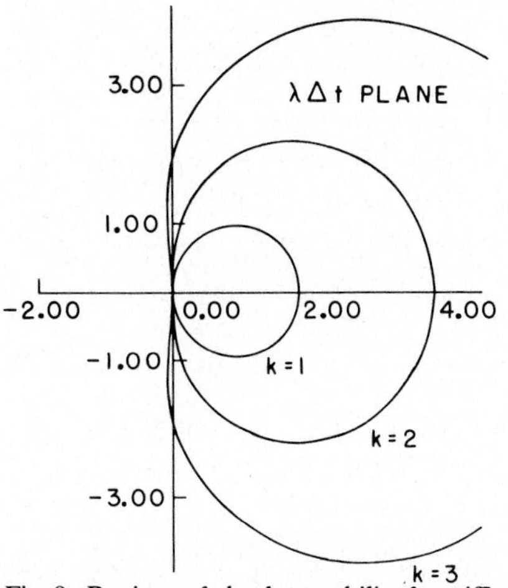

Fig. 9. Regions of absolute stability for stiffly stable methods of orders one through three. Methods are stable outside of closed contours [14].

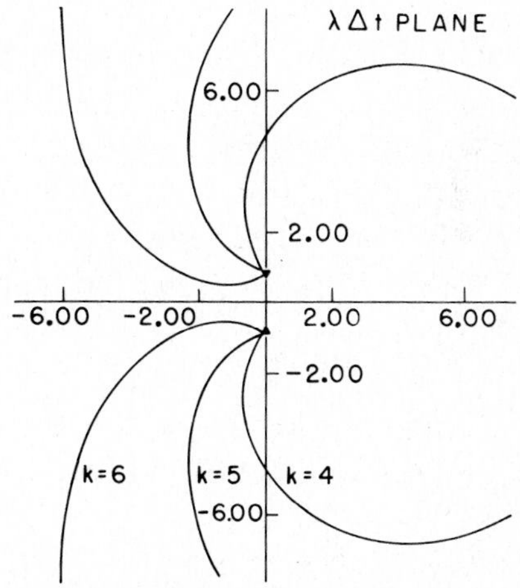

Fig. 10. Regions of absolute stability for stiffly stable methods of orders four through six [14].

B.III.3.d. Park's method

Park [57] has developed a second-order accurate, A-stable algorithm, which retains good accuracy in the low frequencies and strong dissipative characteristics in the high frequencies, by combining Gear's 2-step and 3-step stiffly-stable algorithms. The resultant scheme is defined by

$$\alpha_0 = -1, \quad \alpha_1 = \tfrac{15}{10}, \quad \alpha_2 = -\tfrac{6}{10}, \quad \alpha_3 = \tfrac{1}{10}, \quad \beta_0 = \tfrac{6}{10}\,. \tag{31}$$

Definition 5. An LMS method is called *stiffly A-stable* if it is A-stable and $z_{n+1}/z_n \to 0$ as $\mathrm{Re}(\lambda\Delta t) \to -\infty$.

(This condition is analogous to insisting $\rho(\boldsymbol{A}) \to 0$ as $\Delta t/T \to \infty$ for the cases considered in Section II.)

Gear's methods with $k = 1$ and 2, and Park's method are examples of stiffly A-stable methods.

B.III.3.e. Routh–Hurwitz criterion

Since the determination of stability hinges upon the calculation of roots of polynomials, only the simplest cases can be worked out by hand. For this reason, stability regions are often calculated via numerical procedures. An alternative technique for determining the region of stability, which often has advantages in a hand analysis, consists of applying the transformation $\zeta = (1+\eta)/(1-\eta)$ which maps the unit circle, $|\zeta| = 1$, into the imaginary axis Re $\eta = 0$, and the interior of the circle into the half-plane, Re $\eta < 0$, and then invoking the Routh–Hurwitz criterion which gives necessary and sufficient conditions for the roots of a polynomial to have negative real parts. As an example, suppose the stability polynomial is of third-order, viz.

$$\zeta^3 - 2A_1\zeta^2 + A_2\zeta - A_3 = 0 . \tag{32}$$

If the algorithm had been written in amplification-matrix form, then the A's are the invariants of $\boldsymbol{A}$:

$$A_1 = \tfrac{1}{2}\,\text{trace}\,\boldsymbol{A} , \tag{33}$$

$$A_2 = \text{sum of principal minors of } \boldsymbol{A} , \tag{34}$$

$$A_3 = \det \boldsymbol{A} . \tag{35}$$

Following the procedure given by Lambert [49] we substitute $\zeta = (1+\eta)/(1-\eta)$ into (32), multiply through by $(1-\eta)^3$ and obtain the polynominal equation

$$p_0\eta^3 + p_1\eta^2 + p_2\eta + p_3 = 0 . \tag{36}$$

A necessary and sufficient condition for the roots of this equation to lie in the half plane Re $\eta \le 0$ (i.e., for the roots of (32) to lie within or on the circle $|\zeta| = 1$) is that the following inequalities hold:

$$p_0 = 1 - 2A_1 + A_2 - A_3 \ge 0 , \tag{37}$$

$$p_1 = 3 - 2A_1 - A_2 + 3A_3 \ge 0 , \tag{38}$$

$$p_2 = 3 + 2A_1 - A_2 - 3A_3 \ge 0 , \tag{39}$$

$$p_3 = 1 + 2A_1 + A_2 + A_3 \ge 0 , \tag{40}$$

$$\tfrac{1}{8}(p_1p_2 - p_0p_3) = 1 - A_2 + A_3(2A_1 - A_3) \ge 0 . \tag{41}$$

The above procedure may be employed to establish the stability of Park's method for example.

B.III.4. LMS methods for second-order equations

We consider a system of second-order, linear, ordinary differential equations written in the form:

$$\ddot{\boldsymbol{y}} = \boldsymbol{f}(\boldsymbol{y}, \dot{\boldsymbol{y}}, t) = \boldsymbol{G}_0\boldsymbol{y} + \boldsymbol{G}_1\dot{\boldsymbol{y}} + \boldsymbol{H}(t) \tag{42}$$

where $\boldsymbol{G}_0$ and $\boldsymbol{G}_1$ are given constant matrices. A *k-step, LMS method for* (42) is given by (Geradin [15]):

$$\sum_{i=0}^{k} \{\alpha_i \boldsymbol{y}_{n+1-i} + \Delta t\beta_i \boldsymbol{G}_1 \boldsymbol{y}_{n+1-i} + \Delta t^2\gamma_i[\boldsymbol{G}_0\boldsymbol{y}_{n+1-i} + \boldsymbol{H}(t_{n+1-i})]\} = \boldsymbol{0}\,. \tag{43}$$

The method is defined by the values of α_i, β_i and γ_i, $i = 0, 1, \ldots, k$.

An LMS method of this type is called *explicit* if β_0 and $\gamma_0 = 0$. (If $\boldsymbol{G}_1 = \boldsymbol{0}$, then clearly only γ_0 need be zero.) The method is a *backward-difference method* if β_i and $\gamma_i = 0$, $i \geq 1$. (Likewise, if $\boldsymbol{G}_1 = \boldsymbol{0}$, only the γ_i's, $i \geq 1$, need be zero.)

The equation of motion may be put in the form of (42) simply by multiplying through by $\boldsymbol{M}^{-1}$. The pertinent arrays are then defined as

$$\boldsymbol{y} = \boldsymbol{d}\,, \tag{44}$$

$$\boldsymbol{G}_0 = -\boldsymbol{M}^{-1}\boldsymbol{K}\,, \tag{45}$$

$$\boldsymbol{G}_1 = -\boldsymbol{M}^{-1}\boldsymbol{C}\,, \tag{46}$$

$$\boldsymbol{H}(t) = \boldsymbol{M}^{-1}\boldsymbol{F}(t)\,. \tag{47}$$

By way of (44), we see that (43) is a displacement difference-equation form of the algorithm. The commonly used algorithms of structural dynamics can all be put into this form, although very few are naturally cast this way. An example of how to convert an algorithm written in so-called multi-value form is illustrated by the developments for the Newmark method in Section II, (105)–(109). Equation (109) is the displacement difference-equation form of Newmark's method. (Actually, (109) of Section II is the typical modal equation. Clearly, a similar result may be obtained for the coupled system equations.) In the sense of an LMS method for second-order systems, Newmark's method is a two-step method. LMS methods applied to the first-order form of the equations of motion may likewise be recast as second-order-type LMS methods (e.g., Park's method).

The stability properties of second-order-type LMS methods may be investigated in similar fashion to first-order LMS methods. Briefly, modal reduction may be employed which, for the equation of motion, leads to the following polynomial:

$$\sum_{i=0}^{k} (\alpha_i + 2\xi\omega\Delta t\beta_i + (\omega\Delta t)^2\gamma_i)\zeta^{n+1-i} = 0\,. \tag{48}$$

The stability requirements are $|\zeta| \leq 1$, and multiple ζ's must additionally satisfy $|\zeta| < 1$.

Krieg [47] has proven that an explicit unconditionally stable LMS method for second-order equations does not exist, an analog of part 1 of Dahlquist's theorem. All evidence reported thus far supports the conjecture that the analogs of parts 2 and 3 of Dahlquist's theorem also hold for LMS methods for second-order equations.

B.III.5. Survey of some commonly used algorithms in structural dynamics

This section surveys and compares some algorithms which have been proposed for structural dynamics applications. The algorithms considered are all LMS methods of first-order, or second-order, type, and the comparison is based upon both stability behavior and accuracy. Accuracy characteristics of LMS methods may be studied in similar fashion to the study of accuracy for Newmark's method (see Section II.5). The only complication is that one needs to ignore so-called 'spurious roots', focusing attention on the 'principal roots'. Spurious roots vanish as $\Omega \to 0$, whereas principal roots approach 1 in this limit. The spurious roots are a manifestation of the order of the stability polynomial or, equivalently, the rank of the associated amplification matrix. For example, in consideration of Newmark's method in Section II.5, there were two roots of $\boldsymbol{A}$, and both are principal roots.[6] For methods which have amplification matrices of greater rank, say n, there will be two principal roots and $n-2$ spurious roots. (There are two principal roots for algorithms for second-order systems.) The principal roots are used to define the measures of accuracy (i.e., the algorithmic damping ratio and relative period error) in the same way as in Section II.5 (see eqs. (118)–(120)).

The comparisons which follow are for the physically undamped case (i.e., $\xi = 0$).

B.III.5.a. Houbolt's method

Houbolt's method [26] was one of the earliest employed in structural dynamics computations. It is defined by the following formulas:

$$\boldsymbol{M}\boldsymbol{a}_{n+1} + \boldsymbol{C}\boldsymbol{v}_{n+1} + \boldsymbol{K}\boldsymbol{d}_{n+1} = \boldsymbol{F}_{n+1}\,, \tag{49}$$

$$\boldsymbol{a}_{n+1} = (2\boldsymbol{d}_{n+1} - 5\boldsymbol{d}_n + 4\boldsymbol{d}_{n-1} - \boldsymbol{d}_{n-2})/\Delta t^2\,, \tag{50}$$

$$\boldsymbol{v}_{n+1} = (11\boldsymbol{d}_{n+1} - 18\boldsymbol{d}_n + 9\boldsymbol{d}_{n-1} - 2\boldsymbol{d}_{n-2})/(6\Delta t^2)\,. \tag{51}$$

Combining (49)–(51) leads to a 3-step LMS method of second-order type which is unconditionally stable and second-order accurate. Furthermore, if $\boldsymbol{C} = \boldsymbol{0}$ it is a backward-difference, stiffly A-stable method, and thus has many features in common with Park's method. However, Park's method is significantly more accurate and thus Houbolt's is now more of historical interest.

B.III.5.b. Collocation schemes

Collocation methods generalize and combine aspects of the Newmark method and Wilson-θ method [62]. A systematic analysis of collocation is contained in Hilber and Hughes [21]. The collocation schemes are defined by the following

[6]It has been pointed out by Krieg and Key [48] that Newmark's method is the most general, consistent, rank-2 method for second-order equations. Hilber [20] has classified methods according to the rank of their associated amplification matrix.

equations:

$$Ma_{n+\theta} + Cv_{n+\theta} + Kd_{n+\theta} = F_{n+\theta}\,, \tag{52}$$

$$a_{n+\theta} = (1-\theta)a_n + \theta a_{n+1}\,, \tag{53}$$

$$F_{n+\theta} = (1-\theta)F_n + \theta F_{n+1}\,, \tag{54}$$

$$d_{n+\theta} = d_n + \theta\Delta t v_n + \frac{(\theta\Delta t)^2}{2}\{(1-2\beta)a_n + 2\beta a_{n+\theta}\}\,, \tag{55}$$

$$v_{n+\theta} = v_n + \theta\Delta t\{(1-\gamma)a_n + \gamma a_{n+\theta}\}\,. \tag{56}$$

To (52)–(56) are appended the Newmark formulas (i.e., (5) and (6) of Section II) for purposes of defining d_{n+1} and v_{n+1}, respectively. θ is called the collocation parameter. The collocation schemes can be put into the form of a 3-step LMS method of second-order type. Alternatively, the collocation schemes' model equation can be put in amplification matrix form; see [20]. In this case, the rank of the amplification matrix is 3.

If $\theta = 1$, the scheme reverts to Newmark's method. If $\beta = \frac{1}{6}$ and $\gamma = \frac{1}{2}$, the Wilson-θ methods are obtained. (In the past, the value $\theta = 1.4$, has been often employed.) A necessary and sufficient condition for second-order accuracy is that $\gamma = \frac{1}{2}$. Third-order accuracy can be attained if, in addition, $\beta = \frac{1}{12} - \frac{1}{2}\theta(\theta - 1)$. However, this necessarily entails conditional stability. Unconditionally stable, second-order accurate schemes are defined by (see Figure 11)

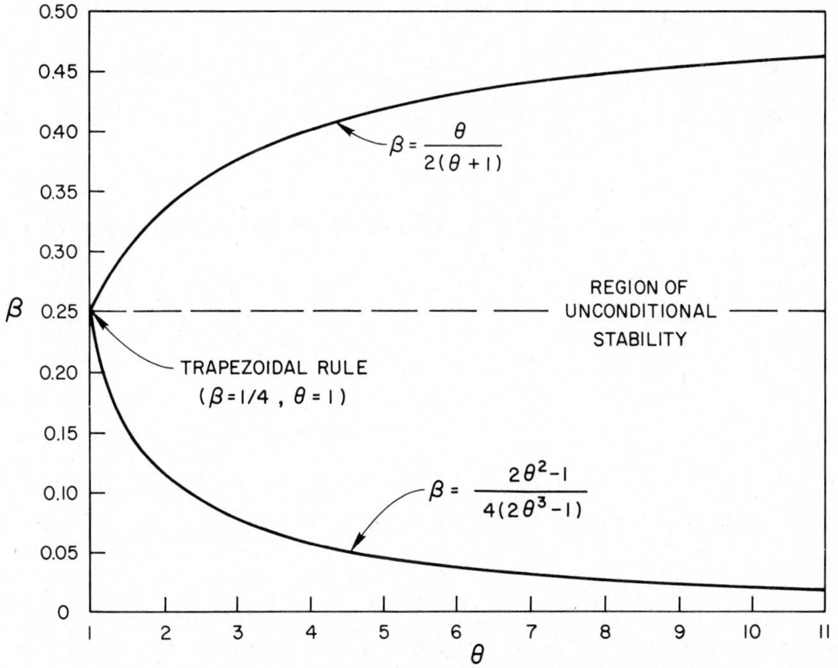

Fig. 11. Region of unconditional stability for collocation schemes [21].

$$\gamma = \tfrac{1}{2}, \qquad \theta \geq 1, \qquad \frac{\theta}{2(\theta+1)} \geq \beta \geq \frac{2\theta^2 - 1}{4(2\theta^3 - 1)} \,. \tag{57}$$

Figure 12 illustrates the behavior of spectral radii versus $\Delta t/T$ for some unconditionally stable members of the Wilson-θ family. The value $\theta = 1.366025$ is the smallest θ within the region of unconditional stability (cf. (57)). For this value $\rho \to 1$ as $\Delta t/T \to \infty$. This behavior, predicted by the stability analysis, is typical of all schemes for which (θ, β) lie upon the lower stability curve of Figure 11, and is highly undesirable since there is no numerical dissipation in the higher modes. Increasing the value of θ for fixed β decreases the limit of ρ as $\Delta t/T \to \infty$ (e.g., the $\theta = 1.4$ curve in Figure 12). The point along the curve at which ρ achieves its minimum marks the bifurcation of the complex conjugate principal roots into real roots. The smallest value of θ for which the principal roots remain complex at all $\Delta t/T$ is found to be $\theta = 1.420815$. Note the vastly superior spectral properties of $\theta = 1.420815$ over $\theta = 1.4$ (cf. [3]). As θ is further increased the limiting value of ρ decreases until it achieves a minimum and then begins to increase again (see for example, $\theta = 2$ in Figure 12). The behavior of ρ in Figure 12 is typical for all $\beta \in (0, \frac{1}{4})$.

As β decreases, θ must increase rapidly to maintain unconditional stability. So as not to incur too large a value of θ, attention has been restricted to the regime $\beta \geq 0.16$. It may also be argued that β may as well as assumed to be less than or equal to $\frac{1}{4}$.

Let $\theta^* = \theta^*(\beta)$ denote the smallest value of θ for which the principal roots remain complex as $\Delta t/T \to \infty$. Values of θ^*, obtained numerically for various

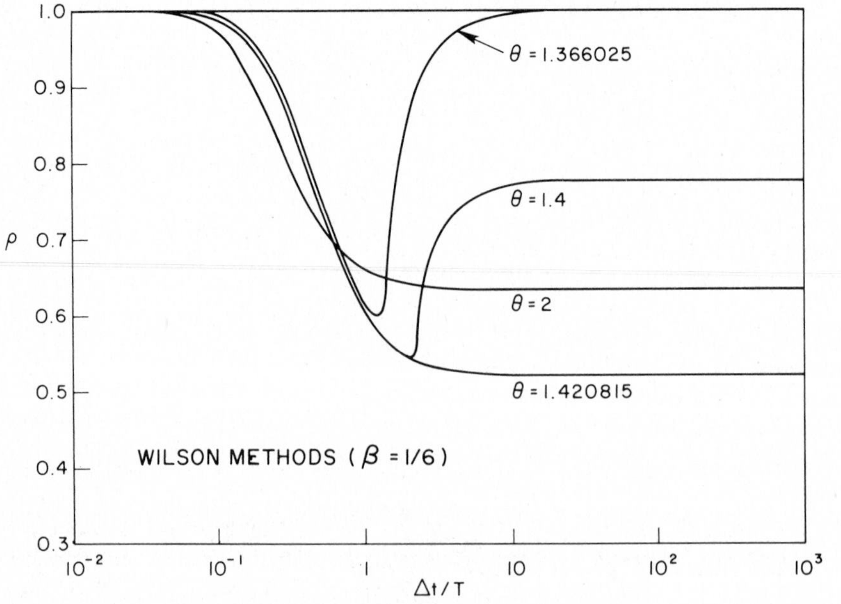

Fig. 12. Spectral radii for Wilson θ methods [21].

Table 7
Smallest collocation parameter which insures complex conjugate principal roots as $\Delta t/T \to \infty$. Corresponding values of algorithmic damping ratio and relative period error for $\Delta t/T = 0.1$ [21]

β	θ^*	$\bar{\xi}$	$(\bar{T}-T)/T$
$\frac{1}{4}$	1	0	0.032
0.24	1.021712	0.60×10^{-4}	0.032
0.23	1.047364	0.27×10^{-3}	0.033
0.22	1.077933	0.70×10^{-3}	0.034
0.21	1.114764	0.14×10^{-2}	0.036
0.20	1.159772	0.27×10^{-2}	0.039
0.19	1.215798	0.46×10^{-2}	0.043
0.18	1.287301	0.77×10^{-2}	0.050
0.17	1.381914	0.13×10^{-1}	0.060
$\frac{1}{6}$	1.420815	0.15×10^{-1}	0.064
0.16	1.514951	0.21×10^{-1}	0.075

values of β, are presented in Table 7, along with the corresponding algorithmic damping ratios and relative period errors for $\Delta t/T = 0.1$. The points (θ^*, β) lie slightly above the lower stability boundary of Figure 11. Note that both the damping ratio and period error increase as β decreases.

In Figure 13 spectral radii of some of the tabulated cases are illustrated. Note

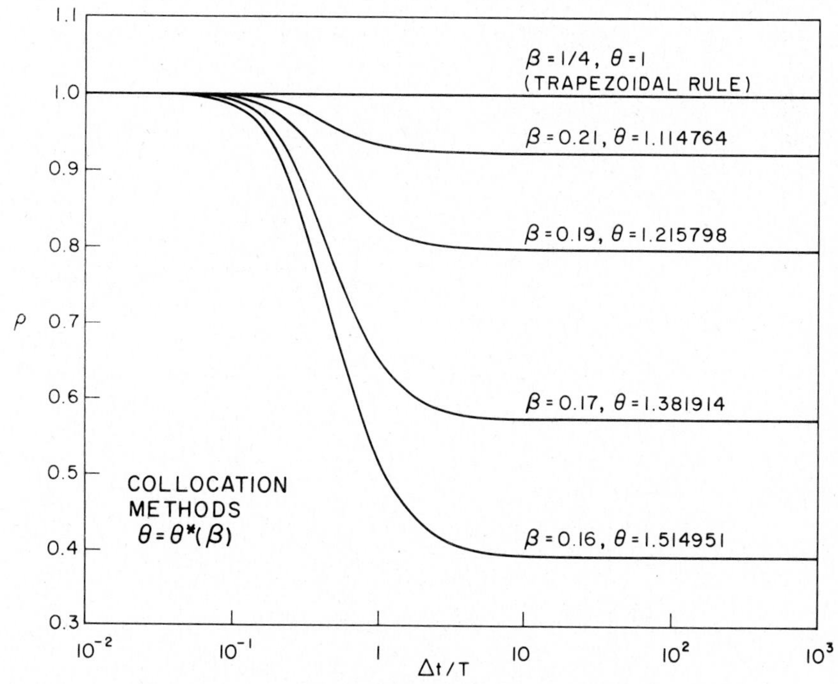

Fig. 13. Spectral radii for optimal collocation schemes [21].

that in each case the minimum value of ρ is obtained as $\Delta t/T \to \infty$, a desirable feature in multidegree-of-freedom applications.

In Figure 14 algorithmic damping ratios are plotted. The continuous control of numerical dissipation in the considered family of algorithms is evident. The damping ratios of the Wilson and Houbolt methods are also depicted in Figure 14 for comparison purposes.

The relative period errors for the various cases are presented in Figure 15. Clearly, for a fixed value of $\Delta t/T$ the relative period errors increase with θ.

As θ is increased beyond θ^* both the damping ratio and period error increase quadratically with θ. On the other hand, for θ fixed and β increasing from the lower unconditional stability limit to the upper one (see (57) and Figure 11), the damping ratio decreases in a linear fashion from its maximum value to zero, whereas the period error increases linearly. Thus it may be concluded: The best unconditionally stable collocation schemes, providing a maximum of dissipation and minimum period error are obtained if $\theta = \theta^*(\beta)$. Such methods are called *optimal collocation methods.* Some pairs $(\beta, \theta^*(\beta))$ are listed in Table 7 and intermediate points can be determined by linear interpolation.

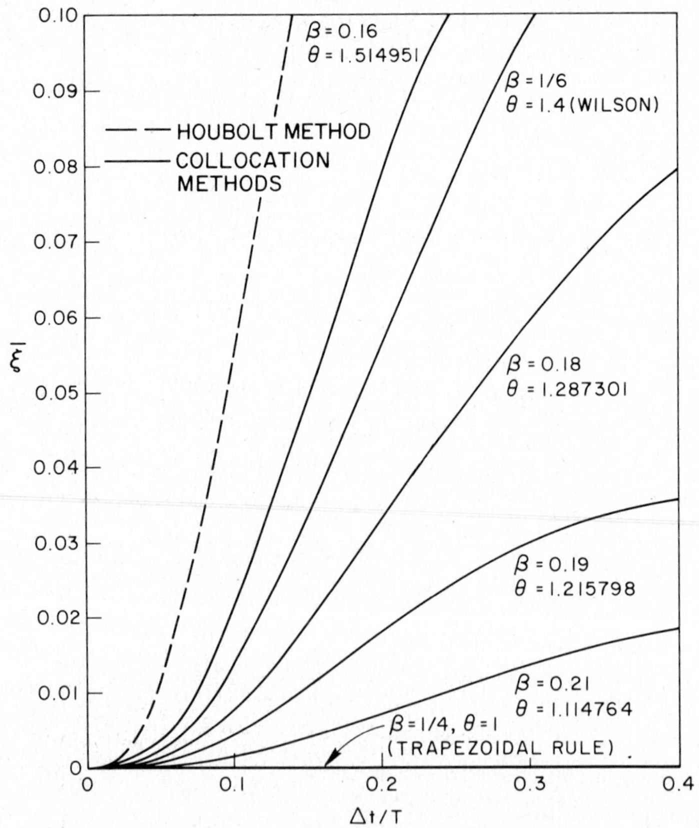

Fig. 14. Algorithmic damping ratios for collocation schemes and Houbolt method [21].

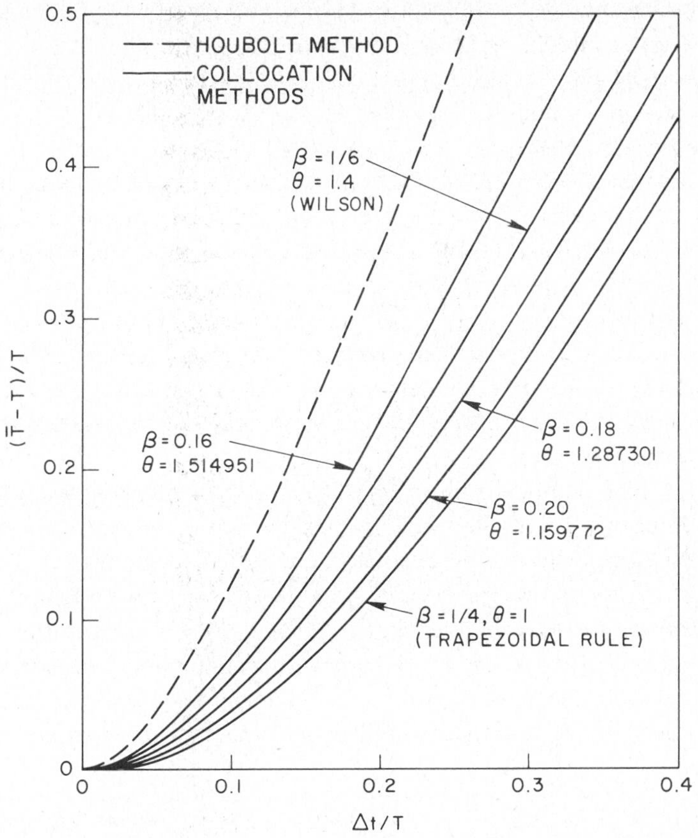

Fig. 15. Relative period errors for collocation schemes and Houbolt method [21].

B.III.5.c. α-method (Hilber, Hughes, Taylor method)

Numerical damping connot be introduced in the Newmark method without degrading the order of accuracy. To improve upon this situation, Hilber, Hughes and Taylor [22] introduced the α-method. In the α-method, the finite difference formulas of the Newmark method are retained (i.e., (5) and (6) of Section II), whereas the time-discrete equation of motion is modified as follows:

$$\boldsymbol{M}\boldsymbol{a}_{n+1} + (1+\alpha)\boldsymbol{C}\boldsymbol{v}_{n+1} - \alpha\boldsymbol{C}\boldsymbol{v}_n + (1+\alpha)\boldsymbol{K}\boldsymbol{d}_{n+1} - \alpha\boldsymbol{K}\boldsymbol{d}_n = (1+\alpha)\boldsymbol{F}_{n+1} - \alpha\boldsymbol{F}_n\,. \tag{58}$$

Clearly, if $\alpha = 0$ we reduce to the Newmark method. If the parameters are picked such that $\alpha \in [-\frac{1}{3}, 0]$, $\gamma = \frac{1}{2}(1-2\alpha)$ and $\beta = \frac{1}{4}(1-\alpha)^2$, an unconditionally stable, second-order accurate scheme results [53]. At $\alpha = 0$, we have the trapezoidal rule. Decreasing α increases the amount of numerical dissipation. The α-method can be put in the form of a 3-step LMS method for second-order equations, or equivalently, in a rank-3 amplification matrix form; see [20].

In Figures 16 and 17 the following cases are compared[7]:

a. Trapezoidal rule ($\alpha = 0$, $\beta = 0.25$, $\gamma = 0.5$).

b. Trapezoidal rule with α-damping ($\alpha = 0.1$, $\beta = 0.25$, $\gamma = 0.5$).

c. Newmark method with γ-damping ($\alpha = 0$, $\beta = 0.3025$, $\gamma = 0.6$).

d. α-method ($\alpha = -0.1$, $\beta = 0.3025$, $\gamma = 0.6$).

The results for case b indicate why α-damping, in itself, is not an effective dissipative mechanism. The effect is seen to be much the same as that for viscous damping (see Section II.4.d). The spectral radii for cases c and d are strictly less than one as $\Delta t/T \to \infty$. This condition insures that the response of higher modes is damped-out. For large $\Delta t/T$, cases c and d are identical. However, for small $\Delta t/T$, the spectral radius for case d is closer to one for a larger range of $\Delta t/T$. This is due to the addition of negative α-dissipation. In fact, it was the observation that combining cases b and c would produce an improved spectral radius graph which led to the α-method.

In Figure 17 the damping ratios versus $\Delta t/T$ are plotted for cases, a, b, c and d. Desirable properties for an algorithmic damping ratio graph to possess are a zero tangent at the origin and subsequently a controlled turn upward. This insures adequate dissipation in the higher modes and at the same time guarantees that the lower modes are not affected too strongly. Notice that for case c the dissipation ratio curve has positive slope at the origin, a manifestation of only first-order accuracy. This is typical for Newmark γ-damping and is the reason why the Newmark family is felt to possess ineffective numerical dissipation. Case b also

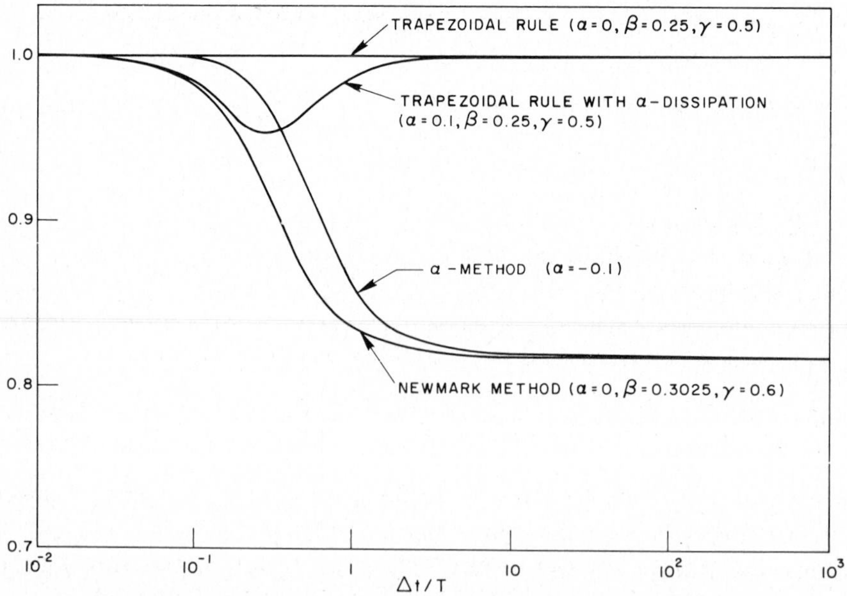

Fig. 16. Spectral radii versus $\Delta t/T$ for α method and Newmark schemes [22].

[7]In all cases $\beta = \frac{1}{4}(\gamma + \frac{1}{2})^2$ which insures unconditional stability.

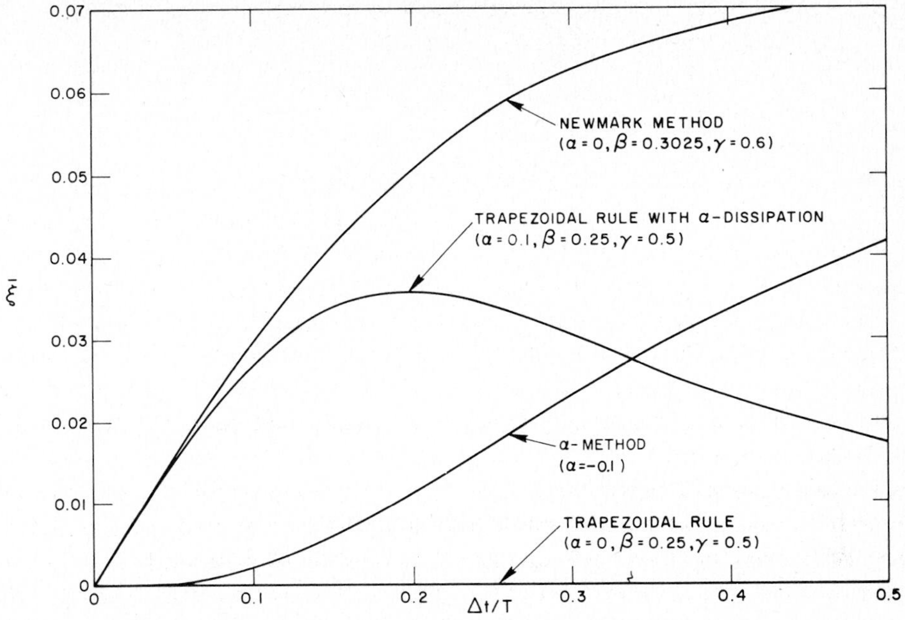

Fig. 17. Damping ratios versus $\Delta t/T$ for α method and Newmark schemes [22].

possesses this property and, in addition, turns downward at $\Delta t/T$ increases, which further emphasizes the ineffectiveness of α-dissipation. On the other hand, the dissipation ratio for case d has a zero slope at the origin and then turns upward.

B.III.5.d. Comparison of algorithms

The algorithms compared in Figures (18)–(20) are:

1. Houbolt's method,
2. Park's method,
3. Optimal collocation methods,
4. α-methods.

These methods have the following features in common: They are implicit, unconditionally stable, second-order accurate and dissipative.

The Houbolt and Park methods do not permit parametric control of the amount of dissipation present, whereas the collocation schemes and α-methods do.

In Figure 18 the spectral radii of the various cases are presented. The spectral radii of the Houbolt and Park methods approach zero as $\Delta t/T \to \infty$, as is typical of backward difference schemes. Collocation schemes and α-methods are seen also to possess strong damping in the high frequency regime. Recall that the effect of $\rho < 1$ is accumulative, e.g., $\rho^n \le e^{-n(1-\rho)}$, which rapidly approaches zero as n is increased.

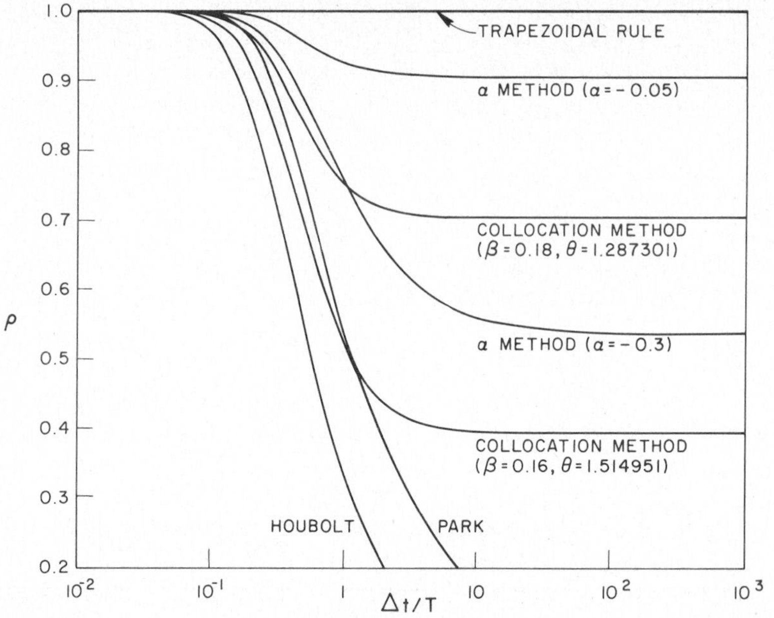

Fig. 18. Spectral radii for α methods, optimal collocation schemes and Houbolt and Park methods [21].

In Figure 19 algorithmic damping ratios are compared. The Houbolt and collocation methods are seen to affect the low modes (i.e., $\Delta t/T \leq 0.1$) too strongly. The inefficiency of the damping in the collocation scheme versus the α-method can be seen by comparing the cases $\beta = 0.18$ and $\alpha = -0.3$, respectively. From Figure 18, $\alpha = -0.3$ is seen to damp the high modes more strongly than $\beta = 0.18$. On the other hand, from Figure 19 the low modes are affected less by $\alpha = -0.3$ than by $\beta = 0.18$.

Relative period errors are compared in Figure 20. The collocation schemes and α-methods have smaller errors than the Houbolt and Park methods.

The following observations summarize the comparisons:

All methods are capable of sufficiently damping out the high modes. The Houbolt method affects the low modes much too strongly, both from the point of view of damping ratio and of period error. Park's method possesses good low mode damping properties, however it's period error is higher than the collocation schemes and α-methods. The collocation schemes damp the low-modes too strongly.

In Figure 21, period error results are presented for undamped Newmark methods ($\gamma = \frac{1}{2}$), and comparison is made with results for the Houbolt and Wilson algorithms. Note that for $\beta < \frac{1}{4}$, the methods presented are conditionally stable, which is made evident by the abrupt increases in period error at finite $\Delta t/T$. Notice also that the central difference method ($\beta = 0$) tends to shorten periods whereas the trapezoidal rule increases periods ($\beta = \frac{1}{4}$).

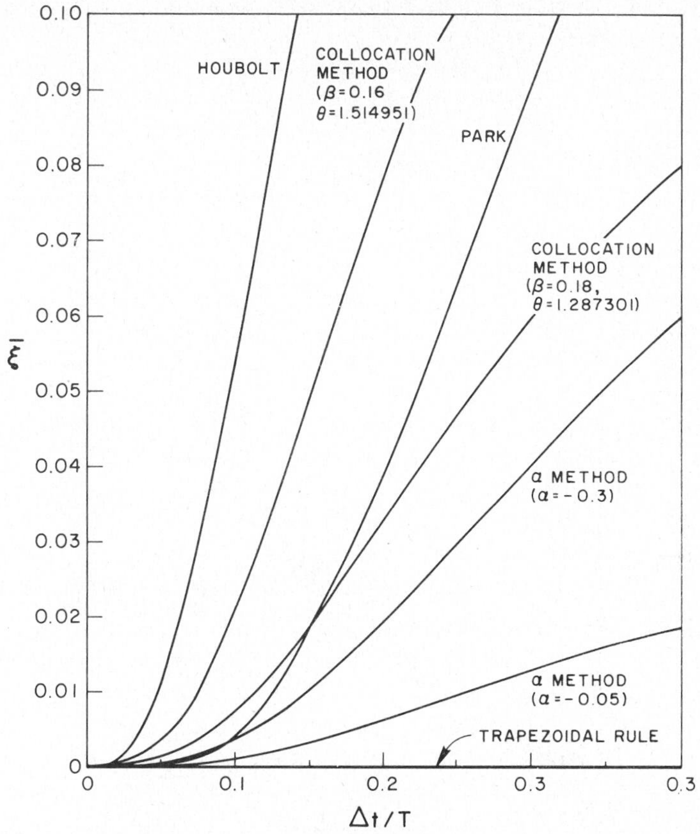

Fig. 19. Algorithmic damping ratios for α-methods, optimal collocation schemes, and Houbolt and Park methods [21].

Another important measure of the performance of unconditionally stable algorithms is called 'overshoot', a tendency to violently oscillate for a short time due to certain initial conditions, or discontinuous loading. It is difficult to characterize overshoot as either a stability or accuracy phenomenon, since it is independent of all stability and accuracy measures considered herein. There is no doubt, however, to its undesirability. Collocation schemes have been shown to behave pathologically with respect to overshoot, whereas the other algorithms considered tend to perform acceptably. In particular, trapezoidal rule and the Houbolt and Park algorithms produce no overshoot whatsoever. A discussion of this phenomenon and numerical results are presented in [21].

B.IV. Algorithms based upon operator and mesh partitions

Although LMS methods are without doubt still the most widely used in structural dynamics, schemes which do not fall within this framework are gaining

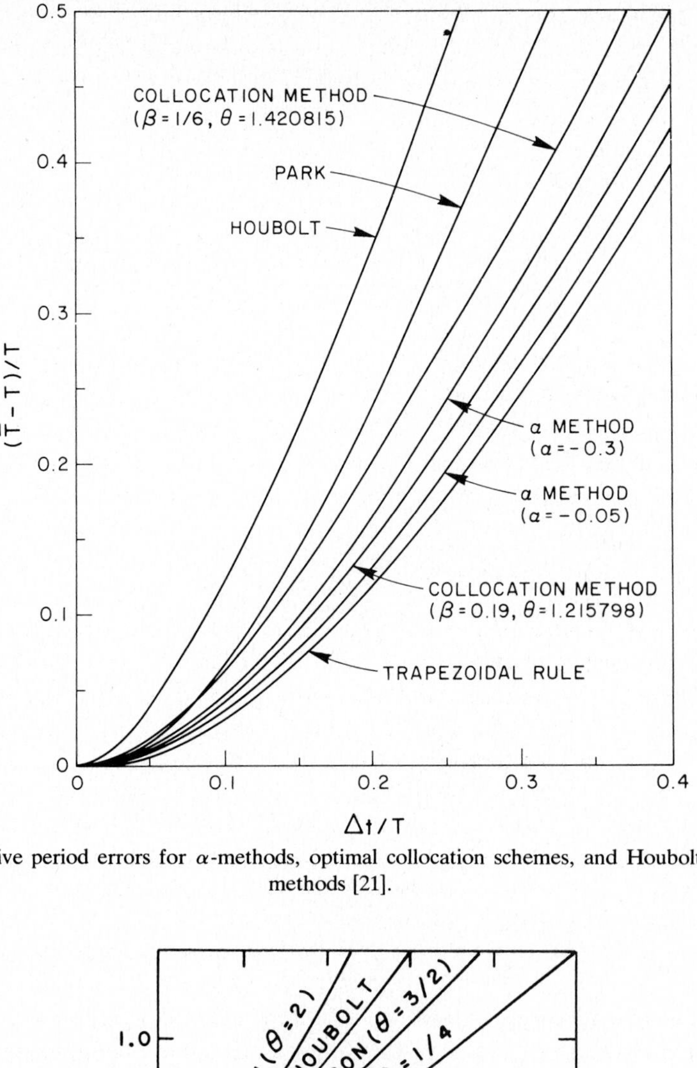

Fig. 20. Relative period errors for α-methods, optimal collocation schemes, and Houbolt and Park methods [21].

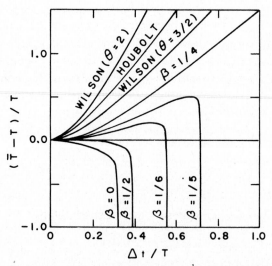

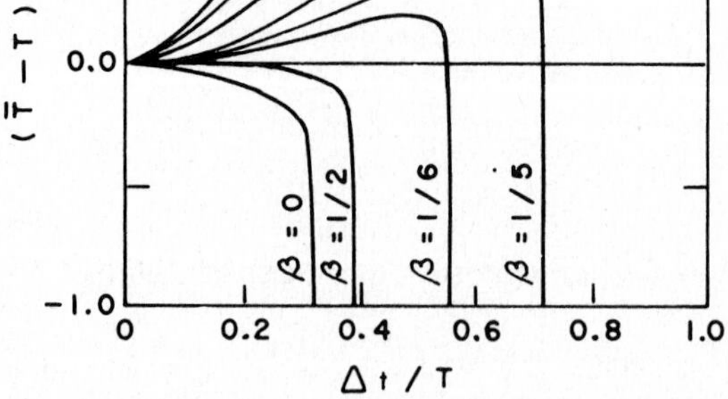

Fig. 21. Period errors for undamped Newmark methods ($\gamma = \frac{1}{2}$) compared with Wilson and Houbolt schemes [16].

popularity. In particular, so called 'partitioned analysis' procedures may be mentioned in this regard. In techniques of this kind, the mesh, or implicit operator, is partitioned in some way to exploit features of the problem under consideration. The chapter by Park and Felippa in this book is devoted to this topic. In this chapter we wish to just briefly touch upon those ideas which are useful for the analysis of methods of this type, and in particular, for ascertaining stability behavior. To this end, we introduce the energy method of stability analysis, which also is important later on in considering nonlinear systems. (The general theory presented in the Park–Felippa chapter is not based upon the energy method.)

As a vehicle for presenting these ideas, we shall employ the so-called implicit-explicit finite element algorithms introduced by Hughes and Liu [35, 36], and subsequently generalized and analyzed by Hughes and co-workers [30, 31, 40, 41]. These algorithms are similar in intent to one introduced earlier by Belytschko and Mullen [4, 5], but possess some implementational advantages.

B.IV.1. Predictor-corrector algorithms

The implicit-explicit algorithms we wish to describe are based upon the Newmark schemes and explicit, predictor-corrector schemes. The predictor-corrector schemes are in one-to-one correspondence with members of the Newmark family, viz.,

$$\boldsymbol{M}\boldsymbol{a}_{n+1} + \boldsymbol{C}\tilde{\boldsymbol{v}}_{n+1} + \boldsymbol{K}\tilde{\boldsymbol{d}}_{n+1} = \boldsymbol{F}_{n+1}\,, \tag{1}$$

$$\tilde{\boldsymbol{d}}_{n+1} = \boldsymbol{d}_n + \Delta t\boldsymbol{v}_n + \frac{\Delta t^2}{2}(1-2\beta)\boldsymbol{a}_n\,, \tag{2}$$

$$\tilde{\boldsymbol{v}}_{n+1} = \boldsymbol{v}_n + \Delta t(1-\gamma)\boldsymbol{a}_n\,, \tag{3}$$

$$\boldsymbol{d}_{n+1} = \tilde{\boldsymbol{d}}_{n+1} + \Delta t^2\beta\boldsymbol{a}_{n+1}\,, \tag{4}$$

$$\boldsymbol{v}_{n+1} = \tilde{\boldsymbol{v}}_{n+1} + \Delta t\gamma\boldsymbol{a}_{n+1}\,, \tag{5}$$

where $\tilde{\boldsymbol{d}}_{n+1}$ and $\tilde{\boldsymbol{v}}_{n+1}$ are 'predictor values.' The corrector equations, (4) and (5), are identical to (5) and (6) of Section II, respectively, for the Newmark method.

B.IV.2. Implicit-explicit algorithms

Consider a finite element model in which the elements are divided into two groups: the implicit elements and the explicit elements. Let I and E superscripts refer to the implicit and explicit groups, respectively. In particular let $\boldsymbol{M}^I$, $\boldsymbol{C}^I$, $\boldsymbol{K}^I$ and $\boldsymbol{F}^{\mathrm{I}}$ (resp., $\boldsymbol{M}^{\mathrm{E}}$, $\boldsymbol{C}^{\mathrm{E}}$, $\boldsymbol{K}^{\mathrm{E}}$ and $\boldsymbol{F}^{\mathrm{E}}$) be the assembled mass, damping, stiffness and load for the implicit (resp., explicit) group. Each of the aforementioned matrices is assumed to be positive semi-definite. The implicit-explicit algorithms, a composite of the implicit Newmark algorithms and explicit, predictor-corrector algorithms, are given by

$$\boldsymbol{M}\boldsymbol{a}_{n+1} + \boldsymbol{C}^{\mathrm{I}}\boldsymbol{v}_{n+1} + \boldsymbol{C}^{\mathrm{E}}\tilde{\boldsymbol{v}}_{n+1} + \boldsymbol{K}^{\mathrm{I}}\boldsymbol{d}_{n+1} + \boldsymbol{K}^{\mathrm{E}}\tilde{\boldsymbol{d}}_{n+1} = \boldsymbol{F}_{n+1} \tag{6}$$

and (2)–(5), where

$$\boldsymbol{M} = \boldsymbol{M}^{\mathrm{I}} + \boldsymbol{M}^{\mathrm{E}}, \tag{7}$$

$$\boldsymbol{C} = \boldsymbol{C}^{\mathrm{I}} + \boldsymbol{C}^{\mathrm{E}}, \tag{8}$$

$$\boldsymbol{K} = \boldsymbol{K}^{\mathrm{I}} + \boldsymbol{K}^{\mathrm{E}}, \tag{9}$$

$$\boldsymbol{F} = \boldsymbol{F}^{\mathrm{I}} + \boldsymbol{F}^{\mathrm{E}}. \tag{10}$$

Many other possibilities fit within this general framework. For example, (7)–(10) may be viewed as general definitions of 'splitting' an operator. In this way, the procedures described in [40] (sometimes referred to as 'semi-implicit schemes') and [61] (i.e., 'triangular splittings') may be subsumed. As observed by Park [58], nodal partitions may also be encompassed by formulations of this type.

B.IV.3. Stability: the energy method

It is obvious from (6), that the implicit-explicit algorithms are not LMS methods. To deduce the stability behavior we shall employ the energy method [6, 59] in which it is attempted to derive bounds on energy-like norms of the discrete solution.

To simplify the subsequent writing, it is helpful to introduce the jump (undivided forward difference) and mean value operators, [ ] and ⟨ ⟩, respectively, which are defined by $[\boldsymbol{x}_n] = \boldsymbol{x}_{n+1} - \boldsymbol{x}_n$ and $\langle \boldsymbol{x}_n \rangle = \frac{1}{2}(\boldsymbol{x}_{n+1} + \boldsymbol{x}_n)$. We shall use the notation $\|\cdot\|$ to denote any vector norm, and also to denote the associated matrix norm (see Section II.3.b). A vector-valued sequence, $\boldsymbol{x}_n$, $n = 0, 1, 2, \ldots$, will be said to be *bounded* if $\|\boldsymbol{x}_n\| \leq c$, $n = 0, 1, 2, \ldots$, where c is a non-negative constant.

The plan for obtaining stability conditions for the implicit-explicit algorithms is to first apply the energy method to the Newmark and predictor-corrector algorithms, and then show that the implicit-explicit algorithms may be reduced to a combination of the previous cases. It suffices to restrict to the case in which $\boldsymbol{F} = \boldsymbol{0}$ for purposes of stability.

B.IV.3.a. Newmark algorithms

Equations (4)–(6) of Section II may be combined to form the following identity:

$$\boldsymbol{a}_{n+1}^{\mathrm{T}}\boldsymbol{A}\boldsymbol{a}_{n+1} + \boldsymbol{v}_{n+1}^{\mathrm{T}}\boldsymbol{K}\boldsymbol{v}_{n+1} = \boldsymbol{a}_n^{\mathrm{T}}\boldsymbol{A}\boldsymbol{a}_n + \boldsymbol{v}_n^{\mathrm{T}}\boldsymbol{K}\boldsymbol{v}_n - (2\gamma - 1)[\boldsymbol{a}_n]^{\mathrm{T}}\boldsymbol{B}[\boldsymbol{a}_n] - 2\Delta t\langle \boldsymbol{a}_n\rangle^{\mathrm{T}}\boldsymbol{C}\langle \boldsymbol{a}_n\rangle \tag{11}$$

where

$$\boldsymbol{B} = \boldsymbol{M} + \Delta t(\gamma - \tfrac{1}{2})\boldsymbol{C} + \Delta t^2(\beta - \tfrac{1}{2}\gamma)\boldsymbol{K}, \tag{12}$$

$$\boldsymbol{A} = \boldsymbol{B} + \Delta t(\gamma - \tfrac{1}{2})\boldsymbol{C}. \tag{13}$$

(There should be no confusion over the present definition of the matrix $\boldsymbol{A}$, and earlier use for denoting an amplification matrix.)

Theorem. *If $\gamma \geq \frac{1}{2}$ and $\boldsymbol{B}$ is positive definite, then $\boldsymbol{a}_n$ and $\boldsymbol{v}_n$ are bounded.*

Proof. The hypotheses imply $\boldsymbol{A}$ is positive definite and

$$\boldsymbol{a}_{n+1}^{\mathrm{T}}\boldsymbol{A}\boldsymbol{a}_{n+1} + \boldsymbol{v}_{n+1}^{\mathrm{T}}\boldsymbol{K}\boldsymbol{v}_{n+1} \leq \boldsymbol{a}_n^{\mathrm{T}}\boldsymbol{A}\boldsymbol{a}_n + \boldsymbol{v}_n^{\mathrm{T}}\boldsymbol{K}\boldsymbol{v}_n \tag{14}$$

which in turn implies

$$\boldsymbol{a}_n^{\mathrm{T}}\boldsymbol{A}\boldsymbol{a}_n + \boldsymbol{v}_n^{\mathrm{T}}\boldsymbol{K}\boldsymbol{v}_n \leq \boldsymbol{a}_0^{\mathrm{T}}\boldsymbol{A}\boldsymbol{a}_0 + \boldsymbol{v}_0^{\mathrm{T}}\boldsymbol{K}\boldsymbol{v}_0, \quad n = 1, 2, 3, \ldots, \tag{15}$$

from which the conclusions follow.

Corrolary: *If $\boldsymbol{K}^{-1}$ exists, then $\boldsymbol{d}_n$ is also bounded.*

Proof. From (4) of Section II,

$$\|\boldsymbol{d}_n\| \leq \|\boldsymbol{K}^{-1}\|(\|\boldsymbol{M}\|\,\|\boldsymbol{a}_n\| + \|\boldsymbol{C}\|\,\|\boldsymbol{v}_n\|)\,. \tag{16}$$

Thus it remains only to ascertain under what conditions $\boldsymbol{B}$ is positive definite to completely determine the stability characteristics of the Newmark algorithms. The results are identical to those obtained previously (see Section II, (86)–(90)).

Remarks. 1. Equation (11) is called a *discrete conservation law*, whereas (14) and (15) are *decay inequalities.*

2. It may be easily shown that any spectrally stable method possesses a discrete conservation law or decay inequality [8].

B.IV.3.b. Predictor-corrector algorithms

Equations (1)–(5) lead to the identity

$$\begin{aligned}\boldsymbol{a}_{n+1}^{\mathrm{T}}\bar{\boldsymbol{A}}\boldsymbol{a}_{n+1} + \boldsymbol{v}_{n+1}^{\mathrm{T}}\boldsymbol{K}\boldsymbol{v}_{n+1} = \boldsymbol{a}_n^{\mathrm{T}}\bar{\boldsymbol{A}}\boldsymbol{a}_n + \boldsymbol{v}_n^{\mathrm{T}}\boldsymbol{K}\boldsymbol{v}_n &- (2\gamma - 1)[\boldsymbol{a}_n]^{\mathrm{T}}\bar{\boldsymbol{B}}[\boldsymbol{a}_n] \\ &- 2\Delta t\langle\boldsymbol{a}_n\rangle^{\mathrm{T}}\boldsymbol{C}\langle\boldsymbol{a}_n\rangle\end{aligned} \tag{17}$$

where

$$\bar{\boldsymbol{B}} = \boldsymbol{B} - \Delta t\gamma\boldsymbol{C} - \Delta t^2\beta\boldsymbol{K}\,, \tag{18}$$

$$\bar{\boldsymbol{A}} = \bar{\boldsymbol{B}} + \Delta t(\gamma - \tfrac{1}{2})\boldsymbol{C}\,. \tag{19}$$

Theorem. *If $\gamma \geq \frac{1}{2}$ and $\bar{\boldsymbol{B}}$ is positive-definite, then $\boldsymbol{a}_n$ and $\boldsymbol{v}_n$ are bounded.*

Proof. The proof is identical to that for the previous theorem, with $\bar{\boldsymbol{A}}$ in place of $\boldsymbol{A}$.

Corollary. *If $\boldsymbol{K}^{-1}$ exists, then $\boldsymbol{d}_n$ is also bounded.*

Proof. From (1), (4) and (5),

$$\|d_n\| \leq \Delta t^2|\beta|\,\|a_n\| + \|K^{-1}\|(\|M\|\,\|a_n\| + \|C\|(\|v_n\| + \Delta t\gamma\|a_n\|))\,. \tag{20}$$

The stability characteristics of the predictor-corrector scnemes are thus determined by the conditions which render $\bar{B}$ positive definite. The requirements for stability are given as follows:

$$\gamma \geq \tfrac{1}{2}, \tag{21}$$

$$\Omega < \Omega_{\text{crit}} := ((\xi^2 + 2\gamma)^{1/2} - \xi)/\gamma\,. \tag{22}$$

Remarks. 1. It is to be expected that there are no unconditionally stable predictor-corrector schemes since they are all explicit linear multistep methods.

2. Spectral radii are presented in Figure 22 for the case $\xi = 0$. The Ω_{crit} obtained from the diagram is consistent with (22). The slope discontinuity in the spectral radius curves corresponds to the point at which the complex conjugate, principal roots of the amplification matrix become real and bifurcate. The value of Ω at which bifurcation occurs is

$$\Omega_{\text{bif}} = 2(1 - \xi)/(\gamma + \tfrac{1}{2}) \tag{23}$$

To insure high-frequency, numerical dissipation when $\xi < 1$, Ω should be kept below Ω_{bif}, resulting in a somewhat smaller time step than that required by Ω_{crit}; cf. (22).

3. Increasing ξ, or γ, *decreases* the critical time step for the predictor-corrector algorithms. Thus an estimate of ξ is essential in determining a stable time step.

4. When viscous damping is absent, $\Omega_{\text{crit}} = (2/\gamma)^{1/2}$. The maximum occurs for $\gamma = \frac{1}{2}$. That is $\Omega_{\text{crit}} = 2$, which is the same result as for the central difference method.

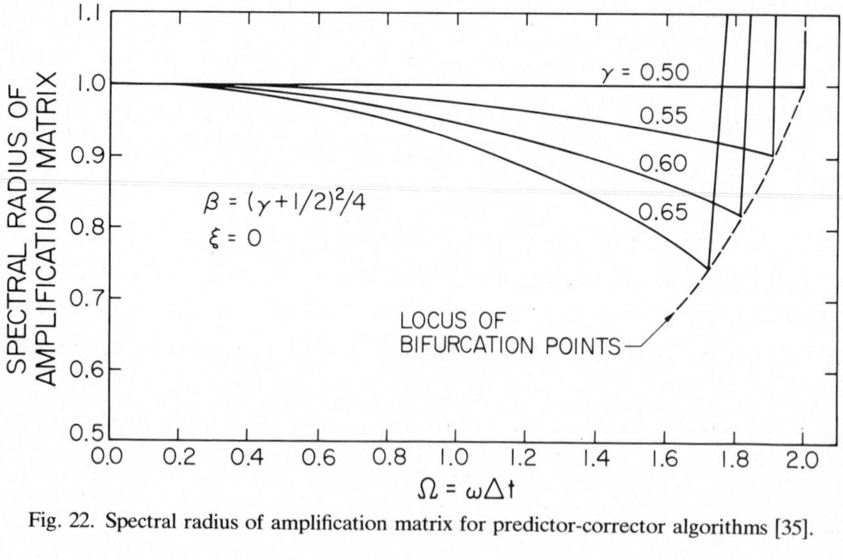

Fig. 22. Spectral radius of amplification matrix for predictor-corrector algorithms [35].

5. A local truncation error analysis of the predictor-corrector schemes indicates that if $\boldsymbol{C}=\boldsymbol{0}$, $\gamma=\frac{1}{2}$ is a necessary and sufficient condition for second-order accuracy. If $\boldsymbol{C}\neq\boldsymbol{0}$, then first-order accuracy is attained for all $\gamma\geq\frac{1}{2}$. (A generalization of the basic scheme, which enables second-order accuracy to be regained under these circumstances, will be discussed subsequently.)

6. Stability conditions for the first-order case (i.e., when $\boldsymbol{K}=\boldsymbol{0}$) are easily deducible from the theorem. In this case, $\bar{\boldsymbol{B}}$ is positive-definite when $\lambda\Delta t<2$, where λ is the maximum eigenvalue of $\boldsymbol{M}^{-1}\boldsymbol{C}$. (Note that $\gamma\geq\frac{1}{2}$ is still required.)

B.IV.3.c. Implicit-explicit algorithms

An identity analogous to those for the previous cases may be derived from (2)–(6):

$$\boldsymbol{a}_{n+1}^{\mathrm{T}}(\boldsymbol{A}^{\mathrm{I}}+\bar{\boldsymbol{A}}^{\mathrm{E}})\boldsymbol{a}_{n+1}+\boldsymbol{v}_{n+1}^{\mathrm{T}}\boldsymbol{K}\boldsymbol{v}_{n+1}=\boldsymbol{a}_{n}^{\mathrm{T}}(\boldsymbol{A}^{\mathrm{I}}+\bar{\boldsymbol{A}}^{\mathrm{E}})\boldsymbol{a}_{n}+\boldsymbol{v}_{n}^{\mathrm{T}}\boldsymbol{K}\boldsymbol{v}_{n}$$
$$-(2\gamma-1)[\boldsymbol{a}_n]^{\mathrm{T}}(\boldsymbol{B}^{\mathrm{I}}+\bar{\boldsymbol{B}}^{\mathrm{E}})[\boldsymbol{a}_n]-2\Delta t\langle\boldsymbol{a}_n\rangle^{\mathrm{T}}\boldsymbol{C}\langle\boldsymbol{a}_n\rangle \tag{24}$$

where

$$\boldsymbol{B}^{\mathrm{I}}=\boldsymbol{M}^{\mathrm{I}}+\Delta t(\gamma-\tfrac{1}{2})\boldsymbol{C}^{\mathrm{I}}+\Delta t^2(\beta-\tfrac{1}{2}\gamma)\boldsymbol{K}^{\mathrm{I}}, \tag{25}$$

$$\boldsymbol{A}^{\mathrm{I}}=\boldsymbol{B}^{\mathrm{I}}+\Delta t(\gamma-\tfrac{1}{2})\boldsymbol{C}^{\mathrm{I}}, \tag{26}$$

$$\bar{\boldsymbol{B}}^{\mathrm{E}}=\boldsymbol{B}^{\mathrm{E}}-\Delta t\gamma\boldsymbol{C}-\Delta t^2\beta\boldsymbol{K}, \tag{27}$$

$$\boldsymbol{B}^{\mathrm{E}}=\boldsymbol{M}^{\mathrm{E}}+\Delta t(\gamma-\tfrac{1}{2})\boldsymbol{C}^{\mathrm{E}}+\Delta t^2(\beta-\tfrac{1}{2}\gamma)\boldsymbol{K}^{\mathrm{E}}, \tag{28}$$

$$\bar{\boldsymbol{A}}^{\mathrm{E}}=\bar{\boldsymbol{B}}^{\mathrm{E}}+\Delta t(\gamma-\tfrac{1}{2})\boldsymbol{C}^{\mathrm{E}}. \tag{29}$$

Theorem. *If $\gamma\geq\frac{1}{2}$ and $\boldsymbol{B}^{\mathrm{I}}+\bar{\boldsymbol{B}}^{\mathrm{E}}$ is positive definite, then $\boldsymbol{a}_n$ and $\boldsymbol{v}_n$ are bounded.*

Proof. Again the proof is identical to that for the first theorem with $\boldsymbol{A}^{\mathrm{I}}+\bar{\boldsymbol{A}}^{\mathrm{E}}$ in place of $\boldsymbol{A}$.

Corollary. *If $\boldsymbol{K}^{-1}$ exists, then $\boldsymbol{d}_n$ is also bounded.*

Proof. From (4), (5) and (18),

$$\|\boldsymbol{d}_n\|\leq\|\boldsymbol{K}^{-1}\|((\|\boldsymbol{M}\|+\Delta t\gamma\|\boldsymbol{C}^{\mathrm{E}}\|+\Delta t^2|\beta|\,\|\boldsymbol{K}^{\mathrm{E}}\|)\|\boldsymbol{a}_n\|+\|\boldsymbol{C}\|\,\|\boldsymbol{v}_n\|)\,. \tag{30}$$

$\boldsymbol{B}^{\mathrm{I}}+\bar{\boldsymbol{B}}^{\mathrm{E}}$ is rendered positive definite, and stability is thereby achieved, when β, γ and Δt satisfy (86) of Section II, or (87)–(90) for that section, and (22). The time step restrictions, (89) and (90) of section II, and (22), pertain only to the implicit and explicit element groups, respectively.

Remark. In practice, one might as well achieve unconditional stability in the implicit element group. Thus for a given value of $\gamma\geq\frac{1}{2}$, β could be selected according to (94) of Section II. The time step restriction then emanates from satisfaction of (22), or (23), in the explicit element group. For example, if $\gamma = \frac{1}{2}$, (94) of Section II yields $\beta = \frac{1}{4}$, resulting in the trapezoidal rule algorithm for the implicit group. In the absence of viscous damping in the explicit element group, (22), or (23), results in $\Omega_{\text{crit}} = 2$. A sufficient condition for stability then is that $\Omega < 2$ for the maximum natural frequency of the explicit elements.

B.IV.4. Convergence

The implicit-explicit algorithms do not fit within the class of LMS methods and the usual convergence arguments, as employed earlier, do not appear to be applicable. In this section an alternative approach is employed to establish convergence.

B.IV.4.a. Preliminaries

A function is said to be *of continuity class* C^k (or C^k*-continuous*) if its derivatives of order i, $0 \leq i \leq k$, are continuous.

A function $\boldsymbol{g}(\boldsymbol{x})$ is said to be *Lipschitz continuous* if there exists a constant C_{L} such that for all $\boldsymbol{y}$ and $\boldsymbol{z}$

$$\|\boldsymbol{g}(\boldsymbol{y}) - \boldsymbol{g}(\boldsymbol{z})\| \leq C_{\text{L}} \|\boldsymbol{y} - \boldsymbol{z}\| . \tag{31}$$

(We may note that functions of class C^k, where $k \geq 1$, are also Lipschitz continuous.)

Assumptions (L). *We assume throughout that* $\boldsymbol{M}$, $\boldsymbol{C}$ *and* $\boldsymbol{K}$ *are constant and symmetric;* $\boldsymbol{M}$ *and* $\boldsymbol{K}$ *are positive-definite;* $\boldsymbol{C}$ *is positive semidefinite and* $\boldsymbol{F}^{\text{ext}} = \boldsymbol{F}^{\text{ext}}(t)$ *is a given continuous function.*

Lemma. *Let* $\boldsymbol{f}(t)$ *be continuous and* $\boldsymbol{g}(\boldsymbol{x}, t)$ *be continuous in* t *and Lipschitz continuous in* $\boldsymbol{x}$, *with Lipschitz constant* C_{L} *(independent of* t*) as in* (31). *Then there exists a unique,* C^1*-continuous function* $\boldsymbol{x}(t)$ *which satisfies*

$$\dot{\boldsymbol{x}}(t) + \boldsymbol{g}(\boldsymbol{x}(t), t) = \boldsymbol{f}(t) , \tag{32}$$

$$\boldsymbol{x}(0) = \boldsymbol{0} . \tag{33}$$

Furthermore

$$\|\boldsymbol{x}(t)\| \leq J(t)\, \mathrm{e}^{C_{\text{L}} t} \tag{34}$$

where

$$J(t) = \int_0^t \|\boldsymbol{f}(\tau)\|\, \mathrm{d}\tau . \tag{35}$$

This is a standard result of ordinary differential equation theory and may be found, for example, in [24].

Corollary 1. *Let $\boldsymbol{A}(t)$ be an invertible, t-continuous, matrix-valued function. Then there exists a unique, C^1-continuous function $\boldsymbol{x}(t)$ which satisfies*

$$\boldsymbol{A}(t)\dot{\boldsymbol{x}}(t) + \boldsymbol{g}(\boldsymbol{x}(t), t) = \boldsymbol{f}(t) , \tag{36}$$

$$\boldsymbol{x}(0) = \boldsymbol{0} , \tag{37}$$

and

$$\|\boldsymbol{x}(t)\| \le J(t)\, \mathrm{e}^{C'_{\mathrm{L}} t} \tag{38}$$

where

$$J(t) = \int_0^t \|\boldsymbol{A}^{-1}(\tau)\boldsymbol{f}(\tau)\|\, \mathrm{d}\tau \tag{39}$$

and C'_{L} is the Lipschitz constant for the function $\mathbf{A}^{-1}(t)\boldsymbol{g}(\boldsymbol{x}, t)$.

Corollary 2. *Let $\boldsymbol{f}_\varepsilon(t)$ be a one-parameter family of t-continuous mappings. For each fixed t, assume $\varepsilon^{-k}\boldsymbol{f}_\varepsilon(t)$ approaches a constant value as $\varepsilon \to 0$. [$f_\varepsilon(t)$ is said to be $\mathrm{O}(\varepsilon^k)$]. Then the solution, $\boldsymbol{x}_\varepsilon(t)$, of*

$$\boldsymbol{A}(t)\dot{\boldsymbol{x}}_\varepsilon(t) + \boldsymbol{g}(\boldsymbol{x}_\varepsilon(t), t) = \boldsymbol{f}_\varepsilon(t) , \tag{40}$$

$$\boldsymbol{x}_\varepsilon(0) = \boldsymbol{0} , \tag{41}$$

is $\mathrm{O}(\varepsilon^k)$.

Proof. Because

$$J_\varepsilon(t) = \int_0^t \|\boldsymbol{A}^{-1}(\tau)\boldsymbol{f}_\varepsilon(\tau)\|\, \mathrm{d}\tau$$

is $\mathrm{O}(\varepsilon^k)$, (38) implies that $\boldsymbol{x}_\varepsilon(t)$ is $\mathrm{O}(\varepsilon^k)$.

B.IV.4.b. Auxiliary problem

Eliminate $\tilde{\boldsymbol{d}}_{n+1}$ and $\tilde{\boldsymbol{v}}_{n+1}$ in (6) by using (4) and (5), respectively. Employing (8) and (9) in the result yields

$$\hat{\boldsymbol{M}}\boldsymbol{a}_{n+1} + \boldsymbol{C}\boldsymbol{v}_{n+1} + \boldsymbol{K}\boldsymbol{d}_{n+1} = \boldsymbol{F}_{n+1} \tag{42}$$

where

$$\hat{\boldsymbol{M}} = \boldsymbol{M} - \Delta t\gamma \boldsymbol{C}^{\mathrm{E}} - \Delta t^2\beta \boldsymbol{K}^{\mathrm{E}} . \tag{43}$$

Equation (42) suggests consideration of the following *auxiliary problem*:

$$\hat{\boldsymbol{M}}\ddot{\boldsymbol{D}} + \boldsymbol{C}\dot{\boldsymbol{D}} + \boldsymbol{K}\boldsymbol{D} = \boldsymbol{F}^{\mathrm{ext}} \tag{44}$$

$$\boldsymbol{D}(0) = \boldsymbol{D}_0 , \tag{45}$$

$$\dot{\boldsymbol{D}}(0) = \boldsymbol{V}_0 . \tag{46}$$

With the aid of (44)–(46), convergence of $\boldsymbol{d}_n$ to $\boldsymbol{d}(t_n)$ may be established by arguing: (i) that $\boldsymbol{d}_n$ converges to $\boldsymbol{D}(t_n)$; and (ii) that $\boldsymbol{D}(t_n)$ converges to $\boldsymbol{d}(t_n)$. The idea is schematically illustrated in Figure 23.

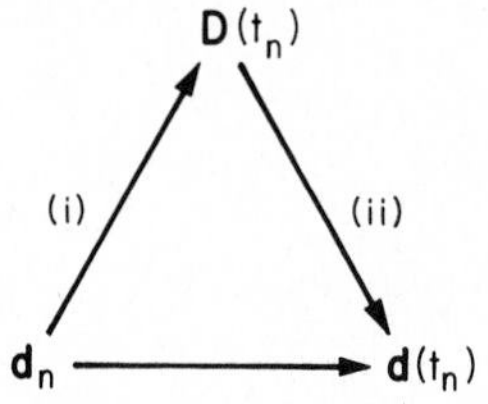

Fig. 23. Schematic of convergence proof [41].

Before proceeding with the convergence proof, we note that by Assumptions (L), there exist unique solutions of (1)–(3) of Section II and (44)–(46) which are of class C^2. In the case of (44)–(46) we need to assume $\hat{\boldsymbol{M}}$ is invertible, which is guaranteed if Δt is taken sufficiently small.

Part (*i*). Observe that the implicit-explicit algorithm becomes the *classical Newmark algorithm* with respect to the auxiliary problem [cf. (2)–(5) and (42) with (44)–(46)]. Consequently, we may apply the results obtained in Section II to the present case, namely

$$\boldsymbol{d}_n - \boldsymbol{D}(t_n) = \mathrm{O}(\Delta t^2) \quad \text{if } \gamma = \tfrac{1}{2}, \tag{47}$$

$$= \mathrm{O}(\Delta t) \quad \text{if } \gamma > \tfrac{1}{2}. \tag{48}$$

Thus to complete the proof it remains only to establish that $\boldsymbol{D}(t_n) \to \boldsymbol{d}(t_n)$.

Part (*ii*). Let $\boldsymbol{e} = \boldsymbol{D} - \boldsymbol{d}$; thus we need to show that $\boldsymbol{e} \to \boldsymbol{0}$ as $\Delta t \to 0$. To this end subtract (44)–(46) from (1)–(3) of Section II, respectively, to obtain

$$\boldsymbol{M}\ddot{\boldsymbol{e}} + \boldsymbol{C}\dot{\boldsymbol{e}} + \boldsymbol{K}\boldsymbol{e} = \boldsymbol{\Lambda}\ddot{\boldsymbol{D}}, \tag{49}$$

$$\boldsymbol{e}(0) = \boldsymbol{0}, \tag{50}$$

$$\dot{\boldsymbol{e}}(0) = \boldsymbol{0}, \tag{51}$$

where

$$\boldsymbol{\Lambda} = \Delta t \gamma \boldsymbol{C}^{\mathrm{E}} + \Delta t^2 \beta \boldsymbol{K}^{\mathrm{E}}. \tag{52}$$

Equations (49)–(51) may be put in the form of (40) and (41) by adopting the following definitions:

$$\boldsymbol{A} = \begin{bmatrix} \boldsymbol{M} & \boldsymbol{0} \\ \boldsymbol{0} & \boldsymbol{I} \end{bmatrix}, \tag{53}$$

$$\boldsymbol{B} = \begin{bmatrix} \boldsymbol{C} & \boldsymbol{K} \\ \boldsymbol{I} & \boldsymbol{0} \end{bmatrix}, \tag{54}$$

$$f_\varepsilon(t) = \begin{Bmatrix} \boldsymbol{\Lambda}\ddot{\boldsymbol{D}}(t) \\ \boldsymbol{0} \end{Bmatrix}, \tag{55}$$

$$\boldsymbol{g}(\boldsymbol{x}_\varepsilon, t) = \boldsymbol{B}\boldsymbol{x}_\varepsilon\,, \tag{56}$$

$$\boldsymbol{x}_\varepsilon = \begin{Bmatrix} \dot{\boldsymbol{e}} \\ \boldsymbol{e} \end{Bmatrix}, \tag{57}$$

$$\varepsilon = \Delta t\,. \tag{58}$$

As remarked previously, $\boldsymbol{D}$ is of class C^2 and consequently $\ddot{\boldsymbol{D}}$ is continuous. Furthermore, from (52), $\boldsymbol{\Lambda}$ is clearly $O(\Delta t)$. Thus $\boldsymbol{f}_\varepsilon(t)$ as defined in (55) satisfies the hypotheses of Corollary 2, and so we conclude $\boldsymbol{e}$ is $O(\Delta t)$.

From (52) and Corollary 2, second-order convergence of $\boldsymbol{e}$ to $\boldsymbol{0}$ is attained if $\boldsymbol{C}^{\mathrm{E}} = \boldsymbol{0}$. By employing a slight extension of the basic implicit-explicit scheme, second-order convergence can be achieved when $\boldsymbol{C}^{\mathrm{E}} \neq \boldsymbol{0}$. This scheme is presented in the following section.

B.IV.5. Predictor/multi-corrector algorithms

Generalization of the previously described schemes to a predictor/multi-corrector format leads to improved characteristics in some circumstances, and is suggestive of further algorithmic generalizations. The implementational sequence is given as follows:

$$i = 0 \quad (i \text{ is the iteration counter}), \tag{59}$$

$$\left.\begin{aligned} \boldsymbol{d}_{n+1}^{(i)} &= \tilde{\boldsymbol{d}}_{n+1} && (60)\\ \boldsymbol{v}_{n+1}^{(i)} &= \tilde{\boldsymbol{v}}_{n+1} && (61)\\ \boldsymbol{a}_{n+1}^{(i)} &= \boldsymbol{0} && (62) \end{aligned}\right\} \quad \text{(predictor phase)},$$

$$\Delta\boldsymbol{F}_{n+1}^{(i)} = \boldsymbol{F}_{n+1} - \boldsymbol{M}\boldsymbol{a}_{n+1}^{(i)} - \boldsymbol{C}\boldsymbol{v}_{n+1}^{(i)} - \boldsymbol{K}\boldsymbol{d}_{n+1}^{(i)}\,, \tag{63}$$

$$\boldsymbol{M}^*\Delta\boldsymbol{a} = \Delta\boldsymbol{F}_{n+1}^{(i)}\,, \tag{64}$$

$$\left.\begin{aligned} \boldsymbol{a}_{n+1}^{(i+1)} &= \boldsymbol{a}_{n+1}^{(i)} + \Delta\boldsymbol{a} && (65)\\ \boldsymbol{v}_{n+1}^{(i+1)} &= \tilde{\boldsymbol{v}}_{n+1} + \Delta t\gamma\boldsymbol{a}_{n+1}^{(i+1)} && (66)\\ \boldsymbol{d}_{n+1}^{(i+1)} &= \tilde{\boldsymbol{d}}_{n+1} + \Delta t^2\beta\boldsymbol{a}_{n+1}^{(i+1)} && (67) \end{aligned}\right\} \quad \text{(corrector phase)}.$$

If additional iterations are to be performed, i is replaced by $i+1$, and calculations resume with (63). When the iterative phase is completed, the solution is defined by the last iterates [i.e., (65)–(67)].

The definition of $\boldsymbol{M}^*$ emanates from the time-discrete equation of motion, which for the predictor/multi-corrector algorithms, allowing for operator and/or mesh partitions, takes the form:

$$\boldsymbol{M}\boldsymbol{a}_{n+1}^{(i+1)} + \boldsymbol{C}^{\mathrm{I}}\boldsymbol{v}_{n+1}^{(i+1)} + \boldsymbol{C}^{\mathrm{E}}\boldsymbol{v}_{n+1}^{(i)} + \boldsymbol{K}^{\mathrm{I}}\boldsymbol{d}_{n+1}^{(i+1)} + \boldsymbol{K}^{\mathrm{E}}\boldsymbol{d}_{n+1}^{(i)} = \boldsymbol{F}_{n+1} . \tag{68}$$

In this case

$$\boldsymbol{M}^* = \boldsymbol{M} + \Delta t\gamma\boldsymbol{C}^{\mathrm{I}} + \Delta t^2\beta\boldsymbol{K}^{\mathrm{I}} . \tag{69}$$

If only one pass is made through (59)–(67), the algorithm is equivalent to the implicit-explicit scheme given earlier. If any part of $\boldsymbol{C}$ is treated explicitly, an additional pass through (63)–(67) can increase accuracy.

To see this, in a simple setting, it suffices to consider the following special case:

$$\boldsymbol{M}\boldsymbol{a}_{n+1}^{(i+1)} + \boldsymbol{C}\boldsymbol{v}_{n+1}^{(i)} = \boldsymbol{0} . \tag{70}$$

It may then be shown that one additional pass through (63)–(67) results in

$$\boldsymbol{v}_{n+1} = (\boldsymbol{I} - \Delta t\boldsymbol{M}^{-1}\boldsymbol{C} + \gamma(\Delta t\boldsymbol{M}^{-1}\boldsymbol{C})^2)\boldsymbol{v}_n \tag{71}$$

which is clearly second-order accurate if $\gamma = \frac{1}{2}$.

The stability condition is easily determined[8] from (71): $\Omega < \Omega_{\mathrm{crit}} = 1/\gamma = 2$, where $\Omega = \lambda\Delta t$ and λ is the maximum eigenvalue of $\boldsymbol{M}^{-1}\boldsymbol{C}$. Thus there is no adverse effect on stability compared with a one-pass formulation, which amounts to the forward-difference algorithm (cf. Section I).

A completely general stability analysis of partitioned predictor/multi-corrector algorithms is not available yet. The unpartitioned case is simpler, however, since all pertinent equations are diagonalizable. Thus one may work in terms of an SDOF model equation. The greater the number of passes, the higher the order of the stability polynomial.

C. Nonlinear symmetric systems

C.1. A simple class of second-order nonlinear problems arising in structural dynamics

Consider the following system of second-order nonlinear ordinary differential equations:

$$\boldsymbol{M}\ddot{\boldsymbol{d}} + \boldsymbol{N}(\boldsymbol{d}, \dot{\boldsymbol{d}}) = \boldsymbol{F} \tag{1}$$

where $\boldsymbol{N}$ is a nonlinear, algebraic function of $\boldsymbol{d}$ and $\dot{\boldsymbol{d}}$. Equation (1) might represent a spatially-discrete system in structural dynamics composed of nonlinear elastic, or a class of nonlinear 'rate-type' viscoelastic, elements. Geometrically nonlinear systems also may be written in the form of (1). Many pertinent situations are, however, not encompassed by (1). For example, plasticity and viscoplasticity may be mentioned in this regard. In these cases, the internal force

[8]This may be seen by noting that the eigenvectors of $\boldsymbol{M}^{-1}\boldsymbol{C}$ diagonalize all positive-integer powers of $\boldsymbol{M}^{-1}\boldsymbol{C}$. That is, if $(\boldsymbol{C} - \lambda\boldsymbol{M})\boldsymbol{\phi} = \boldsymbol{0}$, then $\boldsymbol{\phi}^{\mathrm{T}}(\boldsymbol{M}^{-1}\boldsymbol{C})^j\boldsymbol{\phi} = \lambda^j\boldsymbol{\phi}^{\mathrm{T}}\boldsymbol{\phi}$.

(i.e., $\boldsymbol{N}$) depends on the histories of $\boldsymbol{d}$ and $\dot{\boldsymbol{d}}$ as well as their current values. Very little analysis of a rigorous nature has been performed for systems of this type.

Let

$$\boldsymbol{K}_{\mathrm{T}} = \partial \boldsymbol{N} / \partial \boldsymbol{d} \tag{2}$$

and

$$\boldsymbol{C}_{\mathrm{T}} = \partial \boldsymbol{N} / \partial \dot{\boldsymbol{d}} \tag{3}$$

denote the tangent stiffness and tangent damping matrices, respectively. These are the 'consistent' linearized operators associated to $\boldsymbol{N}$ in the sense of [39]. Each is, generally, a nonlinear function of $\boldsymbol{d}$ and $\dot{\boldsymbol{d}}$.

We shall assume that $\boldsymbol{M}$ is constant; $\boldsymbol{M}$, $\boldsymbol{K}_{\mathrm{T}}$ and $\boldsymbol{C}_{\mathrm{T}}$ are symmetric[9]; $\boldsymbol{M}$ and $\boldsymbol{K}_{\mathrm{T}}$ are positive-definite; and $\boldsymbol{C}_{\mathrm{T}}$ is positive semi-definite.

Various time-stepping algorithms may be applied to (1). We shall consider here the implicit-explicit scheme described in Section B.IV, since it contains as special cases many useful procedures (e.g., Newmark's method). All equations are the same as in Section B.IV.2, except the time-discrete equation of motion which takes on the form:

$$\boldsymbol{M}\boldsymbol{a}_{n+1} + \boldsymbol{N}^{\mathrm{I}}(\boldsymbol{d}_{n+1}, \boldsymbol{v}_{n+1}) + \boldsymbol{N}^{\mathrm{E}}(\tilde{\boldsymbol{d}}_{n+1}, \tilde{\boldsymbol{v}}_{n+1}) = \boldsymbol{F}_{n+1} \tag{4}$$

where $\boldsymbol{N}^{\mathrm{I}}$ and $\boldsymbol{N}^{\mathrm{E}}$ are associated with the implicit and explicit partition, respectively. Clearly, the nonlinear problem which arises in each time step to advance the solution must be solved by an iterative strategy. This is outside the scope of this chapter, but is dealt with elsewhere in this book (see Chapter 9).

C.2. Linearized stability

Linearized stability is without doubt the most widely used stability concept in nonlinear analysis. One must realize, however, that it provides necessary conditions for stability (they may not be sufficient) and so it must be judiciously combined with insight for success in practice.

The first step in determining the linearized stability properties of an algorithm is to define the consistent (in the sense of [39]) linear algorithm. Specifically, the algorithmic equations are viewed as function of $\boldsymbol{d}_n$, $\boldsymbol{v}_n$, $\boldsymbol{a}_n$, $\boldsymbol{d}_{n+1}$, $\boldsymbol{v}_{n+1}$ and $\boldsymbol{a}_{n+1}$, and differentiated with respect to all arguments. The resulting equations are:

$$\boldsymbol{M}\delta\boldsymbol{a}_{n+1} + \boldsymbol{C}_{\mathrm{T}}^{\mathrm{I}}\delta\boldsymbol{v}_{n+1} + \boldsymbol{K}_{\mathrm{T}}^{\mathrm{I}}\delta\boldsymbol{d}_{n+1} + \boldsymbol{C}_{\mathrm{T}}^{\mathrm{E}}\delta\tilde{\boldsymbol{v}}_{n+1} + \boldsymbol{K}_{\mathrm{T}}^{\mathrm{E}}\delta\tilde{\boldsymbol{d}}_{n+1} = \boldsymbol{0}\,, \tag{5}$$

$$\delta\boldsymbol{d}_{n+1} = \delta\tilde{\boldsymbol{d}}_{n+1} + \Delta t^2\beta\delta\boldsymbol{a}_{n+1}\,, \tag{6}$$

$$\delta\boldsymbol{v}_{n+1} = \delta\tilde{\boldsymbol{v}}_{n+1} + \Delta t\gamma\delta\boldsymbol{a}_{n+1}\,, \tag{7}$$

$$\delta\tilde{\boldsymbol{d}}_{n+1} = \delta\boldsymbol{d}_n + \Delta t\delta\boldsymbol{v}_n + \frac{\Delta t^2}{2}(1-2\beta)\delta\boldsymbol{a}_n\,, \tag{8}$$

$$\delta\tilde{\boldsymbol{v}}_{n+1} = \delta\boldsymbol{v}_n + \Delta t(1-\gamma)\delta\boldsymbol{a}_n\,, \tag{9}$$

[9]This is taken to be the definition of symmetry of a system of this type.

where

$$\boldsymbol{K}_{\mathrm{T}}^{\mathrm{E}} = \partial \boldsymbol{N}^{\mathrm{E}}/\partial \boldsymbol{d}\,, \qquad \boldsymbol{K}_{\mathrm{T}}^{\mathrm{I}} = \partial \boldsymbol{N}^{\mathrm{I}}/\partial \boldsymbol{d}\,, \tag{10}$$

$$\boldsymbol{C}_{\mathrm{T}}^{\mathrm{E}} = \partial \boldsymbol{N}^{\mathrm{E}}/\partial \boldsymbol{v}\,, \qquad \boldsymbol{C}_{\mathrm{T}}^{\mathrm{I}} = \partial \boldsymbol{N}^{\mathrm{I}}/\partial \boldsymbol{v}\,, \tag{11}$$

$$\boldsymbol{K}_{\mathrm{T}} = \boldsymbol{K}_{\mathrm{T}}^{\mathrm{I}} + \boldsymbol{K}_{\mathrm{T}}^{\mathrm{E}}\,, \tag{12}$$

$$\boldsymbol{C}_{\mathrm{T}} = \boldsymbol{C}_{\mathrm{T}}^{\mathrm{I}} + \boldsymbol{C}_{\mathrm{T}}^{\mathrm{E}}\,. \tag{13}$$

Equation (5) is sometimes called the 'variational equation'. Physically speaking, (5)–(9) govern the growth/decay properties of small perturbations, or 'variations' (i.e., $\delta \boldsymbol{d}_n$, etc.).

In all subsequent calculations, we *assume* that the arrays appearing in the variational equations (namely: $\boldsymbol{C}_{\mathrm{T}}^{\mathrm{I}}$, $\boldsymbol{C}_{\mathrm{T}}^{\mathrm{E}}$, $\boldsymbol{K}_{\mathrm{T}}^{\mathrm{I}}$ and $\boldsymbol{K}_{\mathrm{T}}^{\mathrm{E}}$) are *constant* and positive semi-definite.

Stability criterion. The stability criterion that we shall require is that $\delta \boldsymbol{d}_n$, $\delta \boldsymbol{v}_n$ and $\delta \boldsymbol{a}_n$, $n = 0, 1, 2, \ldots$, be bounded for arbitrary initial values $\delta \boldsymbol{d}_0$, $\delta \boldsymbol{v}_0$, and $\delta \boldsymbol{a}_0$. (We shall assume that Δt is constant. However, the results remain valid for a finite number of changes in Δt.)

Note that, with the assumption of constancy, the algorithmic equations are identical in form to those for the corresponding linear system. Consequently, application of the energy criterion leads to a discrete conservation law identical to what was previously obtained for the linear case, except here it is in terms of the tangent arrays and governs the behavior of the variations, viz.

Equations (5)–(9) may be combined to form the following identity:

$$\begin{aligned}\delta \boldsymbol{a}_{n+1}^{\mathrm{T}}(\boldsymbol{A}^{\mathrm{I}} + \bar{\boldsymbol{A}}^{\mathrm{E}})\delta \boldsymbol{a}_{n+1} + \delta \boldsymbol{v}_{n+1}^{\mathrm{T}} \boldsymbol{K}_{\mathrm{T}} \delta \boldsymbol{v}_{n+1} &= \delta \boldsymbol{a}_{n}^{\mathrm{T}}(\boldsymbol{A}^{\mathrm{I}} + \bar{\boldsymbol{A}}^{\mathrm{E}})\delta \boldsymbol{a}_{n} + \delta \boldsymbol{v}_{n}^{\mathrm{T}} \boldsymbol{K}_{\mathrm{T}} \delta \boldsymbol{v}_{n} \\ &\quad -(2\gamma - 1)[\delta \boldsymbol{a}_n]^{\mathrm{T}}(\boldsymbol{B}^{\mathrm{I}} + \bar{\boldsymbol{B}}^{\mathrm{E}})[\delta \boldsymbol{a}_n] - 2\Delta t \langle \delta \boldsymbol{a}_n \rangle^{\mathrm{T}} \boldsymbol{C}_{\mathrm{T}} \langle \delta \boldsymbol{a}_n \rangle\end{aligned} \tag{14}$$

where

$$\boldsymbol{B}^{\mathrm{I}} = \boldsymbol{M}^{\mathrm{I}} + \Delta t(\gamma - \tfrac{1}{2})\boldsymbol{C}_{\mathrm{T}}^{\mathrm{I}} + \Delta t^2(\beta - \tfrac{1}{2}\gamma)\boldsymbol{K}_{\mathrm{T}}^{\mathrm{I}}\,, \tag{15}$$

$$\boldsymbol{A}^{\mathrm{I}} = \boldsymbol{B}^{\mathrm{I}} + \Delta t(\gamma - \tfrac{1}{2})\boldsymbol{C}_{\mathrm{T}}^{\mathrm{I}}\,, \tag{16}$$

$$\bar{\boldsymbol{B}}^{\mathrm{E}} = \boldsymbol{B}^{\mathrm{E}} - \Delta t \gamma \boldsymbol{C}_{\mathrm{T}}^{\mathrm{E}} - \Delta t^2 \beta \boldsymbol{K}_{\mathrm{T}}^{\mathrm{E}}\,, \tag{17}$$

$$\boldsymbol{B}^{\mathrm{E}} = \boldsymbol{M}^{\mathrm{E}} + \Delta t(\gamma - \tfrac{1}{2})\boldsymbol{C}_{\mathrm{T}}^{\mathrm{E}} + \Delta t^2(\beta - \gamma/2)\boldsymbol{K}_{\mathrm{T}}^{\mathrm{E}}\,, \tag{18}$$

$$\bar{\boldsymbol{A}}^{\mathrm{E}} = \bar{\boldsymbol{B}}^{\mathrm{E}} + \Delta t(\gamma - \tfrac{1}{2})\boldsymbol{C}_{\mathrm{T}}^{\mathrm{E}}\,. \tag{19}$$

The conditions which lead to stability for the linear case apply here as well except the eigendata must be viewed as those of the local tangent operators (see Section B.IV.3.c).

Note. The concept of stability employed above provides only necessary stability conditions in the nonlinear regime. (In the linear regime they are both necessary and sufficient.) Based upon practical experiences, the following observations may be made: (i) the time step restriction for the explicit elements is sufficient; (ii) if the nonlinear response is accurately discretized by the time step chosen, the stability conditions are also sufficient for the implicit elements; and (iii) if the nonlinear response is *not* accuractely discretized, biased accumulation of truncation errors can occur and may cause pathological growth of the discrete solution [27]. It is debatable whether this phenomenon is a stability or accuracy problem. There is a concurrence, however, that it is unacceptable in practice.

C.3. Convergence

In this section we shall present a convergence proof for the algorithms presented in the previous section.

C.3.a. Preliminaries

Assumptions (N). *We assume that* $\boldsymbol{N}$ *is a* C^1*-continuous function of its arguments and that Assumptions (L) hold with* $\boldsymbol{K}_{\mathrm{T}}$ *and* $\boldsymbol{C}_{\mathrm{T}}$ *in place of* $\boldsymbol{K}$ *and* $\boldsymbol{C}$*, respectively.*

Assumptions (N) guarantee that there exists a unique C^2-continuous solution of the initial-value problem for (1).

The following special cases of Taylor's formula for C^1-continuous functions (see Lang [50]) will be used in the sequel:

$$\boldsymbol{N}(\boldsymbol{d}+\boldsymbol{e}, \dot{\boldsymbol{d}}+\dot{\boldsymbol{e}}) = \boldsymbol{N}(\boldsymbol{d}, \dot{\boldsymbol{d}}) + \bar{\boldsymbol{K}}_{\mathrm{T}}(\boldsymbol{d}, \dot{\boldsymbol{d}}; \boldsymbol{e}, \dot{\boldsymbol{e}})\boldsymbol{e} + \bar{\boldsymbol{C}}_{\mathrm{T}}(\boldsymbol{d}, \dot{\boldsymbol{d}}; \boldsymbol{e}, \dot{\boldsymbol{e}})\dot{\boldsymbol{e}} \tag{20}$$

where

$$\bar{\boldsymbol{K}}_{\mathrm{T}}(\boldsymbol{d}, \dot{\boldsymbol{d}}; \boldsymbol{e}, \dot{\boldsymbol{e}}) = \int_0^1 \boldsymbol{K}_{\mathrm{T}}(\boldsymbol{d}+\varepsilon\boldsymbol{e}, \dot{\boldsymbol{d}}+\varepsilon\dot{\boldsymbol{e}})\,\mathrm{d}\varepsilon\,, \tag{21}$$

$$\bar{\boldsymbol{C}}_{\mathrm{T}}(\boldsymbol{d}, \dot{\boldsymbol{d}}; \boldsymbol{e}, \dot{\boldsymbol{e}}) = \int_0^1 \boldsymbol{C}_{\mathrm{T}}(\boldsymbol{d}+\varepsilon\boldsymbol{e}, \dot{\boldsymbol{d}}+\varepsilon\dot{\boldsymbol{e}})\,\mathrm{d}\varepsilon\,; \tag{22}$$

and

$$\boldsymbol{N}^{\mathrm{E}}(\boldsymbol{d}_{n+1}, \boldsymbol{v}_{n+1}) = \boldsymbol{N}^{\mathrm{E}}(\tilde{\boldsymbol{d}}_{n+1}, \tilde{\boldsymbol{v}}_{n+1}) + \Delta t^2\beta\hat{\boldsymbol{K}}_{\mathrm{T}}^{\mathrm{E}}\boldsymbol{a}_{n+1} + \Delta t\gamma\hat{\boldsymbol{C}}_{\mathrm{T}}^{\mathrm{E}}\boldsymbol{a}_{n+1} \tag{23}$$

where

$$\hat{\boldsymbol{K}}_{\mathrm{T}}^{\mathrm{E}} = \int_0^1 \boldsymbol{K}_{\mathrm{T}}^{\mathrm{E}}(\tilde{\boldsymbol{d}}_{n+1} + \varepsilon\Delta t^2\beta\boldsymbol{a}_{n+1}, \tilde{\boldsymbol{v}}_{n+1} + \varepsilon\Delta t\gamma\boldsymbol{a}_{n+1})\,\mathrm{d}\varepsilon\,, \tag{24}$$

$$\hat{\boldsymbol{C}}_{\mathrm{T}}^{\mathrm{E}} = \int_0^1 \boldsymbol{C}_{\mathrm{T}}^{\mathrm{E}}(\tilde{\boldsymbol{d}}_{n+1} + \varepsilon\Delta t^2\beta\boldsymbol{a}_{n+1}, \tilde{\boldsymbol{v}}_{n+1} + \varepsilon\Delta t\gamma\boldsymbol{a}_{n+1})\,\mathrm{d}\varepsilon\,. \tag{25}$$

Remarks. 1. It may be observed from (20) that

$$\bar{K}_T(d, \dot{d}; e, \dot{e})e + \bar{C}_T(d, \dot{d}; e, \dot{e})\dot{e}$$

is a C^1-continuous function of d, $\dot{d}$, e and $\dot{e}$, and in particular is Lipschitz continuous with respect to e and $\dot{e}$.

2. It will be useful in subsequent developments to extend $\hat{K}_T^E$ and $\hat{C}_T^E$ to smooth functions of t which interpolate at t_{n+1} the values given by (24) and (25), respectively. This may be accomplished by any number of interpolatory schemes (e.g., Lagrange interpolation [11]).

C.3.b. Auxiliary problem

Equations (4) and (23) enable us to write

$$\hat{M}a_{n+1} + N(d_{n+1}, v_{n+1}) = F_{n+1} \tag{26}$$

where

$$\hat{M} = M - \Delta t\gamma \hat{C}_T^E - \Delta t^2 \beta \hat{K}_T^E \,. \tag{27}$$

Thus we are led to consider the *auxiliary problem* consisting of (45) and (46) of Section IV, and

$$\hat{M}\ddot{D} + N(D, \dot{D}) = F \,. \tag{28}$$

By virtue of Remark 2, $\hat{M}$ may be viewed as a smooth function of t. Consequently for sufficiently small Δt, Assumptions (N) imply that there exists a unique C^2-continuous solution of the auxiliary problem.

As in the linear case (see Section IV.4), the convergence proof is broken up into two parts:

Part (*i*). Arguments made for the linear case carry over to the nonlinear case unaltered.

Part (*ii*). Proceeding in analogous fashion to the linear case and using (20), we obtain

$$M\ddot{e} + \bar{C}_T\dot{e} + \bar{K}_T e = \Lambda \ddot{D} \tag{29}$$

where

$$\Lambda = \Delta t\gamma \hat{C}_T^E + \Delta t^2 \beta \hat{K}_T^E \,. \tag{30}$$

Equation (29) may be put in the form of (40), Section IV, by employing the definitions of Section IV, namely, (53), (55)–(58) and in place of (54):

$$B = \begin{bmatrix} \bar{C}_T & \bar{K}_T \\ I & 0 \end{bmatrix}. \tag{31}$$

Observe that the Remark 2, Λ may be viewed as a smooth function of t. This fact and the continuity of $\ddot{\boldsymbol{D}}$ insures that $\boldsymbol{f}_\varepsilon(t)$ is continuous in t. By Remark 1, (56) of Section IV, and (31), $\boldsymbol{g}(\boldsymbol{x}_\varepsilon(t), t)$ is C^1-continuous in t and Lipschitz continuous in $\boldsymbol{x}_\varepsilon(t)$. Furthermore, from (30), $\boldsymbol{\Lambda}$ is clearly $O(\Delta T)$. Thus the hypotheses of Corollary 2 are satisfied, and so $\boldsymbol{e}$ is $O(\Delta t)$.

From (30) and Corollary 2, second-order convergence of $\boldsymbol{e}$ to $\boldsymbol{0}$ is attained if $\boldsymbol{N}^{\mathrm{E}}(\boldsymbol{d}, \dot{\boldsymbol{d}}) = \boldsymbol{N}^{\mathrm{E}}(\boldsymbol{d})$. As in the linear case, a slight generalization of the basic implicit-explicit scheme to a multi-corrector format must be employed to recapture second-order convergence when $\boldsymbol{N}^{\mathrm{E}}$ varies with $\dot{\boldsymbol{d}}$.

C.4. Energy-conserving algorithms

In the linear regime, the notions of stability that we have employed imply that a conservation law or decay inequality imposes limits upon the growth of solutions. The concept of linearized stability does not necessarily imply the same holds true in nonlinear situations. Pathological energy growth has been noted in some situations for algorithms which are unconditionally linearized stable, namely the trapezoidal rule (see [27]). Thus the issue has been raised whether or not the notion of linearized stability is adequate for general nonlinear analysis. Stronger notions may be employed, but require different classes of algorithms than those which are now widely used. The purpose of this section is to present such an algorithm. It is based upon a modification to the trapezoidal algorithm for nonlinear applications, which results in physically appropriate energy growth characteristics [33]. In particular, when external forces are absent, *energy is identically conserved in nonlinear elastodynamics*, a property shared by the exact continuum equations. As a consequence, *unconditional stability is automatically attained.*

The technique used to achieve these ends is the Lagrange multiplier method. The trapezoidal algorithm is written as the Euler–Lagrange equation of a certain functional. The desired energy identity is viewed as a constraint condition on the solution and the variational problem is rendered unconstrained by appending a Lagrange multiplier times the constraint to the original functional.

Throughout this section we assume that $\boldsymbol{N}(\boldsymbol{d}, \dot{\boldsymbol{d}}) = \boldsymbol{N}(\boldsymbol{d})$, which is the case for a nonlinear elastic system, and that $\boldsymbol{N}$ is the gradient of a strain energy function U, i.e.,

$$\boldsymbol{N}(d) = \frac{\partial U}{\partial \boldsymbol{d}}(\boldsymbol{d}) . \tag{32}$$

We further assume $U(\boldsymbol{d})$ is non-negative and that if $U(\boldsymbol{d})$ is bounded, so is $\boldsymbol{d}$. By virtue of (32),

$$\boldsymbol{K}(\boldsymbol{d}) = \frac{\partial \boldsymbol{N}}{\partial \boldsymbol{d}}(\boldsymbol{d}) \tag{33}$$

is necessarily symmetric. Solutions of the initial-value problem satisfy the *fundamental energy identity*:

$$E[\boldsymbol{d}(t), \dot{\boldsymbol{d}}(t)] = \boldsymbol{E}(\boldsymbol{D}_0, \boldsymbol{V}_0) + \int_0^t \dot{\boldsymbol{d}}(\tau)^{\mathrm{T}} \boldsymbol{F}(\tau)\, \mathrm{d}\tau \tag{34}$$

where $\boldsymbol{E}$ is the total energy, defined by

$$\boldsymbol{E}(\boldsymbol{d}, \dot{\boldsymbol{d}}) = \tfrac{1}{2}\dot{\boldsymbol{d}}^{\mathrm{T}} \boldsymbol{M} \dot{\boldsymbol{d}} + U(\boldsymbol{d})\,. \tag{35}$$

In particular, if $\boldsymbol{F} = \boldsymbol{0}$, *total energy is conserved*:

$$E[\boldsymbol{d}(t), \dot{\boldsymbol{d}}(t)] = E(\boldsymbol{D}_0, \boldsymbol{V}_0)\,. \tag{36}$$

C.4.a. Trapezoidal rule: variational formulation

In the present circumstances the trapezoidal rule[10] may be written as

$$\boldsymbol{M}\boldsymbol{a}_{n+1} + \boldsymbol{N}(\boldsymbol{d}_{n+1}) = \boldsymbol{F}_{n+1}\,, \tag{37}$$

$$\boldsymbol{d}_{n+1} = \boldsymbol{d}_n + \frac{\Delta t}{2}(\boldsymbol{v}_n + \boldsymbol{v}_{n+1})\,, \tag{38}$$

$$\boldsymbol{v}_{n+1} = \boldsymbol{v}_n + \frac{\Delta t}{2}(\boldsymbol{a}_n + \boldsymbol{a}_{n+1})\,, \tag{39}$$

$$\boldsymbol{d}_0 = \boldsymbol{D}_0\,, \tag{40}$$

$$\boldsymbol{v}_0 = \boldsymbol{V}_0\,, \tag{41}$$

$$\boldsymbol{a}_0 = \boldsymbol{M}^{-1}[\boldsymbol{F}_0 - \boldsymbol{N}(\boldsymbol{d}_0)]\,. \tag{42}$$

A nonlinear algebraic problem for $\boldsymbol{d}_{n+1}$ may be obtained by eliminating $\boldsymbol{v}_{n+1}$ and $\boldsymbol{a}_{n+1}$ from (37)–(39).

In what follows, it will be convenient to view the nonlinear algebraic problem as the Euler–Lagrange equation of a functional. Specifically let

$$\begin{aligned}\mathscr{F}(\boldsymbol{d}_{n+1}) = {} & \frac{2}{\Delta t^2}\boldsymbol{d}_{n+1}^{\mathrm{T}} \boldsymbol{M} \boldsymbol{d}_{n+1} + U(\boldsymbol{d}_{n+1}) \\ & - \boldsymbol{d}_{n+1}^{\mathrm{T}}\Big\{\boldsymbol{F}_{n+1} + \boldsymbol{M}\Big[\boldsymbol{a}_n + \frac{4}{\Delta t^2}(\boldsymbol{d}_n + \Delta t \boldsymbol{v}_n)\Big]\Big\}\,.\end{aligned} \tag{43}$$

Then

$$\begin{aligned}\delta \boldsymbol{d}_{n+1}^{\mathrm{T}} \frac{\partial \mathscr{F}}{\partial \boldsymbol{d}_{n+1}}(\boldsymbol{d}_{n+1}) = {} & \delta \boldsymbol{d}_{n+1}^{\mathrm{T}}\Big\{\frac{4}{\Delta t^2}\boldsymbol{M}\boldsymbol{d}_{n+1} + \boldsymbol{N}(\boldsymbol{d}_{n+1}) \\ & - \boldsymbol{F}_{n+1} - \boldsymbol{M}\Big[\boldsymbol{a}_n + \frac{4}{\Delta t^2}(\boldsymbol{d}_n + \Delta t \boldsymbol{v}_n)\Big]\Big\}\,.\end{aligned} \tag{44}$$

Thus, the nonlinear algebraic problem is equivalent to (44) being identically zero for arbitrary 'variations', $\delta \boldsymbol{d}_{n+1}$.

[10]Recall that the trapezoidal rule is often referred to as the average acceleration method in structural dynamics.

C.4.b. Modified algorithm

We wish to modify the trapezoidal rule so that the following discrete energy identity is satisfied:

$$E_{n+1} - E_n = \tfrac{1}{2}(\boldsymbol{d}_{n+1} - \boldsymbol{d}_n)^{\mathrm{T}}(\boldsymbol{F}_{n+1} + \boldsymbol{F}_n) \tag{45}$$

where $E_n = E(\boldsymbol{d}_n, \boldsymbol{v}_n)$. When $\boldsymbol{F} = \boldsymbol{0}$, (45) implies that the total energy of the approximate solution is identically conserved.

To achieve the desired end, let

$$\mathscr{G}(\boldsymbol{d}_{n+1}) = E\left[\boldsymbol{d}_{n+1}, \frac{2}{\Delta t}\left(\boldsymbol{d}_{n+1} - \boldsymbol{d}_n - \frac{\Delta t}{2}\boldsymbol{v}_n\right)\right] - E_n - \tfrac{1}{2}(\boldsymbol{d}_{n+1} - \boldsymbol{d}_n)^{\mathrm{T}}(\boldsymbol{F}_{n+1} + \boldsymbol{F}_n)\,. \tag{46}$$

The equality

$$\mathscr{G}(\boldsymbol{d}_{n+1}) = 0\,, \tag{47}$$

when combined with (38), implies (45) is satisfied.

The equations of the modified algorithm are defined to be (38), (39) and the Euler–Lagrange equations of the functional

$$\mathscr{F}(\boldsymbol{d}_{n+1}) + \lambda\mathscr{G}(\boldsymbol{d}_{n+1}) \tag{48}$$

where λ is a Lagrange multiplier. Differentiating (48) yields

$$\delta\boldsymbol{d}_{n+1}^{\mathrm{T}}\left\{\frac{\partial\mathscr{F}}{\partial\boldsymbol{d}_{n+1}}(\boldsymbol{d}_{n+1}) + \lambda\frac{\partial\mathscr{G}}{\partial\boldsymbol{d}_{n+1}}(\boldsymbol{d}_{n+1})\right\} + \delta\lambda\mathscr{G}(\boldsymbol{d}_{n+1})\,. \tag{49}$$

If (49) is to vanish for arbitrary $\delta\boldsymbol{d}_{n+1}$ and $\delta\lambda$, then (47) is satisfied and

$$\begin{aligned} 0 = {} & (1+\lambda)\left\{\frac{4}{\Delta t^2}\boldsymbol{M}\boldsymbol{d}_{n+1} + \boldsymbol{N}(\boldsymbol{d}_{n+1})\right\} - (1+\tfrac{1}{2}\lambda)\boldsymbol{F}_{n+1} - \tfrac{1}{2}\lambda\boldsymbol{F}_n \\ & -\boldsymbol{M}\left\{\boldsymbol{a}_n + \frac{4}{\Delta t^2}[(1+\lambda)\boldsymbol{d}_n + \Delta t(1+\tfrac{1}{2}\lambda)\boldsymbol{v}_n]\right\}. \end{aligned} \tag{50}$$

Equation (50) is the modified equation of motion which accounts for the forces induced by the Lagrange multiplier. The solution of (47) and (50) may be obtained by employing an iterative algorithm. An effective procedure is discussed in [33].

C.4.c. Numerical examples

A materially nonlinear, one-dimensional, small-deformation, elastic rod model was used to compare the behavior of the energy conserving algorithm with trapezoidal rule. Bilinear hardening and softening material laws were employed (see Figure 24). In each case, a rod of length 10, fixed at the left end, is subjected to an initial displacement, varying linearly between 0 on the left end and 0.1 on the right. (This insures significant nonlinear behavior.) The initial velocity and external forces are assumed equal to 0. There are 10 equal-length, 'linear' elements in the model, and each has cross-sectional area 1.0 and density 0.01.

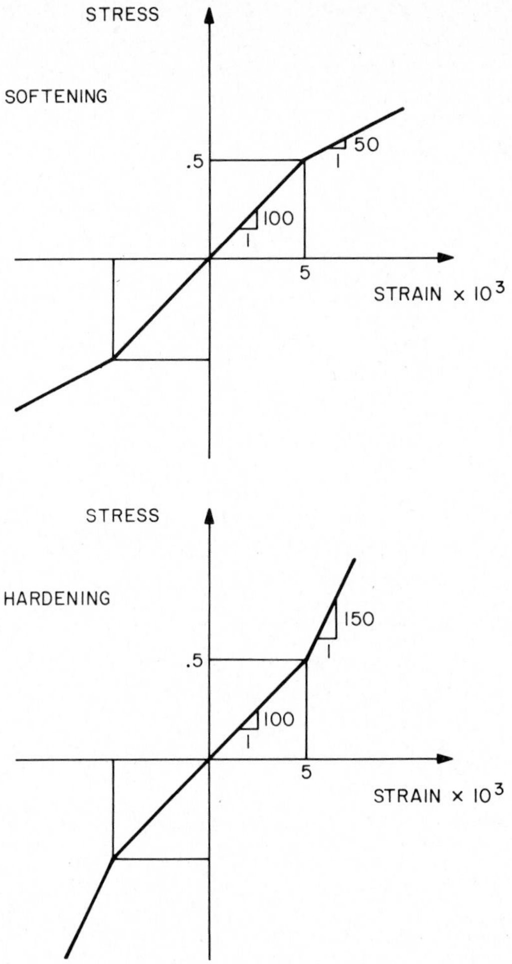

Fig. 24. Bilinear material laws used in the numerical calculations [33].

In Figures 25 and 26, summaries of the growth/decay behavior of total energy are presented. For a time step of 0.04, the energy in the trapezoidal solution is steadily increasing in both the hardening and softening cases. For all time steps considered, the modified algorithm exactly conserves energy.

For small time steps, the modified algorithm and trapezoidal rule yield virtually identical results, as is to be expected. For somewhat larger steps, the energy conserving property of the modified algorithm seems to engender greater accuracy. Sample results, in support of this statement, are presented in Figure 27 (see also [19]).

The idea presented in this section may be viewed as a general methodology for 'stabilization' and applied to any transient algorithm to achieve stability in a physically appropriate sense. Dissipative effects are also easily accounted for. On the negative side, procedures of this kind tend to increase cost somewhat and thus are presently used only in special circumstances (see e.g. [19]).

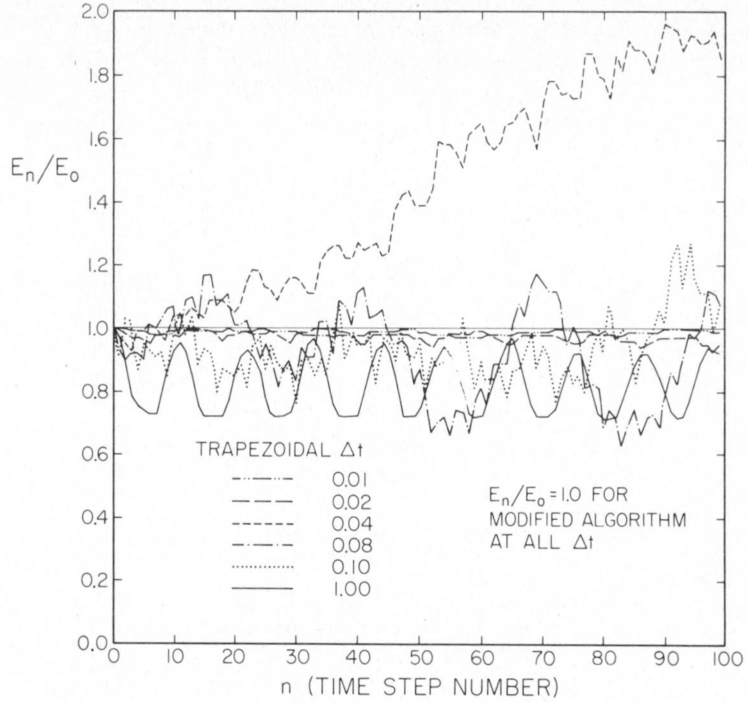

Fig. 25. Energy growth and decay for softening material law [33].

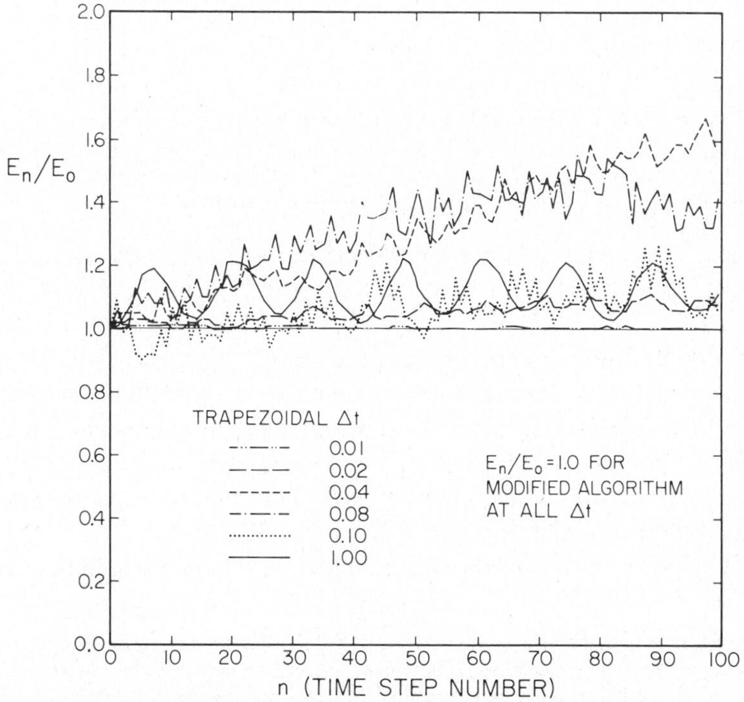

Fig. 26. Energy growth and decay for hardening material law [33].

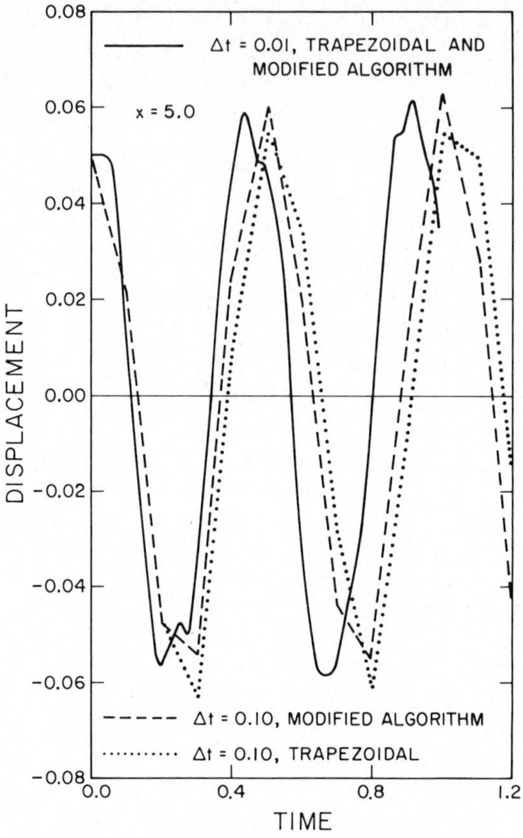

Fig. 27. Displacement response for softening material law [33].

C.5. *Stability of one-step methods for first-order systems*

The class of problems considered is governed by the spatially discrete system

$$\boldsymbol{M}(\boldsymbol{d}, t)\dot{\boldsymbol{d}} + \boldsymbol{K}(\boldsymbol{d}, t)\boldsymbol{d} = \boldsymbol{F}. \tag{51}$$

Equation (51) is the nonlinear generalization of the linear heat conduction equation consider in Section B.I. We assume throughout that $\boldsymbol{M}$ and $\boldsymbol{K}$ are symmetric[11], continuous, matrix-valued functions of their arguments. Furthermore $\boldsymbol{M}$ is assumed positive definite and $\boldsymbol{K}$ is assumed positive semi-definite. In the linear case in which $\boldsymbol{M}$ and $\boldsymbol{K}$ are constants, the stability behavior of one-step methods is well-known and unambiguous (see Section B.I.).

In this section we shall consider the stability of two related families of one-step

[11]Despite the fact that $\boldsymbol{M}$ and $\boldsymbol{K}$ are symmetric, the system (51) is not generally symmetric in the more 'usual sense' that it emanates from a potential, which, by Vainberg's theorem, is equivalent to requiring its consistent linearization be symmetric. This observation may be confirmed by examining the term $\boldsymbol{K}(\boldsymbol{d}, t)\boldsymbol{d}$ alone.

methods for (51), the 'generalized trapezoidal family' and 'generalized midpoint family'. In the linear regime the differences between these two families of algorithms disappear.

Two notions of stability are considered here:

The first is suggested by the criterion used for the linear counterpart of (51), obtained by viewing $\boldsymbol{M}$ and $\boldsymbol{K}$ as constants[12]. This approach employs an SDOF model problem and is *ad hoc* in nature. Results obtained for the trapezoidal family based upon this criterion turn out to be overly conservative.

The second notion of stability is based upon a global energy criterion. We shall present only a special case (i.e., for $\boldsymbol{M}$ constant) but it clearly illustrates that, under some circumstances, the first notion of stability is misleading. These observations were first presented by Hughes [28] and subsequently generalized by Hogge [25].

The *generalized trapezoidal family of methods*, applied to (51), take on the following form:

$$\boldsymbol{M}_{n+1}\boldsymbol{v}_{n+1} + \boldsymbol{K}_{n+1}\boldsymbol{d}_{n+1} = \boldsymbol{F}_{n+1}\,, \tag{52}$$

$$\boldsymbol{d}_{n+1} = \boldsymbol{d}_n + \Delta t\boldsymbol{v}_{n+\alpha}\,, \tag{53}$$

$$\boldsymbol{v}_{n+\alpha} = (1-\alpha)\boldsymbol{v}_n + \alpha\boldsymbol{v}_{n+1}\,, \tag{54}$$

where

$$\boldsymbol{M}_{n+1} = \boldsymbol{M}(\boldsymbol{d}_{n+1}, t_{n+1})\,, \tag{55}$$

$$\boldsymbol{K}_{n+1} = \boldsymbol{K}(\boldsymbol{d}_{n+1}, t_{n+1})\,, \tag{56}$$

$$\boldsymbol{F}_{n+1} = \boldsymbol{F}(t_{n+1})\,. \tag{57}$$

The *generalized midpoint family of methods* also employs (53) and (54), however, the time-discrete heat conduction equation takes on the form

$$\boldsymbol{M}_{n+\alpha}\boldsymbol{v}_{n+\alpha} + \boldsymbol{K}_{n+\alpha}\boldsymbol{d}_{n+\alpha} = \boldsymbol{F}_{n+\alpha} \tag{58}$$

where

$$\boldsymbol{M}_{n+\alpha} = \boldsymbol{M}(\boldsymbol{d}_{n+\alpha}, t_{n+\alpha})\,, \tag{59}$$

$$\boldsymbol{K}_{n+\alpha} = \boldsymbol{K}(\boldsymbol{d}_{n+\alpha}, t_{n+\alpha})\,, \tag{60}$$

$$\boldsymbol{F}_{n+\alpha} = \boldsymbol{F}(t_{n+\alpha})\,, \tag{61}$$

$$\boldsymbol{d}_{n+\alpha} = (1-\alpha)\boldsymbol{d}_n + \alpha\boldsymbol{d}_{n+1}\,, \tag{62}$$

$$t_{n+\alpha} = t_n + \alpha\Delta t\,. \tag{63}$$

Strictly speaking, the trapezoidal and midpoint families are somewhat different,

[12]Note that this is not the same as the *consistently linearized system* which arises from taking the variational derivative of (51) with respect to $\boldsymbol{d}$ and $\dot{\boldsymbol{d}}$. Thus this does not amount to performing a linearized stability analysis as described in Section 2.

even when $\boldsymbol{M}$ and $\boldsymbol{K}$ are constant, due to the different sampling of $\boldsymbol{F}$; cf. (57) and (62). In the context of stability, in which we consider $\boldsymbol{F}$ to be absent, there is no difference when $\boldsymbol{M}$ and $\boldsymbol{K}$ are constant.

C.5.a. Ad hoc SDOF model equation criterion

Concern over the stability of the generalized trapezoidal family has emanated from the following analysis (see Gourlay [17]):

Consider the SDOF model equation

$$\dot{d} + \lambda(d, t)d = 0 \tag{64}$$

in which it is assumed that $\lambda > 0$. Application of the generalized trapezoidal algorithm results in

$$d_{n+1} = A d_n \tag{65}$$

where

$$A = \frac{1 - \Delta t(1-\alpha)\lambda_n}{1 + \Delta t \alpha \lambda_{n+1}} \tag{66}$$

in which

$$\lambda_n = \lambda(d_n, t_n)\,, \tag{67}$$

$$\lambda_{n+1} = \lambda(d_{n+1}, t_{n+1})\,. \tag{68}$$

The stability condition employed is suggested by the linear case, namely

$$|A| \leq 1\,. \tag{69}$$

which must hold for all positive λ_n and λ_{n+1}. For example, if $\alpha = \frac{1}{2}$ (trapezoidal rule or Crank–Nicolson method) and $\lambda_n > \lambda_{n+1}$, then (69) imposes the time step restriction

$$\Delta t \leq 4/(\lambda_n - \lambda_{n+1})\,. \tag{70}$$

Thus the unconditional stability of the trapezoidal rule in linear problems apparently does not carry over to the nonlinear regime.

At this point it is worthwhile to emphasize that the preceding conclusion is in terms of the notion of stability defined by (69). That this result is overly pessimistic will be demonstrated shortly.

An analogous stability analysis of the generalized midpoint family yields

$$A = \frac{1 - \Delta t(1-\alpha)\lambda_{n+\alpha}}{1 + \Delta t \alpha \lambda_{n+\alpha}}\,, \tag{71}$$

where $\lambda_{n+\alpha} = \lambda(d_{n+\alpha}, t_{n+\alpha})$ and condition (69) requires that

$$\lambda_{n+\alpha} \Delta t \leq 2/(1-2\alpha) \tag{72}$$

for $\alpha < \frac{1}{2}$, whereas for $\alpha \geq \frac{1}{2}$ the algorithm in question is unconditionally stable. Thus the midpoint family appears to have the advantage that unconditionally stable methods for linear problems maintain this property in the nonlinear regime.

C.5.b. Global energy criterion

An appropriate notion of stability for the case under consideration may be obtained from an analysis of the governing system. For small enough Δt, it can be shown that the exact solution satisfies the following inequality

$$\boldsymbol{d}(t_{n+1})^{\mathrm{T}}\boldsymbol{M}(\boldsymbol{d}(t_{n+\alpha}), t_{n+\alpha})\boldsymbol{d}(t_{n+1}) \leq \boldsymbol{d}(t_n)^{\mathrm{T}}\boldsymbol{M}(\boldsymbol{d}(t_{n+\alpha}), t_{n+\alpha})\boldsymbol{d}(t_n)\,. \tag{73}$$

Note that decay is measured with respect to a changing norm [i.e., one defined by $\boldsymbol{M}(\boldsymbol{d}(t_{n+\alpha}), t_{n+\alpha})$].

It may be shown [28] that the midpoint family satisfies an essentially identical condition, namely

$$\boldsymbol{d}_{n+1}^{\mathrm{T}}\boldsymbol{M}_{n+\alpha}\boldsymbol{d}_{n+1} \leq \boldsymbol{d}_n^{\mathrm{T}}\boldsymbol{M}_{n+\alpha}\boldsymbol{d}_n \tag{74}$$

whenever the stability conditions above hold for the maximum eigenvalue, $\lambda_{n+\alpha}$, of $\boldsymbol{M}_{n+\alpha}^{-1}\boldsymbol{K}_{n+\alpha}$. In particular, (74) holds for arbitrarily large Δt if $\alpha \geq \frac{1}{2}$. Thus (74), which mimics the behavior of the governing system, reinforces the conclusions gained from the SDOF model problem criterion for the midpoint family.

It may be observed from (74), that if $\boldsymbol{M}$ is constant, then for all n

$$\boldsymbol{d}_n^{\mathrm{T}}\boldsymbol{M}\boldsymbol{d}_n \leq \boldsymbol{d}_0^{\mathrm{T}}\boldsymbol{M}\boldsymbol{d}_0\,. \tag{75}$$

According to the SDOF criterion, trapezoidal rule does not share the above property with midpoint rule unless Δt is appropriately chosen [see (70)]. However, it is known that trapezoidal and midpoint families are identical under a particular transformation of initial data [10]. This result, or a direct calculation, can be shown to yield the following result for the trapezoidal algorithm under the assumption of constant $\boldsymbol{M}$:

$$\begin{aligned}&\boldsymbol{d}_{n+1}^{\mathrm{T}}(\boldsymbol{M} + \Delta t\boldsymbol{K}_{n+1} + \tfrac{1}{4}\Delta t^2\boldsymbol{K}_{n+1}\boldsymbol{M}^{-1}\boldsymbol{K}_{n+1})\boldsymbol{d}_{n+1}\\ &\quad\leq \boldsymbol{d}_n^{\mathrm{T}}(\boldsymbol{M} + \Delta t\boldsymbol{K}_n + \tfrac{1}{4}\Delta t^2\boldsymbol{K}_n\boldsymbol{M}^{-1}\boldsymbol{K}_n)\boldsymbol{d}_n\end{aligned} \tag{76}$$

from which it follows that, for all n,

$$\boldsymbol{d}_n^{\mathrm{T}}\boldsymbol{M}\boldsymbol{d}_n \leq \boldsymbol{d}_0^{\mathrm{T}}(\boldsymbol{M} + \Delta t\boldsymbol{K}_0 + \tfrac{1}{4}\Delta t^2\boldsymbol{K}_0\boldsymbol{M}^{-1}\boldsymbol{K}_0)\boldsymbol{d}_0\,. \tag{77}$$

Thus, the potential growth is bounded by a constant, albeit a somewhat larger one than for the midpoint algorithm. Clearly, (77) is a very strong indicator of stability and demonstrates that the conclusions from the model equation are too stringent. One usually assumes that model equation analysis provides necessary, but not sufficient, stability conditions. The preceding analysis illustrates that the opposite may also be true. That is, the conditions may be sufficient, but not necessary.

Hogge [25] has generalized the preceding argument to all members of the trapezoidal and midpoint families.

The question naturally arises, what is the flaw in the SDOF model problem criterion? The answer is that a modal decomposition argument, such as the ones delineated in Section B.I and B.II, can be used to uncouple the midpoint algorithms and reduce down to the SDOF case [28], whereas the same is *not* true for the trapezoidal family. Thus we see that caution is necessary in drawing conclusions from model equations, unless a precise link can be made with the behavior of the coupled system.

D. Nonsymmetric operators

The thrust of this book is in the areas of solid and structural dynamics, in which symmetric operators predominate. Nevertheless, we wish to give some consideration to fluid dynamical phenomena, especially due to the current interest in problems of fluid-structure interaction. Convection operators, which arise from kinematical descriptions other than Lagrangian are nonsymmetric. This takes us outside the realm of the analyses of the previous sections. The level of understanding of nonsymmetric systems does not seem to be as advanced as that for symmetric systems. Consequently, the presentation devoted to this topic will be less complete than for the symmetric systems considered previously. In particular, our main intent is to summarize a few salient facts about stability. Hopefully, in the ensuing years, much more information will become available.

D.1. Model problem: the linear convection-diffusion equation

To illustrate basic ideas, we shall introduce the following model problem for convective-diffusive phenomena:

$$\boldsymbol{M}\dot{\boldsymbol{v}} + \boldsymbol{C}\boldsymbol{v} + \boldsymbol{Z}\boldsymbol{v} = \boldsymbol{0} \tag{1}$$

where $\boldsymbol{M}$ and $\boldsymbol{C}$ are symmetric and positive definite, and $\boldsymbol{Z}$ is skew-symmetric (i.e., $\boldsymbol{Z}^{\mathrm{T}} = -\boldsymbol{Z}$). Equations of this form arise from Galerkin finite-element and central-difference finite-difference discretizations of typical linear convection-diffusion equations. The $\boldsymbol{C}$-term is to be thought of as arising from diffusion (e.g., viscous effects) and the $\boldsymbol{Z}$-term pertains to convection. In classical-type upwind procedures, the discretized convection operator also includes a positive-definite ('artificial viscosity') term, which may be taken to be subsumed by $\boldsymbol{C}$ in our formalism. Newer, finite-element upwind schemes, based upon Petrov–Galerkin weighted residual methods, (see e.g., [29]) are not included in (1), since, for these cases, all arrays are, ostensibly, non-symmetric. This topic has not reached sufficient maturity at the time of the preparation of this chapter to be adequately considered herein.

The analysis of nonsymmetric systems has so far been treated largely by way of traditional finite-difference stability procedures, such as the von Neumann method [54]. Despite its limitations (the von Neumann method may only be applied to constant-coefficient, linear-difference equations emanating from regular mesh spacing, and does not account for boundary conditions), it presently provides more useful information for conditionally stable methods than any competing technique. However, the energy method has been used to prove a few important results, and considerable potential still exists. We will present a few results deduced from energy criteria, before passing on to a description of the von Neumann technique.

D.1.a. Energy analysis of leap-frog time differencing

Explicit, leap-frog time differencing is known to be conditionally stable for linear skew-symmetric operators by way of von Neumann analysis. However, it is unconditionally unstable for symmetric operators be they positive, or negative-definite, or neither. A potentially stable scheme for (1) would necessarily treat the $\boldsymbol{C}$-term in some other fashion. For example, consider the following algorithm:

$$(\boldsymbol{M}+2\Delta t\gamma\boldsymbol{C})\boldsymbol{v}_{n+1}=(\boldsymbol{M}-2\Delta t(1-\gamma)\boldsymbol{C})\boldsymbol{v}_{n-1}-2\Delta t\boldsymbol{Z}\boldsymbol{v}_n\,. \tag{2}$$

Equation (2) amounts to generalized trapezoidal treatment of the $\boldsymbol{C}$-term, and leap-frog treatment of the $\boldsymbol{Z}$-term. To develop an energy stability analysis of (2), let

$$\boldsymbol{V}_n=\begin{Bmatrix}\boldsymbol{v}_{n+1}\\ \boldsymbol{v}_n\end{Bmatrix}, \tag{3}$$

$$\boldsymbol{A}=\boldsymbol{A}^{\mathrm{T}}=\begin{bmatrix}\boldsymbol{M}+\Delta t(\gamma-\frac{1}{2})\boldsymbol{C} & -\Delta t\boldsymbol{Z}\\ \Delta t\boldsymbol{Z} & \boldsymbol{M}+\Delta t(\gamma-\frac{1}{2})\boldsymbol{C}\end{bmatrix}. \tag{4}$$

Then, (2) may be used to verify that

$$\boldsymbol{V}_{n+1}^{\mathrm{T}}\boldsymbol{A}\boldsymbol{V}_{n+1}=\boldsymbol{V}_n^{\mathrm{T}}\boldsymbol{A}\boldsymbol{V}_n\,. \tag{5}$$

Clearly, stability in 'energy' amounts to $\boldsymbol{A}$ being positive definite. By inspection of $\boldsymbol{A}$, it is immediately apparent that for the $\boldsymbol{C}$-term alone, the usual stability conditions hold (e.g., if $\gamma\geq\frac{1}{2}$, stability is unconditional, etc., see Section B.I). It is also clear that for Δt small enough, the $\boldsymbol{Z}$-term is stabilized by virtue of the positive definiteness of the diagonal blocks. To say more requires a detailed spectral analysis of $\boldsymbol{A}$, which requires consideration of the structure of the constituent arrays. No general result yet exists. (Carey [7] has proposed using the Gershgorin theorem to obtain practical estimates for cases such as these.) However, the present analysis, though not taken to full fruition, encompasses the multi-dimensional, irregular-mesh case, characteristic of finite-element procedures, and thus represents a valuable extension of the classical von Neumann-type analysis.

D.1.b. Energy analysis of the generalized trapezoidal and midpoint methods; nonlinearities

Consider solution of (1) by way of the generalized trapezoidal family of algorithms:

$$\boldsymbol{M}\boldsymbol{a}_{n+1} + \boldsymbol{C}\boldsymbol{v}_{n+1} + \boldsymbol{Z}\boldsymbol{v}_{n+1} = \boldsymbol{0}\,, \tag{6}$$

$$\boldsymbol{v}_{n+1} = \boldsymbol{v}_n + \Delta t\{(1-\gamma)\boldsymbol{a}_n + \gamma\boldsymbol{a}_{n+1}\}\,. \tag{7}$$

Equations (6) and (7) may be manipulated to yield the following identity:

$$\boldsymbol{a}_{n+1}^{\mathrm{T}}\boldsymbol{M}\boldsymbol{a}_{n+1} = \boldsymbol{a}_n^{\mathrm{T}}\boldsymbol{M}\boldsymbol{a}_n - (2\gamma-1)[\boldsymbol{a}_n]^{\mathrm{T}}\boldsymbol{M}[\boldsymbol{a}_n] - [\boldsymbol{v}_n]^{\mathrm{T}}\boldsymbol{C}[\boldsymbol{v}_n]\,. \tag{8}$$

It is immediately clear that (8) implies the unconditional stability of (6) and (7) as long as $\gamma \geq \frac{1}{2}$, the same as when $\boldsymbol{Z}$ is absent. However, no conclusion can be drawn from (8) regarding cases in which $\gamma < \frac{1}{2}$. Thus the presence of $\boldsymbol{Z}$ seems to complicate the energy stability analysis of methods defined by $\gamma < \frac{1}{2}$, but not methods for which $\gamma \geq \frac{1}{2}$. Perhaps an alternative energy analysis will be more revealing. Nevertheless, this gives some indication why other means are still necessary to obtain practical stability conditions for algorithms such as (6)–(7).

It is interesting to note that unconditional stability results may even be derived when the convective forces are nonlinear. To be precise, assume the $\boldsymbol{Z}$-term is replaced by $\boldsymbol{N}(\boldsymbol{v})$, and that the following skew-symmetry property holds:

$$\boldsymbol{v}^{\mathrm{T}}\boldsymbol{N}(\boldsymbol{v}) = 0\,. \tag{9}$$

An example of an algorithm which is unconditionally stable under these circumstances is the midpoint rule, viz.

$$\boldsymbol{M}\boldsymbol{a}_{n+1/2} + \boldsymbol{C}\boldsymbol{v}_{n+1/2} + \boldsymbol{N}(\boldsymbol{v}_{n+1/2}) = \boldsymbol{0}\,, \tag{10}$$

$$\boldsymbol{v}_{n+1} = \boldsymbol{v}_n + \Delta t\boldsymbol{a}_{n+1/2}\,, \tag{11}$$

$$\boldsymbol{a}_{n+1/2} = \tfrac{1}{2}(\boldsymbol{a}_{n+1} + \boldsymbol{a}_n)\,, \quad \text{etc.} \tag{12}$$

Combining (10)–(12) results in

$$\boldsymbol{v}_{n+1}^{\mathrm{T}}\boldsymbol{M}\boldsymbol{v}_{n+1} = \boldsymbol{v}_n^{\mathrm{T}}\boldsymbol{M}\boldsymbol{v}_n - \Delta t\boldsymbol{v}_{n+1/2}^{\mathrm{T}}\boldsymbol{C}\boldsymbol{v}_{n+1/2}\,. \tag{13}$$

which guarantees unconditional stability. This result may be generalized to include other members of the generalized midpoint family, and by the Dahlquist–Lindberg equivalence theorem [10], a related result for trapezoidal algorithms may be also established.

D.2. The von Neumann method

A clear description of this technique, and others used for stability analysis of difference equations, is presented in Mitchell and Griffiths [54]. To illustrate its use, we shall consider only a simple example (the reader interested in further details is urged to consult [54] and references therein).

Example. Consider the one-dimensional heat equation

$$\frac{\partial u}{\partial t} = k\,\frac{\partial^2 u}{\partial x^2} \tag{14}$$

and algorithm

$$u_{n+1}(m) = (1-2r)u_n(m) + r(u_n(m+1) + u_n(m-1)) \tag{15}$$

where $u_n(m) = u_n(x_m)$, etc., and $r = k\Delta t/h^2$, where h is the node spacing. Equation (15) is explicit, and so we would like to know for which values of r is the algorithm stable? The error induced by small perturbations in initial data, round-off, etc., is denoted by $e_n(m)$, and satisfies the algorithmic equation, namely

$$e_{n+1}(m) = (1-2r)e_n(m) + r(e_n(m+1) + e_n(m-1))\,. \tag{16}$$

Furthermore $e_n(m)$ is assumed to take on the following form

$$e_n(m) = \lambda^n \exp(\mathrm{i}\, m\xi), \quad \mathrm{i} = \sqrt{-1}\,. \tag{17}$$

Substitution of (17) into (16) leads to

$$\lambda = 1 - 4r \sin^2(\xi/2)\,. \tag{18}$$

The condition of stability is $|\lambda| \leq 1$ for all ξ, which requires

$$r \leq 1/[2\sin^2(\xi/2)] \tag{19}$$

for all ξ, and so $r \leq \frac{1}{2}$ is the stability result.

D.3. Von Neumann stability results for the convection – diffusion equation

Consider the n-dimensional linear convection-diffusion equation:

$$\frac{\partial u}{\partial t} + v_i\,\frac{\partial u}{\partial x_i} = k\,\frac{\partial^2 u}{\partial x_i \partial x_i} \tag{20}$$

where v_i is the specified flow velocity in the x_i direction (assumed constant). The summation convention is assumed to apply to repeated indices (e.g., $v_i(\partial u/\partial x_i) = v_1(\partial u/\partial x_1) + \cdots + v_n(\partial u/\partial x_n)$).

Assume a finite difference approximation to (20) which is forward in time and centered in space (FTCS):

$$\begin{aligned} u_{n+1}(m) = u_n(m) &+ c_1[u_n(m+1) - u_n(m-1)] \\ &+ \cdots + r_1[u_n(m+1) - 2u_n(m) + u_n(m-1)] + \cdots \end{aligned} \tag{21}$$

where '$+\cdots+$' stands for the missing x_i-direction terms for $i \geq 2$, $c_i = \Delta t v_i/(2h_i)$ and $r_i = \Delta t k/h_i^2$. Precise stability results have been obtained for (21) by Leonard [51] in one dimension and by Hindmarsh and Gresho [23] in multi-dimensions (the latter results were announced in [18]). The stability condition, which is both necessary and sufficient, entails satisfaction of the following two inequalities:

$$\Delta t \leq \frac{1}{2k} \Big/ \sum_{i=1}^{n} (1/h_i^2), \tag{22}$$

$$\Delta t \leq 2k \Big/ \sum_{i=1}^{n} v_i^2 . \tag{23}$$

Remarks. Note that (22) generalizes the result for (14) to the n-dimensional case. As may be concluded from (23), if diffusion is absent, then the FTCS approximation for convection is unconditionally *unstable*, a well-known result. Likewise, in convection-dominated cases, (23) leads to very stringent time-step restrictions. In some practical situations this might preclude use of the FTCS algorithm and thus other alternatives are often resorted to, such as classical 'upwind' differencing [60]. If the upwind effect is introduced via a centered in space 'artificial diffusion' then the preceding stability conditions again apply with k interpreted as the sum of physical and artificial diffusions. For example, assume physical diffusion may be neglected and the artificial diffusion is defined to be

$$1/(2\Delta t)\Big(\sum_{i=1}^{n} (1/h_i^2)\Big) . \tag{24}$$

Then (22) is automatically satisfied and (23) leads to the following Courant-like condition:

$$\Delta t \leq \sqrt{1 \Big/ \Big(\sum_{i=1}^{n} v_i^2\Big)\Big(\sum_{i=1}^{n} (1/h_i^2)\Big)} . \tag{25}$$

The drawback to procedures of this kind is the significant loss in accuracy which is a by-product of the heavy dose of artificial diffusion. In the past, for lack of anything better, this kind of approach has been frequently resorted to. Based on recent developments in the finite element literature, we expect to see more accurate, yet stable, procedures supercede the classical upwind difference technique. There is considerable research going on at this time which appears quite promising (see, e.g., [32] and papers in Hughes [29]). The terminology which has been often applied to these techniques is 'upwind finite element methods', however, their structure and performance represents a dramatic improvement over classical upwind finite differences.

In the case of explicit treatment of convective and diffusive terms in the Navier–Stokes equations, for lack of a more rational alternative, stability results obtained from the linear convection-diffusion equation, such as those above, are often applied (see e.g. [37]). In this case, the kinematic viscosity replaces the diffusivity k. If compressibility effects are accounted for, then Δt must be further restricted by acoustical transit-time conditions.

References

[1] J.H. Argyris, P.C. Dunne and T. Angelopoulos, Dynamic Response by Large Step Integration, *Earthquake Engineering and Structural Dynamics* **2**, 185–203 (1973).

[2] J.H. Argyris, P.C. Dunne and T. Angelopoulos, Non-Linear Oscillations Using the Finite Element Technique, *Computer Methods in Applied Mechanics and Engineering* **2**, 203–250 (1973).

[3] K.J. Bathe and E.L. Wilson, Stability and Accuracy Analysis of Direct Integration Methods, *Earthquake Engineering and Structural Dynamics* **1**, 283–291 (1973).

[4] T. Belytschko and R. Mullen, Mesh Partitions of Explicit-Implicit Time Integration, *Proceedings, U.S.-Germany Symposium on Formulations and Computational Algorithms in Finite Element Analysis*, Massachusetts Institute of Technology, Cambridge, MA (August 1976).

[5] T. Belytschko and R. Mullen, Stability of Explicit-Implicit Mesh Partitions in Time Integration, *International Journal for Numerical Methods in Engineering* **12**(10), 1575–1586 (1978).

[6] T. Belytschko and D.F. Schoeberle, On the Unconditional Stability of an Implicit Algorithm for Nonlinear Structural Dynamics, *Journal of Applied Mechanics* **17**, 865–869 (1975).

[7] G. Carey, An Analysis of Oscillations and Stability in Convection-Diffusion Computations, in: *Finite Element Methods for Convection Dominated Flows*, T.J.R. Hughes, ed., AMD - Vol. 34 (ASME, New York, 1979) pp. 63–71.

[8] A.J. Chorin, T.J.R. Hughes, M.F. McCracken and J.E. Marsden, Product Formulas and Numerical Algorithms, *Communications on Pure and Applied Mathematics* **XXXI**, 205–256 (1978).

[9] G. Dahlquist, A Special Stability Problem for Linear Multistep Methods, *BIT* **3**, 27–43 (1963).

[10] G. Dahlquist and B. Lindberg, On Some Implicit One-Step Methods for Stiff Differential Equations, Report no. TRITA-NA-7302, Department of Information Processing, The Royal Institute of Technology, Stockholm (1973).

[11] P.J. Davis, *Interpolation and Approximation* (Blaisdell, New York, 1963).

[12] D.P. Flanagan and T. Belytschko, A Uniform Strain Hexahedron and Quadrilateral with Orthogonal Hourglass Control, *International Journal for Numerical Methods in Engineering*, **17**, 679–706 (1981).

[13] I. Fried and D.S. Malkus, Finite Element Mass Matrix Lumping by Numerical Integration Without Convergence Rate Loss, *International Journal of Solids and Structures* **11**, 461–466 (1976).

[14] C.W. Gear, *Numerical Initial Value Problems in Ordinary Differential Equations* (Prentice-Hall, Englewood Cliffs, NJ, 1971).

[15] M. Geradin, A Classification and Discussion of Integration Operators for Transient Structural Response, AIAA Paper 74–105, presented at AIAA 12th Aerospace Sciences Meeting, Washington, DC, Jan. 30–Feb. 1 (1974).

[16] G.L. Goudreau and R.L. Taylor, Evaluation of Numerical Methods in Elastodynamics, *Computer Methods in Applied Mechanics and Engineering* **2**, 69–97 (1973).

[17] A.R. Gourlay, A Note on Trapezoidal Methods for the Solution of Initial Value Problems, *Mathematics of Computation* **24**, 629–633 (1970).

[18] P.M. Gresho, S.T.K. Chan, R.L. Lee and C.D. Upson, Solution of the Time-Dependent, Three-Dimensional Incompressible Navier–Stokes Equations via FEM, UCRL Preprint 85337, Lawrence Livermore Laboratory, Livermore, CA (January 1981).

[19] E. Haug, Q.S. Nguyen and A.L. de Rouvray, An Improved Energy Conserving Implicit Time Integration Algorithm for Nonlinear Dynamic Structural Analysis, *Transactions of the 4th International Conference on Structural Mechanics in Reactor Technology*, San Francisco (August 1977).

[20] H.M. Hilber, Analysis and Design of Numerical Integration Methods in Structural Dynamics, EERC Report No. 76–29, Earthquake Engineering Research Center, University of California, Berkeley, CA (November 1976).

[21] H.M. Hilber and T.J.R. Hughes, Collocation, Dissipation and 'Overshoot' for Time Integration Schemes in Structural Dynamics, *Earthquake Engineering and Structural Dynamics* **6**, 99–118 (1978).

[22] H.M. Hilber, T.J.R. Hughes and R.L. Taylor, Improved Numerical Dissipation for Time

Integration Algorithms in Structural Dynamics, *Earthquake Engineering and Structural Dynamics* **5**, 283–292 (1977).

[23] A. Hindmarsh and P. Gresho, The Stability of Explicit Euler Time Integration for Central Finite Difference Approximation of the Multi-dimensional Advection Diffusion Equation, UCRL Report, Lawrence Livermore National Laboratory, December 1982, to appear in *International Journal for Numerical Methods in Fluids.*

[24] M.W. Hirsch and S. Smale, *Differential Equations, Dynamical Systems, and Linear Algebra* (Academic Press, New York, 1974).

[25] M.A. Hogge, Accuracy and Cost of Integration Techniques for Nonlinear Heat Transfer, Second World Congress on Finite Element Methods, Bournemouth, UK, October 23–27 (1978).

[26] J.C. Houbolt, A Recurrence Matrix Solution for the Dynamic Response of Elastic Aircraft, *Journal of the Aeronautical Sciences* **17**, 540–550 (1950).

[27] T.J.R. Hughes, Stability, Convergence and Growth and Decay of Energy of the Average Acceleration Method in Nonlinear Structural Dynamics, *Computers and Structures* **6**, 313–324 (1976).

[28] T.J.R. Hughes, Stability of One-Step Methods in Transient Nonlinear Heat Conduction, *Transactions of the Fourth International Conference on Structural Mechanics in Reactor Technology*, San Francisco, CA (August 1977).

[29] T.J.R. Hughes (editor), *Finite Element Methods for Convection Dominated Flows*, AMD – Vol. 34 (ASME, New York, 1979).

[30] T.J.R. Hughes, Recent Developments in Computer Methods for Structural Analysis, *Nuclear Engineering and Design* **57**(2), 427–439 (1980).

[31] T.J.R. Hughes, Implicit-Explicit Finite Element Techniques for Symmetric and Nonsymmetric Systems, in: *Numerical Methods for Non-Linear Problems, Volume 1*, C. Taylor, E. Hinton and D.R.J. Owen, eds. (Pineridge Press, Swansea, U.K., 1980) pp. 127–139.

[32] T.J.R. Hughes and A. Brooks, A Theoretical Framework for Petrov–Galerkin Methods with Discontinuous Weighting Functions: Application to the Streamline Upwind Procedure, in: *Finite Elements in Fluids, Volume 4*, R.H. Gallagher, ed. (Wiley, London, 1982).

[33] T.J.R. Hughes, T.K. Caughey and W.K. Liu, Finite Element Methods for Nonlinear Elastodynamics Which Conserve Energy, *Journal of Applied Mechanics* **45**, 366–370 (1978).

[34] T.J.R. Hughes, M. Cohen and M. Haroun, Reduced and Selective Integration Techniques in the Finite Element Analysis of Plates, *Nuclear Engineering and Design* **46**(1), 203–222 (1978).

[35] T.J.R. Hughes and W.K. Liu, Implicit-Explicit Finite Elements in Transient Analysis: Stability Theory, *Journal of Applied Mechanics* **45**, 371–374 (1978).

[36] T.J.R. Hughes and W.K. Liu, Implicit-Explicit Finite Elements in Transient Analysis: Implementation and Numerical Examples, *Journal of Applied Mechanics* **45**, 375–378 (1978).

[37] T.J.R. Hughes, W.K. Liu and A. Brooks, Review of Finite Element Analysis of Incompressible Viscous Flows by the Penalty Function Formulation, *Journal of Computational Physics* **30**, 1–60 (1979).

[38] T.J.R. Hughes, W.K. Liu and I. Levit, Nonlinear Dynamic Finite Element Analysis, *Proceedings of the Europe–U.S. Workshop on Finite Elements in Nonlinear Structural Mechanics,*, Bochum, W. Germany (July 1980).

[39] T.J.R. Hughes and K.S. Pister, Consistent Linearization in Mechanics of Solids, *Computers and Structures* **8**, 391–397 (1978).

[40] T.J.R. Hughes, K.S. Pister and R.L. Taylor, Implicit-Explicit Finite Elements in Nonlinear Transient Analysis, *Computer Methods in Applied Mechanics and Engineering* **17/18**, 159–182 (1979).

[41] T.J.R. Hughes and R.S. Stephenson, Convergence of Implicit-Explicit Finite Elements in Nonlinear Transient Analysis, *International Journal of Engineering Science* **19**, 295–302 (1981).

[42] T.J.R. Hughes, R.L. Taylor and W. Kanoknukulchai, A Simple and Efficient Element for Plate Bending, *International Journal for Numerical Methods in Engineering* **11**(10), 1529–1543 (1977).

[43] B.M. Irons, Applications of a Theorem on Eigenvalues to Finite Element Problems, (CR/132/70) University of Wales, Department of Civil Engineering, Swansea (1970).

[44] P.S. Jensen, Transient Analysis of Structures by Stiffly Stable Methods, *Computers and Structures* **4**, 615–626 (1974).

[45] P.S. Jensen, Stiffly Stable Methods for Undamped Second-Order Equations of Motion, *SIAM Journal of Numerical Analysis* **13**(4), 549–563 (1976).

[46] S.W. Key and Z.E. Beisinger, The Transient Dynamic Analysis of Thin Shells by the Finite Element Method, *Proceedings of the 3rd Conference on Matrix Methods in Structural Mechanics*, Wright-Patterson Air Force Base, Ohio (1971).

[47] R.D. Krieg, Unconditional Stability in Numerical Time Integration Methods, *Journals of Applied Mechanics* **40**, 417–421 (1973).

[48] R.D. Krieg and S.W. Key, Transient Shell Response by Numerical Time Integration, *International Journal for Numerical Methods in Engineering* **7**, 273–286 (1973).

[49] J.D. Lambert, *Computational Methods in Ordinary Differential Equations* (John Wiley, London, 1973).

[50] S. Lang, *Differential Manifolds* (Addison Wesley, Reading, MA, 1972).

[51] B.P. Leonard, Note on the Von Neumann Stability of the Explicit FTCS Convective-Diffusion Equation, *Applied Mathematical Modelling* **4**, 401 (1980).

[52] B. Lindberg, On Smoothing and Extrapolation for the Trapezoidal Rule, *BIT* **11**, 29–52 (1971).

[53] G. McVerry, Numerical Integration Schemes for Structural Dynamics, Unpublished Manuscript.

[54] A.R. Mitchell and D.F. Griffiths, *The Finite Difference Method in Partial Differential Equations* (Wiley, London, 1980).

[55] N.M. Newmark, A Method of Computation for Structural Dynamics, *Journal of the Engineering Mechanics Division*, ASCE, 67–94, (1959).

[56] B. Noble, *Applied Linear Algebra* (Prentice-Hall, Englewood Cliffs, NJ, 1969).

[57] K.C. Park, Evaluating Time Integration Methods for Nonlinear Dynamic Analysis, in: *Finite Element Analysis of Transient Non-Linear Behavior*, T. Belytschko, J.R. Osias and P.V. Marcal, eds., Applied Mechanics Symposia Series (ASME, New York, 1975).

[58] K.C. Park, Partioned Transient Analysis Procedures for Coupled-Field Problems, Presented at 2nd Conference on Numerical Methods in Nonlinear Mechanics, TICOM, University of Texas, Austin (March 1979).

[59] D. Richtmyer and K.W. Morton, *Difference Methods for Initial-Value Problems*, Second Edition (Interscience, New York, 1967).

[60] P.J. Roache, *Computational Fluid Dynamics* (Hermosa Publishers, Albuquerque, NM, 1976).

[61] D.M. Trujillo, An Unconditionally Stable Explicit Algorithm for Structural Dynamics, *International Journal for Numerical Methods in Engineering* **11**, 1579–1592 (1977).

[62] E.L. Wilson, A Computer Program for the Dynamic Stress Analysis of Underground Structures, SESM Report No. 68-1, Division Structural Engineering and Structural Mechanics, University of California, Berkeley (1968).

[63] W.L. Wood and R.W. Lewis, A Comparison of Time Marching Schemes for the Transient Heat Conduction Equation, *International Journal for Numerical Methods in Engineering* **9**, 679–689 (1975).

[64] O.C. Zienkiewicz, A New Look at the Newmark, Houbolt and Other Time Stepping Formulas. A Weighted Residual Approach, *Earthquake Engineering and Structural Dynamics* **5**, 413–418 (1977).

CHAPTER 3

Partitioned Analysis of Coupled Systems

K.C. PARK and C.A. FELIPPA

Applied Mechanics Laboratory
Lockheed Palo Alto Research Laboratory
3251 Hanover Street
Palo Alto, CA 94304, USA

Computational Methods for Transient Analysis
Edited by T. Belytschko and T.J.R. Hughes

Abstract: This chapter reviews partitioned analysis procedures for the analysis of coupled-field dynamical problems. These problems involve two or more distinctive subsystems that are tightly coupled. Examples are provided by fluid-structure, soil-structure, thermal-structure and structure-magnetodynamic interactions. The computer analysis of such systems depends on the integration of analysis capabilities for the separate components. Partitioned analysis procedures provide an efficient and modular way of achieving that integration. These procedures advance the solution of the coupled problem in a staged fashion; a process that is naturally implemented through sequential or parallel execution of subsystem (single-field) analysis programs. We review the underlying theory, which is still largely in a formative stage, formulation of the time-advancing process, computer implementation aspects, and some applications.

1. Introduction

This chapter deals with partitioned analysis methods for the computerized treatment of coupled-field dynamic problems. These methods represent a fairly recent but promising development in computational mechanics. Their novelty is reflected in an underlying theory largely in a formative stage, and in a still limited but expanding domain of applications.

1.1. Coupled systems

What are coupled systems? Many engineering problems of current interest require an integrated treatment of *coupled fields*. The key word is *integrated*: tight interaction between the component fields means that the response of the overall system must be calculated *concurrently*.

What component fields? Structures (or substructures), fluid and soil media; thermal, acoustic, electromagnetic or other high-energy fields. Or fields may be selected purely on computational, rather than physical, grounds. A field spatially discretized to a finite number of degrees of freedom (e.g., through a finite element or finite difference method) will be called a *subsystem*.

What specific engineering applications? Fluid-structure interaction in submerged structures, pressure vessels and piping; soil-structure interaction in earthquake engineering; flutter in aerospace structures and turbomachinery; thermal-structure interaction in high-energy equipment.

Considerable progress has been accomplished in the development of computational methods for *single-field problems* such as structures and fluids taken as isolated entities. Because the development of techniques for coupled systems requires multi-disciplinary expertise, it has understandably lagged behind. But more recently, design requirements for efficient use of material resources, emphasis on energy-efficient design, and concerns for equipment safety has boosted the demand for realistic computer analysis of coupled systems.

The computer analysis of most coupled problems tackled to date has been traditionally performed with the "monolithic augmentation" approach. As specific problems arise, large-scale computer programs are expanded to house more interaction effects. An example would be the addition of fluid-volume elements to a finite-element structural analyzer to handle fluid-structure interaction problems.

Experience has indicated that this approach run three risks: uncontrollable complexity growth of the necessary software, inability to accomodate new or improved problem formulations, and lack of flexibility to meet time-critical demands.

First, augmented monolithic codes gradually become unreliable, difficult to use, and overly dependent on developers. Second, rigidity of formulation can be a grave defect; for example, displacement-assumed fluid-volume elements are plainly a poor choice to model irrotational motions, but this grotesque distortion may be forced upon program users by the straightjacket of a direct-stiffness finite-element formulation. Finally, engineering analysis activities are impeded if required analysis capabilities fall outside available programs; panic 'patch jobs' may then be required under tight schedule constraints, and results are obtained too late to intervene in major engineering decisions.

1.2. Partitioned analysis procedures

Partitioned analysis provide an alternative to the traditional approach. Rather than augment and drown in complexity, divide and conquer.

In the partitioned analysis approach, field-state vectors of the coupled equations are processed by separate program modules called *field analyzers*. The solution of the coupled system results from the execution of a set of analyzers synchronized to operate in sequential or parallel fashion.

The analysis of a fluid-structure interaction problem, for example, may be obtained through a 'staggered' solution procedure in which separate fluid and structural analysis programs execute in a strictly sequential fashion and exchange interface-state data such as pressures and velocities at each time step. This approach simplifies the connection of the same fluid analyzer to different structural analysis codes and vice-versa; it thus offers the analyst flexibility in selecting the program(s) that best fit the problem at hand.

The original motivation for developing partitioned analysis procedures was computational efficiency. Mechanical fields often exhibit vastly different response characteristics (in the sense of their natural time scales). This suggests that for efficiency, different solution algorithms and time-marching schemes for each field would be desirable. For example, in the finite element analysis of shells, computational vectors associated with membrane and rotational motions typically exhibit higher frequencies than those associated with bending motions. Thus, an implicit algorithm would be desirable to treat membrane and rotational motions while the bending motion is traced by a more economical explicit scheme.

The two key advantages of partitioned analysis procedures are seen to be computational efficiency and modular implementation. The second aspect is especially significant since the bulk of existing engineering analysis software has been developed for the treatment of single-field problems. Considerable economy may then result by utilizing existing software as modular elements to tackle coupled-field problems.

1.3. History

The decomposition of complex dynamical systems has been exploited in disciplines as diverse as economics, weather prediction, power grid networks, ecosystems, spacecraft control, and computer hardware design. For applications along these lines, the reader is referred to the books [19, 31, 32]. The application of these concepts to structural dynamics is fairly recent, however; as of this writing only a handful of papers have dealt with the subject of partitioned analysis.

Belytschko and Mullen [1, 2, 3] introduced the concept of *implicit-explicit nodal-based* partitions of finite element systems, whereas Hughes and Liu [20, 21] developed a theory of *element-based* implicit-explicit partitions. Park, Felippa and DeRuntz introduced *implicit-implicit* staggered partitions for fluid-structure interaction [24], a subject recently reviewed in [12] in light of the theoretical advances mentioned below. Implicit-implicit partitions have also been studied by Belytschko and coworkers [3, 4, 5] primarily for fluid-solid impact applications, and by Zienkiewicz, Hinton, Leung and Taylor [37] for multiphase soil analysis. Semi-implicit partitions have been studied by Park and Housner [29].

A stability-accuracy theory for general partitions of second-order systems has been recently developed [27, 28]. This theory constitutes the bulk of the material covered in Sections 4 through 7 of this chapter.

1.4. Outline

The outline of this chapter is as follows. Coupled-field equations of motion for various applications are presented and the particular case of a structure submerged in an acoustic medium discussed in some detail to illustrate solution approaches. The equations of motion are time-discretized by introduction of integration formulas of linear multistep type. Partitioning of the resulting difference equations in accordance with problem characteristics yields partitioned analysis formulations.

Formulations are completed through the introduction of specific calculation sequences and predictors. Methods for assessing stability and accuracy of the resulting implementations are described. Then the fluid-structure interaction case is taken upon again in detail, and it is shown that a reformulation of the original equations is needed for stabilization. Application examples pertaining to this particular problem are given. Finally, some views on the present state and projected future of partitioned analysis procedures are offered.

2. Coupled systems

2.1. Governing equations

We shall consider coupled dynamical systems governed by the semi-discrete, second-order matrix equations of motion

$$\boldsymbol{M}\ddot{\boldsymbol{u}} + \boldsymbol{D}\dot{\boldsymbol{u}} + \boldsymbol{K}\boldsymbol{u} = \boldsymbol{f}_A + \boldsymbol{f}_C + \boldsymbol{f}_N \tag{2.1}$$

where $\boldsymbol{u}$ is the state vector of the complete system; $\boldsymbol{M}$, $\boldsymbol{D}$ and $\boldsymbol{K}$ are time-invariant, real-valued matrices with $\boldsymbol{M}$ symmetric; $\boldsymbol{f}_A$, $\boldsymbol{f}_C$ and $\boldsymbol{f}_N$ are vectors of applied, coupling (interaction) and nonlinear forces, respectively; and a dot denotes temporal differentiation.

To make the structure of (2.1) more explicit, consider a *linear* problem involving two coupled fields, X and Y; and let the corresponding state vectors be $\boldsymbol{x}$ and $\boldsymbol{y}$, so that

$$\boldsymbol{u} = \begin{Bmatrix} \boldsymbol{x} \\ \boldsymbol{y} \end{Bmatrix}. \tag{2.2}$$

Then the most general two-field linear system befitting (2.1) is

$$\begin{aligned} \boldsymbol{M}_{xx}\ddot{\boldsymbol{x}} + \boldsymbol{D}_{xx}\dot{\boldsymbol{x}} + \boldsymbol{K}_{xx}\boldsymbol{x} &= \boldsymbol{f}_A^x - \boldsymbol{D}_{xy}\dot{\boldsymbol{y}} - \boldsymbol{K}_{xy}\boldsymbol{y}, \\ \boldsymbol{M}_{yy}\ddot{\boldsymbol{y}} + \boldsymbol{D}_{yy}\dot{\boldsymbol{y}} + \boldsymbol{K}_{yy}\boldsymbol{y} &= \boldsymbol{f}_A^y - \boldsymbol{D}_{yx}\dot{\boldsymbol{x}} - \boldsymbol{K}_{yx}\boldsymbol{x}. \end{aligned} \tag{2.3}$$

Table 2.1 lists several coupled problems of interest in engineering mechanics

Table 2.1
Specialization of Eq. (2.1) to some 'extended structural analysis' problems

(1) Undamped, linear structure–structure interaction	
$\boldsymbol{M}_{xx}\ddot{\boldsymbol{x}} + \boldsymbol{K}_{xx}\boldsymbol{x} = \boldsymbol{f}_A^x - \boldsymbol{K}_{xy}\boldsymbol{y}$	$\boldsymbol{M}_{yy}\ddot{\boldsymbol{y}} + \boldsymbol{K}_{yy}\boldsymbol{y} = \boldsymbol{f}_A^y - \boldsymbol{K}_{yx}\boldsymbol{x}$
(2) Undamped structure subject to nonlinear/nonconservative force field $\boldsymbol{f}_N$	
$\boldsymbol{M}_s\ddot{\boldsymbol{x}} + \boldsymbol{K}_s\boldsymbol{x} = \boldsymbol{f}_A^x + \boldsymbol{y}$	$\boldsymbol{y} = \boldsymbol{f}_N^y(\boldsymbol{x}, \dot{\boldsymbol{x}})$
(3) Linear structure submerged in a DAA_1-fluid medium [9, 14]	
$\boldsymbol{M}_s\ddot{\boldsymbol{x}} + \boldsymbol{D}_s\dot{\boldsymbol{x}} + \boldsymbol{K}_s\boldsymbol{x} = \boldsymbol{f}_A^x - \boldsymbol{T}\boldsymbol{A}_f(\boldsymbol{p}_I + \boldsymbol{p}_S)$	$\boldsymbol{M}_f\dot{\boldsymbol{p}}_S + \rho c\boldsymbol{A}_f\boldsymbol{p}_S = \rho c\boldsymbol{M}_f(\boldsymbol{T}'\ddot{\boldsymbol{x}} - \dot{\boldsymbol{v}}_I)$
(4) Linear structure submerged in a DAA_2-fluid medium [8, 17]	
$\boldsymbol{M}_s\ddot{\boldsymbol{x}} + \boldsymbol{D}_s\dot{\boldsymbol{x}} + \boldsymbol{K}_s\boldsymbol{x} = \boldsymbol{f}_A^x - \boldsymbol{T}\boldsymbol{A}_f(\boldsymbol{p}_I + \boldsymbol{p}_S)$	$\boldsymbol{M}_f\ddot{\boldsymbol{p}}_s + \rho c\boldsymbol{A}_t\dot{\boldsymbol{p}} + \rho c\boldsymbol{\Omega}_f\boldsymbol{A}_f\boldsymbol{p}_S = \rho c[\boldsymbol{M}_f(\boldsymbol{T}'\ddot{\boldsymbol{x}} - \dot{\boldsymbol{v}}_1) + \boldsymbol{\Omega}_f\boldsymbol{M}_f(\boldsymbol{T}'\ddot{\boldsymbol{x}} - \boldsymbol{v}_I)]$
(5) Linear structure buried in a DAA_1-soil medium [34]	
$\boldsymbol{M}_s\ddot{\mathbf{x}} + \boldsymbol{D}_s\dot{\boldsymbol{x}} + \boldsymbol{K}_s\boldsymbol{x} = \boldsymbol{f}_A^x + \boldsymbol{f}_I^x + \boldsymbol{f}_S^x$	$\boldsymbol{f}_S^x = \boldsymbol{T}[\rho c\boldsymbol{A}_m\dot{\boldsymbol{x}} + \mathbf{K}_m\boldsymbol{x} + \boldsymbol{f}_N^m]$

that are governed by matrix differential equations of the form (2.3), possibly augmented with nonlinear force terms.

Before embarking on a general exposition of schemes for the direct time integration of (2.1) in Section 3, we shall review some practical aspects of the derivation of coupled-field equations, as well as three approaches to their solution. This will be done using the third example of Table 2.1 as case study.

2.2. *Submerged structure*

The specific problem used as case study is illustrated in Figure 2.1. A linear or nonlinear three-dimensional structure is submerged in an infinite acoustic fluid. A pressure shock wave propagates through the fluid and impinges the structure. The structure and fluid are spatially discretized through finite-element (FE) and boundary-element (BE) methods, respectively.

Before deriving the governing equations, two practical considerations of relevance to this problem should be mentioned.

First, the structural response (and most especially the structure's survivability) is of primary concern, whereas what happens in the fluid is of little interest.

Second, the FE and BE meshes on the "wet surface" are not necessarily in one-to-one correspondence, as illustrated in the two-dimensional sketch of Figure 2.2. Rather, a "fluid BE" typically overlaps several structural elements. This ties up with the first consideration in the sense that determination of structural deformations and stresses demands a finer subdivision.

2.2.1. *Structural response equations*

The governing matrix equation of motion for the dynamic response of a discrete structure is

$$\boldsymbol{M}_s\ddot{\boldsymbol{x}} + \boldsymbol{D}_s\dot{\boldsymbol{x}} + \boldsymbol{K}_s\boldsymbol{x} = \boldsymbol{f}_C + \boldsymbol{N} \tag{2.4}$$

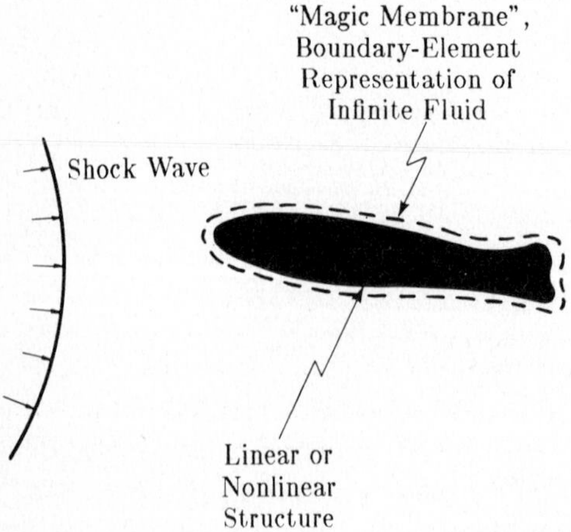

Fig. 2.1. Structure submerged in an acoustic fluid.

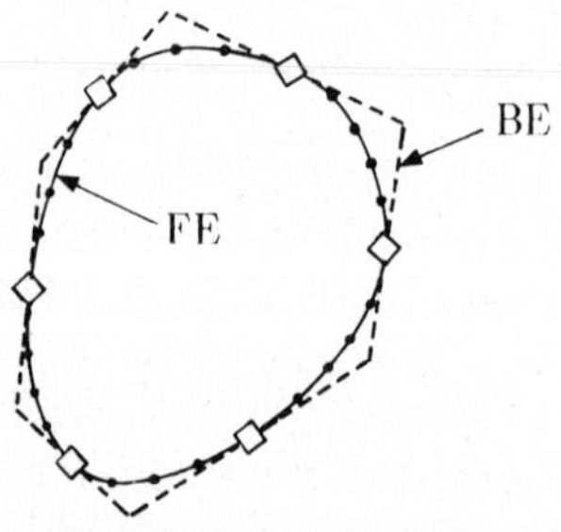

Fig. 2.2. Sketch layout of BE and FE meshes on wet surface (◇ = BE control points; · = FE node points).

where $\boldsymbol{x}=\boldsymbol{x}(t)$ is the structural displacement vector, $\boldsymbol{M}_s$, $\boldsymbol{D}_s$ and $\boldsymbol{K}_s$ are the structural mass, damping and stiffness matrices, respectively; $\boldsymbol{f}_C$ is the interaction force vector, and $\boldsymbol{N}=\boldsymbol{N}(\boldsymbol{x})$ is a nonlinear pseudo-force vector.

For excitation of a submerged structure by an acoustic wave, the interaction force is given by

$$\boldsymbol{f}_C = -\boldsymbol{T}\boldsymbol{A}_f(\boldsymbol{p}_I + \boldsymbol{p}_S) \tag{2.5}$$

where $\boldsymbol{p}_I$ and $\boldsymbol{p}_S$ are nodal pressure vectors for the wet-surface fluid mesh pertaining to the (known) incident wave and the (unknown) scattered wave, respectively, $\boldsymbol{A}_f$ is the diagonal area matrix associated with elements in the fluid mesh, and $\boldsymbol{T}$ is the transformation matrix that relates the structural and fluid nodal forces. Introduction of this matrix takes care of the FE/BE 'mesh-mismatch' illustrated in Figure 2.2.

2.2.2. *Fluid equations*

The response of the fluid is modelled by the first-order doubly asymptotic approximation (DAA_1) of Geers [14, 17]

$$\boldsymbol{M}_f\dot{\boldsymbol{p}}_S + \rho c\boldsymbol{A}_f\boldsymbol{p}_S = \rho c\boldsymbol{M}_f\dot{\boldsymbol{v}} \tag{2.6}$$

where $\boldsymbol{v}$ is the vector of scattered-wave fluid-particle velocity normal to the structure's wet surface, ρ and c are the density and sound velocity of the fluid, respectively, and $\boldsymbol{M}_f$ is the symmetric fluid mass matrix for the wet-surface fluid mesh. This matrix is produced by a boundary-element treatment of Laplace's equation for the irrotational flow generated in an infinite, inviscid, *incompressible* fluid by motions of the structure's wet surface; it is fully populated with non-zero matrix elements. When transformed into structural coordinates, the fluid mass matrix yields the *added mass matrix*, which, when combined with the structural mass matrix, yields the virtual mass matrix for motions of a structure submerged in an incompressible fluid. Details of the calculation procedure may be found in [6].

The approximate pressure-velocity relation (2.6) is called 'doubly asymptotic' because it approaches exactness in both the high-frequency (early-time) and low-frequency (late-time) limits.

For excitation by an incident acoustic wave, $\boldsymbol{v}$ is related to structural response by the kinematic compatibility relation

$$\boldsymbol{T}'\dot{\boldsymbol{x}} = \boldsymbol{v}_I + \boldsymbol{v}_S \tag{2.7}$$

where the prime superscript denotes matrix transposition. Equation (2.7) expresses the constraint that normal fluid-particle velocity match normal structural velocity on the wet surface of the structure. The fact that the transformation matrix relating these velocities is $\boldsymbol{T}'$ follows from the invariance of virtual work with respect to either of the wet-surface coordinate systems.

Generally, $\boldsymbol{T}$ is a rectangular matrix whose height greatly exceeds its width, inasmuch as the number of structural DOF usually exceeds considerably the number of fluid DOF, as noted previously. Typical numbers: 5000 structural DOFs and 160 fluid DOFs.

2.2.3. Interaction equations

The introduction of (2.5) into (2.4) and (2.7) into (2.6) yields the interaction equations

$$\begin{aligned} \boldsymbol{M}_s\ddot{\boldsymbol{x}} + \boldsymbol{D}_s\dot{\boldsymbol{x}} + \boldsymbol{K}_s\boldsymbol{x} &= -\boldsymbol{T}\boldsymbol{A}_f(\boldsymbol{p}_I + \boldsymbol{p}_S) + \boldsymbol{N}, \\ \boldsymbol{M}_f\dot{\boldsymbol{p}}_S + \rho c\boldsymbol{A}_f\boldsymbol{p}_S &= \rho c\boldsymbol{M}_f(\boldsymbol{T}'\ddot{\boldsymbol{x}} - \dot{\boldsymbol{v}}_I). \end{aligned} \tag{2.8}$$

The computational structure of these coupled systems is very different. As can be expected, the FE (structural) system is usually *large but sparse.* The BE (fluid) system is typically *small but dense.* It is therefore of interest to design solution methods that exploit these attributes to maximum advantage.

2.3. Solution approaches

Three approaches to solving the coupled FE/BE system (2.8) are reviewed here. They are presented in chronological order, i.e., in roughly the same sequence as they were tried and evaluated on this particular problem.

2.3.1. Field elimination

As noted previously, the structural response is of primary interest. It is therefore natural to think of eliminating the scattered-pressure vector $\boldsymbol{p}_S$ from the coupled equations of motion (2.8). This yields a third-order system in the structural displacements:

$$\boldsymbol{G}_0\dddot{\boldsymbol{x}} + \boldsymbol{G}_1\ddot{\boldsymbol{x}} + \boldsymbol{G}_2\dot{\boldsymbol{x}} + \boldsymbol{G}_3\boldsymbol{x} = \boldsymbol{r}(\boldsymbol{A}_f, \boldsymbol{M}_f, \boldsymbol{p}_I, \dot{\boldsymbol{v}}_I, \boldsymbol{N}) \tag{2.9}$$

where G_0 through G_3 are complicated matrix functions of the original system matrices and of the generalized inverse of $\boldsymbol{T}$; whereas the right-hand side vector $\boldsymbol{r}$ embodies incident pressure, incident fluid-particle normal velocity, and nonlinear effects. The structural response $\boldsymbol{x}(t)$ can now be determined by numerically integrating (2.9).

This was in fact the first approach tried to tackle the time integration of the coupled system [9]. Although moderately successful for the first problem series (submerged shells of revolution, linear structural behavior), from current perspective it can be properly characterized as a poor strategy that eventually leads to a 'computational horror show' for more general problems. Why?

1. The order of the reduced differential system is raised (in this example, from two to three). The appearance of higher derivatives can be the source of many difficulties, the worst of which is noted next.

2. Proper treatment of initial conditions is complicated by the increased ODE

order. In our case study, it turned out that (2.9) had to be integrated once (yielding an integrodifferential system) so as to regularize the treatment of wavefront-induced singularities. Time integrals of forcing terms had then to be carried along in the calculations – a grievous programming burden.

3. Sparseness and symmetry attributes of the original matrices are adversely affected by the elimination process.

4. The development of specialized software is required. For example, available software for dealing with the uncoupled problems (structural dynamics and acoustic shocks) separately is not likely to be of much use for treating the reduced system (2.9).

2.3.2. Simultaneous integration

In this approach Equations (2.8) are viewed as a single second-order system

$$\begin{bmatrix} \boldsymbol{M}_{\mathrm{s}} & \boldsymbol{0} \\ \boldsymbol{0} & \boldsymbol{0} \end{bmatrix} \begin{Bmatrix} \ddot{\boldsymbol{x}} \\ \ddot{\boldsymbol{p}}_{\mathrm{S}} \end{Bmatrix} + \begin{bmatrix} \boldsymbol{D}_{\mathrm{s}} & \boldsymbol{T}\boldsymbol{A}_{\mathrm{f}} \\ \rho c \boldsymbol{M}_{\mathrm{f}} \boldsymbol{T}' & \boldsymbol{M}_{\mathrm{f}} \end{bmatrix} \begin{Bmatrix} \dot{\boldsymbol{x}} \\ \dot{\boldsymbol{p}}_{\mathrm{S}} \end{Bmatrix} + \begin{bmatrix} \boldsymbol{K}_{\mathrm{s}} & \boldsymbol{0} \\ \boldsymbol{0} & \rho c \boldsymbol{A}_{\mathrm{f}} \end{bmatrix} \begin{Bmatrix} \boldsymbol{x} \\ \boldsymbol{p}_{\mathrm{S}} \end{Bmatrix} = \begin{Bmatrix} -\boldsymbol{T}\boldsymbol{A}_{\mathrm{f}}\boldsymbol{p}_{\mathrm{I}} + \boldsymbol{N} \\ -\rho c \boldsymbol{M}_{\mathrm{f}} \dot{\boldsymbol{v}}_{\mathrm{I}} \end{Bmatrix} \tag{2.10}$$

and a direct integration scheme constructed to advance displacements and pressures *simultaneously* in time.

This approach removes many of the objections raised against the field elimination technique. Inasmuch as the order of the differential equations is not raised, initial-condition difficulties do not arise and better use can be made of existing software for dealing with second-order systems.

Now if these equations are treated by an implicit formula, the assembly and factorization of the implicit coefficient matrix is found to pose enormous computational demands for three-dimensional problems. This is due to the presence of matrix-coupling terms that can extend across thousands of equations. On the other hand, if the equations are treated by an explicit scheme, no particular advantage accrues inasmuch as $\boldsymbol{M}_{\mathrm{f}}$ is a dense matrix.

For example, it was estimated that just the factorization of the implicit coefficient matrix for a problem with 5000 structural degrees of freedom and 150 fluid boundary elements would require roughly 3 wall-clock hours on a Cyber 175 computer. Carrying out a nonlinear transient-response analysis of a realistic structural model was then adjudged infeasible.

2.3.3. Partitioned integration

In the partitioned integration approach, the solution state is advanced over each of the two subsystems: FE structural model and BE fluid model, in a staggered (sequential) fashion. Interaction terms are treated as 'forcing' actions that have to be judiciously extrapolated.

What is now called the *staggered solution procedure* is a specific partitioned-integration scheme originally formulated for the system (2.8) by Park, Felippa and

DeRuntz [24]. A version of this procedure was implemented in a production-level computer program described by DeRuntz et al. [7, 8].

Success of this scheme led to further applications and eventually the development of a general theory of partitioned analysis [27, 28]. A comprehensive review of formulation aspects of staggered solution procedures has recently appeared [12].

The staggered solution procedure was found to offer two important advantages: *enhanced software modularity* and *computational efficiency.*

The first advantage accrues from the fact that comparatively few modifications on programs available for processing the uncoupled systems are required for handling the coupled system. Given current labor-dominated costs in software development, augmentation and maintenance, this is indeed an important virtue of this approach. For our specific problem, a BE fluid analysis module was written, and data-coupled to existing large-scale structural analysis codes such as NASTRAN, SPAR and STAGS.

A key advantage of 'plug-in' modularity is the freedom afforded the analyst as regards the selection of a structural analysis code that best fits the problem at hand; for example, the nonlinear analyzer STAGS when plasticity or finite displacements had to be considered. Moreover, if there is a choice among structural analyzers that can do just about the same thing, the user can select the one he or she is most comfortable with.

As regards computational efficiency, the cost per time step is roughly the same as adding up those incurred in processing the FE and BE models as isolated entities. This is because the overhead introduced by the flow of information, which consists primarily of computational vectors, among the two analysis modules becomes comparatively insignificant in large-scale problems. It follows that the staggered solution procedure appears economically attractive should time stepsizes be dictated only by response-tracing accuracy requirements.

Unfortunately, the latter assumption was not easy to realize in the example problem. The high computational efficiency per time step is counteracted by the fact that satisfactory *numerical stability* properties are hard to achieve; in fact, the practical feasibility of the staggered solution procedure hinges almost entirely on the stability analysis. This topic is taken up again in Section 8, where it is shown that achieving unconditional stability for this problem requires a *reformulation* of the governing equations.

3. Time discretization

The coupled equations of motion (2.1) can be numerically solved directly in terms of the physical coordinates, or indirectly through a dimensionality-reduction transformation from physical to generalized coordinates. In the literature these are commonly referred to as *direct time integration*, and *modal, generalized,* or *global-function integration* approaches, respectively.

We shall deal here only with the direct time integration approach, which enjoys the advantages of generality and physical transparency.

This section reviews computational aspects of direct time integration methods for second-order differential systems. It covers those aspects needed for the presentation of partitioned solution procedures in Sections 4 and 5.

3.1. *Reduction to first-order*

Recall the governing semi-discrete coupled-field equations:

$$\boldsymbol{M}\ddot{\boldsymbol{u}} + \boldsymbol{D}\dot{\boldsymbol{u}} + \boldsymbol{K}\boldsymbol{u} = \boldsymbol{f} \tag{3.1}$$

where $\boldsymbol{f}$ embodies applied and nonlinear forcing terms of (2.1). This second-order system can be reduced to first order through the auxiliary vector $\boldsymbol{v}$ introduced by Jensen [23]

$$\boldsymbol{v} = \boldsymbol{A}\boldsymbol{M}\dot{\boldsymbol{u}} + \boldsymbol{B}\boldsymbol{u} \tag{3.2}$$

where $\boldsymbol{A}$ and $\boldsymbol{B}$ are arbitrary square matrices which may be suitably chosen to reduce operation counts or to achieve various implementation goals. These matrices are independent of time and state although they may be functions of the integration stepsize. Differentiating (3.2) yields

$$\dot{\boldsymbol{v}} = \boldsymbol{A}\boldsymbol{M}\ddot{\boldsymbol{u}} + \boldsymbol{B}\dot{\boldsymbol{u}} = \boldsymbol{A}(\boldsymbol{f} - \boldsymbol{D}\dot{\boldsymbol{u}} - \boldsymbol{K}\boldsymbol{u}) + \boldsymbol{B}\dot{\boldsymbol{u}}\,. \tag{3.3}$$

Equations (3.2) and (3.3) can be presented as a set of first-order equations:

$$\begin{bmatrix} \boldsymbol{A}\boldsymbol{M} & \boldsymbol{0} \\ \boldsymbol{A}\boldsymbol{D} - \boldsymbol{B} & \boldsymbol{I} \end{bmatrix} \begin{Bmatrix} \dot{\boldsymbol{u}} \\ \dot{\boldsymbol{v}} \end{Bmatrix} + \begin{bmatrix} \boldsymbol{B} & -\boldsymbol{I} \\ \boldsymbol{A}\boldsymbol{K} & \boldsymbol{0} \end{bmatrix} \begin{Bmatrix} \boldsymbol{u} \\ \boldsymbol{v} \end{Bmatrix} = \begin{Bmatrix} \boldsymbol{0} \\ \boldsymbol{A}\boldsymbol{f} \end{Bmatrix}. \tag{3.4}$$

3.2. *Integration formulas*

We consider the numerical integration of the first-order system (3.1) by the use of m-step, one-derivative, linear multistep (LMS) formulas. For a constant stepsize h, the formulas used in this chapter are

$$\begin{aligned} \boldsymbol{u}_n + \sum_{j=1}^{m} \alpha_j \boldsymbol{u}_{n-j} &= h \sum_{j=0}^{m} \beta_j \dot{\boldsymbol{u}}_{n-j}\,, \\ \boldsymbol{v}_n + \sum_{j=1}^{m} \alpha_j \boldsymbol{v}_{n-j} &= h \sum_{j=0}^{m} \beta_j \dot{\boldsymbol{v}}_{n-j}\,. \end{aligned} \tag{3.5}$$

Here α_j and β_j are coefficients associated with specific formulas, and $\boldsymbol{u}_k, \boldsymbol{v}_k, \ldots$ denote the vectors $\boldsymbol{u}(t_k)$, $\boldsymbol{v}(t_k)$ computed at sample times t_k.

A more compact representation of (3.5) is

$$\boldsymbol{u}_n = \delta \boldsymbol{u}_n + \boldsymbol{h}_n^{\boldsymbol{u}}\,, \qquad \boldsymbol{v}_n = \delta \boldsymbol{v}_n + \boldsymbol{h}_n^{\boldsymbol{v}}\,, \tag{3.6}$$

in which

$$\delta = \beta_0 h \tag{3.7}$$

is a stepsize-dependent coefficient (sometimes called the 'generalized stepsize'), and $\boldsymbol{h}_n^u$, $\boldsymbol{h}_n^v$ are the *historical vectors*:

$$\boldsymbol{h}_n^u = \sum_{j=1}^{m} (-\alpha_j \boldsymbol{u}_{n-j} + h\beta_j \dot{\boldsymbol{u}}_{n-j}) , \tag{3.8}$$

$$\boldsymbol{h}_n^v = \sum_{j=1}^{m} (-\alpha_j \boldsymbol{v}_{n-j} + h\beta_j \dot{\boldsymbol{v}}_{n-j}) , \tag{3.9}$$

which embody the effect of past solutions.

Note that if $\delta = 0$, then $\boldsymbol{u}_n$ and $\boldsymbol{v}_n$ can be explicitly computed in term of past solutions. Thus formulas (3.5) in which $\beta_0 = 0$ are called *explicit*, and *implicit* otherwise.

Remark. More generally, one can use different integration formulas for $\boldsymbol{u}$ and $\boldsymbol{v}$, see e.g. [10, 11, 12, 26]. A common formula is used in Sections 3 through 7 for simplicity. Extension to the general case is formally straightforward, but complicates the equations.

3.3. Difference equations

We consider only the implicit integration case. Elimination of the time-derivatives $\dot{\boldsymbol{u}}_n$ and $\dot{\boldsymbol{v}}_n$ in (3.2) through (3.4) then yields the algebraic system

$$\begin{bmatrix} \boldsymbol{AM} + \delta\boldsymbol{B} & -\delta\boldsymbol{B} \\ \boldsymbol{AD} - \boldsymbol{B} + \delta\boldsymbol{AK} & \boldsymbol{I} \end{bmatrix} \begin{Bmatrix} \boldsymbol{u}_n \\ \boldsymbol{v}_n \end{Bmatrix} = \begin{Bmatrix} \boldsymbol{AMh}_n^u \\ (\boldsymbol{AD} - \boldsymbol{B})\boldsymbol{h}_n^u + \boldsymbol{h}_n^v + \delta\boldsymbol{Af}_n \end{Bmatrix} . \tag{3.10}$$

Finally, elimination of $\boldsymbol{v}_n$ from (3.10) yields the algebraic system

$$\boldsymbol{Eu}_n = \boldsymbol{g}_n \tag{3.11}$$

where

$$\boldsymbol{E} = \boldsymbol{M} + \delta\boldsymbol{D} + \delta^2\boldsymbol{K} , \tag{3.12}$$

$$\boldsymbol{g}_n = [\boldsymbol{M} + \delta(\boldsymbol{D} - \boldsymbol{A}^{-1}\boldsymbol{B})]\boldsymbol{h}_n^u + \delta\boldsymbol{A}^{-1}\boldsymbol{h}_n^v + \delta^2\boldsymbol{f}_n . \tag{3.13}$$

The order (i.e., the number of equations) of this system is the same as that of the original second-order system (3.1)

3.4. Implementations

Computer implementations of the integration algorithm outlined in Section 3.3 may differ in two respects: the manner in which the auxiliary vector $\boldsymbol{v}_n$ and its time derivative $\dot{\boldsymbol{v}}_n$ are computed in each advancing cycle, and the selection of weighting matrices $\boldsymbol{A}$ and $\boldsymbol{B}$ in the definition (3.2) of $\boldsymbol{v}$.

3.4.1. Computational paths

Three basic computational paths, labeled (0), (1) and (2), may be followed in advancing the discrete solution over a typical time step. The general expression of these paths is flow-charted in References [10, 11, 12].

The path identification index (0, 1 or 2) gives the number of backward-difference operations performed in the determination of $\boldsymbol{u}_n$, $\boldsymbol{v}_n$ and $\dot{\boldsymbol{v}}_n$ in each time step. It is shown in [26] that this index plays an important role in the computational error propagation behavior of corresponding implementations.

Path (0) is consistent with the difference system (3.10); that is, the computed vectors satisfy this system exactly if computational errors are neglected. The original differential equations (3.1) are also satisfied. On the other hand, the computed $\boldsymbol{u}_n$, $\dot{\boldsymbol{u}}_n$, and $\boldsymbol{v}_n$ do not generally satisfy the auxiliary vector definition (3.2), which holds only in the limit of a zero stepsize.

There is a variant of path (0), called (0′), in which the velocity vector $\dot{\boldsymbol{u}}_n$ is recomputed so that (3.2) holds at past time stations. This variant occurs naturally in the conventional choice $\boldsymbol{v} = \dot{\boldsymbol{u}}$ discussed below.

Path (1) enforces both differential expressions (3.1) and (3.2) at the expense of (3.10). Finally, path (2) enforces the auxiliary vector definition (3.2) and the difference system (3.10) at the expense of (3.1).

3.4.2. *Auxiliary vector*

The computational effort per step, and to a lesser extent the storage requirements, can be significantly reduced for certain choices of $\boldsymbol{A}$ and $\boldsymbol{B}$ because some computational steps can be either simplified or bypassed entirely. In [10] the two following selections were studied for linear problems:

$$\boldsymbol{v} = \dot{\boldsymbol{u}} \qquad (\boldsymbol{A} = \boldsymbol{M}^{-1}, \boldsymbol{B} = \boldsymbol{0}), \tag{3.14}$$

$$\boldsymbol{v} = \boldsymbol{M}\dot{\boldsymbol{u}} + \boldsymbol{D}\boldsymbol{u} \qquad (\boldsymbol{A} = \boldsymbol{I}, \boldsymbol{B} = \boldsymbol{D}). \tag{3.15}$$

The choices (3.14) and (3.15) were labeled the conventional (C) and Jensen's (J) formulations, respectively, in [10]. When these two choices are combined with the four computational paths, a total of eight formulations of the advancing step result. The computational sequences associated with six of these are shown in Tables 3.1 and 3.2. A comparative ranking of these formulations for *simultaneous-solution* analysis is given in that reference.

Additional formulations of some interest for *nonlinear* dynamics are obtained if matrices $\boldsymbol{A}$ and $\boldsymbol{B}$ are allowed to be stepsize-dependent. Two possibilities are discussed in Section 3.2 of [11].

3.5. *Operational expressions*

The operational formulation of dynamical systems is useful for concise derivation of general properties. In this subsection we collect some expressions used later in Sections 6 through 8.

In the investigation of stability and accuracy properties of direct time integration, the *indicial* or *generating polynomials* of the integration formulas play

Table 3.1
Computational sequences associated with the conventional (C) formulation $\boldsymbol{v} = \dot{\boldsymbol{u}}$

Form	Step	Calculation
(C0′)	a	$\ddot{\boldsymbol{u}}_{n-1} = \boldsymbol{M}^{-1}(\boldsymbol{f}_{n-1} - \boldsymbol{D}\dot{\boldsymbol{u}}_{n-1} - \boldsymbol{K}\boldsymbol{u}_{n-1})$
	b, b′	$\dot{\boldsymbol{u}}_{n-1} = \boldsymbol{h}^{\dot{u}}_{n-1} + \delta\ddot{\boldsymbol{u}}_{n-1}$
	c	$\boldsymbol{h}^{\dot{u}}_n = \sum_{j=1}^{m}(-\alpha_j\dot{\boldsymbol{u}}_{n-j} + \beta_j\ddot{\boldsymbol{u}}_{n-j})$
	d	$\boldsymbol{h}^{u}_n = \sum_{j=1}^{m}(-\alpha_j\boldsymbol{u}_{n-j} + \beta_j\dot{\boldsymbol{u}}_{n-j})$
	e	$\boldsymbol{g}_n = (\boldsymbol{M} + \delta\boldsymbol{D})\boldsymbol{h}^{u}_n + \delta\boldsymbol{M}\boldsymbol{h}^{\dot{u}}_n + \delta^2\boldsymbol{f}_n$
	f	$\boldsymbol{E} = \boldsymbol{M} + \delta\boldsymbol{D} + \delta^2\boldsymbol{K}$
	g	$\boldsymbol{u}_n = \boldsymbol{E}^{-1}\boldsymbol{g}_n$
	h	$\dot{\boldsymbol{u}}_n = (\boldsymbol{u}_n - \boldsymbol{h}^{u}_n)/\delta$
(C1)	a	$\boldsymbol{M}\ddot{\boldsymbol{u}}_{n-1} = \boldsymbol{f}_{n-1} - \boldsymbol{D}\dot{\boldsymbol{u}}_{n-1} - \boldsymbol{K}\boldsymbol{u}_{n-1}$
	b	$\dot{\boldsymbol{u}}_{n-1} = \dot{\boldsymbol{u}}_{n-1}$ (trivial)
	c	$\boldsymbol{M}\boldsymbol{h}^{\dot{u}} = -\boldsymbol{M}\sum_{j=1}^{m}\alpha_j\dot{\boldsymbol{u}}_{n-j} + \sum_{j=1}^{m}\beta_j(\boldsymbol{M}\ddot{\boldsymbol{u}}_{n-j})$
	d–h	Same as (C0′)
(C2)	a	$\dot{\boldsymbol{u}}_{n-1} = \dot{\boldsymbol{u}}_{n-1}$ (trivial)
	b	$\ddot{\boldsymbol{u}}_{n-1} = (\dot{\boldsymbol{u}}_{n-1} - \boldsymbol{h}^{\dot{u}}_{n-1})/\delta$
	c–g	Same as (C0′)

Table 3.2
Computational sequences associated with the Jensen (J) formulation $\boldsymbol{v} = \boldsymbol{M}\dot{\boldsymbol{u}} + \boldsymbol{D}\boldsymbol{u}$

Form	Step	Calculation
(J0)	a	$\dot{\boldsymbol{v}}_{n-1} = \boldsymbol{f}_{n-1} - \boldsymbol{K}\boldsymbol{u}_{n-1}$
	b	$\boldsymbol{v}_{n-1} = \boldsymbol{h}^{v}_{n-1} + \delta\dot{\boldsymbol{v}}_{n-1}$
	c	$\boldsymbol{h}^{v}_n = \sum_{j=1}^{m}(-\alpha_j\boldsymbol{v}_{n-j} + \beta_j\dot{\boldsymbol{v}}_{n-j})$
	d	$\boldsymbol{h}^{u}_n = \sum_{j=1}^{m}(-\alpha_j\boldsymbol{u}_{n-j} + \beta_j\dot{\boldsymbol{u}}_{n-j})$
	e	$\boldsymbol{g}_n = \boldsymbol{M}\boldsymbol{h}^{u}_n + \delta\boldsymbol{h}^{v}_n + \delta^2\boldsymbol{f}_n$
	f	$\boldsymbol{E} = \boldsymbol{M} + \delta\boldsymbol{D} + \delta^2\boldsymbol{K}$
	g	$\boldsymbol{u}_n = \boldsymbol{E}^{-1}\boldsymbol{g}_n$
	h	$\dot{\boldsymbol{u}}_n = (\boldsymbol{u}_n - \boldsymbol{h}^{u}_n)/\delta$
(J0′)	a–b	Same as (J0)
	b′	$\boldsymbol{M}\dot{\boldsymbol{u}}_{n-1} = \boldsymbol{v}_{n-1} - \boldsymbol{D}\boldsymbol{u}_{n-1}$
	c	Same as (J0)
	d	$\boldsymbol{M}\boldsymbol{h}^{u}_n = -\boldsymbol{M}\sum_{j=1}^{m}\alpha_j\boldsymbol{u}_{n-j} + \sum_{j=1}^{m}\beta_j(\boldsymbol{M}\dot{\boldsymbol{u}}_{n-j})$
	e–h	Same as (J0)
(J2)	a	$\boldsymbol{v}_{n-1} = \boldsymbol{M}\dot{\boldsymbol{u}}_{n-1} + \boldsymbol{D}\boldsymbol{u}_{n-1}$
	b	$\dot{\boldsymbol{v}}_{n-1} = (\boldsymbol{v}_{n-1} - \boldsymbol{h}^{v}_{n-1})/\delta$
	c–h	Same as (J0)

an important role. The polynomials associated with the LMS formulas (3.5) are

$$\rho(\lambda)=\sum_{j=0}^{m}\alpha_j\lambda^{m-j},$$

$$\sigma(\lambda)=\lambda^m+\sum_{j=1}^{m}\hat{\beta}_j\lambda^{m-j},\quad \text{with } \hat{\beta}_j=\beta_j/\beta_0\,. \tag{3.16}$$

Table 3.3 lists the indicial polynomials of three sample integration formulas.

Table 3.3
Indicial polynomials of sample integration formulas

Integration formula	Expressions
Trapezoidal rule ($m=1$)	$\boldsymbol{u}_n=\boldsymbol{u}_{n-1}+0.5h(\dot{\boldsymbol{u}}_n+\dot{\boldsymbol{u}}_{n-1})$ $\rho(\lambda)=\lambda-1$ $\sigma(\lambda)=\lambda+1$ $\delta=0.5h$
Park 3-step ($m=3$)	$\boldsymbol{u}_n=1.5\boldsymbol{u}_{n-1}-0.6\boldsymbol{u}_{n-2}+0.1\boldsymbol{u}_{n-3}+0.6h\dot{\boldsymbol{u}}_n$ $\rho(\lambda)=\lambda^3-1.5\lambda^2+0.6\lambda-0.1$ $\sigma(\lambda)=\lambda^3$ $\delta=0.6h$
Backward Euler ($m=1$)	$\boldsymbol{u}_n=\boldsymbol{u}_{n-1}+h\dot{\boldsymbol{u}}_n$ $\rho(\lambda)=\lambda-1$ $\sigma(\lambda)=\lambda$ $\delta=h$

The argument λ is a complex number called the *amplification factor* on account of its role in stability analysis (Section 6). Two related complex variables often seen in numerical analysis and control-theory literature are the z-transform variable

$$z^j=\lambda^{m-j},\quad j=0,\ldots,m\,, \tag{3.17}$$

and the discrete Laplace-transform variable s

$$s=-\log(z)/h=\log(\lambda)/h,\quad z=\exp(-sh)\,, \tag{3.18}$$

where h is a sampling interval (the integration stepsize in our case). The product sh (often called the complex sampling frequency) plays an important role in the accuracy analysis of Section 7.

Operational counterparts to several of this section's formulas have been compiled in Table 3.4 for ready reference. In this list, expressions paired to the time-continuous equations (3.1) through (3.3) are the conventional Laplace transforms in the time-image variable s (ignoring boundary conditions). On the other hand, for the transforms of time-discrete equations use is made of the polynomials

Table 3.4
Operational expressions

Equation	Operational form
(3.1)	$s^2\boldsymbol{M} + s\boldsymbol{D} + \boldsymbol{K} = \boldsymbol{f}(s)$
(3.2)	$\boldsymbol{v} = (s\boldsymbol{AM} + \boldsymbol{B})\boldsymbol{u}$
(3.3)	$s\boldsymbol{v} = \{\boldsymbol{A}[\boldsymbol{f}(s) - s\boldsymbol{D} - \boldsymbol{K}] + \boldsymbol{B}\}\boldsymbol{u}$
(3.5)	$\rho\boldsymbol{u}_n = \delta\sigma\dot{\boldsymbol{u}}_n, \quad \rho\boldsymbol{v}_n = \delta\sigma\dot{\boldsymbol{v}}_n$
(3.8)	$\boldsymbol{h}^u_n = (\Lambda - \rho)\boldsymbol{u}_n + \delta(\sigma - \Lambda)\dot{\boldsymbol{u}}_n = \begin{cases} [\Lambda - \rho + \delta(\Lambda - \sigma)\boldsymbol{M}^{-1}\boldsymbol{D} \\ \quad + \delta^2(\sigma - \Lambda)\sigma/\rho\boldsymbol{M}^{-1}\boldsymbol{K}]\boldsymbol{u}_n \\ \quad + \delta^2[\sigma - \Lambda]\sigma/\rho\boldsymbol{M}^{-1}\boldsymbol{f}_n \quad \text{for path } (0') \\ (\Lambda - \rho/\sigma)\boldsymbol{u}_n \quad \text{otherwise} \end{cases}$
(3.9)	$\boldsymbol{h}^v_n = (\Lambda - \rho)\boldsymbol{v}_n + \delta(\sigma - \Lambda)\dot{\boldsymbol{v}}_n = \begin{cases} (\boldsymbol{B} - \boldsymbol{AD} + \delta\boldsymbol{K})(\Lambda - \rho/\sigma)\boldsymbol{u}_n \\ \quad + \delta(\Lambda - \rho/\sigma)\boldsymbol{Af}_n \\ \quad \text{for paths } (0') \text{ and } (0) \\ [\rho(\Lambda - \rho)/(\delta\sigma)\boldsymbol{AM} + (\Lambda - \rho/\sigma)\boldsymbol{B} \\ \quad + (\Lambda - \rho)\rho/\sigma\boldsymbol{AD} \\ \quad + \delta(\Lambda - \sigma)\boldsymbol{AK}]\boldsymbol{u}_n + \delta(\Lambda - \sigma)\boldsymbol{Af}_n \\ \quad \text{for path } (1) \\ (\Lambda - \rho/\sigma)[\rho/(\delta\sigma)\boldsymbol{AM} + \boldsymbol{B}]\boldsymbol{u}_n \\ \quad \text{for path } (2) \end{cases}$
(3.14)	$\boldsymbol{g}_n = [a_M\boldsymbol{M} + a_D\boldsymbol{D} + a_K\boldsymbol{K}]\boldsymbol{u}_n + a_f\boldsymbol{f}_n$ where for paths (0′), (0), (1), (2): $a_M = \Lambda - \rho,\ \Lambda - (\rho/\sigma),\ \Lambda - (\rho^2/\sigma),\ \Lambda - (\rho^2/\sigma^2)$ $a_D = \delta(\Lambda - \sigma),\ 0,\ \delta(\Lambda - \rho),\ \delta(\Lambda - \rho/\sigma)$ $a_K = \delta^2[\Lambda - (\sigma^2/\rho)],\ \delta^2(\Lambda - \rho/\sigma),\ \delta^2(\Lambda - \sigma),\ 0$ $a_f = \delta^2\rho/\sigma,\ \delta^2\sigma/\rho,\ \delta^2\sigma,\ \delta^2$

In the above, $\Lambda = \lambda^m$ if ρ and σ are expanded in powers of λ, as in Table 3.3; $\Lambda = 1$ if ρ and σ are expanded in powers of z as defined by (3.17).

(3.16); the arguments of which (λ, z or sh) are left unspecified as a matter of convenience. Path-dependent expressions listed for $\boldsymbol{h}^u_n$, $\boldsymbol{h}^v_n$ and $\boldsymbol{g}_n$ are derived in Section 4 of [12].

4. Partitioned analysis formulations

Two strategies may be followed to formulate partitioned analysis procedures.

Algebraic partitioning. The full coupled system (3.1) is treated by an implicit integration formula as described in the previous section. The resulting time-discrete algebraic system is partitioned in accordance with problem characteristics. Finally, a predictor is applied to the state vector appearing on the right-hand side.

Differential partitioning. Matrices appearing in (3.1) are partitioned in accordance with problem characteristics. The right-hand side state vector is treated with a predictor. Finally, an integration formula is applied to obtain a time-discrete system.

Or, in shorthand:

Algebraic partitioning: integrate–partition–predict,
Differential partitioning: partition–predict–integrate.

It should be noted at the outset that the two strategies do not necessarily lead to the same results, although for certain predictor choices they may be forced to coalesce.

Historically, differential partitioning was the original approach used in the derivation of *staggered solution procedures* (a special implicit-implicit partition) for fluid-structure interaction [24]. More recently, it has been shown [27, 28] that algebraic partitioning provides superior implementation flexibility. On this account, algebraic partitioning will be treated first.

4.1. Algebraic partitioning

Start from the algebraic system (3.11). Partition the $\boldsymbol{K}$ and $\boldsymbol{D}$ matrices as

$$\boldsymbol{D} = \boldsymbol{D}^{\mathrm{I}} + \boldsymbol{D}^{\mathrm{E}}, \qquad \boldsymbol{K} = \boldsymbol{K}^{\mathrm{I}} + \boldsymbol{K}^{\mathrm{E}} \tag{4.1}$$

where superscripts I and E stand for 'implicit part' and 'explicit part', respectively, on account of the interpretation given in Section 5. The induced partition of the coefficient matrix $\boldsymbol{E}$ in the algebraic system (3.11) is

$$\boldsymbol{E} = \boldsymbol{E}^{\mathrm{I}} + \boldsymbol{E}^{\mathrm{E}} \tag{4.2}$$

where

$$\boldsymbol{E}^{\mathrm{I}} = \boldsymbol{M} + \delta\boldsymbol{D}^{\mathrm{I}} + \delta^2\boldsymbol{K}^{\mathrm{I}}, \qquad \boldsymbol{E}^{\mathrm{E}} = \delta\boldsymbol{D}^{\mathrm{E}} + \delta^2\boldsymbol{K}^{\mathrm{E}}. \tag{4.3}$$

Observe that the matrix $\boldsymbol{M}$ is *not* partitioned, which is crucial to the success of these methods.

The term involving $\boldsymbol{E}^{\mathrm{E}}$ is then transferred to the right-hand side, and a predictor is applied:

$$\boldsymbol{E}^{\mathrm{I}}\boldsymbol{u}_n = \boldsymbol{g}_n - \boldsymbol{E}^{\mathrm{E}}\boldsymbol{u}_n^{\mathrm{P}}, \tag{4.4}$$

$$\boldsymbol{g}_n = [\boldsymbol{M} - \delta(\boldsymbol{D} - \boldsymbol{A}^{-1}\boldsymbol{B})]\boldsymbol{h}_n^u + \delta\boldsymbol{A}^{-1}\boldsymbol{h}_n^v + \delta^2\boldsymbol{f}_n \tag{4.5}$$

where $\boldsymbol{u}_n^{\mathrm{P}}$ denotes predicted value extrapolated from past solutions.

To complete the formulation, the block structure of the matrices $\boldsymbol{D}^{\mathrm{I}}$, $\boldsymbol{D}^{\mathrm{E}}$, $\boldsymbol{K}^{\mathrm{I}}$ and $\boldsymbol{K}^{\mathrm{E}}$ has to be specified. This topic is elaborated upon in Section 5.

4.2. Differential partitioning

In this approach the partitions (4.1) of $\boldsymbol{D}$ and $\boldsymbol{K}$ are introduced into the governing differential equation (3.1), and terms involving $\boldsymbol{D}^{\mathrm{E}}$ and $\boldsymbol{K}^{\mathrm{E}}$ transferred to the right-hand side:

$$\boldsymbol{M}\ddot{\boldsymbol{u}} + \boldsymbol{D}^{\mathrm{I}}\dot{\boldsymbol{u}} + \boldsymbol{K}^{\mathrm{I}}\boldsymbol{u} = \boldsymbol{f} - \boldsymbol{D}^{\mathrm{E}}\dot{\boldsymbol{u}} - \boldsymbol{K}^{\mathrm{E}}\boldsymbol{u}\,. \tag{4.6}$$

The next step is to apply a predictor at sample time t_n:

$$\boldsymbol{M}\ddot{\boldsymbol{u}} + \boldsymbol{D}^{\mathrm{I}}\dot{\boldsymbol{u}}_n + \boldsymbol{K}^{\mathrm{I}}\boldsymbol{u}_n = \boldsymbol{f} - \boldsymbol{D}^{\mathrm{E}}\dot{\boldsymbol{u}}_n^{\mathrm{P}} - \boldsymbol{K}^{\mathrm{E}}\boldsymbol{u}_n^{\mathrm{P}}\,. \tag{4.7}$$

Now the right-hand side of (4.7) can be viewed as a pseudo-force vector $\hat{\boldsymbol{f}}_n$:

$$\boldsymbol{M}\ddot{\boldsymbol{u}} + \boldsymbol{D}^{\mathrm{I}}\dot{\boldsymbol{u}}_n + \boldsymbol{K}^{\mathrm{I}}\boldsymbol{u}_n = \hat{\boldsymbol{f}}_n\,. \tag{4.8}$$

Finally, (4.8) is treated with an implicit time integration formula. This step yields the algebraic system

$$\boldsymbol{E}^{\mathrm{I}}\boldsymbol{u}_n = \boldsymbol{g}_n \tag{4.9}$$

where

$$\boldsymbol{E}^{\mathrm{I}} = \boldsymbol{M} + \delta\boldsymbol{D}^{\mathrm{I}} + \delta^2\boldsymbol{K}^{\mathrm{I}}\,, \tag{4.10}$$

$$\boldsymbol{g}_n = [\boldsymbol{M} - \delta(\boldsymbol{D}^{\mathrm{I}} - \boldsymbol{A}^{-1}\boldsymbol{B})]\boldsymbol{h}_n^{u} + \delta\boldsymbol{A}^{-1}\boldsymbol{h}_n^{v} + \delta^2\hat{\boldsymbol{f}}_n\,, \tag{4.11}$$

$$\boldsymbol{v} = \boldsymbol{A}\boldsymbol{M}\dot{\boldsymbol{u}} + \boldsymbol{B}\boldsymbol{u}\,. \tag{3.2}$$

As in the previous case, the formulation is completed by selecting specific partitions.

Remark 1. A distinctive feature of differential partitioning is the need for two predictors, one for $\boldsymbol{u}_n$ and one for $\dot{\boldsymbol{u}}_n$; however, for most application problems only one of these appears.

Remark 2. For certain stability investigations, the intermediate 'predicted pseudo-force' form (4.7) can be viewed as a differential-difference (delayed differential) equation, which has not been yet fully discretized in time. This interpretation is exploited in Section 8.3.

5. Partitions

As mentioned in Section 4, specific partitions result from selecting 'block patterns' for the matrices in (4.1) tailored to the problem at hand (or perhaps to available software). In this section we examine how the structure of these partitions affects the flow of computations, and catalog those so far found most useful in applications.

5.1. Partition-induced computational flow

5.1.1. An illustrative example

To illustrate the effect of matrix structure on the computational process, consider the fully-coupled, linear, three-field system (X, Y, Z) sketched in Figure

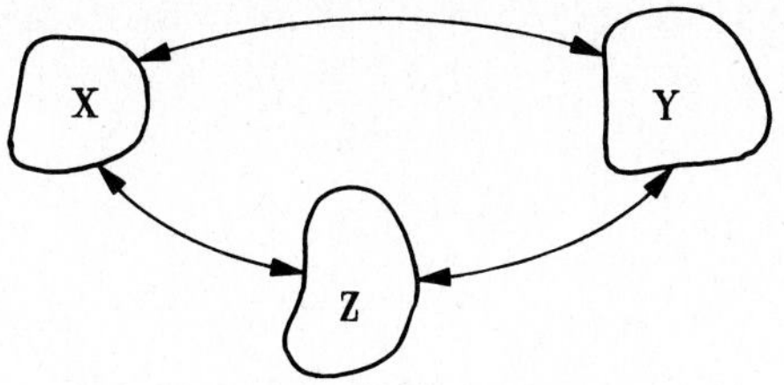

Fig. 5.1. Fully-coupled three-field system.

5.1. The subsystem state vectors are x, y and z. The corresponding block structure for the full-system matrix $\boldsymbol{K}$ is

$$\begin{bmatrix} \boldsymbol{K}_{xx} & \boldsymbol{K}_{xy} & \boldsymbol{K}_{xz} \\ \boldsymbol{K}_{yx} & \boldsymbol{K}_{yy} & \boldsymbol{K}_{yz} \\ \boldsymbol{K}_{zx} & \boldsymbol{K}_{zy} & \boldsymbol{K}_{zz} \end{bmatrix} \tag{5.1}$$

with a similar configuration for matrix $\boldsymbol{D}$. Now consider the partition

$$\begin{aligned} \boldsymbol{K} &= \boldsymbol{K}^{\mathrm{I}} + \boldsymbol{K}^{\mathrm{E}} \\ &= \begin{bmatrix} \boldsymbol{K}_{xx} & \boldsymbol{K}_{xy} & \boldsymbol{K}_{xz} \\ \boldsymbol{0} & \boldsymbol{K}_{yy} & \boldsymbol{K}_{yz} \\ \boldsymbol{0} & \boldsymbol{0} & \boldsymbol{0} \end{bmatrix} + \begin{bmatrix} \boldsymbol{0} & \boldsymbol{0} & \boldsymbol{0} \\ \boldsymbol{K}_{yx} & \boldsymbol{0} & \boldsymbol{0} \\ \boldsymbol{K}_{zx} & \boldsymbol{K}_{zy} & \boldsymbol{K}_{zz} \end{bmatrix} \end{aligned} \tag{5.2}$$

and similarly for matrix $\boldsymbol{D}$. If this partition is inserted in the difference equations (4.4) or (4.9), a natural flow of computations, directed by the presence or absence of zero blocks, emerges. This flow is diagrammed in Figure 5.2, which we now try to explain.

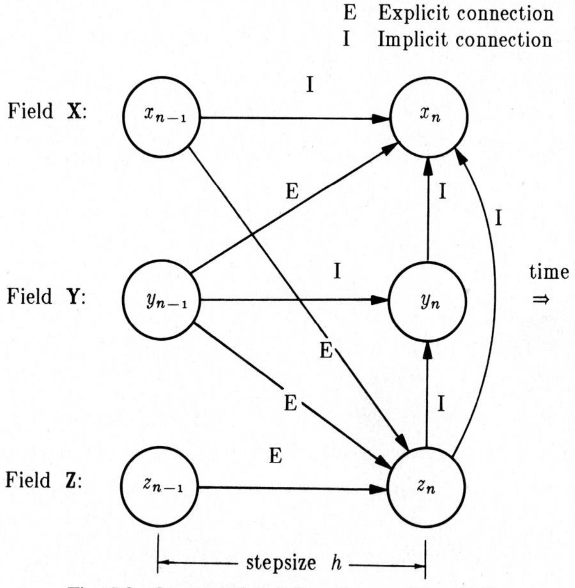

Fig. 5.2. Computational flow for partition (5.2).

All $\boldsymbol{K}$ and $\boldsymbol{D}$ blocks associated with field Z end up in the right-hand side of the difference system. This means that the matrix equation for this field is, for algebraic partitioning,

$$\boldsymbol{M}_z \boldsymbol{z}_n = \boldsymbol{g}_n - \boldsymbol{E}^{\mathrm{E}}_{zx} \boldsymbol{x}^{\mathrm{P}}_n - \boldsymbol{E}^{\mathrm{E}}_{zy} \boldsymbol{y}^{\mathrm{P}}_n - \boldsymbol{E}^{\mathrm{E}}_{zz} \boldsymbol{z}^{\mathrm{P}}_n . \tag{5.3}$$

Now if $\boldsymbol{M}_z$ is diagonal and nonsingular, this equation can be easily solved for z_n. This means that field Z has been treated in a *fully explicit* fashion. One can say that vector z_n is obtained simply by extrapolating from previous solutions.

Conversely, the three first-row matrix blocks have been kept in the left-hand side of (5.2). Thus, field X is treated in a *fully implicit* fashion. After the solution for fields Y and Z has been advanced to the current time, one must solve a system of algebraic equations to obtain $\boldsymbol{x}_n$, but the equation for it *does not* involve a predictor.

How about field Y? This is treated in a mixed fashion. The equation for $\boldsymbol{y}_n$ is implicit in the sense that a system of equations must be processed, but it contains a predictor for $\boldsymbol{x}_n$ on the right-hand side. The presence of this predictor introduces a wisp of explicitness. One can informally say that field Y is treated implicitly in Y and Z, and explicitly in X. To abbreviate, this is often referred to as a *quasi-implicit* treatment.

Note that the sequence in which the fields are processed is dictated by the block structure of the partition (5.2). Fully-explicit fields are processed first, then quasi-implicit fields, finally fully-implicit fields. In this case a sequential process results:

$$\text{Field } Z \rightarrow \text{Field } Y \rightarrow \text{Field } X . \tag{5.4}$$

5.1.2. Terminology and general properties

Some terminology and algebraic properties of partitioned analysis procedures emerge from this example, and are readily generalized to arbitrary partitions.

1. On first cut, fields for which *diagonal blocks* of $\boldsymbol{D}$ and $\boldsymbol{K}$ are kept on the left side (the 'implicit part') will be said to be treated implicitly, and explicitly otherwise.

2. On second cut, implicit fields for which all matrix blocks are kept on the left side will be said to be treated in fully implicit manner, and partly- or quasi-implicitly otherwise.

3. A partition containing at least one explicitly treated field will be called *implicit-explicit* (I–E). If all fields are treated in implicit or quasi-implicit fashion, the partition will be called *implicit-implicit* (I–I).

4. There are *no* explicit-explicit (E–E) partitions within the context of procedures considered in this chapter, because they are trivially equivalent to a fully explicit treatment of the whole system. (E–E partitions deserve attention, however, in multi-stepping or fractional-step schemes [36], in which subsystems are integrated by different stepsizes. These more general schemes are not covered here.)

5. Strictly sequential processing of fields, as in (5.4), is associated with partitions which may be presented in *block triangular form* for a suitable arrangement of field state vectors. Arrangements not reducible to this form permit (at least in principle) *parallel* processing of two or more fields.

6. The example partition (5.2) belongs to the *all-or-nothing* type, in which a matrix block either is retained on the left-hand side, or is moved to the right. There are more general partitions that involve *block splittings*, in which a block contributes nonzero entries to both sides.

Next we examine specific partitions for simple fields couplings. Only the structure of $\boldsymbol{K}^{\mathrm{I}}$ and $\boldsymbol{K}^{\mathrm{E}}$ are shown for brevity.

5.2. *Two-field problems*

The simplest type of coupled system involve two interacting fields (X, Y). These arise naturally in the following contexts:

1. Structure-medium interaction (see examples in Table 2.1).
2. Structures under boundary-state-dependent forces, e.g. aeroelasticity.
3. Volume-volume interactions, as in coupled thermoelasticity.

Even for only two fields, there is a surprisingly large number of all-or-nothing partitions: ten to be exact. All possible $\boldsymbol{K}^{\mathrm{E}}$ configurations are listed below, omitting field subscripts for brevity:

$$\begin{bmatrix} \mathbf{0} & \mathbf{0} \\ \mathbf{0} & \mathbf{0} \end{bmatrix} \begin{bmatrix} \mathbf{0} & \mathbf{0} \\ \mathbf{0} & \boldsymbol{K} \end{bmatrix} \begin{bmatrix} \mathbf{0} & \mathbf{0} \\ \boldsymbol{K} & \mathbf{0} \end{bmatrix} \begin{bmatrix} \mathbf{0} & \mathbf{0} \\ \boldsymbol{K} & \boldsymbol{K} \end{bmatrix} \begin{bmatrix} \mathbf{0} & \boldsymbol{K} \\ \mathbf{0} & \boldsymbol{K} \end{bmatrix}$$
$$\begin{bmatrix} \mathbf{0} & \boldsymbol{K} \\ \boldsymbol{K} & \mathbf{0} \end{bmatrix} \begin{bmatrix} \mathbf{0} & \boldsymbol{K} \\ \boldsymbol{K} & \boldsymbol{K} \end{bmatrix} \begin{bmatrix} \boldsymbol{K} & \mathbf{0} \\ \mathbf{0} & \boldsymbol{K} \end{bmatrix} \begin{bmatrix} \boldsymbol{K} & \mathbf{0} \\ \boldsymbol{K} & \boldsymbol{K} \end{bmatrix} \begin{bmatrix} \boldsymbol{K} & \boldsymbol{K} \\ \boldsymbol{K} & \boldsymbol{K} \end{bmatrix} \tag{5.5}$$

(There are actually $4^2 = 16$ of these, but six correspond to trivial field switchings.)

The first and last of (5.5) correspond to the limit cases of *fully implicit* and *fully explicit* treatment of the problem, respectively. Of the other eight, two (nos. 3 and 6) are implicit-implicit, one (no. 8) is explicit-explicit and thus trivially equivalent to the last one; and the other five are implicit-explicit.

The most practically important is no. 3, the staggered I–I partition, which is studied in detail in Section 8. The other I–I partition (no. 6, sometimes called a block-Jacobi partition) has some applications in implicit dynamic relaxation [12]. The most interesting I–E partition is no. 2, which follows under the purview of the 'DOF-by-DOF' partition examined in Section 5.3.3.

Remark. The number of distinct all-or-nothing partitions actually grows very rapidly: 120 for three fully-coupled fields, as in Figure 5.1; 2520 for four fully-coupled fields.

5.3. *Simplified three-field problem*

As next step up in complexity, consider two fields: X and Y, which interact through a boundary B (Figures 5.3). The boundary is viewed as a separate field.

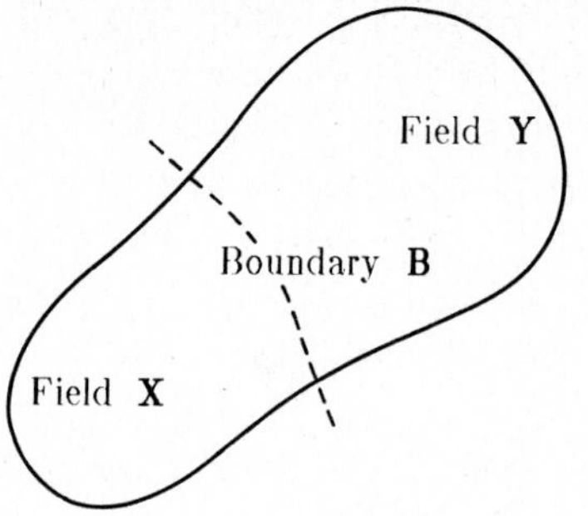

Fig. 5.3. Three-field coupled system consisting of two domain fields (X, Y) separated by boundary B.

Physically this corresponds to the interaction of two domain-type discretizations, e.g. two finite element meshes. For such problems the boundary unknowns may warrant special treatment.

The $\boldsymbol{K}$ matrix for the coupled system has the block structure

$$\begin{bmatrix} \boldsymbol{K}_{xx} & \boldsymbol{K}_{xb} & \boldsymbol{0} \\ \boldsymbol{K}_{bx} & \boldsymbol{K}^{x}_{bb} + \boldsymbol{K}^{y}_{bb} & \boldsymbol{K}_{by} \\ \boldsymbol{0} & \boldsymbol{K}_{by} & \boldsymbol{K}_{yy} \end{bmatrix} \tag{5.6}$$

with a similar structure for $\boldsymbol{D}$.

There are 102 distinct partitions of (5.6), most of which are of little interest. In what follows we examine five practically useful partitions: three are implicit-explicit (I–E) and two are implicit-implicit (I–I).

5.3.1. *Node-by-node I–E partition*

Consider

$$\boldsymbol{K} = \begin{bmatrix} \boldsymbol{K}_{xx} & \boldsymbol{K}_{xb} & \boldsymbol{0} \\ \boldsymbol{K}_{bx} & \boldsymbol{K}_{bb} & \boldsymbol{K}_{by} \\ \boldsymbol{0} & \boldsymbol{0} & \boldsymbol{0} \end{bmatrix} + \begin{bmatrix} \boldsymbol{0} & \boldsymbol{0} & \boldsymbol{0} \\ \boldsymbol{0} & \boldsymbol{0} & \boldsymbol{0} \\ \boldsymbol{0} & \boldsymbol{K}_{by} & \boldsymbol{K}_{yy} \end{bmatrix} \tag{5.7}$$

in which

$$\boldsymbol{K}_{bb} = \boldsymbol{K}^{x}_{bb} + \boldsymbol{K}^{y}_{bb}\,.$$

The state vector for Y is computed explicitly; then the boundary and X-field unknowns are computed by a fully-implicit scheme.

This I–E partition was proposed by Belytschko and Mullen [1, 2], who described it in a more physical context. It is a natural partition for two interacting finite element meshes if *boundary nodes* are identified as part of the implicitly-treated mesh, as illustrated in Figure 5.4. This interpretation motivates its name.

5.3.2. *Element-by-element I–E partition*

The node-by-node partition views boundary unknowns as part of the implicit (X) field; the resulting 'partition anisotropy' is reflected in an unsymmetric $\boldsymbol{K}^{\mathrm{I}}$.

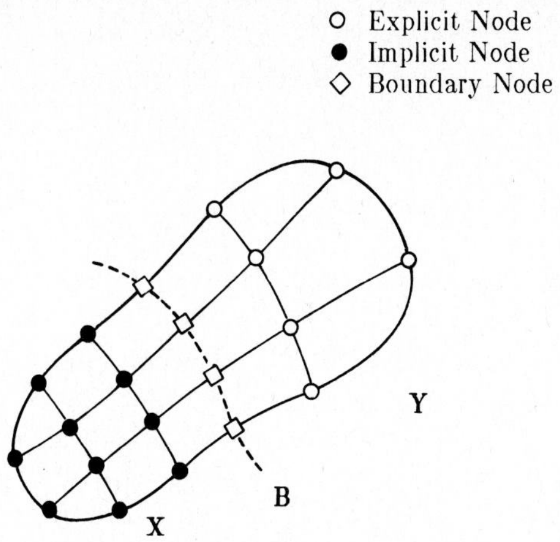

Fig. 5.4. Node-by-node I–E partition.

Now consider the symmetric partition

$$K = \begin{bmatrix} K_{xx} & K_{xb} & 0 \\ K_{bx} & K_{bb}^{x} & 0 \\ 0 & 0 & 0 \end{bmatrix} + \begin{bmatrix} 0 & 0 & 0 \\ 0 & K_{bb}^{y} & K_{by} \\ 0 & K_{by} & K_{yy} \end{bmatrix}. \tag{5.8}$$

Field X and its connection to the boundary is treated implicitly whereas field Y and its connection to the boundary are treated explicitly. Thus, part of the boundary values are obtained explicitly, and the rest implicitly.

This partition is natural for finite element codes in which it is desired to label *elements* as implicit or explicit as proposed by Hughes and Liu [20, 21]. This concept is illustrated in Figure 5.5. The fact that boundary nodes need not be explicitly tagged as such simplifies programming. On the other hand, if the $X-Y$ interface is extensive, this partition entails more computational work than a node-by-node partition.

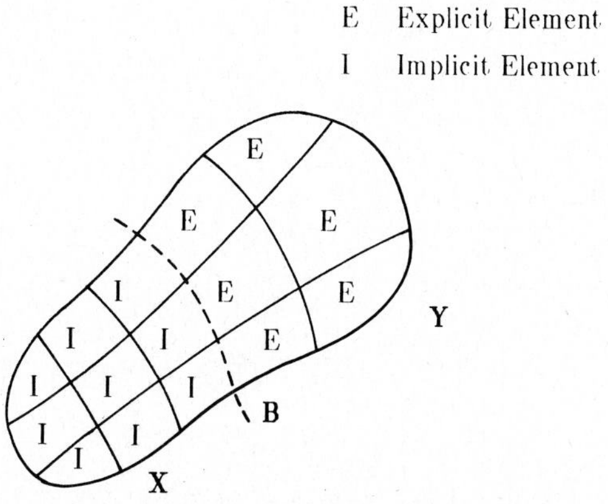

Fig. 5.5. Element-by-element I–E partition.

5.3.3. DOF-by-DOF I–E partition

As final I–E example, consider the symmetric partition

$$\boldsymbol{K} = \begin{bmatrix} \boldsymbol{K}_{xx} & \boldsymbol{K}_{xb} & \boldsymbol{0} \\ \boldsymbol{K}_{bx} & \boldsymbol{K}_{bb} & \boldsymbol{K}_{by} \\ \boldsymbol{0} & \boldsymbol{K}_{by} & \boldsymbol{0} \end{bmatrix} + \begin{bmatrix} \boldsymbol{0} & \boldsymbol{0} & \boldsymbol{0} \\ \boldsymbol{0} & \boldsymbol{0} & \boldsymbol{0} \\ \boldsymbol{0} & \boldsymbol{0} & \boldsymbol{K}_{yy} \end{bmatrix} . \tag{5.9}$$

Here only the Y-field vector is treated in a quasi-explicit fashion; note the difference with regard to (5.7).

The partition (5.9) is useful when different degrees of freedom are to be treated by different algorithms. For example, rotational degrees of freedom associated with beam or shell structures display higher frequencies than translational freedoms. It would be advantageous to treat the former implicitly whereas the latter are treated explicitly.

5.3.4. Staggered I–I partition

Although the staggered partition has been primarily used in two-field coupled problems (cf. Section 8), it can also be formulated for the two-field-plus-boundary problem as

$$\boldsymbol{K} = \begin{bmatrix} \boldsymbol{K}_{xx} & \boldsymbol{K}_{xb} & \boldsymbol{0} \\ \boldsymbol{K}_{bx} & \boldsymbol{K}_{bb} & \boldsymbol{K}_{by} \\ \boldsymbol{0} & \boldsymbol{0} & \boldsymbol{K}_{yy} \end{bmatrix} + \begin{bmatrix} \boldsymbol{0} & \boldsymbol{0} & \boldsymbol{0} \\ \boldsymbol{0} & \boldsymbol{0} & \boldsymbol{0} \\ \boldsymbol{0} & \boldsymbol{K}_{by} & \boldsymbol{0} \end{bmatrix} . \tag{5.10}$$

As all diagonal blocks remain on the left side, the partition (5.10) is implicit-implicit. Only field Y is treated in a quasi-implicit fashion, inasmuch as boundary values have to be predicted for it.

5.3.5. Element-by-element I–I partition

This is an I–I partition designed to overcome certain accuracy problems of the staggered partition (5.10), notably the distortion of rigid body motions discussed in Section 7.3. The partition is structurally similar to (5.8), but both fields are treated implicitly. Two implicit solutions for the boundary unknowns are required at each time step. The time-advancing procedure is briefly outlined in Section 6.5, in connection with the stability analysis, and explained more fully in [27].

5.3.6. Semi-implicit partitions

Another class of solution procedures, known as *semi-implicit*, results if a strict triangular partition is used:

$$\boldsymbol{K} = \begin{bmatrix} \boldsymbol{K}^{\mathrm{L}}_{xx} & \boldsymbol{0} & \boldsymbol{0} \\ \boldsymbol{K}_{bx} & \boldsymbol{K}^{\mathrm{L}}_{bb} & \boldsymbol{0} \\ \boldsymbol{0} & \boldsymbol{K}_{by} & \boldsymbol{K}^{\mathrm{L}}_{yy} \end{bmatrix} + \begin{bmatrix} \boldsymbol{K}^{\mathrm{U}}_{xx} & \boldsymbol{K}_{xb} & \boldsymbol{0} \\ \boldsymbol{0} & \boldsymbol{K}^{\mathrm{U}}_{bb} & \boldsymbol{K}_{by} \\ \boldsymbol{0} & \boldsymbol{0} & \boldsymbol{K}^{\mathrm{U}}_{yy} \end{bmatrix} . \tag{5.11}$$

where L and U superscripts denote lower- and upper-triangular components, respectively, of the diagonal blocks of (5.6). These partitions have been studied by Park and Housner [29].

6. Stability

The numerical stability of the partitioned difference equations (4.4) or (4.9) is governed by four interacting factors:

1. the integration formulas (3.5);
2. the partition (4.1);
3. the predictor formula in (4.4) or (4.7); and
4. the computational path (Section 3.4.1).

The algorithmic properties of the integration formulas (3.5) for fixed stepsize are completely determined by the indicial polynomials $\rho(\lambda)$ and $\sigma(\lambda)$ defined by (3.16). The partition of $\boldsymbol{D}$ and $\boldsymbol{K}$ will be left initially in the general form (4.1). Next we have to study the predictor.

6.1. Predictors

The general expression of a *linear multistep predictor* appropriate for second-order differential systems is

$$\boldsymbol{u}_n^{\mathrm{P}} = \sum_{j=1}^{m} (a_j \boldsymbol{u}_{n-j} + \delta b_j \dot{\boldsymbol{u}}_{n-j} + \delta^2 c_j \ddot{\boldsymbol{u}}_{n-j}) \tag{6.1}$$

where a_j, b_j and c_j are numerical coefficients. The operational version of (6.1) is

$$\begin{aligned} \boldsymbol{u}_n^{\mathrm{P}} &= e_0(\lambda)\boldsymbol{u}_{n-m} + \delta e_1(\lambda)\dot{\boldsymbol{u}}_{n-m} + \delta^2 e_2(\lambda)\ddot{\boldsymbol{u}}_{n-m} \\ &= e(\lambda)\boldsymbol{u}_{n-m} \end{aligned} \tag{6.2}$$

where

$$\begin{aligned} e_0(\lambda) &= \sum_{j=1}^{m} a_j \lambda^{m-j}, \\ e_1(\lambda) &= \sum_{j=1}^{m} b_j \lambda^{m-j}, \\ e_2(\lambda) &= \sum_{j=1}^{m} c_j \lambda^{m-j}. \end{aligned} \tag{6.3}$$

What computed derivatives are to be used in (6.1)? As discussed in Section 3, this depends on the choice of auxiliary vector and on the computational path followed. For example, Table 6.1 lists the expressions for the conventional (C) formulation (3.14).

6.2. Characteristic equations

Start from the difference equation (4.4) of algebraic partitioning. Delete the applied force term, and seek homogeneous solutions of the form

$$\boldsymbol{u}_n = \lambda^m \boldsymbol{u}_{n-m}. \tag{6.4}$$

(Alternatively: take the discrete Laplace transform, and replace $\exp(sh)$ by λ.)

Table 6.1
Path-dependent quantities for stability analysis

Algebraic partition path	Derivative calculations for predictor formula (6.1)	Coefficients in (6.8) c_D c_K
(C0′)	$\ddot{\boldsymbol{u}}_k = \boldsymbol{f}_k - \boldsymbol{D}(\boldsymbol{u}_k - \boldsymbol{h}_k^{u})/\delta - \boldsymbol{K}\boldsymbol{u}_k$ $\dot{\boldsymbol{u}}_k = \boldsymbol{h}_k^{u} + \delta \boldsymbol{u}_k$	ρ
(C0)	$\dot{\boldsymbol{u}}_k = (\boldsymbol{u}_k - \boldsymbol{h}_k^{u})/\delta$ $\ddot{\boldsymbol{u}}_k$ = same as (C0′)	$\rho\sigma/\lambda^m$
(C1)	$\dot{\boldsymbol{u}}_k$ = same as (C0) $\ddot{\boldsymbol{u}}_k$ = not explicitly computed	σ
(C2)	$\dot{\boldsymbol{u}}_k$ = same as (C0) $\ddot{\boldsymbol{u}}_k = (\dot{\boldsymbol{u}}_k - \boldsymbol{h}_k^{\dot{u}})/\delta$	σ^2/λ^m

For differential partitioning $c_D = \rho\sigma/\lambda^m$ and $c_K = \sigma^2/\lambda^m$ regardless of computational path.

Substitute the operational expressions listed in Table 3.1 for the historic vectors, the $\boldsymbol{E}$ matrices by (4.3), and the predicted value by (6.2). The result is the characteristic system

$$\boldsymbol{C}(\lambda) = \boldsymbol{J}(\lambda)\boldsymbol{u}_n = \boldsymbol{0} \tag{6.5}$$

where $\boldsymbol{J}(\lambda)$ is the characteristic matrix

$$\boldsymbol{J}(\lambda) = \boldsymbol{J}^{\mathrm{I}}(\lambda) + [e(\lambda) - \lambda^m]\boldsymbol{J}^{\mathrm{E}}(\lambda), \tag{6.6}$$

$$\boldsymbol{J}^{\mathrm{I}}(\lambda) = \rho^2\boldsymbol{M} + \delta\rho\sigma\boldsymbol{D} + \delta^2\sigma^2\boldsymbol{K}, \tag{6.7}$$

$$\boldsymbol{J}^{\mathrm{E}}(\lambda) = \delta c_D\boldsymbol{D}^{\mathrm{E}} + \delta^2 c_K\boldsymbol{K}^{\mathrm{E}}, \tag{6.8}$$

where c_D and c_K are scalar monomial functions of ρ and σ listed in Table 6.1. Note that these depend on the computational path.

Setting the predictor e to λ^m reduces $\boldsymbol{J}$ to $\boldsymbol{J}^{\mathrm{I}}$. But this is the same as treating the difference equation by a fully implicit method. Thus, $\boldsymbol{J}^{\mathrm{I}}$ is the characteristic matrix for a fully-implicit simultaneous-solution procedure. It is well known that the stability of such procedures is independent of the computational path.

For (6.5) to have a nontrivial solution,

$$\det|\boldsymbol{J}| = 0. \tag{6.9}$$

Equation (6.9) is the characteristic equation. The solutions (6.4) remain bounded if the roots λ_i of (6.9) satisfy the stability condition

$$|\lambda_i| \leq 1. \tag{6.10}$$

Remark. For differential partitioning (Section 4.2) a characteristic matrix of the form (6.6) also results, but it is *independent* of the computational path. For details the interested reader is referred to [12] or [27].

6.3. Stability analysis example

Most comprehensive stability results for second-order partitioned systems have been obtained for *symmetric* system-matrices $\boldsymbol{D}$ and $\boldsymbol{K}$. To illustrate the general technique, select the trapezoidal rule integrator, for which (cf. Table 3.3)

$$\rho(\lambda) = \lambda - 1, \qquad \sigma(\lambda) = \lambda + 1, \qquad \delta = 0.5h, \tag{6.11}$$

computational path (1), and the predictor

$$\boldsymbol{u}_n^{\mathrm{P}} = \boldsymbol{u}_{n-1} + \delta b \dot{\boldsymbol{u}}_{n-1}, \tag{6.12}$$

or

$$e_0 = 1, \qquad e_1 = b, \qquad e_2 = 0, \tag{6.13}$$

where b is a free parameter.

With the foregoing choices, the stability matrix specializes to

$$\begin{aligned} \boldsymbol{J}(\lambda) = {} & (\lambda^2 - 1)^2 \boldsymbol{M} + \delta(\lambda^2 - 1)\boldsymbol{D} + \delta^2(\lambda + 1)^2 \boldsymbol{K} \\ & -[\lambda^2 + b\lambda + (1-b)](\delta \boldsymbol{D}^{\mathrm{E}} + \delta^2 \boldsymbol{K}^{\mathrm{E}}) . \end{aligned} \tag{6.14}$$

It is convenient to transform (6.14) into a Routh polynomial $\boldsymbol{J}(z)$ through the change of variable

$$\lambda = (1+z)/(1-z) \tag{6.15}$$

which maps the stability disk $|\lambda| \leq 1$ onto the half-plane $\mathrm{Re}(z) \leq 0$. The transformed polynomial is

$$\begin{aligned} \boldsymbol{J}(z) = {} & z^2(4\boldsymbol{M} - 2b\delta \boldsymbol{D}^{\mathrm{E}} - 2b\delta^2 \boldsymbol{K}^{\mathrm{E}}) \\ & + 4\delta \boldsymbol{D} + 4\delta^2 \boldsymbol{K} + 2(b-2)[\delta \boldsymbol{D}^{\mathrm{E}} + \delta^2 z \boldsymbol{K}^{\mathrm{E}}] . \end{aligned} \tag{6.16}$$

It is shown in [27] that the optimal choice for b is 2, which corresponds to the predictor

$$\boldsymbol{u}_n^{\mathrm{P}} = \boldsymbol{u}_{n-1} + h\dot{\boldsymbol{u}}_{n-1} . \tag{6.17}$$

With this choice, the last bracketed term in (6.16) vanishes. Call the resultant characteristic matrix $\boldsymbol{J}^*(\lambda)$. It is shown in [27] that the z-roots of $\boldsymbol{J}^*$ have no positive real parts if the 'modified mass matrix'

$$\boldsymbol{M} - \delta \boldsymbol{D}^{\mathrm{E}} - \delta^2 \boldsymbol{K}^{\mathrm{E}} \tag{6.18}$$

has only nonnegative z-roots. This result can be strengthened with further assumptions on $\boldsymbol{D}$ and $\boldsymbol{K}$ for specific partitions.

Implicit-explicit partitions. If $\boldsymbol{D}$ and $\boldsymbol{K}$ are both positive semidefinite, then the resulting I–E procedure is equivalent to integrating the implicit (explicit) partition

Table 6.2
Stable predictors for trapezoidal rule

Algebraic partition computational path	Recommended predictor(s)
(C0′)	$u_n^P = \begin{cases} u_{n-1} \\ u_{n-1} + h\dot{u}_{n-1} + \frac{1}{2}h^2\ddot{u}_{n-1} \end{cases}$
(C1)	$u_n^P = u_{n-1} + h\dot{u}_{n-1}$
(C2)	$u_n^P = u_{n-1} + h\dot{u}_{n-1} + \frac{1}{4}h^2\ddot{u}_{n-1}$
(J0)	$u_n^P = u_{n-1} + \frac{1}{2}h\dot{u}_{n-1} = h_n^u$

Notes: 1. The second predictor listed for (C0′) is recommended only for I–E partitions that do not preserve rigid-body motions, e.g. the DOF-by-DOF partition (Section 7.3). The stability limit is then somewhat reduced [27].

2. In the case of differential partitioning, use the path (2) formula for predicting in u, and the path (0) formula for predicting in $\dot{u}$.

by the trapezoidal rule (central difference) formula. The stability limit then becomes that of the central difference formula applied to the coupled problem.

Implicit-implicit partitions. The resulting staggered solution procedure is unconditionally stable if the symmetric parts of both D^I and K^I are positive semidefinite. The element-by-element I–I partition is not covered by this analysis, however, because it involves additional computational steps; this partition is analyzed in Section 6.5.

6.4. *Computational path compensation*

The previous analysis is valid for the computational sequence (C1). What happens for other paths? If one tries the same predictor, the characteristic equation of algebraic partitioning changes, and stability is generally lost.

But it is possible to adjust the predictor so that the *same characteristic equation results for each path.* This means that if stabilization is achieved for a specific path, it can be stabilized for all others by simply adjusting the predictor. This leads to the notion of *equivalent predictors.* Equivalent predictors recommended for the trapezoidal rule formula are listed in Table 6.2. For general integration rules the reader is referred to Section 6 of [12] or Table 2 of [28].

6.5. *Element-by-element I–I procedure*

The stability analysis of Section 6.3 is applicable to implicit-explicit and staggered implicit-implicit procedures. When the component single-field spatial operators are completely separated (disjoint) and each field has to be treated by implicit integrators on account of computational considerations, then additional computational steps are required beyond solving the partitioned difference equation (4.4). For example, the element-by-element I–I solution for the problem of Figure 5.3 with the partition (5.8) may be effected by

$$E_1^I u_n^{(1)} = g_n - E_1^E u_n^P , \tag{6.19}$$

$$E_2^I u_n^{(2)} = g_n - E_2^E u_n^P , \tag{6.20}$$

where

$$E_{(1,2)}^I = M + \delta D_{(1,2)}^I + \delta^2 K_{(1,2)}^I , \tag{6.21}$$

$$E_{(1,2)}^E = \delta K_{(1,2)}^E + \delta^2 K_{(1,2)}^E , \tag{6.22}$$

$$D = D_1^I + D_2^I, \qquad D_1^I = D_2^E , \tag{6.23}$$

$$K = K_1^I + K_2^I, \qquad K_1^I = K_2^E . \tag{6.24}$$

The characteristic equation for the above I–I procedure can be obtained as before for various computational paths. For example, the (C1) path implementation yields

$$J(\lambda) = J^I(\lambda) + (\lambda^m \sigma - e_0 \sigma - e_1 \rho) L \tag{6.25}$$

where

$$L = \delta^2 D_1^I M^{-1} D_2^I + \delta^3 (K_1^I M^{-1} D_2^I + D_1^I M^{-1} K_2^I) + \delta^4 K_1^I M^{-1} K_2^I \tag{6.26}$$

and $J^I(\lambda)$ is given by (6.7).

For the trapezoidal rule in conjunction with the predictor (6.17), Equation (6.25) reduces to

$$J(\lambda) = (\lambda - 1)^2 (M + L) + \delta(\lambda^2 - 1) D + \delta^2 (\lambda + 1)^2 K \tag{6.27}$$

or, in terms of the Routh polynomial

$$J(z) = 4[(M + L) z^2 + \delta C z + \delta^2 K] . \tag{6.28}$$

It can be shown [27] that $J(z)$ as given by (6.28) is a stable matrix polynomial provided the matrices

$$D_{(1,2)}^I, \qquad K_{(1,2)}^I \tag{6.29}$$

are positive semidefinite and M is nonsingular.

Remark. If solution sequences are *alternated* for implicit-implicit partitions, viz. for the nth step we use

$$E_1^I u_n^{(1)} = g_n - E_1^E u_n^P, \qquad E_2^I u_n = g_n - E_2^E u_n^{(1)} , \tag{6.30}$$

whereas for the following step we switch fields,

$$E_2^I u_{n+1}^{(1)} = g_{n+1} - E_2^E u_{n+1}^P, \qquad E_1^I u_{n+1} = g_{n+1} - E_1^E u_{n+1}^P . \tag{6.31}$$

Then for the (C0′) path the following Routh characteristic matrix polynomial results [27]:

$$J(z) = [4M + 2(L' + L)]z^2 + [4\delta C + 2(L' - L)]z + 4\delta^2 K\,. \tag{6.32}$$

This polynomial is always stable provided the matrices (6.29) are symmetric. Hence, the alternating-solution procedure defined by (6.30) and (6.31) is stable.

6.6. Unsymmetric systems

The stability results of Sections 6.3 and 6.5 rely on matrix symmetry and positiveness attributes. These assumptions cover a wide range of coupled mechanical systems, such as structure-structure interaction.

The case of unsymmetric D and K is not presently well understood, and in fact constitutes one of the open research areas mentioned in Section 10. Of particular concern in the applications are general damping mechanisms, diffusion and convective terms that occur in structure-medium interaction.

For the latter class of problems specialized investigations are presently unavoidable. Although the search for stable solution schemes is simplified by general results (e.g., the concept of equivalent predictors), it nonetheless may require extensive computer-aided numerical experimentation. Insight into causes of instability (physical, integration formula, predictor, partition) can be quite valuable in directing attention to critical components in method design. A specific stabilization process conducted along these guidelines is described in Section 8.

7. Accuracy

In Section 6 we have presented techniques for investigating the stability of the calculations. For a program user's viewpoint, stability means that the solution of an alledgedly physically-stable system can't 'blow up' if stepsize constraints are met. Having been assured of this comforting property, the user may next ponder on the accuracy of the solutions delivered by the computer.

It is fair to say that accuracy aspects of a new computational technique can be a fuzzy, even mystic, subject. A fair assessment of method performance in this regard is frequently clouded by flimsy evidence, because of lack of agreed-upon performance measures. The attending confusion may lead to claims and counterclaims, and to the rediscovery of old methods (central difference, trapezoidal rule, . . .) coated in a new plumage sometimes sprinkled with Greek letter qualifiers.

In this section we try to establish some common grounds for assessing the accuracy of competing stable partitions. In applying numerical techniques to mechanical systems, engineers have replaced, with good reason, the mathematical truncation-error measure by the more physically meaningful amplitude and phase errors, and the error in tracing rigid-body motions as an asymptotic accuracy measure (for which the frequency approaches zero while the stepsize remains finite).

Following these antecedents, we focus our attention on three aspects: partition-induced amplitude and phase errors, interwoven effect of predictor and computational path on numerical damping, and preservation of rigid-body motions under partitioning. We hope that these aspects provide sufficient information for selecting a satisfactory scheme from the combinatorial morass of partitions, integrators, predictors and computational paths.

7.1. Effects of partitioning on accuracy

The conventional Fourier technique for assessing accuracy of integration formulas for second-order systems proceeds as follows (see, e.g. [25]). First, the homogeneous undamped equations are uncoupled by projecting them on normal (modal) coordinates. Second, the *frequency distortion* and *numerical damping* of an uncoupled difference equation driven by steady harmonic input of frequency ω are presented as functions of the normalized sampling frequency ωh. This procedure is not immediately applicable to partitioned difference equations, however, because these are not generally diagonalized by the normal modes of the differential equations (3.1).

We shall instead use a *limit differential equation* approach to assess the frequency-dependent accuracy of various partition schemes, assuming that the solutions can be expanded in Taylor series up to the order of the integration formula.

The limit-differential-equation approach has been extensively used in finite difference hydrodynamic calculations for the evaluation of artificial viscosity effects [30, 35]. But it has not received the attention it deserves in finite-element structural dynamics. Because of this neglect, we outline first the basic concepts for a non-partitioned treatment of the equations of motion.

7.1.1. Limit DE for non-partitioned system

Consider the linear homogeneous, undamped governing equations

$$\boldsymbol{M}\ddot{\boldsymbol{u}} + \boldsymbol{K}\boldsymbol{u} = \boldsymbol{0}\,. \tag{7.1}$$

Treat (7.1) by the LMS integration formulas (3.5), transform the difference equations using the indicial polynomials (3.16), and divide through by $(\delta\sigma)^2$. The result is

$$(\phi^2\boldsymbol{M} + \boldsymbol{K})\boldsymbol{u}_n = \boldsymbol{0} \tag{7.2}$$

in which

$$\phi = \rho/(\delta\sigma)\,. \tag{7.3}$$

Here the indicial polynomials ρ and σ are to be viewed as functions of the complex sampling frequency $sh = -\log(\lambda)$ rather than of λ, because we are interested in the behavior as $sh \to 0$. Expansions of ϕ in sh for three well-known integration formulas are listed in Table 7.1. (Note that ϕ and ϕ^2 may be interpreted as discrete approximations to the differentiation operators $\partial/\partial t$ and $\partial^2/\partial t^2$, respectively; and the series in sh yields all there is to know about their truncation errors.)

Table 7.1
sh expansions of (7.3) for some integration formulas

Integration formula	Expansions
Trapezoidal rule	$\rho = 1 - e^{-sh}, \quad \sigma = 1 + e^{-sh}, \quad \delta = 0.5h$ $\phi = \rho/(\delta\sigma) = s(1 - \frac{1}{12}s^2h^2 + \frac{1}{240}s^4h^4 - \cdots)$ $\phi^2 = s^2(1 - \frac{1}{6}s^2h^2 + \frac{1}{72}s^4h^4 - \cdots)$
Park 3-step	$\sigma = 1 - 1.5\,e^{-sh} + 0.6\,e^{-2sh} - 0.1\,e^{-3sh}$ $\sigma = 1, \quad \delta = 0.6h$ $\phi = \rho/(\delta\sigma) = s(1 - \frac{1}{6}s^2h^2 + \frac{11}{200}s^4h^4 - \cdots)$ $\phi^2 = s^2(1 - \frac{1}{3}s^2h^2 + \frac{7}{120}s^4h^4 - \cdots)$
Backward Euler	$\rho = 1 - e^{-sh}, \quad \sigma = 1, \quad \delta = h$ $\phi = \rho/(\delta\sigma) = s(1 - sh + \frac{1}{2}s^2h^2 + \frac{1}{6}s^3h^3 + \cdots)$ $\phi^2 = s^2(1 - 2sh + 2s^2h^2 - s^3h^3 + \cdots)$

Take now the trapezoidal rule expansion, insert in (7.2), and inverse-transform back to the time domain:

$$\boldsymbol{M}\ddot{\boldsymbol{u}} - \tfrac{1}{6}h^2\boldsymbol{M}\ddot{\ddot{\boldsymbol{u}}} + \tfrac{1}{72}h^4\boldsymbol{M}\ddot{\ddot{\ddot{\boldsymbol{u}}}} - \cdots + \boldsymbol{K}\boldsymbol{u} = \boldsymbol{0}\,. \tag{7.4}$$

Finally, reduce the order of high $\boldsymbol{u}$ derivatives by two from the differentiated (7.1):

$$[\boldsymbol{M} + \tfrac{1}{6}h^2\boldsymbol{K}]\ddot{\boldsymbol{u}} - \tfrac{1}{72}h^4\boldsymbol{K}\ddot{\ddot{\boldsymbol{u}}} + \cdots + \boldsymbol{K}\boldsymbol{u} = \boldsymbol{0}\,. \tag{7.5}$$

Equation (7.5) is the limit DE associated with the trapezoidal rule. It tells us that the primary effect of the trapezoidal rule integrator is a frequency distortion, and that such distortion is a relative elongation of order $\omega^2h^2/6$ for $\omega h \ll 1$. A similar technique can be followed for other integrators.

The key idea is to obtain a *perturbed differential equation* whose exact solution would, asymptotically for small stepsize h, reproduce the computed values at sample times. In numerical analysis, this approach is known as *backward error analysis*: find which changes in the original data would give rise to observed or predicted computational errors.

What are the advantages of this approach? First, no assumptions as to modal decomposition are needed. Second, perturbations of the original problem are often of immediate physical significance. For example, in (7.5) the main perturbations appear as modifications in mass coefficients; which an engineer can readily grasp and directly correlate, for a given h, with expected uncertainties in the physical data.

7.1.2. Limit DE for partitioned system

The counterpart of (7.2) for an optimally stabilized partitioned treatment of (7.1) is (cf. Section 6)

$$[\phi^2(\boldsymbol{M} - \delta^2\boldsymbol{K}^{\mathrm{E}}) + \boldsymbol{K}]\boldsymbol{u}_n = \boldsymbol{0}\,. \tag{7.6}$$

Inserting the expansion for the trapezoidal rule and proceeding as before yields the limit DE

$$[\boldsymbol{M} + \tfrac{1}{12}h^2(2\boldsymbol{K}^{\mathrm{I}} - \boldsymbol{K}^{\mathrm{E}})]\ddot{\boldsymbol{u}} + \cdots + \boldsymbol{K}\boldsymbol{u} = \boldsymbol{0}\,, \tag{7.7}$$

or

$$[\boldsymbol{M} + \tfrac{1}{6}h^2\boldsymbol{K}^{\mathrm{I}} - \tfrac{1}{12}h^2\boldsymbol{K}^{\mathrm{E}}]\ddot{\boldsymbol{u}} + \cdots + \boldsymbol{K}\boldsymbol{u} = \boldsymbol{0}\,, \tag{7.8}$$

which of course reduces to (7.5) if $\boldsymbol{K}^{\mathrm{E}}$ vanishes. Note that the main effect of partitioning is concentrated on the mass matrix, and that no perturbation terms associated with the velocity appear. We therefore conclude that the primary algorithmic error caused by partitioning is manifested in *frequency* (*phase*) *distortion.* This distortion may be calculated from the frequency equation

$$\det|[\boldsymbol{M} + \tfrac{1}{6}h^2\boldsymbol{K}^{\mathrm{I}} - \tfrac{1}{12}h^2\boldsymbol{K}^{\mathrm{E}}]\Omega^2 + \cdots + \boldsymbol{K}| = 0 \tag{7.9}$$

where

$$\Omega = \omega h = -\mathrm{j}sh\,. \tag{7.10}$$

The two limit cases for (7.9) are: $\boldsymbol{K}^{\mathrm{E}} = \boldsymbol{0}$ (fully implicit integration) and $\boldsymbol{K}^{\mathrm{I}} = \boldsymbol{0}$ (fully explicit integration). These correspond in turn to the standard trapezoidal rule and central difference integration formulas, respectively. It is well known that these two formulas introduce no numerical damping. Moreover, the central difference formula shortens the period whereas the trapezoidal rule elongates it by an amount asymptotically twice as big; a feature that can be immediately deduced from (7.9).

That frequency distortion is also the primary algorithmic error due to partitioning can also be shown to hold for other, numerically damped, integration formulas. The demonstration relies again on the limit DE approach: first non-vanishing odd-derivative perturbation terms that occur are of higher order than the integration formula truncation error.

7.1.3. A model system

As noted previously, the frequency equation (7.9) is not generally diagonalizable by the natural modes of the original differential equations. To assess the accuracy performance of specific partitions we can exploit, however, the fact that the frequency distortion due to partitioning emanates from partition boundaries. This enables us to construct a two-degree-of-freedom model system and to assess the partitioning-induced error from it. The governing equations of this system are

$$\begin{bmatrix} 1+\alpha & 0 \\ 0 & 1 \end{bmatrix} \ddot{\boldsymbol{u}} + \begin{bmatrix} 1+(1/\alpha) & -1 \\ -1 & 1 \end{bmatrix} \boldsymbol{u} = \boldsymbol{0}\,. \tag{7.11}$$

Equations (7.11) can be viewed as the normalized axial equations of motion for a fixed-free prismatic bar consisting of two elements whose length ratio is $1:\alpha$ (see Figure 7.1).

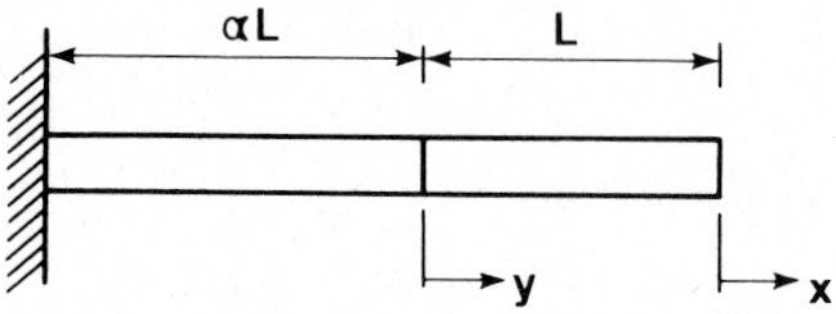

Fig. 7.1. Two-bar-element model system.

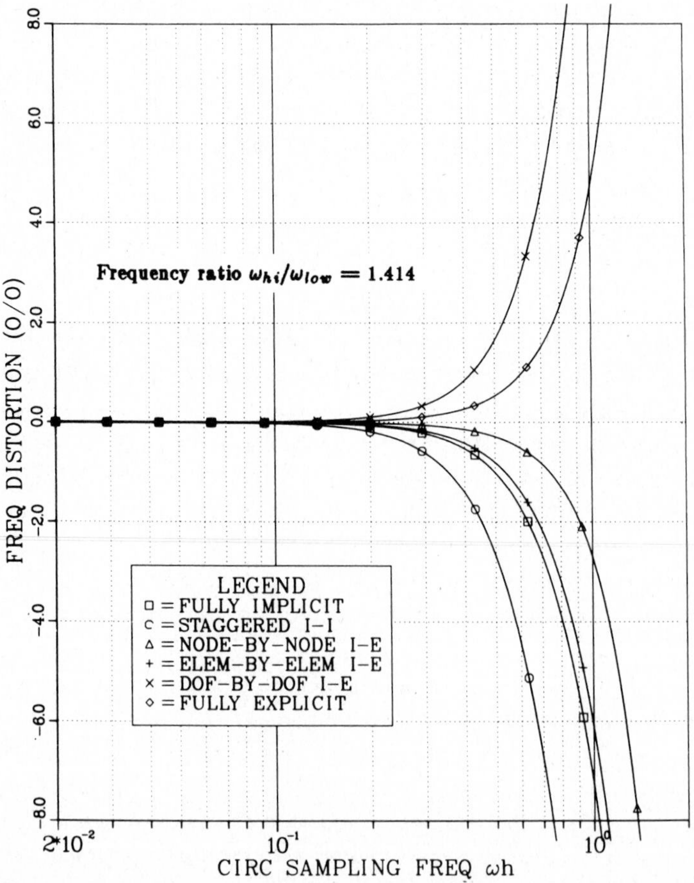

Fig. 7.2. Distortion of low-frequency characteristic root model system as function of sampling frequency, for $\alpha = 1.0$, trapezoidal integration rule with last-solution predictor, and specific partitions.

For the above system the partition-caused frequency distortion can be asymptotically evaluated from (7.9). Alternatively, the distortion can be exactly computed from the characteristic equation (7.7). As the exact calculations are not restricted to $\omega h \ll 1$, we present results obtained by the second approach.

Figures 7.2 and 7.3 show the frequency distortion of the low and high frequency components for the equal length case $\alpha = 1$. For this case the node-by-node implicit-explicit partition is most accurate, followed by the element-by-element, the DOF-by-DOF, and the staggered partitions. As the element length ratio (roughly the square of the frequency ratio) is decreased, however, an intermediate value of α is reached at which the element-by-element partition becomes more accurate than the node-by-node partition, as illustrated by Figure 7.4. If the length ratio is further decreased, the low (high) frequency error

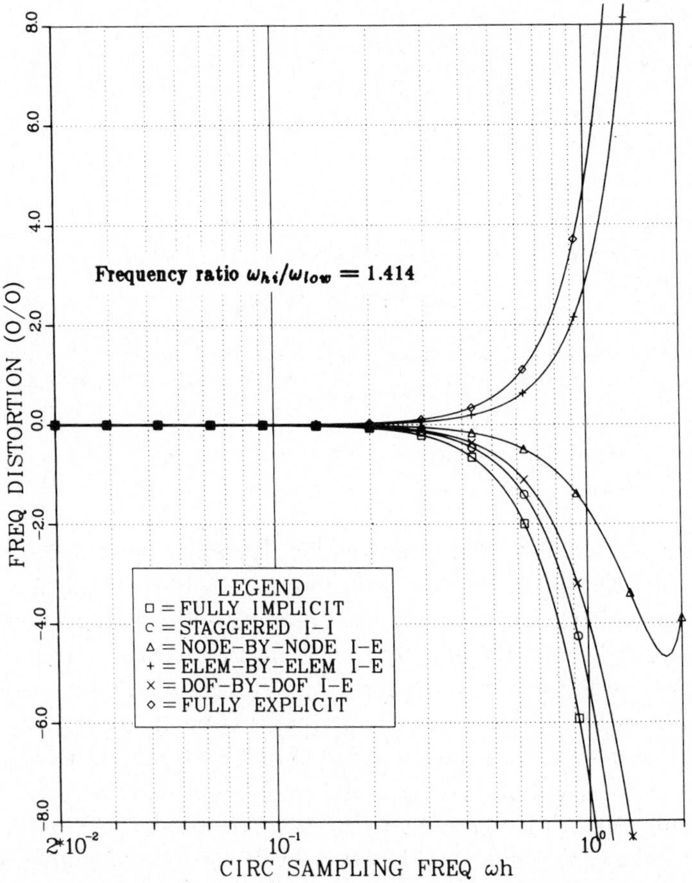

Fig. 7.3. Distortion of high-frequency characteristic root model system as function of sampling frequency, for $\alpha = 1.0$, trapezoidal integration rule with last-solution predictor, and specific partitions.

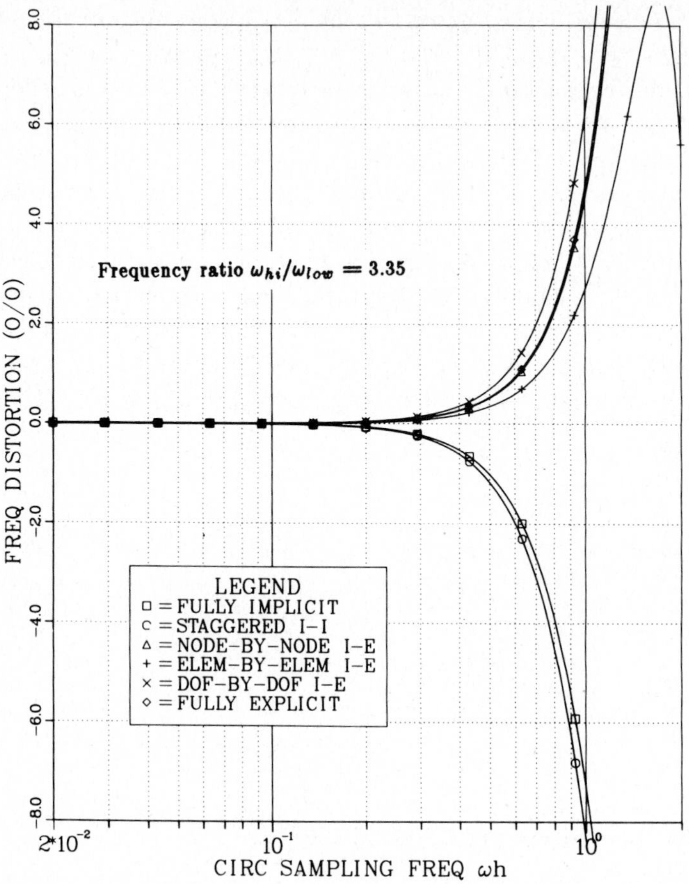

Fig. 7.4. Distortion of low-frequency characteristic root model system as function of sampling frequency, for $\alpha = 0.1$, trapezoidal integration rule with last-solution predictor, and specific partitions.

approaches that of the fully explicit (implicit) formula, as can be observed in Figure 7.5. (Increasing the frequency ratio weakens modal coupling.)

The foregoing analysis shows that accuracy of solution components far from partition boundaries is primarily controlled by the accuracy of the integration formulas used. For solution components at or near partition boundaries, the element-by-element and node-by-node implicit-explicit partitions enjoy a somewhat higher accuracy than either the explicit or implicit formula by itself, in the sense that frequency distortions are less than those associated with the fully explicit or fully implicit integration formulas. On the other hand, for the DOF-by-DOF and the staggered partitions the accuracy of partition-boundary components is somewhat inferior to that provided by fully explicit or implicit formulas.

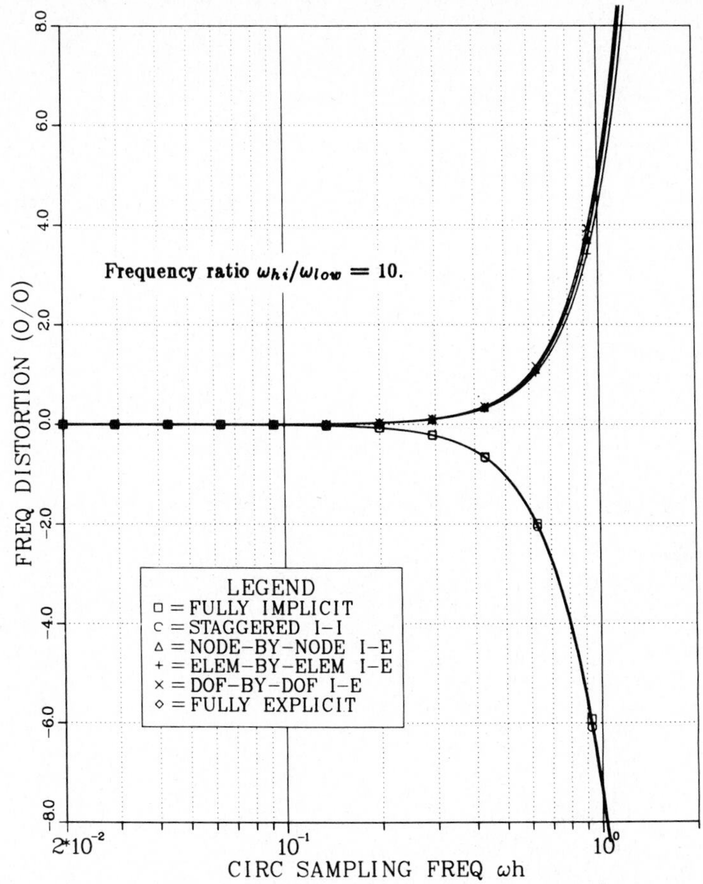

Fig. 7.5. Distortion of low-frequency characteristic root model system as function of sampling frequency, for $\alpha = 0.02$, trapezoidal integration rule with last-solution predictor, and specific partitions.

7.2. *Effect of predictor and computational path on accuracy*

It was shown in Section 6 that different combinations of predictors and computational paths will generally give rise to different characteristic equations, even if all other factors (governing equations, integration formulas, partition) remain the same. This has major implications not only on stability, but also in accuracy (see Table 6.2).

To further illustrate this point, consider again the model system (7.11). Figure 7.6 displays numerical damping ζ versus sampling frequency ωh for the characteristic roots of (7.11) with $\alpha = 1$, when treated by the trapezoidal rule implemented in the four computational paths: (0′), (0), (1), (2), using the last solution

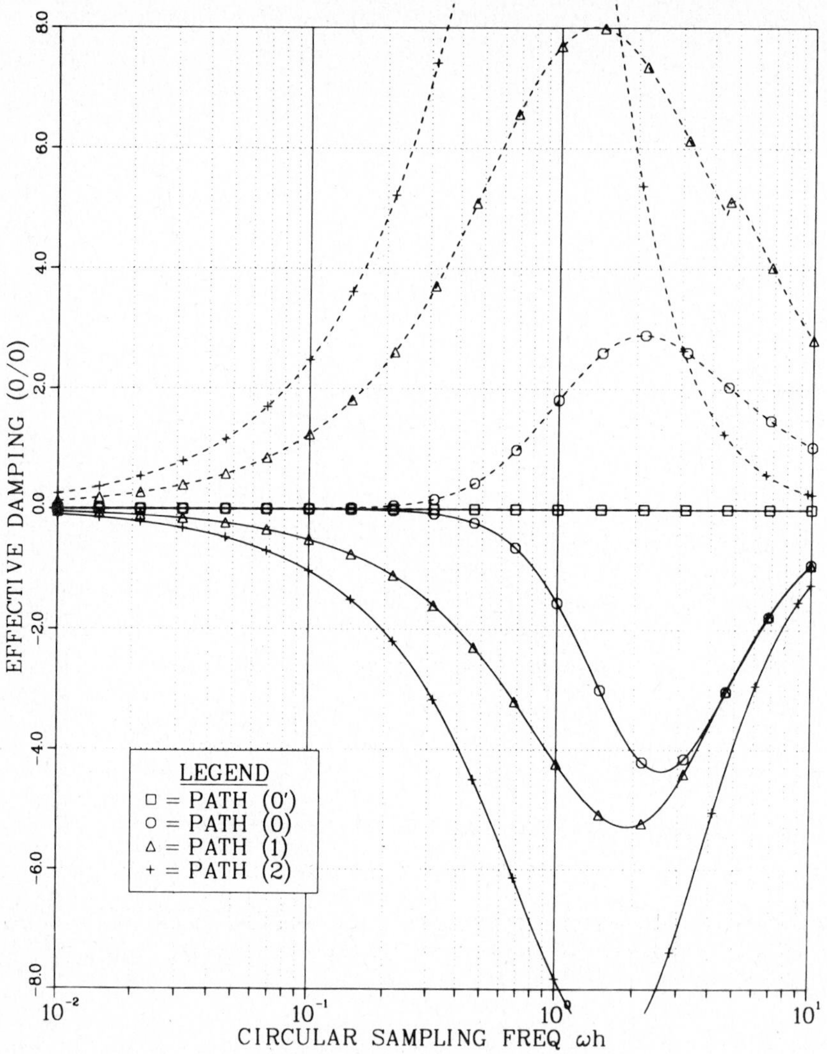

Fig. 7.6. Numeric damping associated with essential characteristic roots for model system (7.9), trapezoidal rule integration with last-solution predictor, four computational paths, and varying sampling rates (solid and dashed curves pertain to high- and low-frequency roots, respectively).

predictor

$$\boldsymbol{u}_n^{\mathrm{P}} = \boldsymbol{u}_{n-1} \tag{7.12}$$

and the staggered partition. [The numerical damping coefficient ζ associated with a characteristic root λ_i is measured as

$$\zeta_i = -\log|\lambda_i|/(\omega_i h) \tag{7.13}$$

where $|\lambda|$ denotes root modulus.]

Note that only path (0′) in conjunction with (7.12) preserves the no-damping algorithmic characteristic of the trapezoidal rule. This verifies the theorems of [27]. For the other three paths the use of (7.12) introduces positive numerical damping on the high-frequency root, and negative damping on the low-frequency one.

This example illustrates that a harmonious assembly of integration formula, predictor and computational path not only stabilizes the solution procedure, but renders it more accurate. There are exceptions to this blanket statement, however, in the treatment of rigid-body motions, which is studied next.

7.3. Preservation of rigid body motions

As final tool for accuracy assessment, consider a mechanical system whose response is dominated by *rigid-body motions*, for example a free-falling or orbiting structure. Assume that the state vector $\boldsymbol{u}$ is a linear combination of such motions, and ignore damping. Then (3.1) collapse to the equations of rigid-body dynamics:

$$\boldsymbol{M}\ddot{\boldsymbol{u}} = \boldsymbol{f}. \tag{7.14}$$

Treating (7.14) as one entity with an integration scheme of second-order accuracy (for example: trapezoidal rule, Gear 2-step, Park 3-step, central difference), yields the exact response if the force vector $\boldsymbol{f}$ does not vary in time. This *constant acceleration case* is experienced, for instance, by a free-falling structure to a good first approximation. Inasmuch as two is the maximum order of accuracy ordinarily used for structural dynamics, there is no point in considering more general time-varying motions.

What happens if a partition is introduced? This question cannot be answered directly with the limit-DE approach utilized in Section 7.1, because we are dealing with zero-frequency motions. Consequently $sh = \mathrm{j}\omega h$ vanishes for all stepsizes, and arguments based on Taylor-series expansions (as in Table 7.1) become vacous. It is necessary to treat this case separately.

The partitioned difference equations (4.4) or (4.8) with $\boldsymbol{D} = \boldsymbol{0}$ and the conventional choice (3.14) read

$$(\boldsymbol{M} + \delta^2\boldsymbol{K}^{\mathrm{I}})\boldsymbol{u}_n = \boldsymbol{M}(\boldsymbol{h}_n^{u} + \delta\boldsymbol{h}_n^{\dot{u}}) + \delta^2\boldsymbol{f}_n - \delta^2\boldsymbol{K}^{\mathrm{E}}\boldsymbol{u}_n^{\mathrm{P}}. \tag{7.15}$$

Since rigid body motions satisfy $\boldsymbol{K}\boldsymbol{u} = \boldsymbol{0}$, then

$$\boldsymbol{K}^{\mathrm{I}}\boldsymbol{u} = -\boldsymbol{K}^{\mathrm{E}}\boldsymbol{u}. \tag{7.16}$$

Considering now only constant-acceleration motions, from (7.14) it follows that $\boldsymbol{f}_n$ in (7.15) may be replaced by $\boldsymbol{M}\ddot{\boldsymbol{u}}_k$ for any k, in particular $k = n - 1$. Taking also (7.16) into account, (7.15) may be rewritten as

$$(\boldsymbol{M} - \delta^2\boldsymbol{K}^{\mathrm{E}})\boldsymbol{u}_n = \boldsymbol{M}(\boldsymbol{h}_n^{u} + \delta\boldsymbol{h}_n^{\dot{u}} + \delta^2\ddot{\boldsymbol{u}}_{n-1}) - \delta^2\boldsymbol{K}^{\mathrm{E}}\boldsymbol{u}_n^{\mathrm{P}}. \tag{7.17}$$

The corresponding difference equation for the non-partitioned case $\boldsymbol{K}^{\mathrm{E}} = \boldsymbol{0}$ is

$$\boldsymbol{M}\boldsymbol{u}_n = \boldsymbol{M}(\boldsymbol{h}_n^{u} + \delta\boldsymbol{h}_n^{\dot{u}} + \delta^2\ddot{\boldsymbol{u}}_{n-1}). \tag{7.18}$$

Comparing (7.18) with (7.17) it follows that exactness of rigid-body representation demands that either

$$\boldsymbol{K}^{\mathrm{E}}\boldsymbol{u} \equiv \boldsymbol{0} \tag{7.19}$$

identically, or that the following predictor be used:

$$\boldsymbol{u}_n^{\mathrm{P}} = \boldsymbol{h}_n^u + \delta \boldsymbol{h}_n^{\dot{u}} + \delta^2 \ddot{\boldsymbol{u}}_{n-1} \, . \tag{7.20}$$

For the trapezoidal rule (cf. Table 7.1), this predictor is

$$\boldsymbol{u}_n^{\mathrm{P}} = \boldsymbol{u}_{n-1} + h\dot{\boldsymbol{u}}_{n-1} + \tfrac{1}{2}h^2 \ddot{\boldsymbol{u}}_{n-1} \, . \tag{7.21}$$

Remark 1. If one insists only on preservation of *constant velocity* rigid-body motions, for which $\ddot{\boldsymbol{u}} \equiv \boldsymbol{0}$, any of the following predictors can be used:

$$\boldsymbol{u}_n^{\mathrm{P}} = \boldsymbol{h}_n^u + \delta \boldsymbol{h}_n^{\dot{u}} + \theta \delta^2 \ddot{\boldsymbol{u}}_{n-1} \tag{7.22}$$

where θ is a free parameter adjustable to meet stability constraints. For example, two predictors of this form for the trapezoidal rule are

$$\boldsymbol{u}_n^{\mathrm{P}} = \boldsymbol{u}_{n-1} + h\dot{\boldsymbol{u}}_{n-1} \, , \tag{7.23}$$

$$\boldsymbol{u}_n^{\mathrm{P}} = \boldsymbol{u}_{n-1} + h\dot{\boldsymbol{u}}_{n-1} + \tfrac{1}{4}h^2 \ddot{\boldsymbol{u}}_{n-1} \, . \tag{7.24}$$

Remark 2. Both the element-by-element and node-by-node implicit-explicit partitions preserve any rigid-body motions because they satisfy the identity (7.19). For these two partitions, the last-solution predictor (7.12) is optimally stable in conjunction with the trapezoidal rule and path (0′), while preserving rigid-body motions.

Remark 3. The last-solution predictor (7.12), although stable in conjunction with the trapezoidal rule and path (0′), distorts constant-velocity and constant-acceleration rigid-body motions for the DOF-by-DOF implicit-explicit and the staggered implicit-implicit partitions treated with the trapezoidal rule. This effect is illustrated in Figure 7.7 on a two-degree-of-freedom free-free bar.

The distortion can be eliminated by iterating each time step to convergence. Alternatively, predictor (7.23) or (7.24) in conjunction with path (1) or (2), respectively, could be used to preserve constant-velocity rigid-body motions.

Remark 4. The accurate predictor (7.21) gives rise to numerical instability in general. Thus stability and accuracy *are not* generally equivalent attributes for partitioned integrations, within the context of rigid-body motion preservation.

The main findings of this study are summarized in Table 7.2. Clearly path (0′) is the choice when accurate tracing of rigid-body motions is important, and a DOF-by-DOF or staggered partition is used. The price paid for this im-

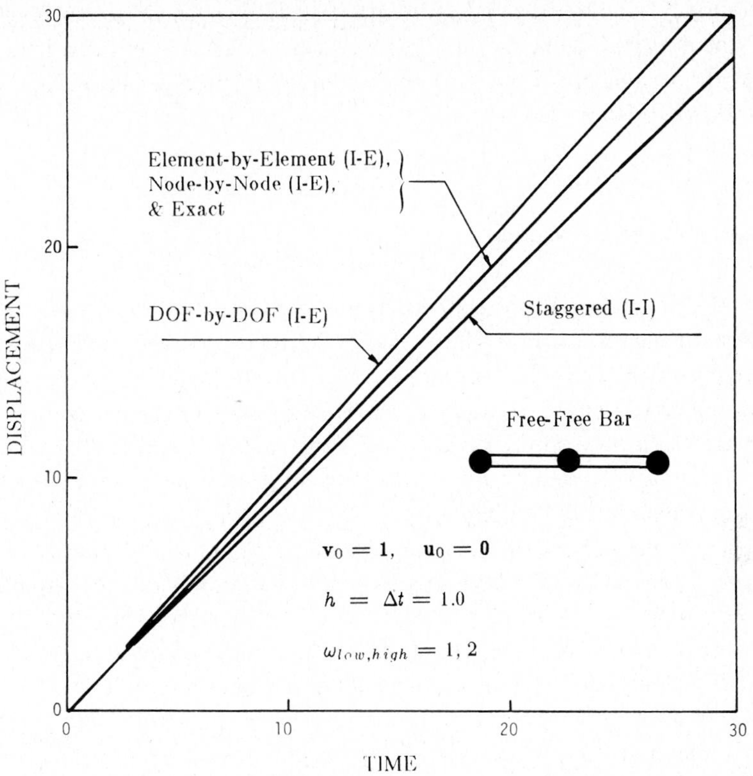

Fig. 7.7. Distortion of constant-velocity rigid-body response of free-free bar for the DOF-by-DOF and staggered partitions.

Table 7.2
Test on rigid-body motion preservation (trapezoidal rule integration)

Procedure	Computational paths			
	(C0′)	(C1)	(C2)	(J0)
Node-by-node implicit-explicit	Any	Any	Any	Any
Element-by-element implicit-explicit	Any	Any	Anv	Anv
DOF-by-DOF implicit-explicit	A	V	V	D
Staggered implicit-implicit	A	V	V	D
Element-by-element implicit-implicit	A	A	A	A

Any = any rigid-body motion; A = preserves constant acceleration (free fall); V = preserves constant velocity; D = distorts all but constant displacement.
For all computational paths, only stable predictors are used.

plementation is the need for additional calculations, such as solving for accelerations, in each advancing step [10]. If the accuracy requirement is relaxed to preservation of constant-velocity rigid-body motions, then either path (1) or (2) is acceptable for those partitions.

8. A case study in stabilization

A key application of partitioned analysis procedures has been to fluid-structure interaction problems, as noted in Section 2. More precisely: the analysis of a three-dimensional structure surrounded by an infinite acoustic medium. The response of this system to a shock wave propagating in the fluid is to be determined. Such a response exhibits early-time, high-frequency-dominated transients that gradually settle down (because of fluid radiation damping) into late-time, low-frequency oscillations.

Efficient tracing of the late-time response by large time steps demands *implicit integration* of the structural equations. A preliminary study of the coupled system revealed a wide overlapping of response characteristics. Thus the use of an implicit-implicit partition appeared attractive, and the staggered solution procedure was in fact developed [24] to meet these computational requirements.

The stability theory presented in Section 6 provides sufficient grounds for stabilizing the partitioned analysis of undamped and Rayleigh-damped coupled mechanical systems. But in the fluid-structure problem (as well as for structure-medium interaction in general) we are faced with more general damping mechanisms. For these problems stabilization of the time integration process may demand a *reformulation* of the original equations. The central purpose of this sample case study is to show that the conventional formulation of coupled field equations is not necessarily the best from a numerical stability viewpoint for constructing implicit-implicit partitions. Investigators faced with similar classes of problems are advised to keep this important fact in mind.

8.1. The model problem

It is shown in [24] that stabilization of the fluid-structure interaction problem can be conveniently studied on the following two-degree-of-freedom homogeneous model problem:

$$m_s\ddot{d} + k_s d = -a\dot{q}, \qquad m_f\dot{q} + \rho c a_f q = \rho c m_f \dot{d}, \tag{8.1}$$

where m_s, k_s, a_f and m_f are generalized quantities resulting from the projection of the structural mass $\boldsymbol{M}_s$, structural stiffness $\boldsymbol{K}_s$, 'wet surface' area $\boldsymbol{A}_f$, and fluid mass $\boldsymbol{M}_f$, respectively, on normal coordinates d and q that diagonalize the left-hand side of the coupled matrix equations (2.8); with the fluid equations *integrated once in time* for computational convenience. Thus d and q in (8.1) are state variables representing generalized structural displacement and generalized scattered fluid pressure-integral, respectively. We note that the first of (8.1) represents a pressure-excited undamped mechanical oscillator ($p = \dot{q}$ is pressure) whereas the second represents a velocity-excited pressure-integral decay equation.

The system (8.1) can be further reduced to the *non-dimensional* form

$$\xi \ddot{x} + \omega^2 x = -\dot{y}, \qquad \dot{y} + \mu y = \dot{x}, \tag{8.2}$$

through introduction of the dimensionless variables

$$\begin{aligned}
&x = d/l, \qquad y = q/(\rho c l), \\
&\xi = m_s/(\rho l a_f), \qquad \omega^2 = k_s l^2/(\xi m_s c^2), \\
&\mu = m_s/(\xi m_f) = \rho l a_f/m_f, \\
&(\dot{\ }) \equiv \partial/\partial\tau, \qquad \tau = ct/l.
\end{aligned} \tag{8.3}$$

Here, l denotes a characteristic length of the problem, e.g., the radius of a submerged cylinder or sphere; ξ is a 'buoyancy ratio' (structural mass divided by displaced fluid mass); ω is a reduced vibration frequency, and μ is a generalized-pressure decay exponent. Note that the dot superscript has been redefined to denote differentiation with respect to the dimensionless time τ, rather than physical time t.

The numerical integration of the system (8.2) brings four dimensionless parameters into play: ξ, μ, ω and the stepsize $h = \nabla\tau$. The first three embody spatial characteristics of the response whereas h introduces the effect of the time integrator stepsize. It is of interest to the forthcoming discussions on the applicability of various implementations of the staggered procedure, to exhibit typical ranges assumed by such quantities. This information is collected in Table 8.1.

For ξ, μ and ω, two limit conditions and an illustrative case are shown. The *cavity condition* is the limit of modal motions heavily dominated by the inertia of the fluid, viz. low-frequency motions of a very thin submerged shell. The *dry mode condition* is realized by structural motions that do not interact with the fluid, such as torsional modes of structures of revolution. The illustrative problem of a submerged spherical shell is of interest because this is one of the few geometries amenable to exact analytical modal treatment [24, Appendix B]. Ranges quoted for the dimensionless time increment $h = \nabla\tau = c\nabla t/l$ are tabulated according to the temporal characteristics of the excitation, which determines the energy-spectrum characteristics of the response.

Next we examine various staggered solution procedures for the model system, roughly following the chronological order in which they were formulated.

8.2. Pressure extrapolation

The simplest formulation of the staggered scheme suggested by the partition (5.10) applied to (8.2) is the *pressure extrapolation* (PE) formulation, which may be

Table 8.1
Range of non-dimensional parameters

Parameter	Limit cases		Illustrative case
	Cavity $m_s \to 0,\ k_s \to 0$	Dry structural mode $m_f \to 0,\ a \to 0$	Submerged spherical shell[a]
ω	0[b]	≥ 0	0 to $\sim \bar{\rho}\bar{c}e^{1/2}n^2$
ξ	0	∞	$\sim \bar{\rho}e$ to $\sim \bar{\rho}en^2$
μ	1	indet.	1 to $n+1$
	Shock-excited problems[c] Early-time[d] response	Shock-excited problems[c] Late-time[e] response	Structural dynamics problems[f]
h	0.01–0.1	1–100	$\gg 1$

[a] $\bar{\rho} = \rho_s/\rho$ = ratio of shell and fluid densities, $\bar{c} = c_s/c$ = ratio of sound speeds in shell and fluid, e = thickness-to-radius ratio, n = highest circumferential wave number retained.

[b] k_s tends to zero faster than m_s.

[c] Problems characterized by wave propagation effects.

[d] Period during which shock wave starts to envelop the structure ($\tau < 1$); characterized by high-frequency structural motions, high radiation damping, relatively small hydrodynamic inertia forces.

[e] Period characterized by low-frequency structural motions, dominant hydrodynamic inertia, and low radiation damping (usually $\tau \lesssim 10$).

[f] Problems characterized by low-frequency motions throughout.

described as follows. Assume that solutions up to the $(n-1)^{\text{th}}$ time station are known. We first predict the pressure $\dot{y}_n$ in the first of (8.2), solve for the structural displacement x_n, obtain the velocity $\dot{x}_n$ from (8.4), insert this into the second of (8.2), and finally solve for the pressure $\dot{y}_n$. This corrected value may be used in an iterative setting if desired. We proceed now to examine various aspects of the PE formulation.

8.2.1. Time-discrete system

Introducing the pair of integration formulas

$$x_n = \delta_x \dot{x}_n + h_n^x, \qquad \delta_x = \beta_x h\,,$$
$$y_n = \delta_y \dot{y}_n + h_n^y, \qquad \delta_y = \beta_y h\,, \tag{8.4}$$

to treat (8.2) yields the algebraic system

$$(\xi + \delta_x^2 \omega^2) x_n = -\delta_x^2 \dot{y}_n + \xi(h_n^x + \delta_x h_n^{\dot{x}})\,,$$
$$(1 + \delta_y \mu)\dot{y}_n = \dot{x}_n - \mu h_n^y\,. \tag{8.5}$$

The secondary state variables: structural velocity $\dot{x}_n$ and pressure integral y_n, are calculated from the integration formulas:

$$\dot{x}_n = (x_n - h_n^x)/\delta_x, \qquad y_n = h_n^y + \delta_y \dot{y}_n\,. \tag{8.6}$$

8.2.2. Iteration convergence

First, does the pressure iteration alluded above converge? Using k as an iteration cycle index, the iterated PE scheme may be written

$$E_x x_n^{(k)} = -\beta_x h \chi \dot{y}_n^{(k-1)} + b_x, \qquad k = 1, 2, \ldots, \tag{8.7}$$
$$E_y \dot{y}_n^{(k)} = (x_n^{(k)} - h_n^x)/\delta_x + b_y,$$

where

$$\begin{aligned} E_x &= 1 + \beta_x^2 \Omega^2, & \Omega^2 &= \omega^2 h^2/\xi, \\ E_y &= 1 + \beta_y \Psi, & \Psi &= \mu h, \\ \chi &= h/\xi, & b_x &= h_n^x + \delta_x h_n^{\dot{x}}, \qquad b_y = -\mu h_n^y. \end{aligned} \tag{8.8}$$

The initial $\dot{y}_n^{(0)} = \dot{y}_n^{\mathrm{P}}$ may be obtained from a predictor such as

$$\dot{y}_n^{(0)} = (1+\gamma)\dot{y}_{n-1} - \gamma \dot{y}_{n-2} \tag{8.9}$$

where γ is an extrapolation parameter.

The scheme (8.7) fits the standard stationary-iteration form

$$\boldsymbol{u}^{(k)} = \boldsymbol{R}\boldsymbol{u}^{(k-1)} + \boldsymbol{b}, \qquad \boldsymbol{u} = \begin{Bmatrix} x \\ \dot{y} \end{Bmatrix}. \tag{8.10}$$

The iteration converges if and only if the spectral radius κ of $\boldsymbol{R}$ (modulus of its largest eigenvalue) is less than one. Since $\kappa(\boldsymbol{R})$ is easily found to be $\beta_x \chi/(E_x E_y)$, the convergence condition is

$$\chi = h/\xi \leq (1 + \beta_x^2 \Omega^2)(1 + \beta_y \Psi)/\beta_x. \tag{8.11}$$

The right-hand side of (8.11) is minimized for modal motions with $\Omega \to 0$, $\Psi \to 0$, in which case (8.11) reads

$$\chi \leq 1/\beta_x. \tag{8.12}$$

The structural equations will normally be treated with an A-stable integration formula (8.4a). But for all such formulas of the type (8.4a), β lies in the range 0.5 to 1.0. Consequently,

$$\chi = h/\xi \leq 2.0 \tag{8.13}$$

is the best that can be achieved by using the trapezoidal rule ($\beta_x = \frac{1}{2}$). Note that this is the only parameter in (8.4) that matters.

8.2.3. Stability

How about numerical stability? When one sees a result such as (8.13), a similar stepsize limitation for numerical stability can be usually expected. Indeed this is the case. The stability of the PE formulation was studied for a variety of time integration formulas, assuming the two-step predictor (8.9) and a fixed number (k) of iterations per time step. No details of the analysis will be given here, inasmuch as the staggered PE procedure is not recommended for practical use. Only one illustrative example is shown.

The stability region of the PE formulation is χ-bounded, i.e., physical parameters ω and μ play no role (in more precise terms, the worst combination is $\omega = \mu = 0$), and neither does the fluid integration formula (8.2b). For the trapezoidal formula and the one-parameter predictor (8.9) the stability regions can be conveniently displayed in the (γ, χ) plane as shown in Figure 8.1. The largest stable χ is 4, which can be attained for the single-pass solution ($k = 1$) if $\gamma = -\frac{1}{2}$ (the predicted pressure value is the mean of the last two values). If the process is iterated, the peak of the stability region is reduced; and as k goes to infinity the stability limit approaches the predictor-independent value (8.13).

It should be stressed that a time increment limited by an h/ξ of order 2 to 4 is unacceptable for a general-purpose integration package. The constraint can be practically met only in a limited class of problems, such as early-time shock

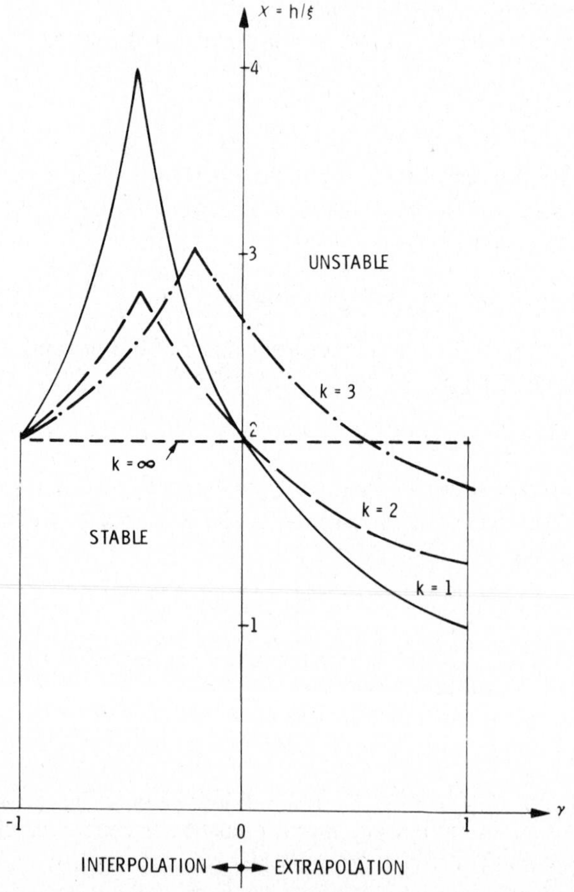

Fig. 8.1. Stability limit of the PE formulation (8.4) treated by the trapezoidal rule and the two-step predictor (8.9) as a function of the number of passes (k) per time step.

response analysis (cf. Table 8.1). If the response is dominated by low-frequency structural motions, however, the limited time increment is intolerably small; so small, in fact, that computational error accumulation and computation time become critical.

Why the detailed convergence study of the iterated PE formulation? Suppose we had found that the iteration (8.7) converges for the entire range (or at least a wide range) of parameters. The *converged solution* as $k \to \infty$ would then be identical to that furnished by the *fully-implicit* scheme resulting from the simultaneous solution approach. Under such assumptions, the stability region of the iterated PE formulation must approach that of the fully implicit scheme. The latter can be made unconditionally stable by selecting a suitable A-stable linear multistep method (e.g., the trapezoidal rule). Iteration would then provide a simple strategy for transmuting the single-pass, conditionally stable staggered solution procedure into an unconditionally stable scheme of greater accuracy; this strategy might have led to a potentially favorable tradeoff between iteration cost and the ability to utilize larger time steps.

As increasing k does not actually improve stability to any significant extent, it appears that there is no point in iterating at all should the PE formulation be used. (In linear problems, an iteration cycle costs essentially the same as one advancing step, but higher accuracy can be generally achieved through equivalent reductions in the stepsize.)

8.2.4. Velocity extrapolation

An alternative formulation of the staggered solution procedure can be based on velocity extrapolation (VE) applied to the fluid equation (8.2b). The algorithmic properties of this formulation are identical to those of the PE formulation, and need not be discussed further.

8.3. Sources of instability

What causes the stability limitations of the PE and VE formulations of the staggered scheme? Can it be blamed on the physical system, the partitioning process, or the integration formula? The answer to this question provides direct insight into techniques for stabilization of the staggered solution procedure at the *governing-equation level*; such techniques are exploited in Section 8.4.

8.3.1. Characteristics of the fully implicit system

We obtain the homogeneous form of the fully-implicit model system by transferring the coupling terms in (8.1) to the left-hand side, which yields

$$\xi \ddot{x} + \dot{y} + \omega^2 x = 0\,, \qquad \dot{y} - \dot{x} + \mu y = 0\,. \tag{8.14}$$

Laplace transformation of these equations without consideration of initial conditions then yields

$$\begin{bmatrix} \xi s^2 + \omega^2 & s \\ -s & s+\mu \end{bmatrix} \begin{Bmatrix} \hat{x}(s) \\ \hat{y}(s) \end{Bmatrix} = \mathbf{0} \tag{8.15}$$

in which $\hat{x}(s)$ and $\hat{y}(s)$ denote the transforms of $x(t)$ and $y(t)$, respectively. The associate characteristic equation is

$$(\xi s^2 + \omega^2)(s+\mu) + s^2 = 0\,. \tag{8.16}$$

The root-locus diagram for (8.16) is shown in Figure 8.2. Because all physically relevant branches are on the left-hand side $\mathrm{Re}(s) < 0$, (8.14) are inherently stable equations. Consequently, a stable numerical solution is guaranteed if the time integrator applied to (8.14) is A-stable. This also shows that the source of the PE/VE stability limitations cannot be traced to the governing equations.

8.3.2. Characteristics of the PE formulation

A differential-difference (DD) equation for the PE formulation can be obtained by expressing the two-step predictor (8.9) in a *delayed continuous* form,

$$\dot{y}^{\mathrm{P}}(\tau) = (1+\gamma)\dot{y}(\tau - h) - \gamma\dot{y}(\tau - 2h) \tag{8.17}$$

where $h = \nabla\tau$. This form is applied to the model system (8.2) to obtain

$$\xi\ddot{x} + \omega^2 x = -(1+\gamma)\dot{y}(\tau-h) + \gamma\dot{y}(\tau - 2h)\,, \qquad \dot{y} + \mu y = \dot{x}\,, \tag{8.18}$$

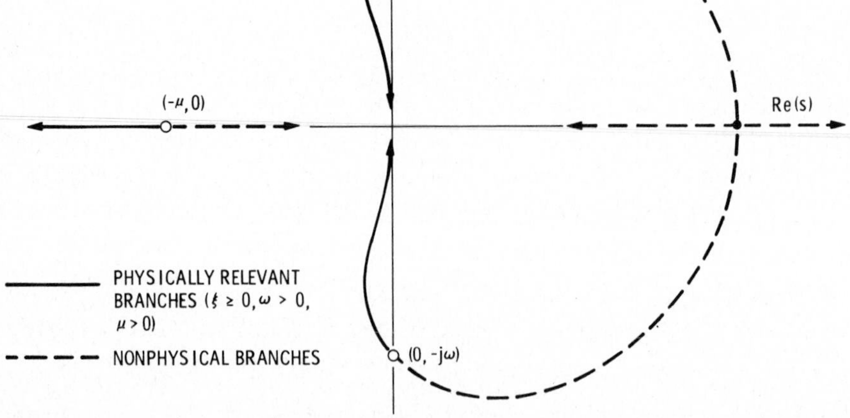

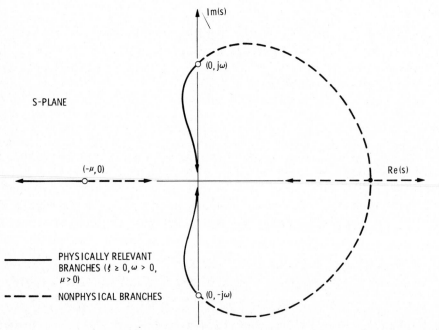

Fig. 8.2. Root locus diagram of the fully-implicit system (8.14) in the Laplace-transform variable plane (s-plane).

which is the *differential-difference form* associated with the single-pass PE formulation. Note that (8.18) is independent of any time integration scheme. Its characteristic equation in the Laplace-transform variable s is obtained following a procedure identical to that used to derive (8.16):

$$(\xi s^2 + w^2)(s+\mu) + s^2[(1+\gamma)\,e^{-sh} - \gamma\, e^{-2sh}] = 0\,. \tag{8.19}$$

In control theory, the identifier *dead-time element* is often used for terms such as $\exp(-sh)$ and $\exp(-2sh)$, which result from Laplace-transforming time-delay terms. It is generally acknowledged that the occurrence of such terms in control loops has a destabilizing effect.

In the limit of vanishing stepsize h, (8.19) approaches the fully implicit characteristic equation (8.16), which is stable. For a finite h, it can be shown that the critical combination of parameters pertaining to the stability of (8.19) is $\omega = \mu = 0$, in which case (8.19) reduces to

$$\xi s^2 + (1+\gamma)\,e^{-sh} - \gamma\, e^{-2sh} = 0\,. \tag{8.20}$$

Application of Nyquist's stability criterion [33] to (8.20) yields the stability condition

$$\begin{aligned}&(1+\gamma)\cos \mathrm{j}sh - \gamma \cos 2\mathrm{j}sh = 0\,,\\ &\xi\, \mathrm{j}sh \geq [(1+\gamma)\sin \mathrm{j}sh - \gamma \sin 2\mathrm{j}sh]h\,,\end{aligned} \tag{8.21}$$

in which $\mathrm{j} = \sqrt{-1}$. Figure 8.3 shows the stability limit in the (γ, χ) plane as

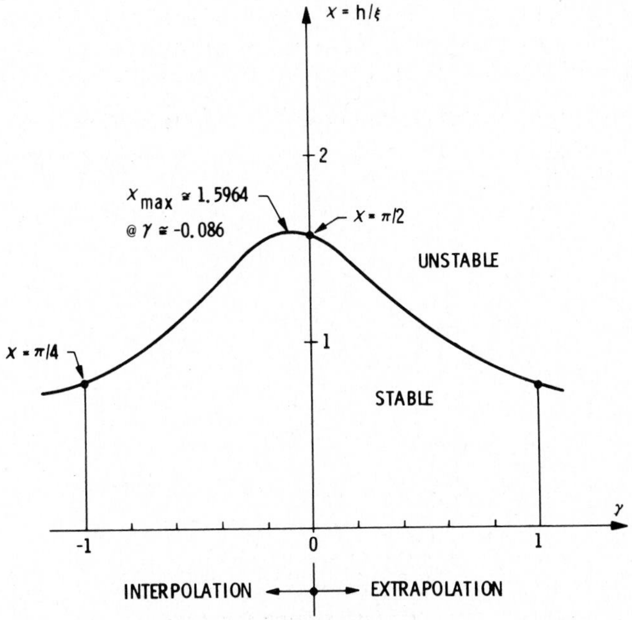

Fig. 8.3. Stability limit of the differential-difference form (8.18) of the PE formulation with the two-step predictor (8.9).

calculated from (8.21). For the special case $\gamma = 0$, i.e., using the last pressure solution as the predicted value, (8.21) gives the stability limit

$$\chi = h/\xi = \pi/2\,. \tag{8.22}$$

The destabilizing effect of the dead-time element is now apparent. It must be stressed that the stability boundary shown in Figure 8.3 applies only if an integration scheme of infinite accuracy (such as analytical integration or a non-truncated Taylor-series algorithm) were used to solve the DD system (8.16), and that the boundary is perturbed by the intrusion of a difference time integrator. For example, the use of the trapezoidal rule as the structural integrator increases somewhat the stability domain, as can be deduced from a comparison of Figures 8.1 and 8.3; on the other hand, the use of the backward Euler method (not shown here) has a detrimental effect on stability.

8.3.2. Summary of findings

The χ-bounded instability of the PE formulation is due to the dead-time (delayed) pressure feedback (8.17) into the structural equations (8.18a). In control theory terms, this process can be viewed as a sampled-feedback system, in which the pressure feedback is delayed by a 'hold' (dead-time interval) equivalent to the stepsize h. This representation is illustrated in the block diagram of Figure 8.4, which also includes nonhomogeneous (input source) terms.

The most commonly used strategy for stabilizing a system with dead-time components consists of introducing a series of compensating elements that reduce the bandwidth of the system's frequency response (see, e.g., [18], p. 118). These stabilization techniques essentially amount to the introduction of damping into the feedback loop.

In the fully-implicit treatment, damping is inherently present by virtue of the energy radiation term $\dot{y}$ in the model system (8.14) on the left-hand side. The

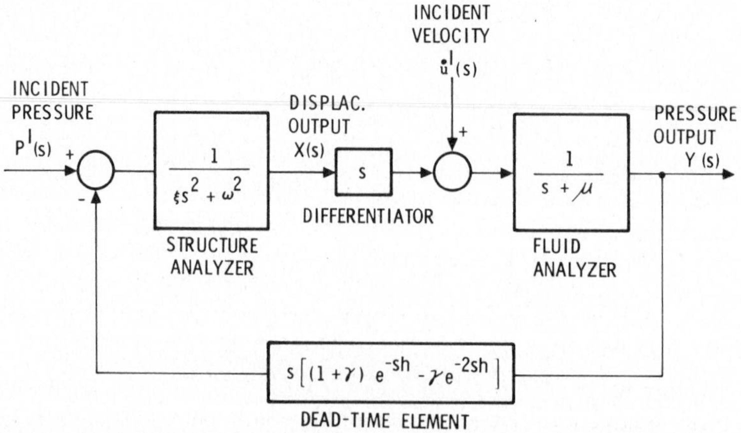

Fig. 8.4. Control loop representation of the pressure extrapolation (PE) procedure.

effect of the radiation damping on the structural response roots can be readily appreciated by examining Figure 8.2. The structural equation (8.18a) of the PE formulation, however, contains no homogeneous damping term. We are therefore motivated to attempt the restoration of sufficient damping to achieve overall stability *without* adding extraneous artificial damping. This leads to the technique of 'stabilization by reformulation' described next.

8.4. Stabilization

The foregoing results clearly suggest that the staggered solution may be stabilized by the addition of damping terms into the left-hand side of the governing equations of motion. However, the use of artificial damping is *ruled out by accuracy considerations.* Therefore, the governing equations must be tailored in such a way that homogeneous damping terms either appear in the structural equations, or be added to the fluid equations, or both. We now proceed two stabilized formulations generated through such 'augmentation' techniques.

8.4.1. Pressure-integral extrapolation (PIE) formulation

The structural equation for the PE formulation (8.18a) is excited by a time-lagged velocity feedback; a destabilizing effect occurs because of the delayed energy dissipation in the feedback loop (cf. Figure 8.4). We can correct this situation by eliminating the velocity feedback. To do this, solve the second of (8.2) for $\dot{y}$:

$$\dot{y} = \dot{x} - \mu y , \tag{8.23}$$

replace this in the first equation, and transfer the damping term $\dot{x}$ to the left-hand side, to yield the *augmented structural equation*:

$$\xi \ddot{x} + \dot{x} + \omega^2 x = \mu y . \tag{8.24}$$

The staggered solution procedure based upon (8.24) along with the original pressure equation (8.2b) will be called the *pressure-integral extrapolation* (PIE) formulation, inasmuch as the pressure-integral y is involved in the prediction process.

8.4.2. Displacement extrapolation (DE) formulation

Stabilization can also be achieved by augmentation of the fluid equation so as to increase the pressure decay rate in the homogeneous system. Qualitatively speaking, this strategy works as long as the stability 'margin' of the modified pressure equation overwhelms the destabilizing effect of the structural equations. The modified pressure equation is obtained as follows. Differentiate the second of (8.2) with respect to time and multiply through by ξ:

$$\xi \ddot{y} + \xi \mu \dot{y} = \xi \ddot{x} . \tag{8.25}$$

Solve for $\xi \ddot{x}$ from the first of (8.2), insert in the right-hand side of the above equation, and transfer the $\dot{y}$ term to the left. The resulting *augmented fluid equation* is

$$\xi\ddot{y} + (1 + \xi\mu)\dot{y} = -\omega^2 x\,. \tag{8.26}$$

The staggered solution procedure based upon (8.26) along with the original structural equation (8.2a) will be called the *displacement extrapolation* (DE) formulation.

It turns out that more complex formulations may be constructed by modifying *both* equations. Such forms will not be exhibited, as the process of constructing augmented equations should be by now fairly obvious to the reader.

8.4.3. *Stability properties*

It is shown in [24] that both the PIE and DE formulations enjoy considerable stability advantages over the PE formulation. More specifically:

1. Unconditional stability for single-pass staggered solution schemes may be attained for both formulations if the predictor and time integrator are appropriately chosen.

2. The corrector iteration for fixed time step converges over the entire feasible domain of the parameters ξ, ω and μ.

The first result indicates that the two augmented formulations are capable of delivering stable single-pass solutions provided the predictor accuracy is suitably restricted. Such restrictions are, in general, affected by the integration method. For example, application of the trapezoidal rule to the DE formulation limits the extrapolator accuracy to first order [24, p. 113] whereas the application of an A-stable backward-difference scheme such as the Park 3-step method allows the use of somewhat more accurate predictors.

The significance of the second result is: both stabilized formulations can, if *iterated to convergence*, attain the unconditional stability and accuracy of the fully-implicit solution procedure, regardless of the choice of predictors. This observation raises again the issue of finding a cost-minimization compromise between a single-pass solution process with small time increments and an iterated solution process with larger time increments. To be sure, the issue is more important in nonlinear problems than in linear ones.

Finally, it should be mentioned that stability analyses of the differential-difference (DD) equations pertaining to the single-pass PIE and DE formulations were also performed using integral-transform techniques similar to those used in Section 8.3.2 for the PE formulation. These analysis provided valuable insight into the sensitivity of the stability characteristics to the various parameters appearing in the model formulations. The results will not be reported here, however, because of the length and complexity of the algebraic manipulations involved.

8.4.4. *Control theory interpretation*

From the viewpoint of control theory, the destabilizing effect of the dead-time element of Figure 8.4 can often be overcome by introducing *compensating elements.* Appropriate selection of such elements reduces the bandwidth of the frequency response of the system. A compensated control loop for the original PE formulation can take the fairly general form illustrated in Figure 8.5. The four compensating elements should be chosen so as to stabilize the staggered solution procedure while minimizing the deviation of the response of the compensated system from that of the original system. In the following we interpret the equation-augmentation process of Sections 8.4.1 and 8.4.2 using control theory terminology.

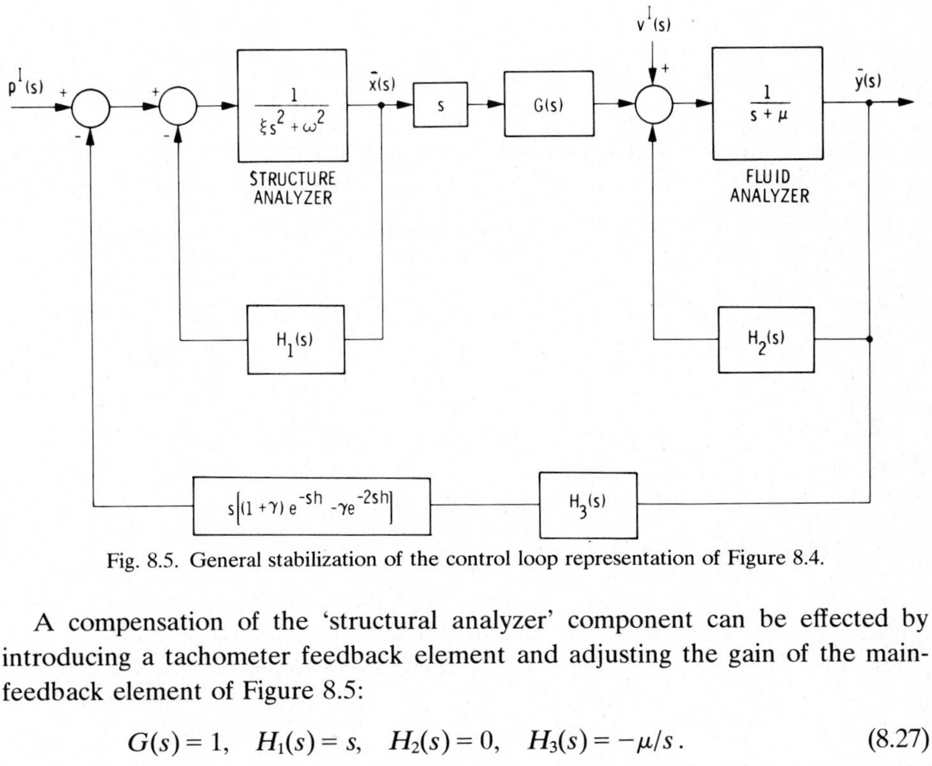

Fig. 8.5. General stabilization of the control loop representation of Figure 8.4.

A compensation of the 'structural analyzer' component can be effected by introducing a tachometer feedback element and adjusting the gain of the main-feedback element of Figure 8.5:

$$G(s)=1,\quad H_1(s)=s,\quad H_2(s)=0,\quad H_3(s)=-\mu/s\,. \tag{8.27}$$

The resulting process, shown in Figure 8.6, corresponds to the PIE formulation of Section 8.4.1.

Alternatively, the 'fluid analyzer' may be compensated by selecting

$$\begin{aligned}&G(s)=(\omega/\xi)\exp(-sh)/s,\quad H_1(s)=0,\quad H_2(s)=1/\xi\,,\\&H_3(s)=1/\{s[(1+\gamma)\exp(-sh)-\gamma\exp(-2sh)]\}\,,\end{aligned} \tag{8.28}$$

The resulting process, shown in Figure 8.7, corresponds to the DE formulation of Section 8.4.2.

One can also compensate on both structural and fluid analyzers while adjusting the main feedback element to minimize response distortions.

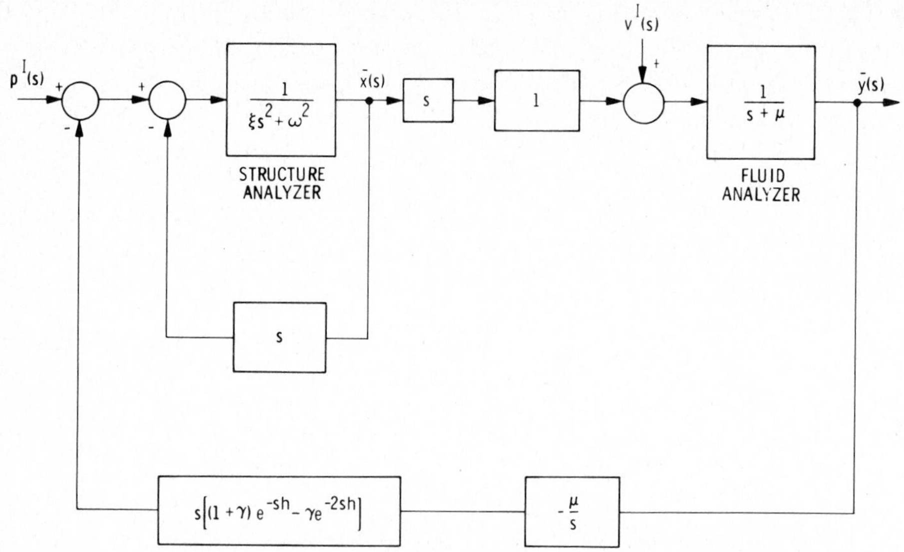

Fig. 8.6. Control loop representation of the pressure integral extrapolation (PIE) procedure.

It should be emphasized again that stabilization is possible for system (8.2) *without introducing distortions* in the response. That is, no artificial damping or stiffness is required to stabilize the staggered solution procedure. This fortunate circumstance may not always be the case for more general coupled-field problems, however; then a compromise between stability and response distortion has to be worked out.

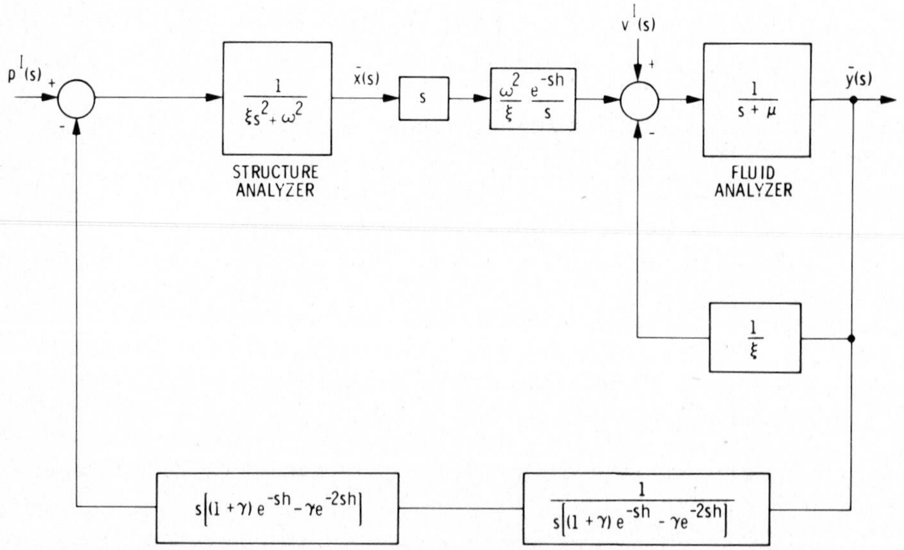

Fig. 8.7. Control loop representation of the displacement extrapolation (DE) procedure.

8.5. Implementation considerations

Thus far we have succeeded in developing several stabilized forms of the staggered partition of the two degree-of-freedom system (8.1). Application of the two formulations of greatest interest to the original matrix system (2.8) provides the equations shown in Table 8.2. Before selecting one of these formulations for implementation in a fluid-structure analysis code, it is necessary to discuss the practical implications associated with such a decision.

Table 8.2
Matrix implementation forms for stabilized procedures

Formulation	Implementation
Pressure integral extrapolation (PIE)	$\boldsymbol{M}\ddot{\boldsymbol{u}} + (\boldsymbol{D} + \rho c\boldsymbol{TAT}')\dot{\boldsymbol{u}} + \boldsymbol{Ku}$ $= \boldsymbol{f}_s + \boldsymbol{r} - \boldsymbol{TA}(\boldsymbol{p}^{\mathrm{I}} + \rho c\dot{\boldsymbol{u}}^{\mathrm{I}}) + \rho c\boldsymbol{TAM}_{\mathrm{f}}^{-1}\boldsymbol{Aq}$ $\boldsymbol{A}\dot{\boldsymbol{q}} + \rho c\boldsymbol{AM}_{\mathrm{f}}^{-1}\boldsymbol{Aq} = \rho c\boldsymbol{A}(\boldsymbol{T}'\dot{\boldsymbol{u}} - \dot{\boldsymbol{u}}^{\mathrm{I}})$
Displacement extrapolation (DE)	$\boldsymbol{M}\ddot{\boldsymbol{u}} + \boldsymbol{D}\dot{\boldsymbol{u}} + \boldsymbol{Ku} = \boldsymbol{f}_s + \boldsymbol{r} - \boldsymbol{TA}(\dot{\boldsymbol{q}} + \boldsymbol{p}^{\mathrm{I}})$ $\boldsymbol{A}\ddot{\boldsymbol{q}} + \rho c(\boldsymbol{AM}_{\mathrm{f}}^{-1}\boldsymbol{A} + \boldsymbol{AM}^{-\mathrm{G}}\boldsymbol{A})\dot{\boldsymbol{q}}$ $= -\rho c\boldsymbol{A}\ddot{\boldsymbol{u}}^{\mathrm{I}} - \rho c\boldsymbol{AM}^{-\mathrm{G}}\boldsymbol{Ap}^{\mathrm{I}}$ $+ \rho c\boldsymbol{AM}^{-\mathrm{G}}\boldsymbol{C}^{-1}\boldsymbol{T}'(\boldsymbol{f}_s + \boldsymbol{r} - \boldsymbol{D}\dot{\boldsymbol{u}} - \boldsymbol{Ku})$ where $\boldsymbol{M}^{-\mathrm{G}} = \boldsymbol{C}'(\boldsymbol{T}'\boldsymbol{MT})^{-1}\boldsymbol{C}, \quad \boldsymbol{C} = \boldsymbol{T}'\boldsymbol{T}$

If there were no *a-priori* constraints with regard to software development, i.e., the program developer had complete freedom to construct both the structural and fluid analyzers from scratch, the overriding selection factor would probably be the accuracy characteristics displayed by each formulation. As it is, there now exist many large-scale linear and nonlinear structural analyzers that incorporate capabilities for transient response analysis.

Inspection of Table 8.2 reveals that only the DE formulation has no impact upon the structural equations of motion, while the PIE formulation (as well as the other, more complicated, formulations not shown there) requires the addition of a damping term, $\rho c\boldsymbol{TAT}'$, to the left-hand side of the structural equation. This term can be expected to have an adverse effect upon the sparseness characteristics of the coefficient matrix $\boldsymbol{E}_x$, because each fluid boundary element on the contact or 'wet' surface will normally overlap several structural elements. Additional nonzero coefficients then appear outside the 'profile' of the structural stiffness matrix in entry positions pertaining to the wet structural displacements. As matrix connectivity characteristics are generally controlled by the structural grid information, extensive and expensive software modifications would be required to include such a damping term in existing 'dry-structure' analyzers.

For the analysis of *linear* structures, a general-purpose, stand-alone integration package could be built for the PIE and other formulations, which would accept, as input data, preprocessed matrices assembled by existing linear structural analysis codes (with, perhaps, some restrictions on sparse-matrix storage formats). For structures exhibiting *nonlinear* behavior, however, such an undertaking would necessitate the coupling of the nonlinear solver with the dedicated integration package to provide continuously updated information in the form of factored matrices, pseudo-force vector, and the like. In this case, the driving consideration would certainly be preservation of the autonomy of the structural analyzer, a restriction that precludes the use of either the PIE or other forms that augment the structural equations.

An additional argument that reinforces this conclusion is that the fluid matrix equation (2.8b) is only the lowest-order member (DAA_1) of a family of doubly-asymptotic surface interaction approximations [17]. The next member of that family (DAA_2) is characterized by a *second-order* matrix differential equation, shown in Table 2.1, which would replace (2.8b). The inclusion of a higher derivative of the scattered pressure has the effect of adding widely-connected matrix terms to both the damping *and* stiffness matrix in the structural equations if the equivalents of the PIE and other formulations are adopted.

In conclusion, it seems clear that the displacement-extrapolation formulation is the only stabilized procedure that satisfies the practical requirements of modularity and avoids the proliferation of special-purpose versions of existing structural analyzers.

9. Application examples

This section presents selective results generated by a large-scale fluid-structure interaction code for two idealized configurations illustrating underwater and free-surface shock problems. The USA Code, written by J.A. DeRuntz [7, 8], solves the fluid-structure interaction problem discussed in Section 2.2 using a staggered solution procedure stabilized through the displacement-extrapolation reformulation described in Section 8.4.

9.1. Sample configurations

The structure involved in these problems is a linear-elastic, isotropic, infinite, circular cylindrical shell of monocoque wall construction. The shell is excited by a transverse, plane stepwave of unit incident pressure propagating through an infinite and a semi-infinite fluid. These fairly simple configurations are used as checkout and demonstration problems because they are two of the very few cases amenable to exact analytical treatment of the fluid by residual potential techniques [13, 14, 15]. Availability of these exact solutions allows direct assessment of the accuracy of the computed solutions.

Note that there are two principal error sources: that due to discretization (spatial and temporal), and that arising from the use of doubly asymptotic approximations for the fluid interaction equations.

9.2. Submerged shell

In the first configuration the cylinder is submerged in an *infinite* fluid. The shell structure was modeled by 36 square finite elements uniformly spaced around the circumference. Only one element is used longitudinally; appropriate kinematic constraints were applied to enforce an infinite-shell plane-strain condition. The structural stiffness and mass matrices were generated by the structural analyzer SPAR. The fluid model consists of 36 equally-spaced boundary elements; the

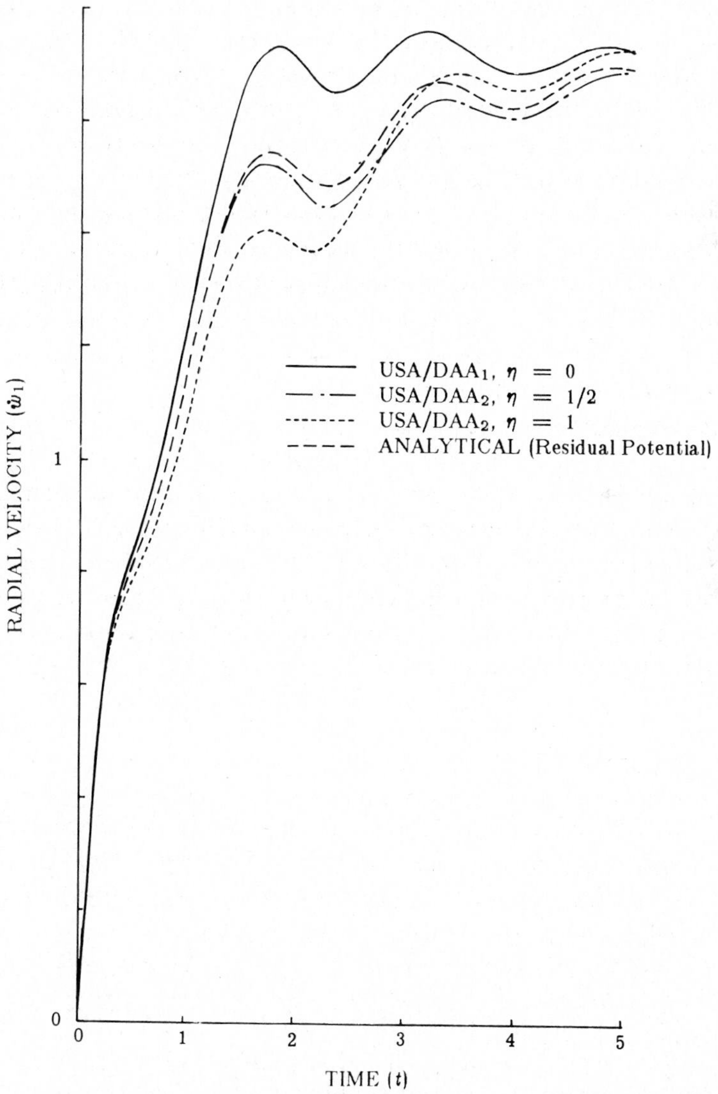

Fig. 9.1. $n = 1$ radial velocity response of plane stepwave-excited submerged cylinder [7]: computed DAA_1 and DAA_2 time histories vs. exact (residual potential) history.

two-dimensional nature of the infinite shell geometry was simulated by setting a 'fictitious element' options in the fluid analyzer.

The material properties used correspond to a steel shell immersed in water. The input data is normalized so that density and speed of sound for the fluid both equal unity; hence, the density, Young's modulus, and Poisson's ratio for the structural material are taken as 7.85, 98.125, and 0.3, respectively. The radius and wall thickness of the cylinder are 1.0 and 0.01, respectively.

The computed responses were postprocessed by a circumferential Fourier analyzer to generate $n = 0$, 1, and 2 modal response results, n being the circumferential harmonic number. This decomposition facilitates orderly comparison with exact solutions, which are naturally obtained on a mode-by-mode basis. The primary response variables of interest are radial displacements for $n = 0$ (breathing mode), radial and tangential velocities for $n = 1$ (a translational rigid-body mode), and radial and tangential displacements for $n = 2$ (ovalization mode).

A time step of 0.025 (1/40 of the envelopment period) was used up to $t = 1$; it was increased to 0.05 for t between 1 and 2, and to 0.1 for t greater than 2.

The $n = 1$ tangential velocity results are shown in Figure 9.1 in which the analytical residual-potential response is compared with computed DAA_1 and DAA_2 results. The latter approximation embodies an adjustable coefficient, η, which varies between zero and one [17]; for $\eta = 0$ the DAA_2 reduces to the DAA_1. Best results are obtained with $\eta = 0.5$.

A comparison of the DAA_1 responses computed through the staggered solution procedure with an exact (analytical) treatment of the DAA_1 [15, 16] shows maximum errors of the order of 1%, cf. Figure 9.2. Thus for this simple configuration the error due to the approximate treatment of the fluid interaction is seen to dominate the computational (discretization) error. The reverse is true, however, for complex structures of arbitrary geometry with internal components, and most especially if the structure exhibits localized nonlinear behavior such as plasticity; in such problems the structural discretization governs response accuracy.

9.3. *Semi-submerged shell*

In the second configuration, the axis of the infinite cylindrical shell lies in the plane of the *free surface* of a *semi-infinite* acoustic medium. Analytical solutions for this case may be readily obtained by the method of imaging, as explained in Section 4.3 of [8].

For the numerical treatment of this problem, a 19-element SPAR model was constructed with one 5 deg. plate element, seventeen 10 deg. plate elements and a second 5 deg. plate element. The arriving shock front forms a 45 deg. angle with the surface. Time steps were selected as in the infinite fluid case.

Computed $n = 0$ and $n = 1$ circumferential harmonic responses generally agree within 1% of the exact DAA_1 results. The $n = 1$ radial velocity response is shown in Figure 9.3.

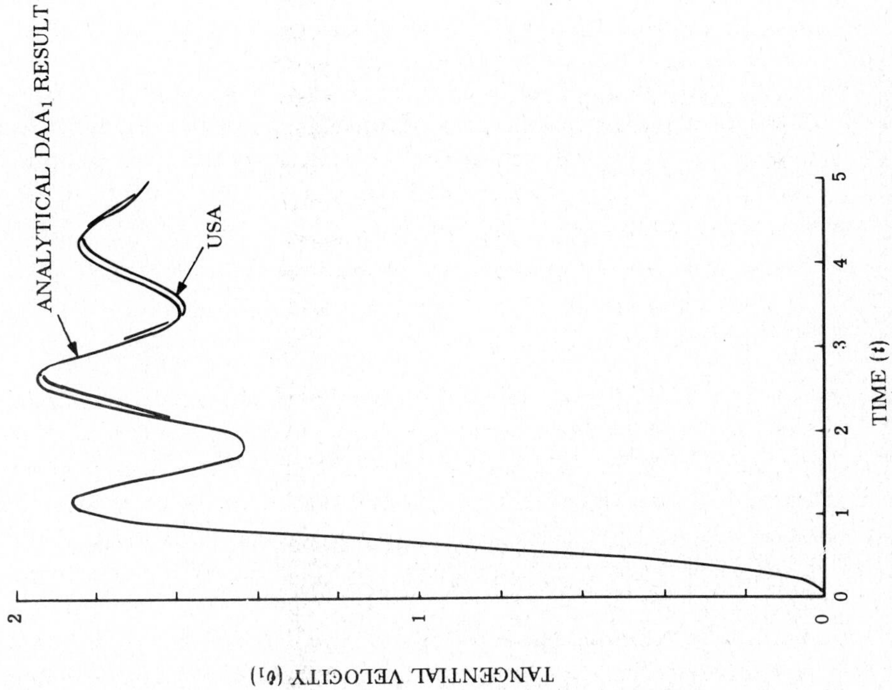

Fig. 9.2. $n = 1$ radial velocity response of plane step-wave-excited submerged cylinder [8]: computed DAA_1 history vs. analytical DAA_1 history.

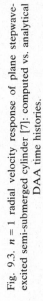

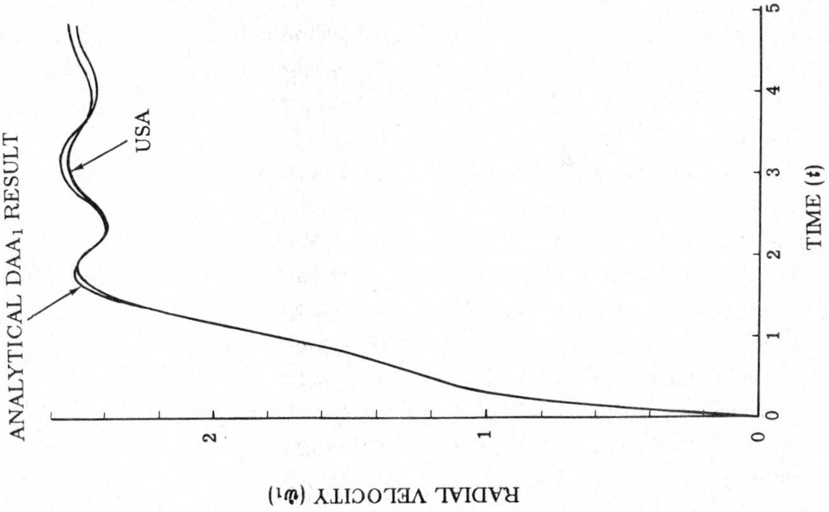

Fig. 9.3. $n = 1$ radial velocity response of plane stepwave-excited semi-submerged cylinder [7]: computed vs. analytical DAA time histories.

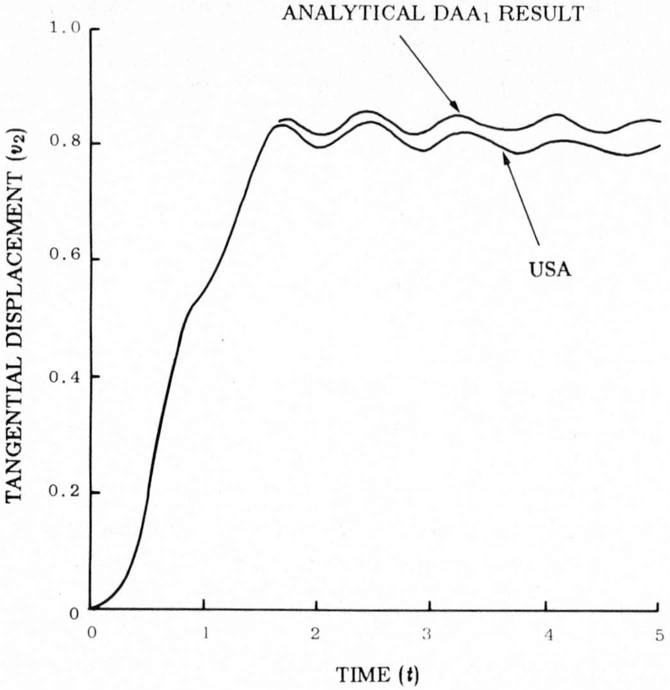

Fig. 9.4. $n = 2$ tangential displacement response of plane stepwave-excited semi-submerged cylinder [8]: computed vs. analytical DAA histories.

On the other hand, the $n = 2$ post-envelopment solution during the final phase of shell envelopment is more sensitive to temporal and spatial discretization details. This sensitivity is reflected in Figure 9.4, which shows the $n = 2$ tangential velocity response, and is associated with the numerical representation of the pressure discontinuity at the stepwave front.

10. Concluding remarks

In this exposition we have endeavored to present motivating forces, underlying theory, algorithmic properties and applications of partitioned analysis procedures for coupled-field problems. The material covers coupled dynamical systems governed by second-order differential equations when the time step used in the direct numerical integration is the same for every subsystem. Particular attention is given to first-order acoustic-fluid equations coupled to the second-order equations of structural dynamics, since the bulk of applications of partitioned analysis has so far centered on the fluid-structure interaction problem. There are, however, other potentially rewarding areas of research and applications that remain unexplored.

When fluid-dynamics formulations acquire convective terms, as in the classical Navier-Stokes equations, resulting matrix operators are not symmetric. Interaction of such flows with other dynamical systems is a virgin topic in which progress is presently hampered by the lack of a partitioned-analysis stability theory that accounts for unsymmetry and highly nonlinear terms.

In problems dealing with highly 'layered' or 'anisotropic' systems, (e.g., weather prediction, boundary layers, stratified multiphase flows) the use of different subsystem time steps may be computationally attractive. Some 'fractional step' schemes have been proposed [36] for the classical equations of mathematical physics, but no systematic study as to their implementation and potential advantages is presently available.

Existing formulations of coupled-field problems have been traditionally built in the formulate-discretize-implement-interface sequence. Field interface conditions appear as a byproduct of the other steps. But this sequence is not necessarily the optimal one for partitioned analysis. For many problems it would be desirable to reverse the order, and start by seeking interface conditions consistent with the philosophy of partitioned analysis. Single-field formulation and discretization would then become subordinates to interfacing requirements.

Finally, a critical assessment of, when is partitioned analysis beneficial? remains a future task. One key difficulty here is that theory and methods are now ahead of software. It was noted in the Introduction that huge, monolithic programs and partitioned analysis do not mix well. Only when modular software tools configured to fully exploit the operational features of partitioned analysis become widely available to engineering users can this evalution phase begin seriously.

Acknowledgements

The theoretical development and computer implementation of partitioned analysis procedures has been supported by the Independent Research Program of Lockheed Missiles and Space Company, Inc., the Office of Naval Research, and the Defense Nuclear Agency. Preparation of this chapter has been supported by LMSC's Independent Research Program. We are indebted to John DeRuntz for supplying us the application examples of Section 9 from the USA Code Reference Manual.

Note added in proof

The doctoral dissertation by D.K. Paul, Efficient Dynamical Solutions for Single and Coupled Multiple Field Problems, School of Engineering, University of Swansea, Wales, U.K. (1982), contains applications of staggered solution procedures to a wide range of coupled mechanical systems. Some of Paul's results appear on pp.

321–334 of the book *Numerical Methods for Coupled Problems*, coedited by E. Hinton, D.J.R. Owen and P. Bettess, Pineridge Press, Swansea, U.K. (1981). Multi-stepping procedures, which allow different sybsystem time increments, have been recently studied by W.K. Liu and T. Belytschko, *Computers and Structures* **15**, 445–450 (1982).

References

[1] T. Belytschko and R. Mullen, Mesh Partitions of Explicit-Implicit Time Integration, in: *Formulations and Computational Algorithms in Finite Element Analysis*, K.J. Bathe, J.T. Oden and W. Wunderlich, eds. (MIT Press, Cambridge, MA, 1976) pp. 673–690.

[2] T. Belytschko and R. Mullen, Stability of Explicit-Implicit Mesh Partitions in Time Integration, *International Journal of Numerical Methods in Engineering* **12**, 1575–1586 (1978).

[3] T. Belytschko, H.J. Yen and R. Mullen, Mixed Methods for Time Integration, *Computer Methods in Applied Mechanics and Engineering* **17/18**, 259–275 (1979).

[4] T. Belytschko, Fluid-Structure Interaction, *Computers and Structures*, **12**, 459–470 (1980).

[5] T. Belytschko and R. Mullen, Two-Dimensional Fluid-Structure Impact Computations with Regularization, *Computer Methods in Applied Mechanics and Engineering* **27**, 139–154 (1981).

[6] J.A. DeRuntz and T.L. Geers, Added Mass Computation by the Boundary Integral Method, *Internat. Journal of Numerical Methods in Engineering* **12**, 531–550 (1978).

[7] J.A. DeRuntz, T.L. Geers and C.A. Felippa, The Underwater Shock Analysis (USA) Code; A Reference Manual, Report DNA-4524F to Defense Nuclear Agency, Lockheed Palo Alto Research Lab., Palo Alto, CA (1978).

[8] J.A. DeRuntz, T.L. Geers and C.A. Felippa, The Underwater Shock Analysis (USA-Version 3) Code: A Reference Manual, Interim Report to Defense Nuclear Agency, Lockheed Palo Alto Research Lab., Palo Alto, CA (1980).

[9] C.A Felippa, T.L. Geers and J.A. DeRuntz, Response of a Ring-Stiffened Cylindrical Shell to a Transient Acoustic Wave, Report LMSC-D403671 to Office of Naval Research, Structural Mechanics Laboratory, Lockheed Palo Alto Research Lab., Palo Alto, CA (1974).

[10] C.A. Felippa and K.C. Park, Computational Aspects of Time Integration Procedures in Structural Dynamics, Part I: Implementation, *Journal of Applied Mechanics* **45**, 595–602 (1978).

[11] C.A. Felippa and K.C. Park, Direct Time Integration Methods in Nonlinear Structural Dynamics, *Computer Methods in Applied Mechanics and Engineering* **24**, 61–111 (1980).

[12] C.A. Felippa and K.C. Park, Staggered Solution Transient Analysis Procedures for Coupled Mechanical Systems: Formulation, *Computer Methods in Applied Mechanics and Engineering* **24**, 61–111 (1980).

[13] T.L. Geers, Excitation of an Elastic Cylindrical Shell by a Transient Acoustic Wave, *Journal of Applied Mechanics* **36**, 459–469 (1969).

[14] T.L. Geers, Residual Potential and Approximate Methods for Three-Dimensional Fluid-Structure Interaction, *Journal of the Acoustical Society of America* **45**, 1505–1510 (1971).

[15] T.L. Geers, Scattering of a Transient Acoustic Wave by an Elastic Cylindrical Shell, *Journal Acoustical Society of America* **51**, 5 (Part 2), 1640–1651 (1972).

[16] T.L. Geers, Shock Response Analysis of Submerged Structures, *Shock and Vibration Bulletin* **44**, Supp. 3, 17–32 (1974).

[17] T.L. Geers, Doubly Asymptotic Approximations for Transient Motions of Submerged Structures, *Journal of the Acoustical Society of America* **64**, 1500–1508 (1980).

[18] J.E. Gibson, *Nonlinear Automatic Control* (McGraw-Hill, New York, 1963).

[19] D.M. Himmelblau (editor), *Decomposition of Large-Scale Problems* (North-Holland, Amsterdam–New York, 1978).

[20] T.J.R. Hughes and W.K. Liu, Implicit-Explicit Finite Elements in Transient Analysis: Stability Theory, *Journal of Applied Mechanics* **45**, 371–374 (1978).

[21] T.J.R. Hughes and W.K. Liu, Implicit-Explicit Finite Elements in Transient Analysis: Implementation and Numerical Examples, *Journal of Applied Mechanics* **45**, 375–378 (1978).
[22] T.J.R. Hughes, K.S. Pister and R.L. Taylor, Implicit-Explicit Finite Elements in Nonlinear Transient Analysis, *Computer Methods in Applied Mechanics and Engineering* **17/18**, 159–182 (1979).
[23] P.S. Jensen, Transient Analysis of Structures by Stiffly Stable Methods, *Computers and Structures* **4**, 615–626 (1974).
[24] K.C. Park, C.A. Felippa and J.A. DeRuntz, Stabilization of Staggered Solution Procedures for Fluid-Structure Interaction Analysis, in: *Computational Methods for Fluid-Structure Interaction Problems*, T. Belytschko and T.L. Geers, eds., ASME Applied Mechanics Symposia Series, AMD - Vol. 26 (1977) pp. 94–124.
[25] K.C. Park, Practical Aspects of Numerical Time Integration, *Computers and Structures* **7**, 343–353 (1977).
[26] K.C. Park and C.A. Felippa, Computational Aspects of Time Integration Procedures in Structural Dynamics, Part II: Error Propagation, *Journal of Applied Mechanics* **45**, 603–611 (1978).
[27] K.C. Park, Partitioned Transient Analysis Procedures for Coupled-Field Problems: Stability Analysis, *Journal of Applied Mechanics* **47**, 370–376 (1980).
[28] K.C. Park and C.A. Felippa, Partitioned Transient Analysis Procedures for Coupled-Field Problems: Accuracy Analysis, *Journal of Applied Mechancis* 47, 919–926 (1980).
[29] K.C. Park and J.M. Housner, Semi-Implicit Transient Analysis Procedures for Structural Dynamics Analysis, *Internat. Journal for Numerical Methods in Engineering* **18**, 609–622 (1982).
[30] P.J. Roache, On Artificial Viscosity, *Journal of Computational Physics* **10**, 169–184 (1972).
[31] A.P. Sage, *Methodology for Large Scale Systems* (McGraw-Hill, New York, 1977).
[32] D.D. Siljak, *Large Scale Dynamic Systems: Stability and Structure* (North-Holland, Amsterdam–New York, 1978).
[33] T. Takahashi, *Mathematics of Automatic Control* (Holt, Rinehart and Winston, New York, 1966).
[34] P.G. Underwood and T.L. Geers, Doubly-Asymptotic, Boundary-Element Analysis of Nonlinear Soil-Structure Interaction, in: *Proc. Second International Symposium on Innovative Numerical Analysis in Applied Engineering Sciences, Montreal, June 16–20, 1980*, R. Shaw et al., eds. (University Press of Virginia, Charlottesville, VA, 1980) pp. 413–422.
[35] J. von Neumann and R.D. Richtmyer, A. Method for the Numerical Calculation of Hydrodynamic Shocks, *Journal of Applied Physics* **21**, 232–257 (1950).
[36] N.N. Yanenko, *The Method of Fractional Steps* (Springer, New York, 1971).
[37] O.C. Zienkiewicz, E. Hinton, K.H. Leung and R.L. Taylor, Staggered Time Marching Schemes in Dynamic Soil Analysis and a Selective Explicit Extrapolaton Algorithm, in: *Proc. Second International Symposium on Innovative Numerical Analysis in Applied Engineering Sciences, Montreal, June 16–20, 1980*, R. Shaw et al., eds. (University Press of Virginia, Charlottesville, VA, 1980) pp. 525–530.

CHAPTER 4

Boundary-Element Methods for Transient Response Analysis

Thomas L. GEERS

Senior Staff Scientist
Lockheed Palo Alto Research Laboratory
Palo Alto, California

Computational Methods for Transient Analysis
Edited by T. Belytschko and T.J.R. Hughes

1. Introduction

The term ‘Boundary Element Method’ (BEM) consititutes a coalescence of the terms ‘Boundary Integral Equations’ (BIE) and ‘Finite Element Method’ (FEM). The name is apt, for BEM combines the primary advantage of BIE, viz., reduction in problem dimension, with the primary advantage of FEM, viz., versatility of spatial discretization.

1.1. *Illustrative problem*

A simple one-dimensional problem that demonstrates the attractiveness of BEM is shown in Figure 1. The governing equations for this problem may be written

$$\begin{gathered} m\ddot{w}(t)+b\dot{w}(t)+kw(t)=q(t)-p(0,t)\,, \\ w(0)=w_0, \qquad \dot{w}(0)=\dot{w}_0, \qquad \dot{w}(t)=u(0,t), \qquad p(x,t)=\rho\dot{\phi}(x,t), \qquad (1) \\ u(x,t)=-\phi_{,x}(x,t)\,, \qquad \ddot{\phi}(x,t)=c^2\phi_{,xx}(x,t), \qquad \phi(x,0)=0\,, \end{gathered}$$

where the familiar overhead dot and subscript comma denote temporal and

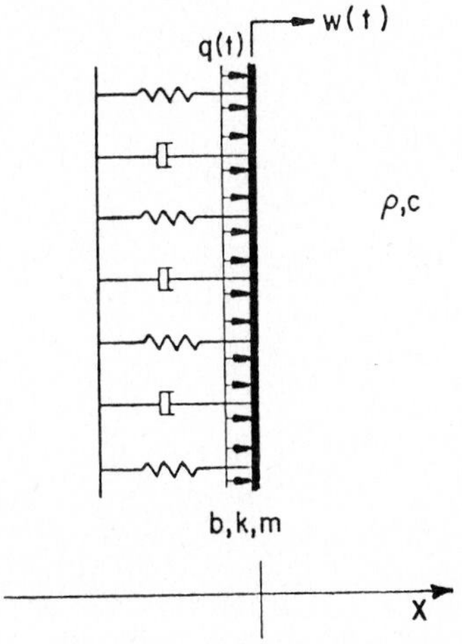

Fig. 1. Infinite plate-oscillator bounding a semi-infinite acoustic medium and excited by a given internal pressure loading $q(t)$ [m, b and k are oscillator mass, damping-constant and spring-constant (per unit plate area), respectively].

spatial differentiation, respectively. The first of (1) is the equation of motion for the plate-oscillator, the second and third are initial conditions for the plate, the fourth enforces compatibility at the fluid-plate interface, the fifth and sixth define fluid pressure and particle velocity in terms of the velocity potential $\phi(x, t)$ (ρ and c are the fluid density and speed of sound, respectively), the seventh states that $\phi(x, t)$ satisfies the wave equation, and the last specifies an initially quiescent fluid.

Strict, i.e., naive application of FEM in order to determine the response of the plate-oscillator would involve one-dimensional discretization of the semi-infinite fluid medium out to a point sufficiently far to the right that waves reflected from the end of the mesh would not return to the surface of the plate-oscillator during the time span of interest. The computational inefficiency of such a procedure is apparent, especially for a broad-spectrum excitation, which would require a refined mesh to accomodate high-frequency components, as well as an extended mesh to prevent premature arrival of the waves reflected from the end of the mesh.

1.2. Indirect BEM approach

Application of BEM to this problem can follow either of two approaches. The indirect approach involves replacing the surface of the plate with acoustic sources that generate a plane acoustic wave that travels to the right. Such a wave has the form $\sigma(t - x/c)$, hence one may express the (one-sided) Laplace transform of $\phi(x, t)$ as

$$\tilde{\phi}(x, s) = \tilde{\sigma}(s)\, e^{-sx/c} . \tag{2}$$

From the fifth and sixth of (1), then $\tilde{p}(x, s) = \rho s\tilde{\sigma}(s)\, e^{-sx/c}$ and $\tilde{u}(x, s) = (s/c)\tilde{\sigma}(s)\, e^{-sx/c}$. The elimination of $\tilde{\sigma}(s)$ from these equations, followed by inverse transformation, then yields $p(x, t) = \rho c u(x, t)$. Most importantly, the pressure-velocity relationship at the boundary is

$$p(0, t) = \rho c u(0, t) . \tag{3}$$

which, when used in conjunction with the first and fourth of (1) yields

$$m\ddot{w} + (b + \rho c)\dot{w} + kw = q(t) . \tag{4}$$

This equation may be regarded as the equation of motion for the submerged plate-oscillator under internal pressure-loading.

1.3. Direct BEM approach

The second BEM approach to the problem of Figure 1 is the direct approach, which is based on Green's second identity [1], written here in terms of the Laplace transforms of two scalar functions ϕ and ψ as

$$\int_V (\tilde{\phi}\nabla^2\tilde{\psi} - \tilde{\psi}\nabla^2\tilde{\phi})\,\mathrm{d}V = -\int_S (\tilde{\phi}\tilde{\psi}_{,n} - \tilde{\psi}\tilde{\phi}_{,n})\,\mathrm{d}S \tag{5}$$

where $\tilde{\phi}$ and $\tilde{\psi}$ possess continuous second derivatives, the volume V is bounded completely by the piece-wise continuous surface S and n is the normal to S pointing *into* V. The application of (5) to the present problem yields

$$\int_0^\infty (\tilde{\phi}\tilde{\psi}_{,xx} - \tilde{\psi}\tilde{\phi}_{,xx})\,\mathrm{d}x = -[\tilde{\phi}\tilde{\psi}_{,x} - \tilde{\psi}\tilde{\phi}_{,x}]_0^\infty . \tag{6}$$

At this point, $\tilde{\psi}(x, s)$ is taken as the fundamental solution for rightward wave propagation, viz., $\tilde{\psi}(x, s) = \mathrm{e}^{-sx/c}$ [cf. (2)]. Hence (6) becomes

$$\int_0^\infty [\tilde{\phi}(s/c)^2\,\mathrm{e}^{-sx/c} - \mathrm{e}^{-sx/c}\tilde{\phi}_{,xx}]\,\mathrm{d}x = [\tilde{\phi}(s/c)\,\mathrm{e}^{-sx/c} + \mathrm{e}^{-sx/c}\tilde{\phi}_{,x}]_0^\infty . \tag{7}$$

But, from the seventh of (1), $\tilde{\phi}_{,xx} = (s/c)^2\tilde{\phi}$, so the integral on the left side of (7) vanishes; all that remains of (7) is then

$$\frac{s}{c}\,\tilde{\phi}(0, s) = -\tilde{\phi}_{,x}(0, s) . \tag{8}$$

Inverse transformation, followed by the use of the fifth and sixth of (1) then yields (3).

1.4. Discussion

It is interesting to note that the direct approach, in contrast to the indirect approach, makes no assumptions regarding the nature of ϕ in the domain other than it satisfy the wave equation and possess continuous second derivatives. Accordingly, it provides no solution for ϕ in the domain. Another interesting point is the smoothing power of Laplace transformation. For example, $\tilde{\psi}(x, s) = \mathrm{e}^{-sx/c}$ corresponds to $\psi(x, t) = \delta(t - x/c)$, i.e., a propagating Dirac delta-function. This function certainly lacks the continuity characteristics stated after (5), whereas $\mathrm{e}^{-sx/c}$ clearly possesses the required smoothness. Similarly, derivative-discontinuities that might appear in $\phi(x, t)$ are nicely smoothed by Laplace transformation, so that the desired pressure-velocity relation given by (3) is readily obtained. In contrast, discontinuous ϕ-derivatives may present problems in the FEM application described earlier.

The preceding applications of BEM to a one-dimensional problem have been so effective that the resulting equation of motion for the submerged plate-oscillator, i.e., (4), is exact! The BEM mesh consists of a single element with uniform distribution of pressure and velocity.

2. BEM in three dimensions

From the preceding, it is clear that the application of BEM requires the determination of a fundamental solution of the field equation in the domain. Usually that solution pertains to an infinite domain, which leads to an integral equation over the entire surface bounding the actual domain. Fundamental solutions have been determined for relatively few field equations. Those of greatest interest to structural dynamicists pertain to Laplace's equation, the wave equation of acoustics, and the equations of elasto-dynamics.

As with FEM, BEM has enjoyed a wider application to static and steady-state problems than to transient ones. Problems in potential flow, elastostatics, and acoustics currently tend to dominate the extensive literature on the integral equations of mechanics in general, and on BEM in particular (see, e.g., [2]–[10]). When proposed for application to problems in bounded domains, BEM is in direct competition with FEM, whose power of local approximation is hard to beat. The argument that BEM possesses computational-efficiency advantages over FEM in some problems do not impress this writer, among others [11]. The argument that BEM can accomodate discontinuities and singularities better than FEM, as seen in the illustrative problem above, is persuasive [12, 13].

For problems in unbounded domains, however, BEM really comes into its own, by virtue of its capacity to reduce problem dimensionality. (Even here, however, FEM is challenging BEM with 'infinite elements' [14] and 'non-reflecting boundaries' [15, 16].) For this reason, the present paper concentrates on the application of BEM to infinite-domain problems. Also, in the interest of simplicity, clarity and brevity, the expository portion of the paper (Section 2) focuses on three-dimensional problems involving the wave equation of acoustics.

2.1. *Singular fundamental solutions – indirect approach*

Almost all of the extant formulations of BEM are based on the use of free-space Green's functions as fundamental solutions [1]. For the Laplace-transformed wave equation of acoustics, viz.

$$c^2\nabla^2\tilde{\phi} = s^2\tilde{\phi}\,, \tag{9}$$

the free-space (monopole) Green's function is [17]

$$\tilde{\psi}(r, s) = \frac{1}{r(p, q)}\,e^{-(s/c)r(p,q)} \tag{10}$$

where $r(p, q)$ is the magnitude of the vector from the 'receiving point' p to the 'transmitting point' q, i.e., $r = |\overrightarrow{pq}|$. It is clear that $\tilde{\psi}(r, s)$ becomes singular as q approaches p, i.e., as $r \to 0$.

In the indirect approach to the use of (10) for BEM in transient acoustics, a distribution of monopole and dipole sources in an infinite medium replaces the fluid boundary such that

$$\tilde{\phi}(P, s) = \int_S [\tilde{\sigma}(Q, s)\tilde{\psi}(R, s) + \tilde{\mu}(Q, s)\tilde{\psi}_{,n_Q}(R, s)]\, dS,$$
$$\tilde{\phi}_{,n_P}(P, s) = \int_S [\tilde{\sigma}(Q, s)\tilde{\psi}_{,n_P}(R, s) + \tilde{\mu}(Q, s)\tilde{\psi}_{,n_P n_Q}(R, s)]\, dS, \tag{11}$$

where P and Q are receiver and transmitter points, respectively, on the boundary S, $R = |\overrightarrow{PQ}|$, and n_P and n_Q are the normals to S at P and Q, respectively, going into V (Figure 2). As $R \to 0$, the singular functions $\tilde{\psi}(R, s)$, $\tilde{\psi}_{,n_P}(R, s)$ and $\tilde{\psi}_{,n_Q}(R, s)$ remain integrable but $\tilde{\psi}_{,n_P n_Q}$ does not. However, a regularization procedure has been developed [18] for the numerical treatment of $\tilde{\psi}_{,n_P n_Q}$.

The discretization of (11) is readily effected through the use of standard finite-element procedures [19] as

$$\tilde{\sigma}(Q, s) \simeq \sum_j N_j(Q)\tilde{\sigma}_j(s), \qquad \tilde{\mu}(Q, s) \simeq \sum_j M_j(Q)\tilde{\mu}_j(s), \tag{12}$$

where the $\tilde{\sigma}_j$ and $\tilde{\mu}_j$ are source values at the nodes of the finite-element model of S, and where the N and M are narrowly based shape functions. Hence (11) lead to

$$\tilde{\phi}(P, s) \simeq \sum_j [\tilde{b}_j(P, s)\tilde{\sigma}_j(s) + \tilde{d}_j(P, s)\tilde{\mu}_j(s)],$$
$$\tilde{\phi}_{,n_P}(P, s) \simeq -\sum_j [\tilde{c}_j(P, s)\tilde{\sigma}_j(s) + \tilde{e}_j(P, s)\tilde{\mu}_j(s)], \tag{13}$$

in which the $\tilde{b}_j(P, s)$, $\tilde{d}_j(P, s)$, $\tilde{c}_j(P, s)$ and $\tilde{e}_j(P, s)$ follow from (11) and (12). BEM solutions based on (13) may then be sought by means of standard collocation, i.e., enforcement of (13) at the boundary nodes. This yields the matrix equations

$$\tilde{\boldsymbol{\phi}}(s) = \tilde{\boldsymbol{B}}(s)\tilde{\boldsymbol{\sigma}}(s) + \tilde{\boldsymbol{D}}(s)\tilde{\boldsymbol{\mu}}(s), \qquad \tilde{\boldsymbol{\phi}}_{,n_P}(s) = -\tilde{\boldsymbol{C}}(s)\tilde{\boldsymbol{\sigma}}(s) - \tilde{\boldsymbol{E}}(s)\tilde{\boldsymbol{\mu}}(s) \tag{14}$$

for which the matrix elements of $\tilde{\boldsymbol{B}}(s)$, $\tilde{\boldsymbol{D}}(s)$, $\tilde{\boldsymbol{C}}(s)$ and $\tilde{\boldsymbol{E}}(s)$ are given by

$$\tilde{b}_{ij}(s) = \int_S N_j(Q)\tilde{\psi}(R_i, s)\, dS, \qquad \tilde{d}_{ij}(s) = \int_S M_j(Q)\tilde{\psi}_{,n_Q}(R_i, s)\, dS,$$
$$\tilde{c}_{ij}(s) = \int_S N_j(Q)\tilde{\psi}_{,n_P}(R_i, s)\, dS, \qquad \tilde{e}_{ij}(s) = \int_S M_j(Q)\tilde{\psi}_{,n_P n_Q}(R_i, s)\, dS, \tag{15}$$

where $R_i = |\overrightarrow{P_iQ}|$.

Now the objective in the derivation of (14) has been to obtain an approximate, but direct, relationship between $\phi(P, t)$ and $\phi_{,n_P}(P, t)$. This is clearly not possible if $\tilde{\sigma}(Q, s)$ and $\tilde{\mu}(Q, s)$ are independent. In a recent study [20], they have been related as

$$\tilde{\sigma}(Q, s) = \alpha\tilde{\nu}(Q, s), \qquad \tilde{\mu}(Q, s) = \beta\tilde{\nu}(Q, s), \tag{16}$$

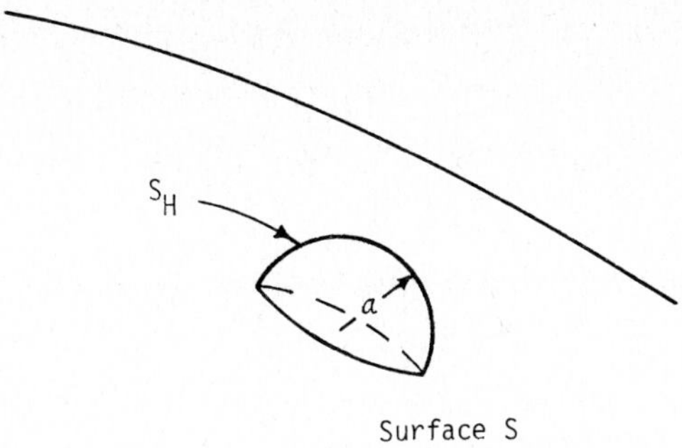

Fig. 2. Evaluation of integrals containing singularities by the construction of a hemispherical dimple on the boundary.

so that (14) become

$$\tilde{\boldsymbol{\phi}}(s)=[\alpha\tilde{\boldsymbol{B}}(s)+\beta\tilde{\boldsymbol{D}}(s)]\tilde{\boldsymbol{\nu}}(s),\qquad \tilde{\boldsymbol{\phi}}_{,n_P}(s)=-[\alpha\tilde{\boldsymbol{C}}(s)+\beta\tilde{\boldsymbol{E}}(s)]\tilde{\boldsymbol{\nu}}(s)\,. \tag{17}$$

The desired relationship between $\phi(P,t)$ and $\phi_{,n_P}(P,t)$ may then be obtained through solution of the second of (17) for $\tilde{\nu}(s)$, substitution of the result in the first of (17) and inversion of the resulting Laplace-transform relation.

The difficulty with the procedure just described, aside from the selection of α and β, is that inversion of the Laplace-transform relation between $\tilde{\boldsymbol{\phi}}(s)$ and $\tilde{\boldsymbol{\phi}}_{,n}(s)$ can only be done numerically. While this is not a crippling difficulty for problems involving simple boundary conditions, it can be deadly for problems in which S is the surface of an FEM model of a complex, even nonlinear, structure. Hence, a better choice of $\tilde{\sigma}(Q,s)$ and $\tilde{\mu}(Q,s)$ is needed.

2.2. *Singular fundamental solutions – direct approach*

What constitutes an excellent choice of $\tilde{\sigma}(Q,s)$ and $\tilde{\mu}(Q,s)$ is provided by the direct approach to the use of (10) for BEM in transient acoustics. First, the use of (9) in (5) yields, for a fixed surface-receiver point P [see (10)],

$$\int_V \{\tilde{\phi}(q,s)[\nabla_q^2\tilde{\psi}(r,s)]_{p=P}-[\tilde{\psi}(r,s)]_{p=P}(s/c)^2\tilde{\phi}(q,s)\}\,\mathrm{d}V_q$$
$$=-\int_S [\tilde{\phi}(Q,s)\tilde{\psi}_{,n_Q}(R,s)-\tilde{\psi}(R,s)\tilde{\phi}_{,n_Q}(Q,s)]\,\mathrm{d}S_Q\,. \tag{18}$$

But, from (10) for $r\neq 0$, $\nabla_q^2\tilde{\psi}(r,s)=(s/c)^2\tilde{\psi}(r,s)$; hence the left side of (10) vanishes. On the right, the functions $\tilde{\psi}(R,s)$ and $\tilde{\psi}_{,n_Q}(R,s)$ are singular as $R\to 0$. By considering the surface S as possessing a small hemispherical dimple with its center at P, however, one may obtain, for continuous $\tilde{\phi}(Q,s)$, (see Figure 2)

$$\int_{S_H}(\tilde{\phi}\tilde{\psi}_{,n_Q}-\tilde{\psi}\tilde{\phi}_{,n_Q})\,\mathrm{d}S_Q$$
$$\simeq -2\pi a^2\left[\tilde{\phi}(P,s)\left(\frac{1}{a^2}+\frac{s}{ac}\right)+\frac{1}{a}\,\tilde{\phi}_{,a}(P,s)\right]\mathrm{e}^{-sa/c}\,. \tag{19}$$

Hence, with the left side of (18) vanishing and a approaching zero, (18) and (19) yield

$$2\pi\tilde{\phi}(P, s) = \int\!\!\!\!-_{S} [\tilde{\phi}(Q, s)\tilde{\psi}_{,n_Q}(R, s) - \tilde{\psi}(R, s)\tilde{\phi}_{,n_Q}(Q, s)]\, \mathrm{d}S_Q \tag{20}$$

where the line through the integral sign indicates omission of P in the integration. Comparison of this equation with the first of (11) reveals that the excellent choices for $\tilde{\sigma}(Q, s)$ and $\tilde{\mu}(Q, s)$ are

$$\tilde{\sigma}(Q, s) = -\frac{1}{2\pi}\phi_{,n_Q}(Q, s), \qquad \tilde{\mu}(Q, s) = \frac{1}{2\pi}\phi(Q, s). \tag{21}$$

Now the preceding development can be extended to accomodate edges and corners in S, and to consider a receiver point off the surface S, either inside or outside the volume V [21]. The result is [cf. (20)]

$$4\pi\varepsilon(p)\tilde{\phi}(p, s) = \int_{S} [\tilde{\phi}(Q, s)\tilde{\psi}_{,n_Q}(r, s) - \tilde{\psi}(r, s)\tilde{\phi}_{,n_Q}(Q, s)]\, \mathrm{d}S_Q \tag{22}$$

where $\varepsilon(p) = 1$ for p in V, $\varepsilon(p) = 0$ for p outside of V and not on S, $\varepsilon(p) = \frac{1}{2}$ for p ($\equiv P$) on smooth portions of S [yielding (20)], and $\varepsilon(p) = \Omega_p/4\pi$ for p ($\equiv P$) on an edge or corner of S, in which Ω_P is the solid angle subtended by V at P.

Fortunately, (22) may immediately be inverse-transformed to yield, using (10),

$$\begin{aligned} 4\pi\varepsilon(p)\phi(p, t) = -\int_{S} \Big[&\frac{1}{r}\phi_{,n_Q}(Q, \langle t - r/c\rangle) + \left(\frac{1}{r^2}\right)\frac{\partial r}{\partial n_Q}\phi(Q, \langle t - r/c\rangle) \\ &+ \left(\frac{1}{rc}\right)\frac{\partial r}{\partial n_Q}\dot{\phi}(Q, \langle t - r/c\rangle)\Big]\, \mathrm{d}S_Q \end{aligned} \tag{23}$$

where the pointed brackets act as parentheses when $t \geq r/c$, but act to zero their dependent functions when $t < r/c$. The key relation (23) is Kirchhoff's retarded-potential integral equation [1, 17].

It is interesting to note that here, in contrast to Section 1.3, the direct approach provides a solution for ϕ in the domain, provided one knows the solution on the boundary. This result is due to the singular behavior of $\tilde{\psi}_{,n_Q}$ as $r \to 0$. One seeks the solution on the boundary by considering receiver points only on S, i.e., by solving (23) with $p = P$ and $r = R$.

BEM solutions to (23) with $p = P$ and $r = R$ may be obtained through boundary discretization of ϕ and $\phi_{,n_Q}$ as

$$\phi(Q, t) = \sum_j N_j(Q)\phi_j(t), \qquad \phi_{,n_Q}(Q, t) = \sum_j M_j(Q)\phi_j'(t) \tag{24}$$

followed by collocation, as described earlier. This yields the matrix equation

$$4\pi\boldsymbol{\varepsilon}\boldsymbol{\phi}(t) = \boldsymbol{F}\boldsymbol{u}\langle t - \bar{\boldsymbol{R}}/c\rangle - \boldsymbol{G}\boldsymbol{\phi}\langle t - \bar{\boldsymbol{R}}/c\rangle - \boldsymbol{H}\dot{\boldsymbol{\phi}}\langle t - \bar{\boldsymbol{R}}/c\rangle \tag{25}$$

where $\boldsymbol{\varepsilon}$ is a diagonal matrix with non-zero elements $\varepsilon(P_i)$, $\boldsymbol{u} = -\boldsymbol{\phi}'$, the matrix elements of $\boldsymbol{F}$, $\boldsymbol{G}$ and $\boldsymbol{H}$ are given by

$$f_{ij} = \int_S \frac{M_j(Q)}{R_i(Q)} \, dS_Q ,$$

$$g_{ij} = \int_S \frac{N_j(Q)}{R_i^2(Q)} \left(\frac{\partial R_i}{\partial n_Q}\right) dS_Q , \tag{26}$$

$$h_{ij} = \frac{1}{c} \int_S \frac{N_j(Q)}{R_i^2(Q)} \left(\frac{\partial R_i}{\partial n_Q}\right) dS_Q ,$$

and the matrix element $\bar{R}_{ij}$ of $\bar{\boldsymbol{R}}$ is merely the distance between the centroids of the ith and j^{th} elements.

Inasmuch as $\boldsymbol{p}(t) = \rho\dot{\boldsymbol{\phi}}$, (25) has accomplished the desired objective of providing a BEM matrix equation explicitly relating $\boldsymbol{p}(t)$ and $\boldsymbol{u}(t)$. Although the numerical solution of (25) by means of time-stepping techniques may appear straightforward, some pitfalls lie in the path. First, the elements f_{ii}, g_{ii} and h_{ii} must be evaluated with care (see, e.g., [22, 23]). Second, if $\phi(t)$ is discontinuous, as in the case of acoustic shockwaves, the differentiation of (25) to obtain surface pressures generates a line-integral contribution by Liebniz' rule [24–26].

2.3. *Regular fundamental solutions*

A formulation of BEM that has not been widely recognized derives from the 'transition matrix' method of Waterman [27]. Originally developed for steady-state acoustic scattering, the method has been extended to the treatment of both acoustic and elastic scattering (see, e.g., [28, 29]).

The foundation of this new method for BEM analysis is the use of regular, as opposed to singular, fundamental solutions of the field equations. For the 3-D, Laplace-transformed equations of acoustics, such fundamental solutions are

$$\tilde{\psi}_{lm}(r, \theta, \varphi, s) = k_m\left(\frac{rs}{c}\right) P^l_m(\cos\theta)\, e^{jl\varphi} \tag{27}$$

where $j = \sqrt{-1}$, $k_m(z)$ is the modified spherical Bessel function of the third kind of order m [30], $P_m(z)$ is the Legendre function of order m and degree l [30], and r, θ and φ are the usual spherical coordinates[1].

The first step in the development of this method consists of expressing $\tilde{\phi}$ as a series in the fundamental solutions as

$$\tilde{\phi}(r, \theta, \varphi, s) = \sum_{j=0}^{\infty} \sum_{i=0}^{j} \tilde{\phi}_{ij}(s)\tilde{\psi}_{ij}(r, \theta, \varphi) \tag{28}$$

and to take $\tilde{\psi} = \tilde{\psi}_{lm}$. Under these conditions, it can be shown that (5) yields (see,

[1]In this method, r is the distance from the origin of the spherical coordinate system (not in the fluid domain V), *not* the distance between receiver and transmitter points, as in previous sections. In the present section, r is never zero.

e.g., [28])

$$\int_S (\tilde{\phi}\tilde{\psi}_{lm,n} - \tilde{\psi}_{lm}\tilde{\phi}_{,n})\,\mathrm{d}S = 0\,. \tag{29}$$

The second step consists of noting that

$$\tilde{\psi}_{lm,n} = \frac{\partial\tilde{\psi}_{mn}}{\partial r}\frac{\partial r}{\partial n} = \frac{s}{c}\frac{\partial r}{\partial n}\,k'_m\left(\frac{rs}{c}\right)P^l_m(\cos\theta)\,\mathrm{e}^{\mathrm{j}l\varphi} \tag{30}$$

where k'_m is the derivative of k_m with respect to argument, and that

$$k_m(z) = \mathrm{e}^{-z}\sum_{i=1}^{m+1}\frac{a_i^{(m)}}{z^i}, \qquad k'_m(z) = \mathrm{e}^{-z}\sum_{i=1}^{m+2}\frac{b_i^{(m)}}{z^i}, \tag{31}$$

where the $a_i^{(m)}$ and $b_i^{(m)}$ are known constants [30]. The third step involves the introduction of (27) and (30) into (29) to obtain

$$\int_S \left[\frac{s}{c}\tilde{\phi}k'_m\left(\frac{rs}{c}\right)\frac{\partial r}{\partial n} - \tilde{\phi}_{,n}k_m\left(\frac{rs}{c}\right)\right]P^l_m(\cos\theta)\,\mathrm{e}^{\mathrm{j}l\varphi}\,\mathrm{d}S = 0 \tag{32}$$

which, following the introduction of (31) and inverse transformation with $p = \rho\dot{\phi}$ and $u = -\phi_{,n}$, yields the time-dependent equation

$$\begin{aligned}\int_S \Bigg[&\left(\frac{b_1^{(m)}}{\rho r}\overset{*}{p}\left\langle t-\frac{r}{c}\right\rangle + \frac{cb_2^{(m)}}{\rho r^2}\overset{**}{p}\left\langle t-\frac{r}{c}\right\rangle + \cdots (m+2\text{ terms})\cdots\right)\frac{\partial r}{\partial n}\\ &+\frac{ca_1^{(m)}}{r}\overset{*}{u}\left\langle t-\frac{r}{c}\right\rangle + \frac{c^2a_2^{(m)}}{r^2}\overset{**}{u}\left\langle t-\frac{r}{c}\right\rangle + \cdots (m+1\text{ terms})\cdots\Bigg]\\ &\times P^l_m(\cos\theta)\,\mathrm{e}^{\mathrm{j}l\varphi}\,\mathrm{d}S = 0\end{aligned} \tag{33}$$

where an overhead asterisk denotes a time integration of the pertinent quantity.

Spatial discretization of $p(Q, t)$ and $u(Q, t)$ may be effected as indicated in (24). This, coupled with $(m+2)$-fold differentiation of (33), yields

$$\begin{aligned}&\sum_{k=1}^{K}\left[\beta_{klm}^{(1)}p_k^{(m+1)}\left\langle t-\frac{\bar{r}_k}{c}\right\rangle + \beta_{klm}^{(2)}p_k^{(m)}\left\langle t-\frac{\bar{r}_k}{c}\right\rangle + \cdots (m+2\text{ terms})\ldots\right]\\ &\quad = -\sum_{k=1}^{K}\left[\alpha_{klm}^{(1)}u_k^{(m+1)}\left\langle t-\frac{\bar{r}_k}{c}\right\rangle + \alpha_{klm}^{(2)}u_k^{(m)}\left\langle t-\frac{\bar{r}_k}{c}\right\rangle + \cdots (m+1\text{ terms})\cdots\right]\end{aligned} \tag{34}$$

where $\bar{r}_k$ is the value of r for the kth node, $p_k^{(m+1)} = \mathrm{d}^{m+1}p/\mathrm{d}t^{m+1}$, etc., and

$$\begin{aligned}\alpha_{klm}^{(i)} &= a_i^{(m)}c^i\int_{S_k}\frac{M_k(Q)}{r^i(Q)}P^l_m(\cos\theta_Q)\,\mathrm{e}^{\mathrm{j}l\varphi_Q}\,\mathrm{d}S\,.\\ \beta_{klm}^{(i)} &= \frac{b_i^{(m)}}{\rho}c^{i-1}\int_{S_k}\frac{N_k(Q)}{r^i(Q)}\left(\frac{\partial r}{\partial n}\right)_Q P^l_m(\cos\theta_Q)\,\mathrm{e}^{\mathrm{j}l\varphi_Q}\,\mathrm{d}S\,,\end{aligned} \tag{35}$$

Now (34) constitutes, for fixed l and m, a single equation relating K variables p_k (and their temporal derivatives) and K variables $\dot{u}_k$ (and their temporal derivatives). The system may be 'squared', of course, by the selection of an appropriate number of combinations of l and m. Hence, as K increases, the maximum values of l and m are increased, and the analysis becomes increasingly refined.

The system of equations given by (34) for various values of l and m appears to be considerably more cumbersome to solve than the singularity-based system given by (25). It avoids, however, the singularity-evaluation and line-integral complications of the latter system mentioned at the end of Section 2.2. Furthermore, the regular-function formulation has exhibited distinct advantages over the singular-function formulation in steady-state applications [31]. Finally, it should be mentioned that the preceding development can be expanded to provide solutions in V as well as on S [27–29].

2.4. Doubly asymptotic approximations

In an attempt to secure simpler and computationally more efficient treatments of boundary interaction problems, numerous investigators have introduced a variety of boundary interaction approximations (see, e.g., [32, 33]). Perhaps the most successful of these for unbounded domains have been the doubly asymptotic approximations (DAA) for acoustic and elastic media [34–38]. The primary attribute of DAA is that they constitute simplified BEM formulations that are asymptotically exact for both low- and high-frequency motions.

The most straightforward derivation of a DAA for transient acoustics proceeds as follows. Consider the pressure/normal-velocity relationship for a boundary-element mesh. In the high-frequency limit, where the acoustic wavelength is much smaller than the characteristic wavelength for boundary motion, this relationship is

$$\boldsymbol{p} = \rho c \boldsymbol{u} \tag{36}$$

where $\boldsymbol{p}$ is the pressure vector and $\boldsymbol{u}$ is the normal-velocity vector. In the hydrodynamic low-frequency limit, where the acoustic wavelength is much larger than a characteristic boundary wavelength, the relationship is

$$\boldsymbol{A}\boldsymbol{p} = \boldsymbol{M}\dot{\boldsymbol{u}} \tag{37}$$

where $\boldsymbol{A}$ is the diagonal area matrix for the boundary-element mesh and $\boldsymbol{M}$ is the frequency-independent fluid-mass matrix for that mesh [23]. From (36) and (37), the simplest combined relationship between $\boldsymbol{p}$ and $\boldsymbol{u}$ that approaches both the high- and low-frequency asymptotic relationships is

$$\boldsymbol{M}\dot{\boldsymbol{p}} + \rho c \boldsymbol{A}\boldsymbol{p} = \rho c \boldsymbol{M}\dot{\boldsymbol{u}} . \tag{38}$$

The simplicity of this expression contrasts sharply with the complexity of the exact relationships given in previous sections.

The question that immediately arises is: how accurate are the DAA? Numerous comparisons between first- and second-order DAA solutions and their exact counterparts for simple structure-medium interaction problems have shown them to be surprisingly accurate for linear media [32, 33, 35, 37] and disappointing for nonlinear media [38]. With regard to the latter, the exact formulations of Sections 2.1–2.3 appear to be strictly limited to the treatment of linear media.

2.5. Concluding remarks for Section 2

This section has focused on BEM in three dimensions. The application of BEM to two-dimensional transient response problems is, of course, possible, but is generally more restricted and often more difficult. For example, the free-space Green's function for the 2-D Laplace-transformed wave equation of acoustics is

$$\tilde{\psi}(r, s) = K_0\left(\frac{rs}{c}\right) \tag{39}$$

where $K_0(z)$ is the modified cylindrical Bessel function of order zero. This function does not lend itself well to formal inversion, in that $K_0(z)$ has a branch point at $z = 0$ [30]. Hence one is usually constrained to seek solutions by direct numerical inversion, as, e.g., in [39, 40], or by treatment of 2-D problems as 3-D problems for cylindrical surfaces (see, e.g., [41]). The same is true for approaches based on regular fundamental solutions. For example, the 2-D counterpart to (27) is

$$\tilde{\psi}_m(r, \theta, s) = K_m\left(\frac{rs}{c}\right) e^{jm\theta} . \tag{40}$$

Even in three dimensions, the exact BEM formulations must be considered complex. If manifested in the form of highly reliable computer codes, this may not be a critical problem, however. On the other hand, the construction of such codes is a complicated process, which provides the necessary motivation to pursue DAA-based codes (see, e.g., [42–44]).

It should be remembered that the ability to calculate solutions out in the domain, following the generation of appropriate boundary solutions, is possessed by both singular and regular BEM formulations. In the case of the singular BEM formulation, such calculations can be difficult and costly for derivative quantities (such as strains), however, because of the highly singular nature of the associated Green's functions. Hence recourse is sometimes made to finite-difference calculations of derivative quantities [38, 45].

Finally, it should be reiterated that the primary objective of the preceding developments has been to obtain a relationship (preferably in the time domain) between appropriate conjugate mechanical quantities, such as force and displacement, pressure and velocity, etc., at the boundary of a domain of interest. With respect to these developments, perhaps the best way to illustrate the motivation behind the objective is to expand on the illustrative problem of Section 1, as follows.

Consider a three-dimensional structure submerged in an infinite acoustic medium. The equations of motion for a finite-element model of that structure may be written in matrix form as

$$\boldsymbol{M}_s\ddot{\boldsymbol{x}} + \boldsymbol{C}_s\dot{\boldsymbol{x}} + \boldsymbol{K}_s\boldsymbol{x} = \boldsymbol{f} - \boldsymbol{G}\boldsymbol{A}\boldsymbol{p} \tag{41}$$

where $\boldsymbol{x}$ is the structural displacement vector, $\boldsymbol{M}_s$, $\boldsymbol{C}_s$ and $\boldsymbol{K}_s$ are the structural mass, damping and stiffness matrices, respectively, $\boldsymbol{f}$ is a known external force vector, $\boldsymbol{G}$ is the transformation matrix that relates the structural (finite-element) and fluid (boundary-element) nodal forces at the wet surface of the structure, $\boldsymbol{A}$ is the diagonal wet-surface area matrix, and $\boldsymbol{p}$ is the wet-surface fluid-pressure vector. These equations are complemented by a normal-velocity compatibility statement at the wet surface of the structure, viz.

$$\boldsymbol{G}^{\mathrm{T}}\dot{\boldsymbol{x}} = \boldsymbol{u}\,, \tag{42}$$

where $\dot{x}$ pertains to the finite-element mesh and $\boldsymbol{u}$ pertains to the boundary-element mesh, and where the superscript T denotes matrix transposition. The appearance of $\boldsymbol{G}^{\mathrm{T}}$ in (42) follows from the invariance of virtual work with respect to either the wet-surface portion of the finite-element mesh or the 'completely wet' boundary-element mesh. Generally, $\boldsymbol{G}$ is a rectangular matrix whose height greatly exceeds its width, inasmuch as the number of structural (finite-element) degrees of freedom usually exceeds considerably the number of fluid (boundary-element) degrees of freedom.

If no excitation sources are present in the fluid, $\boldsymbol{p}$ and $\boldsymbol{u}$ pertain to waves radiated outward by the structure. Then the establishment of a BEM-based relationship between $\boldsymbol{p}$ and $\boldsymbol{u}$ 'closes the loop', thereby making possible a numerical solution of the problem. If prescribed excitation sources are present in the fluid, $\boldsymbol{p}$ and $\boldsymbol{u}$ pertain to the combined effects of known incident waves and unknown scattered waves, i.e.

$$\boldsymbol{p} = \boldsymbol{p}_{\mathrm{I}} + \boldsymbol{p}_{\mathrm{S}}, \qquad \boldsymbol{u} = \boldsymbol{u}_{\mathrm{I}} + \boldsymbol{u}_{\mathrm{S}}\,. \tag{43}$$

Then it is the establishment of a BEM-based relationship between $\boldsymbol{p}_{\mathrm{S}}$ and $\boldsymbol{u}_{\mathrm{S}}$ that closes the loop.

3. BEM solutions

This section describes solutions to specific problems, both simple and complex, obtained through the use of BEM in transient response analysis. The survey focuses on incompressible flow problems (Laplace's equation), acoustics problems (the wave equation), and problems in dynamic elasticity (equations of elastodynamics).

3.1. *Incompressible flow*

The use of BEM to treat added-mass effects associated with transient, inviscid, irrotational, incompressible flow about a submerged body or within a container is relatively straightforward [23, 46, 47]. A recent application of a dipole formulation of BEM to a complex fluid-structure interaction problem in the nuclear reactor industry is described in [48]. Figure 3 (taken from [48]) illustrates the complexity of the problem and the fluid-boundary discretization used.

Recently, BEM have been applied to incompressible flow problems in un-

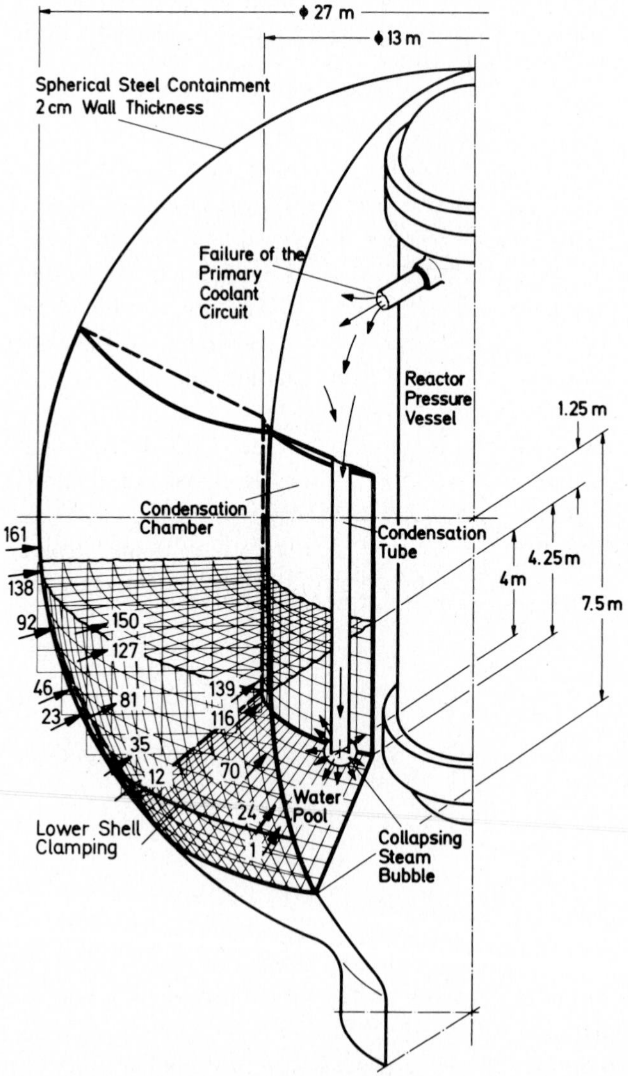

Fig. 3. BEM model of a pressure-suppression pool for a boiling water reactor (arrows indicate nodal points of particular interest in [48]).

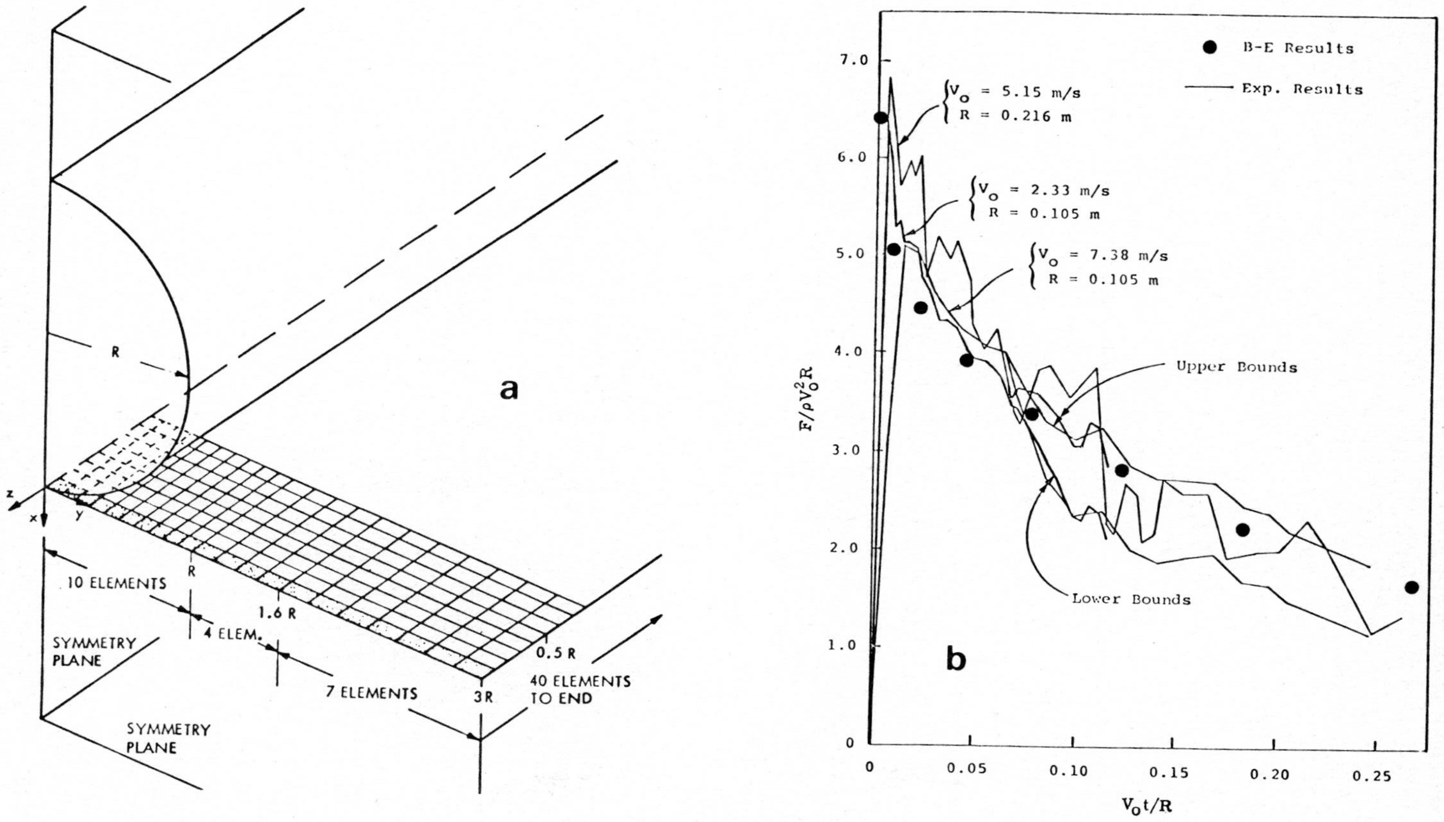

Fig. 4. BEM calculation of the force on a rigid cylinder impacting the free surface on an incompressible fluid at constant velocity [50]. (a) Computational quarter-model, (b) Normalized force per unit length.

bounded domains with a free surface. One of these is fluid-structure impact characteristic of ship slamming [49, 50]. Figure 4a shows a computational quarter-model of an infinite, rigid, circular cylinder suffering normal impact on the surface of an infinite, incompressible fluid; computational and experimental results are compared in Figure 4b. Another such problem is the transient scattering of ocean waves by an island [51]. In [51], the fluid domain in the immediate vicinity of the island, which is generally characterized by a variable depth, is treated by finite-differences. The remainder of the fluid domain, which is considered to possess a fixed depth, is treated by BEM. Figure 5a shows the geometry for a circular island; Figure 5b shows some computational results for that case. Actually, the effects of gravity in this second problem establishes the wave equation, not Laplace's equation, as the governing equation; hence Figure 5 pertains both to this section and to the one following.

3.2. Transient acoustics

A considerable number of BEM solutions for transient acoustic problems have appeared in the literature [24–26, 52–58]. Most of these solutions pertain to idealized boundary conditions for very simple boundary geometries. An exception is the analysis of [26], which deals with the excitation of a submerged spherical shell by a transient acoustic wave.

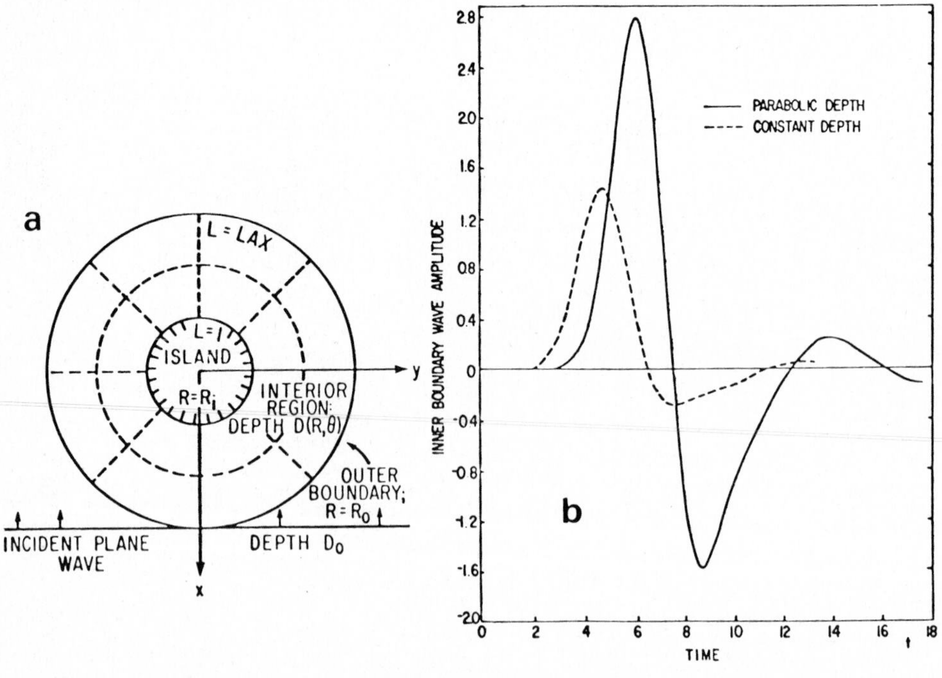

Fig. 5. BEM boundary for a finite-difference treatment of a Tsunami problem [51]. (a) Geometry for a circular island. (b) Wave-height history for a Gaussian incident wave.

The problem of [26] is illustrated in Figure 6a, where the incident wave is taken as a step-wave. Velocity response histories produced by BEM for the fluid and FEM for the shell are compared with corresponding histories produced by a classical separation-of-variables approach in Figure 6b. The comparisons shown in this figure [59] exhibit better agreement than that exhibited in [26], thus indicating steady improvement in the technology.

As mentioned above, a number of DAA-based, general-purpose response codes have been built for the analysis of structures submerged in an acoustic medium. Most of these use a BEM treatment of the wet surface [42, 44, 60, 61], but some do not [43, 62]. BEM-generated DAA results for an infinite, circular, cylindrical shell excited by a transverse, plane, acoustic step-wave are compared with their analytical-DAA and analytical-exact counterparts in Figure 7. Satisfactory agreement is observed.

3.3. Dynamic elasticity

Relatively few BEM solutions for elasto-dynamic problems have appeared in the literature, and these pertain to simple boundary conditions and boundary geometries [40, 63, 64]. One such problem involves determination of the dynamic

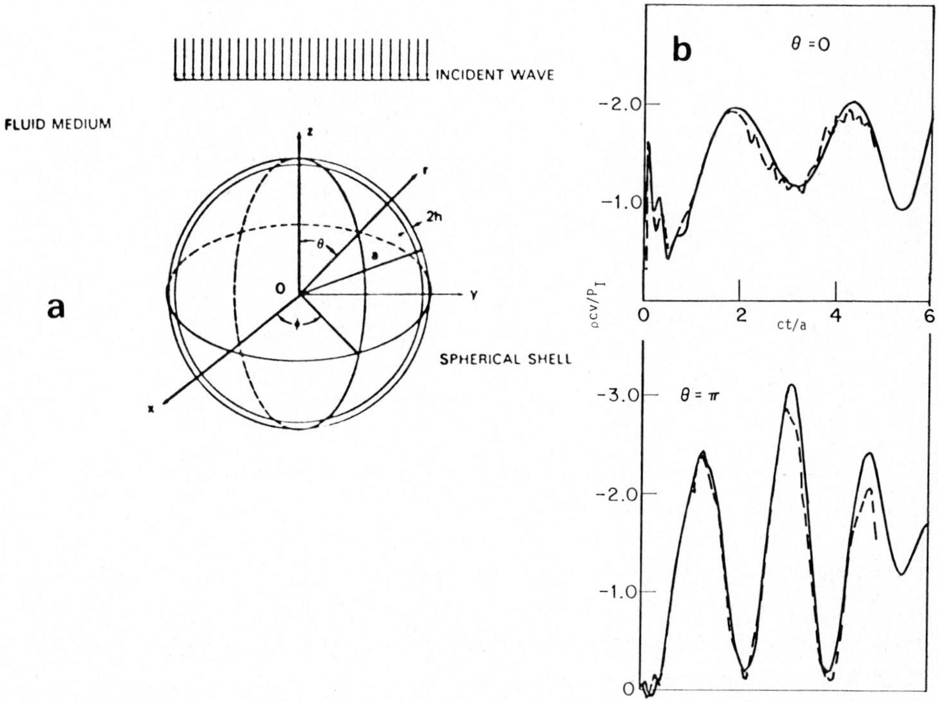

Fig. 6. BEM/FEM analysis of a step-wave-excited, submerged spherical shell [59]. (a) Geometry of problem, (b) Velocity response histories (solid line: separation of variables solution; dashed line: BEM/FEM solution).

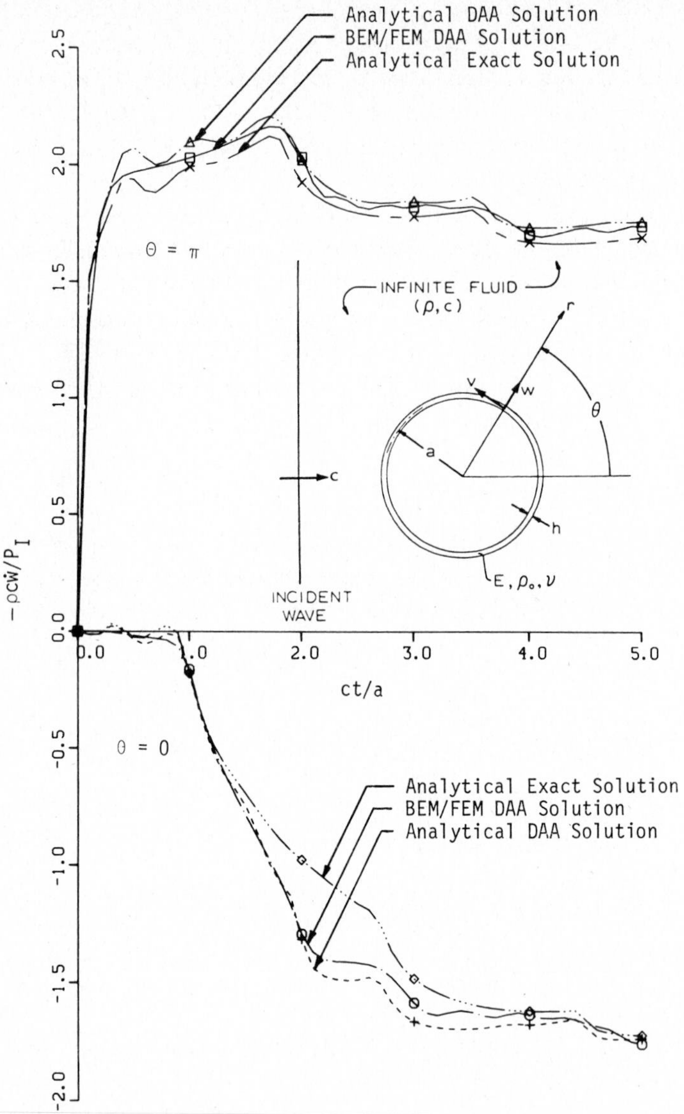

Fig. 7. Radial-velocity response histories for a step-wave-excited, infinite cylindrical shell submerged in an infinite acoustic medium.

stress-concentration factor for a circular hole in an infinite plate excited by a dilatational step-wave. A BEM solution for this problem is compared with circumferential Fourier Series solutions in Figure 8, which is adapted from [64]. Satisfactory agreement is observed.

DAA-based BEM solutions for some simple problems are presented in [37]. Figure 9, which is taken from that paper, pertains to the two-dimensional problem of an infinite cylindrical shell embedded in an infinite, elastic medium that is

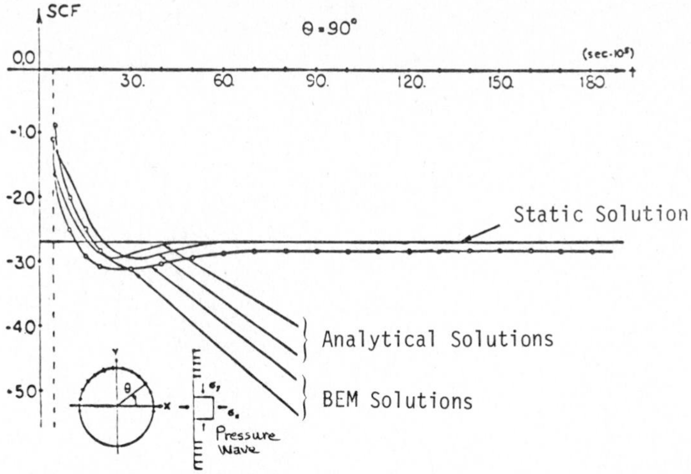

Fig. 8. Dynamic stress-concentration factor for a circular hole in an infinite plate [64].

excited by a plane, transverse, dilatational step-wave propagating through the medium. In the figure, stress-response histories produced by a DAA/BEM treatment of the structure-medium interaction along with an FEM treatment of the shell are compared with histories based on circumferential Fourier Series solutions pertaining to both an exact and a DAA treatment of the structure-medium interaction [65]. It is seen that the primary error committed by the DAA is ommission of the modest 'under-shoot' exhibited by the exact response histories.

4. Conclusion

Although the boundary-element method has been used to treat a host of static and steady-state problems, it has enjoyed only limited application to transient problems. General-purpose BEM codes for transient problems apparently exist only for ideal fluids or DAA treatments of acoustic media. This leaves much room for further development of the technology.

Acknowledgement

The author expresses his appreciation to his colleagues, Dr. J.A. DeRuntz (for generating Figure 7 of this paper), and Dr. C.A. Felippa and Mr. P.G. Underwood (for providing reference material from their libraries). He also wishes to thank the authors and publishers of [37], [48], [50], [51], [59] and [64] for permitting reproduction of their figures in this paper.
The preparation of this paper has been sponsored by the Independent Research Program of Lockheed Missiles and Space Company.

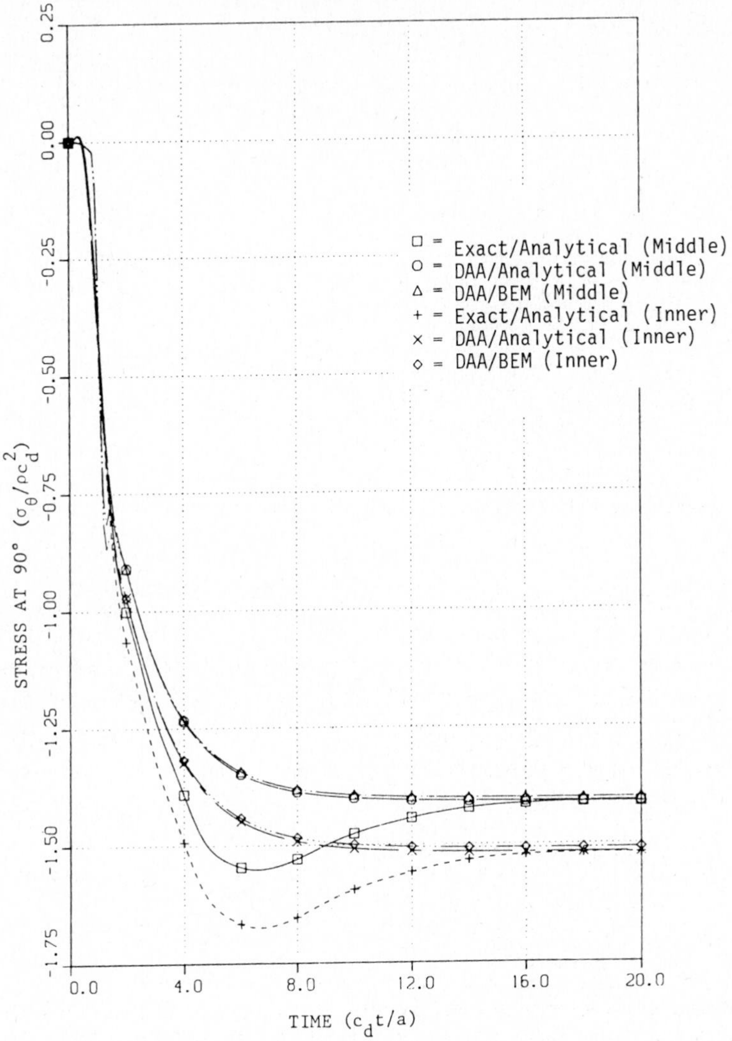

Fig. 9. Stress response in the middle and inner fibers of an infinite cylindrical shell, embedded in an infinite elastic medium, to an incident dilatational step-wave [37].

References

[1] S.L. Sobolev, *Partial Differential Equations of Mathematical Physics, Trans*, by E.R. Dawson (Pergamon Press, Oxford, 1964).

[2] C.A. Brebbia (editor), *Proc. 3rd Int. Seminar on Recent Advances in Boundary Element Methods* (1981).

[3] R.P. Shaw et al. (editors), *Innovative Numerical Analysis for the Engineering Sciences* (University Press of Virginia, Charlottesville, 1980).

[4] R.S. Anderssen et al. (editors), *The Application and Numerical Solution of Integral Equations* (Sijthoff and Noordhoff, Alphen aan den Rijn. The Netherlands, 1980).

[5] B. Hunt, (editor), *Numerical Methods and Applied Fluid Dynamics* (Academic Press, London, 1980).
[6] P.K. Banerjee and R. Butterfield (editors), *Developments in Boundary Element Methods – I* (Applied Science Publishers, London, 1980).
[7] T.A. Cruse et al. (editors), *Proc. Int. Symp. on Innovative Numerical Analysis in Applied Engineering Science*, Versailles (1977).
[8] M.A. Jaswon and G.T. Symm, *Integral Equation Methods in Potential Theory and Elastostatics* (Academic Press, London, 1977).
[9] O.C. Zienkiewicz et al., The Coupling of the Finite Element Method and Boundary Solution Procedures, *Int. J. Num. Meth. Eng.* **11**, 355–375 (1977).
[10] T.A. Cruse and F.J. Rizzo (editors), *Boundary Integral Equation Method: Computational Applications in Applied Mechanics*, AMD – Vol. 11 (Amer. Soc. Mech. Eng., New York, 1975).
[11] P. Bettess, Operation Counts for Boundary Integral and Finite Element Methods, *Int. J. Num. Meth. Eng.* **17**, 306–308 (1981).
[12] T.A. Cruse, Application of the Boundary Integral Equation Method to Three-Dimensional Stress Analysis, *Comp. and Struc.* **3**, 509–527 (1973).
[13] T.A. Cruse and R.B. Wilson, Advanced Applications of Boundary-Integral Equation Methods, *Nucl. Eng. and Design* **46**, 223–234 (1978).
[14] P. Bettess, Infinite Elements, *Int. J. Num. Meth. Eng.* **11**, 53–64 (1977).
[15] W.D. Smith, A Non-Reflecting Plane Boundary for Wave Propagation Problems, *J. Comp. Phys.* **15**, 492–503 (1974).
[16] P.A. Cundall et al., Solution of Infinite Dynamic Problems by Finite Modeling in the Time Domain, *Proc. 2nd Int. Conf. Appl. Num. Modeling*, Madrid (1978).
[17] B.B. Baker and E.T. Copson, *The Mathematical Theory of Huygens' Principle* (Oxford University Press, Oxford, 1953).
[18] A.J. Burton and G.F. Miller, The Application of Integral Equation Methods to the Numerical Solution of Some Exterior Boundary Value Problems, *Proc. Royal Soc.* **323**, 201–210 (1971).
[19] O.C. Zienkiewicz, *The Finite Element Method* (McGraw Hill, London, 1977).
[20] P. Filippi, Layer Potentials and Acoustic Diffraction, *J. Sound and Vib.* **54**, 473–500 (1977).
[21] V.D. Kupradze, Dynamical Problems in Elasticity, in: *Progress in Solid Mechanics, Vol. 3*, I.N. Sneddon and R. Hill, eds. (North Holland, Amsterdam, 1963).
[22] J.A. Lachat and J.O. Watson, Effective Numerical Treatment of Boundary Integral Equations, *Int. J. Num. Meth. Eng.* **10**, 991–1005 (1976).
[23] J.A. DeRuntz and T.L. Geers, Added Mass Computation by the Boundary Integral Method, *Int. J. Meth. Engr.* **12**, 531–549 (1978).
[24] M.B. Friedman and R. Shaw, Diffraction of Pulses by Cylindrical Obstacles of Arbitrary Cross-Section, *J. Appl. Mech.* **29**, 40–46 (1962).
[25] R.P. Shaw, Diffraction of Pulses by Obstacles of Arbitrary Shape with an Impedance Boundary Condition, *J. Acoust. Soc. Amer.* **44**, 1962–1968 (1968).
[26] H. Huang et al., Retarded Potential Techniques for the Analysis of Submerged Structures Impinged by Weak Shock Waves, in: *Computational Methods for Fluid-Structure Interaction Problems*, T. Belytschko and T.L. Geers, eds., AMD – Vol. 26 (Amer. Soc. Mech. Eng., New York, 1977) pp. 83–93.
[27] P.C. Waterman, New Formulation of Acoustic Scattering, *J. Acoust. Soc. Amer.* **45**, 1417–1429 (1969).
[28] Y.H. Pao, The Transition Matrix for the Scattering of Acoustic Waves and for Elastic Waves, *Proc. IUTAM Symp. on Modern Problems in Elastic Wave Propagation*, J. Miklowitz and J. Achenbach, eds. (Wiley Interscience, New York, 1978).
[29] Y.H. Pao, Bettie's Identity and Transition Matrix for Elastic Waves, *J. Acoust. Soc. Amer.* **64**, 302–310 (1978).
[30] M. Abramowitz and I.A. Stegun, *Handbook of Mathematical Functions*, Natl. Bur. of Stds. Appl. Math. Series 55 (U.S. Dept. of Commerce, 1964).
[31] J.C. Bolomey and A. Wirgin, Numerical Comparison of the Green's Function and the Waterman and Rayleigh Theories of Scattering from a Cylinder with Arbitrary Cross-Section, *Proc. Inst. Elec. Eng.* **12**, 794–804 (1974).
[32] T.L. Geers, Shock Response Analysis of Submerged Structures, *Shock and Vib. Bull.* **44**, Suppl. 3, 17–32 (1974).

[33] T.L. Geers, Transient Response Analysis of Submerged Structures, in: *Finite Element Analysis of Transient Nonlinear Structural Behavior*, T. Belytschko et al., eds., AMD – Vol. 14 (Amer. Soc. Mech. Eng., New York, 1975) pp. 59–84.
[34] T.L. Geers, Residual Potential and Approximate Methods for Three-Dimensional Fluid-Structure Interaction Problems, *J. Acoust. Soc. Amer.* **49**, 1505–1510 (1971).
[35] T.L Geers, Doubly Asymptotic Approximations for Transient Motions of Submerged Structures, *J. Acoust. Soc. Amer.* **64**, 1500–1508 (1978).
[36] C.A. Felippa, Top-Down Derivation of Doubly Asymptotic Approximations for Structure Fluid Interaction Analysis, pp. 79–88 in [3].
[37] P.G. Underwood and T.L. Geers, Doubly-Asymptotic Boundary-Element Analysis of Dynamic Soil-Structure Interaction, *Int. J. Solids and Struct.* **17**, 687–697 (1981).
[38] P.G. Underwood and T.L. Geers, Doubly-Asymptotic Boundary-Element Analysis of Nonlinear Soil-Structure Interaction, pp. 413–422 in [3].
[39] T.A. Cruse and F.J. Rizzo, A Direct Formulation and Numerical Solution of the General Transient Elastodynamic Problem – I, *J. Math. Anal. Appl.* **22**, 244–259 (1968).
[40] T.A. Cruse, A Direct Formulation and Numerical Solution of the General Transient Elastodynamic Problem – II, *J. Math. Anal. Appl.* **22**, 341–355 (1968).
[41] R.P. Shaw, Boundary Integral Equation Methods Applied to Wave Problems, pp. 121–153 in [6].
[42] J.A. DeRuntz et al., The Underwater Shock Analysis Code (USA-Version 3), a Reference Manual, DNA 5615, Defense Nuclear Agency, Washington, DC (Sept. 1980).
[43] D. Ranlet et al., Elastic Response of Submerged Shells with Internally Attached Structures to Shock Loading, *Comp. and Struc.* **7**, 355–364 (1978).
[44] R.F. Jones et al., Transient Response of Submerged Shell Structures, in: *Computational Methods for Offshore Structures*, H. Armen and S. Stiansen, eds., AMD – Vol. 37 (Amer. Soc. Mech. Eng., New York, 1980) pp. 1–7.
[45] D.N. Cathie and P.K. Banerjee, Numerical Solutions in Axisymmetric Elastoplasticity by the Boundary Element Method, pp. 331–339 in [3].
[46] G.R. Khabbaz, Dynamic Behavior of Liquids in Elastic Tanks, *AIAA J.* **9**, 1985–1990 (1971).
[47] Y. Ousset, Hydroelastic Vibrations of Tanks: Added Mass Computation with an Integral Equation Method, pp. 107–118 in [3].
[48] R. Krieg et al., A Boundary Integral Method for Highly Transient Internal Flow Problems Coupled with Structural Dynamics, pp. 89–98 in [3].
[49] T.L. Geers et al., Boundary Element Analysis of Fluid-Solid Impact, in: *Computational Methods for Fluid-Structure Interaction Problems*, T. Belytschko and T.L. Geers, eds., AMD – Vol. 26 (Amer. Soc. Mech. Eng., New York, 1977) pp. 125–138.
[50] T.L. Geers, A Boundary Element Method for Slamming Analysis, *J. Ship Res.* **26**, 117–124 (1981).
[51] R.P. Shaw, An Outer Boundary Integral Equation Applied to Transient Wave Scattering in an Inhomogeneous Medium, *J. Appl. Mech.* **42**, 147–152 (1975).
[52] R.P. Shaw, Diffraction of Acoustic Pulses by Obstacles of Arbitrary Shape with a Robin Boundary Condition, *J. Acoust. Soc. Amer.* **41**, Part A, 855–859 (1967).
[53] K.M. Mitzner, Numerical Solution for Transient Scattering from a Hard Surface of Arbitrary Shape – Retarded Potential Technique, *J. Acoust. Soc. Amer.* **42**, 391–397 (1967).
[54] C.L.S. Farn and H. Huang, Transient Acoustic Fields Generated by a Body of Arbitrary Shape, *J. Acoust. Soc. Amer.* **43**, 252–257 (1968).
[55] R.P. Shaw and D.F. Courtine, Diffraction of a Plane Acoustic Pulse by a Free Orthogonal Trihedron, *J. Acoust. Soc. Amer.* **46**, 1382–1384 (1969).
[56] R.P. Shaw and J.A. English, Transient Acoustic Scattering by a Free Sphere, *J. Sound and Vib.* **20**, 321–331 (1972).
[57] R.P. Shaw, Transient Scattering by a Circular Cylinder, *J. Sound and Vib.* **42**, 295–304 (1975).
[58] Y.P. Lu, The Application of Retarded Potential Techniques to Submerged Dynamic Structural Systems, pp. 59–68 in [3].
[59] H. Huang, Private communication (June 1981).
[60] J.A. DeRuntz and F.A. Brogan, Underwater Shock Analysis of Nonlinear Structures, A Reference Manual for the USA-STAGS Code (Version 3), DNA 5545, Defense Nuclear Agency, Washington, DC (1980).

[61] G.C. Everstine, A NASTRAN Implementation of the Doubly Asymptotic Approximation for Underwater Shock Response, NASTRAN: Users' Experiences, NASA TM X3428 (1976) pp. 207–228.

[62] R. Atkatsh and R.P. Daddazio, Dynamic Elastoplastic Response of Shells in an Acoustic Medium: Users Manual for the EPSA Code, Technical Rpt. No. 27, Weidlinger Associates, New York (1980).

[63] R.P. Shaw, Retarded Potential Approach to the Scattering of Elastic Pulses by Rigid Obstacles of Arbitrary Shape, *J. Acoust. Soc. Amer.* **44**, 745–748 (1968).

[64] G.G. Manolis and D.E. Beskos, Dynamic Stress Concentration Studies by the Boundary Integral Equation Method, pp. 459–463 in [3].

[65] T.L. Geers and C.-L. Yen, Doubly Asymptotic Methods for Transient Motions of Buried Structures, in: *Theoretical and Applied Mechanics*, Proc. XIV IUTAM Congress, W.T. Koiter, ed. (North-Holland, Amsterdam, 1977).

CHAPTER 5

Dynamic Relaxation

Philip UNDERWOOD

Applied Mechanics Laboratory
Lockheed Palo Alto Research Laboratory
3251 Hanover Street
Palo Alto, CA 94304, USA

Computational Methods for Transient Analysis
Edited by T. Belytschko and T.J.R. Hughes

1. Introduction

In this chapter a review of dynamic relaxation (DR), an explicit iterative method for the static solution of structural mechanics problems, is presented. DR is based on the fact that the static solution is the steady state part of the transient response for a temporal-step load. Hence for structural mechanics problems the familiar structural dynamics equations are the governing equations. This review of DR will be presented from a structural dynamics viewpoint in keeping with the subject matter of this book: transient response.

The DR method is especially attractive for problems with highly nonlinear geometric and material behavior, which include limit points or regions of very soft stiffness characteristics. The explicit nature of the method makes it highly suitable for computations because all quantities may be treated as vectors, resulting in an easily programmed method with low storage requirements. In many cases the number of iterations to obtain convergence may be quite large, but the DR method pursues its goal with a tenacity seldom found in many computational methods. This combination of simplicity and tenacity result in an efficient solution method for nonlinear problems. Here efficiency is assumed to be based on time from start to finish of an analysis.

This chapter begins with a section on the development of the DR method presented from a structural dynamics viewpoint. This development will mostly parallel the historical path that was followed so that the reader may see how the current usage of DR evolved. The next section contains a brief presentation of a recently developed adaptive DR method for nonlinear structural analysis. Two numerical examples are shown which illustrate the DR method. The chapter concludes with some remarks on what the next improvements in DR may be.

2. Development

This section on the development of DR will primarily focus on the linear problem. A few aspects of the nonlinear problem will be introduced but the nonlinear problem will not be considered until Section 3. In the first portion of this section background on the history of DR, formulation of the equations for structural analysis, and the central difference time integration method are presented. In the second portion of this section the DR algorithm properties and some methods of applying DR are given.

2.1. *History*

The name 'dynamic relaxation' appears to have been coined by Otter [1–3] or Day [4] in the mid-1960's. These papers [1–4] represent the beginning of the engineer's interest in DR and introduce the idea of obtaining a static solution from a dynamic transient analysis method. The DR method originates from the 2nd order Richardson method developed by Frankel [5] in 1950. Frankel states that the formal equivalence of the Richardson algorithm [6] to first order time dependent equations suggests the extension to a solution algorithm equivalent to a second order time dependent equation. From this it would appear Frankel first made the connection with dynamics. However, the first exploitation of computational structural dynamics methods with regard to improving DR appears to have been made by either Welsh [7] or Cassell et al. [8], who introduced the idea of fictitious density (mass). Rushton [9] appears to have made the first application of DR to a nonlinear problem. Another early paper on DR by Wood [10] compares DR with other iterative methods. Since 1970 DR literature has expanded considerably. The papers by Cassell [11], Ruston [1], Alwar [13, 14], Frieze [15], Pica [16], and Key [17] are mainly concerned with applications of DR to certain structural analyses. Papers by Brew [18], Wood [19], Bunce [20], Alwar [21], Cassell [22] and Papadrakakis [23] are concerned with understanding and improving the DR method. The paper by Papadrakakis [23] is of special interest in that it presents a method for the automatic selection of DR iteration parameters. Also, the author will present some previously unpublished work [24] on an adaptive DR method in Section 3.

2.2. *Structural analysis formulations*

In this presentation the equations governing the structural behavior are assumed to have been derived from a finite difference [25] or finite element [26] discretization method. This discrete equation is considered to be in the form

$$\boldsymbol{P}(\boldsymbol{q}) = \boldsymbol{f} \tag{2.1}$$

where $\boldsymbol{P}$ is the vector of internal forces, $\boldsymbol{q}$ is the vector of dependent discrete variables (the solution vector), and $\boldsymbol{f}$ is the vector of external forces. In general $\boldsymbol{P}$ is obtained from variational principles, so

$$\boldsymbol{P}(\boldsymbol{q}) = \frac{\partial E(\boldsymbol{q})}{\partial \boldsymbol{q}} \tag{2.2}$$

where E is the internal energy. For the linear problem $\boldsymbol{P}$ is commonly in the form

$$\boldsymbol{P}(\boldsymbol{q}) = \boldsymbol{K}\boldsymbol{q}\,, \tag{2.3}$$

where $\boldsymbol{K}$ is the stiffness matrix. For the nonlinear problem $\boldsymbol{P}$ is commonly in the incremental form

$$\Delta P(\boldsymbol{q}) = \boldsymbol{K}(q)\Delta q\,. \tag{2.4}$$

where $\boldsymbol{K}(\boldsymbol{q})$ is the tangent stiffness matrix obtained from

$$\boldsymbol{K}(\boldsymbol{q}) = \frac{\partial \boldsymbol{P}(\boldsymbol{q})}{\partial \boldsymbol{q}} = \frac{\partial^2 E(\boldsymbol{q})}{\partial \boldsymbol{q} \partial \boldsymbol{q}} . \tag{2.5}$$

Another form for the nonlinear problem is the pseudo-force representation [27],

$$\boldsymbol{P}(\boldsymbol{q}) = \boldsymbol{K}\boldsymbol{q} + \boldsymbol{Q}(\boldsymbol{q}) , \tag{2.6}$$

where $\boldsymbol{K}$ is a linear stiffness matrix and $\boldsymbol{Q}$ (the pseudo-force) accounts for the nonlinear behavior.

For nonlinear static analysis the incremental form; (2.4) is most common. However, in nonlinear structural dynamic analysis by an explicit method the internal forces are most efficiently computed in the vector form, (2.2). The pseudo-force form is commonly used for implicit methods and this $\boldsymbol{P}$ is identical in value to the vector form if derived from the same internal energy functional. Note that, $\boldsymbol{P}$ can also be obtained by integrating (2.4). This is not a trivial integration, and the integration must be performed with great care [28, 29].

2.3. *Explicit, transient structural dynamics and DR*

Since the DR method is based on an equivalence to the second order time dependent equation, the equation of motion governing structural dynamic response is the appropriate equation for developing the DR method for structural analysis. For the nth time increment this equation is given by

$$\boldsymbol{M}\ddot{\boldsymbol{q}}^n + \boldsymbol{C}\dot{\boldsymbol{q}}^n + \boldsymbol{P}(\boldsymbol{q}^n) = \boldsymbol{f}(t^n) \tag{2.7}$$

where $\boldsymbol{M}$ is the mass matrix, $\boldsymbol{C}$ is the damping matrix, t is time, n indicates the nth time increment, a superimposed dot indicates a temporal derivative, and the other terms are defined in Section 2.2. To obtain the DR algorithm (2nd order Richardson) developed by Frankel [5], the following central difference expressions are used for the temporal derivatives

$$\dot{\boldsymbol{q}}^{n-1/2} = (-\boldsymbol{q}^{n-1} + \boldsymbol{q}^n)/h, \qquad \ddot{\boldsymbol{q}}^n = (-\dot{\boldsymbol{q}}^{n-1/2} + \dot{\boldsymbol{q}}^{n+1/2})/h , \tag{2.8}$$

where h is a fixed time increment. The expression for $\dot{\boldsymbol{q}}^n$ is obtained by the average value:

$$\dot{\boldsymbol{q}}^n = \tfrac{1}{2}(\dot{\boldsymbol{q}}^{n-1/2} + \dot{\boldsymbol{q}}^{n+1/2}) . \tag{2.9}$$

Substituting (2.8) and (2.9) into (2.7) gives the pair of equations used to advance to the next velocity and displacement as

$$\begin{aligned} \dot{\boldsymbol{q}}^{n+1/2} &= \frac{(\boldsymbol{M}/h - \frac{1}{2}\boldsymbol{C})}{(\boldsymbol{M}/h + \frac{1}{2}\boldsymbol{C})}\dot{\boldsymbol{q}}^{n-1/2} + \frac{(\boldsymbol{f}^n - \boldsymbol{P}^n)}{(\boldsymbol{M}/h + \frac{1}{2}\boldsymbol{C})}, \\ \boldsymbol{q}^{n+1} &= \boldsymbol{q}^n + h\dot{\boldsymbol{q}}^{n+1/2} , \end{aligned} \tag{2.10}$$

where $\boldsymbol{f}^n = \boldsymbol{f}(t^n)$ and $\boldsymbol{P}^n = \boldsymbol{P}(\boldsymbol{q}^n)$.

To preserve the explicit form of the central difference integrator $\boldsymbol{M}$ must be diagonal and to obtain the form used for DR, $\boldsymbol{C}$ has the form

$$\boldsymbol{C} = c\boldsymbol{M}. \tag{2.11}$$

Substituting (2.11) into (2.10) gives

$$\begin{aligned} \dot{\boldsymbol{q}}^{n+1/2} &= \frac{(2-ch)}{(2+ch)}\dot{\boldsymbol{q}}^{n-1/2} + 2h\boldsymbol{M}^{-1}(\boldsymbol{f}^n - \boldsymbol{p}^n)/(2+ch), \\ \boldsymbol{q}^{n+1} &= \boldsymbol{q}^n + h\dot{\boldsymbol{q}}^{n+1/2}, \end{aligned} \tag{2.12}$$

where $\boldsymbol{M}^{-1}$ indicates the inverse of $\boldsymbol{M}$. Since $\boldsymbol{M}$ is diagonal, (2.12) is algebraic. That is, each solution vector component may be computed individually from

$$\begin{aligned} \dot{q}_i^{n+1/2} &= \frac{(2-ch)}{(2+ch)}\dot{q}_i^{n-1/2} + 2h(f_i^n - P_i^n)/[m_{ii}(2+ch)], \\ q_i^{n+1} &= q_i^n + h\dot{q}_i^{n+1/2}, \end{aligned} \tag{2.13}$$

where the subscript i indicates the ith vector component and m_{ii} is the ith diagonal element of $\boldsymbol{M}$. This time integration algorithm is seen to be very simple.

The first expression of (2.12) or (2.13) cannot be used to start the integration because the velocity is not known at $t^{-1/2}$, only at t^0. For the DR algorithm the starting conditions are of the form

$$\boldsymbol{q}^0 \neq 0; \quad \dot{\boldsymbol{q}}^0 = 0. \tag{2.14}$$

Using (2.9) and the second of (2.14) gives

$$\dot{\boldsymbol{q}}^{-1/2} = -\dot{\boldsymbol{q}}^{1/2}. \tag{2.15}$$

So, for the first time increment the first expression of (2.12) becomes

$$\dot{\boldsymbol{q}}^{1/2} = h\boldsymbol{M}^{-1}(\boldsymbol{f}^0 - \boldsymbol{P}^0)/2. \tag{2.16}$$

Note that the value of the damping coefficient, c, does not enter into the starting procedure. The central difference time integrator for diagonal mass and mass proportional damping is therefore given by

$$\begin{aligned} &\text{if } n = 0; \quad \dot{\boldsymbol{q}}^{1/2} = h\boldsymbol{M}(\boldsymbol{f}^0 - \boldsymbol{P}^0)/2, \\ &\text{if } n \neq 0; \quad \dot{\boldsymbol{q}}^{n+1/2} = \frac{(2-ch)}{(2+ch)}\dot{\boldsymbol{q}}^{n-1/2} + 2h\boldsymbol{M}^{-1}(\boldsymbol{f}^n - \boldsymbol{P}^n)/(2+ch), \\ &\text{for all } n: \quad \boldsymbol{q}^{n+1} = \boldsymbol{q}^n + h\dot{\boldsymbol{q}}^{n+1/2}. \end{aligned} \tag{2.17}$$

To obtain the static solution from the transient response equation one selects the damping coefficient c, the time increment h, and the mass matrix $\boldsymbol{M}$ to obtain the fastest convergence or some measure of convergence to determine $\boldsymbol{q}$ such that $\boldsymbol{P}(\boldsymbol{q}) = \boldsymbol{f}$. Physically, the damping, time increment, and mass are selected so that the transient response is attenuated leaving the steady state or static solution for the applied load $\boldsymbol{f}$. Note that only $\boldsymbol{P}$ and $\boldsymbol{f}$ must represent the physical problem, and c and $\boldsymbol{M}$ do not need to represent the physical structure. Also, h is dependent on $\boldsymbol{M}$.

Formally, the DR algorithm may be written [11, 20]:

$$
\begin{aligned}
&\text{(a) choose } v \ (v = ch) \text{ and } \boldsymbol{M}; \quad \boldsymbol{q}^0 \text{ given}; \quad \dot{\boldsymbol{q}}^0 = \boldsymbol{0},\\
&\text{(b) } \boldsymbol{r}^n = \boldsymbol{f}^n - \boldsymbol{P}(\boldsymbol{q}^n),\\
&\text{(c) if } \boldsymbol{r}^n \approx \boldsymbol{0} \text{ stop, otherwise continue},\\
&\text{(d) } n = 0; \quad \dot{\boldsymbol{q}}^{1/2} = h\boldsymbol{M}^{-1}\boldsymbol{r}^0/2\,,\\
&\qquad n \neq 0; \quad \dot{\boldsymbol{q}}^{n+1/2} = \frac{(2-v)}{(2+v)}\dot{\boldsymbol{q}}^{n-1/2} + 2h\boldsymbol{M}^{-1}\boldsymbol{r}^n/(2+v),\\
&\text{(e) } \boldsymbol{q}^{n+1} = \boldsymbol{q}^n + h\dot{\boldsymbol{q}}^{n+1/2},\\
&\text{(f) } n = n+1; \text{ return to (b).}
\end{aligned}
\tag{2.18}
$$

Steps (d) and (e) in the DR algorithm (2.18) are identical to the central difference time integrator, i.e. (2.17). The only difference is that in DR, v and $\boldsymbol{M}$ are fictitious values chosen so that the static solution $\boldsymbol{r} = \boldsymbol{0}$ is obtained in a minimum number of steps. Also, h is a pseudo-time increment which must be chosen to ensure *stability* and *accuracy* of the iterations.

2.4. DR algorithm properties

In this subsection a theoretical basis for the DR algorithm is presented. This basis is developed from: (1) characterization of the mode-by-mode convergence rate in terms of the spectral radius of the iterative error equations, (2) determination of the optimum convergence rate, (3) the diagonal damping coefficient, and (4) the stable time increment. The presentation follows closely the development of Frankel [5]; here it is specialized to linear structural mechanics equations. For the linear problem the residual $\boldsymbol{r}$ is

$$\boldsymbol{r} = \boldsymbol{f} - \boldsymbol{K}\boldsymbol{q} \tag{2.19}$$

where (2.3) is used for $\boldsymbol{P}$. Substituting (2.19) into (2.18d) and using (2.18e) and the first expression of (2.8) gives the equation to advance to the new iteration in terms of the displacement.

$$
\begin{aligned}
&\boldsymbol{q}^{n+1} = \boldsymbol{q}^n + \beta(\boldsymbol{q}^n - \boldsymbol{q}^{n-1}) - \alpha\boldsymbol{A}\boldsymbol{q}^n + \alpha\boldsymbol{b}^n\,;\\
&\alpha = 2h^2/(2+v), \qquad \beta = (2-v)/(2+v)\,,
\end{aligned}
\tag{2.20}
$$

where $\boldsymbol{A} = \boldsymbol{M}^{-1}\boldsymbol{K}$, $\boldsymbol{b} = \boldsymbol{M}^{-1}\boldsymbol{f}$ and only the general equation for $n \neq 0$ is considered. To study the convergence, the error in the iteration at the nth step is defined by

$$\boldsymbol{e}^n = \boldsymbol{q}^n - \boldsymbol{q}^* \tag{2.21}$$

where $\boldsymbol{q}^*$ is the solution for $\boldsymbol{r} = \boldsymbol{0}$, (2.19). Substituting (2.21) into (2.20) gives the error equation:

$$\boldsymbol{e}^{n+1} = \boldsymbol{e}^n - \alpha\boldsymbol{A}\boldsymbol{e}^n + \beta(\boldsymbol{e}^n - \boldsymbol{e}^{n-1})\,. \tag{2.22}$$

A solution of (2.22) may be obtained by assuming

$$e^{n+1} = \kappa e^n \tag{2.23}$$

where $|\kappa| = \rho$ is the spectral radius [30]. Note that the most rapid convergence is obtained for the smallest possible $\rho < 1$. Substituting (2.23) into (2.22) gives

$$\kappa^2 - (1 + \beta - \alpha A)\kappa + \beta = 0 \tag{2.24}$$

where A denotes any eigenvalue of $\boldsymbol{A}$. From (2.23) it is seen that convergence is obtained for $|\kappa| < 1$ and for *optimum* convergence κ would be the minimum κ that produces uniform convergence over the entire range of eigenvalues $A_0 \leq A \leq A_m$. This optimum convergence condition, κ^*, is obtained for

$$(1 + \beta - \alpha A) = \pm 2\beta^{1/2}, \tag{2.25}$$

which gives

$$|\kappa^*| = \rho^* = \beta^{1/2}. \tag{2.26}$$

A plot of the spectral radius versus the eigenvalues of $\boldsymbol{A}$ for the optimum condition given by (2.25) and any other choice of condition satisfying (2.24) and $|\kappa| < 1$ is shown in Figure 2.1. This figure clearly illustrates that (2.25) is the

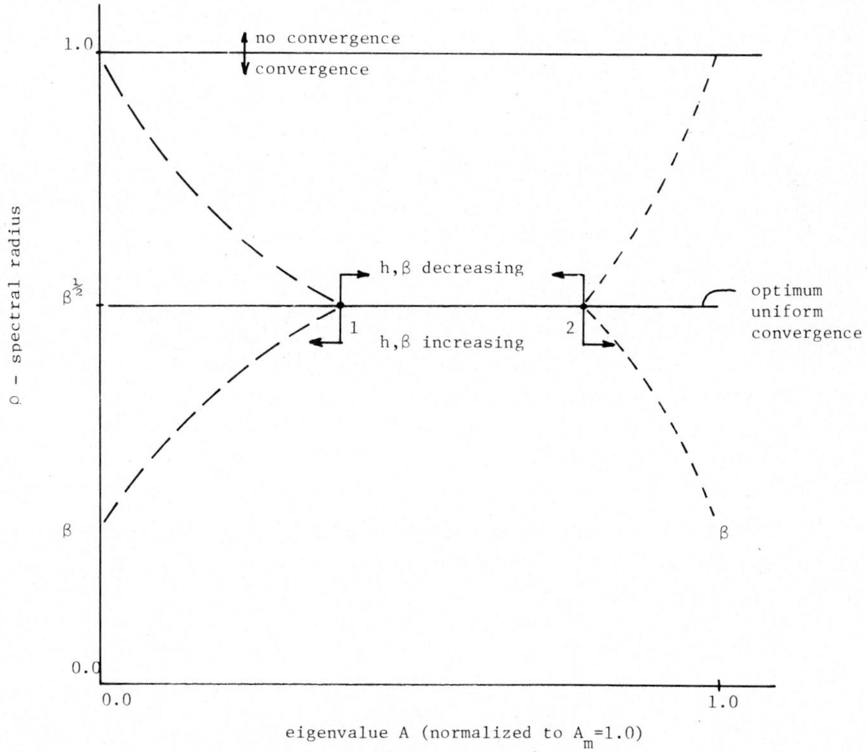

Fig. 2.1. Spectral radius vs. eigenvalues of $\boldsymbol{A}$.

optimum convergence condition. For any other choice two distinct roots are obtained and the value of ρ for one of the roots is always larger than ρ^* over some portion of the eigenvalues; hence the convergence rate for the other ρ's would be slower than for ρ^*. The points 1 and 2 shown in Figure 2.1 move along the ρ^* line as indicated for decreasing and increasing values of h and β. Points 1 and 2 are not necessarily symmetrically located. For $A = 0$ and $A = A_m = 1.0$ the maximum and minimum values of the two roots of ρ are 1.0 and β, as shown. Papadrakakis [23] also presents a good discussion of this topic.

Since (2.25) holds for all possible eigenvalues of $\boldsymbol{A}$, the two equations that satisfy the condition for ρ^* are

$$1 + \beta - \alpha A_0 = 2\beta^{1/2} \tag{2.27}$$

and

$$1 + \beta - \alpha A_m = -2\beta^{1/2}\,. \tag{2.28}$$

Adding (2.27) and (2.28) gives

$$\alpha(A_0 + A_m) = 2(1 + \beta) \tag{2.29}$$

and rewriting (2.27) gives

$$\alpha A_0 = (\beta^{1/2} - 1)^2\,. \tag{2.30}$$

So, combining (2.29) and (2.30) gives an expression for ρ^* in terms of A_0 and A_m:

$$\rho^* = \beta^{1/2} \approx \left|1 - 2\sqrt{\frac{A_0}{A_m}}\right|\,, \tag{2.31}$$

where $A_0 \ll A_m$ has been assumed. From (2.31) it is seen that scaling $\boldsymbol{A}$ to maximize the ratio A_0/A_m produces more rapid convergence, i.e. ρ^* is a minimum.

It now remains to determine the values of h and c to be used in the DR iteration, (2.18), that satisfy the condition for optimum convergence, (2.25). Combining (2.11), (2.30), and the definitions of α and β, (2.20), gives

$$h \leq 2/\sqrt{A_m} = 2/\omega_{\max} \tag{2.32}$$

and

$$c \approx 2\sqrt{A_0} = 2\omega_0\,, \tag{2.33}$$

where the approximation $A_0 \ll A_m$ has been used and ω_0 and $\omega_{\max}$ are the lowest and highest circular frequencies of the undamped equation (2.7). Also, in terms of ω_0 and $\omega_{\max}$ (2.31) becomes

$$\rho^* \approx \left|1 - 2\left(\frac{\omega_0}{\omega_{\max}}\right)\right|\,. \tag{2.34}$$

To minimize ρ^* by maximizing the ratio $\omega_0/\omega_{\max}$ through a judicious choice for $\boldsymbol{M}$ is the same as scaling $\boldsymbol{A}$ to maximize the ratio A_0/A_m.

Relating the above to structural dynamics, it is seen that (2.32) is the expression for the stability limit for the central difference time integrator [31, 32] and that (2.33) is the expression for critical damping of lowest eigenvalue [33]. Note that, these observations are not original here. The relationship of the stability limit to the transit time for information between two adjacent nodes in the discrete elements (the Courant–Friedricks–Lewy condition [34]) has been exploited to develop the idea of 'fictitious mass' [7, 11, 17], which minimizes ρ^* while retaining stability. The critical damping property has been used by many, e.g., [9, 11, 20]. Use of the relationship of DR to structural dynamics computational methods will be made in Section 3.

2.5. DR methodology

Now that the criteria for choosing $\boldsymbol{M}$, h and c are established there remains the task of how these terms are actually determined for a DR analysis. The selection of the fictitious mass matrix $\boldsymbol{M}$ and the pseudo-time increment h are covered first, followed by the selection of the damping matrix coefficient c.

The most common method for determining the elements of $\boldsymbol{M}$ is to choose m_{ii} such that the transit times for information transfer for degree of freedom i to adjacent and like degrees of freedoms is a constant. For convenience this constant is typically chosen, such that $h = 1$, i.e. the constant is slightly greater than 1 so that stability is ensured. For example, consider the axial motion of a one-dimensional bar with an elastic modulus E, cross-sectional area A and a fixed distance of $\Delta 1$ between nodes. Then for finite-difference and low-order finite-element formulations the transit time T is

$$T = \Delta 1/\sqrt{EA/m} = h_{\text{critical}} \tag{2.35}$$

which gives

$$m \geq h^2 EA/(\Delta 1)^2 , \tag{2.36}$$

where m is an element of the fictitious mass matrix $\boldsymbol{M}$. The author has found that evaluating m for $h = 1.1$ and iterating with $h = 1$ provides a sufficient margin to ensure stability. The reader has probably already made the connection between this approach and the Courant–Friedricks–Lewy condition [34]. Key [17] discusses this method for one finite element. For finite difference, Cassell [11, 22] presents expressions like (2.36) for several structural elements, such as beams, plates and shells. The expressions are derived by determining the highest frequency for the discretized structural element. Reference [22] illustrates the use of this method for a nonlinear problem. This approach for determining $\boldsymbol{M}$ appears to have been proposed by Welsh [7]. The author feels that this approach is good as long as the structural model is composed of only one element type, but this approach is very cumbersome for models containing many element types.

An attractive alternative [11, 20] to the above approach is through an application of Greschgorin's theorem. This theorem states [30]:

"*Every eigenvalue of* $\boldsymbol{A}$ *lies in at least one of the circles* $C_1, \ldots, C_n$, *where* C_i *has its center at the diagonal entry* a_{ii} *and its radius* $r_i = \sum_{j \neq 1} |a_{ij}|$ *equal to the absolute sum along the rest of the row.*"

If every row is scaled such that the absolute sum along every row is identical, then Gerschgorin's theorem tells us that all the circles are coincident for equal mesh spacing and they will be nearly coincident for unequal mesh spacing. In addition, it guarantees the largest eigenvalue to be less than or equal to the absolute sum along a row. Therefore, a good estimate of the highest frequency ω_{max} is obtained. For the axial bar problem considered above the elements of a row in the $\boldsymbol{A}$ matrix ($\boldsymbol{A} = \boldsymbol{M}^{-1}\boldsymbol{K}$) are

$$EA/m(\Delta 1)^2, \quad -2EA/m(\Delta 1)^2, \quad EA/m(\Delta 1)^2 . \tag{2.37}$$

Hence Gerschgorin's theorem gives

$$A_m \leq 4EA/m(\Delta 1)^2 \tag{2.38}$$

and using (2.32) gives

$$m \geq h^2 EA/(\Delta 1)^2 \tag{2.39}$$

which is identical to (2.36).

Physically, Gerschgorin's theorem is equivalent to assuming a highest eigenvector, that is, a sequence of 1, −1, 1, −1, . . . values. This eigenvector is identical to the highest eigenvector for the bar, ignoring boundary conditions; hence the reason equations (2.36) and (2.39) are identical. Using (2.32) this approach gives the following general expression for m_{ii}:

$$m_{ii} \geq \frac{1}{4} h^2 \sum_j |K_{ij}| \tag{2.40}$$

where K_{ij} are the elements of the stiffness matrix $\boldsymbol{K}$. Since for an explicit method $\boldsymbol{K}$ is not explicitly required, only the internal force $\boldsymbol{P}(\boldsymbol{q})$, evaluating (2.40) could be a burden, but fortunately it is not. The author has successfully used numerical differentiation of $\boldsymbol{P}(\boldsymbol{q})$ to obtain $|K_{ij}|$. In fact, the differentiation does not have to be super accurate as only an estimate is being sought. Also, this approach is not hindered by a variety of elements in the structural model.

Selection of m_{ii} using (2.40) or using the first approach produces a scaling that in the least does not reduce the ratio ω_0/ω_{max} and in general will increase the ratio ω_0/ω_{max} for faster convergence. This can be shown as follows. Consider two eigenvalue problems. The first is the unscaled problem

$$\boldsymbol{Kx} = \omega^2 \alpha \boldsymbol{Mx} \tag{2.41}$$

where α is chosen such that $\omega_{max}^2 \simeq 4$, i.e.

$$\alpha m_{ii} \geq \tfrac{1}{4} \sum_j |K_{ij}|$$

where the equality holds for at least one *i*. The second is the scaled problem

$$\boldsymbol{K}\boldsymbol{x} = {}_{s}\omega^2 \tfrac{1}{4}\Big[\sum_j |K_{ij}|\Big]\boldsymbol{x} \tag{2.42}$$

where ${}_{s}\omega^2_{max} \simeq 4$ from Gerschgorin's theorem. So for both problems the maximum eigenvalue is essentially identical. The lowest frequency for both problems is estimated for Rayleigh's quotient [35], which gives

$$\omega_0^2 \simeq \frac{\boldsymbol{u}^T\boldsymbol{K}\boldsymbol{u}}{\boldsymbol{u}^T\boldsymbol{M}\boldsymbol{u}} \tag{2.43}$$

and

$${}_{s}\omega_0^2 \simeq \frac{4\boldsymbol{v}^T\boldsymbol{K}\boldsymbol{v}}{\boldsymbol{v}^T[\Sigma_j |K_{ij}|]\boldsymbol{v}} \tag{2.44}$$

where $\boldsymbol{u}$ and $\boldsymbol{v}$ are the eigenvectors associated with the lowest eigenvalue for each eigenvalue problem. Since $\boldsymbol{u} \simeq \boldsymbol{v}$, because they are both the solution mode [35] equations (2.43) and (2.44) give

$$\omega_0^2 \leq {}_{s}\omega_0^2 \tag{2.45}$$

based on the choice of the mass matrix in (2.41) and (2.42). The equality in (2.45) would hold for the case $\boldsymbol{M} \simeq \boldsymbol{I}$, where $\boldsymbol{I}$ is the identity matrix. Therefore, (2.45) gives

$$\frac{\omega_0^2}{\omega_{max}^2} \leq \frac{{}_{s}\omega_0^2}{{}_{s}\omega_{max}^2}, \tag{2.46}$$

which illustrates the assertion that the ratio ω_0/ω_{max} is at least preserved and generally increased.

A third approach has been proposed by Papadrakakis [23]. He assumes

$$m_{ii} \sim k_{ii} \tag{2.47}$$

and then determines *h* for stability from Gerschgorin's theorem. This approach is very simple and should be considered by those using DR.

Now the determination of the damping matrix coefficient *c* will be considered. The majority of investigators using DR determine *c* from a numerical experiment, e.g. [8, 9, 11, 15]. This experiment consists of forming the fictitious mass matrix by either the first or second approach described above and then with $c = 0$ computing the response for a number of iterations. The number of iterations must be sufficient to observe the lowest frequency ω_0 and then *c* is computed from (2.33). It is easy to see that these iterations are unproductive with regard to advancing the solution vector. Rushton [9], Frieze [15] and others also determine a distinct damping coefficient for each intrinsic coordinate. For example, a beam with stretching requires in plane and lateral displacements fields and a distinct damping coefficient is determined for both the in plane and lateral motion. A second interesting approach to determining c based on Rayleigh's quotient, is proposed by Bunce [20]. This is still an unproductive iteration approach, but it is more deterministic than the first approach. In Section 3 this approach is used in an adaptive DR method.

A third approach is proposed by Alwar [21]. In this approach a small value is chosen for c, so that the solution will oscillate about the true solution but the oscillations will be decaying. The true solution is estimated from the envelope of the decaying oscillations.

A fourth approach, that is most interesting, is proposed by Papadrakakis [23]. He calculates a series of approximations to the dominant eigenvalue λ_{DR} from

$$\lambda_{DR} = \frac{\|\boldsymbol{q}^{n+1} - \boldsymbol{q}^n\|}{\|\boldsymbol{q}^n - \boldsymbol{q}^{n-1}\|}. \tag{2.48}$$

When λ_{DR} given by (2.48) has converged to an almost constant value, then this is the minimum eigenvalue needed to determine c for the optimum iteration parameters. This is a very attractive approach as it automatically computes the iteration parameters. A similar approach developed by Underwood [24] will be presented in Section 3.

3. An adaptive DR method

In this section the author would like to present briefly a DR method which he developed [24], but has not previously appeared in the literature. Although this is not in the spirit of a review, the author hopes the reader will find this section worthwhile.

Here the emphasis is on the nonlinear problem and specifically the problem of snap-through buckling and plastic deformation are of concern. Therefore, the DR method must not overshoot the true solution, but must always approach the solution from below. By putting together many of the pieces of Section 2. with a few extensions it is possible to develop an adaptive DR method that successfully solves the nonlinear problem. Here adaptive means that information generated during the computations is used to modify system parameters in order to maintain optimum convergence.

For this adaptive method the fictitious mass is computed from (2.40). To provide a safety margin for stability, (2.40) is evaluated for $h \simeq 1.1$ and the iterations are performed with $h = 1.0$. For the nonlinear problem K_{ij} must represent the tangent stiffness, (2.5), so that the stability requirements of the central difference integrator are maintained [36]. Numerical differentiation of the internal force $\boldsymbol{P}$ is used to determine K_{ij}. The summation is performed with the differentiation so provisions for storing a matrix are not required.

For the nonlinear problem the initial fictitious mass matrix may not satisfy the conditions for stability throughout the analysis due to increases in the stiffness. To determine when the fictitious mass matrix should be updated the 'perturbed apparent frequency' error measure [37–39] is used. This error, ε, is computed as [38]:

$$\varepsilon = \max(\varepsilon_1^n, \ldots, \varepsilon_{\max\text{ D.O.F.}}^n) \tag{3.1}$$

where

$$\varepsilon_i^n = \frac{h_n^2}{4}\frac{a_i^n}{b_i^n}, \qquad a_i^n = |\ddot{q}_i^n - \ddot{q}_i^{n-1}|, \qquad b_i^n = |q_i^n - q_i^{n-1}|\,.$$

If ε is greater than 1.0 then the fictitious mass matrix must be reformed or a smaller h must be used. The error measure given by (3.1) could also be used to indicate that the fictitious mass matrix should be reformed to obtain optimum convergence for regions of very soft stiffness, but this has not been done.

For this adaptive DR method the damping matrix coefficient c is computed at each iteration from Rayleigh's quotient as

$$c_n = 2\sqrt{(\boldsymbol{q}^n)^{\mathrm{T}}\,{}^{\mathrm{l}}\boldsymbol{K}^n\boldsymbol{q}^n/(\boldsymbol{q}^n)^{\mathrm{T}}\boldsymbol{M}\boldsymbol{q}^n} \tag{3.2}$$

where superscript T indicates the transpose and the elements of the diagonal matrix ${}^{\mathrm{l}}\boldsymbol{K}^n$ (a 'local' stiffness matrix) are given by

$${}^{\mathrm{l}}K_{ii}^n = [-P_i(\boldsymbol{q}^{n-1}) + P_i(\boldsymbol{q}^n)]/h\dot{q}_i^{n-1/2}\,. \tag{3.3}$$

For the linear problem ${}^{\mathrm{l}}\boldsymbol{K}^n\boldsymbol{q}^n$ would be replaced by $\boldsymbol{P}(\boldsymbol{q}^n)$. In (3.2) the square root term would give ω_0 for $\boldsymbol{q}^n$ equal to the eigenvector associated with the lowest frequency and with ${}^{\mathrm{l}}\boldsymbol{K}^n = \boldsymbol{K}$, the linear stiffness matrix. In reality, (3.2) gives an estimate of the critical damping for the current deformation mode, $\boldsymbol{q}^n$, based on an estimate for the local ('tangent') stiffness, (3.3). During the iterative DR procedure the high frequencies are damped out first and hence no overshoot of the solution occurs. Eventually (3.2) gives the critical damping based on the final solution. This value of the critical damping may or may not be equal to the value based on the lowest frequency of the system. This is important in that if the solution is better represented by the third mode then the DR damping would be based on the third mode when (3.2) is used. Use of (3.2) has proven to be very effective and it requires no unproductive iterations. This method was successfully used by Key [17] for some large complex nonlinear finite element analyses using DR. Formally, this adaptive DR method may be written as:

(a) $\boldsymbol{q}^0$ given; $\dot{\boldsymbol{q}}^0 = \boldsymbol{0}$; $n = 0$,
(b) compute $\boldsymbol{M}$ from (2.40) with $h = h^*$, where $h^* > h$ and K_{ij} is determined from (2.5),
(c) $\boldsymbol{r}^n = \boldsymbol{f}^n - \boldsymbol{P}(\boldsymbol{q}^n)$,
(d) if $\boldsymbol{r}^n \simeq \boldsymbol{0}$ stop, otherwise continue,
(e) $n = 0$; $\dot{\boldsymbol{q}}^{1/2} = h\boldsymbol{M}^{-1}\boldsymbol{r}^0/2$,

$$n \neq 0;\ \dot{\boldsymbol{q}}^{n+1/2} = \frac{(2 - \upsilon_n)}{(2 + \upsilon_n)}\dot{\boldsymbol{q}}^{n-1/2} + 2h\boldsymbol{M}^{-1}\boldsymbol{r}^n/(2 + \upsilon_n)\,, \tag{3.4}$$

(f) $\boldsymbol{q}^{n+1} = \boldsymbol{q}^n + h\dot{\boldsymbol{q}}^{n+1/2}$,
(g) evaluate error from (3.1)
and reform $\boldsymbol{M}$ if necessary; repeat (c)–(f),
(h) $n = n + 1$,
(i) $v_n = 2h\sqrt{(\boldsymbol{q}^n)^{\mathrm{T}}\,{}^{\mathrm{I}}\boldsymbol{K}^n\boldsymbol{q}^n/(\boldsymbol{q}^n)^{\mathrm{T}}\boldsymbol{M}\boldsymbol{q}^n}$,
(j) return to (c).

In step (i) if the argument of the square root is not positive, v_n is set to zero. This is required for problems which traverse an unstable region; see Subsection 4.2. In addition, step (b) gives $\omega_{\max} \leq 2$; so if the square root in step (i) is greater than 2, then it is set to a value less than 2 (typically 1.9). This is necessary as ${}^{\mathrm{I}}\boldsymbol{K}^n$ is an estimate that does not always satisfy the physics of the problem.

If (3.4) is compared to (2.18) it is seen that the adaptive method requires a few additional computations, but because it is adaptive the robustness of the DR method is increased considerably. To solve a nonlinear material problem these effects are included in $\boldsymbol{P}(\boldsymbol{q}^n)$ at each iteration just as in a dynamic transient analysis [17]. The method (3.4) has consistently produced good results with no overshoot of the solution. It can also be used for the linear problem; here step (g) is not required and step (i) is much simpler to evaluate.

4. Numerical examples

Two numerical examples have been chosen to illustrate the behavior of the DR method. These relatively simple problems are used because they make it easier to focus on the characteristics of the DR method. The first example is a nonlinear truss-spring and the second example concerns the snap-through buckling of a uniformly loaded arch.

4.1. Truss-spring example

The truss-spring model being considered [40] is shown in Figure 4.1. This problem has been solved by other methods: Newton–Raphson, self-correcting and initial value [27, 40, 41] so a comparison with other methods is possible.

From [40] the internal force $P(q)$ and the tangent stiffness $K(q)$ are given by

$$P(q) = \tfrac{1}{2}AE\cos^2\phi(q/l)^2[3\sin\phi + \cos^2\phi(q/l)] + K_s q + \frac{AE}{l}\sin^2\phi q \quad (4.1)$$

and

$$K(q) = \tfrac{1}{2}AE(3/l)\cos^2(q/l)[2\sin\phi + \cos^2\phi(q/l)] + K_s + \frac{AE}{l}\sin^2\phi\,. \quad (4.2)$$

Since this is a single degree of freedom system, the fictitious mass must be chosen not only to ensure stability, but to obtain accuracy. A constant mass (M) of 30.0 was chosen to satisfy stability and accuracy for $h = 1$ and an assumed maximum q

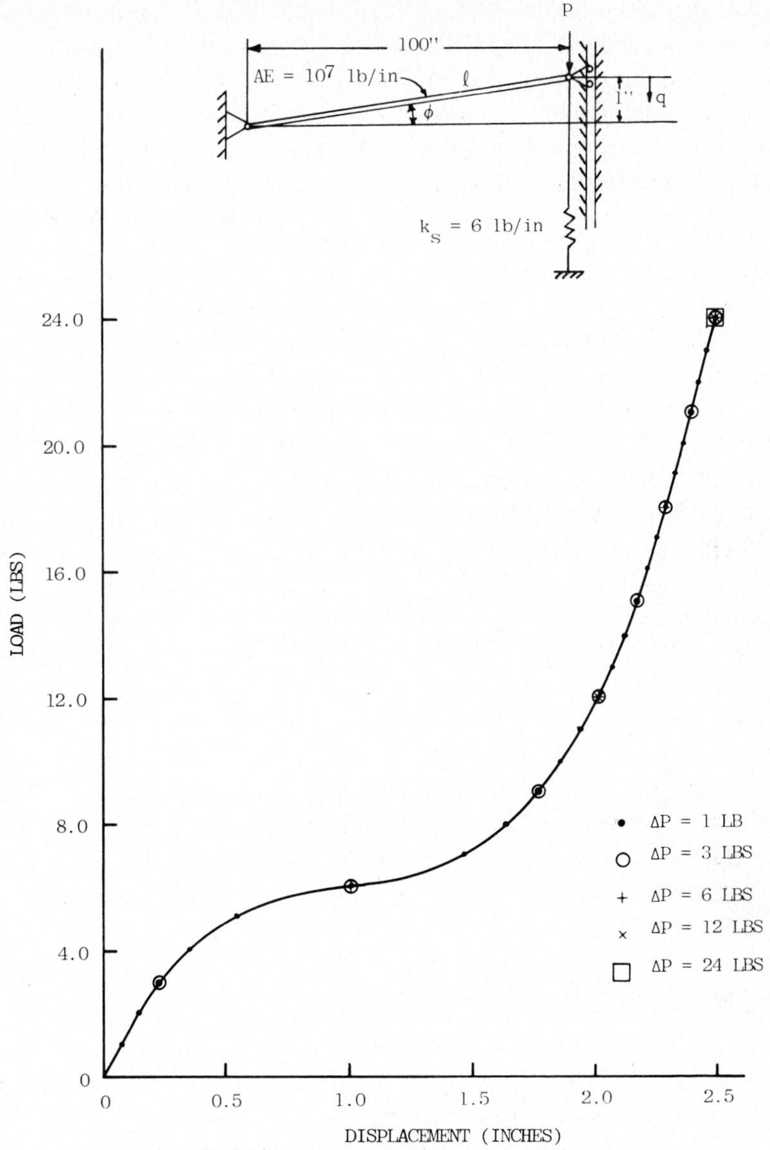

Fig. 4.1. Nonlinear truss-spring, load-displacement.

of 2.5. The critical damping parameter is computed directly from

$$v = 2h\sqrt{K(q)/M} \tag{4.3}$$

where $K(q)$ is computed from (4.2).

The load deflection curve shown in Figure 4.1 and Table 4.1 was computed for various load increments ($\Delta P = 1$, 3, 6, 12, and 24 lbs) up to a peak of 24 lbs. It is seen that the DR method gives accurate results at all loads and most important is that the DR method converges for any load step. This is in contrast to the results given in references [27, 40, 41] which show that very small load steps are required to obtain accuracy for most of the other methods and none would converge for a large load step. The number of iterations for convergence ($|r| \leq 10^{-6}$) is shown in parentheses in Table 4.1. These iteration counts show that convergence in a region of near zero slope in the load deflection curve requires more iterations as the frequency of the system is low. In this case using a variable mass would probably speed convergence.

Table 4.1
Results for nonlinear truss-spring example

Load (lbs)	Exact deflection (ins)	$\Delta P = 1$ lb deflection (ins)	$\Delta P = 3$ lb deflection (ins)	$\Delta P = 6$ lb deflection (ins)	$\Delta P = 12$ lb deflection (ins)	$\Delta P = 24$ lb deflection (ins)
1.0		0.06657 (27)				
2.0		0.1434 (29)				
3.0	0.2354	0.2354 (31)	0.2354 (33)			
4.0	0.3532	0.3532 (36)				
5.0	0.5275	0.5275 (45)				
6.0	1.000	0.9995 (89)	0.9995 (92)	0.9995 (94)		
7.0	1.472	1.472 (46)				
8.0	1.647	1.647 (37)				
9.0	1.765	1.765 (32)	1.765 (35)			
10.0	1.857	1.857 (29)				
11.0	1.934	1.934 (27)				
12.0	2.000	2.000 (26)	2.000 (28)	2.000 (30)	2.000 (30)	
13.0		2.059 (24)				
14.0		2.113 (23)				
15.0	2.162	2.162 (23)	2.162 (25)			
16.0		2.207 (22)				
17.0		2.250 (21)				
18.0	2.289	2.289 (21)	2.289 (23)	2.289 (24)		
19.0		2.327 (20)				
20.0		2.362 (20)				
21.0		2.396 (19)	2.396 (21)			
22.0		2.429 (19)				
23.0		2.460 (19)				
24.0		2.489 (18)	2.489 (20)	2.489 (21)	2.489 (22)	2.489 (24)

4.2. Uniformly loaded arch

The problem considered here is an arch composed of a thin circular segment clamped at both ends. The radius of the circular segment (R) is 57.3 inches and the half angle (β) of the circular segment span is 5° giving a length of 10 inches. The arch has a thickness of 0.04367 inches and a unit width. The elastic modulus is 10^7 psi. The loading is uniformly distributed and acts toward the center of the circle defining the arch. For this particular geometry snap-through buckling occurs in an asymmetric mode. That is, the inplane load developed in the arch reaches the inplane load required to buckle into the second mode before the arch snaps through [42].

The discrete model for the arch is based on the finite difference method applied to a reduction of Sanders' shell equations [43] to give the equations for the arch. Both inplane and lateral displacements are included. A very coarse mesh comprising 10 elements along the arch was chosen. The problem was solved by the DR method given by (3.4). The fictitious mass was computed for $h = 1.05$ and the iterations were performed with $h = 1$.

The displacement of the arch for the three load cases considered is shown in Figure 4.2. The loading $f = 0.976$ lb/in is just before snap-through buckling. Note that at this state the lowest eigenvalue is less than zero and the DR method successfully converged for this very unstable condition. It required 116 iterations to reach convergence for 0.5 lb/in, another 232 iterations to reach 0.976 lb/in, and another 639 iterations to reach 1.1 lb/in. Figure 4.3 illustrates the displacement of the crown of the arch and the damping parameter versus the number of iterations. Note that the displacement never overshoots and the damping parameter is mostly well behaved. The region in which snap-through is occurring is indicated by the zero damping value. The occasional spikes that occur in the damping history appear to be due to waves traveling in the inplace direction in the arch and sometimes the spikes are due to the poor numerical estimate for ${}^{1}\boldsymbol{K}$ obtained from (3.3). Since the duration of the spikes last for only one or two iterations they do not seem to degard the convergence properties.

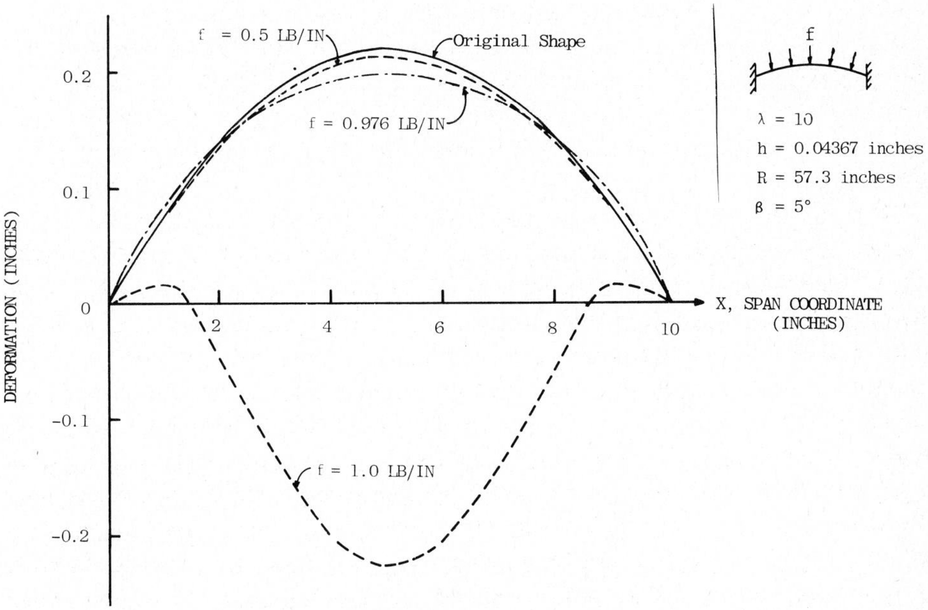

Fig. 4.2. Load-deformation pattern for arch.

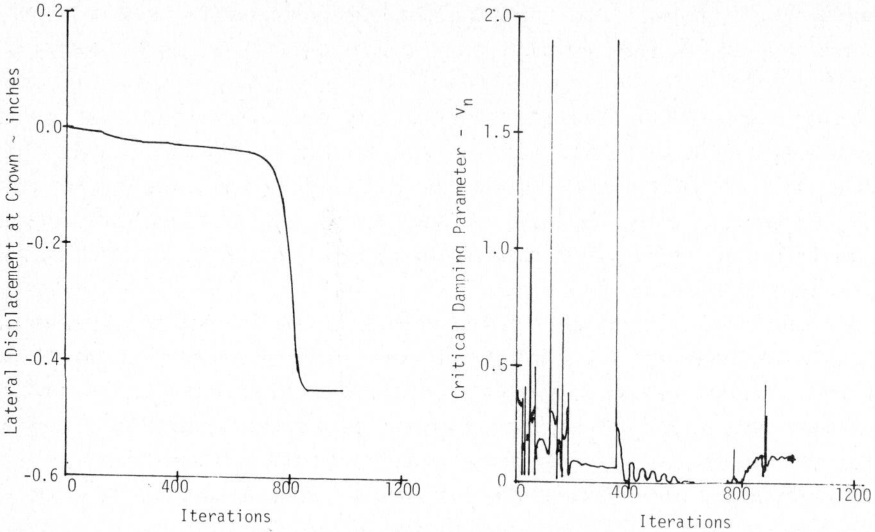

Fig. 4.3. Arch snap-through buckling.

5. Remarks

This section contains some remarks that did not seem to fit into the other sections, but are necessary to tie up some loose ends. Besides, it gives the author a chance to make some comments on the future, that is, the future of DR.

The most successful applications of DR have been based on discrete structural models that were developed from finite differences, not finite elements. The reason for this seems to lie in ratio A_0/A_m. Since most finite element discretizations include rotations as degrees of freedom and finite difference discretizations do not, the A_0/A_m ratio is generally smaller for finite elements than for finite differences. Hence the convergence rate is slower, viz. (2.31), for finite element discretizations.

This problem with slow convergence for DR actually exists for all refined discrete element models and it leads one to think of methods to increase the ratio A_0/A_m. One method that may be promising is to consider an integrator other than the central difference integrator. Other explicit integrators do not seem to offer much because the central difference integrator is hard to beat [36]. Implicit integrators are not attractive because they require matrix operations and the simplicity of the DR approach is destroyed. This leaves semi-explicit or semi-implicit methods, e.g. [44, 45]. These methods are attractive in that the critical time increment, (2.32) can be increased. This increased critical time increment effectively increases the ratio A_0/A_m. The only problem with the semi-explicit methods appears to be retaining enough accuracy to obtain an acceptable solution [46]. Someday, some bright person will sort out this dilemma. Along this same line of thought is the paper by Park [47] which considers many solution methods for nonlinear analysis within the framework of the transient response equations. Even the Newton-like methods can be considered as transient response methods [47].

Although the DR method has been applied to linear problems by many people the real forte of DR is in the nonlinear problem, e.g. [15, 17]. Other iterative methods [48] are available for linear problems and in general they are more suitable than DR. The report by Crisfield [49] has an excellent comparison of iterative methods for both linear and nonlinear problems. In this report the DR method was found to be inferior to many other methods, so this report should be required reading for an honest evaluation of DR. Also, if the solution of nonlinear structural analysis problems is the goal of the reader, he should be aware of the recent extensions of the Newton-family of solution methods [50, 51]. These methods are very attractive for geometrical nonlinearities and they should be considered.

And if you read only one other paper on DR, other than this one, read the paper by Papadrakakis [23].

References

[1] J.R.H. Otter, Computations for Prestressed Concrete Reactor Pressure Vessels Using Dynamic Relaxation, *Nucl. Struct. Engng.* **1**, 61–75 (1965).

[2] J.R.H. Otter, Dynamic Relaxation, *Proc. Inst. Civ. Engrs.* **35**, 633–656 (1966).

[3] J.R.H. Otter, E. Cassel and R.E. Hobbs, Dynamic Relaxation, *Proc. Inst. Civ. Engrs.* **35**, 723–750 (1966).

[4] A.S. Day, An Introduction to Dynamic Relaxation, *The Engineer* **219**, 218–221 (1965).

[5] S.P. Frankel, Convergence Rates of Iterative Treatments of Partial Differential Equations, *Mathl. Tabl. Natn. Res. Coun.*, Washington, **4**, 65–75 (1950).

[6] L.F. Richardson, The Approximate Arithmetical Solution by Finite Differences of Physical Problems Involving Differential Equations, with an Application to the Stresses in a Masonary Dam, *R. Soc. London Phil. Trans A* **210**, 307–357 (1911).

[7] A.K. Welsh, Discussion on Dynamic Relaxation, *Proc. Inst. Civ. Engrs.* **37**, 723–750 (Aug. 1967).

[8] A.C. Cassell et al., Cylindrical Shell Analysis by Dynamic Relaxation, *Proc. Inst. Civ. Engrs.* **39**, 75–84 (Jan. 1968).

[9] K.R. Rushton, Large Deflexion of Variable-Thickness Plates, *Int. J. Mech. Sci.* **10**, 723–735 (1968).

[10] W.L. Wood, Comparison of Dynamic Relaxation with Three Other Iterative Methods, *Engineer* **224**, 683–687 (1967).

[11] A.C. Cassell, Shells of Revolution Under Arbitrary Loading and the Use of Fictitious Densities in Dynamic Relaxation, *Proc. Inst. Civ. Engrs.* **45**, 65–78 (Jan. 1970).

[12] K.R. Rushton, Buckling of Laterally Loaded Plates Having Initial Curvature, *Int. J. Mech. Sci.* **14**, 667–680 (1972).

[13] R.S. Alwar and N.R. Rao, Nonlinear Analysis of Orthotropic Skew Plates, *AIAA J.* **11**(4), 495–498 (April 1973).

[14] R.S. Alwar and N.R. Rao, Large Elastic Deformations of Clamped Skewed Plates by Dynamic Relaxation, *Computers and Structures* **4**, 381–398 (1974).

[15] P.A. Frieze, R.E. Hobbs and P.J. Dowling, Application of Dynamic Relaxation to the Large Deflection Elasto-Plastic Analysis of Plates, *Computers and Structures* **8**, 301–310 (1978).

[16] A. Pica and E. Hinton, Transient and Pseudo-Transient Analysis of Mindlin Plates, *Int. J. for Num. Meth. in Engrg.* **15**, 189–208 (1980).

[17] S.W. Key, C.M. Stone and R.D. Kreig, A Solution Strategy for the Quasi-Static, Large Deformation, Inelastic Response of Axisymmetric Solids, in: *Nonlinear Finite Element Analysis in Structural Mechanics,* W. Wunderlich et al., eds. (Springer, Berlin–New York, 1981) pp. 585–620.
[18] J.S. Brew and D.M. Brotton, Nonlinear Structural Analysis by Dynamic Relaxation, *Int. J. for Num. Meth. in Engrg.* **3**, 145–147 (1971).
[19] W.L. Wood, Note on Dynamic Relaxation, *Int. J. for Num. Meth. in Engrg.* **3**, 145–147 (1971).
[20] J.W. Bunce, A Note on the Estimation of Critical Damping in Dynamic Relaxation, *Int. J. for Num. Meth. in Engrg.* **4**, 301–304 (1972).
[21] R.S. Alwar, N.R. Rao and M.S. Rao, An Alternative Procedure in Dynamic Relaxation, *Computers and Structures* **5**, 271–274 (1975).
[22] A.C. Cassell and R.E. Hobbs, Numerical Stability of Dynamic Relaxation Analysis of Non-Linear Structures, *Int. J. for Num. Meth. in Engrg.* **10**, 1407–1410 (1976).
[23] M. Papadrakakis, A Method for the Automated Evaluation of the Dynamic Relaxation Parameters, *Computer Methods in Applied Mechanics and Engineering* **25**, 35–48 (1981).
[24] P.G. Underwood, An adaptive Dynamic Relaxation Technique for Nonlinear Structural Analysis, Lockheed Palo Alto Research Lab, LMSC-D678265 (1979).
[25] G.E. Forsythe and W.R. Wasow, *Finite-Difference Methods for Partial Differential Equations Computer Methods in Applied Mechanics and Engineering* **25**, 35–48 (1981).
[26] O.C. Zienkiewicz, *The Finite Element Method in Engineering Science* (McGraw-Hill, London, 1971).
[27] J.A. Stricklin and W.E. Haisler, Formulations and Solution Procedures for Nonlinear Structural Analysis, *Computers and Structures* **7**, 125–136 (1977).
[28] S. Rajasekasan and D.W. Murray, Incremental Finite Element Matrices, Proc. ASCE, *J. of the Struc. Div. St.* **12**, 2423–2437 (Dec. 1973).
[29] C.A. Felippa, Discussion of [28], Proc. ASCE, *J. of the Struc. Div. St.* **12**, 2521–2523 (Dec. 1973).
[30] G. Strang, *Linear Algebra and Its Applications* (Academic Press, New York, 1976).
[31] J.W. Leech, Stability of Finite-Difference Equations for the Transient Response of a Flat Plate, *AIAA J.* **3**, (9) 1772–1773 (Sept. 1965).
[32] G.G. O'Brien, M.A. Hyman and S. Kaplan, A Study of the Numerical Solution of Partial Differential Equations, *J. Math. and Phys.* **29**, 223–251 (1951).
[33] W.T. Thompson, *Vibration Theory and Applications* (Prentice-Hall, Englewood Cliffs, NJ, 1965).
[34] R. Courant, K. Friedrichs and H. Lewy, On the Partial Difference Equations of Mathematical Physics, *Mathematische Annalen* **100**, 32–74 (1928); English translation in *IBM J.* 215–234 (Mar. 1967).
[35] L. Meirovitch, *Analytical Methods in Vibrations* (MacMillan, New York, 1967).
[36] K.C. Park, Practical Aspects of Numerical Time Integration, *Computers and Structures* **7**, 343–353 (1977).
[37] K.C. Park and P.G. Underwood, A Variable-Step Central Difference Method for Structural Dynamic Analysis – Part 1. Theoretical Aspects, *Computer Methods in Applied Mechanics and Engineering* **22**, 241–258 (1980).
[38] P.G. Underwood and K.C. Park, A Variable-Step Central Difference Method for Structural Dynamic Analysis – Part 2. Implementation and Performance Evaluation, *Computer Methods in Applied Mechanics and Engineering* **23**, 259–279 (1980).
[39] P. Underwood and K.C. Park, STINT/CD: A Stand-Alone Explicit time Integration Package for Structural Dynamics Analysis, *Int. J. for Num. Meth. in Engrg.* **18**, 609–622 (1982).
[40] W.E. Haisler, Development and Evaluation of Solution Procedures for Nonlinear Structural Analysis, Ph.D. Dissertation, Texas A&M University (Dec. 1970).
[41] J.R. Tillerson, J.A. Stricklin and W.E. Haisler, Numerical Method for the Solution of Nonlinear Problems in Structural Analysis, in: *Numerical Solution of Nonlinear Structural Problems,* R.F. Hartung, ed., AMD – Vol. 6 (ASME, 1973) pp. 67–101.
[42] H.L. Schreyer and E.F. Masur, Buckling of Shallow Arches, Proc. ASCE. *Journal of the Engr. Mech. Div.* EM4, **92**, 1–19 (Aug. 1966).
[43] J.L. Sanders, Jr., Nonlinear Theories for Thin Shells, *Q. Appl. Math.* **21**, 21–36 (1963).

[44] D.M. Trujillo, An Unconditionally Stable Explicit Algorithm for Structural Dynamics, *Int. J. for Num. Meth. Engr.* **11**, 1579–1592 (1977).

[45] K.C. Park and J.M. Housner, Semi-Implicit Transient Analysis Procedures for Structural Dynamics Analysis, *Int. J. for Num. Meth. in Engrg.* **18**, 609–622 (1982).

[46] O.C. Zienkiewicz, E. Hinton, K.H. Leung and R.L. Taylor, Staggered Time Marching Schemes in Dynamics Soil Analysis and a Selective Explicit Extrapolation Algorithm, in: *Proceedings of the Second Intl. Symp. on Innovative Numerical Analysis in Applied Engineering Science*, Montreal (The University Press of Virginia, 1980) pp. 525–530.

[47] K.C. Park, A Family of Solution Algorithms for Nonlinear Structural Analysis Based on Relaxation Equations, in: *New Concepts in Finite Element Methods*, ASME Applied Mechanics Division Symposia Series, Vol. 44 (1981).

[48] A. Jennings and G.M. Malik, The Solution of Sparse Linear Equations by the Conjugate Gradient Method, *Int. J. for Num. Meth. Engr.* **12**, 141–158 (1978).

[49] M.A. Crisfield, Iterative Solution Procedures for Linear and Non-Linear Structural Analysis, TRRL Report 900, Transport and Road Research Laboratory, Crowthorne, Berkshire (1979).

[50] E. Riks, An Incremental Approach to the Solution of Snapping and Buckling Problems, *Int. J. Solids and Structures* **15**, 529–551 (1979).

[51] M.A. Crisfield, A Fast Incremental/Iterative Solution Procedure that Handles Snap-Through, *Computers and Structures* **13**, 55–62 (1981).

CHAPTER 6

Dispersion of Semidiscretized and Fully Discretized Systems

Howard L. SCHREYER

Department of Mechanical Engineering
University of New Mexico
Albuquerque, NM 87131, USA

Computational Methods for Transient Analysis
Edited by T. Belytschko and T.J.R. Hughes
© Elsevier Science Publishers B.V. (1983) 267–299

1. Introduction

Dynamic motion is a broad scientific subject with a rich history that includes the development of mathematical concepts and techniques. The phenomena that are considered are so broad that analytical solutions exist for just the simplest cases and only approximate solutions are feasible for the others. One method for obtaining approximate solutions to the governing equations of motion is to use numerical schemes which, unfortunately, often introduce phenomena that are not present in the physical system. Consequently, to properly interpret numerical results, an analyst must have an extensive knowledge of the features of both the solution scheme and the governing differential equation. It is fortunate that equations of motion appear at all technical levels so that exact solutions of the simplest governing equations provide an excellent foundation for studying the numerical schemes that must be used for the more complex systems.

The subject of this chapter is primarily concerned with the phenomenon of dispersion that is introduced by the finite difference and finite element schemes that are commonly used in engineering analysis. Dispersion associated with physical characteristics such as geometry, boundaries and material properties will not be discussed. Furthermore, only a restricted class of numerical schemes will be treated since the potential number of approaches is limitless.

The objective of this presentation is to present the basic concepts associated with numerical dispersion in a manner that would be useful to someone new in the field. However, it is presumed that the reader has some background in basic integration algorithms. This development is an assimilation of approaches outlined in several journal articles. The reference list should be useful as a guide for those who want to study the basic ideas in more depth. Some of the referenced work has been utilized directly; it is hoped that the use of material in this manner will be looked upon as an expression of appreciation and recognition for the contribution that is described. Clarity and brevity were also factors in the selection of the method in which the theory is presented.

Inevitably, in any discussion involving characteristics of numerical dispersion, other concepts such as numerical stability, order of accuracy, cut-off frequencies, overshoot and dissipation arise. The relationships among these aspects of numerical integration will be mentioned briefly since the primary focus will be on dispersion. However, in many cases more than one of these features must be considered simultaneously and it is important to realize this from the beginning.

The prototype for dispersive motion is based on a type of solution rather than a type of equation [1]. A dispersive system is any system which admits solutions of the form

$$u = A e^{i(kx - \omega t)}, \quad A = A(\omega), \tag{1.1}$$

for spatial and temporal variables x and t, respectively. The frequency ω is a real function of the wave number k and the relation

$$\omega = W(k) \tag{1.2}$$

is determined by the particular system. A is the amplitude of the wave and $i = \sqrt{-1}$. The phase speed (which is not the wave speed, in general) is

$$c = \frac{\omega}{k} \tag{1.3}$$

and the waves are said to be dispersive if c depends on k. The reason for considering several waves is that a more general solution will consist of a superposition of several modes such as those of (1.1) or in the most general case, a Fourier integral. If c is not the same for all k, the modes with different k will propagate at different speeds; i.e., they will disperse. A conventional definition is to say that the function of (1.1) is dispersive if $W''(k) \neq 0$.

In general, there can be more than one relation given by (1.2) for a system in which case the corresponding solutions are also referred to as modes. For clarity these solutions will be called 'branch' modes and for those modes not eliminated on physical grounds, they can be superposed to form the complete solution for linear problems.

The quantity

$$\theta = kx - \omega t \tag{1.4}$$

is the phase, the wavelength is

$$\lambda = \frac{2\pi}{k} \tag{1.5}$$

and the period is

$$T = \frac{2\pi}{\omega}. \tag{1.6}$$

For a nondispersive system the phase speed is a constant, c_0, which is called the wave speed, and (1.1) and (1.2) become

$$u = A e^{ik(x - c_0 t)}, \tag{1.7}$$

$$\omega = c_0 k, \tag{1.8}$$

which represents a solution to the one-dimensional equation (hyperbolic) for plane waves

$$u_{,tt} = c_0^2 u_{,xx} \tag{1.9}$$

where commas represent partial derivatives for the given variables. This equation is chosen as the model problem for illustrating in a simple manner each of the features associated with dispersion. It is easily verified that the general solution to (1.9) is

$$u = f(x - c_0 t) + g(x + c_0 t) \tag{1.10}$$

in which the functions f and g represent waves traveling to the right and left, respectively. The function of (1.7) is a particular form for f and, in fact, f can be represented formally by the Fourier integral

$$f(x - c_0 t) = \int_{-\infty}^{\infty} F(k)\, \mathrm{e}^{ik(x - c_0 t)}\, \mathrm{d}k \tag{1.11}$$

where F is chosen to fit initial or boundary data. As a side light, it is seen that the phase speed is also the wave speed for a nondispersive system.

To illustrate the effect of dispersion with a simple example, consider the solution composed of a sum of terms (modes) as follows:

$$u = \tfrac{1}{2} - \frac{2}{\pi} \sum_{n=1}^{\infty} \frac{(-1)^n}{(2n-1)} \cos(k_n x - \omega t)\,. \tag{1.12}$$

If we choose $k_n = 2n - 1$ and $\omega = k_n$, (1.12) is a Fourier series representation of the square wave function with period 2π;

$$f(\theta) = \begin{cases} 1 & \text{for } 0 < |\theta| < \tfrac{1}{2}\pi\,, \\ 0 & \text{for } \tfrac{1}{2}\pi \le |\theta| \le \pi\,. \end{cases} \tag{1.13}$$

Consider an initial condition consisting of the Fourier series representation of the square wave pattern which is shown in the vicinity of the origin of x and t in Figures 1a and 1b, respectively. Since this wave pattern which is labeled as $\varepsilon = 0$, is not dispersive ($c = c_0 = 1$), the deviation of the 20-term approximation from the exact wave form exists because of the attempt to represent the discontinuity (Gibbs' phenomenon).

Now suppose $\omega = k^{\varepsilon+1}$ or $c = k^{\varepsilon}$ for $|\varepsilon| \neq 0$. Then from the definition given above, the modes propagate at different velocities and the solution is dispersive. Examples for positive ($c > c_0$ or $\varepsilon > 0$) and negative ($c < c_0$ or $\varepsilon < 0$) dispersion are also displayed in Fig. 1 for an elapsed time of 1 or a displacement of 1 for the principal mode. The choice of unity for time and displacement was dictated by the desire to avoid the display of interaction between consecutive waves. All curves have been shifted to a common origin for detailed comparison. For the highest mode retained ($n = 20$ or $k = 39$) the phase velocities are 1.44 and 0.69 for $\varepsilon = 0.1$ and $\varepsilon = -0.1$, respectively. The differences between the curves ($\varepsilon \neq 0$) and the

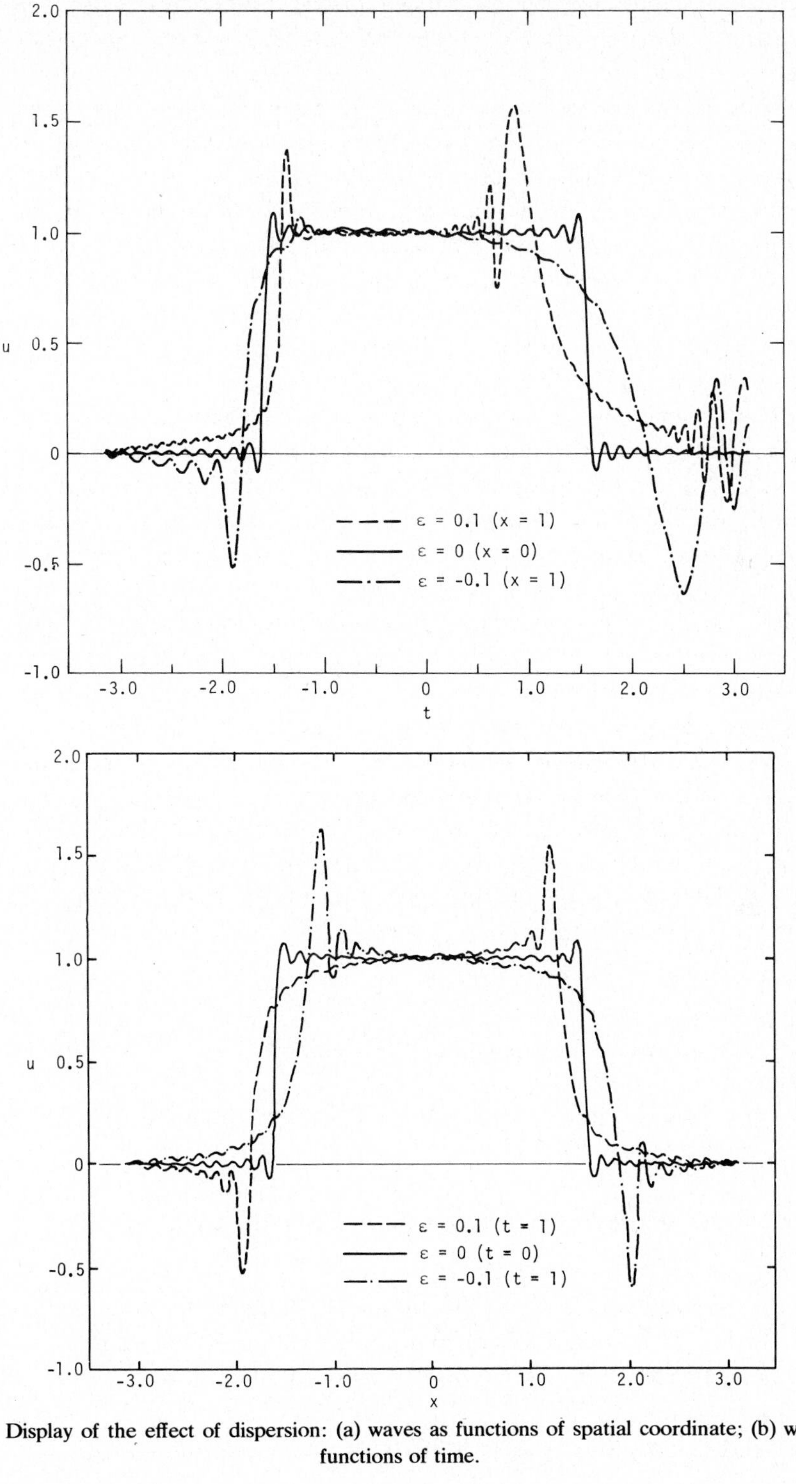

Fig. 1. Display of the effect of dispersion: (a) waves as functions of spatial coordinate; (b) waves as functions of time.

Fourier series representation ($\varepsilon = 0$) are attributable to dispersion. Since the coefficients of the trigonometric terms do not depend on x or t, there is no dissipation.

In Figure 1a, the waves formed in the presence of dispersion show a bias to the right and left for positive and negative dispersion, respectively. Positive dispersion displays outrunning which can have significant implications with regard to picking up arrival times. Negative dispersion results in a similar phenomenon at the trailing edge of the wave. Also the pulse width is severely distorted. Figure 1b shows the same phenomena from the perspective of an observer who watches the wave pass at $x = 1$. Overall, these results show that severe overshoot and wave modification can occur, and if the dispersion is not based on physical considerations, important features could be easily obscured.

For most engineering applications, the governing equation is discretized with respect to both spatial and temporal variables. As will be seen, each discretization introduces a pattern of dispersion but it must be emphasized that for the final numerical solution of the governing partial differential equation, dispersion from both sources must be simultaneously considered. Initially, for the sake of clarity in this introductory exposition, each feature will be discussed separately. Only dispersion associated with temporal discretizations (time integrators) as applied to the model problem of (1.9) is considered in the next section. The following section will illustrate dispersion associated with spatial discretizations while Section 4 presents typical cases where the effects are combined. Also included is a brief discussion on dispersion associated with other cases such as two-dimensional wave propagation, beam flexural waves and reduced systems. From the relatively simple examples considered and the fact that many of the most significant contributions have been made only recently, the conclusions can be made that this is indeed a fruitful area for further research. The last section also suggests possible areas of work that should be investigated.

2. Dispersion associated with temporal discretization[1]

For the model problem defined by (1.9), consider a solution of the form

$$u = \bar{u}(t)\,\mathrm{e}^{ikx} \tag{2.1}$$

so that the time dependent function is governed by the equation

$$\ddot{\bar{u}} + \omega^2\bar{u} = 0, \quad 0 \le t \le \tau > 0\,, \tag{2.2}$$

where a superposed dot denotes differentiation with respect to time. Approximate solutions to (2.2) for initial values of $\bar{u}$ and $\dot{\bar{u}}$ can be obtained by a variety of difference methods. For a time step of duration s and with $t_n = ns$, the ap-

[1]The material in References [2] and [3] has been used extensively for this section.

proximations to $\bar{u}(t_n)$, $\dot{\bar{u}}(t_n)$, $\ddot{\bar{u}}(t_n)$ are $\bar{u}_n$, $\bar{v}_n$ and $\bar{a}_n$, respectively. We restrict our analysis to those algorithms which are given symbolically by the integration scheme (three root)

$$\bar{u}_{n+1} = F_1(\bar{u}_n, \bar{v}_n, \bar{a}_n, \bar{a}_{n+1}, s)\,, \tag{2.3}$$

$$\bar{v}_{n+1} = F_2(\bar{v}_n, \bar{a}_n, \bar{a}_{n+1}, s)\,, \tag{2.4}$$

and the finite difference form used for (2.2) is assumed to be of the type

$$\bar{a}_{n+1} = F_3(\bar{u}_n, \bar{u}_{n+1}, \bar{a}_n, s) \tag{2.5}$$

for $n = 0, \ldots, N-1$ and

$$N = \tau/s\,. \tag{2.6}$$

Let

$$X_n = (\bar{u}_n, s\bar{v}_n, s^2\bar{a}_n) \tag{2.7}$$

denote the vector of these approximations. For the sake of definiteness, consider only those algorithms for which (2.3), (2.4) and (2.5) can be collectively written in the recursive form

$$X_{n+1} = AX_n\,. \tag{2.8}$$

The matrix A, which is called the amplification matrix, is central to the theory since general properties of the integration scheme can be related to its eigenvalues. The characteristic equation for A is

$$-\det(A - \lambda I) = \lambda^3 - A_1\lambda^2 + A_2\lambda - A_3 = 0 \tag{2.9}$$

where I is the identity matrix. If the eigenvalues of A are denoted by λ_1, λ_2 and λ_3, then

$$\begin{aligned} A_1 &= \text{trace } A = \lambda_1 + \lambda_2 + \lambda_3\,, \\ A_2 &= \tfrac{1}{2}(A_1^2 - \text{trace } A^2) = \lambda_1\lambda_2 + \lambda_2\lambda_3 + \lambda_3\lambda_1\,, \\ A_3 &= \text{determinant } A = \lambda_1\lambda_2\lambda_3\,, \end{aligned} \tag{2.10}$$

are three independent invariants of A. The spectral radius is $\rho = \max|\lambda_i|$.

If the velocities and accelerations are eliminated by repeated use of (2.8), then the resulting difference equation in terms of displacements is

$$\bar{u}_{n+1} - A_1\bar{u}_n + A_2\bar{u}_{n-1} - A_3\bar{u}_{n-2} = 0, \quad 2 \le n \le N-1\,. \tag{2.11}$$

For distinct eigenvalues, a comparison of (2.9) and (2.11) indicates that the discrete solution has the representation

$$\bar{u}_n = \sum_{i=1}^{3} c_i\lambda_i^n \tag{2.12}$$

where the coefficients c_1, c_2 and c_3 are determined from initial data. If $\lambda_1 = \lambda_2$, then the general solution of (2.11) is

$$\bar{u}_n = (c_1 + nc_2)\lambda_1^n + c_3\lambda_3^n \tag{2.13}$$

whereas if $\lambda_1 = \lambda_2 = \lambda_3$,

$$\bar{u}_n = (c_1 + nc_2 + n^2c_3)\lambda_1^n . \tag{2.14}$$

The restriction of the time integration algorithm to one involving a maximum of three roots results in fairly simple algebraic expressions for the general solution. Also, specific relations can be given for the local trunction error and the stability limit, s_c, on the time step. The related notions of consistency, stability, and convergence are discussed thoroughly in Chapter 2 of this book. Here, we explore in some detail the dispersive properties of three-root algorithms by noting that an explicit solution for the eigenvalues of A is available [4]. Let

$$\hat{\lambda} = \lambda - \tfrac{1}{3}A_1 . \tag{2.15}$$

Then (2.9) becomes

$$\hat{\lambda}^3 + \hat{A}_2\hat{\lambda} - \hat{A}_3 = 0 \tag{2.16}$$

where

$$\hat{A}_2 = A_2 - \tfrac{1}{3}A_1^2, \qquad \hat{A}_3 = \tfrac{1}{27}(2A_1^3 - 9A_1A_2 + 27A_3) . \tag{2.17}$$

With

$$D = (\tfrac{1}{3}\hat{A}_2) + (\tfrac{1}{2}\hat{A}_3)^2 \tag{2.18}$$

assumed to be positive, let

$$u = [\tfrac{1}{2}\hat{A}_3 + \sqrt{D}]^{1/3}, \qquad v = [\tfrac{1}{2}\hat{A}_3 - \sqrt{D}]^{1/3} . \tag{2.19}$$

Then the solutions to (2.16) exist as complex conjugates and a real root

$$\hat{\lambda}_1, \hat{\lambda}_2 = \frac{-(u+v)}{2} \pm \tfrac{3}{2}(u-v)\,\mathrm{i}, \qquad \hat{\lambda}_3 = u + v . \tag{2.20}$$

It follows from (2.15) that, for a stable algorithm, the solution of (2.12) can be written in the form

$$\bar{u}_n = \exp(-\bar{\zeta}\bar{\omega}t_n)(c_1 \cos \bar{\omega}t_n + c_2 \sin \bar{\omega}t_n) + c_3\lambda_3^n \tag{2.21}$$

where λ_3 is called a spurious root with $|\lambda_3| < |\lambda_1, \lambda_2| \le 1$ and

$$\begin{aligned} &\lambda_1, \lambda_2 = A \pm B\mathrm{i} = \exp[\bar{\Omega}(-\bar{\zeta} \pm \mathrm{i})], \\ &A = \tfrac{1}{3}A_1 - \tfrac{1}{2}(u+v), \qquad B = \tfrac{1}{2}\sqrt{3}(u-v), \\ &\bar{\omega} = \bar{\Omega}/s, \qquad \bar{\Omega} = \arctan(B/A), \qquad \bar{\zeta} = -\ln(A^2 + B^2)/(2\bar{\Omega}). \end{aligned} \tag{2.22}$$

A measure of numerical dissipation, as noted in Chapter 2 is the algorithmic damping ratio $\bar{\zeta}$ which is defined in terms of the two principal roots. From the definition of the spectral radius it follows that

$$\bar{\zeta} = -\ln \rho \Big/ \left(\arctan \frac{B}{A}\right). \tag{2.23}$$

For many structural problems, it is desirable to have the high frequency terms damp out in which case a non-zero value for $\bar{\zeta}$ is desirable. However, for the analysis of waves over a long period of time, $\bar{\zeta} = 0$ is a necessary condition. The role of the spurious root in characterizing a numerical solution has not been thoroughly studied.

If the spurious root is not considered, then the form of the complete solution to the numerical scheme from (2.1) and (2.21) is given by

$$u = e^{-\bar{\zeta}\bar{\omega}t_n}\, e^{i(kx - \bar{\omega}t_n)} \tag{2.24}$$

which falls into the category of a dispersive system defined by (1.1) and (1.2). One measure of the dispersion for the time integration algorithm is

$$e_d = \frac{c}{c_0} = \frac{\bar{\omega}}{\omega} \tag{2.25}$$

since the wave numbers are identical for the exact solution to the model problem and the solution to the time discretized problem. In terms of periods of the two systems

$$T = \frac{2\pi}{\omega}, \qquad \bar{T} = \frac{2\pi}{\bar{\omega}}, \tag{2.26}$$

an alternate measure of dispersion is the relative period error

$$e_t = (\bar{T} - T)/T \tag{2.27}$$

which is the form used in Chapter 2. Since

$$e_d = \frac{1}{1 + e_t}, \tag{2.28}$$

the integration algorithm is nondispersive if $e_d = 1$ or $e_t = 0$.

To illustrate the amount of dispersion that might be expected in practice, consider the α-method family of algorithms defined by the following relations [2] (see (58) in Chapter 2, Section B.III):

$$\begin{aligned} M\bar{a}_{n+1} + (1+\alpha)K\bar{u}_{n+1} - \alpha K\bar{u}_n &= 0\,, \\ \bar{u}_{n+1} = \bar{u}_n + s\bar{v}_n + s^2[(\tfrac{1}{2} - \beta)\bar{a}_n &+ \beta\bar{a}_{n+1}]\,, \\ \bar{v}_{n+1} = \bar{v}_n + s[(1-\gamma)\bar{a}_n &+ \gamma\bar{a}_{n+1}]\,, \end{aligned} \tag{2.29}$$

with $\bar{u}_0$ and $\bar{v}_0$ prescribed. These equations represent a specific example of (2.3), (2.4) and (2.5). The time integration procedure is of the Newmark family with an additional parameter α used to control dissipation. We use these equations to show how the theory in this section can be used to evaluate dispersive properties. Other algorithms could be analysed similarly.

By direct substitution, it can be shown that the amplification matrix of (2.8) is

$$A = \frac{1}{D}\begin{bmatrix} 1+\alpha\beta\Omega^2 & 1 & \frac{1}{2}-\beta \\ -\gamma\Omega^2 & 1+(1+\alpha)(\beta-\gamma)\Omega^2 & (1-\gamma)+(1+\alpha)\left(\beta-\frac{\gamma}{2}\right)\Omega^2 \\ -\Omega^2 & -(1+\alpha)\Omega^2 & -(1+\alpha)(\frac{1}{2}-\beta)\Omega^2 \end{bmatrix} \tag{2.30}$$

in which

$$\omega = (K/M)^{1/2}, \qquad \Omega = s\omega, \qquad D = 1+(1+\alpha)\beta\Omega^2 . \tag{2.31}$$

It follows from (2.10) that

$$A_1 = \frac{1}{D}[2+\Omega^2\{\beta(3\alpha+2)-(1+\alpha)(\gamma+\tfrac{1}{2})\}],$$

$$A_2 = \frac{1}{D}[1+\Omega^2\{\alpha(3\beta-2\gamma)+\beta-\gamma+\tfrac{1}{2}\}], \tag{2.32}$$

$$A_3 = \frac{1}{D}[\alpha(\tfrac{1}{2}+\beta-\gamma)\Omega^2] .$$

The free parameters α, β and γ govern the stability, dissipative and dispersive properties of the system. Although dissipation has been the primary focus of research on the effect of these parameters [2], we choose to use this algorithm to illustrate the effect that these parameters have on dispersion. We note in passing that the algorithm is second order accurate if $\gamma = \frac{1}{2}-\alpha$ and $\beta = \frac{1}{4}(1-\alpha)^2$ but these limitations are not imposed in the sequel.

If $\alpha = 0$ this family of algorithms reduces to the Newmark subset. Of this subset, if $2\beta \geq \gamma \geq \frac{1}{2}$ the algorithm is unconditionally stable. The trapezoidal rule ($\beta = 0.25$, $\gamma = 0.5$) is one of the most commonly used implicit time integrators while the central difference method ($\beta = 0$, $\gamma = \frac{1}{2}$), which is conditionally stable with $s_c = T/\pi$, is the most popular explicit method.

We observe from (2.32) that $A_3 = 0$ if $\alpha = 0$ or $\frac{1}{2}+\beta-\gamma = 0$ and there are only two non-trivial eigenvalues of A as follows:

$$\lambda_1, \lambda_2 = \tfrac{1}{2}[A_1 \pm (A_1^2 - 4A_2)^{1/2}] . \tag{2.33}$$

If $A_1^2 - 4A_2 < 0$ the eigenvalues are complex conjugate and

$$\bar{\Omega} = \arctan\left(\frac{4A_2 - A_1^2}{A_1^2}\right)^{1/2}, \qquad \rho = A_2^{1/2} . \tag{2.34}$$

Additional simplifications can be made for the central difference algorithm in which case

$$A_1 = 2-\Omega^2, \qquad A_2 = 1 \tag{2.35}$$

and

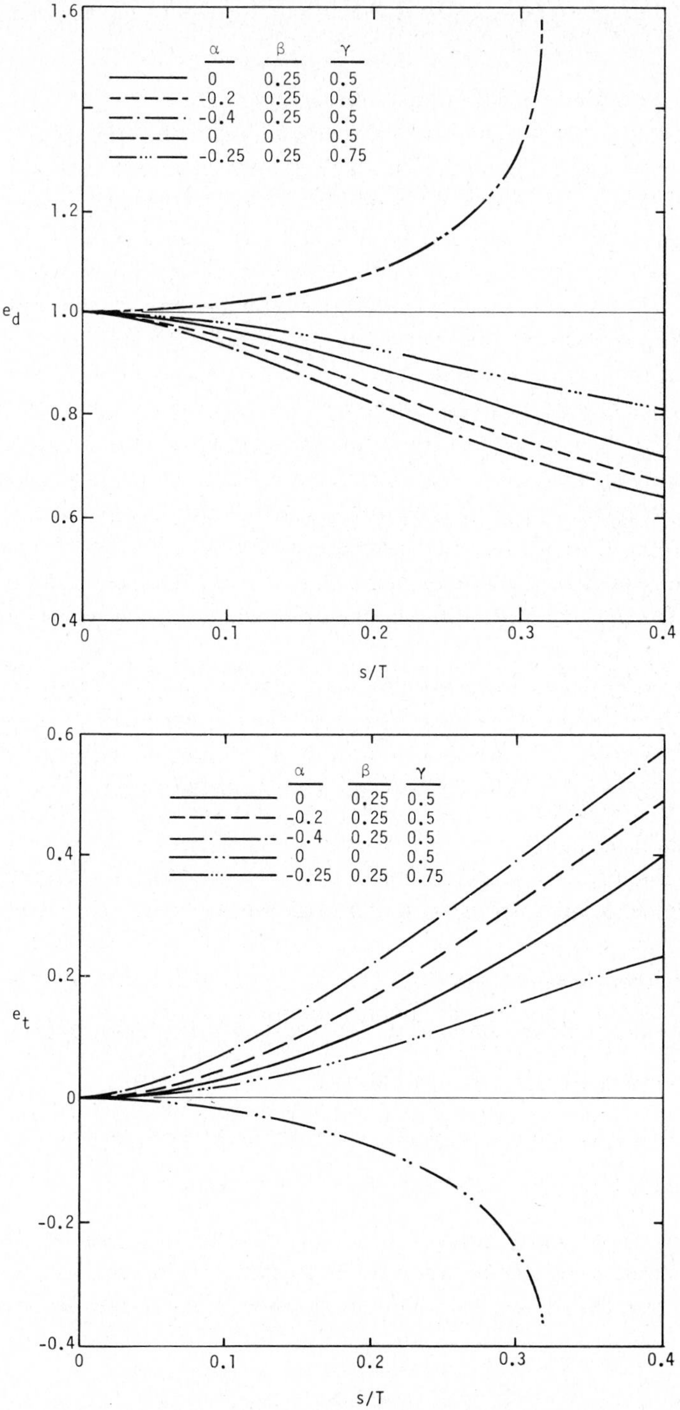

Fig. 2. Dispersion for various time integrators: (a) dispersion in terms of phase speed; (b) dispersion in terms of relative period error.

$$\bar{\Omega} = \arctan\left[\frac{\Omega^2(4-\Omega^2)}{(2-\Omega^2)^2}\right]^{1/2}, \qquad \rho = 1\,, \tag{2.36}$$

so that this particular algorithm is non-dissipative.

Dispersion curves associated with the general algorithm for specific choices of α, β and γ are given in Figure 2a. The same information expressed in terms of the relative period error is given in Figure 2b. Curves associated with the central difference algorithm are terminated at the stability limit, $s/T = 1/\pi$.

The effect of the parameter α for the trapezoidal rule is displayed by the first three curves. In general the effect is relatively minor which is encouraging since the primary purpose for the parameter is to control dissipation. The fourth (central-difference explicit) and fifth (implicit) curves display dispersion for schemes that contain no spurious root ($\alpha = \frac{1}{2} + \beta$) with α chosen such that $\gamma = \frac{1}{2} - \alpha$; i.e., $\alpha = -\beta$. The central difference scheme shows dispersion with a sign opposite to that displayed by the others. The latter curve indicates that a study might be appropriate to optimize the scheme with regard to dispersion.

Relative period errors for the collocation method and Houbolt's and Park's schemes are also presented in Figures 15 and 20 of Chapter 2. In general, the Houbolt and Park methods have larger errors than optimized collocation and α-methods.

In this section, basic concepts and the relationship of certain properties to dispersion have been introduced. Representative dispersion curves for some common time integrators are shown. However, nothing has been said about the effect on dispersion of spatial discretization which is discussed in the next section.

3. Dispersion associated with spatial discretization

3.1. *Basic formulation*

For the model problem defined by (1.9), consider a solution of the form

$$u = \hat{u}(x)\, e^{-i\omega t} \tag{3.1}$$

so that the space dependent function is governed by the equation

$$\hat{u}_{,xx} + k^2\hat{u} = 0, \quad 0 \le x \le L\,, \tag{3.2}$$

over an appropriate region for x. An alternate formulation based on the principle of virtual displacements is obtained by multiplying each term in (3.2) by $\delta\hat{u}$ and integrating over the region. After use of boundary conditions on $\hat{u}$ and $\hat{u}_{,x}$, the result is

$$k^2 \int_0^L \hat{u}\delta\hat{u}\, dx - \int_0^L \hat{u}_{,x}\delta\hat{u}_{,x}\, dx = 0\,. \tag{3.3}$$

Let $\hat{u}_m$ represent the approximation to $\hat{u}(mh)$ for $m = 0, \ldots, M$ where the number of spatial steps, each of length h, is

$$M = L/h\,. \tag{3.4}$$

Solution values for $\hat{u}_m$ are obtained by utilizing a finite difference scheme associated with (3.2) or finite element shape functions in (3.3). Both approaches result in a matrix equation of the form

$$[k^2h^2M - K]\{\hat{u}\} = 0 \tag{3.5}$$

in which M and K are dimensionless mass and stiffness matrices, respectively, and $\{\hat{u}\} = \langle \hat{u}_0, \ldots, \hat{u}_M\rangle^{\mathrm{T}}$. The mass matrix is diagonal for the finite difference approach but, in general, it is not diagonal for the finite element method (consistent mass matrix). However, the use of a diagonal mass matrix is so convenient for dynamic calculations that for the latter case, the consistent mass matrix is often replaced with a diagonal mass matrix M_{d}. There are various approaches for choosing such a matrix. Perhaps the most common way is to define the diagonal component of M_{d} as the sum of the components in the corresponding row of the consistent mass matrix. This is appropriate only if all terms are dimensionally consistent; otherwise alternate methods are required.

For many discretization methods, the solution to (3.5) for a typical nodal variable may be given symbolically as follows:

$$\hat{u}_m = G(kh;\, \hat{u}_{m-1}, \hat{u}_{m+1}, \hat{u}_{m-2}, \hat{u}_{m+2}, \ldots)\,. \tag{3.6}$$

This equation holds provided all nodes are weighted equally. The function G depends on the particular scheme. If symmetry exists for every other node rather than equal weighting, (3.6) is not valid and instead an alternate form governs:

$$\begin{aligned}\hat{u}_{2m} &= G_1(kh;\, \hat{u}_{2m-1}, \hat{u}_{2m+1}, \hat{u}_{2m-2}, \hat{u}_{2m+2}, \ldots)\,,\\ \hat{u}_{2m+1} &= G_2(kh;\, \hat{u}_{2m}, \hat{u}_{2m+2}, \hat{u}_{2m-1}, \hat{u}_{2m+3}, \ldots)\,.\end{aligned} \tag{3.7}$$

For (3.6), a solution of the form

$$\hat{u}_m = A_m\, \mathrm{e}^{\mathrm{i}\hat{k}mh} \tag{3.8}$$

may be assumed where A_m denotes a constant. From (1.1) and (3.1) it is seen that such a solution falls in the category of a dispersive system.

For convenience define dimensionless wave numbers

$$\kappa = kh, \qquad \hat{\kappa} = \hat{k}h\,. \tag{3.9}$$

Then the result of substituting (3.8) in (3.5) is

$$A_m\, \mathrm{e}^{\mathrm{i}m\hat{\kappa}} F(\hat{\kappa}, \kappa) = 0\,, \tag{3.10}$$

from which it follows that a necessary and sufficient condition for such a solution to exist is that the characteristic function F be identically zero. The solution can be given as a plot of $\hat{\kappa}$ versus κ which is called a characteristic curve but it can also be considered a mode from (1.2) since

$$\kappa = \omega h / c_0 . \tag{3.11}$$

The same information may be presented in the form of a dispersion curve since

$$\hat{\kappa} = \omega h / c \tag{3.12}$$

and (2.24) yields

$$e_{\mathrm{d}} = \frac{c}{c_0} = \frac{\kappa}{\hat{\kappa}} . \tag{3.13}$$

A characteristic curve will display a maximum value for κ which we denote as κ_{c}. From (1.5) it follows that the corresponding wave length is

$$\lambda_{\mathrm{c}} = \frac{2\pi h}{\kappa_{\mathrm{c}}} \tag{3.14}$$

with a logical interpretation that the spatial grid cannot transmit waves with a wave length shorter than λ_{c}. Alternately, (1.8) and (3.9) can be used to show that the maximum (cutoff) frequency that can be transmitted without an exponentially decaying amplitude is

$$\omega_{\mathrm{c}} = c_0 \kappa_{\mathrm{c}} / h . \tag{3.15}$$

If a particular formulation falls in the category of (3.7) rather than (3.6), a solution of the form

$$\hat{u}_{2m} = A_{2m}\, \mathrm{e}^{\mathrm{i}2m\hat{\kappa}}, \qquad \hat{u}_{2m+1} = A_{2m+1}\, \mathrm{e}^{\mathrm{i}(2m+1)\hat{\kappa}} \tag{3.16}$$

yields a characteristic function that displays two characteristic curves or 'branch' modes. If the maximum value of κ for the first curve ($\kappa_{1\mathrm{c}}$) is less than the minimum value of κ associated with the second curve (κ_{2L}), then the system acts like a band filter where frequencies in the range

$$\frac{c_0 \kappa_{1\mathrm{c}}}{h} < \omega < \frac{c_0 \kappa_{2L}}{h} \tag{3.17}$$

decay exponentially, just as frequencies above the cutoff. The ramifications of such systems have not been completely explored although Belytschko and Mullen [5] and Brillouin [6] discuss the topic to some extent.

For a large class of systems, the dispersion function is periodic with respect to $\hat{\kappa}$ with a half period denoted by $\hat{\kappa}_0$. To compare the dispersive properties of such systems it is convenient to plot e_{d} as a function of normalized wave number $\hat{\kappa}/\hat{\kappa}_0$ which has the range [0, 1]. Also, dispersion can be expressed in terms of the change in wave length so that an alternative function to e_{d} is

$$e_{\lambda} = \frac{\lambda - \hat{\lambda}}{\lambda} = 1 - e_{\mathrm{d}} . \tag{3.18}$$

Finite elements and finite differences are the two dominant methods for spatial discretization. We choose to present primarily discretizations based on the finite element procedure since there is a variety of approaches associated with the method of which some are directly analogous to finite difference formulations. Explicit formulations are given for a few cases to illustrate the procedure with other results summarized for brevity.

For many physical systems with finite dimensions, solutions to the semidiscretized systems have identical wavelengths to corresponding solutions of the exact differential equation. Thus, for these systems, the corresponding frequencies are different which is analogous to the situation in the previous section. For wave propagation in an infinite medium we use (3.13) as our interpretation of dispersion; for a medium of finite domain the appropriate interpretation is $e_d = \bar{\omega}/\omega$. Also for the latter case, the mode and wave number are identical so that $\hat{\kappa}/\hat{\kappa}_0 = n/N$ where n is the mode and N denotes the number of degrees of freedom for the discretized system.

3.2. *Finite element semidiscretizations*

(a) *Constant strain element.* Consider elements in which the displacement varies linearly so that the strain is constant. The stiffness, and consistent and diagonal mass matrices for this element are

$$K^e = \begin{bmatrix} 1 & -1 \\ -1 & 1 \end{bmatrix}, \qquad M^e = \tfrac{1}{6}\begin{bmatrix} 2 & 1 \\ 1 & 2 \end{bmatrix}, \qquad M^e_D = \tfrac{1}{2}\begin{bmatrix} 1 & 0 \\ 0 & 1 \end{bmatrix}. \tag{3.19}$$

Each mass matrix is associated with certain desirable characteristics so it has been proposed [7] that a linear combination

$$M^e_L = \eta M^e + (1-\eta) M^e_D \tag{3.20}$$

may, in a certain sense, provide optimal properties. The assembly of the resulting element equations yields the system described in (3.5) from which a typical component equation is

$$\eta \hat{u}_{m+1} + (6-2\eta)\hat{u}_m + \eta \hat{u}_{m-1} + \frac{6}{\kappa^2}(\hat{u}_{m+1} - 2\hat{u}_m + \hat{u}_{m-1}) = 0\,. \tag{3.21}$$

The lumped mass equation ($\eta = 0$) is identical to a second order central difference approximation to (3.2). Also, the equation describes the motion of a one-dimensional lattice of point masses interconnected by springs and has served as a model for sound wave propagation among others [6].

After some algebraic manipulations it can be shown that the characteristic equation is (see, for example, Goudreau [8])

$$\kappa^2(3 - 2\eta \sin^2 \tfrac{1}{2}\hat{\kappa}) - 12 \sin^2 \tfrac{1}{2}\hat{\kappa} = 0 \tag{3.22}$$

from which it follows that

$$\kappa_c = 2/(1 - \tfrac{2}{3}\eta)^{1/2}, \qquad \hat{\kappa}_0 = \pi\,. \tag{3.23}$$

Dispersion curves for the lumped mass ($\eta = 0$), consistent mass ($\eta = 1$) and combined mass ($\eta = \frac{3}{4}$) cases are given in Figure 3. These results show that the dispersive velocity is always greater (less) than the actual velocity for the consistent (diagonal) mass case. Also, a linear combination of the two mass matrices displays improved dispersive properties [7].

(b) *Linear strain elements.* For the quadratic displacement, linear strain element, the element stiffness and mass matrices are [5]

$$K^e = \frac{1}{3}\begin{bmatrix} 7 & -8 & 1 \\ -8 & 16 & -8 \\ 1 & -8 & 7 \end{bmatrix}, \qquad M^e = \frac{1}{30}\begin{bmatrix} 4 & 2 & -1 \\ 2 & 16 & 2 \\ -1 & 2 & 4 \end{bmatrix},$$

$$M_D^e = \frac{1}{6}\begin{bmatrix} 1 & 0 & 0 \\ 0 & 4 & 0 \\ 0 & 0 & 1 \end{bmatrix}. \tag{3.24}$$

This formulation falls in the category of (3.6). For the diagonal mass case the characteristic equation is

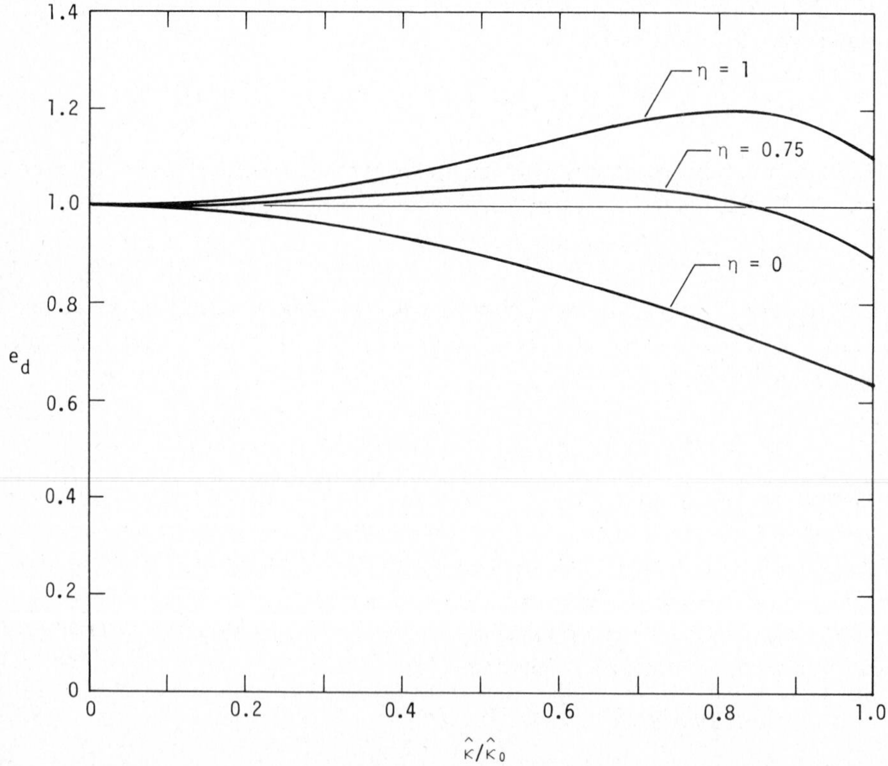

Fig. 3. Dispersion for constant strain element with a linear combination of consistent and diagonal mass matrices.

$$2\kappa^2 - 6 + \sin^2\hat{\kappa} \pm (36 - 36\sin^2\hat{\kappa} + \sin^4\hat{\kappa})^{1/2} = 0 \tag{3.25}$$

which displays the two branches shown in Figure 4. It is easily shown that

$$\hat{\kappa}_0 = \tfrac{1}{2}\pi, \qquad \kappa_{1c} = \sqrt{2}, \qquad \kappa_{2L} = \sqrt{3}. \tag{3.26}$$

and frequencies in the range described by (3.17) are damped out. These results show that this element exhibits relatively little dispersion for the lower branch (acoustical) and considerable dispersion for the upper branch (optical). This information is also given in [5] together with corresponding curves for the consistent mass matrix in terms of the frequency spectrum (ω vs. $\hat{\kappa}$).

(c) *Consistent diagonal mass elements.* For the previous two cases, the lumped mass could be considered as the sum of the terms in the corresponding row of the consistent mass matrix. In reference [9], it is shown that, with the use of orthogonal (self-conjugate) polynomials and a mixed variational procedure, a diagonal mass matrix can be obtained directly from the expression for kinetic energy.

For this approach where masses are considered to be lumped at element nodes, the even- and odd-numbered nodes decouple to a 'leap-frog' scheme and a typical

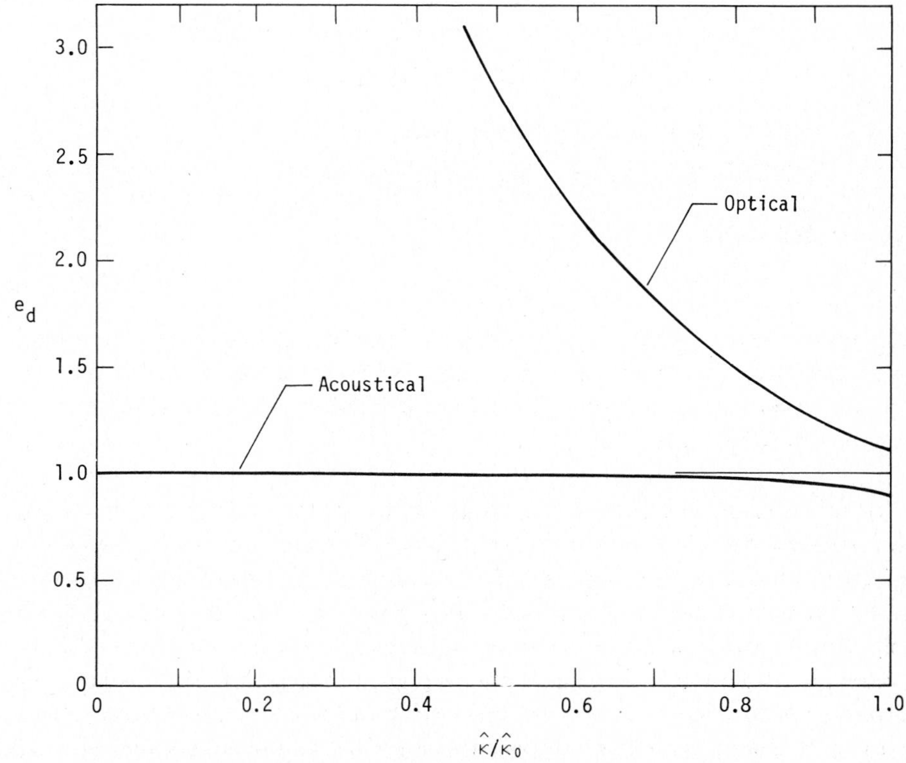

Fig. 4. Dispersion for linear strain element with diagonal mass matrix.

component equation is

$$4\kappa^2\hat{u}_m = \hat{u}_{m+1} - 2\hat{u}_m + \hat{u}_{m-1}\,. \tag{3.27}$$

The characteristic equation is

$$\kappa = \sin\hat{\kappa} \tag{3.28}$$

so that this discretization is identical to the lumped-mass constant-strain element provided h is replaced with $2h$, and the dispersive properties for the two elements are identical.

It can be shown from rigid body mechanics that if mass is lumped uniformly at all nodes, or if mass is lumped in any fashion at nodes on the periphery of an element, then the expression for kinetic energy under rigid body rotation of such a lumped-mass element cannot be correct. In reference [10], an element is described in which two-thirds of the element mass is lumped at the centroid and self-conjugate shape functions are used with the result that the correct rigid body kinetic energy is obtained (kinetic energy element). This system also falls in the category of (3.6) with the specific form given as follows:

$$\begin{aligned} 8\kappa^2\hat{u}_{2m} &= 4(\hat{u}_{2m+2} - 2\hat{u}_{2m} + \hat{u}_{2m-2}) \\ &\quad -(\hat{u}_{2m+3} - \hat{u}_{2m+1} - \hat{u}_{2m-1} + \hat{u}_{2m-3})\,, \\ 16\kappa^2\hat{u}_{2m+1} &= (\hat{u}_{2m+5} + 8\hat{u}_{2m+3} - 18\hat{u}_{2m+1} + 8\hat{u}_{2m-1} + \hat{u}_{2m-3}) \\ &\quad -4(\hat{u}_{2m+4} - \hat{u}_{2m+2} - \hat{u}_{2m} + \hat{u}_{2m-2})\,. \end{aligned} \tag{3.29}$$

The associated two-branch characteristic equation is

$$2\kappa^2 - \sin^2\hat{\kappa}\,[5 - \sin^2\hat{\kappa} \pm \{(9 - \sin^2\hat{\kappa})\cos^2\hat{\kappa}\}^{1/2}] = 0\,. \tag{3.30}$$

It can be shown that

$$\hat{\kappa}_0 = \tfrac{1}{2}\pi, \qquad \kappa_{1c} = \tfrac{3}{2}, \qquad \kappa_{2c} = \sqrt{2}\,, \tag{3.31}$$

and based on the graphical representation of Figure 5 it is seen that the optical dispersion curve is considerably different from that associated with the conventional linear strain element. It is not known whether or not this difference is of any practical significance.

(d) *Beam elements.* Flexural wave propagation in beams is different from uniaxial wave propagation in that dispersion is a physical phenomenon. Thus, a measure of the error introduced by spatial semidiscretization is the difference in dispersion between approximate and exact solutions. The conventional Euler–Bernouilli beam theory yields a linear relation between dispersion and the radius to wavelength ratio for circular beams. For a sufficiently large value of this ratio, this relation is in significant error in comparison to an exact solution based on the theory of elasticity. However, dispersion based on Timoshenko's theory which includes corrections to the Euler–Bernouilli theory for shear and rotatory inertia

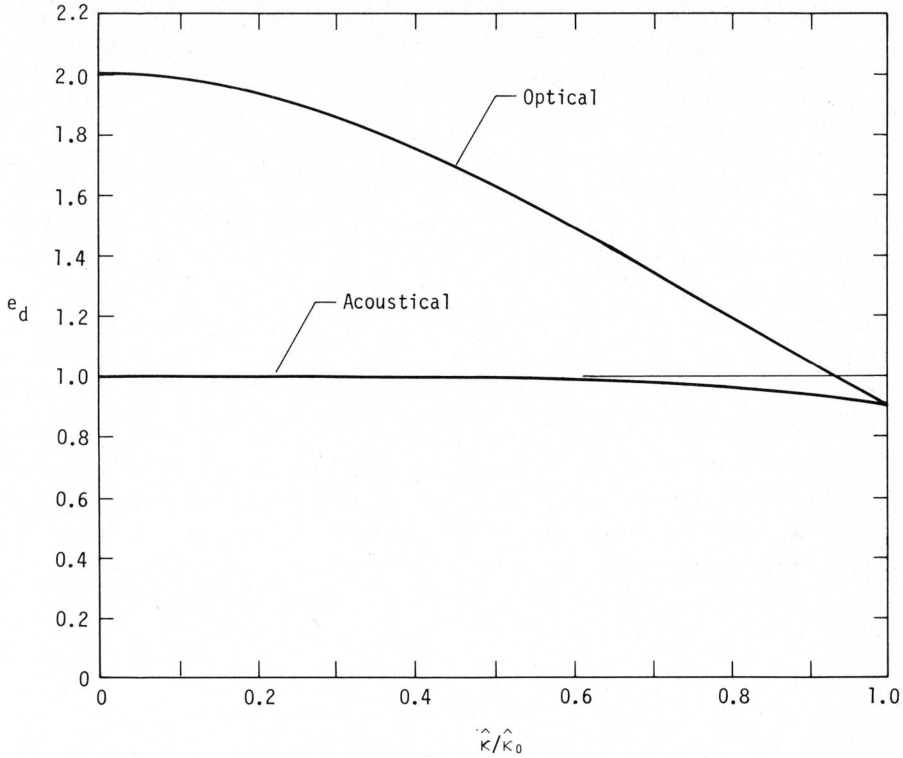

Fig. 5. Dispersion for kinetic energy element.

agrees remarkably well with the exact theory. Thus, Timoshenko's theory forms an appropriate base from which to obtain dispersion error due to spatial semidiscretization. Archer [11] has developed an element based on this theory and the element was used by Belytschko and Mindle [12] for their study. For a consistent mass matrix an illustrative example given in Figure 6 shows that dispersion is always greater than the exact solution. For increasing values of the radius to element length ratio, r/h, predicted values of dispersion monotonically converge to the exact value for a given ratio of radius to wave length, r/Λ. Also, in conformity with previous illustrative results, it is shown in [12] that dispersion associated with a diagonal mass matrix generally falls below the exact curve.

(e) *Two-dimensional problems.* In what appears to be the first investigation of dispersion associated with two-dimensional semidiscretizations, Goudreau [8] considered a clamped rectangular membrane. An ordinary five-point Laplacian finite difference operator and a bilinear quadrilateral finite element expansion with consistent and diagonal mass matrices were studied. Dispersion is shown in Fig. 7 for the full range of n/N and discrete values of m/M where n and m represent mode numbers for the two orthogonal coordinates and the corresponding maximum mode numbers are N and M, respectively.

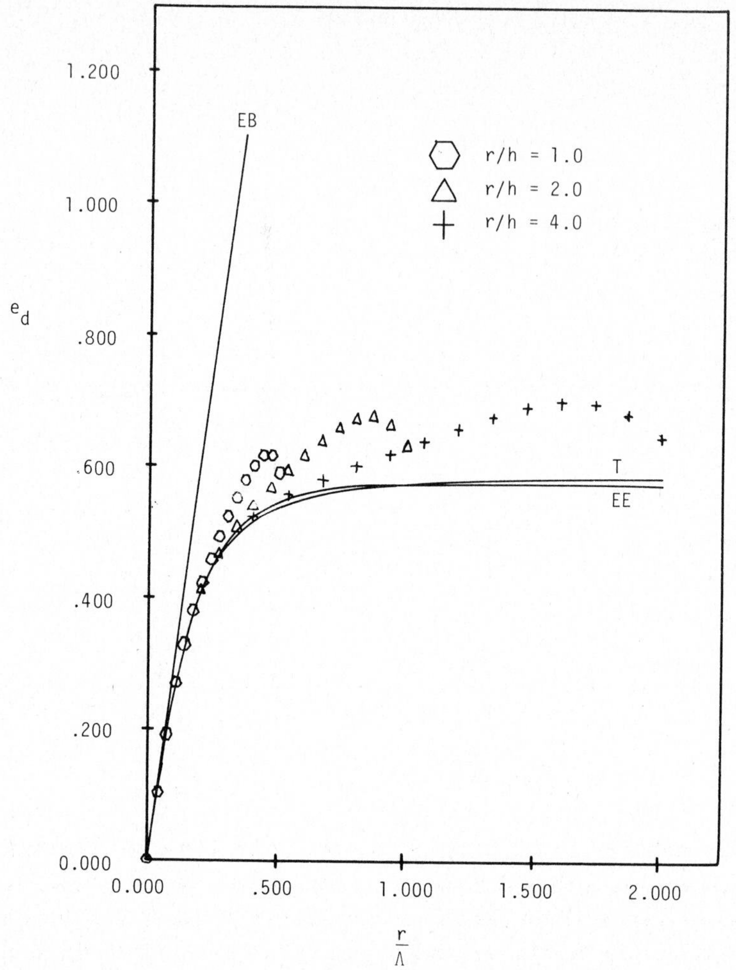

Fig. 6. Dispersion for consistent mass archer element (EB – Euler–Bernoulli exact; T – Timoshenko exact; EE – exact elasticity) [12].

For the two-dimensional wave equation, both the angle of wave propagation and the scheme for arranging elements can have an effect on dispersion. In general, Mullen and Belytschko [13] claim that quadrilateral elements have superior dispersion properties to any arrangement of constant strain triangles. Consistent mass matrices are preferable to diagonal matrices as shown by Figures 8 and 9 in which θ denotes the angle of wave propagation with respect to the x-axis. Here $\hat{\kappa}/\hat{\kappa}_0 = \kappa h/\pi$ for the square elements under consideration. The use of lumped masses introduces severe dispersion for directions of propagation that do not coincide with mesh lines, and Figures 10 shows that this property is exacerbated by the use of reduced integration. In general, the quadrilateral element with consistent mass matrices had dispersive properties that are superior to those associated with the five-point Laplacian finite difference operator. Explicit

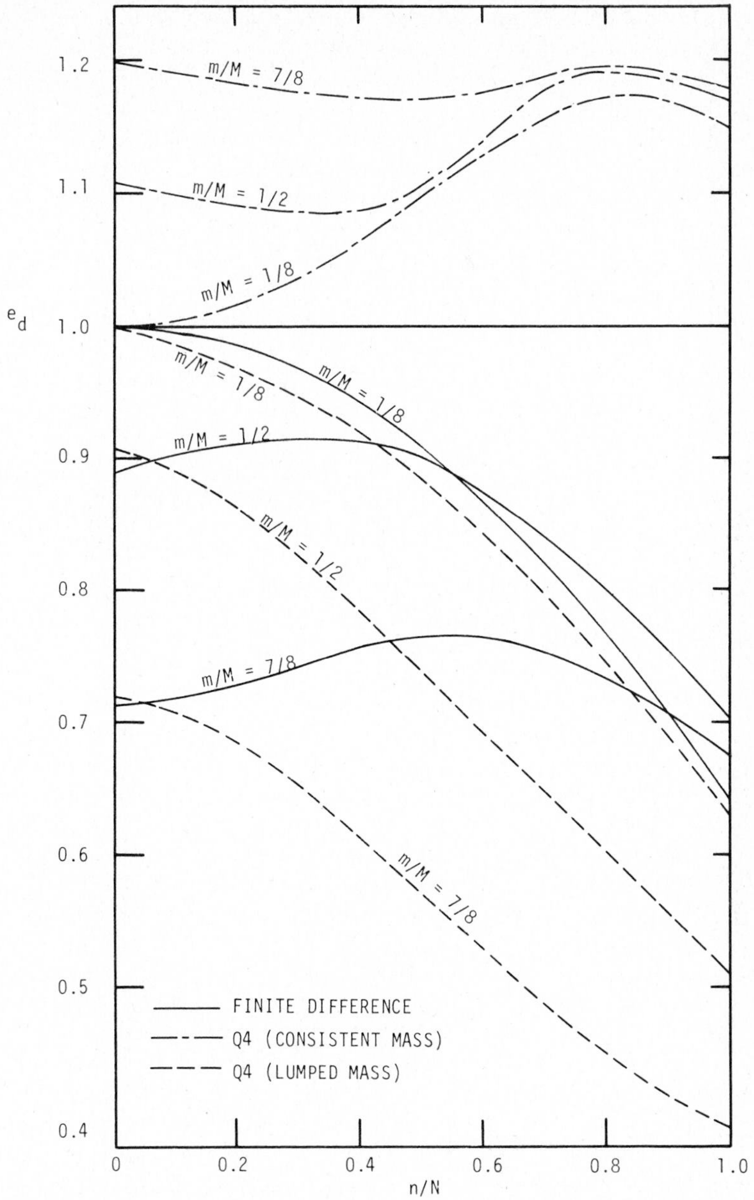

Fig. 7. Dispersion as a function of mode numbers for a vibrating membrane [8].

expressions for dispersion of rectangular elements as well as various arrangements of triangular elements are also given in [13] which represents the most comprehensive two-dimensional study to this date.

(f) *Reduced systems.* A scheme for reducing the number of degrees of freedom for finite element systems has been proposed by Hughes et al. [14]. The basic idea

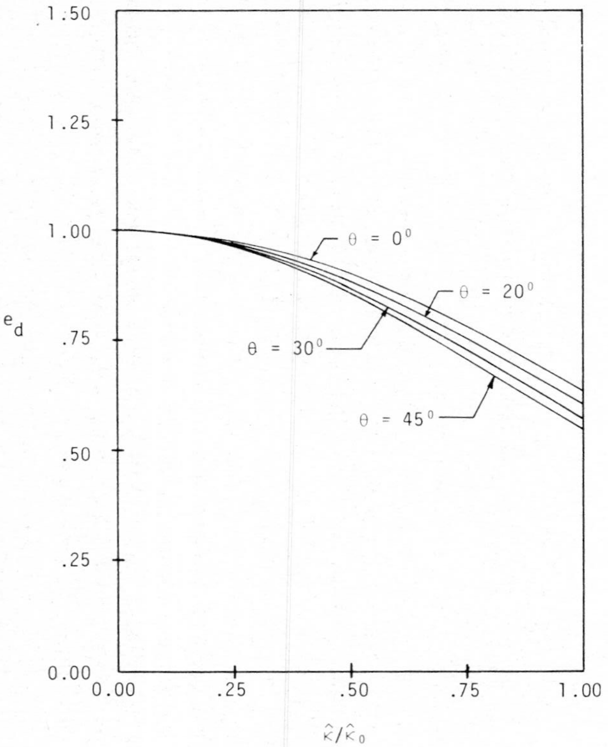

Fig. 8. Dispersion of bilinear quadrilateral element with consistent mass as a function of relative wave number for various propagation directions [13].

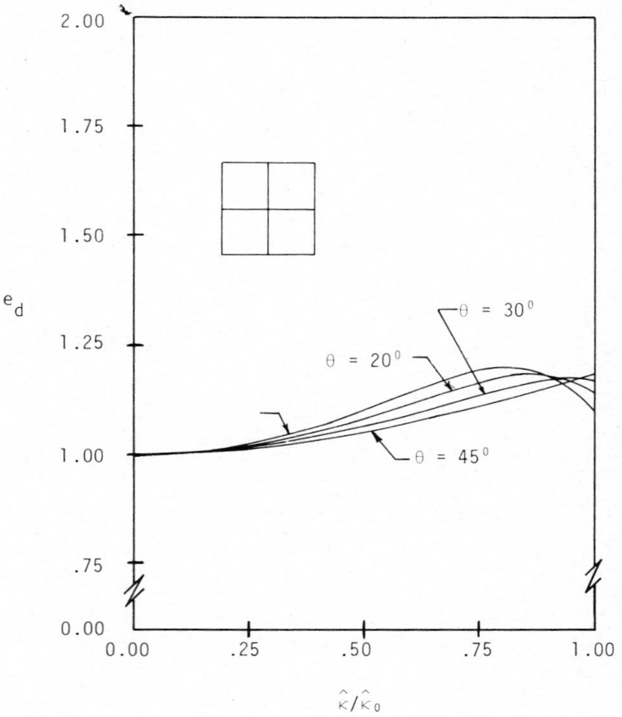

Fig. 9. Dispersion of bilinear quadrilateral element with lumped mass as a function of relative wave number for various propagation directions [13].

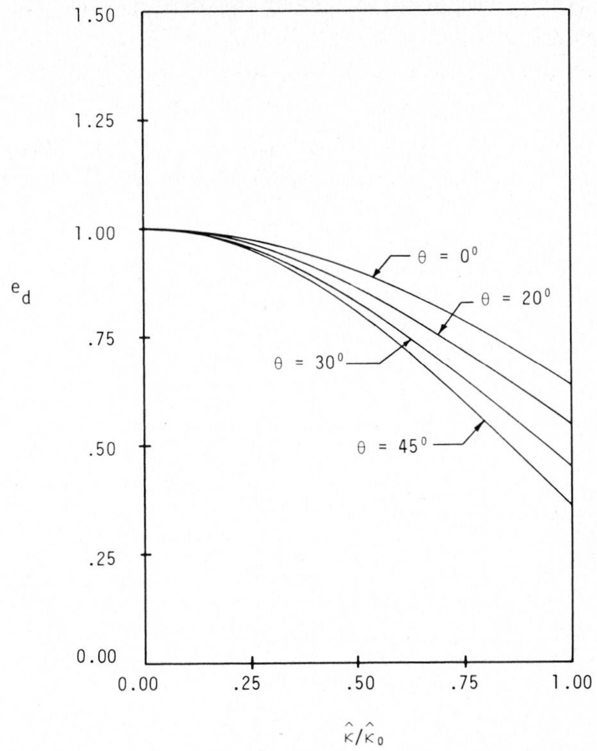

Fig. 10. Dispersion of underintegrated bilinear quadrilateral element with lumped mass as a function of relative wave number for various propagation directions [13].

is to assume different interpolation functions for the displacement and velocity fields. If the number of velocity degrees of freedom is less than the number of displacement degrees of freedom, then the combined finite element system is reduced from the system obtained by the conventional approach of using the same number of degrees of freedom for both field variables. It is shown that the rate of convergence of the conventional consistent mass approach is maintained provided $\hat{k} \geq k - m$ where $2m$ is the order of the spatial differential operator, and k and $\hat{k}$ are the degrees of the complete polynomial contained in the interpolation assumptions for the displacement and velocity fields, respectively. In addition to the convergence study, dispersion properties of certain elements were obtained and are summarized below.

The differential equation for the longitudinal displacement, u, of a rod is given by $u'' + \omega^2 u = 0$ where ω is the natural frequency; thus $m = 1$. Fixed-free and fixed-fixed boundary conditions were studied. Within each element the displacement field is assumed to vary quadratically ($k = 2$) with the displacements of the end points and the midpoint as the finite element variables (linear strain element discussed previously). Results are presented for three cases: a conventional mass matrix, a diagonal mass matrix in which $\frac{1}{6}$ the mass is lumped at each end-point and $\frac{2}{3}$ at the midpoint, and a reduced system involving a linear velocity approximation for each element ($\hat{k} = 1$).

Dispersion for the fixed-fixed case is presented in Figure 11 where n is the mode number and N is the number of elements. The curves corrobate the upper bound properties of the consistent and reduced system. Also, the maximum relative error for the reduced system (approximately 15 percent) is less than that for either of the other cases.

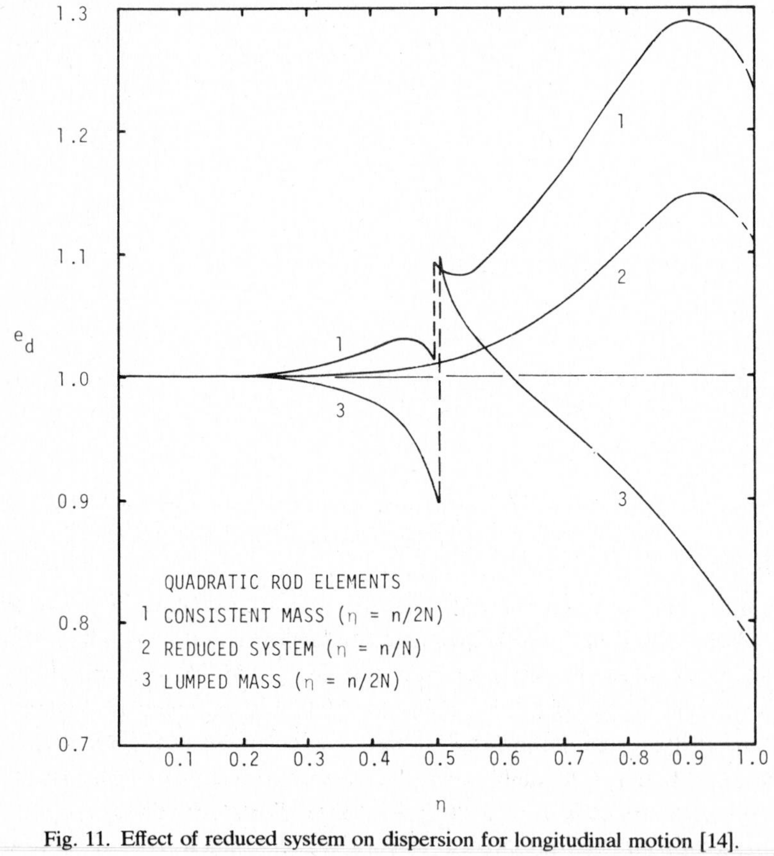

Fig. 11. Effect of reduced system on dispersion for longitudinal motion [14].

The model beam equation $u^{IV} - \omega^2 u = 0$ where u denotes the transverse displacement is used with simply-supported boundary conditions to illustrate the case $m = 2$. The traditional cubic displacement function is employed ($k = 3$). Cases considered included a consistent mass matrix, a diagonal mass matrix with $\frac{1}{2}$ the mass lumped at each translatory degree of freedom and no mass associated with the rotatory degrees of freedom, and a reduced system involving a linear velocity approximation for each element ($\hat{k} = 1$); thus $\hat{k} = k - m$. The rate of convergence is quartic for all cases. Dispersive properties are summarized in Figure 12 with conclusions similar to those made for the rod element.

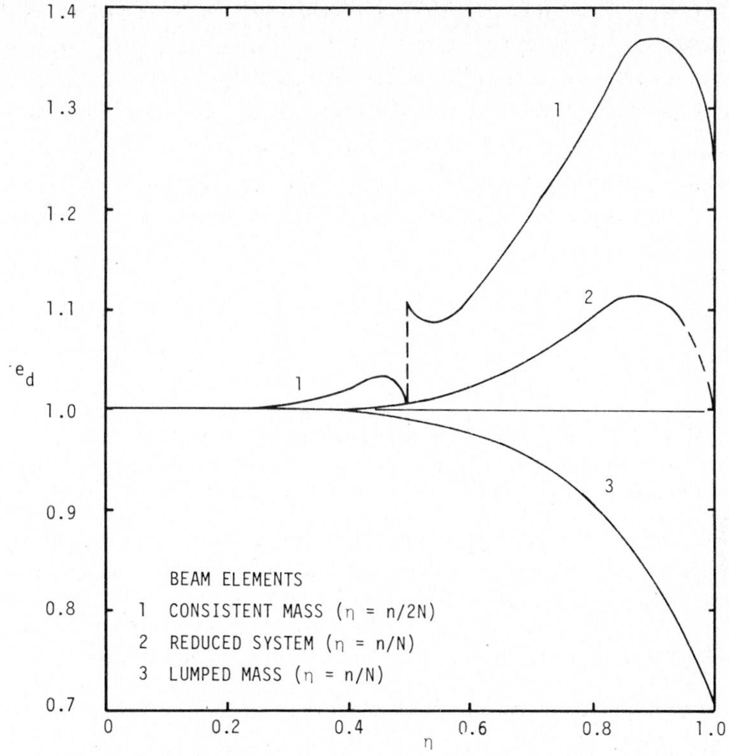

Fig. 12. Effect of reduced system on dispersion for flexural motion [14].

A simply-supported square plate was also analyzed in this study with an element which contains a complete cubic displacement function. A bilinear velocity assumption was made for the reduced case and the resulting dispersive properties were consistent with those obtained for rod and beam elements.

In this section, dispersion associated with spatial semidiscretizations based on a representative set of finite elements has been analyzed. A comparison with dispersion associated with time integration algorithms indicates that with certain combinations, the two sources of dispersion may be cumulative whereas for other cases, there will be a certain amount of compensation. A brief look at dispersion associated with combined spatial and temporal discretization is presented next.

4. Dispersion associated with complete discretization

The model problem of (1.9) may be restated in a variational form and for the sake of explicitness, we choose the principle of virtual displacements in which case the governing equation is

$$\int_0^L [u_{,tt}\delta u + c_0^2 u_{,x}\delta u_{,x}]\,\mathrm{d}x = 0\,. \tag{4.1}$$

Let u_n^m represent the approximation to $u(mh, ns)$ for $m = 0, \ldots, M$ and $n = 0, \ldots, N$. Solution values for u_n^m may be obtained by a variety of time integration and finite element or finite difference schemes. To illustrate the degree of dispersion associated with certain specific combinations, consider the time integration scheme of (2.29) and the constant strain element formulation of (3.19) which incorporates both consistent and diagonal mass matrices. The utilization of (4.1) to obtain appropriate element mass and stiffness matrices yields the following representative equation:

$$\tfrac{1}{6}[\eta a_{n+1}^{m+1} + (6-2\eta)a_{n+1}^{m} + \eta a_{n+1}^{m-1}] + (1+\alpha)\frac{c_0^2}{h^2}[u_{n+1}^{m+1} - 2u_{n+1}^{m} + u_{n+1}^{m-1}]$$
$$-\alpha\frac{c_0^2}{h^2}[u_n^{m+1} - 2u_n^m + u_n^{m-1}] = 0\,,$$
$$u_{n+1}^m = u_n^m + s v_n^m + s^2[(\tfrac{1}{2}-\beta)a_n^m + \beta a_{n+1}^m]\,,$$
$$v_{n+1}^m = v_n^m + s[(1-\gamma)a_n^m + \gamma a_{n+1}^m]\,, \tag{4.2}$$

in which a_n^m and v_n^m denote the approximations to $u_{,tt}(mh, ns)$ and $u_{,t}(mh, ns)$, respectively, and the linear combination of mass matrices given in (3.20) is used.

For conciseness, we introduce the operators

$$\nabla_x u_n^m = u_n^{m+1} - 2u_n^m + u_n^{m-1}, \qquad \nabla_t u_n^m = u_{n+1}^m - 2u_n^m + u_{n-1}^m\,. \tag{4.3}$$

Then, by eliminating the velocity terms in (4.2_2) and (4.2_3) it follows that

$$\nabla_t u_n^m = s^2[\tfrac{1}{2}(a_n^m + a_{n-1}^m) + \gamma(a_n^m - a_{n-1}^m) + \beta\nabla_t a_n^m]\,. \tag{4.4}$$

If each term in (4.2_1) is operated on by ∇_t and each term in (4.4) by ∇_x, then (4.4) can be used to eliminate the displacement terms in (4.2_1). The resulting governing equation in terms of accelerations only is

$$\eta\nabla_t\nabla_x a_n^m + 6\nabla_t a_n^m = 6\sigma^2[\nabla_x a_n^m + \beta\nabla_x\nabla_t a_n^m + (\gamma - \tfrac{1}{2} + \alpha)\nabla_x(a_n^m - a_{n-1}^m)$$
$$+\alpha\nabla_x\nabla_t\{(\gamma - \tfrac{1}{2})a_{n-1}^m + \beta(a_n^m - a_{n-1}^m)\}] \tag{4.5}$$

where σ, the Courant number, is defined by

$$\sigma = \frac{c_0 s}{h}\,. \tag{4.6}$$

Assume a solution of the following type:

$$a_n^m = A\,e^{i\phi_n^m}, \quad \phi_n^m = m\hat{\kappa} - n\bar{\Omega}\,. \tag{4.7}$$

The substitution of (4.7) in (4.5) will yield imaginary terms which implies that the assumed solution is not correct and that in actual fact dissipation exists. Dissipation may be an undesirable feature and is caused by the presence of the term a_{n-1}^m in (4.5). This term can be eliminated if the parameters γ and β are chosen such that

$$\gamma = \tfrac{1}{2} \quad \text{and} \quad \alpha = 0 \tag{4.8_1}$$

or

$$\gamma = \tfrac{1}{2} - \alpha \quad \text{and} \quad \beta = -\alpha\,. \tag{4.8$_2$}$$

It is interesting to note that the first equation yields a time integrator that is second order accurate [3]. Dispersion curves associated with (4.8$_1$) have been provided by Belytschko and Mullen [5]. We follow their procedure, but with the general condition associated with (4.8$_2$), to illustrate the approach and to provide information on dispersion for a time integrator that was introduced to provide good dissipation properties. For this case, it was shown [3] that the most useful range for α is $[-\frac{1}{3}, 0]$. From (4.8$_2$) these values also encompass the values of β that are most frequently used; e.g., $\beta = 0$ yields the central difference operator and $\beta = \frac{1}{4}$ yields the trapezoidal rule. In the following part of this section, we limit ourselves to the study of the two-parameter family (η, α) defined by (4.5) and (4.8$_2$).

The substitution of Eqs. (4.7) and (4.8$_2$) in (4.5) followed by the use of trigonometric identities yields the following characteristic equation:

$$3\sigma^2 \sin^2 \tfrac{1}{2}\hat{\kappa} = \sin^2 \tfrac{1}{2}\bar{\Omega}[3 - 2\{\eta + 6\alpha(1+\alpha)\sigma^2\}\sin^2 \tfrac{1}{2}\hat{\kappa}]\,. \tag{4.9}$$

Special cases are readily obtained by assigning particular values to the free parameters; e.g., lumped mass ($\eta = 0$) or consistent mass ($\eta = 1$).

In plotting dispersion curves, it appears from (4.9) that too many variables $(\sigma, \bar{\Omega}, \hat{\kappa})$ are available. However, we first note that

$$\frac{\bar{\Omega}}{\hat{\kappa}\sigma} = \frac{(s\bar{\omega})}{(\omega h/c)(c_0 s/h)} = \left(\frac{c}{c_0}\right)^2, \tag{4.10}$$

so that the appropriate dispersion parameter is

$$e_d = \left(\frac{\bar{\Omega}}{\hat{\kappa}\sigma}\right)^{1/2}. \tag{4.11}$$

By inspection $\hat{\kappa}_0 = \pi$, and $\partial\bar{\Omega}/\partial\hat{\kappa} = 0$ for $\hat{\kappa} = 0, \pi$. Thus, the cutoff frequency is obtained by setting $\hat{\kappa} = \pi$ in (4.9) from which it follows that

$$\sin^2 \tfrac{1}{2}\bar{\Omega} = \frac{3\sigma^2}{[3 - 2\eta - 12\alpha(1+\alpha)\sigma^2]} \tag{4.12}$$

and in the limit for small time step

$$\lim_{s\to 0} \bar{\omega}_c = \omega_c = \frac{2c_0/h}{(1-\frac{2}{3}\eta)^{1/2}} \tag{4.13}$$

which is identical to (3.23). The maximum value of σ over the admissible range of $\bar{\Omega}_c$ yields the dimensionless critical time step

$$\sigma_c = \left[\frac{1-\frac{2}{3}\eta}{1+4\alpha(1+\alpha)}\right]^{1/2}. \tag{4.14}$$

For the central difference algorithm ($\beta = 0$) with the lumped mass matrix ($\eta = 0$), $\sigma \leq 1$ for numerical stability and for the limiting case of $\sigma = 1$, it follows from (4.9) and (4.11) that $\hat{\kappa} = \bar{\Omega}$ and $e_d = 1$, i.e., there is no dispersion.

Following Krieg and Key [15] we see that there is actually a family of discretizations that can be considered optimal with respect to dispersion in the sense that there is no dispersion for $\sigma = 1$ if η is chosen such that

$$\eta = -6\alpha(1+\alpha)\,. \tag{4.15}$$

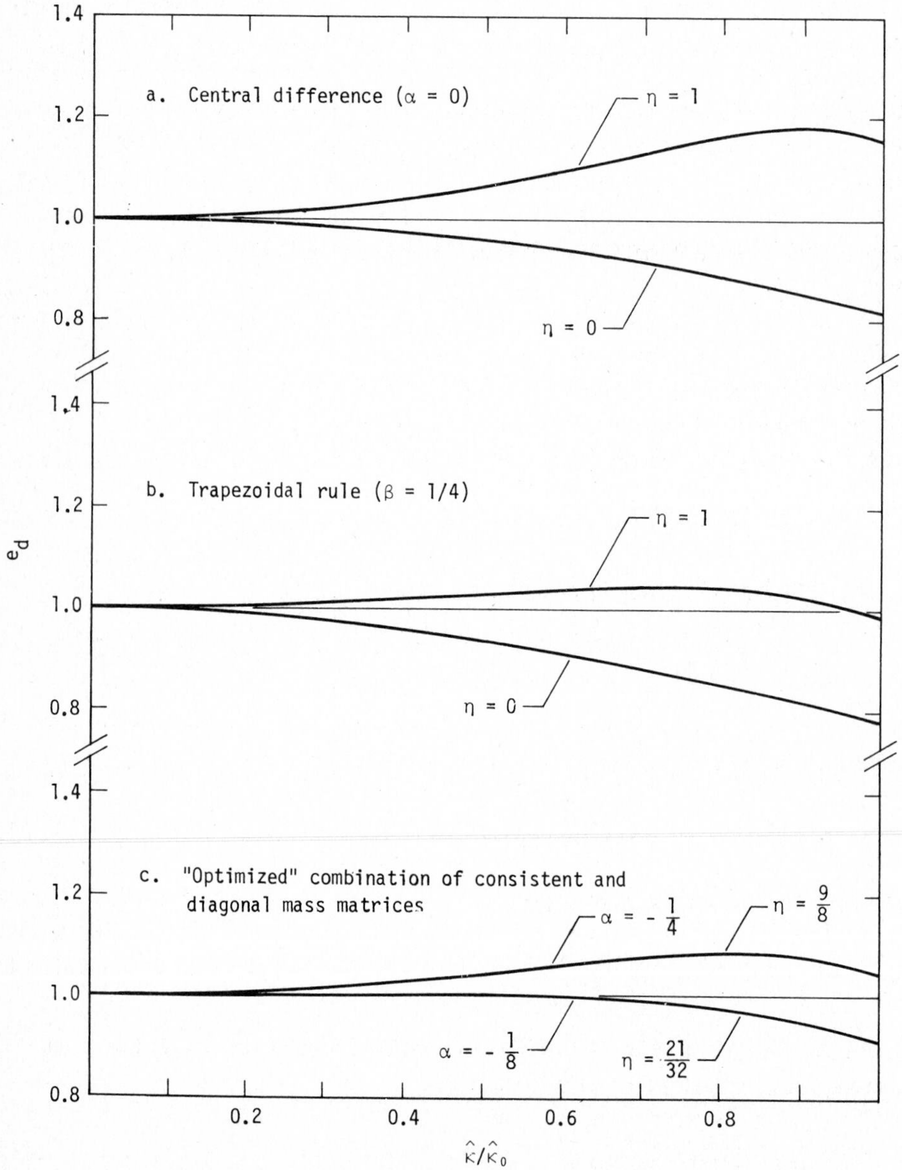

Fig. 13. Representative dispersion curves for complete discretization ($\sigma = 0.5$).

For example, if $\alpha = -\frac{1}{4}$, then $\eta = \frac{9}{8}$ which implies that for the trapezoidal rule, an 'overcompensated' consistent mass matrix would be optimal. However, this property is not that important since any practical engineering problem contains variable mesh sizes so that σ will not be one for all elements. Furthermore, for the extreme case of $\eta = \frac{3}{2}$, (4.14) indicates that the critical time step is zero as was shown in [16].

There are several other interesting consequences to the requirement of $(\ddot{4}.8_2)$. For example, the trapezoidal rule which is an unconditionally stable algorithm for temporally discretized systems becomes conditionally stable and (4.14) yields $\sigma_c = 2(1 - \frac{2}{3}\eta)^{1/2}$ for $\alpha = -\frac{1}{4}$. The algorithm is unconditionally stable if $1 + 4\alpha(1 + \alpha) = 0$ or $\alpha = -\frac{1}{2}$.

To illustrate the effect that the parameters α and η have on dispersion a few characteristic curves are shown in Figure 13 for $\sigma = \frac{1}{2}$. The first pair ($\alpha = 0$) represent the central difference algorithm with conventional diagonal mass ($\eta = 0$) and consistent mass ($\eta = 1$) approaches. The second pair were obtained for the same values of η but represent dispersion for the trapezoidal rule ($\alpha = -\frac{1}{4}$). The third pair represent values for α and η of $(-\frac{1}{8}, \frac{21}{32})$ and $(-\frac{1}{4}, \frac{9}{8})$, respectively, that satisfy (4.15).

The set of curves shown in Figure 14 represent the unconditionally stable case of $\alpha = -\frac{1}{2}$ for the consistent mass case ($\eta = 1$). A set of three typical values of dimensionless time step were chosen to show a realistic range of dispersion that might be expected. The results are disconcerting in that an unconditionally stable scheme displays significantly more dispersion than any of the others. Also, a typical

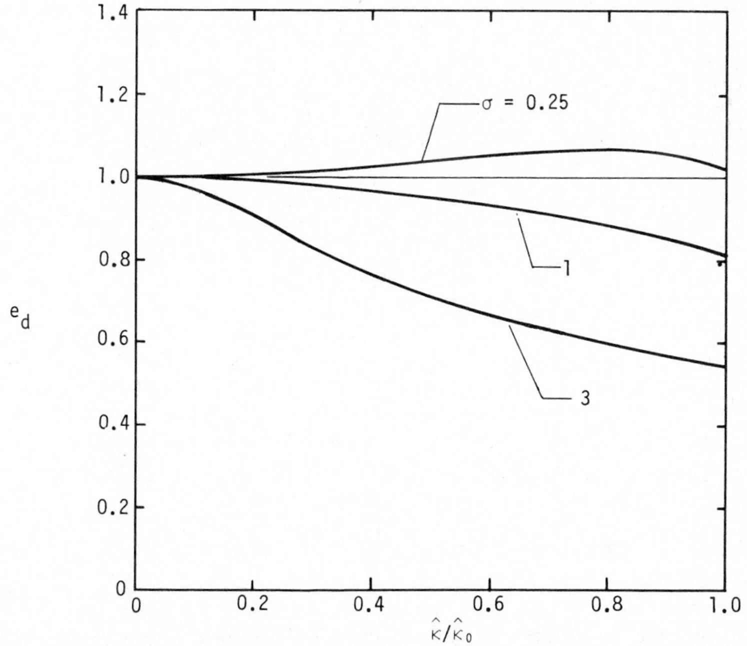

Fig. 14. Dispersion for unconditionally stable ($\alpha = -0.5$), consistent mass ($\eta = 1$) case.

problem will normally require the use of several values of σ so that the dispersion will not necessarily be of one type (positive or negative).

The result of combined temporal and spatial discretizations yield characteristic functions that are not intuitive based on a knowledge of dispersion associated with each discretization considered separately. The combined situation is the one of practical interest and unless several simplifying assumptions are made, characteristic curves are not catalogued as easily as the ones shown here for the constant strain element. In addition, other factors such as those associated with bending complicate the situation even further. Krieg and Key [15] have performed such a study for a beam element with a cubic displacement field. Dispersion as a function of mode number is shown in Figure 15 for various combinations of time integrators and mass matrices for the simply-supported case. The diagonal mass matrix is obtained by equal lumping for the mass and by appropriately scaling the corresponding diagonal component of the consistent mass matrix for rotational inertia. The curves are labeled according to the combination of consistent mass (CM) or diagonal mass (DM) and central difference (CD) or trapezoidal rule (TR).

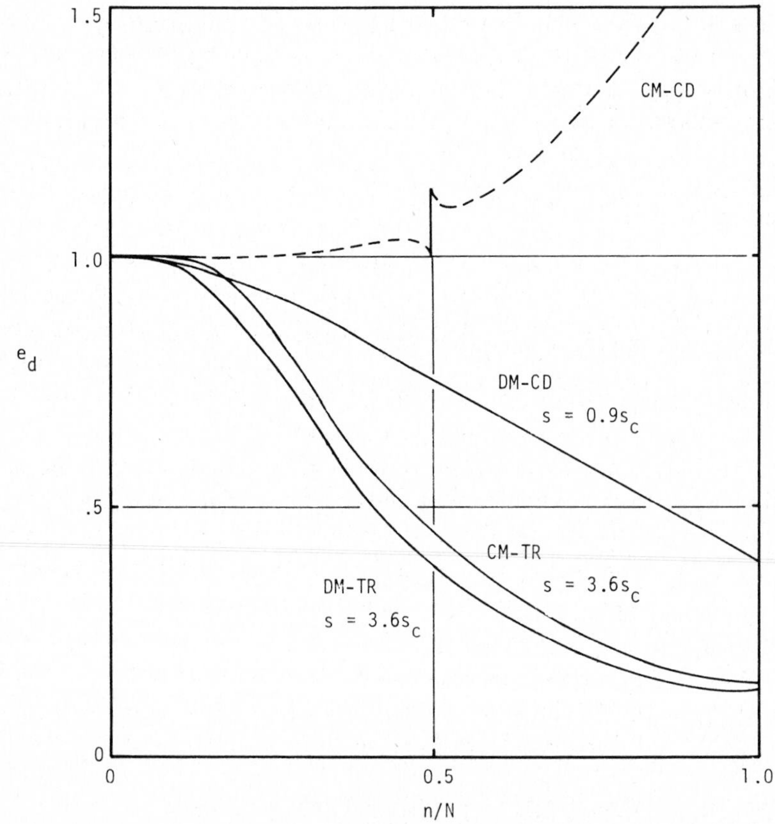

Fig. 15. Dispersion as a function of mode number for the complete discretization of the beam equation [15].

Solid curves imply that time steps were chosen such that these results are on an equal basis with regard to computational work. The critical time step is based on the diagonal mass-central difference combination. The consistent mass-central difference combination had a different critical time step so the computational effort for this curve is fifteen times that for any of the other curves. The conclusions based on these results is that the diagonal mass-central difference combination is superior. Of course, the actual dispersive nature of the Euler–Bernouilli theory must be kept in mind.

5. Conclusion

In Section 2 we outlined a conventional approach for studying dispersion associated with temporal semidiscretizations. The theory is neat and succinct, at least for the two-root time integrators that are commonly in use. For algorithms that involve a spurious root, definitive statements concerning dispersion cannot be made.

Conclusions with regard to dispersion associated with spatial semidiscretizations are even harder to summarize because of the variety of approaches that can be adopted. Higher order finite element and finite difference schemes can introduce branch modes and the resulting difference between numerical and exact solutions is loosely collected under the term of 'noise'. Actually this is just a reflection of our lack of understanding of dispersion, dissipation and Gibb's phenomenon [17, 18].

A method for obtaining characteristic (dispersion) curves was presented in Section 3 and examples of these curves for a few elements were also given. Here, the situation is complicated by the fact that different forms for an element mass matrix are in common use, and the choice of a particular form often depends on factors other than dispersive properties. A variety of element base functions also leads to the requirement of multiple investigations.

The literature is just beginning to reflect systematic investigations of dispersion associated with many of the numerical algorithms that are commonly used by engineers. These studies show that significant advances in understanding dispersion due to spatial discretizations are being made but still there are several difficult problems that have not been analyzed. An obvious area is the effect of changes in mesh size in both one [16, 19] and two-dimensions. Changes in density and material properties can also be studied in the infinite domain as can flexural motion in plates and shells. Dispersion due to the modeling of finite boundaries including voids and rigid inclusions would appear to be a formidable research area.

Ultimately, the dispersive properties of completely discretized systems are of the most value to analysts. As indicated by the results of Section 4, several parameters exist even for the simplest algorithms and it is not a trivial task to display the combined effect in a simple manner. Furthermore, Gibb's phenomenon and dissipation, in addition to dispersion, are important properties that must be considered simultaneously. Although dispersion due to semidiscretizations may be considered a classical subject, it appears that this is still a potentially rewarding research field. Each of the areas that have been mentioned with regard to semi discretizations have scarcely been touched as far as totally discretized systems are concerned. As indicated by Engquist and Kreiss [20], dispersion may also be introduced by transmitting or absorbing boundaries and different time scales inherent in a physical problem. Changes in time steps should also be mentioned in this context.

Overall, extensive developments are currently being reported but much remains to be done.

References

[1] G.B. Whitham, *Linear and Nonlinear Waves* (Wiley, New York, 1974).

[2] H.M. Hilber, T.J.R. Hughes and R.L. Taylor, Improved Numerical Dissipation for Time Integration Algorithms in Structural Dynamics, *Earthquake Engineering and Structural Dynamics* **5**, 283–292 (1977).

[3] H.M. Hilber and T.J.R. Hughes, Collocation, Dissipation and 'Overshoot' for Time Integration Schemes in Structural Dynamics, *Earthquake Engineering and Structural Dynamics* **6**, 99–117 (1978).

[4] J.J. Tuma, *Engineering Mathematics Handbook* (McGraw-Hill, New York, 1970).

[5] T. Belytschko and R. Mullen, On Dispersive Properties of Finite Element Solutions, *Modern Problems in Elastic Wave Propagation*, International Union of Theoretical and Applied Mechanics Symposium held at Northwestern University, Evanston, IL, 12–15 September 1977, J. Miklowitz and J.D. Achenbach, eds. (Wiley, New York, 1978).

[6] L. Brillouin, *Wave Propagation in Periodic Structures* (Dover, New York, 1946).

[7] R.H. MacNeal (editor), *The NASTRAN Theoretical Manual*, NASA SP-221 (1970).

[8] G.L. Goudreau, Evaluation of Discrete Methods for the Linear Dynamic Response of Elastic and Viscoelastic Solids, Report No. 69-15, Dept. of Civil Engineering, University of California, Berkeley, CA (June 1970).

[9] H.L. Schreyer, Consistent Diagonal Mass Matrices and Finite Element Equations for One-Dimensional Problems, *Int. J. for Numerical Methods in Engineering* **12**(7), 1171–1184 (1978).

[10] H.L. Schreyer and J.J. Fedock, Orthogonal Base Functions and Consistent Diagonal Mass Matrices for Two-Dimensional Elements, *Int. J. for Numerical Methods in Engineering* **14**(9), 1279–1398 (1979).

[11] J.S. Archer, Consistent Matrix Formulations for Structural Analysis Using Finite Element Techniques, *Journal of the American Inst. of Aeronatucis and Astronautics* **3**, 1910–1918 (1965).

[12] T. Belytschko and W.L. Mindle, Flexural Wave Propagation Behavior of Lumped Mass Approximations, *Journal of Computers and Structures* **12**, 805–812 (1980).

[13] R. Mullen and T. Belytschko, Dispersion Analysis of Finite Element Semidiscretizations of the Two-Dimensional Wave Equation, *Int. J. for Numerical Methods in Engineering* **18**, 11–29 (1982).

[14] T.J.R. Hughes, H.M. Hilber and R.L. Taylor, A Reduction Scheme for Problems of Structural Dynamics, *Int. J. Solids Structures* **12**, 749–767 (1976).

[15] R.D. Krieg and S.W. Key, Transient Shell Response by Numerical Time Integration, *International Journal for Numerical Methods in Engineering* **7**, 273–286 (1973).

[16] Z.P. Bazant, Spurious Reflection of Elastic Waves in Nonuniform Finite Element Grids, *Computer Methods in Applied Mechanics and Engineering* **16**, 91–100 (1978).

[17] H.S. Carslaw, *Introduction to the Theory of Fourier's Series and Integrals*, 3rd ed. (MacMillan, London, 1930).

[18] R.C.Y. Chin, Dispersion and Gibbs Phenomenon Associated with Difference Approximations to Initial Boundary-Value Problems for Hyperbolic Equation, *J. of Computational Physics* **18**, 233–247 (1975).

[19] I. Fried, Accuracy of String Element Mass Matrix, *Computer Methods in Applied Mechanics and Engineering* **20**, 317–321 (1979).

[20] B. Engquist and H.O. Kreiss, Difference and Finite Element Methods for Hyperbolic Differential Equations, *Computer Methods in Applied Mechanics and Engineering* **17/18**, 581–596 (1979).

CHAPTER 7

Silent Boundary Methods for Transient Analysis

Martin COHEN

Mobil Oil Company, Dallas, Texas, USA

and

Paul C. JENNINGS

California Institute of Technology
Pasadena, California, USA

Computational Methods for Transient Analysis
Edited by T. Belytschko and T.J.R. Hughes

Abstract: This chapter analyzes and compares several dynamic models which are designed to absorb waves radiating toward infinity in a finite computational grid. The analysis is directed primarily toward the problem of soil-structure interaction, where energy propagates from a region near a structure, outwardly toward the boundaries. One proposed method, called the extended-paraxial boundary, was derived originally from one-directional, wave theories. In this chapter, the theory is presented from a more general viewpoint, and its stability properties are studied. The boundary's implementation for finite element calculations is also presented and discussed.

The extended-paraxial boundary is then compared, both analytically and numerically, with two other transmitting (or silent) boundaries currently available – the standard-viscous and unified-viscous methods. The analytical results indicate that the extended-paraxial boundary enjoys advantage in cancelling wave reflections; however, actual numerical tests revealed only a small superiority over the viscous approaches.

1. Introduction and review of literature

1.1. Introduction

The use of finite element and finite difference methods has been extensive, and they have proven their versatility in solving a wide range of practical problems. One of their limitations, however, arises when they are employed for the modeling of an infinite domain, in which energy radiates from a source outwardly toward infinity. In numerical calculations, only a finite region of the medium is analyzed. Unless something is done to prevent outwardly radiating waves from reflecting from the region's boundaries, errors are introduced into the results. The present chapter concerns itself with the study of such effects, using the finite-element method, and an artificial, transmitting boundary at the edge of the computational grid. (Most of the discussion applies as well to finite differences.) The reasons for studying infinitely-radiating waves, and for using this particular approach to a solution, have been documented elsewhere. A background to the study of silent boundaries is included now in order to clarify our objectives.

Several different methods for the treatment of absorbing boundaries have been proposed and employed with varying success. In all cases, the object of the work has been to make the artificial boundary behave, as nearly as possible, as if the mesh extended to infinity. In particular, since economy dictates that the boundary be near the central field of the mesh, the methods all try to avoid large, direct reflections of energy. The resulting techniques are variously known as transmitting boundaries, absorbing boundaries, or silent boundaries. These terms are used interchangeably in this chapter.

Lysmer and Kuhlemeyer [68] conceived of using viscous damping forces, which act along the boundary, as a means of absorbing, rather than reflecting, the radiated energy. The method, being directly analogous to the use of viscous dashpots, is relatively easy to implement, and it appears to treat both dilatational and shear waves with acceptable accuracy in many applications. The viscous forces, or dashpots, enjoy a third advantage, in that they do not depend upon the frequencies of the transmitted waves. The technique is thus suitable for transient analysis.

It is commonly believed ([27], [51], [53], [65], [68]) that one drawback to the viscous boundary is its inability to transmit Rayleigh waves as effectively as it does body waves. A special viscous boundary for Rayleigh waves was devised [68] in which the dashpots have coefficients that depend upon the frequency of the waves. The accuracy of this Rayleigh-wave boundary is not well established. It has been noted [65] that in order to avoid inaccuracies, the computational mesh may have to be especially refined near the ground surface. This is because at one point, a parameter of the dashpot goes to infinity. In addition, there have been few comparisons between the standard- and Rayleigh-viscous boundaries, except for one axisymmetric problem which was discussed in [68]. The use of a standard-viscous boundary for problems which involve Rayleigh waves should not necessarily be ruled out. Unlike the Rayleigh boundary, it is independent of frequency and is much easier to implement. For example, Haupt [31], using the standard-viscous boundary along with some of his own boundary innovations, achieved a good, steady-state, Rayleigh-wave solution. Another Rayleigh-wave example is presented in Section 4.

White, Valliappan, and Lee [78] attempted to improve upon Lysmer and Kuhlemeyer's scheme, and also to broaden the theory to include anisotropic materials. To do this, they selected, and then minimized with respect to C_{ij}, a certain norm, I_{ij}, where

$$I_{ij} = \int_{\theta} (B_{ij} + C_{ij})^2 \cos^2 \theta \, \mathrm{d}\theta \; ; \tag{1.1}$$

i = the number of stress components and j = the number of displacement components; θ is the angle of incidence of the wave. C_{ij} is the desired damping matrix that is used to cancel the stresses, $[\boldsymbol{B}]\{\boldsymbol{u}\}$, of the incoming waves. Therefore, I_{ij} is a measure of residual stress, or energy, that is not removed by the boundary terms, and will cause reflected waves.

After the minimization of I_{ij} to find C_{ij} is completed, this C_{ij} is used as a starting point to iterate for more 'energy efficient' values of C_{ij}. This second minimization is performed on the energy ratio, $E_{\text{reflected}}/E_{\text{incident}}$, which is calculated by using harmonic wave forms and the previous values of C_{ij}.

The benefit to this approach is that anisotropic materials can be modeled. The authors have not shown, however, just how efficient this boundary is for such materials. For the isotropic case, the method offers virtually no improvement upon the Lysmer–Kuhlemeyer boundary, and is more complicated to implement.

Claerbout [1] devised the idea of creating equations which transmit waves in only one direction. He derived these equations, termed paraxial approximations, for the two-dimensional, scalar-wave case. Clayton and Engquist [59] later expanded Claerbout's ideas to include elastic waves, and conceived the notion of applying it as an energy-absorbing boundary. In their approach, one takes the triple Fourier transform (two spatial and one temporal) of the following two-dimensional, elasticity equations in plane strain.

$$
\begin{aligned}
&u_{tt} - c_d^2 u_{xx} - (c_d^2 - c_s^2) w_{xz} - c_s^2 u_{zz} = 0\,, \\
&w_{tt} - c_s^2 w_{xx} - (c_d^2 - c_s^2) u_{xz} - c_d^2 w_{zz} = 0\,,
\end{aligned}
\tag{1.2}
$$

where c_d = the dilatational wave speed, c_s = the shear wave speed, and u and w are the respective horizontal and vertical displacements; x and z represent the corresponding spatial coordinates, and t = time.

The authors use the scalar-wave, paraxial equations "to provide a hint as to the general form of the paraxial approximation" for elasticity. They then take the Fourier transform of this general paraxial form, and "fit the coefficients by matching to the full, elastic wave equation". Using these derived coefficients, the general paraxial form becomes the governing equation for the boundary.

The authors implemented these equations using a finite-difference numerical technique and, in the cases they presented, they obtained good body-wave transmission. The method, however, suffers from several difficulties. First, the technique, as formulated, does not directly lend itself to finite-element utilization. A straightforward, finite-element analysis divides the domain into two parts: the interior, where the regular equations of elasticity hold, and the boundary, where the paraxial equations are valid. The interface between the two discretized regions (elastic and paraxial) does not permit smooth wave transmission, and waves which do arrive in the paraxial area do not propagate correctly. A significant part of the wave energy is reflected into the elastic medium.

The effectiveness of the finite-element paraxial boundary was notably improved by our utilization of some upwinding techniques suggested by Hughes [15], and by our introduction of an interface element [60]. The details of this approach are presented in Section 3, along with some comparisons between this boundary and others. Another major difficulty with the earlier paraxial technique is that when Poisson's ratio is greater than $\frac{1}{3}$, a negative stiffness term is introduced into the paraxial equations. This term clearly leads to instabilities [60]; the boundary erroneously causes the displacements and stresses to grow in time. This problem is reviewed in Section 2.

In a series of papers (Lysmer [69], Lysmer and Waas [70], and Lysmer and Drake [43]), a boundary was developed in order to transmit either Love waves or Rayleigh waves. The boundary was especially designed for a layered medium. The method first assumes that a wave of a certain frequency is propagating in a certain layer. If another finite element having width h were present beyond the boundary, then its displacements would be e^{-ikh} (k equals the wave number) times those of the last element at the boundary. They then calculate the stiffness contribution of the supposed elements and put them into the equations of motion for the lumped masses at the boundary. Therefore,

$$\boldsymbol{M}\ddot{\boldsymbol{u}} + \boldsymbol{K}\boldsymbol{u} - k^2\boldsymbol{K}^*\boldsymbol{u} = 0\,. \tag{1.3}$$

If the frequency, ω, is known, and having set $\boldsymbol{C} = \omega^2\boldsymbol{M} + \boldsymbol{K}$, equation (1.3) becomes

$$[\boldsymbol{C} - k^2\boldsymbol{K}^*]\boldsymbol{u} = \boldsymbol{0}\,. \tag{1.4}$$

k, the wave number, is different for each of the various layers, so the eigenvalue problem (1.4) must be solved. The impinging wave (shear or Rayleigh) causes stresses at the boundary. The idea is to apply oppositional forces to effectively nullify them. For the shear wave example, these stresses are proportional to both the displacements at the boundary and to the wave numbers (eigenvalues) solved for in equation (1.4). This enables the authors to find the matrix, $\boldsymbol{R}$, which relates the nullifying forces to be applied at the boundary to the displacements at the boundary. $\boldsymbol{Ru}$ serves as the boundary force contribution to the finite-element equations of the interior.

Although suitable for transmitting periodic, surface waves, this method is highly restrictive. First, the boundary terms are frequency dependent, meaning that one cannot generally perform a transient analysis in the time domain. If the governing equations of the interior region are *linear,* one can perform a transient analysis in the frequency domain, but this may be more costly. Also the method is inapplicable if the interior equations are nonlinear. This technique is also restrictive in that only shear waves or Rayleigh waves can be transmitted. Other boundary methods are broader in scope. Finally, this transmitting boundary is more difficult to implement than most other boundary schemes.

Smith [76] proposed the adding together of wave solutions having different boundary conditions, in order to eliminate reflections. In a one-dimensional example, the reflection of a wave striking a free boundary cancels that of a wave striking a fixed boundary, when they are added to each other.

Two lubricated-rigid boundary conditions are imposed for two dimensions:

$$u_1 = 0,\ \ \tau = 0 \quad \text{and} \quad u_2 = 0,\ \ \sigma = 0\,. \tag{1.5}$$

u_1 and u_2 are the respective normal and tangential displacements at the boundary, and σ and τ are the respective normal and tangential stresses there. For the three-dimensional case, the plane of the boundary is either lubricated or fixed, and the normal displacement is conversely fixed or left free.

Smith demonstrates that this boundary method eliminates all reflections, regardless of frequency or angle of incidence. It also handles all types of waves, including body, Rayleigh, or Love waves. The only drawback to this method, and it is an inescapable one, is that two solutions are required for each possible wave reflection. For example, with a two-dimensional corner, two solutions for each boundary side are needed. The problem must be solved four times to cancel the wave reflections. Likewise, if there is enough time for a wave to reflect from one boundary, strike another, and then return, then the number of calculations must be doubled. Therefore, 2^n is equal to the number of complete solutions, where n is the number of possible reflections. If one performs the calculations over a long period of time, the number of required solutions increases very rapidly. This method does not appear as attractive as other approaches, except for one-dimensional problems and problems with very short characteristic times.

Cundall et al. [61] have recently introduced a cost-saving scheme that attempts to retain the advantages of the Smith boundary. Their idea is to set up a small boundary region, in which equations are formed and solved for each of the above-mentioned boundary conditions. Smith's idea of adding the two solutions is implemented at every fourth time step. Thus, the boundary area, which is four elements deep, requires two solutions at each step, while the interior region needs only one solution. Cundall et al. encountered some practical difficulties with this incremental approach. They solved them by using two pairs of constant stress and constant velocity conditions at the boundary, instead of the fixed and free conditions that Smith presented. Wave-reflection theory clearly shows how the reflections are eliminated in the Smith model by adding the 'fixed' and 'free' solutions. With constant velocity, and stress boundary conditions, it is not obvious how the reflections are controlled. Cundall et al., however, have observed that these conditions perform well and reduce the "numerical shocks" that are caused by the adding of the fixed and free boundary solutions. The main question about this incremental ('superposition') approach is its practicability. It effectively adds a layer of eight elements, which have to be formed and calculated at each step. In addition, the constant velocity and constant stress conditions at the boundary must be stored for all times. It would be cheaper, and less cumbersome, to simply add a layer of eight elastic elements and to employ either the viscous, or paraxial boundary conditions. However, the comparative accuracy of the superposition method and the latter schemes is unknown.

Tseng and Robinson [75] and Robinson [72] investigated wave propagation using another transmitting boundary proposal. This method relies on the separation of S and P waves by the potentials, ϕ and ψ. First, the transmitting condition for two-dimensional, plane waves is obtained. For example,

$$\phi_t + c_d\phi_x = 0\,. \tag{1.6}$$

Then a correction is added for cylindrical waves. Although equations such as (1.6) are written in terms of potentials, they appear to be similar to the equations of the first paraxial approximation [63, 1] shown below

$$u_t + c_d u_x = 0\,. \tag{1.7}$$

Equations (1.6) and (1.7) are derived for waves oriented in the positive-x direction. Robinson and Tseng demonstrate the method's superior transmission of cylindrical waves but, so far, the benefit does not appear to be significant enough to clearly justify this boundary's use over other techniques.

Still another idea was proposed by Isenberg [33] and later by Zhen-peng, Pai-puo and Yi-fan [79]. At the nth time step, they use the known displacements of all points on or near the boundary to predict the boundary's motion at the $(n+1)^{\text{th}}$ step. The rest of the displacements in the interior are then found. In order to develop a predictor for the boundary, one needs the frequency and wave number of the impinging wave. In this way, this method could be effective for steady-state problems. One would probably have to resort to Fourier analysis to model transient waves.

Isenberg [66] suggested an alternate frequency-independent, predictor method. In this proposal, the preliminary step would be to apply unit forces at the nodes adjacent to the boundary, and to calculate the boundary's reaction to each of these loads. This information would be stored for use during the main calculation. The effects of the various nodal loadings are scaled and then superimposed on each other, in order to predict the boundary's response for the next time step. No one has, as yet, implemented this idea, so its feasibility and accuracy are unknown. It appears that a significant effort to set up such a boundary is required, but the concept may have the potential to solve infinite-domain problems.

Researchers [62, 77] have also investigated the boundary integral method, where the interior displacements of a region are found by evaluating integrals along the region's boundary. The boundary is discretized into segments, and the integrations are performed numerically. As yet, the technique has not been completely satisfactory. Not only does taking integrals along the boundary depend upon the linearity of the interior equations, it also leads to nonsymmetric matrices. For large problems, it could be computationally expensive.

In still another attempt to simulate the effects of wave radiation, analysts [41] have incorporated material damping into their models. Alternatively, one could employ numerical damping [12] to account for the transmitted energy. While these damping procedures are rather easily implemented, how one could practically employ them is not clear. Just how much damping should be put into the system, and where should it be applied? How can this damping discriminate the effects of wave radiation from the actual physical dissipation taking place within the model?

A systematic approach to the usage of damping in various systems is not available. Luco et al. [41] demonstrated some of the problems that can occur. They compared analytical solutions for wave propagation to calculations from a finite element model which used 'plausible' damping estimates. In general, the material damping did a poor job of duplicating the radiation effect. At this stage, the proper application of damping to account for radiation effects seems to be more of an art than a science.

Another relatively simple idea, which was proposed by Haupt [31], can be applied for repetitive analyses of certain systems which can be split into interior and exterior parts. The interior is altered for each analysis (e.g., each interior geometry or load history could be different), but the exterior region remains constant. One initially sets up an extensive mesh of the entire system, but then condenses those degrees-of-freedom that are in the outlying region. Each successive problem could be solved by utilizing just the small interior mesh and the force contribution from the condensed equations. This substructuring method reduces the computational expense for these special cases.

Finally, investigators [25, 28, 35] have experimented with extensive meshes, and have determined where the boundary should be placed in order to produce acceptably small reflections. Day [28] found that by successively increasing the size of outlying elements by a factor of 1.1, he could prevent undesirable reflections. This growth factor of 1.1 provides some help in reducing the number of required elements, but the computational costs remain high, prohibitively so for three-dimensional calculations.

In summary, each of the proposed transmitting boundary schemes has been shown to be effective for selected wave problems. The basic criticism of all these methods is that, to one degree or another, none has been extensively verified. For example, there are available few published comparisons between the results of using a truncated mesh having a transmitting boundary and of calculations with an extended mesh. Some of these ideas, such as the viscous damping mechanism, have been incorporated into programs for other purposes [27, 44, 47, 53]. This has generated a degree of qualitative confidence in their use, but their accuracy is not well known.

1.2. Organization

The object of this presentation is to analyze transmitting boundaries which can be applied to transient analysis in the time domain. It is also desirable that these boundaries be applicable to situations where nonlinear behavior is important.

In Section 2, theories of some absorbing boundaries are outlined. Since the boundaries are supposed to remove energy from the system, we discuss several of the forms which these mechanisms might take. Next, several ideas for transmitting boundaries are analyzed. The basic paraxial idea [1] is presented in a different light, employing features that improve its energy-absorbing character [60]. The viscous-boundary proposal and the unified version of the viscous boundary are also presented. The various boundary methods are then compared, using standard, wave-reflection analysis. We also discuss the boundaries' ability to transmit Rayleigh waves and suggest some special, frequency-independent, Rayleigh-wave boundaries.

Section 3 primarily deals with the numerical implementation of the boundary schemes discussed in Section 2. We point out some of the practical considerations of implementing a silent boundary. The boundary methods should be designed so that the limitations, such as the time-step size, are no greater than are those which other considerations impose.

We then outline some procedures which aid the paraxial method's application to finite elements. These include the 'upwinding' of certain paraxial terms and a nodal assembly procedure that eliminates a finite-element 'interfacing' effect. Next, limits of numerical instability are determined for the various boundaries.

Section 4 contains results and comparisons of the boundary methods. Examples include horizontal- and vertical-pulse loadings and a Rayleigh-wave excitation. These problems illustrate the primary strengths and drawbacks of the various methods and suggest improvements that could be made.

In the final section, we present conclusions and offer recommendations for the use of the silent boundaries.

2. Theory of paraxial viscous silent boundaries

2.1. Introduction

In this and the following two sections, we examine silent-boundary methods suitable for nonlinear, transient, finite-element analysis. These are the paraxial and viscous boundary methods. It is important in this case that a transmitting boundary be independent of frequency, and that it employ information only from nearby regions of the mesh. These requirements allow the interior mesh equations to be nonlinear; only the outer-region equations (those near the silent boundary) need remain linear. In addition, it is desirable to have the method be easily implemented and understood. We would also hope to establish a stability criterion and a measure of the boundary's accuracy. Lastly, the boundary should be capable of handling all types of incoming waves, and of proving its reliability over a wide range of different conditions. For example, it should place no significant restriction on the material properties which are being modeled.

2.2. Theory of paraxial boundaries

Herein, we present the paraxial boundary from a different viewpoint than did Clayton and Engquist [59], who originally applied the paraxial equations to a silent boundary. We believe that our alternative approach gives a better insight into the method. Also, a stability analysis which is presented indicates that the paraxial equations must be modified from their original form.

In the derivation here, the paraxial boundary idea is best introduced by means of the one-dimensional wave equation,

$$u_{tt} - c^2 u_{xx} = 0, \quad -\infty < x < \infty,\ 0 \le t < \infty . \tag{2.1}$$

It has the solution

$$u = p(x - ct) + q(x + ct), \tag{2.2}$$

where $p(x-ct)$ and $q(x+ct)$ represent arbitrarily-shaped waves moving in the positive and negative x-directions, respectively. For the paraxial boundary, we seek a similar partial differential equation whose solution only allows waves to travel in the positive x-direction. This equation will then govern a boundary region which only transmits waves in the positive x-direction.

Several partial differential equations produce the appropriate solution. One of these was presented by Clayton and Engquist [59]:

$$u_t + cu_x = 0\,. \tag{2.3}$$

However, we choose to analyze another, related equation:

$$u_{tt} + cu_{tx} = 0\,. \tag{2.4}$$

Equations (2.3) and (2.4) both admit the solution $p(x-ct)$, but equation (2.4) is dimensionally consistent with the original wave equation. It also more closely resembles the paraxial equations that can be used for two- and three-dimensional problems [60].

For the one-dimensional example, all solutions to the differential equation are available, so we can select the correct paraxial equation. Unfortunately, this is not the case for multidimensional equations. This necessitates a different approach, whereby the equations of the boundary regions are constructed to accommodate *unidirectional, plane, harmonic waves.* For purposes of simplicity, this different method is illustrated with the one-dimensional example.

Consider solutions of the form:

$$u = A\exp[\mathrm{i}(k_x x - \omega t)]\,, \tag{2.5}$$

in which k_x is the wave number and ω is the frequency of the wave.

Substituting equation (2.5) into (2.1) leads to

$$\frac{k_x}{\omega} = \pm\frac{1}{c}\,. \tag{2.6}$$

In equation (2.5), if k_x/ω is positive, then u represents only positive x-traveling waves. We now seek a differential equation which, upon substitution of the desired solution represented by equation (2.5), produces only

$$\frac{k_x}{\omega} = +\frac{1}{c}\,. \tag{2.7}$$

It happens that both equations (2.3) and (2.4) meet this condition: They each possess the harmonic solution (2.5) and satisfy the positive k_x/ω requirement. This can guide the development of differential equations of higher dimensions which also permit only positive traveling waves.

2.3. *Two-dimensional scalar wave equation*

As an introduction to the development of a silent boundary for the equations of elasticity, we present an analysis of the two-dimensional, scalar-wave equation:

$$\begin{aligned} &u_{tt} - c^2(u_{xx} + u_{zz}) = 0\,, \\ &u = f(x, z) \quad \text{at } t = 0\,, \\ &u_t = g(x, z) \quad \text{at } t = 0\,. \end{aligned} \tag{2.8}$$

Once again, the goal is to find a similar differential equation which allows only positively moving waves.

Consider solutions of the form:

$$u = A \exp[\mathrm{i}(k_x x + k_z z - \omega t)]\,, \tag{2.9}$$

Combining (2.8) and (2.9) yields

$$\omega^2 - c^2 k_x^2 - c^2 k_z^2 = 0\,. \tag{2.10}$$

Factoring equation (2.10),

$$\left[ck_x - \omega\sqrt{1 - \left(\frac{ck_z}{\omega}\right)^2}\,\right]\left[ck_x + \omega\sqrt{1 - \left(\frac{ck_z}{\omega}\right)^2}\,\right] = 0\,. \tag{2.11}$$

Equation (2.11) furnishes the two roots of ck_x/ω in equation (2.10); $(ck_x/\omega)_1$ is positive and $(ck_x/\omega)_2$ is negative (with $ck_z/\omega < 1$). If we substitute the positive ck_x/ω root into (2.9), then this equation represents waves traveling in the positive x-direction.

Beginning with the positive ck_x/ω relation.

$$\frac{ck_x}{\omega} - \sqrt{1 - \left(\frac{ck_z}{\omega}\right)^2} = 0\,, \tag{2.12}$$

we expand the square root term producing

$$\frac{ck_x}{\omega} - \left[1 - \frac{1}{2}\left(\frac{ck_z}{\omega}\right)^2 + \mathrm{O}\left(\frac{ck_z}{\omega}\right)^4\right] = 0\,. \tag{2.13}$$

We can now determine the differential equation with corresponds to the first three terms of equation (2.13) by inspection of (2.9) and its derivatives:

$$u_{tt} + cu_{tx} - \tfrac{1}{2}c^2 u_{zz} = 0\,. \tag{2.14}$$

Equation (2.14) has a solution in the form of equation (2.9), but with ck_x/ω defined by the first three terms of (2.13). Further expansions of (2.13) to higher orders in (ck_z/ω) have been carried out by Clayton and Engquist [59].

Equation (2.14) represents an approximation of an equation which would only transmit positive x-directed waves. The nature of this approximation is shown with the help of Figure 2.1. The plane harmonic wave illustrated in this figure has the form:

$$u = A \exp\left[\mathrm{i}\left(\frac{\omega}{c}\cos\alpha\, x + \frac{\omega}{c}\sin\alpha\, z - \omega t\right)\right]. \tag{2.15}$$

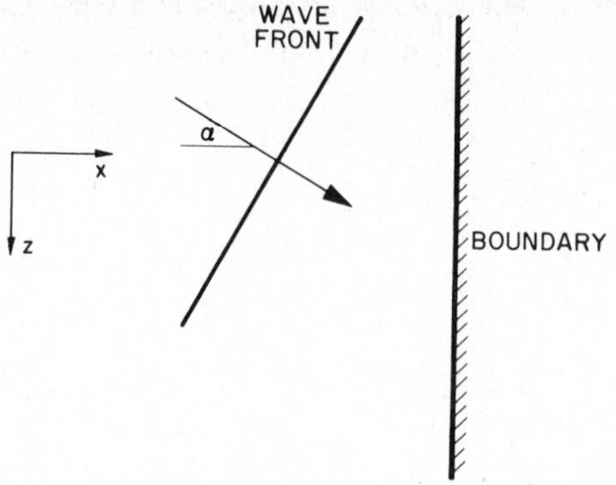

Fig. 2.1. Orientation of wave used in the scalar-wave and elasticity analyses.

In this case, $ck_z/\omega = \sin\alpha$, where α is the angle of incidence. The first three terms of equation (2.13) best approximate (2.12) when ck_z/ω is small. Hence, we expect the differential equation (2.14) to perform best with small values of α, that is, with waves which are nearly normal to the boundary.

2.4. *Boundary approximations for linear elasticity*

The development using the elasticity equations follows along lines similar to the previous cases. We first determine the appropriate paraxial boundary equations for the plane strain equations of elasticity,

$$\boldsymbol{u}_{tt} - \boldsymbol{E}_{11}c^2\boldsymbol{u}_{xx} - \boldsymbol{E}_{12}c^2\boldsymbol{u}_{xz} - \boldsymbol{E}_{22}c^2\boldsymbol{u}_{zz} = \boldsymbol{0}\,, \tag{2.16}$$

where

$$\boldsymbol{u} = \begin{Bmatrix} u \\ w \end{Bmatrix}, \quad \boldsymbol{E}_{11} = \frac{1}{c^2}\begin{bmatrix} c_{\rm d}^2 & 0 \\ 0 & c_{\rm s}^2 \end{bmatrix},$$
$$\boldsymbol{E}_{12} = \frac{1}{c^2}(c_{\rm d}^2 - c_{\rm s}^2)\begin{bmatrix} 0 & 1 \\ 1 & 0 \end{bmatrix}, \quad \boldsymbol{E}_{22} = \frac{1}{c^2}\begin{bmatrix} c_{\rm s}^2 & 0 \\ 0 & c_{\rm d}^2 \end{bmatrix}; \tag{2.17}$$

u and w are the displacements which act in the respective x and z coordinate directions; $c_{\rm d}$ is the dilatational velocity, and $c_{\rm s}$ is the shear velocity; c, which has the dimensions of a wave speed, is included in equation (2.16) in order to render several quantities dimensionless. Equation (2.16) and its solutions are clearly independent of c.

Solutions to (2.16) are assumed to be of the form:

$$u = U\exp[\mathrm{i}(k_x x + k_z z - \omega t)]\,,$$
$$w = W\exp[\mathrm{i}(k_x x + k_z z - \omega t)]\,. \tag{2.18}$$

Substituting (2.18) into (2.16), we find that

$$[\boldsymbol{I}\omega^2 - \boldsymbol{E}_{11}k_x^2c^2 - \boldsymbol{E}_{12}k_xk_zc^2 - \boldsymbol{E}_{22}k_z^2c^2]\begin{Bmatrix} U \\ W \end{Bmatrix} = 0\,. \tag{2.19}$$

Equations (2.18) and (2.19) describe the propagation of plane waves. Guided by the scalar-wave example, we examine the terms obtained by approximate factoring of (2.19).

$$\left[I\frac{k_x c}{\omega} - B_1 - B_2\frac{k_z c}{\omega} - B_3\left(\frac{k_z c}{\omega}\right)^2\right] \times ,$$

$$\left[I\frac{k_x c}{\omega} - B_4 - B_5\frac{k_z c}{\omega} - B_6\left(\frac{k_z c}{\omega}\right)^2\right]\begin{Bmatrix} U \\ W \end{Bmatrix} = \mathbf{0}\,. \tag{2.20}$$

We are solving for the matrices B_j. If we substitute the first root,

$$I\left(\frac{k_x c}{\omega}\right) = B_1 + B_2\frac{k_z c}{\omega} + B_3\left(\frac{k_z c}{\omega}\right)^2, \tag{2.21}$$

into (2.19), we obtain

$$\begin{aligned}&[I - E_{11}B_1^2]\left(\frac{ck_z}{\omega}\right)^0 + [-E_{12}B_1 - E_{11}B_1B_2 - E_{11}B_2B_1]\left(\frac{ck_z}{\omega}\right)^1 \\ &+[-E_{22} + E_{11}B_2 - E_{11}B_1B_3 - E_{11}B_3B_1 - E_{11}B_2^2]\left(\frac{ck_z}{\omega}\right)^2 \\ &+[-E_{12}B_3 - E_{11}B_2B_3 - E_{11}B_3B_2]\left(\frac{ck_z}{\omega}\right)^3 + [-E_{11}B_3^2]\left(\frac{ck_z}{\omega}\right)^4 = 0\,.\end{aligned} \tag{2.22}$$

Setting the coefficients of $(ck_z/\omega)^0$ equal to zero,

$$B_1 = \pm\begin{bmatrix} \dfrac{1}{c_d} & 0 \\ 0 & \dfrac{1}{c_s} \end{bmatrix} c\,. \tag{2.23}$$

The positive root of B_1 leads to a positive x-direction, paraxial boundary. Now setting the $(ck_z/\omega)^1$ and $(ck_z/\omega)^2$ coefficients equal to zero,

$$B_2 = -(c_d - c_s)\begin{bmatrix} 0 & \dfrac{1}{c_d} \\ \dfrac{1}{c_s} & 0 \end{bmatrix}, \tag{2.24}$$

$$B_3 = -\begin{bmatrix} c_s - \frac{1}{2}c_d & 0 \\ 0 & c_d - \frac{1}{2}c_s \end{bmatrix}\frac{1}{c}\,. \tag{2.25}$$

The substitution of the second root of equation (2.20) into (2.19) produces

$$B_4 = -B_1, \qquad B_5 = B_2, \qquad B_6 = -B_3\,. \tag{2.26}$$

If one introduces these matrices into equation (2.20) and multiplies the terms, (2.19) and some additional third and fourth order terms in (ck_z/ω) result. The approximate factoring given by (2.20) is most accurate for small values of $ck_z/\omega = \sin\alpha$. From this analysis, it can also be seen that B_1 is the zeroth order term and, therefore, is the most important part of the paraxial approximation.

The inclusion of $\boldsymbol{B}_2$ and $\boldsymbol{B}_3$ attains closer agreement between the elasticity equations (2.19) and the factored paraxial equations (2.20).

If we use equation (2.20) with the calculated $\boldsymbol{B}$ matrices, the result is

$$\left[-\begin{bmatrix}1 & 0\\ 0 & 1\end{bmatrix}\omega^2+\begin{bmatrix}c_\mathrm{d} & 0\\ 0 & c_\mathrm{s}\end{bmatrix}k_x\omega+(c_\mathrm{d}-c_\mathrm{s})\begin{bmatrix}0 & 1\\ 1 & 0\end{bmatrix}k_z\omega\right.$$
$$\left.+\begin{bmatrix}c_\mathrm{d}(c_\mathrm{s}-\frac{1}{2}c_\mathrm{d}) & 0\\ 0 & c_\mathrm{s}(c_\mathrm{d}-\frac{1}{2}c_\mathrm{s})\end{bmatrix}k_z^2\right]\begin{Bmatrix}U\\ W\end{Bmatrix}=\mathbf{0}\,. \tag{2.27}$$

In analogy to the scalar-wave case, the paraxial differential equations can now be derived. If the harmonic solution (2.18) is substituted into the desired differential equation, then (2.27) is the product. By inverse reasoning, the desired equation is:

$$\begin{bmatrix}1 & 0\\ 0 & 1\end{bmatrix}\boldsymbol{u}_{tt}+\begin{bmatrix}c_\mathrm{d} & 0\\ 0 & c_\mathrm{s}\end{bmatrix}\boldsymbol{u}_{tx}+(c_\mathrm{d}-c_\mathrm{s})\begin{bmatrix}0 & 1\\ 1 & 0\end{bmatrix}\boldsymbol{u}_{tz}$$
$$-\begin{bmatrix}c_\mathrm{d}(c_\mathrm{s}-\frac{1}{2}c_\mathrm{d}) & 0\\ 0 & c_\mathrm{s}(c_\mathrm{d}-\frac{1}{2}c_\mathrm{s})\end{bmatrix}\boldsymbol{u}_{zz}=0\,. \tag{2.28}$$

Equation (2.28) produces approximately the same harmonic wave solution in the positive x-direction as the elasticity equations do. Similarly, one employs equation (2.26), the paraxial equations, for the negative x-direction:

$$\begin{bmatrix}1 & 0\\ 0 & 1\end{bmatrix}\boldsymbol{u}_{tt}-\begin{bmatrix}c_\mathrm{d} & 0\\ 0 & c_\mathrm{s}\end{bmatrix}\boldsymbol{u}_{tx}-(c_\mathrm{d}-c_\mathrm{s})\begin{bmatrix}0 & 1\\ 1 & 0\end{bmatrix}\boldsymbol{u}_{tz}$$
$$-\begin{bmatrix}c_\mathrm{d}(c_\mathrm{s}-\frac{1}{2}c_\mathrm{d}) & 0\\ 0 & c_\mathrm{s}(c_\mathrm{d}-\frac{1}{2}c_\mathrm{s})\end{bmatrix}\boldsymbol{u}_{zz}=0\,. \tag{2.29}$$

Clayton and Engquist [59] derived equations which were the same as those above, except for a change of sign in their $\boldsymbol{B}_2$ and $\boldsymbol{B}_4$ matrices. They did this through a method based on Fourier analysis.

We can immediately identify some characteristics of (2.28) and (2.29). These equations have the same mass term as the original elasticity equations. The other paraxial terms, however, bear little resemblance to those of their parent equations. $\boldsymbol{u}_{tx}$ and $\boldsymbol{u}_{tz}$ resemble damping terms, but an energy analysis reveals that these terms are not necessarily dissipative. The $\boldsymbol{u}_{zz}$ term constitutes a transverse stiffness which can be negative if

$$c_\mathrm{s}<\tfrac{1}{2}c_\mathrm{d}\,. \tag{2.30}$$

The potentially negative stiffness term raises the question of whether the paraxial equations are stable.

In three dimensions, the paraxial equations are derived by the same method. The three-dimensional elasticity equations are

$$\boldsymbol{u}_{tt}-\boldsymbol{E}_{11}c^2\boldsymbol{u}_{xx}-\boldsymbol{E}_{22}c^2\boldsymbol{u}_{yy}-\boldsymbol{E}_{33}c^2\boldsymbol{u}_{zz}-\boldsymbol{E}_{12}c^2\boldsymbol{u}_{xy}$$
$$-\boldsymbol{E}_{13}c^2\boldsymbol{u}_{xz}-\boldsymbol{E}_{23}c^2\boldsymbol{u}_{yz}=\mathbf{0}\,, \tag{2.31}$$

where

$$
\boldsymbol{E}_{11}=\frac{1}{c^2}\begin{bmatrix} c_d^2 & 0 & 0\\ 0 & c_s^2 & 0\\ 0 & 0 & c_s^2\end{bmatrix},\qquad
\boldsymbol{E}_{22}=\frac{1}{c^2}\begin{bmatrix} c_s^2 & 0 & 0\\ 0 & c_d^2 & 0\\ 0 & 0 & c_s^2\end{bmatrix},
$$

$$
\boldsymbol{E}_{33}=\frac{1}{c^2}\begin{bmatrix} c_s^2 & 0 & 0\\ 0 & c_s^2 & 0\\ 0 & 0 & c_d^2\end{bmatrix},\qquad
\boldsymbol{E}_{12}=\frac{(c_d^2-c_s^2)}{c^2}\begin{bmatrix} 0 & 1 & 0\\ 1 & 0 & 0\\ 0 & 0 & 0\end{bmatrix},\tag{2.32}
$$

$$
\boldsymbol{E}_{13}=\frac{c_d^2-c_s^2}{c^2}\begin{bmatrix} 0 & 0 & 1\\ 0 & 0 & 0\\ 1 & 0 & 0\end{bmatrix},\qquad
\boldsymbol{E}_{23}=\frac{(c_d^2-c_s^2)}{c^2}\begin{bmatrix} 0 & 0 & 0\\ 0 & 0 & 1\\ 0 & 1 & 0\end{bmatrix}.
$$

When, as before, we assume harmonic solutions:

$$
\begin{aligned}
u &= A\exp[\mathrm{i}(k_x x + k_y y + k_z z - \omega t)]\,,\\
v &= B\exp[\mathrm{i}(k_x x + k_y y + k_z z - \omega t)]\,,\\
w &= C\exp[\mathrm{i}(k_x x + k_y y + k_z z - \omega t)]\,.
\end{aligned}\tag{2.33}
$$

Then, by substituting into equation (2.31) and solving for ck_x/ω, we find

$$
\begin{aligned}
&\begin{bmatrix} 1 & 0 & 0\\ 0 & 1 & 0\\ 0 & 0 & 1\end{bmatrix}\boldsymbol{u}_{tt}+\begin{bmatrix} c_d & 0 & 0\\ 0 & c_s & 0\\ 0 & 0 & c_s\end{bmatrix}\boldsymbol{u}_{tx}+(c_d-c_s)\begin{bmatrix} 0 & 0 & 1\\ 0 & 0 & 0\\ 1 & 0 & 0\end{bmatrix}\boldsymbol{u}_{tz}\\
&+(c_d-c_s)\begin{bmatrix} 0 & 1 & 0\\ 1 & 0 & 0\\ 0 & 0 & 0\end{bmatrix}\boldsymbol{u}_{ty}-(c_d-c_s)\begin{bmatrix} 0 & 0 & 0\\ 0 & 0 & 1\\ 0 & 1 & 0\end{bmatrix}\boldsymbol{u}_{yz}\\
&-\begin{bmatrix} c_d(c_s-\frac{1}{2}c_d) & 0 & 0\\ 0 & c_s(c_d-\frac{1}{2}c_s) & 0\\ 0 & 0 & \frac{1}{2}c_s\end{bmatrix}\boldsymbol{u}_{yy}\\
&-\begin{bmatrix} c_d(c_s-\frac{1}{2}c_d) & 0 & 0\\ 0 & \frac{1}{2}c_s & 0\\ 0 & 0 & c_s(c_d-\frac{1}{2}c_s)\end{bmatrix}\boldsymbol{u}_{zz}=0\,.
\end{aligned}\tag{2.34}
$$

Equation (2.34) governs a medium which transmits waves only in the positive x-direction. A rotational transformation on the finite-element boundary terms is all that is required for orienting the boundary in other directions.

The overall behavior of (2.34) appears to be the same as that of the two-dimensional paraxial equations. If c_s is less than one-half c_d, then $\boldsymbol{u}_{yy}$ and $\boldsymbol{u}_{zz}$ have coefficients indicating negative stiffnesses. The other paraxial terms in the three-dimensional version are also similar to their two-dimensional counterparts: $\boldsymbol{u}_{tx}$ represents the zeroth-order term; $\boldsymbol{u}_{ty}$ and $\boldsymbol{u}_{tz}$ are the first-order terms; and $\boldsymbol{u}_{yz}$, $\boldsymbol{u}_{yy}$, and $\boldsymbol{u}_{zz}$ are the higher-order contributions. The order of the terms refers to the powers of k_y/ω or k_z/ω that result when (2.33) is substituted into (2.31).

2.5. Stability analysis

One way to evaluate the stability of the paraxial equations is to study the result of a harmonic wave impinging on the boundary (refer to Figure 2.1). For the stability analysis, the wave is conveniently represented by:

$$u = U \exp[\mathrm{i}\kappa(\cos \alpha x + \sin \alpha z - ct)] ,$$
$$w = W \exp[\mathrm{i}\, \kappa(\cos \alpha x + \sin \alpha z - ct)] , \tag{2.35}$$

where κ is the wave number, which is assumed to be real. If c in (2.35) has a negative, imaginary component, then u and w grow exponentially in time and therefore are unstable. The substitution of (2.35) into the paraxial equations (2.28) gives:

$$\left[\begin{array}{c|c} c^2 - c(c_\mathrm{d} \cos \alpha) + c_\mathrm{d}(\tfrac{1}{2}c_\mathrm{d} - c_\mathrm{s}) \sin^2 \alpha & -c(c_\mathrm{d} - c_\mathrm{s}) \sin \alpha \\ \hline -c(c_\mathrm{d} - c_\mathrm{s}) \sin \alpha & c^2 - c(c_\mathrm{s} \cos \alpha) + c_\mathrm{s}(\tfrac{1}{2}c_\mathrm{s} - c_\mathrm{d}) \sin^2 \alpha \end{array}\right] \begin{Bmatrix} U \\ W \end{Bmatrix} = 0 , \tag{2.36}$$

or

$$\bar{\boldsymbol{D}}\boldsymbol{U} = \boldsymbol{0} . \tag{2.37}$$

For a nontrivial solution of (2.37), the determinant of the matrix $\bar{\boldsymbol{D}}$, must vanish, which results in a quartic equation for c. The computed solutions of this quartic, for various angles of incidence and Poisson's ratios, are plotted in Figure 2.2.

One can see that for Poisson's ratios of less than $\frac{1}{3}(c_\mathrm{s}/c_\mathrm{d} < \frac{1}{2})$, the paraxial equations are always stable. They are also stable for waves which impinge almost normally on the boundary. There exists, however, a large unstable zone for Poisson's ratios greater than $\frac{1}{3}$. We conducted several numerical tests using points inside the unstable region, and our results confirmed the existence of the instability. Guided by the approximate nature of the paraxial equations, a practical solution to this problem is to set the negative stiffness term, $c_\mathrm{d}(c_\mathrm{s} - \frac{1}{2}c_\mathrm{d})$, equal to zero if Poisson's ratio is greater than $\frac{1}{3}$. The justification for this simple solution is that the stiffness terms are the least important element of the paraxial approximation, as was noted in the derivation of the paraxial equations. A similar stability analysis of the revised paraxial equations revealed no instabilities.

The set of equations, then, which we have used for paraxial boundaries is [60]:

$$\begin{aligned} &\rho[u_{tt} + c_\mathrm{d}u_{tx} + (c_\mathrm{d} - c_\mathrm{s})w_{tz} - c_\mathrm{d}(c_\mathrm{s} - \tfrac{1}{2}c_\mathrm{d})u_{zz}] = 0 , \\ &\rho[w_{tt} + c_\mathrm{s}w_{tx} + (c_\mathrm{d} - c_\mathrm{s})u_{tz} - c_\mathrm{s}(c_\mathrm{d} - \tfrac{1}{2}c_\mathrm{s})w_{zz}] = 0 , \\ &\qquad \text{for } \nu \le \tfrac{1}{3} ; \\ &\rho[u_{tt} + c_\mathrm{d}u_{tx} + (c_\mathrm{d} - c_\mathrm{s})w_{tz}] = 0 , \\ &\rho[w_{tt} + c_\mathrm{s}w_{tx} + (c_\mathrm{d} - c_\mathrm{s})u_{tz} - c_\mathrm{s}(c_\mathrm{d} - \tfrac{1}{2}c_\mathrm{s})w_{zz}] = 0 , \\ &\qquad \text{for } \nu > \tfrac{1}{3} . \end{aligned} \tag{2.38}$$

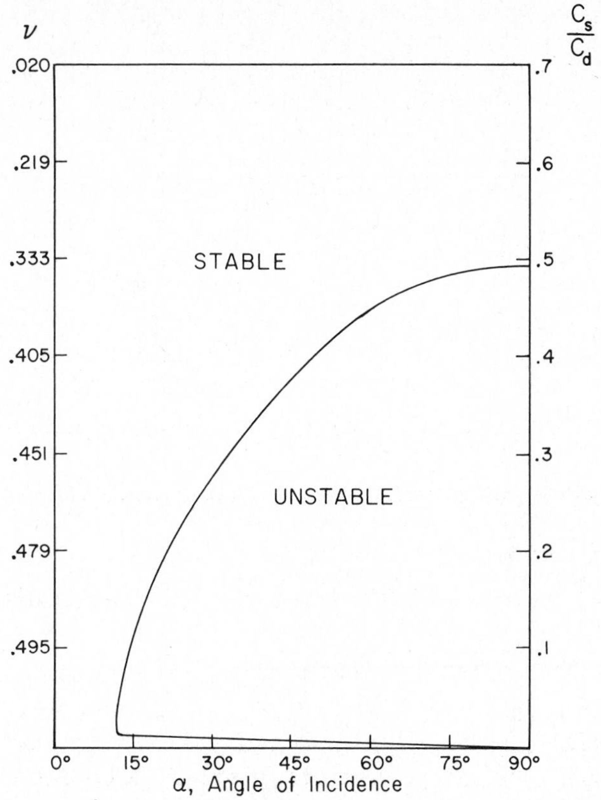

Fig. 2.2. Stability regions for the paraxial equations for different material properties and angles of incidence.

The second set of equations (2.38) could serve for all Poisson's ratios, but when $\nu < \frac{1}{3}$, the first set is theoretically slightly more accurate.

2.6. *Theoretical comparisons of paraxial and viscous boundaries*

2.6.A. *The viscous boundary*

The basic idea of a viscous boundary was proposed by Lysmer and Kuhlemeyer [68], and is illustrated in Figure 2.3 for plane strain. One applies boundary stresses, σ and τ, to an otherwise free boundary. These cancel the stresses which are produced at the boundary by incoming waves, or

$$\sigma_{\text{in}} + \sigma_{\text{bd}} \simeq 0, \qquad \tau_{\text{in}} + \tau_{\text{bd}} \simeq 0\,, \tag{2.39}$$

in which σ_{in} and τ_{in} are the incident stresses, and σ_{bd} and τ_{bd} are the applied boundary stresses. Thus, the zero-traction condition at the free boundary,

$$(\sigma_{\text{in}} + \sigma_{\text{bd}}) + \sigma_{rf} = 0, \qquad (\tau_{\text{in}} + \tau_{\text{bd}}) + \tau_{rf} = 0\,, \tag{2.40}$$

causes the reflected wave stresses, σ_{rf} and τ_{rf}, to be zero.

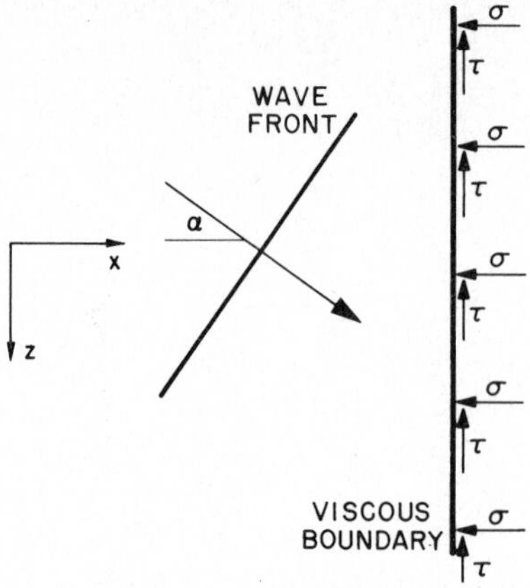

Fig. 2.3. Schematic representation of a viscous boundary.

One set of applied boundary stresses is defined by (2.41):

$$\sigma_{xx} = -\rho c_{\mathrm{d}} u_t, \qquad \tau_{xz} = -\rho c_{\mathrm{s}} w_t\,, \tag{2.41}$$

as was proposed by Lysmer and Kuhlemeyer [68]. These applied stresses are clearly dissipative.

White, Valliappan and Lee [78] proposed a 'unified' viscous boundary, where

$$\sigma_{xx} = -a\rho c_{\mathrm{d}} u_t, \qquad \tau_{xz} = -b\rho c_{\mathrm{s}} w_t\,. \tag{2.42}$$

The parameters, a and b, vary according to the material properties of the medium. The authors performed two minimization processes in order to obtain the optimum values of a and b.

The concepts of the paraxial and viscous boundaries can be related through a heuristic analysis presented in reference [60]. In this analysis a linear material is postulated in which Hooke's law is replaced by a condition that the normal and shear stresses are proportional to the velocities perpendicular and parallel to the boundary. For this postulated medium, the equations of equilibrium are the same as the zeroth order paraxial equations. This correspondence indicates that the viscous and paraxial boundaries use the nodal velocities in a similar manner and suggests that the paraxial technique can be viewed as one way to extend the idea of a viscous boundary. With the capability of rationally introducing higher order terms, the paraxial technique potentially could lead to more efficient energy-absorbing boundaries.

2.6.B. Analysis of wave reflections

In order to compare the theoretical performance of the paraxial and viscous boundaries, we next perform a standard analysis of wave-reflection coefficients. For more details of this procedure, see Miklowitz [4]. We assume that plane-harmonic, elastic waves are impinging upon a boundary strip, as is shown in Figure 2.4. The problem is again one of plane strain.

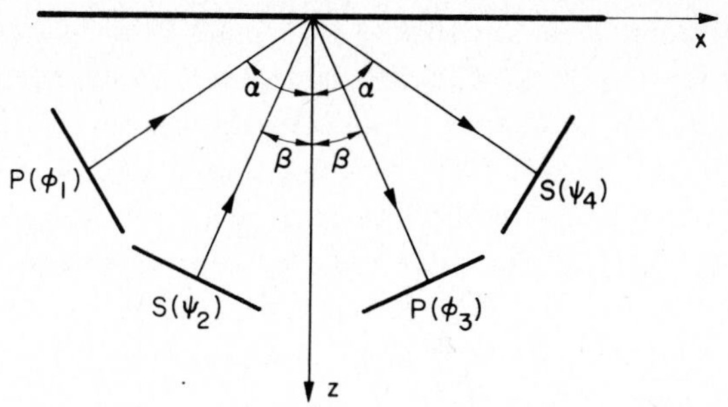

Fig. 2.4. Schematic illustration of dilatational and shear waves reflecting from a boundary.

The wave potentials, which satisfy the governing equations of motion, are:

$$\begin{aligned}
\phi_1 &= \exp\left[\mathrm{i}\frac{\omega}{c_\mathrm{d}}(\sin\alpha\, x - \cos\alpha\, z - c_\mathrm{d}t)\right],\\
\phi_2 &= \exp\left[\mathrm{i}\frac{\omega}{c_\mathrm{d}}(\sin\alpha\, x + \cos\alpha\, z - c_\mathrm{d}t)\right],\\
\psi_3 &= \exp\left[\mathrm{i}\frac{\omega}{c_\mathrm{s}}(\sin\beta\, x - \cos\beta\, z - c_\mathrm{s}t)\right],\\
\psi_4 &= \exp\left[\mathrm{i}\frac{\omega}{c_\mathrm{s}}(\sin\beta\, x + \cos\beta\, z - c_\mathrm{s}t)\right],\\
\phi &= I_\mathrm{p}\phi_1 + A_\mathrm{p}\phi_2, \qquad \psi = I_\mathrm{s}\psi_3 + A_\mathrm{s}\psi_4,
\end{aligned} \tag{2.43}$$

where $I_\mathrm{p} = 1$ and $I_\mathrm{s} = 0$ for an incident P-wave, $I_\mathrm{p} = 0$ and $I_\mathrm{s} = 1$ for an incident S-wave, and A_p and A_s are the amplitudes of reflected P- and S-waves, respectively.

We can now calculate the reflection coefficients for the paraxial boundary. The elastic wave is described by Lamé's solution,

$$u = \frac{\partial\phi}{\partial x} - \frac{\partial\psi}{\partial z}, \qquad w = \frac{\partial\phi}{\partial z} + \frac{\partial\psi}{\partial x}. \tag{2.44}$$

Substitution of the potentials in (2.43) into (2.44) produces the displacements in the elastic region. This wave strikes the paraxial-boundary region, which is governed by (2.38). After substituting the elastic-displacement solution into equations (2.38), we have two equations for the solution of the reflected-wave amplitudes.

In the viscous-boundary scheme the stresses at the boundary are set equal to the viscous stresses. Again, we assume that we have potential solutions which lead to:

$$\begin{aligned}
\sigma_{zz} &= \frac{\lambda}{c_\mathrm{d}^2}\phi_{tt} + 2\mu(\phi_{zz} + \psi_{xz}) = -a\rho c_\mathrm{d} w_t,\\
\tau_{zx} &= \mu[\psi_{tt}/c_\mathrm{s}^2 + 2(\phi_{xz} - \psi_{zz})] = -b\rho c_\mathrm{s} u_t,
\end{aligned} \tag{2.45}$$

in which λ and μ are Lamé's constants. The velocities, μ_t and w_t, are determined by using Lamé's solution, (2.44). The constants, a and b, are positive in (2.45). In the standard viscous boundary, a and b are set to unity. For the unified viscous boundary, however, a and b take values as set forth by White, Valliappan and Lee [78]. The wave-reflection amplitudes, which may be complex, are computed for the various angles of incidence. The absolute values of the amplitudes, for different values of Poisson's ratio, are plotted in Figures 2.5 through 2.8. The three different silent boundaries are labeled as follows:

V – Standard viscous boundary,
O – Unified ('optimized') viscous boundary,
P – Modified paraxial boundary.

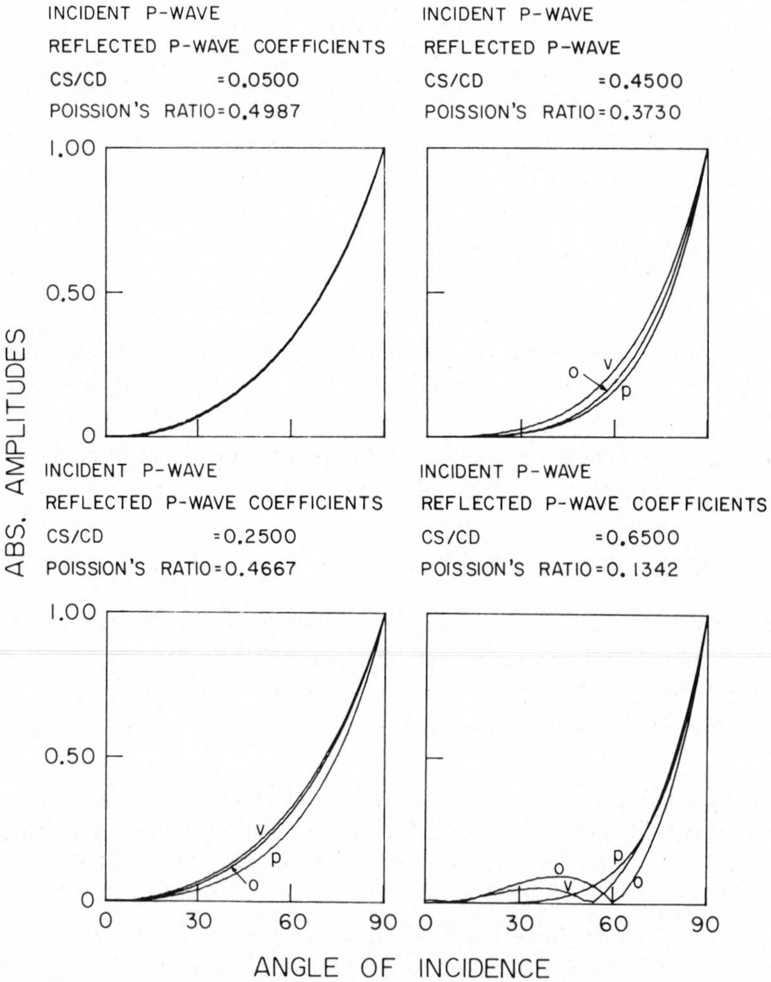

Fig. 2.5. Absolute amplitudes of reflected waves for various angles of incidence (plane strain).

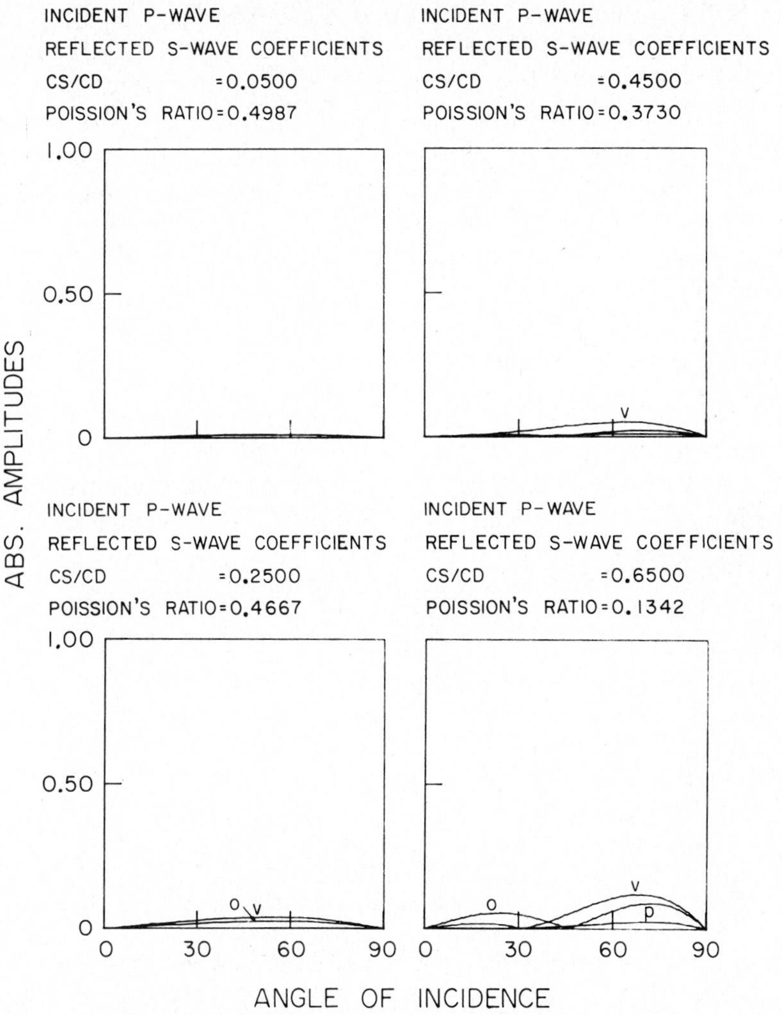

Fig. 2.6. Absolute amplitudes of reflected waves for various angles of incidence (plane strain).

There are four sets of curves:

(1) P-reflections from an incident P-wave,
(2) S-reflections from an incident P-wave,
(3) P-reflections from an incident S-wave,
(4) S-reflections from an incident S-wave.

We should note a few points before interpreting the wave-reflection figures. First, the reflection amplitudes for low angles of incidence are more important than are those for high angles of incidence. Waves which strike a silent boundary at high angles will usually hit one or more boundaries before returning to the interior. In addition, one usually knows in advance the source of wave radiations, and consequently can orient the silent boundaries toward that source. Therefore, the fact that most of the incoming energy is reflected when incident angles are

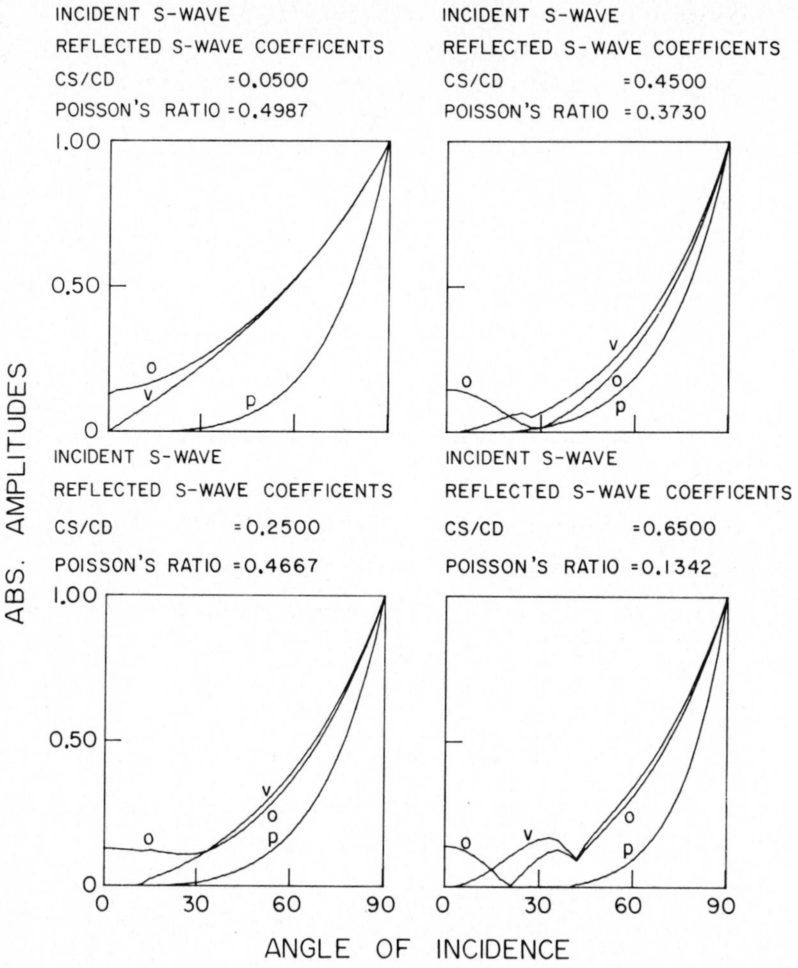

Fig. 2.7. Absolute amplitudes of reflected waves for various angles of incidence (plane strain).

nearly equal to 90°, does not necessarily reduce the boundary's efficiency by a significant amount.

Second, the assumptions which govern the wave-reflection calculations are not strictly the same as those that govern a finite-element representation of the boundary. The finite-element method spatially discretizes the boundary equations, and the boundary contributions are calculated for the outermost set of nodes. In contrast, the above theoretical analysis employs an infinitesimal boundary strip, from which the boundary effects are calculated.

When we consider the first set of curves, the reflected P-wave amplitudes caused by incident P-waves (Figure 2.5), we can see that all of the boundaries are nearly equal in their reflection amplitudes. Nearly perfect absorption is attained for those incident waves which are almost normal to the boundary. Conversely,

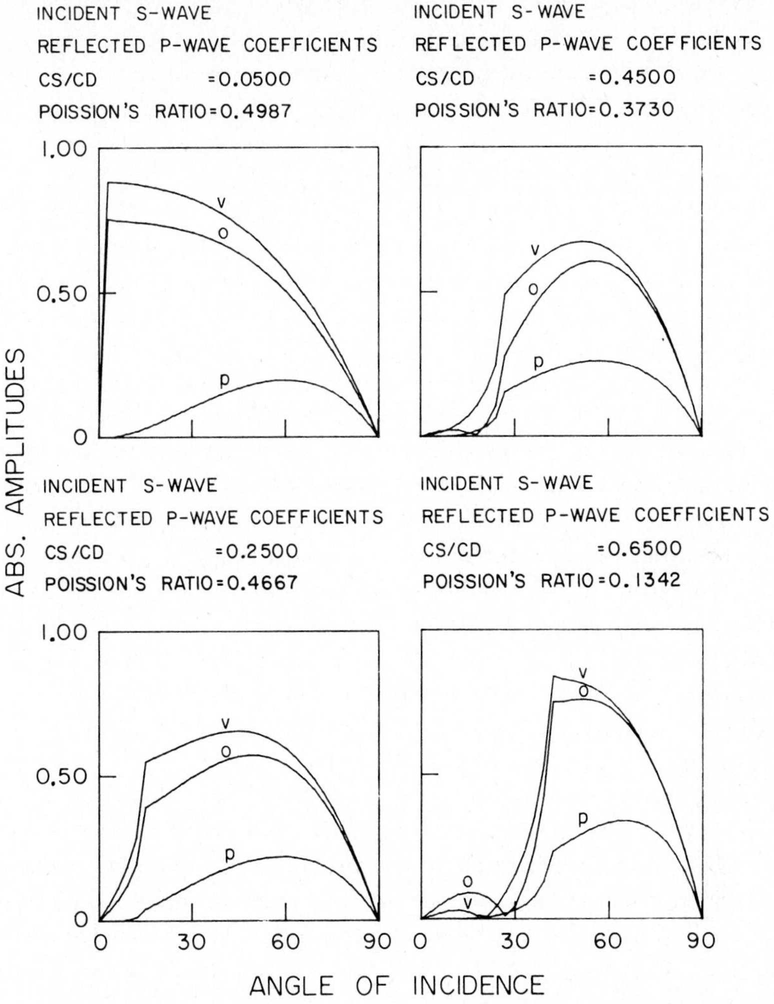

Fig. 2.8. Absolute amplitudes of reflected waves for various angles of incidence (plane strain).

total reflection occurs for the waves which impinge tangentially. As Poisson's ratio decreases, the absorption characteristics of all three silent boundaries improve. The one exception to this trend is the unified viscous boundary when Poisson's ratio is nearly zero, which performs less effectively.

The viscous boundary curves inflect near 60° for $c_s/c_d \geq 0.55$. These inflections are due to changes in the signs of the reflected amplitudes. The reflected, paraxial amplitudes retain the same sign for all angles of incidence, and therefore display no inflections.

The reflected S-wave amplitudes, caused by incident P-waves, are shown in Figure 2.6. All three boundary methods show negligible reflections when Poisson's ratio is near one-half. The modified-paraxial magnitudes are so small that they all are nearly zero. In addition, the boundaries are almost as effective for lower Poisson's ratios as they are for higher ones.

The illustrations of reflected S-wave magnitudes arising from incident S-waves, Figure 2.7, depict larger differences among the three boundary schemes than do the previous figures. In this test, the modified-paraxial boundary clearly is superior. For all Poisson's ratios, the unified viscous boundary produces 10–15 percent reflections of normally-incident waves. It should also be noted, however, that with high Poisson's ratio, the incident S-wave curves are somewhat misleading. In this case, the shear-wave propagation speed is only a small fraction of the dilatational speed. The S-wave reflections will be traveling slowly, and therefore, may not significantly influence the response in the interior region.

The last set of comparisons, Figure 2.8, describes P-wave reflections due to incident S-waves. These curves are influenced by the critical angle phenomenon. If the incident, shear-wave angle is larger than a certain angle called the critical angle, then the reflected, dilatational wave becomes a surface wave which travels along the boundary. In each of the plots in Figure 2.8 one can detect the critical angles by the abrupt changes in the slopes of the reflection curves. This occurs in the region of from 0° to 45°, depending on the Poisson ratio. As the shear-wave speed decreases (i.e., Poisson's ratio increases) the critical angle decreases.

In this analysis, we are dealing with an incident, plane, harmonic wave which extends infinitely in the $x-z$ plane. Likewise, the reflected body waves extend infinitely. The reflected surface wave, however, which is created when the shear-wave angle is greater than the critical angle, is confined to a region near the boundary and contains only a small fraction of the total energy. Therefore, in Figure 2.8, the fact that P-wave amplitudes are large for those incidence angles which are greater than the critical angle, is probably not seriously detrimental. In fact, previous authors [68, 78] eliminated these amplitudes from comparison when they multiplied them by the wave speed times the cosine of the angle of incidence. This new quantity measured the energy flux at the boundary; energy propagating along the boundary was assumed to be confined there.

In general, by considering all of the wave-reflection curves, we conclude that each of the boundary schemes produces acceptable results. All of them perform well when the incident angles are less than 40 degrees. With high Poisson's ratios, however, shear waves may cause some difficulties, particularly for the viscous boundaries.

The modified-paraxial boundary appears to be generally superior to the viscous schemes. It consistently outperforms the others and it demonstrates significant improvement for those reflected, shear-wave amplitudes which are produced by incident shear waves. It must be noted, however, that this superiority of the paraxial boundary diminishes in actual numerical simulations (see Section 4).

As it was first formulated [59], the paraxial boundary includes potentially negative stiffness terms. It has been shown that when these terms are negative ($\nu > \frac{1}{3}$), reflections in similar analyses are greater than unity [60]. This result supports that of the stability analysis presented in Section 2.5.

2.7. *Rayleigh waves*

In a typical case, the energy radiating toward a boundary could be simultaneously composed of Rayleigh, Love, and body waves propagating at different speeds. It is desirable, of course, to install a transmitting boundary which can handle all of these different wave motions.

The previous, wave-reflection analysis, however, is not valid in the case of Rayleigh waves. Therefore, it is difficult to assess the various boundaries' effectiveness in transmitting these waves. In reference [60] an argument is set forth which indicates that the modified-paraxial equations largely absorb Rayleigh waves.

The efficacy of the standard viscous boundary in the transmission of Rayleigh waves is largely untested. Lysmer and Kuhlmeyer [68] pointed out that because of the waves' exponential decay in the z-direction, the formerly constant parameters, a and b, should also be functions of z. These author's plots of the required coefficients, as a function of the normalized depth κz, are reproduced in Figure 2.9.

As can be seen from Figure 2.9, these viscous coefficients vary with κ and, therefore, are frequency dependent. One must know in advance the frequency of the incoming waves and implement the correct values for $a(\kappa z)$ and $b(\kappa z)$ accordingly. Thus, the boundary may be applied for steady-state problems, but it appears unsuitable for transient analysis.

In a practical sense, it is not clear that the use of the variable coefficients would improve the absorption of Rayleigh waves. It would be extremely difficult, in employing a finite element mesh, to approximate the variation in $a(\kappa z)$ from $-\infty$ to $+\infty$ (Figure 2.9). Further, as will be seen in Section 4, the behavior of the viscous boundaries is relatively insensitive to changes in a and b.

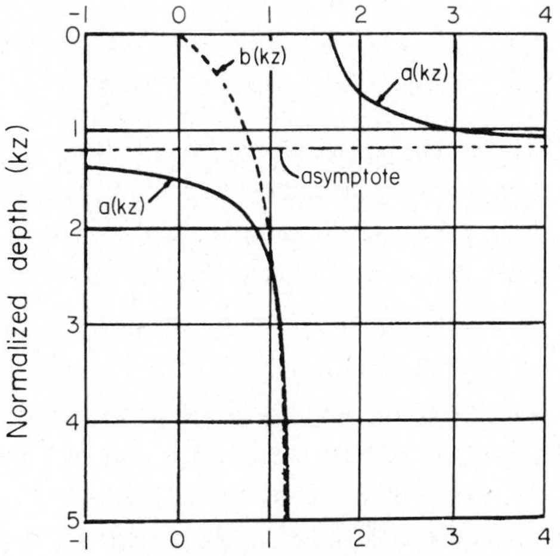

(b) The functions a and b

Fig. 2.9. The functions $a(\kappa z)$ and $b(\kappa z)$ used for Rayleigh waves, as proposed by Lysmer and Kuhlemeyer [68].

2.8. Other silent-boundary applications

2.8.A. Spherically symmetric case

The paraxial and viscous boundaries can be easily derived for this one-dimensional situation. The spherically symmetric equation of elasticity is:

$$u_{rr} + \frac{2u_r}{r} - \frac{2u}{r^2} = \frac{u_{tt}}{c_d^2}, \tag{2.46}$$

wherein u is the radial displacement, and r is the radial coordinate. The solution to (2.46), expressed in terms of a potential ϕ, is [4]:

$$\phi = \frac{1}{r} f(r - c_d t) + \frac{1}{r} g(r + c_d t), \tag{2.47}$$

$$u = \frac{\partial \phi}{\partial r}. \tag{2.48}$$

If we consider only outwardly radiating waves,

$$\begin{aligned} \phi &= \frac{1}{r} f(r - c_d t), \\ u &= -\frac{1}{r^2} f(r - c_d t) + \frac{1}{r} f'(r - c_d t). \end{aligned} \tag{2.49}$$

One partial differential equation which has (2.49) as its approximate solution is:

$$u_{tt} + c_d u_{tr} + \frac{c_d}{r} u_t = 0. \tag{2.50}$$

The substitution of (2.49) into (2.50) produces a residual term,

$$\frac{-f'(r - c_d) c_d^2}{r^3},$$

which grows smaller as r increases (assuming that $f'(r - c_d t)$ is bounded for large r). Equation (2.50) appears to be a suitable paraxial approximation for outwardly radiating waves, and it may be useful for spherically symmetric problems.

The corresponding viscous boundary employs the boundary stress,

$$\sigma_{rr} = -\rho c_d u_t. \tag{2.51}$$

Castellani [58] discusses this version of the viscous boundary. He offers a method for evaluating its effectiveness, which we believe can be adapted to the paraxial equation (2.50). It compares stresses not cancelled by the viscous boundary to those stresses created by the incident wave. Applying this approach indicates that both the paraxial and viscous boundaries are frequency dependent. For a given r, they behave better for high frequency motion.

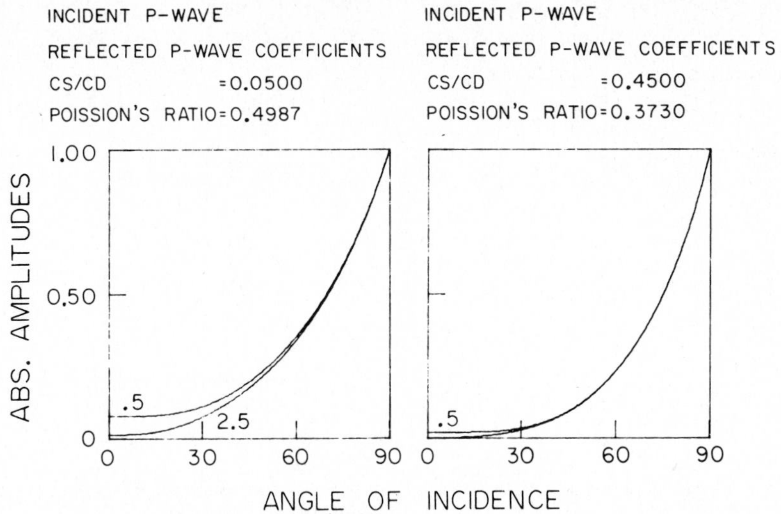

Fig. 2.10. Absolute amplitudes of reflected waves for various angles of incidence (axial symmetry).

2.8.B. Axially symmetric waves

In [60] a preliminary investigation of the viscous boundary's effectiveness for the axially symmetric case is presented. Selected results are summarized in Figures 2.10 through 2.13. The results for an incident P-wave are illustrated in Figures 2.10 and 2.11; the S-wave reflections are shown in Figures 2.12 and 2.13. The numbers '0.5' and '2.5' in the figures denote the reflection coefficients calculated with $R/L_d = 0.5$ and $R/L_d = 2.5$, respectively. In this normalization, R is the distance from the axis of symmetry to the viscous boundary and L_d is c_d times the wavelength, divided by the wave speed.

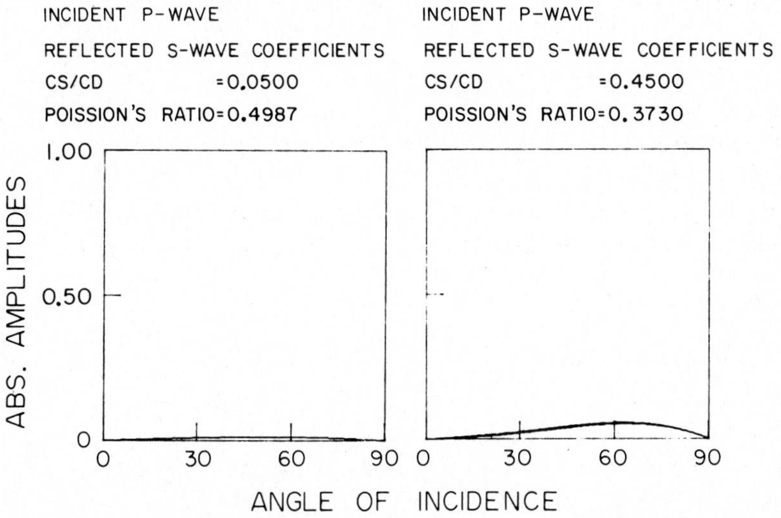

Fig. 2.11. Absolute amplitudes of reflected waves for various angles of incidence (axial symmetry).

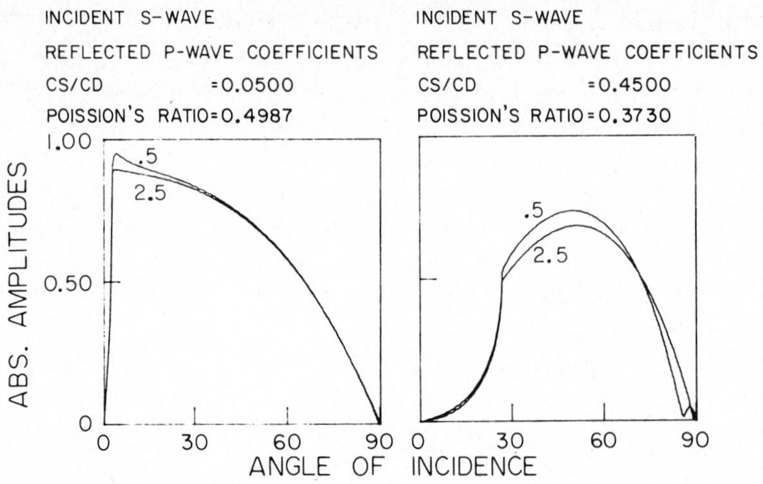

Fig. 2.12. Absolute amplitudes of reflected waves for various angles of incidence (axial symmetry).

Several preliminary conclusions can be gathered from these results. Unlike the plane-strain case, the axially symmetric, viscous boundary depends on R/L_d, and is therefore frequency dependent. Although the boundary's accuracy is greater for higher frequency waves, it does not significantly downgrade in the range of R/L_d shown. This result would seem to be corroborated in [78].

One may also notice that the axially symmetric amplitudes, with $R/L_d = 2.5$, are nearly identical to the corresponding plane-strain results in Section 2.6. As R/L_d is decreased, the differences become more perceptible. The curves with $R/L_d = 0.5$ intimate a trend of increased reflections for smaller ratios of R/L_d. The fact that some reflection amplitudes exceed one, when α is close to 90°, and that errors are introduced when R/L_d is small, is discussed in [60].

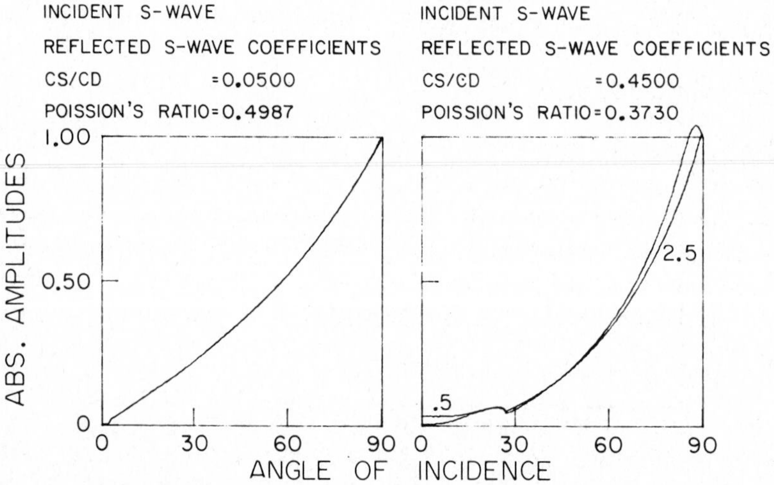

Fig. 2.13. Absolute amplitudes of reflected waves for various angles of incidence (axial symmetry).

Some authors have implemented the viscous boundary for axially symmetric waves and have attained mixed results. Lysmer and Kuhlemeyer [68], having discovered the energy-absorbing potential of the standard viscous boundary for plane waves, proceeded to apply the boundary to axially symmetric problems. The boundary was situated at $R/L_R = 0.75$ and $R/L_R = 1.50$, which correspond to $R/L_d = 0.35$ and $R/L_d = 0.69$, respectively. (L_R = the Rayleigh wavelength, Poisson's ratio $= \frac{1}{3}$). These authors' results indicate good agreement with analytical solutions when $R/L_d = 0.69$, but the boundary with $R/L_d = 0.35$ acquits itself only fairly. These test results appear to be consistent with the reflection curves in Figures 2.10 through 2.13.

Baladi [57] also scrutinized the standard viscous boundary in an axially symmetric setting. His boundary was aligned to absorb waves travelling in the z-direction. He found generally satisfactory agreement between the silent boundary results and those produced with an extended mesh.

In summary, the limited data available indicate that the standard viscous boundary can be adapted to treat axially symmetric waves.

3. Implementation

3.1. *Introduction*

The analyses of Section 2 suggest that the previously discussed boundary schemes could reproduce most of the effects of an infinite domain. There are, however, a number of practical limitations to their usage which one must consider. For example, we need to appraise the boundary's *numerical* stability. If a silent boundary demands a smaller critical time step than does the interior region, it may engender larger computer costs. Also, problems in accuracy may arise in the numerical treatment of the boundary terms.

3.2. *Finite-element procedures*

3.2.A. *Implicit-explicit algorithm*

In the work reported herein, we employ the implicit-explicit method which was developed by Hughes and Liu [18]. The basic procedure is outlined in references [19] and [20]. The algorithm enjoys considerable versatility in coping with transient problems. One can effectively divide the domain of analysis into 'implicit' and 'explicit' regions, thus capitalizing on the advantages of each scheme.

3.2.B. *Finite-element implementation of the paraxial boundary*

1. *General considerations.* We recall the one-dimensional, paraxial equation

$$u_{tt} + c u_{tx} = 0 . \tag{3.1}$$

It is easy to show how the application of standard finite-element procedures to the u_{tx} term leads to a nonsymmetric matrix. The matrix equation of motion, including both interior and boundary contributions, is:

$$\boldsymbol{M}\boldsymbol{d}_{tt} + \boldsymbol{C}\boldsymbol{d}_t + \boldsymbol{K}\boldsymbol{d} + \boldsymbol{N}(\boldsymbol{d}) = \boldsymbol{F}. \tag{3.2}$$

$\boldsymbol{C}$ is the matrix which is derived from the boundary term, u_{tx}; $\boldsymbol{N}$ is a nonlinear operator in the interior. Equation (3.2) may be solved with a variety of numerical algorithms which lead to an equation of the form

$$\boldsymbol{K}^{*(j)}\boldsymbol{d}_{n+1}^{(j+1)} = \boldsymbol{F}^{*(j)}, \tag{3.3}$$

where n denotes the time step number, and j is an iteration counter. $\boldsymbol{K}^*$ is sometimes called the effective stiffness, and $\boldsymbol{F}^*$ is referred to as the residual force. $\boldsymbol{K}^*$ may include components derived from $\boldsymbol{M}$, $\boldsymbol{C}$, $\boldsymbol{K}$, and $\boldsymbol{N}$.

If the contributions from $\boldsymbol{M}$, $\boldsymbol{K}$, and $\boldsymbol{N}$ are symmetric, then the use of a nonsymmetric boundary matrix, $\boldsymbol{C}$, would require an increased storage for $\boldsymbol{K}^*$. It would also demand a nonsymmetric equation solver and involve a concomitant increase in computational effort. An explicit, paraxial boundary scheme does not suffer from the above difficulties and was therefore chosen. The explicit algorithm, however, does limit the allowable time step.

2. *Upwinding.* As was noted above, the finite-element treatment of the u_{tx} term leads to a nonsymmetric matrix, $\boldsymbol{C}$. In other problems which generate nonsymmetric matrices (e.g., fluid mechanics), it has been observed that spurious oscillations can occur. In many respects, the paraxial equations have features in common with the equations of fluid mechanics. In our case, u_{tx} serves as a convective mechanism for transporting energy.

Special techniques are required in a finite implementation to capture the physical behavior of the convective terms appropriately. In the present work the 'quadrature scheme' of Hughes [15] was adopted. This leads to a so-called 'upwind formulation' which is an area of much interest in computational fluid dynamics. In the present circumstances, it corrects the tendency of the standard finite element implementation to produce spurious oscillations. The result is a dissipative procedure, which is in keeping with the purpose of a silent-boundary method.

A brief sketch of the essential features of the procedure is facilitated with the aid of Figure 3.1. In calculating the element contributions of convection terms, a nonsymmetric quadrature rule is used in which the evaluation points are located as shown in Figure 3.1. The reasons for this are discussed at length in Hughes [15] and Cohen [60]. It is, in fact, only essential to use the nonsymmetric rule on $\boldsymbol{u}_{tx}$; $\boldsymbol{u}_{tz}$ could be integrated with the standard 2×2 or with 1-point Gaussian quadrature. The two-point integration method, however, seems to provide slightly better accuracy. The paraxial stiffness term, $\boldsymbol{u}_{zz}$, can be integrated using standard 2×2 quadrature. The quadrature method used for $\boldsymbol{u}_{tz}$ and $\boldsymbol{u}_{zz}$, however, does not

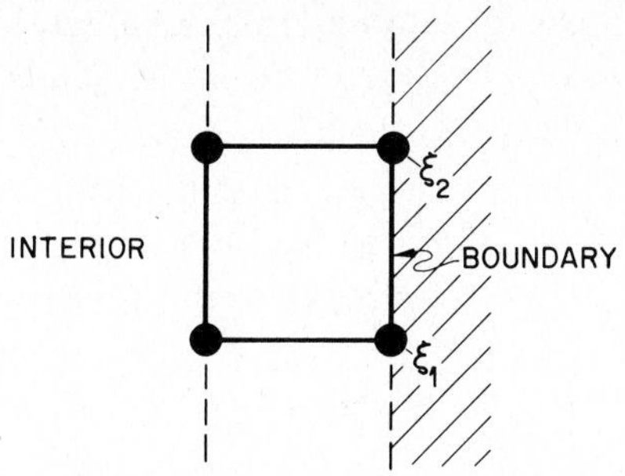

Fig. 3.1. Location of the integration points, ξ_1 and ξ_2, with respect to the interior domain.

significantly affect the boundary's behavior because they are terms in the lower-order paraxial sense.

3. *Assembly procedure.* The algorithm employed in our transient calculations performs well for waves in either the elastic medium or the paraxial medium. We found, however, that an interfacing effect inhibits elastic waves for smoothly proceeding into the paraxial area. The method developed to circumvent this problem, an element-assembly procedure, is depicted in Figure 3.2.

The finite-element equations for node 2 are written as if there were interior elements present on each side, while the equations for node 3 are formulated as if it were bracketed by boundary elements. In other words, the ith element, shown in the picture, contributes regular wave terms to node 2 and boundary terms to node 3. The mass matrix is unaffected because it remains uniform for all of the elements.

The reason for adopting this procedure is the apparent success of the finite difference solution, used by Clayton and Engquist [59]. The finite-element method of assembling contributions on an element basis simply did not work at the interface. Therefore, we attempted to 'finite difference' these interface nodes. The latter procedure allows the elastic and paraxial nodal equations to be assembled independently. In the former approach, elastic and paraxial contributions were being added together in the *same* set of nodal equations. This led to the difficulty.

In two and three dimensions, the interface nodes are assembled in the same manner as described above.

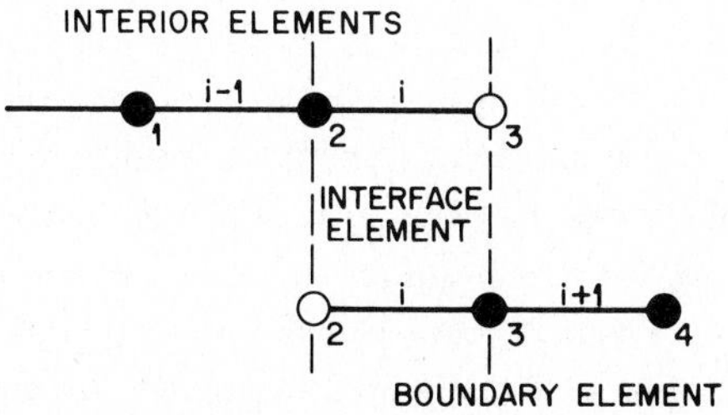

Fig. 3.2. Schematic diagram of the interface element, paraxial boundary.

3.2.C. Numerical implementation of the viscous boundary

The viscous stresses are applied continuously along the boundary, as depicted in Section 2, Figure 2.3. Their contributions are assembled at the nodes through the use of one-dimensional, finite-element, shape functions (as shown in Figure 3.3, wherein linear shape functions are employed)

$$N_1 = N_4 = 0\,, \tag{3.4}$$

$$N_2 = \tfrac{1}{2}(1-\zeta), \qquad N_3 = \tfrac{1}{2}(1+\zeta)\,, \tag{3.5}$$

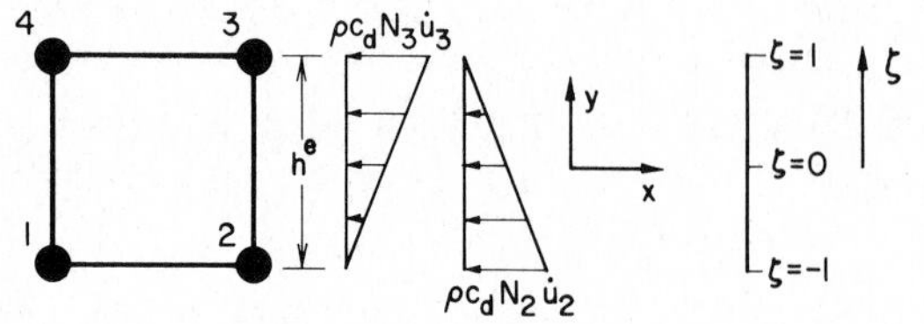

Fig. 3.3. Distribution of the viscous stresses applied to a boundary element, viscous boundary.

where ζ is the natural coordinate $(-1 \le \zeta \le 1)$, and $\dot{u}_2$ and $\dot{u}_3$ are the nodal velocities. The boundary force, F_a, acting at node a is defined as:

$$F_a = \int_{y_2}^{y_3} N_a \sigma(y)\,\mathrm{d}y$$

$$= \int_{-1}^{1} -N_a \rho c_d (N_2 \dot{u}_2 + N_3 \dot{u}_3) \frac{\mathrm{d}y}{\mathrm{d}\zeta}\,\mathrm{d}\zeta, \quad \text{where } a = 2, 3\,. \tag{3.6}$$

If we substitute into (3.6) both $\mathrm{d}y/\mathrm{d}\zeta = \tfrac{1}{2}h^e$ and the above defined shape function equations, N_a, we find that:

$$\begin{Bmatrix} F_2 \\ F_3 \end{Bmatrix} = \frac{-\rho c_d h^e}{6} \begin{bmatrix} 2 & 1 \\ 1 & 2 \end{bmatrix} \begin{Bmatrix} \dot{u}_2 \\ \dot{u}_3 \end{Bmatrix} = -\boldsymbol{C}\dot{\boldsymbol{u}}\,. \tag{3.7}$$

Though we chose not to, it appears that one could 'lump' the C matrix in the same way that mass matrices are lumped. That is,

$$\boldsymbol{C}_\ell = \frac{\rho c_d h^e}{2} \begin{bmatrix} 1 & 0 \\ 0 & 1 \end{bmatrix}. \tag{3.8}$$

The boundary shear stresses are applied in the same manner as the normal stresses. When boundary parameters, a and b, different from unity are used they multiply the appropriate terms in (3.7) and its counterpart for shearing stresses.

One advantage of the viscous boundary is that equations (3.7) and (3.8) are symmetric. Thus, one can convert this boundary to fit either an implicit or an explicit algorithm without any difficulty. For an implicit algorithm, the symmetric $\boldsymbol{C}$ matrix enables us to easily determine the boundary's numerical stability. The boundary's simplicity is another major advantage. The C matrix is explicitly defined by (3.7) or (3.8), so there is little additional cost in making these contributions to the boundary elements.

3.3. Numerical stability of the extended-paraxial and viscous boundaries

A numerical stability analysis for the paraxial equations is presented in [60]. It was determined that explicit, paraxial elements largely share the same stability properties as do those of regular elastic elements.

The stability characteristics of the viscous boundary proceed from the stability theorem in [18]. An *implicit*, viscous boundary is unconditionally stable if algorithmic parameters are appropriately chosen. The numerical stability limits for the explicit, viscous boundary are not as specific, but they can be evaluated for each separate case (see [60] for further discussion).

4. Numerical procedures and results

4.1. Introduction

In order to assess the relative merits of the extended-paraxial, standard-viscous, and unified-viscous boundaries, we present in this section summaries of our numerical investigations. The investigations were focused on the abilities of the three boundaries to treat high frequency waves, to handle materials wherein Poisson's ratio is near $\frac{1}{2}$, and to transmit Rayleigh waves.

4.1.A. Description of the numerical procedure

All of the examples in this section were solved by using the silent boundaries described in Section 3. Both the elastic and the boundary regions were discretized with four-noded elements which employ bilinear, isoparametric shape functions. A selective integration scheme is used to develop the element stiffness, as described in [14]. This enables satisfactory treatment of situations involving near incompressibility in which Poisson's ratio approaches $\frac{1}{2}$.

The captions to the figures in this section specify whether the 'interior' region is solved implicitly or explicitly. The Newmark parameters, γ and β [18, 19], and the time step, DT, are also defined in the figure captions. The viscous boundaries are always treated implicitly, in order to utilize the implicit algorithm's unconditional stability. The explicit method, however, is employed in the extended-paraxial domain, for the reasons mentioned in Section 3. To facilitate the calculations of energy, the mass matrices are lumped, regardless of the chosen algorithm.

4.1.B. Selection of wave problems

The meshes used in this chapter are significantly coarser than those which one would normally use for practical engineering problems. In fact, most of the loadings of the following systems are applied over one or two elements and for only one time step. This input, similar to a delta function, generates much high-frequency motion in the solutions. We selected this approach because it subjects the silent boundaries to relatively severe test conditions. Since the boundaries are designed to absorb smooth pulses, this transient, high-frequency excitation should challenge the limits of their capabilities.

The purpose of these tests, then, is to evaluate and compare the silent boundaries using the above described input. Because of the deliberate selection of coarse meshes, the resulting motion may not duplicate the correct solution in all of its details, but that is not really the concern here. The accuracy of the finite-element method, and its dependence upon the fineness of the mesh, is well understood. For our purposes the 'correct' response is that which is produced by a mesh extensive enough to prevent the boundary reflections from reaching the 'interior' zone. In all of these problems, the silent boundaries were implemented as indicated in Figure 4.1. To depict a silent boundary, we employ a thick black strip, which is illustrated on the left side of the figure.

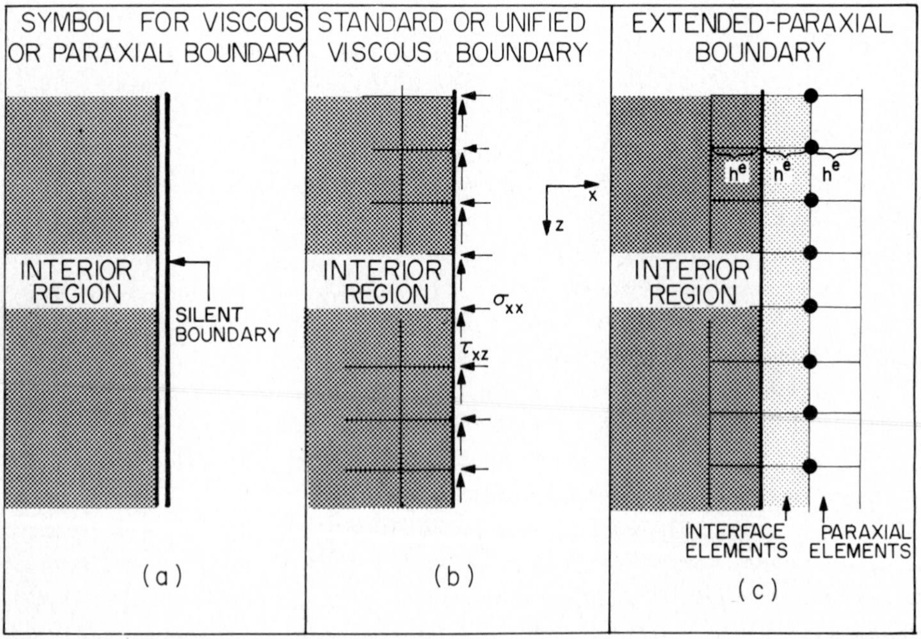

Fig. 4.1. Schematic drawings of the silent boundaries: (a) thick black strip which symbolizes a silent boundary; (b) the standard-, or unified-viscous boundary, as applied in all our examples; (c) the extended-paraxial boundary.

4.2. *Direct incidence of dilatational waves*

Radial, dilatational pulses which directly impinge on the boundaries were tested in [60]. In these examples, where Poisson's ratio ranged from 0.33 to 0.48, all the boundaries demonstrated nearly perfect absorption.

4.3. *Pulse loadings – general discussion*

With the following two-dimensional examples of pulse loadings, we compare the results produced by a small mesh having absorbing boundaries to those generated by an extended mesh. The meshes are shown in Figure 4.2. Time histories of the response are recorded at three points, labeled A, B, and C, in each of the meshes.

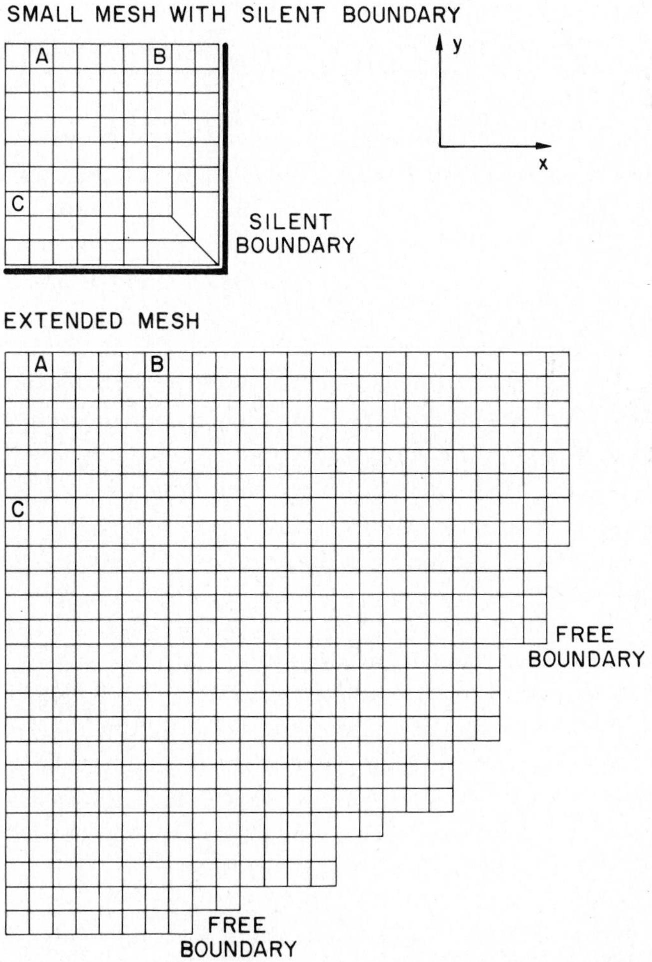

Fig. 4.2. Finite-element meshes used for pulse loadings: (top) mesh with absorbing boundary; (bottom) extended mesh with free boundary.

The two excitations which are considered are a vertical pulse that generates mainly dilatational waves, and a horizontal pulse that generates primarily shear waves. These are applied to the top surface near point A (see Figure 4.3). These loadings were selected because of their simplicity and their relevance to the vertical and horizontal loadings that occur in soil-structure interaction problems.

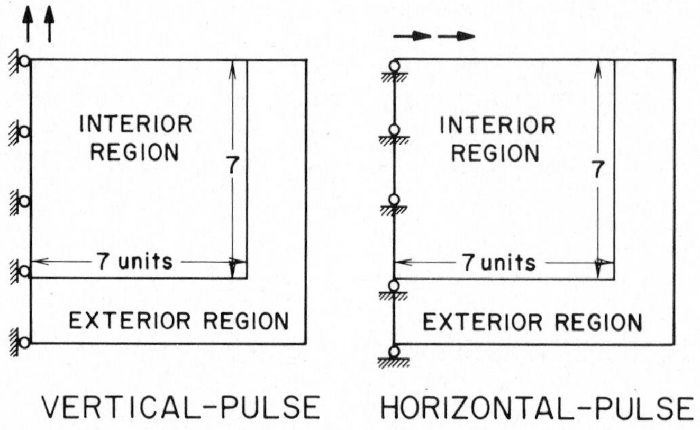

Fig. 4.3. Schematic drawing of the horizontal- and vertical-pulse loadings.

In the vertical-pulse loading, the x degree-of-freedom is fixed along the left face of the mesh. These are the 'symmetrical' boundary conditions which allows us to analyze a half space with a quarter mesh. To simulate the pulse, nodal forces at two nodes are applied vertically during the first time step only.

Figure 4.3 also illustrates the loading the boundary conditions for the horizontal-pulse loading. With this applied-traction problem, we are not analyzing a half space because it does not meet the requirements of symmetry. The conditions of Figure 4.3, however, ensure that shearing waves impinge upon the horizontal boundary.

4.4. *Horizontal-pulse loading*

4.4.a. *Comparisons of the boundary methods*

For the first horizontal-pulse-loading calculation, $c_s = 0.5345$ units/sec, $c_d =$ 1 unit/sec, $\rho =$ density $= 1$ (Poisson's ratio $= \frac{1}{3}$), DT $= 0.85$ sec, where 1 unit is the width of one of the square elements. With these values, the dilatational waves reach the boundary of the smaller mesh in 9 seconds; the shear waves arrive at the same point in 16 seconds. No material damping is present in the system. The interior region was solved using the explicit algorithm.

The horizontal displacements, the main components of the motion, are plotted in Figures 4.4 through 4.7. Figures 4.4 and 4.5 report those displacements recorded near the side boundary (point B in Figure 4.2); the arrival of the main pulse is evident in all the figures. Wave reflections caused by a free boundary are

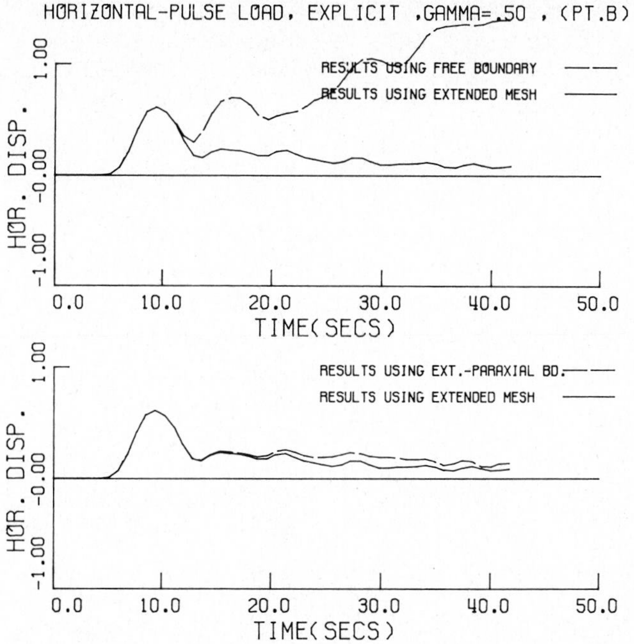

Fig. 4.4. Horizontal displacement at point *B* as a function of time (horizontal-pulse loading, undamped case).

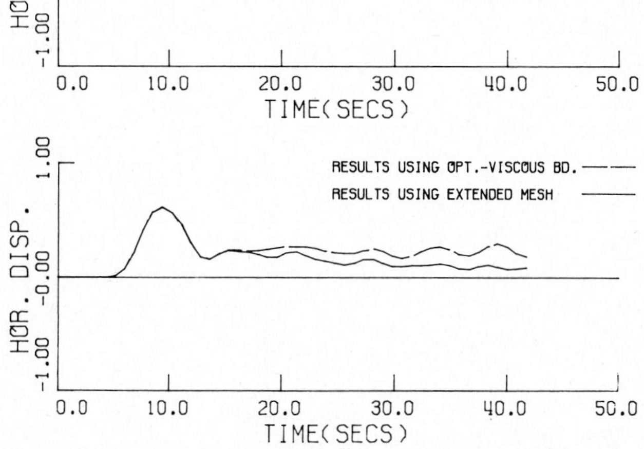

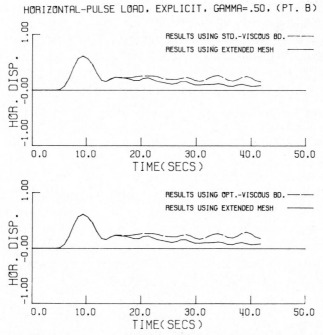

Fig. 4.5. Horizontal displacements at point B as a function of time (horizontal-pulse loading, undamped case).

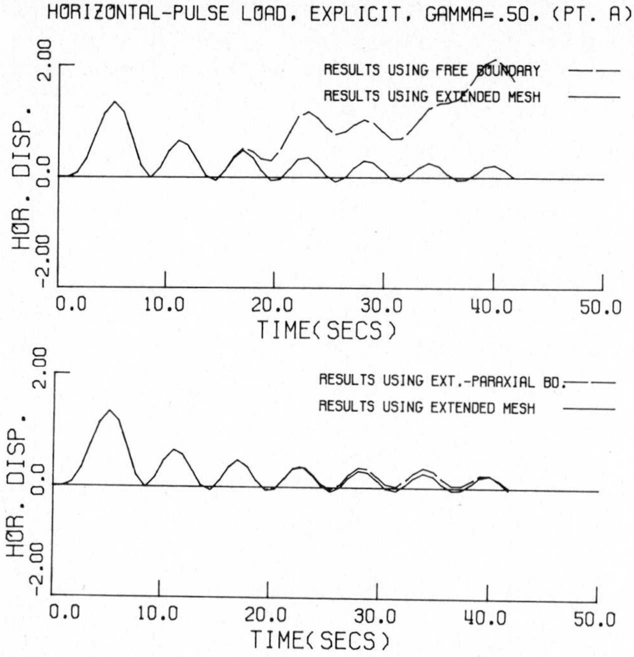

Fig. 4.6. Horizontal displacements at point A as a function of time (horizontal-pulse losding, undamped case).

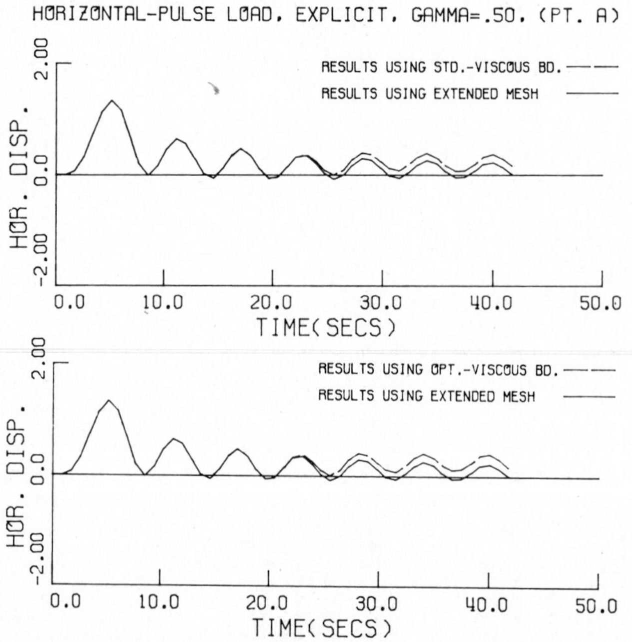

Fig. 4.7. Horizontal displacements at point A as a function of time (horizontal-pulse loading, undamped case).

clearly seen in Figure 4.4, while the paraxial and viscous boundaries largely succeed in eliminating this reflection. The extended-paraxial boundary is slightly more accurate than the other transmitting boundaries. On the other hand, both the standard-, and the unified-viscous boundaries produce nearly the same response (see Figure 4.5).

Figures 4.6 and 4.7 demonstrate how well the silent boundaries simulate the infinite domain at point A in Figure 4.2. Figure 4.6 shows the results for the extended-paraxial boundary, and Figure 4.7 illustrates the behavior of the viscous boundaries. These figures corroborate the slight edge in the paraxial boundary's performance as seen earlier.

Oscillations of the horizontal displacement in Figures 4.6 and 4.7 arise in all of the calculations, including the extended mesh. Each of the transmitting boun-

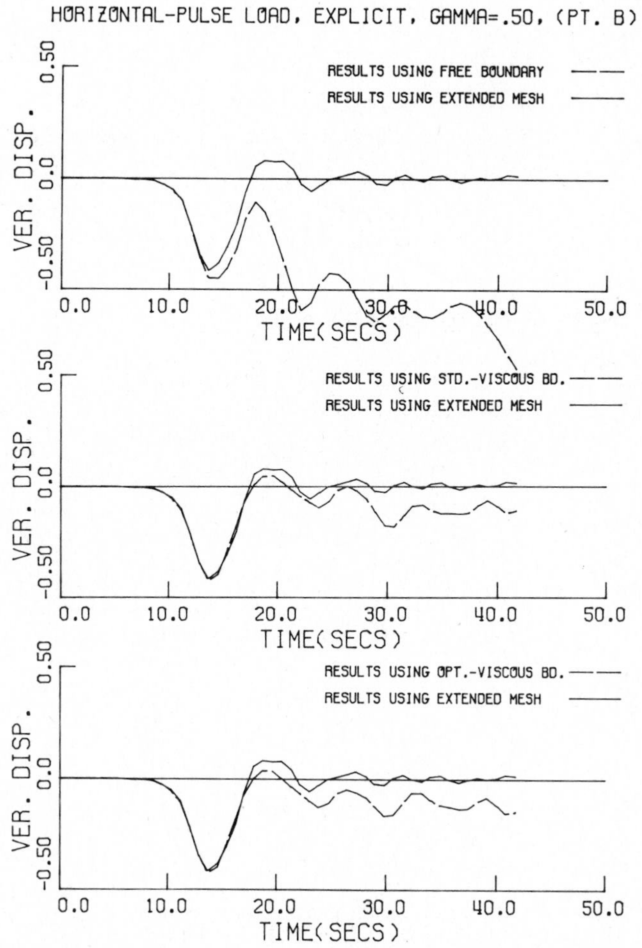

Fig. 4.8. Vertical displacements at point B as a function of time (horizontal-pulse loading, undamped case).

daries preserves the period of these high frequency motions, which arise from the coarseness of the mesh and the character of the loading.

In the calculations of the displacements, as illustrated in Figures 4.5 and 4.7, the unified-viscous boundary almost duplicates those results produced by the standard-viscous boundary. The largest quantitative differences between the two methods appear in Figure 4.8. The vertical displacements that are recorded next to the side boundary (point B) indicate that the standard viscous approach may be, to a small extent, the better alternative. The result for the paraxial boundary, corresponding to Figure 4.8, indicates a slightly better agreement with the extended mesh than do the viscous boundaries.

In Figure 4.8 it may also be noted that the vertical wave arrives later than does the horizontal pulse shown in Figure 4.4. This later arrival seems to be composed mainly of a shear wave, which travels more slowly than does the horizontal, dilatational pulse.

4.4.B. High-frequency waves

One weakness of the silent boundaries is highlighted in Figure 4.9. The numerical noise following the arrival of the main stress pulses (between 9 and 14 sec) is the predominant feature seen in this figure. Both the extended and the small meshes, which provide reasonable solutions to the high-frequency displacements, are too coarse to determine the stresses accurately; these latter

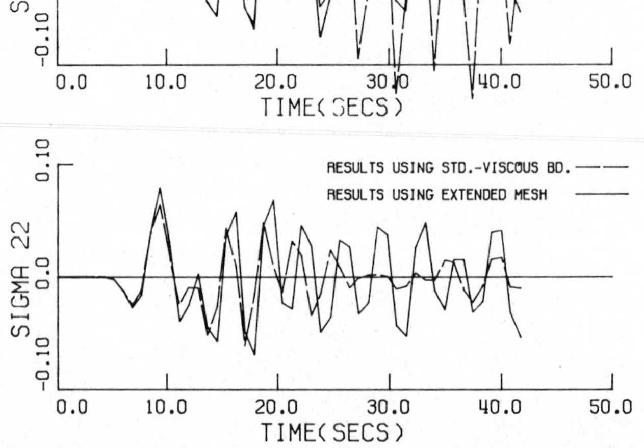

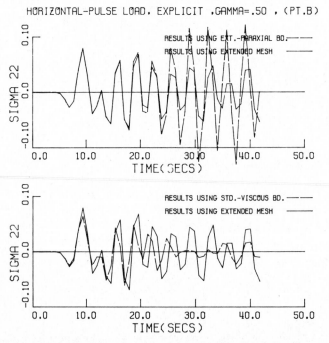

Fig. 4.9. Stresses, σ_{22}, at point B as a function of time (horizontal-pulse loading, undamped case).

quantities are evaluated from the derivatives of the displacements. It is, therefore, understandable that none of the boundary schemes produce the same high-frequency oscillations of the stress, σ_{22}, that the extended mesh does. The paraxial boundary actually magnifies this numerical noise.

These oscillations are also evident in the energy plots of Figure 4.10. The total energy is the sum of the kinetic and strain energies, which are contained in the 'interior' region of Figure 4.3. For each time step, we computed this value by integrating the energy terms over the region (using one-point Gaussian quadra-

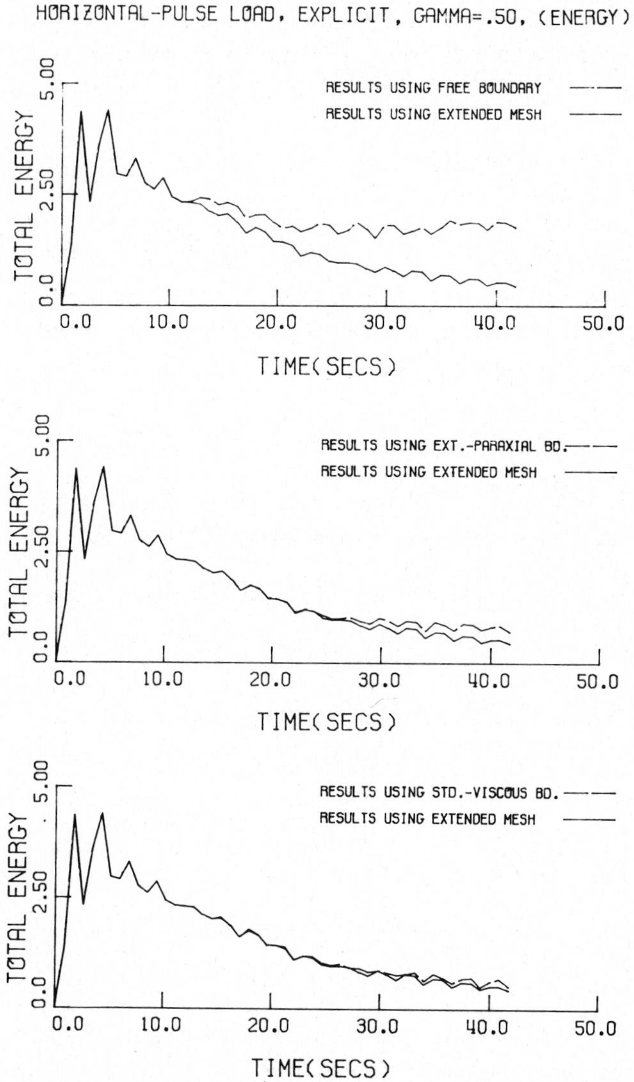

Fig. 4.10. Total energy in the interior region as a function of time (horizontal-pulse loading, undamped case).

ture on each element.[1] The effects of this reflected, high-frequency energy for the case of the extended-paraxial boundary become apparent after $t = 30$ sec. The paraxial boundary is responsible for a slightly larger error than is the viscous mechanism.

One technique for filtering high-frequency noise is to apply some numerical damping, which selectively attenuates the high-frequency motion of the mesh. We implemented this idea in order to eliminate the noise introduced by the silent boundaries. In Figures 4.11 to 4.13, γ is set equal to 0.55, with $\beta = 0.276$. These

[1]We should clarify why the energy level, even with the use of free boundary, declines in Figure 4.10. Some of the energy resides in a narrow band of elements, the 'exterior' zone shown in Figure 4.3. Also, if numerical damping is present, then the high-frequency energy is reduced. In Figure 4.10 where no damping is used, the total energy in the free-boundary system is conserved after $t = 20$ sec.

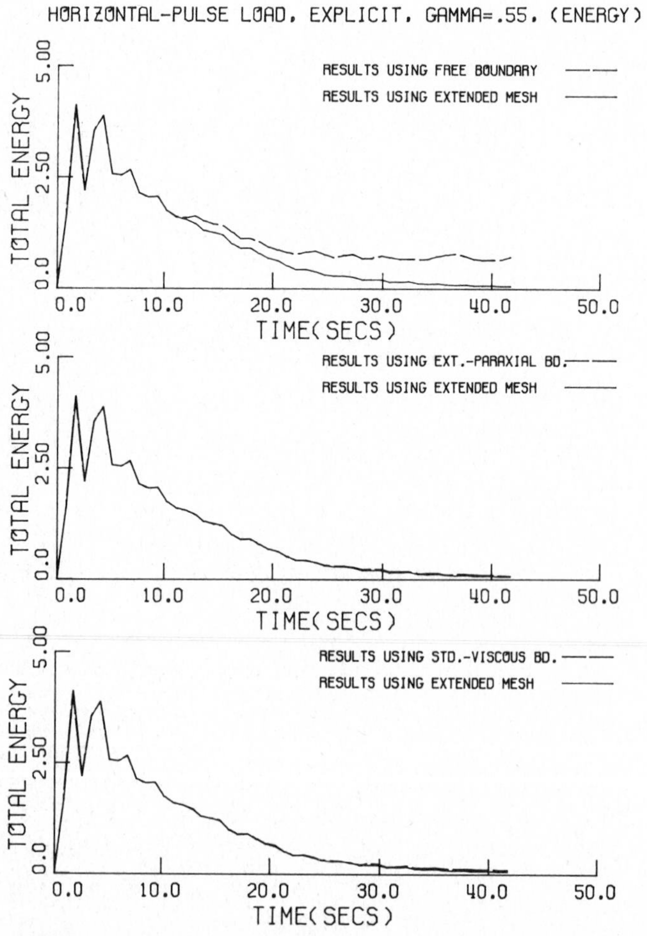

Fig. 4.11. Total energy in the interior region as a function of time (horizontal-pulse loading, damped case).

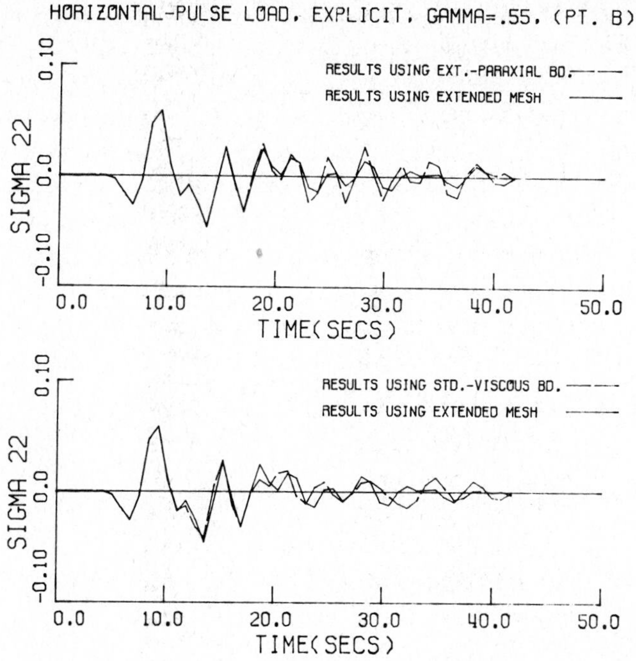

Figure 4.12. Stresses, σ_{22}, at point B as a function of time (horizontal-pulse loading, damped case).

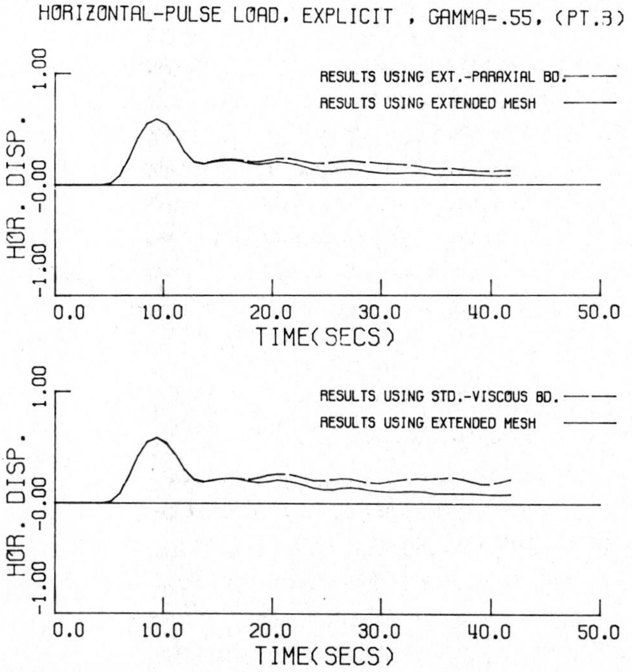

Fig. 4.13. Horizontal displacements at point B as a function of time (horizontal-pulse loading, damped case).

parameters introduce some algorithm damping, which mainly affects the higher modes. Except for this change, Figure 4.11 depicts the same energy graphs as in Figure 4.10. The beneficial effect of the damping is clearly evident. It not only reduces the total energy for long times, it completely removes the high-frequency errors in the total energy, including those associated with reflections from the silent boundaries.

On a local level, the effect of numerical damping is demonstrated in Figure 4.12. The spurious, high-frequency oscillations are significantly reduced compared to those of Figure 4.9. The numerical damping is especially effective in reducing the high-frequency reflections from the paraxial boundary.

For the lower modes, that is, the longer wavelength components of the response, the algorithmic damping we applied has little effect, and the conclusions for the undamped case are also valid here. For example, in Figure 4.13, we plot the horizontal displacements near the side boundary (point B). As a function of time, the displacement curve appears to be a little smoother than does that for the undamped case in Figures 4.4 and 4.5, but the amplitudes are almost identical. The extended-paraxial boundary again enjoys a slight advantage over the viscous boundary.

4.4.C. Shear waves

Shearing motion dominates the solution at point *C*, near the bottom of the mesh. Figures 4.14 and 4.15 present the undamped shearing stresses and horizon-

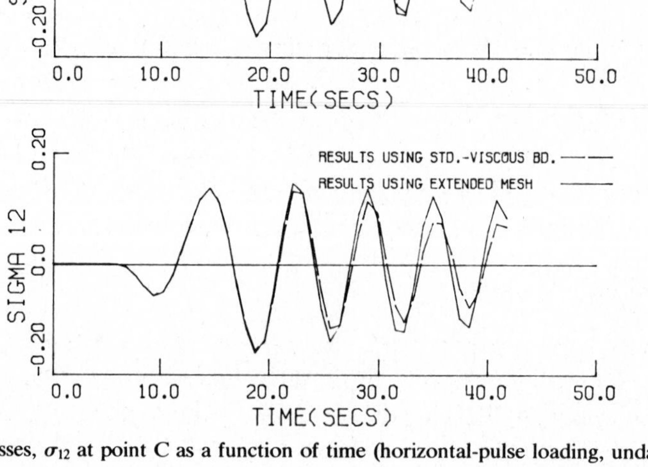

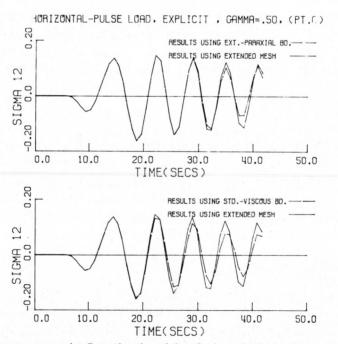

Fig. 4.14. Stresses, σ_{12} at point C as a function of time (horizontal-pulse loading, undamped case).

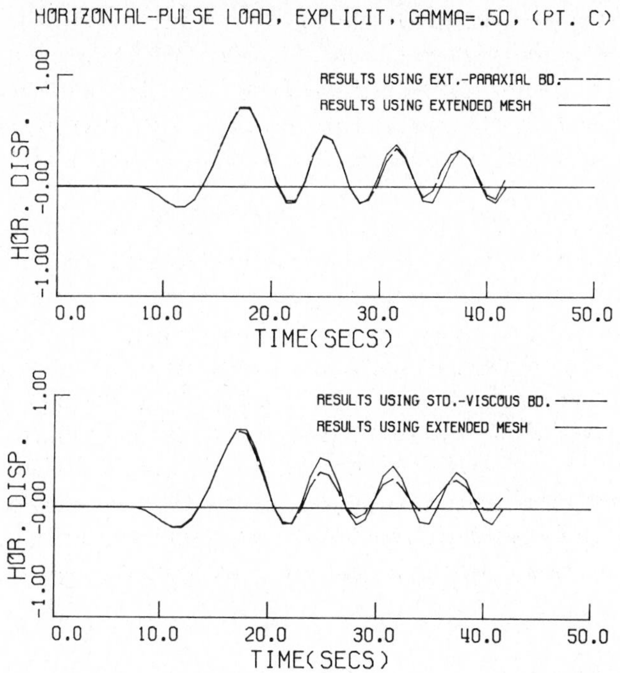

Fig. 4.15. Horizontal displacements at point C as a function of time (vertical-pulse loading, undamped case).

tal displacements at the lower measuring point. The figures indicate that the viscous boundary causes a greater reduction in the shearing amplitudes than is shown by the extended mesh. These results, however, do provide evidence of the boundaries' ability to transmit shear waves, as well as dilatational waves.

4.5. *Vertical-pulse loading*

The computations of the response to the vertical loading were conducted without using numerical or material damping. The interior equations were solved implicitly, and DT = 0.9 sec. The material constants and the method of loading are the same as those above. These results, which are presented in [60], confirm the conclusions drawn from the horizontal loading case. The fact that we used the implicit algorithm in this case did not seem to significantly affect the behavior of the silent boundaries.

4.6. *Vertical half-sine pulse*

Figures 4.16 and 4.17 illustrate the progress of a half-sine pulse applied over 5 elements in the quarter mesh. The load is applied for one step and is then removed. $c_s = 0.5345$ units/sec, $c_d = 1$ unit/sec, DT = 0.9 sec, $\gamma = 0.55$, and $\beta = 0.276$. The quarter mesh is drawn in Figure 4.2 and the boundary conditions are

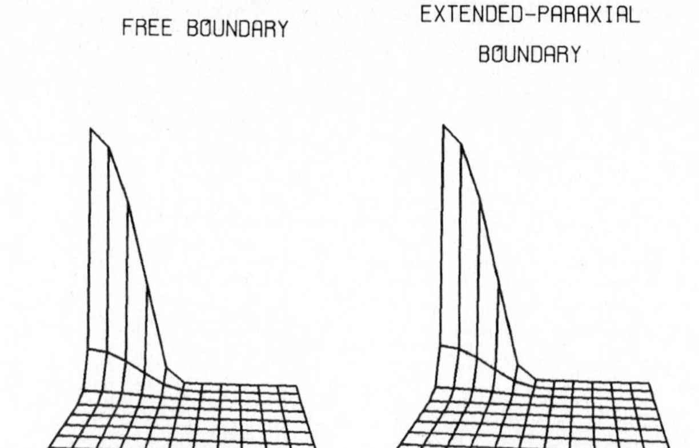

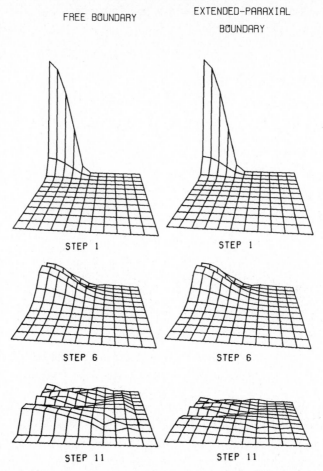

Fig. 4.16. Absolute amplitudes of the velocities plotted as ordinates on a planar, two-dimensional mesh. Vertical half-sine pulse. (Continued on Figure 4.17).

shown in Figure 4.3. This problem is identical to the previous, vertical-pulse examples, except for the half-sine loading. The purpose of this calculation is to show, qualitatively, the benefits of using the silent boundaries. These are used under conditions of relative mesh size which are more representative of practical applications. However, the pulse is 'rapidly' applied, generating significant high frequency motion.

In Figures 4.16 and 4.17 the velocity amplitudes, v_a, are plotted over the two-dimensional mesh; v_a is defined by

$$v_a = \sqrt{v_{1a}^2 + v_{2a}^2}\,, \tag{4.1}$$

where v_{1a} and v_{2a} are the velocity components of node a. For the free boundary, there is added in the plots an outer ring of undeformed elements to clarify the

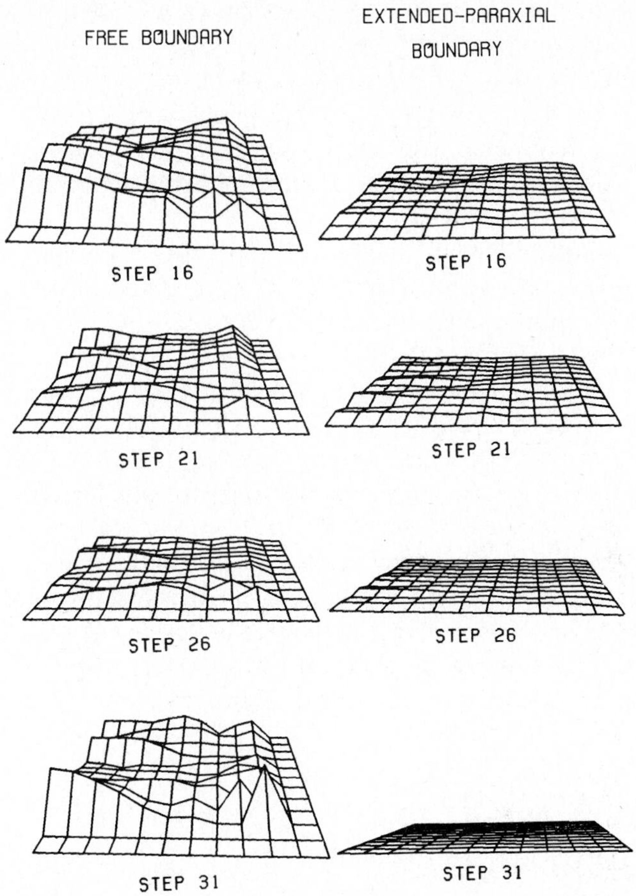

Fig. 4.17. Absolute amplitudes of the velocities plotted as ordinates on a planar, two-dimensional mesh. Vertical half-sine pulse.

portrayal of the amplitudes. These three-dimensional plots illuminate the qualitative behavior of the free and extended-paraxial boundaries. The waves effectively propagate out of the system with the silent boundary, whereas energy is trapped within the mesh when it has only a free boundary.

4.7. *Discussion of nonlinear waves*

The analyses of Section 2 and the above results are strictly valid only for waves in linear, elastic media. The interior of the mesh may be governed by nonlinear equations, but the region adjacent to the silent boundary must be linear. This assumption, that the governing equations are linear on the outer fringes of the computational mesh, is appropriate for many different problems. For example, in soil-structure interaction analysis the soil's strongly nonlinear behavior is mainly confined to an area near the structure. The wave motion which emanates from the interaction zone and propagates to the outer boundaries often can be represented adequately by linear models of the soil.

In some problems, however, the nonlinearity of the wave motion at locations 'far' from the wave source cannot be ignored. In these cases a linear wave may be followed by slower-traveling, nonlinear waves. The silent boundaries, which are designed for only one set of wave speeds, may not effectively transmit the slower-moving waves.

The silent boundaries' ability to absorb waves depends on the parameters, ac_d and bc_s, in equation (2.42) of Section 2. The standard-viscous and extended-paraxial boundaries set both a and b equal to unity. The wave-reflection theory in Section 2 and the numerical examples in this chapter both indicate that $a = b = 1$ is the foremost choice for linearly-elastic waves. For waves traveling at a slower speed, a or b should be somewhat less than 1.

Our experience with the unified viscous boundary indicates, however, that the viscous boundaries are relatively insensitive to the parameters a and b. For the unified boundary, a ranges from 0.959 to 1.011, b lies between 0.740 and 0.773. The numerical results, using either the standard- or the unified-viscous boundary, are nearly identical. This finding implies that the standard-viscous boundary might effectively absorb waves traveling at widely varying speeds. In fact, in numerical tests with a varying from 0.6 to 1.3, we found that the use of the unified-viscous boundaries instead of the standard-viscous boundary produces virtually no differences in the stresses and vertical displacements [60].

4.8. *Rayleigh-wave example*

The previous examples illustrate the boundary schemes' ability to transmit high-frequency, body-wave pulses. In this section a Rayleigh wave is used to excite the system. The purpose of this test is twofold. We wish, first, to subject the boundary to a steady-state motion. A certain amount of reflection was observed for the pulse loadings, and this raises the question of whether the errors accumulate as the loading continues. Second, the ability of the silent boundaries to transmit Rayleigh waves need to be evaluated. The analyses of Section 2.6 indicate that the extended-paraxial boundary may be effective in this case. The efficiency of the standard-viscous boundary is uncertain.

It often happens in soil-structure interaction analyses that some of the energy propagating from the interaction zone is in transition from body waves to Rayleigh waves. Wave motions, including shear and dilatational components, are superimposed, gradually forming Rayleigh waves as they move out along the surface. In most practical instances a computational mesh will not extend far enough for Rayleigh waves to form completely. Therefore, the measuring of the reflections of this transitory motion is of interest. Similar comments apply to Love waves, where layered media are considered.

Figures 4.18 and 4.19 illustrate the test problem in which plane-strain elasticity

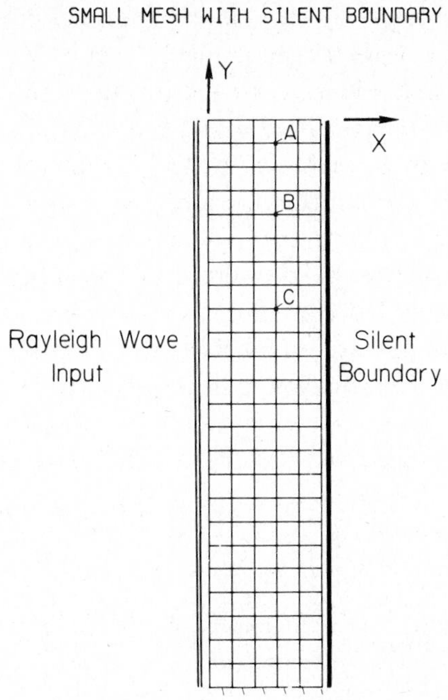

Fig. 4.18. Finite-element mesh with absorbing boundary used for the Rayleigh-wave loading.

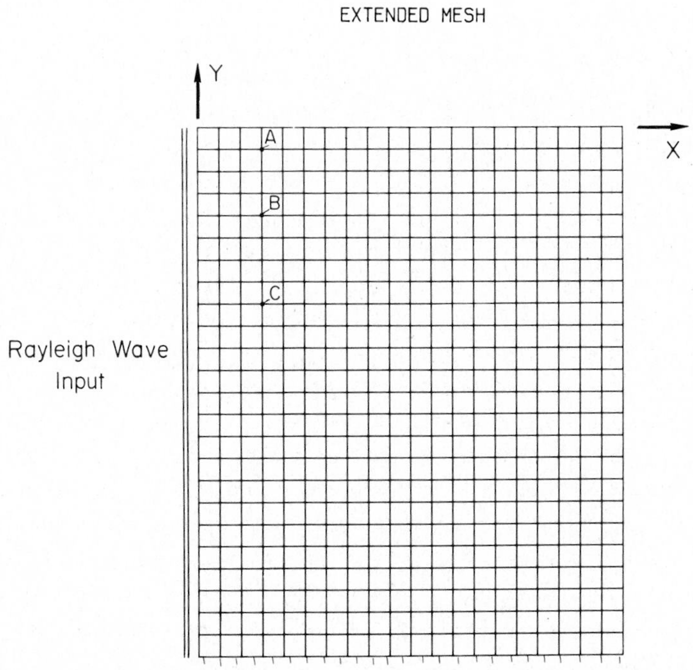

Fig. 4.19. Extended, finite-element mesh with free boundary used for the Rayleigh-wave loading.

is used. The mesh is initially at rest. The horizontal and vertical displacements along the left side of the mesh are prescribed, for all subsequent time, according to a known Rayleigh-wave solution. The mesh is two wavelengths deep, and the bottom nodes are fixed. Due to reflections from the bottom boundary, it has been recommended [56] that the mesh depth extend to three or four wavelengths. In our case these reflections will propagate to the outer areas of the extended mesh, so that the region adjacent to the input will not be significantly affected. For the small mesh in Figure 4.18, energy from the bottom can propagate to the surface, however, if it reflects from the right side.

The excitation on the left generates transient waves at first, then the motion approaches steady state. The displacements on the left are prescribed as:

$$u = D[\exp(0.8475\,\kappa_R y) - 0.5773\exp(0.3933\,\kappa_R y)]\sin(\kappa_R c_R t)\,,$$

$$w = D[-0.8475\exp(0.8475\,\kappa_R y) + 1.4678\exp(0.3933\,\kappa_R y)]\cos(\kappa_R c_R t)\,. \tag{4.2}$$

Equations (4.2) represent the Rayleigh-wave solution for $\nu = 0.25$ [2], $\omega = 0.2781$ radians/sec, L_R = Rayleigh wavelength = 12 units, $\kappa_R = 0.5236$, c_R = Rayleigh-wave speed = 0.5312 units/sec, $c_s = 0.5774$ units/sec, $c_d = 1$ unit/sec, and

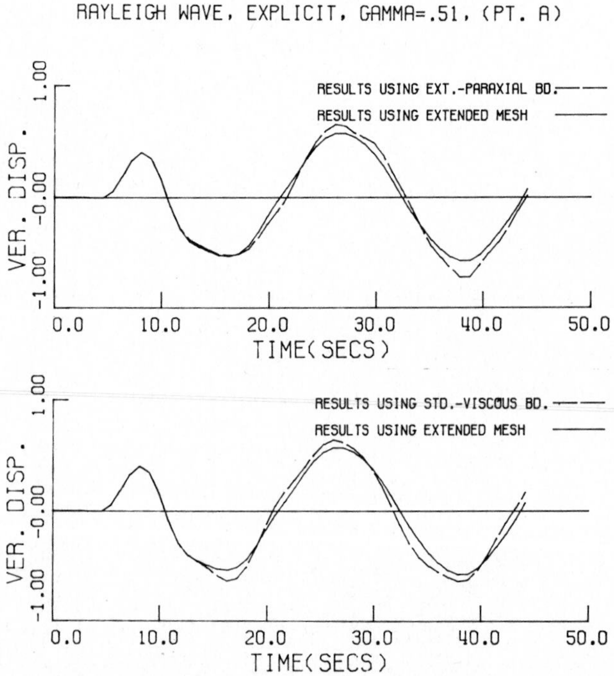

Fig. 4.20. Vertical displacements at point A as a function of time (Rayleigh-wave loading, slight damping).

1 unit = length of one element. $\gamma = 0.51$ and $\beta = 0.255$, so there is a negligible amount of numerical damping present in the system. The equations were solved explicitly with DT = 0.9 sec. Energy first strikes the right boundary of the small mesh at $t = 6$ sec; Rayleigh-wave components follow shortly after.

We represent, in Figures 4.20 and 4.21, several comparisons among the various boundaries. Initial transient motion is evident in the first 10 seconds, and then the response becomes more nearly periodic. It has not quite reached steady state.

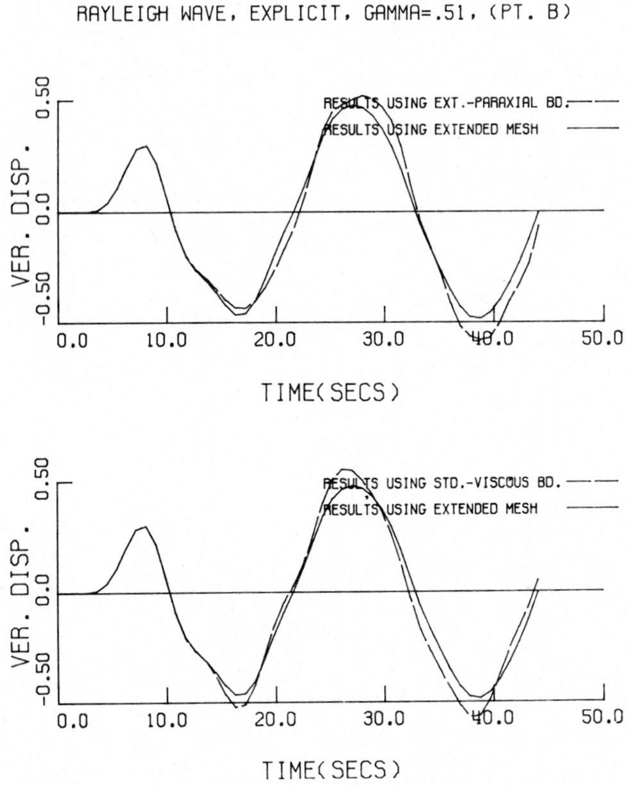

Fig. 4.21. Vertical displacements at point B as a function of time (Rayleigh-wave loading, slight damping).

The silent boundaries generally prevent the reflection of energy. The vertical displacements in these figures exhibit a fairly close agreement between the extended-mesh results and those from using a silent boundary. Other measurements of stresses and displacements, which were taken at points A, B, and C but are not displayed here, suggest that the same or better agreement exists than that found in Figures 4.20 and 4.21. The distortions caused by a free boundary are most visible at point A near the surface.

Figures 4.20 and 4.21 illustrate most of the poorest agreement between the extended-paraxial boundary and the extended mesh. The largest discrepancies caused by the standard-viscous boundary are shown in Figure 4.22, where the phase of the response appears to have been shifted.

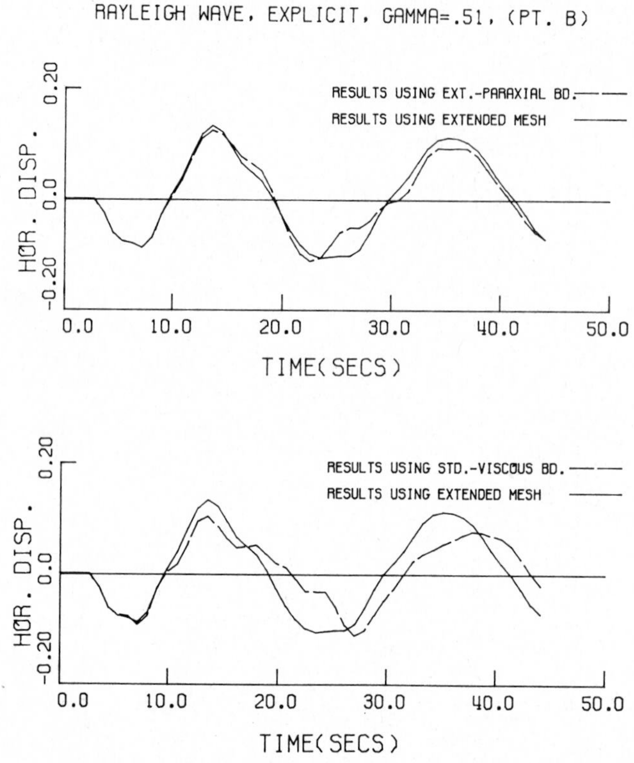

Fig. 4.22. Horizontal displacements at point B as a function of time (Rayleigh-wave loading, slight damping).

The Rayleigh mode shapes can also be used to estimate the accuracy of the transmitting boundaries. In these graphs presented in Figures 4.23 and 4.24, the solid lines represent the Rayleigh mode shape that excites the system. The dashed-line curves are the resulting displacements at $x = 3$ when $t = 27$ sec. These profiles were calculated with four conditions at the right side of the mesh: (a) extended mesh, (b) free boundary, (c) extended-paraxial boundary, and (d) standard-viscous boundary.

In the ideal case the profile of the displacements at $x = 3$ will duplicate the input motion at $x = 0$. However, due to the presence of transient waves and discretization errors, the displacement profiles differ somewhat. As can be perceived in Figures 4.23 and 4.24, the extended mesh and the silent boundaries each generate displacement configurations that are similar to the input.

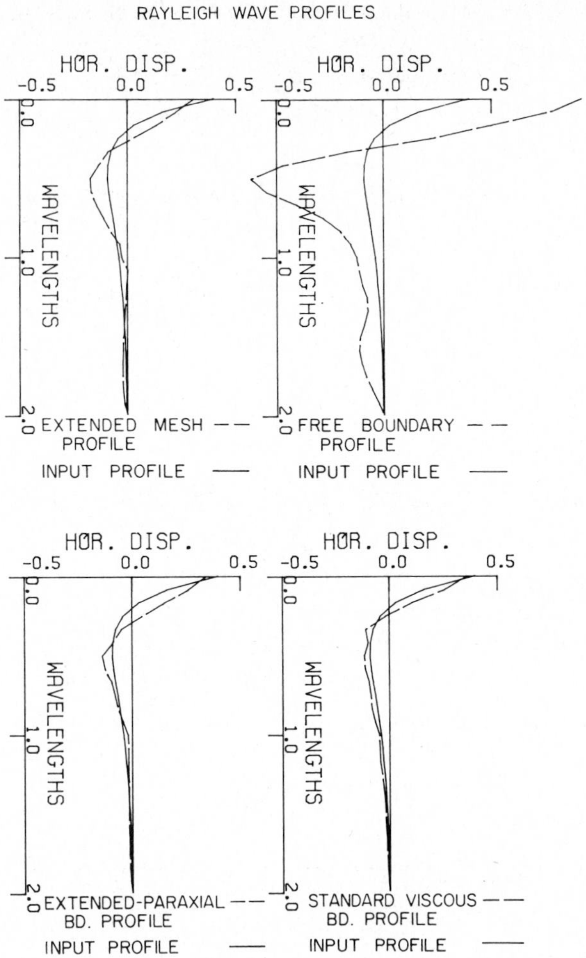

Fig. 4.23. Horizontal displacements as a function of depth ($x = 3$ units, $t = 27$ sec, Rayleigh-wave input).

Overall, the boundary methods do not induce large distortions in the response, such as those that are observed in the free boundary case. The silent-boundary errors are of the same magnitude as those introduced by the discretization and by the transient motion. The two methods, unified-paraxial and standard-viscous, are again comparable in accuracy. If we consider all the data accumulated in this problem, including deformed mesh plots not illustrated here, the paraxial results are slightly closer to those of the extended mesh.

As we mentioned in Section 1, several authors have pointed out that the viscous boundary may be ineffective in the case of Rayleigh waves. Our experience, at least with this example, is different; the viscous boundary appeared to function for Rayleigh waves about as well as it does for body waves.

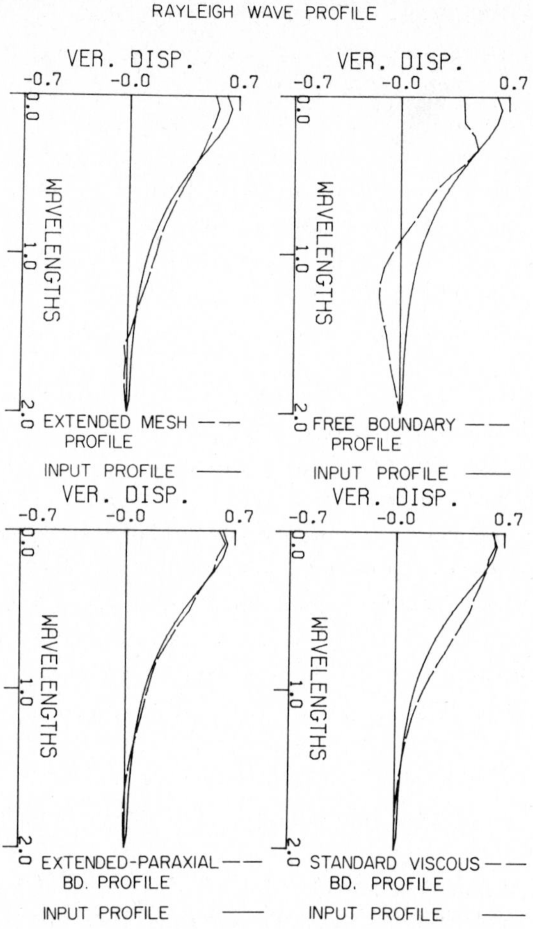

Fig. 4.24. Vertical displacements as a function of depth ($x = 3$ units, $t = 27$ sec, Rayleigh-wave input).

5. Concluding remarks

We have restricted this presentation to the more easily implemented silent boundaries which can be directly employed in transient, time-domain analysis, and which would be serviceable for many problems involving nonlinear materials. Two of the boundary methods, paraxial and viscous, which have these qualities, were analyzed in detail.

The major emphasis centered upon the adaptation, development, and testing of the various boundary approaches as they apply to finite element calculations. As the first result, general derivation of the paraxial equations for plane, harmonic waves led to questions of stability and to a modification of the equations' original form. The revised paraxial equations were then compared to two viscous-boundary proposals. It was noted that the standard-viscous boundary, in essence, embodies the first-order terms of the paraxial equations. A wave-reflection analysis indicated a distinct potential superiority of the paraxial approach. This analysis implies that the use of the unified parameters as suggested in [78] does not improve the efficiency of the boundary. From this theoretical standpoint, the modified-paraxial boundary reflects waves of the smallest amplitudes, but the standard- and unified-viscous boundaries are also fairly effective.

In Section 3 it was indicated that a straightforward, finite-element implementation of the paraxial equations does not lead to a practicable transmitting boundary. Certain alterations in the numerical procedures are needed to upgrade the boundary's accuracy.

The test examples presented in Section 4 furnish most of the data for an evaluation and comparison of the various boundaries. In the problems we studied, all of the silent boundaries produced from adequate to very accurate results. There are a number of features which should be emphasized. First, the viscous boundaries cause a permanent, 'residual' rigid displacement of the mesh. This motion is, perhaps, 10% of that which results from the use of a free boundary. The extended-paraxial boundary largely eliminates this residual. On the other hand, the viscous boundaries do effectively absorb the correct amount of energy from the system. In most cases, the energy curves which result from extended-mesh calculations coincide exactly with those resulting from a viscous boundary. The extended-paraxial boundary is also competent in removing energy from the system, with the exception of eradicating noise from the solution.

A second feature of all the silent boundaries is their relative inability to remove high-frequency, node-to-node oscillations. The extended-paraxial boundary is worse in this capacity because it aggravates this noise and sends it to the interior. One solution to this problem is to apply slight numerical damping to the algorithm. In the problem studied it was found that the setting of $\gamma = 0.55$ eliminates most of the numerical oscillations, without distorting the solution at lower frequencies, and allows the silent boundaries to operate more effectively. Our experience indicates that $\gamma = 0.51$ does not induce enough damping into the system ($\gamma = 0.50$ is the undamped case).

A third conclusion is that the standard- and unified-viscous boundaries perform almost identically in all ways. Therefore, it would seem unnecessary to modify the parameters, a and b, in order to improve the boundary's behavior (the two boundaries differ only in the selected values of a and b). In fact, in one example we tested, the doubling of the parameter a causes little change in the absorption character of the viscous boundary. This insensitivity of the viscous boundary suggests that it may be suitable for nonlinear waves.

Fourth, the performance of the silent boundaries varies with the parameters being measured. By and large, the extended-paraxial boundary hold a slight advantage in accuracy over the viscous schemes. Thus, as implemented in our studies [60], the advantages in accuracy of the paraxial approach indicated by the analysis of wave reflections were reduced when comparisons were made for example calculations.

Finally, the implementation of the two silent boundaries, as considered, is inexpensive. Both the viscous and paraxial damping matrices can be easily calculated by hand, in terms of the element dimensions, h^x and h^z. Hence, the generation of these expressions entails virtually no costs on the computer. A slight expense is incurred in the formation and assembly of the paraxial stiffness, and in the solution of the paraxial degrees of freedom. The cost of this latter computation, always being done explicitly, is not significant. Overall, the extended-paraxial scheme is slightly more expensive to use than the viscous boundary.

In summary, the silent boundaries studied herein possess the following characteristics:

- The extended-paraxial boundary is founded upon a mathematical theory which indicates its capability to transmit wave energy. In this context the viscous boundary can be considered as a first-order paraxial boundary.
- The boundary schemes are easily implemented and exact a minimal computational expense.
- The extended-paraxial boundary sometimes projects numerical noise into the system, but this tendency is controlled by a slight amount of numerical damping.
- The extended-paraxial boundary enjoys some advantage in accuracy over the viscous boundaries as measured by analysis of wave reflections. In our implementation, however, we were not able to fully mobilize this potential advantage.
- They offer a broad scope in that they can be applied to Rayleigh waves and, presumably, to anisotropic materials.
- They can be applied to many problems in which nonlinear, material effects are important.
- They do not adversely affect the numerical stability properties of the family of algorithms that we tested.

In the course of our investigation we ascertained that the viscous boundary performs reasonably well and that it would be a suitable transmitting boundary in many applications. It is widely believed that this boundary acquits itself poorly when confronted by Rayleigh waves, but in the one example that we studied, the boundary seemed to absorb the waves to an acceptable degree, although it was somewhat less accurate than the extended-paraxial approach.

Acknowledgments

The results presented in this paper are abstracted from the senior author's doctoral thesis [60]. The financial support of the National Science Foundation and the California Institute of Technology Earthquake Research Affiliates during the course of the thesis work is gratefully acknowledged. The authors also appreciate the assistance to Projessor T.J.R. Hughes in the conduct of these studies and the preparation of this chapter.

References

General references

[1] J.F. Claerbout, *Fundamentals of Geophysical Data Processing* (McGraw-Hill, New York, 1976).
[2] W.M. Ewing, W.S., Jardetsky and F. Press, *Elastic Waves in Layered Media* (McGraw-Hill, New York, 1957).
[3] H. Lamb, *Philosophical Transactions of the Royal Society* (*London*) **203**, 1–42 (1904).
[4] J. Miklowitz, *The Theory of Elastic Waves and Waveguides* (North-Holland, Amsterdam–New York–Oxford, 1978).
[5] B. Noble, *Applied Linear Algebra* (Prentice-Hall, Englewood Cliffs, NJ, 1969).
[6] C.C. Mow and Y.H. Pao, The Diffraction of Elastic Waves and Stress Concentrations, RAND Report, R-482-PR (1971).
[7] I.A. Viktorov, *Rayleigh and Lamb Waves* (Plenum Press, New York, 1967).

Finite Element Methods

[8] A. Brooks and T.J.R. Hughes, Streamline-Upwind/Petrov–Galerkin Methods for Advection Dominated Flows, *Proc. Third Internat. Conf. Finite Element Methods in Fluid Flow*, Banff, Canada (June 1980).
[9] I. Christie, D.F. Griffiths, A.R. Mitchell and O.C. Zienkiewicz, Finite Element Methods for Second Order Differential Equations with Significant First Derivatives, *Internat. J. Numerical Methods in Eng.* **10**, 1389–1396 (1976).
[10] D.F. Griffiths and J. Lorenz, An Analysis of the Petrov–Galerkin Finite Element Method, *Computer Methods in Applied Mechanics and Eng.* **14**, 39–64 (1978).
[11] J.C. Heinrich, F.S. Huyakorn, O.C. Zienkiewicz and A.R. Mitchell, An 'Upwind' Finite-Element Scheme for Two-Dimensional Convective Transport Equation, *Internat. J. Numerical Methods in Eng.* **11**, 131–143 (1977).
[12] H.M. Hilber, T.J.R. Hughes and R.L. Taylor, Improved Numerical Dissipation for Time Integration Algorithms in Structural Dynamics, *Earthquake Eng. and Structural Dynamics* **6**, 283–292 (1977).
[13] T.J.R. Hughes, Stability, Convergence, and Growth and Decay of Energy at the Average Acceleration Method in Nonlinear Structural Dynamics, *Computers and Structures* **6**, 313–324 (1976).
[14] T.J.R. Hughes, Equivalence of Finite Elements for Nearly-Incompressible Elasticity, *J. Appl. Mechanics* **44**, Series E, 181 (1977).
[15] T.J.R. Hughes, A Simple Scheme for Developing 'Upwind' Finite Elements, *Internat. J. Numerical Methods in Eng.* **12**, 1359–1369 (1978).
[16] T.J.R. Hughes and A. Brooks, A Multidimensional Upwind Scheme with no Crosswind Diffusion, in AMD **34**, *Finite Element Methods for Convection Dominated Flows*, T.J.R. Hughes, ed. (ASME, New York, 1979).
[17] T.J.R. Hughes, M. Cohen and M. Haroun, Reduced and Selective Integration Techniques in the Finite Element Analysis of Plates, *Nuclear Eng. and Design* **46**, 203–222 (1978).
[18] T.J.R. Hughes and W.K. Liu, Implicit-Explicit Finite Elements in Transient Analysis: Stability Theory, *J. Appl. Mechanics* **45**, 371–374 (1978).
[19] T.J.R. Hughes and W.K. Liu, Implicit-Explicit Finite Elements in Transient Analysis: Implementation and Numerical Examples, *J. Appl. Mechanics* **45**, 375–378 (1978).

[20] T.J.R. Hughes, K.S. Pister and R.L. Taylor, Implicit-Explicit Finite Elements in Nonlinear Transient Analysis, *Computer Methods in Applied Mechanics and Eng.* **17/18**, 159–182 (1979).

[21] R.D. Krieg and S.W. Key, Transient Shell Response by Numerical Time Integration, *Internat. J. Numerical Methods in Eng.* **7**, 273–286 (1973).

[22] R.L. Kuhlemeyer and J. Lysmer, Finite Element Method Accuracy for Wave Propagation Problems, *Journal of the Soil Mechanics and Foundations Div. ASCE*, 421–426 (May 1973).

[23] R. Mullen and T. Belytschko, Dispersion Analysis of Finite Element Semidiscretizations of the Two-Dimensional Wave Equation, Preprint, Depth. of Civil Engineering, The Technological Institute, Northwestern University, Evanston, IL 60201 (1980).

[24] O.C. Zienkiewicz, *The Finite Element Method*, 3rd ed. (McGraw-Hill, London, 1977).

Solid-structure interaction analysis – finite element approach

[25] J.C. Anderson, Seismic Response Effects on Embedded Structures, *Bull. Seismological Soc. of Amer.* **62**, 177–194 (1972).

[26] W.W.H. Chen, and M. Chatterjee, Response of Foundation Embedded in Layered Media, *Seventh Conference on Electronic Computation*, St. Louis, August 6–8 (1979).

[27] S.P. Dasgupta and N.S.V. Kameswara Rao, Dynamic Response of Strip Footing on an Elastic Halfspace, *Internat. J. Numerical Methods in Eng.* **14**, 1597–1612 (1979).

[28] S.M. Day, Finite Element Analysis of Seismic Scattering Problems, Ph.D. Thesis, University of California, San Diego (1977).

[29] A. Gomez-Masso, J. Lysmer, J.C. Chen and H.B. Seed, Soil-Structure Interaction in Different Seismic Environments, Earthquake Engineering Research Center, Univ. of California, Berkeley, Report No. UCB/EERC-79/18 (August 1979).

[30] A.H. Hadjian, J. Luco and N.C. Tsai, Soil-Structure Interaction: Continuum or Finite Element, *Nuclear Eng. and Design* **31**, 151–167 (1975).

[31] W.A. Haupt, Surface-Waves in Nonhomogeneous Half-Space, *Dynamical Methods in Soil and Rock Mechanics Proceedings*, Karlsruhe, Germany, September 5–16 (1977).

[32] I.M. Idriss and K. Sadigh, Seismic SSI of Nuclear Power Plant Structures, *J. Geotechnical Div. ASCE* **102** (1977).

[33] J. Isenberg, Interaction between Soil and Nuclear Reactor Foundations during Earthquakes, proposed for The Research Foundation of the University of Toledo, Contract No. AT(40-1)-3822, June 10 (1970).

[34] J. Isenberg and S.A. Adham, Interaction of Soil and Power Plants in Earthquakes, *J. Power Div. ASCE* **98**, 273–291 (1972).

[35] J. Isenberg, D.K. Vaughan and I. Sandler, Nonlinear Soil-Structure Interaction, report prepared by Weidlinger Associated for the Electric Power Research Institute, (Dec. 1978).

[36] J.J. Johnson and R.P. Kennedy, Earthquake Response of Nuclear Power Facilities, *J. Energy Div. ASCE* **105**, 15–32 (1979).

[37] W.B. Joyner, A Method for Calculating Nonlinear Seismic Response in Two Dimensions, *Bull. Seismological Soc. of Amer.* **65**, 1337–1357 (1975).

[38] W.B. Joyner and A.T.F. Chen, Calculation of Nonlinear Ground Response in Earthquakes, *Bull. Seismological Soc. of Amer.* **65**, 1315–1336 (1975).

[39] E. Kausel, J.M. Roesset and G. Waas, Dynamic analysis of Footings on Layered Media, *J. Eng. Mechanics Div. ASCE* **101**, 679–693 (1975).

[40] V.C. Liang, Dynamic Response of Structures in Layered Solids, Ph.D. Thesis, MIT, (Jan. 1974).

[41] J.E. Luco, A.H. Hadjian and H.D. Bos, The Dynamic Modeling of the Half-Plane by Finite Elements, *Nuclear Eng. and Design* **31**, 184–194 (1975).

[42] J.E. Luco and A.H. Hadjian, Two-Dimensional Approximations to the Three-Dimensional Soil-Structure Interaction Problem, *Nuclear Eng. and Design* **31**, 195–203 (1975).

[43] J. Lysmer and L.A. Drake, A Finite Element Method for Seismology, in: *Methods of Computational Physics*, Vol. 11, Chapter 6 (Academic Press, New York, 1972).

[44] J. Lysmer, T. Udaka, C-F. Tsai and H.B. Seed, FLUSH – A Computer Program for Approximate 3-D Analysis of Soil-Structure Interaction Problems, Earthquake Engineering Research Center, University of California, Berkeley, Report No. EERC 75-30 (Nov. 1975).

[45] J. Lysmer, Analytical Procedures in Soil Dynamics, Earthquake Engineering Research Center, University of California, Berkeley, Report No. EERC 78-29 (Dec. 1978).
[46] G.A. Nestor, Soil-Structure Interaction Analyses Using the Finite Element Method, prepared for the National Science Foundation, University of California, Los Angeles (June 1977).
[47] W.S. Tseng and N.C. Tsai, Soil-Structure Interaction for Transient Loads due to Safety Relief Valve Discharges, *Nuclear Eng. and Design* **45**, 251–259 (1978).
[48] P.B. Schnabel, J. Lysmer and H.B. Seed, SHAKE – A Computer Program for Earthquake Response Analysis of Horizontally Layered Sites, Earthquake Engineering Research Center, Report No. EERC 72-12, University of California, Berkeley (Dec. 1972).
[49] H.B. Seed and J. Lysmer, Soil-Structure Interaction Analysis by Finite Elements – State-of-Art, *Nuclear Eng. and Design* **46**, 349–365 (1978).
[50] K. Ukaji, Analysis of Soil-Foundation-Structure Interaction During Earthquakes, Ph.D. Thesis, Stanford University (1975).
[51] C.M. Urlich and R.L. Kuhlemeyer, Coupled Rocking and Lateral Vibrations of Embedded Footings, *Canadian Geotechnical Jour.* **10**, 145–160 (1973).
[52] A.K. Vaish and A.K. Chopra, Earthquake Finite Element Analysis of Structure-Foundation Systems, *J. Eng. Mechanics Div. ASCE* **100**, 1101–1116 (1974).
[53] S. Valliappan, J.J. Favaloro and W. White, Dynamic Analysis of Embedded Footings, *J. Geotechnical Div. ASCE* **103**, 129–133 (1977).
[54] G. Waas and J. Lysmer, Vibrations of Footings Embedded in Layered Media, Proc. Water-Ways Experimental Station, Symposium on Applications of the Finite Element Method in Geotechnical Engineering, U.S. Army Engineers, Waterways Experimental Station, Vicksburg, MS (1972).
[55] W.J. Weaver, G.E. Brandow and K. Hoeg, Three Dimensional Soil-Structure Response to Earthquakes, *Bull. Seismological Soc. of Amer.* **63**, 1041–1056 (1973).
[56] G. Wojcik, Personal Communication, Weidlinger Associates, 3000 Sand Hill Rd., Bldg. 4, Suite 245, Menlo Park, CA 94025.

Silent boundaries

[57] G.Y. Baladi, Ground Shock Calculation Parameter Study; Effect of Various Bottom Boundary Conditions, Tech. Report S-71-4, Report 2, U.S. Army Engineer Waterways Experimental Station, CE, Vicksburg, MS (1972).
[58] A. Castellani, Boundary Conditions to Simulate an Infinite Space, *Meccanicca* **9**, 199–205 (1974).
[59] J. Clayton and B. Engquist, Absorbing Boundary Conditions for Acoustic and Elastic WAve Equations, *Bull. Seismological Soc. of Amer.* **67**, 1529–1541 (1977).
[60] M. Cohen, Silent Boundary Methods for Transient Wave Analysis, Earthquake Engineering Laboratories Report No. EERL 80 (Sept. 1980).
[61] P.A. Cundall, R.R. Kunar, P.C. Carpenter and J. Marti, Solution of Infinite Dynamic Problems by Finite Modelling in the Time Domain, *Proc. 2nd Internat. Conf. on Applied Numerical Modelling*, Madrid, Spain (1978).
[62] J. Dominguez, Dynamic Stiffness of Rectangular Foundations, MIT Report, Publication No. R78-20, Order No. 626 (Aug. 1978).
[63] B. Engquist and A. Majda, Absorbing Boundary Conditions for the Numerical Simulation of Waves, *Mathematics of Computation* **31**, 629–651 (1977).
[64] B. Engquist and A. Majda, Radiation Boundary Condition for Acoustic and Elastic Wave Calculations, *Communications on Pure and Appl. Math.* **32**, 313–357 (1979).
[65] W.A. Haupt, Numerical Methods for the Computation of Steady-State Harmonic Wave Fields, *Dynamical Methods in Soil and Rock Mechanics Proceedings*, Karlsruhe, Germany, Sept. 5–16 (1977).
[66] J. Isenberg, Personal Communication, Weidlinger Associates, 3000 Sand Hill Rd., Bldg. 4, Suite 245, Menlo Park, CA 94025.
[67] K.K. Kunar and Rodriguez-Ovejero, Model with Nonreflecting Boundaries for Use in Explicit Soil-Structure Interaction Analyses, *Earthquake Eng. and Structural Dynamics* **8**, 361–374 (1980).
[68] J. Lysmer and R.L. Kuhlemeyer, Finite Dynamic Model for Infinite Media, *J. Eng. Mechanics Div. ASCE*, 859–877 (August 1969).

[69] J. Lysmer, Lumped Mass Method for Rayleigh Waves, *Bull. Seismological Soc. of Amer.* **60**, 89–104 (1970).

[70] J. Lysmer and G. Waas, Shear Waves in Plane Infinite Structures, *J. Eng. Mechanics* Div. ASCE, 85–105 (Feb. 1972).

[71] A.T. Matthews, Effects of Transmitting Boundaries in Ground Shock Computation, Tech. Report S-71-8, U.S. Army Engineer Waterways Experimental Station, Vicksburg, MS (Sept. 1971).

[72] A.R. Robinson, The Transmitting Boundary-Again, in: *Structural and Geotechnical Mechanics*, W.J. Hall, ed. (Prentice-Hall, Englewood Cliffs, NJ, 1976).

[73] J.M. Roesset and M. Ettouney, Transmitting Boundaries: A Comparison, *Internat. J. for Numerical and Analytical Methods in Geomechanics* **1**(2), (1977).

[74] J.M. Roesset and H. Scaletti, Boundary Matrices for Semi-Infinite Problems, *Third Engineering Mechanics Division Specialty Conference*, University of Texas at Austin, Sept. 17–19 (1979).

[75] M.N. Tseng and A.R. Robinson, A Transmitting Boundary for Finite-Difference Analysis of WAve Propagation in Solids, Ph.D. Thesis, University of Illinois at Urbana-Champaign (Nov. 1975).

[76] W.D. Smith, A Nonreflecting Plane Boundary for Wave Propagation Problems, *J. Computational Physics* **15**, 492–503 (1974).

[77] P.G. Underwood and T.L. Geers, Doubly Asymptotic Boundary-Element Analysis of Dynamic Soil Structure Interaction, DNA Report 4512T, Defense Nuclear Agency (SPSS), Washington, DC 20305 (March 1978).

[78] W. White, S. Valliappan and I.K. Lee, Unified Boundary for Finite Dynamic Models, *J. Eng. Mechanics Div. ASCE*, 949–964 (Oct. 1977).

[79] L. Zhen-peng, Y. Pai-puo and Y. Yi-fan, Feedback Effect of Low-Rise Buildings on Vertical Earthquake Ground Motion and Application of Transmitting Boundaries for Transient Wave Analysis, Institute of Engineering Mechanica, Academia Sinica, Harbin, China (Oct. 1978).

CHAPTER 8

Explicit Lagrangian Finite-Difference Methods*

W. HERRMANN and L.D. BERTHOLF

*Sandia National Laboratories***
Albuquerque, NM 87185, USA

*This work was supported by the US Department of Energy under Contract DE-AC04-76-DP00789.
**A US Department of Energy Facility.

Computational Methods for Transient Analysis
Edited by T. Belytschko and T.J.R. Hughes

1. Introduction

Exigencies of the Manhattan project generated demands for solutions to fluid flow and stress wave problems through the necessity to predict performance of designs which could not readily be tested. During the early 1950s, computer programs were developed at Los Alamos Scientific Laboratory and the Lawrence Livermore Laboratory which were capable of solving transient compressible fluid flow problems. These were based on much earlier theoretical works, which had as their main objective the construction of ordered sequences of approximations for the investigation of the properties of partial differential equations.

By the late 1950s, the ability to handle stress waves in elastic-plastic materials had been developed. Stress wave problems arising in a very wide variety of applications are routinely handled today by computer programs based directly on these methods. Typical programs of this type are HEMP [43], TOODY [35], PISCES [32], STEALTH [33], REXCO [29] and many others derived from them. The methods underlying them may be characterized as direct, explicit, second-order methods. While the basic finite-difference methods have undergone little change in the last two decades, important ancilliary developments have occurred which have greatly enhanced the usefulness of these programs.

All of the programs we discuss are Lagrangian, in which the computational grid distorts with the material. While distorted grids were considered from the beginning, Lagrangian codes are severely limited in handling problems of very large distortion, since the computational grid then becomes tangled and disordered. Two important developments of the 1960s alleviated this limitation; the development of slide lines, representing slip surfaces along which materials on either side may slide relative to each other, and the development of rezone methods, by means of which a grid which is becoming tangled may be remapped into a less distorted grid in order to allow the computation to proceed. These features are fundamental to the usefulness of Langrangian methods in production calculations of a wide variety of applications.

A second major line of advance has concerned the development of constitutive equations to realistically represent the dynamic behavior of real materials. Means of describing viscoelastic and viscoplastic materials, and of integrating the resultant equations, were developed during the 1960s. A major thrust of the 1970s has been the development of constitutive equations to describe phase changes, chemical reactions, dissociation, the accumulation of distributed damage and many other rate-dependent physical processes.

Although the basic finite-difference methods were developed a quarter of a century ago, many tests have shown that they are still among the most efficient in terms of computer resources used to provide a solution of given accuracy, as well as the most flexible in terms of handling complicated initial value problems.

The purpose of this discussion is to describe the methods underlying the class of dynamic Lagrangian computer programs which have been identified above. Since much of the recent work has involved the development of constitutive equations, a general tensor treatment of this topic is included. After preliminaries in which the mass, momentum and energy equations are summarized for reference and to establish the notation, the constitutive equations for a general material with internal state variables are presented and placed into a form suited to numerical integration. The governing equations are then expanded into physical components, leading to the full set of governing differential equations. Since very close relationships exist with explicit Eulerian codes and coupled thermal-stress codes, convective and heat flux terms are retained in these equations, although they are only used to show how the Lagrangian non-conductive methods may be extended to include them.

Finite-difference expressions for the differential equations are then developed. These involve replacing spatial partial derivatives by difference expressions, and then finding a suitable method to perform the integration in time. The discussion is limited to the methods used by the particular class of computer programs under consideration, although the treatment of the constitutive descriptions is quite general, and goes beyond most common implementations. For simplicity, the discussion is restricted to two-dimensional rectangular and cylindrical coordinates, although extension to three dimensions is obvious, and three-dimensional computer programs have been developed using them.

All of the programs we consider use an artificial viscosity to render first and second order discontinuities, that is, the ubiquitous shock and acceleration waves that are a distinguishing feature of non-linear stress wave propagation, into steep but continuous waves. The various forms of artificial viscosity and damping which are in common use are discussed. Finally a discussion is given of sliding interface and rezone treatments in common use.

2. Conservation laws

In order to establish the notation, the governing partial differential equations are given. They include the equations expressing the conservation of mass, momentum and energy. In direct tensor notation[1], they are [1]

[1]For second-order tensors, the notation $\boldsymbol{\sigma} \cdot \boldsymbol{D} = \mathrm{tr}(\boldsymbol{\sigma D})$ implies, in Cartesian indicial notation, $\sigma_{ij}D_{ji}$.

$$\dot{\rho} = -\rho \operatorname{div} \boldsymbol{u}\,, \tag{2.1}$$

$$\dot{\boldsymbol{u}} = \frac{1}{\rho} \operatorname{div} \boldsymbol{\sigma} + \boldsymbol{b}\,, \tag{2.2}$$

$$\dot{\mathscr{E}} = \frac{1}{\rho} \boldsymbol{\sigma} \cdot \boldsymbol{D} + \frac{1}{\rho} \operatorname{div} \boldsymbol{h} + r\,, \tag{2.3}$$

where ρ is the mass density, $\boldsymbol{u}$ the material velocity vector, $\boldsymbol{\sigma}$ the symmetric Cauchy stress tensor, $\boldsymbol{b}$ the specific body force vector, $\mathscr{E}$ the specific internal energy, $\boldsymbol{h}$ the heat flux vector and r the specific heat source strength. The velocity strain, or stretching tensor $\boldsymbol{D}$ and spin tensor $\boldsymbol{W}$ are defined by

$$\boldsymbol{D} = \tfrac{1}{2}(\boldsymbol{L} + \boldsymbol{L}^{\mathrm{T}})\,, \qquad \boldsymbol{W} = \tfrac{1}{2}(\boldsymbol{L} - \boldsymbol{L}^{\mathrm{T}})\,, \tag{2.4}$$

where $\boldsymbol{L}$ is the velocity gradient, defined by

$$\boldsymbol{L} = \mathbf{grad}\, \boldsymbol{u}\,. \tag{2.5}$$

As usual, the superposed dot represents the material time derivative while the gradient and divergence are taken with respect to the spatial coordinates $\boldsymbol{x}$.

Deformation is measured relative to a reference configuration. If $\boldsymbol{X}$ refers to the positions of material particles in the reference configuration, the entire motion is described by a function $\boldsymbol{x} = \hat{\boldsymbol{x}}(\boldsymbol{X}, t)$. The deformation gradient tensor is defined by

$$\boldsymbol{F} = \mathbf{Grad}\, \hat{\boldsymbol{x}}\,, \tag{2.6}$$

the majuscule indicating that the gradient is taken with respect to the reference configuration $\boldsymbol{X}$. For continuity $\boldsymbol{F}$ is nonsingular, that is, the determinant of $\boldsymbol{F}$, denoted J, is finite and non-vanishing. It is taken to be positive. Differentiating (2.6) and using (2.5)

$$\dot{\boldsymbol{F}} = \mathbf{Grad}\, \boldsymbol{u} = \boldsymbol{L}\boldsymbol{F}\,. \tag{2.7}$$

Since $\boldsymbol{F}$ is nonsingular, it may be decomposed into a symmetric positive-definite part $\boldsymbol{U}$ representing a pure stretch, and an orthogonal part $\boldsymbol{R}$ representing a pure rotation

$$\boldsymbol{F} = \boldsymbol{R}\boldsymbol{U}, \qquad \boldsymbol{U}^2 = \boldsymbol{F}^{\mathrm{T}}\boldsymbol{F}, \qquad \boldsymbol{R} = \boldsymbol{F}\boldsymbol{U}^{-1}\,. \tag{2.8}$$

While $\boldsymbol{U}$ may be used as a measure of strain, a more convenient measure is Green's strain tensor

$$\boldsymbol{E} = \tfrac{1}{2}(\boldsymbol{U}^2 - \mathbf{1})\,. \tag{2.9}$$

By differentiating (2.8), using $(2.7)_2$ and taking the symmetric part of the result

$$\dot{\boldsymbol{E}} = \boldsymbol{F}^{\mathrm{T}}\boldsymbol{D}\boldsymbol{F}\,. \tag{2.10}$$

It is also useful to define a stress tensor in terms of directions and areas in the reference configuration. A convenient measure is the second Piola–Kirchoff stress tensor, defined from the Cauchy stress $\boldsymbol{\sigma}$ by

$$\boldsymbol{\Sigma} = J\boldsymbol{F}^{-1}\boldsymbol{\sigma}\boldsymbol{F}^{-\mathrm{T}} . \tag{2.11}$$

Note that $\boldsymbol{\Sigma}$ is symmetric. Differentiating and using (2.7)

$$\dot{\boldsymbol{\Sigma}} = J\boldsymbol{F}^{-1}\mathring{\boldsymbol{\sigma}}\boldsymbol{F}^{-\mathrm{T}} \tag{2.12}$$

where $\mathring{\boldsymbol{\sigma}}$ is a convected stress rate

$$\mathring{\boldsymbol{\sigma}} = \dot{\boldsymbol{\sigma}} - \boldsymbol{L}\boldsymbol{\sigma} - \boldsymbol{\sigma}\boldsymbol{L}^{\mathrm{T}} + \boldsymbol{\sigma}(\mathrm{tr}\,\boldsymbol{L}) . \tag{2.13}$$

The conservation laws can be rewritten in terms of $\boldsymbol{E}$ and $\boldsymbol{\Sigma}$, converting independent variables from $\boldsymbol{x}$ to $\boldsymbol{X}$ in the process. The resultant referential forms of the equations expressing momentum and energy conservation are

$$\dot{\boldsymbol{u}} = \frac{1}{\rho^{\mathrm{R}}}\,\mathrm{Div}(\boldsymbol{\Sigma}\boldsymbol{F}^{\mathrm{T}}) + \boldsymbol{b} , \tag{2.14}$$

$$\dot{\mathscr{E}} = \frac{1}{\rho^{\mathrm{R}}}\,\boldsymbol{\Sigma}\cdot\dot{\boldsymbol{E}} + \frac{1}{\rho^{\mathrm{R}}}\,\mathrm{Div}\,\boldsymbol{H} + r , \tag{2.15}$$

where $\rho^{\mathrm{R}} = J\rho$ is the density in the reference configuration and $\boldsymbol{H}$ is the heat flux referred to area in the reference configuration

$$\boldsymbol{H} = J\boldsymbol{F}^{-1}\boldsymbol{h} . \tag{2.16}$$

Note that the majuscule indicates that the divergence is taken with respect to $\boldsymbol{X}$.

3. Constitutive equations

Much of the early work in finite-difference methods relied on simple fluid equations of state, whence derives the still common appelation ‘hydrocodes’ describing computer programs based thereon. Major advances in the field in the last decade relate to the development of constitutive equations, describing various aspects of real material behavior, and their numerical solution. We take as an example the constitutive equations for a material with internal state variables [2, 3]. This theory has been studied in some detail with regard to the types of material behavior which it predicts [4–6], and it and its simplifications have been used quite successfully to represent viscoelastic and viscoplastic materials, materials undergoing polymorphic, melting and vaporization phase changes, chemical reactions, ionization and dissociation, and materials undergoing damage such as crack accumulation [7]. Fluid, elastic and plastic material descriptions arise naturally as special cases. Not all materials can be described in terms of this theory, notable exceptions include diffusive mixtures in which components have relative velocities, and materials which have internal degrees of freedom in which inertial effects accompany microstructural rearrangements [8, 9].

We first note that the constitutive equations should be objective [1], that is, that the material behavior should not be affected by rigid motions relative to the fixed inertial coordinate frame $\boldsymbol{x}$. In other words, they should be invariant under transformations of the form

$$\bar{\boldsymbol{x}} = \boldsymbol{a}(t) + \boldsymbol{Q}(t)\boldsymbol{x} \tag{3.1}$$

where $\boldsymbol{a}$ is a vector representing a time-dependent translation and $\boldsymbol{Q}$ is an orthogonal tensor representing a time-dependent rotation. It is easily verified [1] that $\boldsymbol{E}$, $\boldsymbol{\Sigma}$ and $\boldsymbol{H}$ are invariant under this transformation, as are scalars like the density. It is particularly convenient to construct constitutive equations in terms of such invariant quantities to ensure objectivity.

A further property desired of the constitutive equations is that they be such that the conservation laws (2.1), (2.2) and (2.3) be obeyed for all physically possible motions. Furthermore, it is desired that the second law of thermodynamics embodying the concept of irreversibility be obeyed for all such motions. A set of constitutive equations satisfying all of these requirements has been developed by Coleman and Gurtin [2] and Bowen [3]. It is most convenient to write their results in terms of the Helmholtz free energy potential, defined by

$$A = \mathscr{E} - TS \tag{3.2}$$

where T is the absolute temperature and S is the entropy. They find that the constitutive equations reduce to the forms

$$A = \hat{A}(\boldsymbol{E}, T, \boldsymbol{\alpha}), \qquad \boldsymbol{\Sigma}_{\mathrm{e}} = \rho^{\mathrm{R}} \frac{\partial}{\partial \boldsymbol{E}} \hat{A}(\boldsymbol{E}, T, \boldsymbol{\alpha}), \qquad S = -\frac{\partial}{\partial T} \hat{A}(\boldsymbol{E}, T, \boldsymbol{\alpha}), \tag{3.3}$$

where $\boldsymbol{\alpha}$ is a k-tuple $\{\alpha_1, \alpha_2, \ldots, \alpha_k\}$ representing a finite number of internal state variables whose evolution is governed by equations of the form[2]

$$\dot{\boldsymbol{\alpha}} = \hat{f}(\boldsymbol{E}, T, \boldsymbol{\alpha}). \tag{3.4}$$

The second-order tensor $\boldsymbol{\Sigma}_{\mathrm{e}}$-represents an elastic stress related to the total stress $\boldsymbol{\Sigma}$ and a dissipative viscous stress $\boldsymbol{\Sigma}_{\mathrm{v}}$ by

$$\boldsymbol{\Sigma} = \boldsymbol{\Sigma}_{\mathrm{e}} + \boldsymbol{\Sigma}_{\mathrm{v}}. \tag{3.5}$$

The viscous stress $\boldsymbol{\Sigma}_{\mathrm{v}}$ and heat flux $\boldsymbol{H}$ will here be taken in the simple forms[2]

$$\boldsymbol{\Sigma}_{\mathrm{v}} = \hat{\boldsymbol{\Sigma}}_{\mathrm{v}}(\boldsymbol{E}, \dot{\boldsymbol{E}}, T), \qquad \boldsymbol{H} = \hat{\boldsymbol{H}}(\boldsymbol{E}, \boldsymbol{G}, T), \tag{3.6}$$

where $\boldsymbol{G}$ is the material temperature gradient

$$\boldsymbol{G} = \mathbf{Grad}\, T = \boldsymbol{F}^{\mathrm{T}}\boldsymbol{g} \tag{3.7}$$

[2]More general dependencies on $(\boldsymbol{E}, \dot{\boldsymbol{E}}, T, \boldsymbol{G}, \alpha)$ can be considered [2, 3].

where, in turn, $\boldsymbol{g}$ is the spatial temperature gradient

$$\boldsymbol{g} = \mathbf{grad}\, T. \tag{3.8}$$

As before, **Grad** is the gradient with respect to $\boldsymbol{X}$, while **grad** is the gradient with respect to $\boldsymbol{x}$. Newtonian viscosity and Fourier heat conduction follow if $(3.6)_1$ and $(3.6)_2$ can be represented by their first (linear) terms of a Taylor series expansion about mechanical equilibrium $\dot{\boldsymbol{E}} = \mathbf{0}$ and thermal equilibrium $\boldsymbol{G} = \mathbf{0}$ respectively

$$\boldsymbol{\Sigma}_{\mathrm{v}} = \boldsymbol{M}^{\mathrm{R}}[\dot{\boldsymbol{E}}], \qquad \boldsymbol{H} = -\mathcal{K}^{\mathrm{R}}\boldsymbol{G}, \tag{3.9}$$

where $\boldsymbol{M}^{\mathrm{R}}$ is the fourth-order tensor of viscosity coefficients[4] and $\mathcal{K}^{\mathrm{R}}$ is the second-order tensor of heat conduction coefficients

$$\boldsymbol{M}^{\mathrm{R}} = \rho^{\mathrm{R}} \frac{\partial}{\partial \dot{\boldsymbol{E}}} \hat{\boldsymbol{\Sigma}}_{\mathrm{v}}(\boldsymbol{E}, \mathbf{0}, T), \qquad \mathcal{K}^{\mathrm{R}} = -\frac{\partial}{\partial \boldsymbol{G}} \hat{\boldsymbol{H}}(\boldsymbol{E}, \mathbf{0}, T). \tag{3.10}$$

The material is in thermodynamic equilibrium at given strain $\boldsymbol{E}$ and temperature T when the $\boldsymbol{\alpha}$ are such that

$$\hat{\boldsymbol{f}}(\boldsymbol{E}, T, \boldsymbol{\alpha}) = \mathbf{0}, \tag{3.11}$$

that is, that the rates of change of the internal state variables vanish.

The material is said to be in frozen equilibrium if the motion is so rapid, relative to rates of change of the internal state variables, that the $\dot{\boldsymbol{\alpha}}$ are negligibly small. Since the $\boldsymbol{\alpha}$ are then effectively fixed, they may be dropped as independent variables in (3.3). The theory reduces to that for a thermoelastic material with viscosity and heat conduction considered by Coleman and Mizel [10].

At the other extreme, the motion may be so slow relative to rates of change of the internal state variables that the material is always effectively in a state of relaxed equilibrium, that is, the $\boldsymbol{\alpha}$ effectively satisfy (3.11) throughout the motion. The $\boldsymbol{\alpha}$ are determined implicitly as functions of $(\boldsymbol{E}, T)$ by (3.11). They may be eliminated between (3.3) and (3.11) to once again achieve equations of form identical to those of a thermoelastic material. It is not often possible to carry this out explicitly, however, and (3.11) serves as an implicit constraint on the constitutive equations (3.3) in this case.

Plasticity theories arise quite naturally when there is a finite region in strain space, for given values of $(T, \boldsymbol{\alpha})$, upon which the equilibrium condition (3.11) is satisfied. For motions in which the state remains within this region, the $\boldsymbol{\alpha}$ do not change and the material is thermoelastic. Existence of the elastic stress function $(3.4)_2$ implies that the boundary of the thermoelastic region can be described in terms of the stress by an equation of the form

$$\psi(\boldsymbol{\Sigma}_{\mathrm{e}}, T, \boldsymbol{\alpha}) = 0 \tag{3.12}$$

which is called the yield surface. The way in which the material relaxes to

[4]The fourth-order tensor operation $\boldsymbol{M}^{\mathrm{R}}[\dot{\boldsymbol{E}}]$ implies, in Cartesian indicial notation, $M^{\mathrm{R}}_{ijkl}\dot{E}_{kl}$.

equilibrium on the yield surface, including the flow rule, is embodied in the functions $\hat{f}$.

The thermodynamic quantities A, T and S have been introduced which do not appear in the conservation laws. The description can be simplified as follows. Existence of the Helmholtz free energy function (3.3), together with appropriate smoothness and convexity assumptions, implies through (3.2) the existence of an internal energy function

$$\mathscr{E} = \tilde{\mathscr{E}}(\boldsymbol{E}, S, \boldsymbol{\alpha}), \qquad \boldsymbol{\Sigma}_{\mathrm{e}} = \rho^{\mathrm{R}} \frac{\partial}{\partial \boldsymbol{E}} \tilde{\mathscr{E}}(\boldsymbol{E}, S, \boldsymbol{\alpha}), \qquad T = \frac{\partial}{\partial S} \tilde{\mathscr{E}}(\boldsymbol{E}, S, \boldsymbol{\alpha}). \tag{3.13}$$

Since the absolute temperature T is non-vanishing, the third of these implies that the first is invertible in S. Inserting the result into the other two allows elimination of the entropy

$$\boldsymbol{\Sigma}_{\mathrm{e}} = \bar{\boldsymbol{\Sigma}}_{\mathrm{e}}(\boldsymbol{E}, \mathscr{E}, \boldsymbol{\alpha}), \qquad T = \bar{T}(\boldsymbol{E}, \mathscr{E}, \boldsymbol{\alpha}), \tag{3.14}$$

which may be used in place of (3.3). Similarly, eliminating S from (3.4)

$$\dot{\boldsymbol{\alpha}} = \bar{\boldsymbol{f}}(\boldsymbol{E}, \mathscr{E}, \boldsymbol{\alpha}). \tag{3.15}$$

It is usually more convenient to use (3.14) in differential form

$$\begin{aligned} \dot{\boldsymbol{\Sigma}}_{\mathrm{e}} &= \boldsymbol{K}^{\mathrm{R}}_{\mathscr{E}}[\dot{\boldsymbol{E}}] + \rho^{\mathrm{R}} \boldsymbol{\Gamma}^{\mathrm{R}} \dot{\mathscr{E}} + \boldsymbol{\Phi}^{\mathrm{R}}, \\ \dot{T} &= \frac{\boldsymbol{Y}^{\mathrm{R}}}{\rho^{\mathrm{R}}} \cdot \dot{\boldsymbol{E}} + \frac{1}{C_E} \dot{\mathscr{E}} + \Psi, \end{aligned} \tag{3.16}$$

where the coefficients are defined as

$$\begin{aligned} \boldsymbol{K}^{\mathrm{R}}_{\mathscr{E}} &= \frac{\partial}{\partial \boldsymbol{E}} \bar{\boldsymbol{\Sigma}}_{\mathrm{e}}(\boldsymbol{E}, \mathscr{E}, \boldsymbol{\alpha}), \quad \boldsymbol{Y}^{\mathrm{R}} = \rho^{\mathrm{R}} \frac{\partial}{\partial \boldsymbol{E}} \bar{T}(\boldsymbol{E}, \mathscr{E}, \boldsymbol{\alpha}), \\ \boldsymbol{\Gamma}^{\mathrm{R}} &= \frac{1}{\rho^{\mathrm{R}}} \frac{\partial}{\partial \mathscr{E}} \bar{\boldsymbol{\Sigma}}_{\mathrm{e}}(\boldsymbol{E}, \mathscr{E}, \boldsymbol{\alpha}), \quad \frac{1}{C_E} = \frac{\partial}{\partial \mathscr{E}} \bar{T}(\boldsymbol{E}, \mathscr{E}, \boldsymbol{\alpha}), \\ \boldsymbol{\Phi}^{\mathrm{R}} &= \frac{\partial}{\partial \boldsymbol{\alpha}} \bar{\boldsymbol{\Sigma}}_{\mathrm{e}}(\boldsymbol{E}, \mathscr{E}, \boldsymbol{\alpha}) \cdot \bar{\boldsymbol{f}}(\boldsymbol{E}, \mathscr{E}, \boldsymbol{\alpha}), \quad \Psi = \frac{\partial}{\partial \boldsymbol{\alpha}} \bar{T}(\boldsymbol{E}, \mathscr{E}, \boldsymbol{\alpha}) \cdot \bar{\boldsymbol{f}}(\boldsymbol{E}, \mathscr{E}, \boldsymbol{\alpha}). \end{aligned} \tag{3.17}$$

The fourth-order tensor $\boldsymbol{K}^{\mathrm{R}}_{\mathscr{E}}$ represents elastic moduli at constant internal energy, the second-order tensor $\boldsymbol{\Gamma}^{\mathrm{R}}$ represents Grueneisen coefficients, C_E is the specific heat at constant strain, and $\boldsymbol{\Phi}^{\mathrm{R}}$ and Ψ are stress and temperature relaxation functions respectively. Note that, since $\boldsymbol{\Sigma}$ is invariant under the transformation (3.1), so is $\dot{\boldsymbol{\Sigma}}$ and (3.16) is objective.

The energy conservation law (2.15) may be used to eliminate $\dot{\mathscr{E}}$ from (3.16) whence

$$\begin{aligned} \dot{\boldsymbol{\Sigma}}_{\mathrm{e}} &= \boldsymbol{K}^{\mathrm{R}}_{\mathrm{S}}[\dot{\boldsymbol{E}}] + \boldsymbol{\Gamma}^{\mathrm{R}}(\boldsymbol{\Sigma}_{\mathrm{v}} \cdot \dot{\boldsymbol{E}} + \operatorname{Div} \boldsymbol{H} + \rho^{\mathrm{R}} r) + \boldsymbol{\Phi}^{\mathrm{R}}, \\ \dot{T} &= T \boldsymbol{\Gamma}^{\mathrm{R}} \cdot \dot{\boldsymbol{E}} + \frac{1}{\rho^{\mathrm{R}} C_E} (\boldsymbol{\Sigma}_{\mathrm{v}} \cdot \dot{\boldsymbol{E}} + \operatorname{Div} \boldsymbol{H} + \rho^{\mathrm{R}} r) + \Psi, \end{aligned} \tag{3.18}$$

where thermodynamic identities relate the isentropic moduli $\boldsymbol{K}_S^R$ and Grueneisen tensor $\boldsymbol{\Gamma}^R$ to previously defined derivatives[5]

$$\boldsymbol{K}_S^R[\dot{\boldsymbol{E}}] = \boldsymbol{K}_{\mathscr{E}}^R[\dot{\boldsymbol{E}}] + \boldsymbol{\Gamma}^R(\boldsymbol{\Sigma}_e \cdot \dot{\boldsymbol{E}}), \qquad T\boldsymbol{\Gamma}^R = \frac{\boldsymbol{Y}^R}{\rho^R} - \frac{\boldsymbol{\Sigma}_e}{\rho^R C_E}. \tag{3.19}$$

The equations in referential form, (3.18), are most suitable for treating problems involving arbitrarily large deformations of solid materials. However, they are seldom used. The constitutive equations can be cast into spatial forms by use of $(2.10)_1$ and (2.12), whence

$$\begin{aligned} \mathring{\boldsymbol{\sigma}}_e &= \boldsymbol{K}_S[\boldsymbol{D}] + \boldsymbol{\Gamma}(\boldsymbol{\sigma}_v \cdot \boldsymbol{D} + \operatorname{div} \boldsymbol{h} + \rho r) + \boldsymbol{\Phi}, \\ \dot{T} &= T\boldsymbol{\Gamma} \cdot \boldsymbol{D} + \frac{1}{\rho C_E}(\boldsymbol{\sigma}_v \cdot \boldsymbol{D} + \operatorname{div} \boldsymbol{h} + \rho r) + \Psi, \end{aligned} \tag{3.20}$$

where[6]

$$\boldsymbol{K}_S[\boldsymbol{D}] = \frac{1}{J}\boldsymbol{F}\boldsymbol{K}_S^R[\boldsymbol{F}^T\boldsymbol{D}\boldsymbol{F}]\boldsymbol{F}^T, \qquad \boldsymbol{\Gamma} = \frac{1}{J}\boldsymbol{F}\boldsymbol{\Gamma}^R\boldsymbol{F}^T, \qquad \boldsymbol{\Phi} = \frac{1}{J}\boldsymbol{F}\boldsymbol{\Phi}^R\boldsymbol{F}^T. \tag{3.21}$$

Equivalently, for the viscous stress and heat conduction equations (3.9)

$$\boldsymbol{\sigma}_v = \boldsymbol{M}[\boldsymbol{D}], \qquad \boldsymbol{h} = -\mathscr{K}\boldsymbol{g}, \tag{3.22}$$

where

$$\boldsymbol{M}[\boldsymbol{D}] = \frac{1}{J}\boldsymbol{F}\boldsymbol{M}^R[\boldsymbol{F}^T\boldsymbol{D}\boldsymbol{F}]\boldsymbol{F}^T, \qquad \mathscr{K} = \frac{1}{J}\boldsymbol{F}\mathscr{K}^R\boldsymbol{F}^T. \tag{3.23}$$

Note that the convected stress rate (2.13) appears in (3.20). It is simpler to use the corotational stress rate defined by

$$\overset{\triangledown}{\boldsymbol{\sigma}}_e = \dot{\boldsymbol{\sigma}}_e - \boldsymbol{W}\boldsymbol{\sigma}_e - \boldsymbol{\sigma}_e\boldsymbol{W}^T. \tag{3.24}$$

Using (2.4), $(2.20)_1$ may be simplified to give

$$\overset{\triangledown}{\boldsymbol{\sigma}}_e = \boldsymbol{K}[\boldsymbol{D}] + \boldsymbol{\Gamma}(\boldsymbol{\sigma}_v \cdot \boldsymbol{D} + \operatorname{div} \boldsymbol{h} + \rho r) + \boldsymbol{\Phi}, \tag{3.25}$$

where[7]

$$\boldsymbol{K}[\boldsymbol{D}] = \boldsymbol{K}_S[\boldsymbol{D}] + \boldsymbol{D}\boldsymbol{\sigma}_e + \boldsymbol{\sigma}_e\boldsymbol{D} - \boldsymbol{\sigma}_e(\operatorname{tr} \boldsymbol{D}). \tag{3.26}$$

We note that there are a number of restrictions on the constitutive functions arising from stability considerations. These have only been cursorily explored for the general theory given here. Nature provides a wealth of examples of in-

[5]In Cartesian indicial notation $(K_S^R)_{ijkl} = (K_{\mathscr{E}}^R)_{ijkl} + (\Gamma^R)_{ij}(\Sigma_e)_{kl}$.

[6]In Cartesian indicial notation $K_{ijkl} = J^{-1}F_{im}F_{jn}F_{kr}F_{ls}K^R_{mnrs}$.

[7]In Cartesian indical notation $K_{ijkl} = (K_S)_{ijkl} + \sigma_{ik}\delta_{jl} + \sigma_{jk}\delta_{il} - \sigma_{ij}\delta_{kl}$.

stabilities, bifurcations and non-uniqueness to suggest that the theory of stability of materials with internal state variables is likely to be a rich one. We only note a few results in passing.

Irreversibility demands that the viscous stress and heat flux vanish in mechanical and thermal equilibrium $\dot{\boldsymbol{E}} = \mathbf{0}$, $\boldsymbol{G} = \mathbf{0}$ respectively, while the viscous coefficient tensor $\boldsymbol{M}^{\mathrm{R}}$ and heat conductivity tensor $\mathscr{K}^{\mathrm{R}}$ are positive semi-definite there [2, 3]. Coleman and Noll have postulated that the elastic modulus tensor $\boldsymbol{K}_{\mathrm{S}}^{\mathrm{R}}$ is positive definite while the specific heat C_E is positive, generalizing the inequalities of Gibbs [11]. Coleman and Gurtin [2] and Bowen [3] have considered the stability properties of the evolution equations (3.4) at constant strain and temperature, but little has been done to consider stability in general motions.

No restriction can be placed on the Grueneisen coefficients. If $\boldsymbol{\Gamma}^{\mathrm{R}} = \mathbf{0}$, it can be seen from (3.18) that mechanical and thermal processes are then uncoupled, except insofar as the coefficients in the stress equation may depend on internal energy or temperature, while the viscous stress work appears in the temperature equation and its coefficients may depend on strain.

If the coefficients in (3.18) are independent of $\boldsymbol{\alpha}$, the equations reduce to generalizations of the constitutive equations of a Maxwellian material [12].

Material symmetries are most easily represented in the referential description. For example, if the material is isotropic in its reference configuration, then (3.18) and (3.9) become isotropic tensor functions. Forms for all the point-symmetry classes are easily developed [11]. The spatial forms (3.20) and (3.22) do not, in general, share these symmetries because of strain-induced anisotropies entering through (3.21) and (3.23). However, isotropy is retained in (3.20) and (3.22) if the deformation differs only infinitesimally from a pure volume change and rotation. Consider an infinitesimal displacement $\boldsymbol{a} = \boldsymbol{x} - \boldsymbol{X}$ where $\sup\{|a_i|\} \leq \varepsilon$, ε being a small number. The displacement gradient is given by

$$\boldsymbol{H} = \mathbf{Grad}\,\boldsymbol{a} = \boldsymbol{F} - \mathbf{1}\,, \tag{3.27}$$

while the infinitesimal strain tensor is defined by

$$\boldsymbol{e} = \tfrac{1}{2}(\boldsymbol{H} + \boldsymbol{H}^{\mathrm{T}})\,. \tag{3.28}$$

Then it may be shown that

$$\dot{\boldsymbol{e}} = \boldsymbol{D} + \mathrm{O}(\varepsilon^2)\,. \tag{3.29}$$

Now consider a deformation composed of an infinitesimal deformation, superimposed upon a large rotation $\boldsymbol{R}$ and large dilatation θ. Then the deformation gradient for the combined deformation is

$$\boldsymbol{F} = (\mathbf{1} + \boldsymbol{H})\boldsymbol{R}\theta\,. \tag{3.30}$$

Without loss of generality, the total volume strain may be lumped into θ so that we may take $\mathrm{tr}\,\boldsymbol{e} = 0$ and $\boldsymbol{e}$ is a deviatoric tensor. Under this deformation, the coefficients in (3.20) and (3.25) are found to remain isotropic, and are found to reduce to[8]

$$K[D] = \lambda(\text{tr}\, D)\mathbf{1} + 2\mu D, \qquad \Gamma = \Gamma\mathbf{1}, \qquad C_E = C_v \tag{3.31}$$

to within terms of $O(\varepsilon^2)$, where the Lamé elasticity coefficients λ and μ, the Grueneisen coefficient Γ and specific heat at constant volume C_v may be taken to be functions of the density instead of the full strain tensor. The evolution functions $\boldsymbol{f}$ and stress relation function $\boldsymbol{\Phi}$ become isotropic tensor functions of the strain $\boldsymbol{e}'$, as well as of the density. For example,

$$\boldsymbol{\Phi}(\boldsymbol{E}, T, \boldsymbol{\alpha}) = \Phi_b(\rho, T, \boldsymbol{\alpha})\mathbf{1} + \Phi_d(\rho, T, \boldsymbol{\alpha})\boldsymbol{e}' \tag{3.32}$$

to within terms of $O(\varepsilon^2)$.

A fluid cannot support elastic deviator stresses, $\boldsymbol{\sigma}_e = -p\mathbf{1}$ where p is the pressure. In this case

$$K[D] = K(\text{tr}\, D)\mathbf{1}, \qquad \Phi = \Phi_b \mathbf{1}, \tag{3.33}$$

where K is the bulk modulus given by

$$K = \lambda + \tfrac{2}{3}\mu\,. \tag{3.34}$$

Under similar deformations, the viscous stress and heat conduction equations become isotropic. In their linear approximations

$$\boldsymbol{\sigma}_v = \zeta(\text{tr}\, D)\mathbf{1} + 2\eta D, \qquad \boldsymbol{h} = -\mathcal{K}\boldsymbol{g}, \tag{3.35}$$

where the bulk and shear viscosity coefficients ζ and η and the heat conduction coefficient $\mathcal{K}$ are functions of density instead of the full strain tensor.

While the referential forms are the most suitable for arbitrary deformations of solid materials, the spatial forms have been used almost exclusively. The governing equations will be gathered here for convenience. They are the mass and momentum conservation equations (2.1) and (2.2), stress equation (3.25), evolution equations (3.15) and either the energy or temperature equations (2.3) or $(3.20)_2$. Leaving out the latter for the moment, they may be written most conveniently in terms of deviator stresses defined by

$$\boldsymbol{\sigma}_e' = \boldsymbol{\sigma}_e + p\mathbf{1}, \qquad \boldsymbol{\sigma}_v' = \boldsymbol{\sigma}_v + q\mathbf{1}, \tag{3.36}$$

where q is a viscous pressure. The full set of equations are

$$\begin{aligned}
&\dot{\boldsymbol{u}} = \frac{1}{\rho}\,\text{div}\, \boldsymbol{\sigma}_e' - \frac{1}{\rho}\,\textbf{grad}\, p + \frac{1}{\rho}\,\text{div}\, \boldsymbol{\sigma}_v' - \frac{1}{\rho}\,\textbf{grad}\, q + \boldsymbol{b}\,,\\
&\dot{\boldsymbol{x}} = \boldsymbol{u}, \qquad \dot{\rho} = -\rho(\text{tr}\, D), \qquad \dot{\boldsymbol{e}}' = D'\,,\\
&\dot{p} = -K(\text{tr}\, D) - \Gamma(\boldsymbol{\sigma}_v' \cdot D - q\, \text{tr}\, D + \text{div}\, \boldsymbol{h} + \rho r) - \Phi_b\,,\\
&\dot{\boldsymbol{\sigma}}_e' = W\boldsymbol{\sigma}_e' + \boldsymbol{\sigma}_e' W^{\mathrm{T}} + 2\mu D' + \Phi_d \boldsymbol{e}'\,,\\
&\dot{\boldsymbol{\alpha}} = \boldsymbol{f}\,,
\end{aligned} \tag{3.37}$$

where $D' = D - \frac{1}{3}(\text{tr}\, D)$ is the deviator of the stretching tensor.

[8]In Cartesian indicial notation $K_{ijkl} = \lambda\delta_{ij}\delta_{kl} + \mu(\delta_{ik}\delta_{jl} + \delta_{il}\delta_{jk})$.

When heat conduction is absent $\boldsymbol{h} = \boldsymbol{0}$, the temperature is not needed explicitly, and the energy equation (2.3) may be used.

$$\dot{\mathscr{E}} = \frac{1}{\rho}\boldsymbol{\sigma}_{\mathrm{e}}' \cdot \boldsymbol{D} - \frac{p}{\rho}\operatorname{tr}\boldsymbol{D} + \frac{1}{\rho}\boldsymbol{\sigma}_{\mathrm{v}}' \cdot \boldsymbol{D} - \frac{q}{\rho}\operatorname{tr}\boldsymbol{D} + \frac{1}{\rho}\operatorname{div}\boldsymbol{h} + r\,. \tag{3.38}$$

The coefficients K, μ, Γ, Φ_{b} and Φ_{d} are functions of $(\rho, \mathscr{E}, \boldsymbol{\alpha})$, $\boldsymbol{\sigma}_{\mathrm{v}}'$ and q are nonlinear tensor functions of $(\rho, \boldsymbol{D}, \mathscr{E})$, and b and r are prescribed functions of $(\boldsymbol{x}, t)$.

When heat conduction is present, the temperature is needed explicitly, and the most direct procedure is to replace the energy equation by the temperature equation $(3.20)_2$

$$\dot{T} = T\Gamma(\operatorname{tr}\boldsymbol{D}) + \frac{1}{\rho C_{\mathrm{v}}}(\boldsymbol{\sigma}_{\mathrm{v}}' \cdot \boldsymbol{D} - q\operatorname{tr}\boldsymbol{D} + \operatorname{div}\boldsymbol{h} + \rho r) + \Psi \tag{3.39}$$

where the coefficients K, μ, Γ, C_{v}, Φ_{b}, Φ_{d} and Ψ are now reexpressed as functions of $(\rho, T, \boldsymbol{\alpha})$, $\boldsymbol{\sigma}_{\mathrm{v}}'$ and q are functions of $(\rho, \boldsymbol{D}, T)$, h is a function of $(\rho, \boldsymbol{g}, T)$, and b and r are once again prescribed functions of $(\boldsymbol{x}, t)$. Of course, the temperature description may also be used when heat conduction is absent.

Equations paralleling these for more general situations and for the referential description are easily extracted from the foregoing.

The constitutive equations of this section have seldom been used in their full generality. However, they include as special cases many of the forms which have been used. Applications to reactive mixtures was envisioned in the original development of the theory [2, 3], the internal state variables in this case representing extents of reaction among phases or species. A particular example has been worked out, for example, by Hayes [13] for phase transitions and by Nunziato and Kipp [14] for explosives. In applications to viscoelastic or viscoplastic materials, the internal state variables represent degrees of relaxation. A particular example has been worked out, for example, by Nunziato et al. [15]. In applications to materials undergoing damage, for example finely distributed cracking, the internal state variables represent one or more damage measures related to the extent of cracking [16]. Specific applications have been worked out by Seaman et al. [17] and Grady and Kipp [18].

The Maxwell description, in which internal state variables do not appear explicitly, has been used very widely. Specialization of the present theory leads to results similar to the viscoplastic theory of Perzyna [19] which has been used in a particular application, for example, by Herrmann [20]. Porous materials have also been handled by equations of this type, for example, by Butcher [21]. Applications of equilibrium fluid, elastic and elastic-plastic specializations are too numerous to mention, but include most of the common rate-independent constitutive equations in use.

4. Differential equations

The differential equations governing the motion may be obtained by expanding the tensor equations of the last two sections into physical components for the coordinate system of interest. This is a straightforward exercise both for the spatial and referential forms of the equations. When general anisotropic materials are considered in general three-dimensional motions, the differential equations can be quite lengthy. In order to illustrate the results and to allow discussion of finite-difference forms, only the spatial forms for an isotropic material will be given for rectangular and cylindrical coordinates in two dimensions.

Expanding (3.37) through (3.39), the equations are given in full in the Appendix. For rectangular coordinates x and y, the final terms in square brackets in each equation, arising from Christoffel symbols for cylindrical coordinates, are absent. For cylindrical coordinates, x denotes the radial direction while y denotes the axial direction. A more direct approach for cylindrical coordinates involves redefinition of the dependent variables so that the terms arising from Christoffel symbols vanish [22], but this has not been used in the methods under consideration.

We recall that, when heat conduction is absent, the temperature is not needed explicitly, and the internal energy equation (A14a) can be used. The elasticity coefficients K and μ, Grueneisen's coefficient Γ, the bulk and deviator stress relaxation functions Φ_b and Φ_d are known constitutive functions of $(\rho, \mathscr{E}, \boldsymbol{\alpha})$, the evolution functions of the internal state variables are known functions of $(\rho, \boldsymbol{e}, \mathscr{E}, \boldsymbol{\alpha})$, the viscous stress $\boldsymbol{\sigma}_v$ is a known constitutive function of $(\rho, \partial \boldsymbol{u}/\partial \boldsymbol{x}, \mathscr{E})$ and the body force $\boldsymbol{b}$ and heat source strength r are prescribed functions of $(\boldsymbol{x}, t)$.

When heat conduction is present, the temperature is needed explicitly, and the temperature equation (A14b) may be used in place of the internal energy equation (A14a). In this case, the internal energy is not calculated explicitly. The constitutive functions are now to be reexpressed as functions of the temperature T instead of the internal energy $\mathscr{E}$, and the heat flux $\boldsymbol{h}$ is a function of $(\rho, \partial T/\partial \boldsymbol{x}, T)$.

Construction of specific constitutive functions for particular applications is beyond the scope of this discussion. We note only that K, Γ and C_v may be related, via the usual equations among thermodynamic derivatives, to derivatives of any of the thermodynamic potential functions, for example the Helmholtz free energy, which is most useful when constitutive equations are obtained from theoretical partition functions, or the Gibbs free enthalpy, which is most useful for treating phase equilibria. The viscous stress function will be considered here chiefly for the introduction of artificial viscosity. Heat conduction will not be considered in detail.

Convective terms arising from the material derivatives appear as the first two terms in each equation. The spatial description has been used, and the independent variables are implicitly understood to be functions of $(\boldsymbol{x}, t)$. The full equations are therefore Eulerian, in the sense that stationary coordinates are used.

The description may also be referred to coordinates which are convected with the material. The spatial positions of material points can be calculated by integrating the velocities, consequently, gradients with respect to $\boldsymbol{x}$ can be found even when material points are being followed. However, since material points are now being followed, the material derivatives reduce to partial derivatives with respect to time and the convective terms vanish. Such a description is termed Lagrangian.

The equations of the Appendix may be represented by the general form

$$\frac{\partial \boldsymbol{U}}{\partial t} = \mathscr{L}[\mathscr{D}\boldsymbol{U}] + \boldsymbol{B} \tag{4.1}$$

where $\boldsymbol{U}$ is the solution vector with, in the most general case considered here, $(13+k)$ components, where k is the number of the internal state variables, given by

$$\boldsymbol{U} = \{u_x, u_y, x, y, \rho, e'_{xx}, e'_{xy}, e'_{yy}, p, \sigma^{e\prime}_{xx}, \sigma^{e\prime}_{xy}, \sigma^{e\prime}_{yy}, \boldsymbol{\alpha}, \mathscr{E} \text{ or } T\} \tag{4.2}$$

and $\boldsymbol{B}$ is a vector with $(13+k)$ components. In (4.1), $\mathscr{L}[\cdot]$ represents a linear combination of differential operators $\mathscr{D}$. Most of these operators are first order partial derivatives. However, note that the viscous stress and heat flux depend on velocity and temperature gradients respectively. Consequently, the differential operators include second partial derivatives with respect to $\boldsymbol{x}$ as well as non-linear terms in the first partial derivatives. When viscous stresses and heat conduction are absent, these nonlinearities vanish and the equations are quasi-linear.

5. Difference equations

The theory of finite differences is well known [23, 24]. Only a sketch of the specific methods involved in the particular computer programs under consideration will be given here.

The governing differential equations of the Appendix have been reduced to the general form (4.1). In the finite difference method, the differential operators with respect to $\boldsymbol{x}$ on the right are first expanded into difference operators, using Taylor series expansions or similar devices. Specific difference operators will be considered subsequently. When this is done, the equations reduce to ordinary differential equations of form

$$\frac{\partial \boldsymbol{U}}{\partial t} = \mathscr{L}[\Delta \boldsymbol{U}] + \boldsymbol{B} \tag{5.1}$$

where the Δ represent difference operators, and the $\boldsymbol{U}$ have been discretized. In general, there are m sets of equations, one for each discrete location at which the $\boldsymbol{U}$ are calculated, and these sets are coupled. Note that each term in (4.1) leads to a corresponding term in (5.1).

The $(13+k)\times m$ differential equations (5.1) must now be integrated in time. The simple forward difference scheme

$$\boldsymbol{U}^{n+1} = \boldsymbol{U}^n + \Delta t\{\mathscr{L}[\Delta \boldsymbol{U}^n] + \boldsymbol{B}\} \tag{5.2}$$

where n is the time step index and Δt is the time step size, provides explicit equations for the U^{n+1}. On the other hand, the centered difference scheme

$$U^{n+1} = U^n + \Delta t\{\mathscr{L}[\Delta U^{n+1/2}] + B\}, \tag{5.3}$$

where

$$U^{n+1/2} = \tfrac{1}{2}(U^{n+1} + U^n), \tag{5.4}$$

leads to $(13 + k) \times m$ coupled implicit equations. While the forward-difference explicit method (5.2) is only first-order accurate, the centered-difference implicit method (5.3) is second-order accurate in Δt for uniform time steps. (For non-uniform time steps, errors of order Δt times the difference in successive time steps are present).

A variety of integration methods have been discussed which avoid the inaccuracies of the first-order explicit method (5.2) without resorting to the necessity of solving $(13 + k) \times m$ coupled implicit equations. All of the programs under consideration use a leapfrog time integration utilized by von Neuman and Richtmyer [25] in a similar but highly simplified one-dimensional application. We will give a general discussion.

Omitting convective terms, which are discussed briefly below, the velocities u are first advanced one time step, using difference forms of (A1) and (A2) with an explicit forward time difference. The results are then used to advance positions x and other dependent variables one time step, again using an explicit forward time difference in discrete forms of the remaining equations of the Appendix. In effect, the velocities are half a time step out of phase with the remaining dependent variables. Defining

$$U_c = \{\rho,\ e'_{xx},\ e'_{xy},\ e'_{yy},\ p,\ \sigma^{e\prime}_{xx},\ \sigma^{e\prime}_{xy},\ \sigma^{e\prime}_{yy},\ \alpha,\ \mathscr{E} \text{ or } T\}, \tag{5.5}$$

the equations for the velocities u and positions x can be written

$$u^{n+1/2} = u^{n+1/2} + \tfrac{1}{2}(\Delta t^{n+1/2} + \Delta t^{n-1/2})\{\mathscr{L}[\Delta(x^n, U^n_c)] + B\}, \tag{5.6}$$

$$x^{n+1} = x^n + \Delta t^{n+1} u^{n+1/2},$$

while the remaining equations take the form

$$U^{n+1}_c = U^n_c + \Delta t^{n+1/2}\{\mathscr{L}[\Delta(x^{n+1/2}, u^{n+1/2}, U^n_c)] + B\}. \tag{5.7}$$

The equations (5.6) are second-order centered time differences. Strictly, calculations should begin and end with a half time step advance of the velocities.

Equations (5.7) still involve first-order forward time differences in the U_c. The problem of finding a higher-order accurate time integration method may be simplified by a suitable choice of centering in space. Space is covered by a computational grid, not necessarily regular. Positions x and velocities u are centered at nodes of the grid, while all other dependent variables are centered in the meshes of the grid. If heat conduction is absent, the equations for U_c can be seen from the Appendix to involve gradients of the velocity, but not of other dependent variables. Once velocities and positions $u^{n+1/2}$, $x^{n+1/2}$ have been calculated for all of the nodes, the velocity gradients can be found, centered in the meshes, by finite-difference expressions to be discussed later. Consequently, the equations (5.7) for the $\boldsymbol{U}_c$ for each mesh are uncoupled from the equations for the $\boldsymbol{U}_c$ for every other mesh. Heat conduction is considered separately below.

Reverting to differential form, and suppressing the dependence on $\boldsymbol{u}$ and $\boldsymbol{x}$, (5.7) take the general form

$$\frac{\partial \boldsymbol{U}_c}{\partial t} = \mathcal{F}(\boldsymbol{U}_c) \tag{5.8}$$

where $\mathcal{F}$ is a non-linear vector-valued function of the $\boldsymbol{U}_c$. The $(9+k)$ nonlinear equations for each mesh are coupled to each other, but are uncoupled from those of every other mesh. A number of integration techniques have been used.

Under certain conditions of linearity, the equations resulting when centered time differences are used in (5.8) can be solved explicitly for the $\boldsymbol{U}_c^{n+1}$. For example the mass equation for the density (A5), omitting convective terms, is linear in ρ, and a centered time difference may be solved explicitly for ρ^{n+1}. It may also be noted that centered time differences for the infinitesimal strain deviators are explicit. The density and strains at $n+1$ can therefore always be calculated explicitly. The remaining equations for the pressure, stress deviators, internal state variables and internal energy or temperature become explicit only with especially simple constitutive functions. A less restrictive second-order method for their integration results from the two-step procedure

$$\begin{aligned} \boldsymbol{U}_c^{n+1/2} &= \boldsymbol{U}_c^n + \tfrac{1}{2}\Delta t^{n+1/2}\mathcal{F}(\boldsymbol{U}_c^n)\,, \\ \boldsymbol{U}_c^{n+1} &= \boldsymbol{U}_c^n + \Delta t^{n+1/2}\mathcal{F}(\boldsymbol{U}_c^{n+1/2})\,, \end{aligned} \tag{5.9}$$

which has also been widely used.

It frequently happens that some of the constitutive functions vary rapidly, relative to the motion, causing the differential equations to be stiff. Very little theoretical work has been done for these cases, but a number of devices have been resorted to in order to obtain solutions. For example, some of the equations have been integrated using a sub-multiple of the time step [20] (subcycling). More elegantly, higher order and variable step Runge–Kutta, Adams or other integration methods have been used [14, 26, 27]. It is not necessarily prohibitive to use implicit time integration, since the equations for each mesh are uncoupled from those of other meshes, and integration techniques useful for stiff ordinary differential equations could also be used.

The explicit time integration scheme is conditionally stable. The stable time step is limited by the spatial mesh size, and also depends upon the constitutive functions. Convergence and stability investigations are sparse, particularly for the rate-dependent internal state variable material description given here, but are urgently needed. In view of the expected richness of the differential equations in terms of bifurcation and nonuniqueness of solutions, such investigation is expected to be non-trivial.

Note that the equations (5.8) essentially include all of the constitutive functions, while (5.6) do not, (with the exception of the viscous stress which is dealt with below). Consequently the integration of (5.8) may be implemented in a well-defined sub-program, which is furnished values of the velocity gradients and advances the $\boldsymbol{U}_c$ for one mesh by one time step. The remainder of the program does not require knowledge of the constitutive functions. Different sub-programs may be prepared for different special material descriptions. In fact, the strains e'_{xx}, e'_{xy}, e'_{yy}, internal state variables $\boldsymbol{\alpha}$ and the thermodynamic variable $\mathscr{E}$ or T are entirely internal to the constitutive procedure, and some or all of them may be left out for specially simple material descriptions, without affecting the remainder of the program.

We now return to a brief discussion of the convective, heat conduction and viscous terms, which have been ignored so far. Many of the foregoing remarks do not apply to these terms. Rather than provide a full discussion of them, we note that they may be dealt with by an application of concepts of operator splitting.

The linear sum in (5.1) may be trivially grouped into two parts. The result may be written as

$$\boldsymbol{U}^{n+1} = (\mathbf{1} + \Delta t\boldsymbol{m}_1 + \Delta t\boldsymbol{m}_2)\boldsymbol{U}^n + \Delta t\boldsymbol{B}. \tag{5.10}$$

This is the same, within terms of $O(\Delta t^2)$, as

$$\boldsymbol{U}^{n+1} = (\mathbf{1} + \Delta t\boldsymbol{m}_1)(\mathbf{1} + \Delta t\boldsymbol{m}_2)\boldsymbol{U}^n + \Delta t\boldsymbol{B}. \tag{5.11}$$

If time steps are uniform and the order of the operators is alternated, then under proper conditions the scheme is accurate through second order. In words, this implies that the solution vector may first be advanced one step in time using only some of the difference operators. The result is then advanced one step in time using the remaining difference operators. The result is expected to be equivalent, within the order of accuracy of the difference operators, to a calculation in which the solution vector is advanced one step in time using all of the difference operators simultaneously. Alternating direction methods are a familiar application in which a multidimensional calculation is split into a sequence of one-dimensional calculations in each coordinate direction. For second-order difference operators, Gottlieb [28], for example, has shown how the sequence must be alternated in order to preserve accuracy when the operator is split into more than two parts.

Operator splitting has been used quite widely for other purposes, although usually without the benefit of convergence, stability and accuracy proofs. The convective terms in the difference equations, arising from their counterparts in the Appendix, may be split from the remaining terms which might be termed Lagrangian. In an Eulerian calculation, the solution vector might first be advanced by one time step using the Lagrangian terms only. The result might then be advanced using the convective terms to produce an Eulerian calculation. Most explicit Eulerian computer programs in effect do this [34]. In principle, an

Eulerian program could be constructed by adding a convective overlay to a suitably structured Lagrangian code. The convective terms will not be considered further here, and the remainder of the discussion will be devoted to Lagrangian methods.

The heat conduction terms may be split off in a similar way. In this case the solution vector is first advanced omitting heat conduction terms. The result is then advanced using only the heat conduction terms. The overall result is then expected to be equivalent to a fully coupled thermal-stress calculation. The heat conduction terms of the Appendix may be recognized as the conventional equations of thermal diffusion. In practice, different types of time integration are usually used for the stress and heat conduction calculations. The hyperbolic dynamic stress calculation is usually accomplished by an explicit time integration scheme, as discussed in the preceding. However, the parabolic heat diffusion calculation is usually best solved by an implicit time integration method. Different step sizes are often used, one being a submultiple of the other. Most stress wave codes which include heat diffusion in effect use this type of splitting [34]. It is again, in principle, possible to alternate stress calculations with heat diffusion calculations, using overlays consisting of software implementing the separate processes to perform fully coupled thermal stress calculations. The diffusion equations are beyond the scope of this discussion, and heat conduction will not be considered further.

It may be noted that the viscous stress terms are also parabolic. These terms may also be split off, and are recognizable as the incompressible but non-linear Navier–Stokes equations. The methods under consideration all use artificial viscosity to smooth discontinuities, and the discussion of the viscous terms will be limited to that context. Since the artificial viscosity is not meant to represent a real physical phenomenon, it is considered sufficient to use a simple explicit forward-difference time integration for the viscous terms. They are not usually split from the remainder of the calculation. Particular forms of the artificial viscous stress function are considered in a later section.

No restrictions have been placed, so far, on the form of the computational grid, and meshes with an arbitrary number of sides and arbitrary connectivity may be considered. Specific finite-difference expressions for spatial gradients will be considered in the next few sections. While they will be given for quadrilateral meshes, they may be generalized to triangular to other meshes, and no restriction will be placed on mesh connectivities. For consistency with the second-order time integration scheme which has been discussed, second-order spatial difference operators will be given. (The operators are, in fact, second-order accurate only on a uniformly spaced grid, but involve errors of order mesh size times the difference in size between adjacent meshes when used with a non-uniform grid.)

Dynamic problems are distinguished by the presence of propagating waves which entail high gradients. In order to resolve them, a fine computational grid is usually required. The corresponding discretized data base is therefore usually very large for non-trivial problems, and data base management becomes important to overall computational efficiency. Highly ordered quadrilateral grids have been found to provide the best compromise between resolution, accuracy and efficiency for problems of this type. Non-quadrilateral meshes and special connectivities can be used in isolated locations which are not easily covered by an isomorph of a rectangular grid.

6. Gradient approximations

The usual method of generating finite-difference expressions for spatial gradients uses Taylor series expansions to approximate first derivatives. In two dimensions, consider four neighboring points labelled 1, 2, 3, 4 at small finite distances from a point 0 under consideration as shown in Figure 6.1. Provided that a function $\varphi(x, y)$ is sufficiently smooth, Taylor series expansions can be written for φ about the point 0 resulting in four equations of type

$$\begin{aligned}\varphi_1 = \varphi_0 &+ (x_1 - x_0)\left(\frac{\partial\varphi}{\partial x}\right)_0 + (y_1 - y_0)\left(\frac{\partial\varphi}{\partial y}\right)_0 \\ &+ \tfrac{1}{2}(x_1 - x_0)^2\left(\frac{\partial^2\varphi}{\partial x^2}\right)_0 + (x_1 - x_0)(y_1 - y_0)\left(\frac{\partial^2\varphi}{\partial x\partial y}\right)_0 \\ &+ \tfrac{1}{2}(y_1 - y_0)^2\left(\frac{\partial^2\varphi}{\partial y^2}\right)_0 + O(\varepsilon^3),\end{aligned} \tag{6.1}$$

where $\sup\{|x_1 - x_0|, |y_1 - y_0|\} \leq \varepsilon$, ε being a small number. The four equations may be solved for the first derivatives by first subtracting those for points 1 and 3, and those for points 2 and 4 to obtain

$$\begin{aligned}(\varphi_1 - \varphi_3) = (x_1 - x_3)\left(\frac{\partial\varphi}{\partial x}\right)_0 &+ (y_1 - y_3)\left(\frac{\partial\varphi}{\partial y}\right)_0 \\ &+ (x_1 - x_3)\chi_{13}\left(\frac{\partial^2\varphi}{\partial x^2}\right)_0 + [(x_1 - x_3)\xi_{13} + (y_1 - y_3)\chi_{13}]\left(\frac{\partial^2\varphi}{\partial x\partial y}\right)_0 \\ &+ (y_1 - y_3)\xi_{13}\left(\frac{\partial^2\varphi}{\partial y^2}\right)_0 + O(\varepsilon^3), \\ (\varphi_2 - \varphi_4) = (x_2 - x_4)\left(\frac{\partial\varphi}{\partial x}\right)_0 &+ (y_2 - y_4)\left(\frac{\partial\varphi}{\partial y}\right)_0 \\ &+ (x_2 - x_4)\chi_{24}\left(\frac{\partial^2\varphi}{\partial x^2}\right)_0 + [(x_2 - x_4)\xi_{24} + (y_2 - y_4)\chi_{24}]\left(\frac{\partial^2\varphi}{\partial x\partial y}\right)_0 \\ &+ (y_2 - y_4)\xi_{24}\left(\frac{\partial^2\varphi}{\partial y^2}\right)_0 + O(\varepsilon^3),\end{aligned} \tag{6.2}$$

where

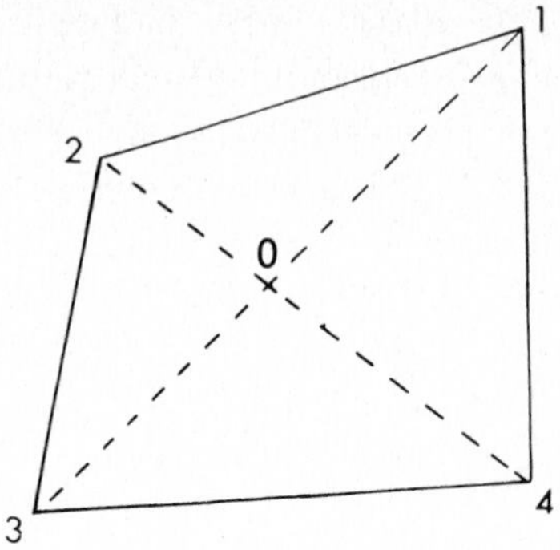

Fig. 6.1. Typical mesh.

$$\chi_{13} = \tfrac{1}{2}(x_1 + x_3) - x_0\,, \qquad \chi_{24} = \tfrac{1}{2}(x_2 + x_4) - x_0\,,$$
$$\xi_{13} = \tfrac{1}{2}(y_1 + y_3) - y_0\,, \qquad \xi_{24} = \tfrac{1}{2}(y_2 + y_4) - y_0\,, \tag{6.3}$$

are measures of mesh distortion. If they are of order ε^2, then

$$\left(\frac{\partial\varphi}{\partial x}\right)_0 = \frac{1}{2A}\{(y_2 - y_4)(\varphi_1 - \varphi_3) - (y_1 - y_3)(\varphi_2 - \varphi_4)\} + \mathrm{O}(\varepsilon^3)\,,$$
$$\left(\frac{\partial\varphi}{\partial y}\right)_0 = \frac{1}{2A}\{(\varphi_2 - \varphi_4)(x_1 - x_3) - (\varphi_1 - \varphi_3)(x_2 - x_4)\} + \mathrm{O}(\varepsilon^3)\,, \tag{6.4}$$

where

$$A = \tfrac{1}{2}\{(y_2 - y_4)(x_1 - x_3) - (y_1 - y_3)(x_2 - x_4)\} \tag{6.5}$$

is the area of quadrilateral 1234.

The same result can be obtained in another way. Green's theorem relates area integrals of gradients to line integrals. In two dimensions

$$\int_A \frac{\partial\varphi}{\partial x}\,\mathrm{d}a = \oint_S \varphi\,\mathrm{d}y\,, \qquad \int_A \frac{\partial\varphi}{\partial y}\,\mathrm{d}a = -\oint_S \varphi\,\mathrm{d}x\,. \tag{6.6}$$

For the quadrilateral 1234, representing integrands by averages

$$\left(\frac{\partial\varphi}{\partial x}\right)_0 = \frac{1}{A}\{\bar\varphi_{43}(y_4 - y_3) + \bar\varphi_{32}(y_3 - y_2) + \bar\varphi_{21}(y_2 - y_1) + \bar\varphi_{14}(y_1 - y_4)\}\,,$$
$$\left(\frac{\partial\varphi}{\partial y}\right)_0 = \frac{1}{A}\{\bar\varphi_{12}(x_1 - x_2) + \bar\varphi_{23}(x_2 - x_3) + \bar\varphi_{34}(x_3 - x_4) + \bar\varphi_{41}(x_4 - x_1)\}\,, \tag{6.7}$$

where $\bar\varphi_{12} = \bar\varphi_{21}$ is the average value of φ over the side 12 etc. If the average is taken to be the average of φ_1 and φ_2,

$$\varphi_{12} = \tfrac{1}{2}(\varphi_1 + \varphi_2) \tag{6.8}$$

etc., then (6.7) reduce immediately to (6.4). It is obvious how this may be extended to non-quadrilateral meshes. Extension to three dimensions is also straightforward.

It should be emphasized that the error terms which have been neglected are of order mesh size times the mesh distortion parameters (6.3). When the mesh distortions are very small, the difference expressions involve error terms which are third order in the mesh size. However, when mesh distortions are of the same order as the mesh size, then the error terms are of second order in the mesh size. This is usually the case.

7. Stress gradients; momentum conservation

Stress gradients appear in the momentum equations for the velocity components given in the Appendix. Coordinate positions are centered at node points, while stresses are centered in the meshes. Consequently, interpolation is required in using the finite difference approximations of Section 6. Consider first the rectangular Cartesian case. Two types of interpolation have been widely used. The labelling system shown in Figure 7.1 will be used for a typical segment of a quadrilateral grid.

In the so-called 'midpoint method', the difference expressions (6.4) are applied to the midpoints labelled 5, 6, 7 and 8 in Figure 7.1. Linear interpolations are used for positions and stresses. Denoting any one of the stress components by φ,

$$x_5 = \tfrac{1}{2}(x_A + x_O), \qquad \varphi_5 = \tfrac{1}{2}(\varphi_1 + \varphi_2)\,, \tag{7.1}$$

etc. When these are used in (6.4) the results are

$$\begin{aligned}\left(\frac{1}{\rho}\frac{\partial\varphi}{\partial x}\right)_0 = \frac{-1}{4A\rho}\{&(\varphi_1 - \varphi_3)(y_D - y_A + y_C - y_B) \\ &- (\varphi_2 - \varphi_4)(y_B - y_A + y_C - y_D)\}\,,\end{aligned} \tag{7.2}$$

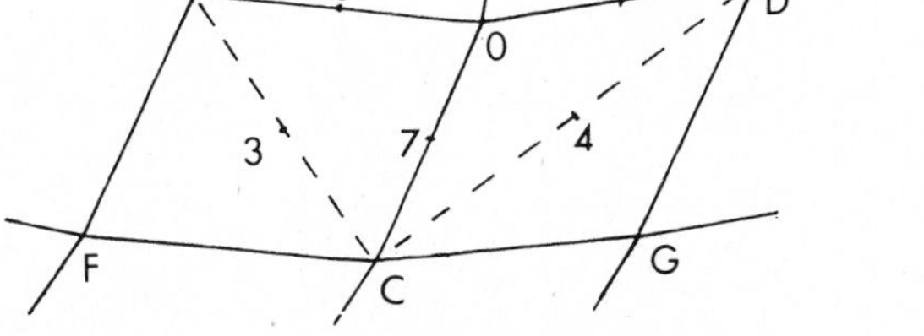

Fig. 7.1. Typical segment of grid.

$$\left(\frac{1}{\rho}\frac{\partial\varphi}{\partial y}\right)_0 = \frac{1}{4A\rho}\{(\varphi_1 - \varphi_3)(x_D - x_A + x_C - x_B) - (\varphi_2 - \varphi_4)(x_B - x_A + x_C - x_D)\},$$

where A is one half the area of quarilateral $ABCD$

$$A = \tfrac{1}{4}\{(y_B - y_D)(x_A - x_C) - (x_B - x_D)(y_A - y_C)\}. \tag{7.3}$$

The density has been introduced into (6.10). The density is centered in meshes and also requires interpolation. This is most conveniently accomplished by writing

$$A\rho = \tfrac{1}{4}(A_1\rho_1 + A_2\rho_2 + A_3\rho_3 + A_4\rho_4) \tag{7.4}$$

where A_1 is the area of quadrilateral $AODH$

$$A_1 = \tfrac{1}{2}\{(y_A - y_D)(x_H - x_O) - (x_A - x_D)(y_H - y_O)\} \tag{7.5}$$

etc. Note that $A_1\rho_1 = M_1$ is the mass per unit thickness of quadrilateral $AODH$ in rectangular coordinates. Consequently $A\rho$ represents one quarter of the mass of the meshes surrounding node O, which we denote M_0. Continuity demands that this mass remain constant. The midpoint method is used in the REXCO code [29] and its derivatives, but dates to much earlier codes used at the Los Alamos Scientific Laboratory.

An alternate method is due to Wilkins [30]. The expressions (6.7) are applied directly to quadrilateral $ABCD$ of Figure 7.1, interpreting the averages of stresses as the mesh centered values.

$$\left(\frac{1}{\rho}\frac{\partial\varphi}{\partial x}\right)_0 = \frac{-1}{2A\rho}\{\varphi_1(y_D - y_A) + \varphi_2(y_A - y_B) + \varphi_3(y_B - y_C) + \varphi_4(y_C - y_D)\},$$

$$\left(\frac{1}{\rho}\frac{\partial\varphi}{\partial y}\right)_0 = \frac{1}{2A\rho}\{\varphi_1(x_D - x_A) + \varphi_2(x_A - x_B) + \varphi_3(x_B - x_C) + \varphi_4(x_C - x_D)\}, \tag{7.6}$$

where the density has again been introduced, and $A\rho$ is given by (7.4) as before. This so-called Green's theorem method has been used in the HEMP [30], TOODY [35], PISCES [32], STEALTH [33] codes and their derivatives.

The Lagrangian terms of the momentum equations may be written in terms of nodal masses, accelerations and forces [36, 37, 38]. For node O

$$A\rho\left(\frac{\partial u_x}{\partial t}\right)_O = F_{x1} + F_{x2} + F_{x3} + F_{x4} + A\rho(b_x)_O,$$

$$A\rho\left(\frac{\partial u_y}{\partial t}\right)_O = F_{y1} + F_{y2} + F_{y3} + F_{y4} + A\rho(b_y)_O, \tag{7.7}$$

where F_{x1}, F_{y1} are the x and y components of the force exerted on node 0 by mesh 1 etc., $(b_x)_O$ and $(b_y)_0$ are the x and y components of the external body force per unit mass at node O, and $A\rho$ is given by (7.4), and represents the constant nodal mass M_O in rectangular coordinates.

For the Green's theorem method, the nodal forces become simply

$$\begin{aligned} F_{x1} &= \tfrac{1}{2}\{(\sigma^{e\prime}_{xx} - p + \sigma^{v\prime}_{xx} - q)(y_A - y_D) - (\sigma^{e\prime}_{xy} + \sigma^{v\prime}_{xy})(x_A - x_D)\}, \\ F_{y1} &= \tfrac{1}{2}\{(\sigma^{e\prime}_{xy} + \sigma^{v\prime}_{xy})(y_A - y_D) - (\sigma^{e\prime}_{yy} - p + \sigma^{v\prime}_{yy} - q)(x_A - x_D)\}, \end{aligned} \tag{7.8}$$

where all the stresses refer to mesh 1, and similarly for the forces due to other meshes.

The midpoint method (7.2) may also be cast into a form to provide nodal forces. The forces at node O due to mesh 1 are

$$\begin{aligned} F_{x1} &= \tfrac{1}{4}\{(\sigma^{e\prime}_{xx} - p + \sigma^{v\prime}_{xx} - q)(y_A - y_D + y_B - y_C) \\ &\quad - (\sigma^{e\prime}_{xy} + \sigma^{v\prime}_{xy})(x_A - x_D + x_B - x_C)\}, \\ F_{y1} &= \tfrac{1}{4}\{(\sigma^{e\prime}_{xy} + \sigma^{v\prime}_{xy})(y_A - y_D + y_B - y_C) \\ &\quad - (\sigma^{e\prime}_{yy} - p + \sigma^{v\prime}_{yy} - q)(x_A - x_D + x_B - x_C)\}. \end{aligned} \tag{7.9}$$

These are not quite as convenient as (7.8) since coordinates of nodes B and C appear. Similar expressions apply to the forces due to other meshes.

A number of other difference expressions have been developed and used. Most are less convenient and involve more computation than the midpoint and Green's theorem methods described above. Simple tests do not reveal significant advantages for any of them [39], and the Green's theorem method is to be preferred because of its simplicity.

In cylindrical coordinates, the mass associated with a node is no longer given by (7.4). In fact the volume V associated with mesh 1, for example, can be found exactly by dividing the quadrilateral mesh into two triangles, which are then rotated about the axis. For mesh 1 of Figure 7.1 for example, the mass per unit radian is given by

$$M_1 = \rho_1(\bar{x}_u A_u + \bar{x}_\ell A_\ell) \tag{7.10}$$

where

$$\bar{x}_u = \tfrac{1}{3}(x_A + x_H + x_D), \qquad \bar{x}_\ell = \tfrac{1}{3}(x_A + x_O + x_D), \tag{7.11}$$

and

$$\begin{aligned} A_u &= \tfrac{1}{2}\{(x_A - x_H)(y_D - y_H) - (y_A - y_H)(x_D - x_H)\}, \\ A_\ell &= \tfrac{1}{2}\{(x_D - x_O)(y_A - y_O) - (y_D - y_O)(x_A - x_O)\}. \end{aligned} \tag{7.12}$$

This does not give the same result as

$$M_1 = \rho_1 \bar{x}_1 A_1 \tag{7.13}$$

if $\bar{x}_1$ is approximated by x_1 defined as

$$x_1 = \tfrac{1}{4}(x_A + x_O + x_D + x_H). \tag{7.14}$$

In fact the effective radius of the mesh $\bar{x}_1$ can be defined by (7.13). While $\bar{x}_1$ can be expressed explicitly in terms of the node coordinates, the result cannot be written in a compact form.

Wilkins [30] has therefore chosen simply to use (7.6) to calculate stress gradients, obtaining $A\rho$ explicitly from node positions and densities via (7.4) and (7.5) etc. at each time step. In each case the terms arising from the Christoffel symbols are approximated by

$$\frac{\varphi}{\rho x}=\frac{1}{4}\left(\frac{\varphi_1}{\rho_1\bar{x}_1}+\frac{\varphi_2}{\rho_2\bar{x}_2}+\frac{\varphi_3}{\rho_3\bar{x}_3}+\frac{\varphi_4}{\rho_4\bar{x}_4}\right) \tag{7.15}$$

where the $\bar{x}$ are calculated from (7.13). The same thing may be done when the midpoint method is used, merely substituting (7.2) for (7.6). These approximations are widely used in the computer programs under consideration. However, the convenience of a constant nodal mass is thereby lost. It might be regained by multiplying both sides of (7.7) by an effective radius. Doing this individually for each mesh contribution would result in

$$\begin{aligned} M_O\left(\frac{\partial u_x}{\partial t}\right)_O &= F_{x1}+F_{x2}+F_{x3}+F_{x4}+M_0(b_x)_O\,,\\ M_O\left(\frac{\partial u_x}{\partial t}\right)_0 &= F_{y1}+F_{y2}+F_{y3}+F_{y4}+M_O(b_y)_O\,, \end{aligned} \tag{7.16}$$

where M_O is one-quarter of the masses of the four surrounding meshes calculated from (7.10), etc., the nodal forces being given by

$$\begin{aligned} F_{x1} &= \tfrac{1}{2}\bar{x}_1\{(\sigma^{e\prime}_{xx}-p+\sigma^{v\prime}_{xx}-q)(y_A-y_D)-(\sigma^{e\prime}_{xy}+\sigma^{v\prime}_{xy})(x_A-x_D)\}\\ &\quad+\tfrac{1}{4}A_1(2\sigma^{e\prime}_{xx}+\sigma^{e\prime}_{yy}+2\sigma^{v\prime}_{xx}+\sigma^{v\prime}_{yy})\,,\\ F_{y1} &= \tfrac{1}{2}\bar{x}_1\{(\sigma^{e\prime}_{xy}+\sigma^{v\prime}_{xy})(y_A-y_D)-(\sigma^{e\prime}_{yy}-p+\sigma^{v\prime}_{yy}-q)(x_A-x_D)\}\\ &\quad+\tfrac{1}{4}A_1(\sigma^{e\prime}_{xy}+\sigma^{v\prime}_{xy})\,, \end{aligned} \tag{7.17}$$

where $\bar{x}_1$ is given by (7.13), and similarly for the other meshes. Hancock [36] has skewed the nodal force contributions to satisfy boundary conditions on the axis $x=0$ without introducing the necessary boundary tractions. Use of the proper boundary conditions together with (7.17) provides the proper behavior on the axis.

8. Velocity gradients; mass conservation

Velocity gradients appear in the mass equation for the density as well as other equations given in the Appendix. Both the velocities and coordinate positions are centered at node points, so that the difference expressions (6.4) are directly applicable. However, the velocities and positions are centered a half time step apart, so that interpolation in time is necessary. Denoting either one of the velocity components by φ, using linear interpolation in time for positions, and applying (6.4) to mesh 1, the results are

$$\begin{aligned}\left(\frac{\partial\varphi}{\partial x}\right)_1^{n+1/2} &= \frac{1}{2(A_1^{n+1}+A_1^n)}\{(\varphi_A^{n+1/2}-\varphi_D^{n+1/2})(y_O^{n+1}-y_H^{n+1}+y_O^n-y_H^n) \\ &\quad -(\varphi_O^{n+1/2}-\varphi_H^{n+1/2})(y_A^{n+1}-y_D^{n+1}+y_A^n-y_D^n\}, \\ \left(\frac{\partial\varphi}{\varphi y}\right)_1^{n+1/2} &= \frac{-1}{2(A_1^{n+1}+A_1^n)}\{(\varphi_A^{n+1/2}-\varphi_D^{n+1/2})(x_O^{n+1}-x_H^{n+1}+x_O^n-x_H^n) \\ &\quad -(\varphi_O^{n+1/2}-\varphi_H^{n+1/2})(x_A^{n+1}-x_D^{n+1}+x_A^n-x_D^n)\},\end{aligned} \tag{8.1}$$

where A_1 is the area of quadrilateral *AODH* given by (7.5).

In a rectangular Lagrangian coordinate system, the differential equation for the density, arising from conservation mass is, from (A5)

$$\frac{1}{\rho}\left(\frac{\partial\rho}{\partial t}\right) = -\left(\frac{\partial u_x}{\partial x}+\frac{\partial u_y}{\partial y}\right). \tag{8.2}$$

Since $A\rho$ is a constant in a rectangular system,

$$\frac{1}{\rho}\left(\frac{\partial\rho}{\partial t}\right) = -\frac{1}{A}\left(\frac{\partial A}{\partial t}\right). \tag{8.3}$$

This may be written in finite difference form as

$$\frac{2}{\Delta t^{n+1/2}}\left(\frac{\rho_1^{n+1}-\rho_1^n}{\rho_1^{n+1}+\rho_1^n}\right) = -\frac{2}{\Delta t^{n+1/2}}\left(\frac{A_1^{n+1}-A_1^n}{A_1^{n+1}+A_1^n}\right). \tag{8.4}$$

Using (7.5) for A_1, the right hand side of (8.4) becomes, after some algebra, identical with the right hand side of (8.2) when (8.1) are used for the velocity gradients together with the centered difference expressions

$$u_x^{n+1/2} = \frac{x^{n+1}-x^n}{\Delta t^{n+1/2}}, \qquad u_y^{n+1/2} = \frac{y^{n+1}-y^n}{\Delta t^{n+1/2}}, \tag{8.5}$$

at each node. Consequently, the three alternatives for calculating the density, given by

$$\rho_1^{n+1} = \frac{M_1}{A_1^{n+1}} \tag{8.6}$$

where M_1 is the constant mass of mesh 1, and

$$\rho_1^{n+1} = \rho_1^n\left\{\frac{1-\frac{1}{2}\Delta t^{n+1/2}Q}{1+\frac{1}{2}\Delta t^{n+1/2}Q}\right\} \tag{8.7}$$

with Q given either by

$$Q = \frac{\partial u_x}{\partial x}+\frac{\partial u_y}{\partial y} \tag{8.8}$$

where the gradients are calculated using (8.1), or by

$$Q = \frac{1}{A}\frac{\partial A}{\partial t} \tag{8.9}$$

where the right-hand side is calculated as in (8.4), are equivalent. Note that each finite-difference form of the mass equation is explicit. The simplest expression (8.6) is usually used although the individual mesh masses must then be stored. If nodal masses are stored, then (8.7) with (8.8) is preferable. The finite-difference expressions (8.1) have been universally used for the velocity gradients required in the remaining equations of the Appendix.

In cylindrical coordinates, a suitably interpolated expression for (u_x/x) appearing in the terms arising from the Christoffel symbols is needed. Wilkins [30] introduced the approximation

$$\left(\frac{u_x}{x}\right)_1^{n+1/2} = 2\frac{(A_u^{n+1}+A_u^n)(\bar{u}_x)_u^{n+1/2}+(A_\ell^{n+1}+A_\ell^n)(\bar{u}_x)_\ell^{n+1/2}}{(A_u^{n+1}+A_u^n)(\bar{x}_u^{n+1}+\bar{x}_u^n)+(A_\ell^{n+1}+A_\ell^n)(\bar{x}_\ell^{n+1}+\bar{x}_\ell^n)} \tag{8.10}$$

where $\bar{x}_u$, $\bar{x}_\ell$, A_u and A_ℓ are given by (7.11) and (7.12) and $(\bar{u}_x)_u$ and $(\bar{u}_x)_\ell$ are given by

$$\begin{aligned}(\bar{u}_x)_u &= \tfrac{1}{3}\{(u_x)_A + (u_x)_H + (u_x)_D\}\,,\\ (\bar{u}_x)_\ell &= \tfrac{1}{3}\{(u_x)_A + (u_x)_O + (u_x)_D\}\,.\end{aligned} \tag{8.11}$$

When (8.10) is used in the mass equation (8.7) with

$$Q = \frac{\partial u_x}{\partial x} + \frac{\partial u_y}{\partial y} + \frac{u_x}{x}\,, \tag{8.12}$$

the result is not identical with the exact expression

$$\rho_1^{n+1} = \frac{M_1}{\bar{x}_u^{n+1}A_u^{n+1} + \bar{x}_\ell^{n+1}A_\ell^{n+1}} \tag{8.13}$$

which is almost always used. In fact, if the density is calculated from (8.13), then the value of (u_x/x) can be calculated from

$$\left(\frac{u_x}{x}\right)_1^{n+1/2} = \left(\frac{\partial u_x}{\partial x} + \frac{\partial u_y}{\partial y}\right) - \frac{2}{\Delta t}\left(\frac{\rho_1^{n+1} - \rho_1^n}{\rho_1^{n+1} + \rho_1^n}\right) \tag{8.14}$$

where (8.1) are used for the velocity gradients. Once again, mesh masses must be stored but (8.14) is preferable to (8.10).

9. Artificial viscosity and damping

Solutions of the inviscid nonlinear partial differential equations commonly exhibit first-order discontinuities, or shock waves, and second-order discontinuities, or acceleration waves, which cause difficulties in finite-difference methods. Such discontinuities arise not only from discontinuities in initial and boundary data, but, as is well known, can be generated by the steepening of continuous waves due to nonlinearities in material behavior. Typical problems often involve many such waves, which may multiply very rapidly by reflections at boundaries and interactions among themselves. It is usually impossible to treat all of them by means of discrete descriptions of moving surfaces of discontinuity.

Discontinuities cause difficulties in the finite-difference representation since, as is seen from Section 6, finite-difference approximations inherently assume the existence of bounded derivatives. On surfaces of discontinuity, derivatives become unbounded and the approximations are invalid. Von Neumann and Richtmyer [25] solved this difficulty by introducing an artificial viscosity, which makes all solutions to the differential equations smooth, discontinuities in the inviscid solution being replaced by steep but continuous steady propagating waves in the viscous solution.

The linear viscous stress function for an isotropic material has been given in (3.35). More generally, the viscous stress may be a non-linear isotropic tensor function of the stretching tensor $\boldsymbol{D}$, which takes the general form [1]

$$\boldsymbol{\sigma}_{\mathrm{v}} = \alpha_0 \mathbf{1} + \alpha_1 \boldsymbol{D} + \alpha_2 \boldsymbol{D}^2 \tag{9.1}$$

where the scalar coefficients α_0, α_1 and α_2 may be nonlinear functions of density, temperature and the invariants of $\boldsymbol{D}$.

Von Neumann and Richtmyer introduced a scalar viscosity which is quadratic in the first invariant of $\boldsymbol{D}$,

$$\alpha_0 = b_{\mathrm{q}}^2 \rho |\mathrm{tr}\, \boldsymbol{D}| (\mathrm{tr}\, \boldsymbol{D}) \tag{9.2}$$

where ρ has been introduced to make the coefficient b_{q} have the dimension of length. Note that (9.2) has been written so that the viscous stress has the correct sign for both expansions and compressions. Von Neumann and Richtmyer showed that the thickness of a continuous steady propagating wave in a simple nonlinearly compressible fluid with the above viscosity is proportional to b_{q}. In order that the thickness of such waves be a fixed multiple of the mesh size, they chose to make b_{q} proportional to mesh size. In two dimensions, the mesh size may be characterized simply as the square root of the mesh area, or is often variously taken to be the length of the smallest mesh side, smallest mesh diagonal, or area of the mesh divided by the largest diagonal, etc. Denoting this characteristic dimension by $\Delta\bar{x}$, we set $b_{\mathrm{q}} = \beta_{\mathrm{q}} \Delta\bar{x}$ where β_{q} is a dimensionless constant.

Introduction of the viscosity (9.2) may be viewed as introducing errors of order $\Delta\bar{x}^2$ relative to the inviscid solution. It is customary to set $\alpha_0 = 0$ on expansion $\mathrm{tr}\, \boldsymbol{D} > 0$, since smoothing is usually needed for compressive shock waves. Velocity gradients away from shock waves are generally small, so that the quadratic viscosity is usually negligible elsewhere. It therefore does no harm to calculate the viscosity on expansions. This has the advantage of providing extra smoothing of numerical oscillations in the solution which may be introduced at shock waves if the viscosity coefficient is too small, as well as at acceleration waves, boundaries, mesh size discontinuities and other sources of truncation errors.

Landshoff [40] noted that the quadratic viscosity did not provide sufficient damping for low amplitude waves, and proposed an additional term which is linear in the first invariant of $\boldsymbol{D}$,

$$\alpha_0 = b_\ell \rho c \operatorname{tr} \boldsymbol{D} \tag{9.3}$$

where $c = \sqrt{K/\rho}$ is the bulk sound speed, which has been introduced to make b_ℓ have the dimension of length. For a nonlinearly compressible fluid, it may also be shown [41] that the thickness of a continuous steady propagating wave is proportional to b_ℓ, and b_ℓ is chosen to be proportional to the characteristic mesh size $b_\ell = \beta_\ell \Delta\bar{x}$, where β_ℓ is nondimensional. The linear viscosity may be viewed as introducing errors of order $\Delta\bar{x}$ relative to the inviscid solution.

Note that the formation of steady propagating waves depends upon material nonlinearities. For very low amplitude waves, where the material is nearly linear, the steady wave thickness is very large and its evolution time is long, so that compressive waves will broaden as they propagate for times which may be longer than the total time of interest in a stress wave problem. Viscosity may also significantly broaden expansive waves. In a linear material, of course, viscosity will always introduce wave dispersion.

It should be noted that the constitutive equations of Section 3, entail dissipation in material relaxation due to changes in the internal state variables. Exact solutions of the differential equations without artificial viscosity may still contain discontinuities [6], however, which may need to be smoothed by artificial viscosity, but the values of the viscous coefficients may need to be adjusted to prevent excessive smoothing of the numerical solution. An analysis of steady propagating waves for the particular constitutive equations to be used may be very helpful in assessing the effects of artificial viscosity, and in choosing reasonable values for the coefficients. Note that artificial viscosity, as well as material relaxation affect the stability limit on the time step size.

Combining (9.1), (9.2) and (9.3), the viscous pressure corresponding to linear and quadratic artificial viscosity is

$$q = -\beta_q^2 \Delta\bar{x}^2 \rho |\operatorname{tr} \boldsymbol{D}| (\operatorname{tr} \boldsymbol{D}) + \beta_\ell \Delta\bar{x} \rho c (\operatorname{tr} \boldsymbol{D}) . \tag{9.4}$$

The right-hand side involves velocity gradients since in two dimensions

$$\operatorname{tr} D = \frac{\partial u_x}{\partial x} + \frac{\partial u_y}{\partial u} + \left[\frac{u_x}{x}\right] \tag{9.5}$$

where the term in square brackets applies for cylindrical coordinates and is absent in rectangular coordinates. The velocity gradients may be calculated as shown in Section 8. Use of centered time differences in the momentum equations for the velocities (A1) and (A2) would render the finite difference forms of these equations implicit. Since the artificial viscosity is not meant to represent a real physical phenomenon, a first-order forward difference is used,

$$\boldsymbol{u}^{n+1/2} = \boldsymbol{u}^{n-1/2} + \tfrac{1}{2}(\Delta t^{n+1/2} + \Delta t^{n-1/2})\{\mathcal{L}[\Delta(\boldsymbol{x}^n, \boldsymbol{U}_c^n, \boldsymbol{u}^{n-1/2})] + B\}\,, \tag{9.6}$$

cf. (5.6) in order to render these equations explicit.

As has been mentioned in Section 8, the term (u_x/x) causes difficulties in the case of cylindrical coordinates. It is preferable to use the mass equation (2.1), which may be written

$$\operatorname{tr} \boldsymbol{D} = -\frac{\dot{\rho}}{\rho} \tag{9.7}$$

to remove the velocity gradients, representing the right hand side of (9.7) by the finite difference expression on the left-hand side of (8.4) for both rectangular and cylindrical symmetry.

In some cases equivoluminal shear oscillations may be present. These may be damped by a shear viscosity. A linear deviatoric tensor viscosity may be used for this purpose. An isotropic linear viscous stress function was given in (3.35). The spherical part is, of course, just the Landshoff linear viscosity when the coefficient $(\zeta + \frac{2}{3}\eta)$ is made proportional to the mesh size. Maenchen and Sack [42] used a linear deviatoric viscosity by making the coefficient η proportional to $\Delta\bar{x}$,

$$\boldsymbol{\sigma}'_{\mathrm{v}} = \beta_{\mathrm{d}}\rho c \Delta\bar{x}\boldsymbol{D}'\,. \tag{9.8}$$

Forward time differences are again used to render the momentum equations explicit. Of course (3.35) may also be used to introduce a physical linear viscosity, although the first-order forward time difference scheme required to make the momentum equations explicit does not provide a very accurate treatment.

A quadratic shear viscosity was introduced by Herrmann [31]. A properly invariant formulation may be derived from the third term of (9.1). In order to eliminate the spherical part, one might use

$$\boldsymbol{\sigma}'_{\mathrm{v}} = \beta_{\mathrm{s}}^2\Delta\bar{x}^2\rho\{\boldsymbol{D}^2 - \tfrac{1}{3}\mathrm{tr}(\boldsymbol{D}^2)\} \tag{9.9}$$

with proper attention to signs.

The spherical and deviator viscous stresses which have been introduced above may be used to damp bulk and shear deformation modes of the meshes. However, there are additional deformation modes of quadrilateral meshes which are not damped by these viscous stresses. An example is the keystone or hourglass deformation modes, illustrated in Figure 9.1, in which alternate nodes have equal velocities which are opposite those of their neighbors. It is readily seen that the finite-difference forms (8.1) yield zero velocity gradients for these modes. Not only are these modes undamped, but they generate no elastic stresses.

The difficulty is not with the finite-difference forms (8.1), but rather with the fact that mesh centered averages of stretchings and stresses are used. The keystone modes involve the gradient of stretching across the meshes which are not accounted for in this description. In some problems, keystone deformations tend

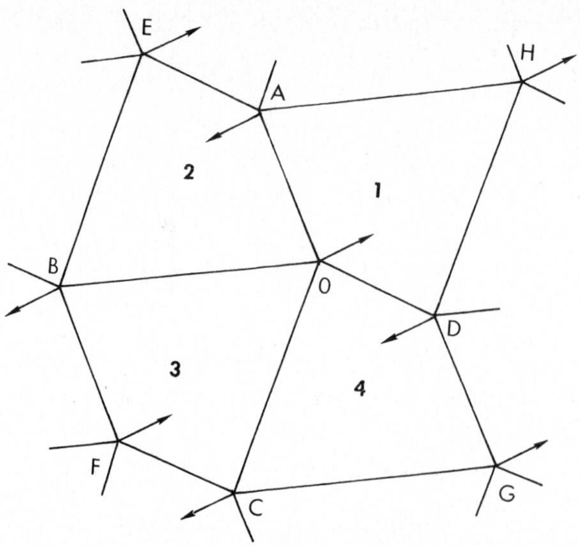

Fig. 9.1. Keystone mode of deformation.

to be initiated, especially at boundary load discontinuities. Since these deformations generate no stresses, they will continue to grow once they are initiated. The problem can be alleviated by rezoning, that is, by replacing a deformed mesh which is developing keystone deformations by a more regular mesh, as described later. The problem may also be delayed by introducing suitable damping.

Maenchen and Sack [42] introduced a keystone viscosity which depends on the difference in rates of rotation of opposite sides of the mesh. Consider mesh *AODH* in Figure 9.1. The rate of rotation of side *AH* is

$$\omega_{AH} = \frac{\{(u_y)_A - (u_y)_H\}\{x_A - x_H\} - \{(u_x)_A - (u_x)_H\}\{y_A - y_H\}}{(x_A - x_H)^2 + (y_A - y_H)^2} \tag{9.10}$$

and similarly for the other four sides. A force is introduced at node O which resists the relative rotation of opposite sides of the mesh. In rectangular coordinates, keystone forces

$$\begin{aligned} F^{\mathrm{k}}_{x1} &= -\tfrac{1}{2}\beta_{\mathrm{k}}\rho_1 c_1 \Delta\bar{x}_1\{(\omega_{AO} - \omega_{HD})(y_A - y_O) + (\omega_{DO} - \omega_{HA})(y_D - y_O)\}\,, \\ F^{\mathrm{k}}_{y1} &= \tfrac{1}{2}\beta_{\mathrm{k}}\rho_1 c_1 \Delta\bar{x}_1\{(\omega_{AO} - \omega_{HD})(x_A - x_O) + (\omega_{DO} - \omega_{HA})(x_D - x_O)\}\,, \end{aligned} \tag{9.11}$$

etc. are added to the stress forces of (7.7). Here β_{k} is a dimensionless constant, $\Delta\bar{x}$ is a characteristic mesh dimension and ρ and c have been introduced to render β_{k} dimensionless as before. The work done by the keystone forces must be included with the dissipative viscous stress work in the energy equation (3.38) and hence also in the stress and temperature equations in (3.37) and (3.39).

It might be possible to retain the convenience of constant nodal masses in cylindrical coordinates by multiplying the nodal force contributions in each mesh by an effective radius, using choices of effective radius similar to those in (7.16). Hancock [36] has used a much simpler means to damp keystone modes. Referring to mesh $AODH$ in Figure 9.1, it may be seen that in dilatational and shear modes of deformation, the midpoints of diagonals AD and OH move at the same velocities. In keystone deformations, however, they move at different velocities. Hancock defines their relative velocity by

$$\begin{aligned} v_{x1} &= \tfrac{1}{4}\{(u_x)_H - (u_x)_A + (u_x)_O - (u_x)_D\}, \\ v_{y1} &= \tfrac{1}{4}\{(u_{yH} - (u_y)_A + (u_y)_O - (u_y)_D\}. \end{aligned} \tag{9.12}$$

In rectangular coordinates, the nodal force contributions

$$F^{\mathrm{k}}_{x1} = -\tfrac{1}{4}\beta_{\mathrm{h}} \frac{M_1 v_{x1}^{n+1/2}}{\Delta t^{n-1/2}}, \qquad F^{\mathrm{k}}_{y1} = -\tfrac{1}{4}\beta_{\mathrm{h}} \frac{M_1 v_{y1}^{n+1/2}}{\Delta t^{n-1/2}} \tag{9.13}$$

are added to the stress forces of (7.7). Here β_{h} is a dimensionless constant and M_1 is the mass of mesh 1. Similar nodal force contributions arise from the other meshes surrounding the node. In cylindrical coordinates, Hancock weights the radial nodal force contribution by

$$F^{\mathrm{k}}_{x1} = -\left(\frac{x_O}{x_O + x_H}\right)^n \tfrac{1}{4}\beta_{\mathrm{h}} \frac{M_1 v_{x1}^{n-1/2}}{\Delta t^{n-1/2}} \tag{9.14}$$

in order for the radial forces to vanish on the axis of symmetry.

Other types of keystone damping have been proposed [30, 35] which, however, also affect dilatational and shear modes of deformation. It is desirable that the damping terms for the various deformational modes be controlled separately, and the two schemes described above are preferable.

It should be noted that undesirable keystone deformations are a global phenomenon, in the sense that adjacent meshes undergo complementary deformations, while the damping forces described above are local, in the sense that they are calculated from the deformation of individual meshes. Meshes may undergo keystone type deformations in motions which do not represent keystoning. For example, the bending of a beam, of which one row of quadrilateral meshes is shown in Figure 9.2, involves relative rotation of opposite sides of the meshes. This is not keystoning, yet the keystone damping described above will

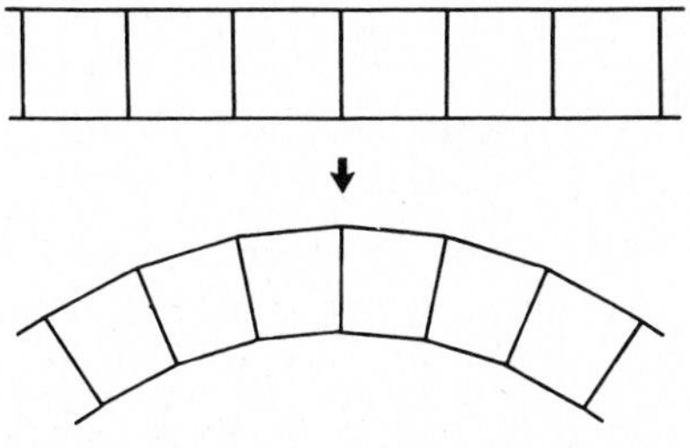

Fig. 9.2. Beam bending deformation mode.

affect this mode of deformation. No global keystone damping schemes have been developed.

It might be noted in passing that the bending of a beam represented by a single row of quadrilateral meshes generates no stresses if the meshes do not change volume. Hence, such a representation of a beam has no bending stiffness. At least two rows of meshes are needed to generate bending stiffness.

10. Boundaries and interfaces

Boundary conditions involve the application of externally prescribed boundary tractions and/or velocities. The boundary conditions are applied by simple modifications in the use of the finite-difference forms of the momentum equations in the methods under consideration.

In order to simplify the logic of the implementation software, phantom meshes are often provided outside the boundary, which carry appropriate densities and stresses derived from the boundary conditions, so that the calculations of the velocities and positions of a boundary node become identical to those for an interior node. If *AOC* in Figure 10.1 is a boundary with adjacent meshes 2 and 3, then phantom meshes 1 and 4 may be constructed so that the equations of Section 7 apply directly.

For example, if *AOC* is a free surface, the stresses and densities of meshes 1 and 4 are set to zero. If *AOC* has a prescribed boundary stress, then appropriate

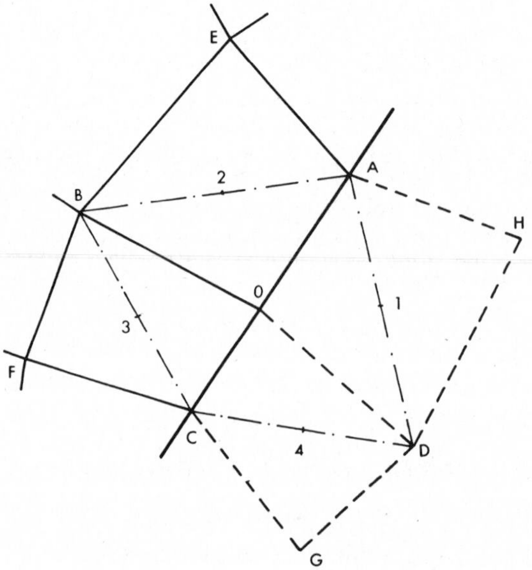

Fig. 10.1. Phantom meshes at a boundary.

average values of the boundary stress acting over AO and OC are set into phantom meshes 1 and 4 respectively, the densities still being set to zero.

When the midpoint method is used, some care is necessary in the choice of location of phantom node D to preserve correct force contributions on node O, but when the Green's theorem method is used, node D is set coincident with node O. In rectangular coordinates, the nodal force contributions due to the boundary stress on AO become simply

$$\begin{aligned} F_x^{\mathrm{b}} &= \tfrac{1}{2}\{\sigma_{xx}^{\mathrm{b}}(y_A - y_O) - \sigma_{xy}^{\mathrm{b}}(x_A - x_O)\}, \\ F_y^{\mathrm{b}} &= \tfrac{1}{2}\{\sigma_{xy}^{\mathrm{b}}(y_A - y_O) - \sigma_{yy}^{\mathrm{b}}(x_A - x_O)\}, \end{aligned} \tag{10.1}$$

which replace the nodal force contributions due to mesh 1 given in (7.8). These can be used directly in cylindrical coordinates if (7.7) is used, or they may be multiplied by $\tfrac{1}{2}(x_A + x_O)$ if the constant nodal mass form (7.16) is used. Similar expressions apply for OC.

Velocity boundary conditions are, of course, straightforward. Instead of calculating nodal velocities from the momentum equations, the values specified by the boundary conditions are used directly. Mixed stress-velocity boundary conditions require more care. In order to discuss them, it is necessary to resolve surface tractions into components parallel and perpendicular to the boundary.

Consider a segment of the boundary AO which makes an angle θ with the x axis. The unit normal vector $\boldsymbol{n}$ and unit tangent vector $\boldsymbol{m}$ have components in two dimensions given by

$$\begin{aligned} n_x &= m_y = \sin\theta = \frac{1}{L}(y_A - y_O), \\ n_y &= -m_x = -\cos\theta = -\frac{1}{L}(x_A - x_O), \end{aligned} \tag{10.2}$$

where $\boldsymbol{n}$ is directed outward from the boundary and $\boldsymbol{m}$ is positive in a counter clockwise sense around the boundary. The length L of the segment is given by

$$L = \{(x_A - x_O)^2 + (y_A - y_O)^2\}^{1/2}. \tag{10.3}$$

The surface traction per unit area $\boldsymbol{f}$ due to a stress $\boldsymbol{\sigma}$ on AO is

$$\boldsymbol{f} = \boldsymbol{\sigma}\boldsymbol{n}, \tag{10.4}$$

or in component form

$$f_x = \sigma_{xx} m_y - \sigma_{xy} m_x, \qquad f_y = \sigma_{xy} m_y - \sigma_{yy} m_x. \tag{10.5}$$

The magnitudes of the normal and tangential components of $\boldsymbol{f}$ on AO, denoted f_{n} and f_{t} respectively, are

$$\begin{aligned} f_{\mathrm{n}} &= f_x m_y - f_y m_x = \sigma_{xx} m_y^2 - 2\sigma_{xy} m_x m_y + \sigma_{yy} m_x^2, \\ f_{\mathrm{t}} &= f_x m_x + f_y m_y = \sigma_{xx} m_x m_y - \sigma_{xy}(m_x^2 - m_y^2) - \sigma_{yy} m_x m_y. \end{aligned} \tag{10.6}$$

Finally the x and y components of the normal and tangential forces $\boldsymbol{F}_{\mathrm{n}}^{\mathrm{b}} = Lf_{\mathrm{n}}\boldsymbol{n}$ and $F_{\mathrm{t}}^{\mathrm{b}} = Lf_{\mathrm{t}}\boldsymbol{m}$ on AO are

$$F_{\mathrm{n}x}^{\mathrm{b}} = f_{\mathrm{n}}(y_A - y_O)\,, \qquad F_{\mathrm{n}y}^{\mathrm{b}} = -f_{\mathrm{n}}(x_A - x_O)\,,$$
$$F_{\mathrm{t}x}^{\mathrm{b}} = f_{\mathrm{t}}(x_A - x_O)\,, \qquad F_{\mathrm{t}y}^{\mathrm{b}} = f_{\mathrm{t}}(y_A - y_O)\,, \tag{10.7}$$

where (10.2) has been used for the components of $\boldsymbol{n}$ and $\boldsymbol{m}$. Note that, in terms of these, the stress boundary conditions (10.1) become simply

$$F_x^{\mathrm{b}} = \tfrac{1}{2}(F_{\mathrm{n}x}^{\mathrm{b}} + F_{\mathrm{t}x}^{\mathrm{b}})\,,$$
$$F_y^{\mathrm{b}} = \tfrac{1}{2}(F_{\mathrm{n}y}^{\mathrm{b}} + F_{\mathrm{t}y}^{\mathrm{b}})\,; \tag{10.8}$$

one half the force on AO being ascribed to node O and one half to node A.

Consider first a symmetry plane. A symmetry plane may be described by providing phantom meshes outside the boundary which are precise reflections of their neighbors inside the boundary. They carry the same mass, and the stresses are such that their nodal force contributions are reflections of those of their neighbors across the plane of symmetry. While the stresses and nodal positions of the phantom meshes outside the boundary may be calculated explicitly, this is unweildy. Rather, the nodal force contributions of the meshes inside the boundary are resolved into components normal and tangential to the boundary plane, using (10.6) and (10.7). The components for mesh 1 in Figure 10.1, for example, are reflections of those for mesh 2 inside the boundary

$$F_{\mathrm{n}x1} = -F_{\mathrm{n}x2}, \qquad F_{\mathrm{n}y1} = -F_{\mathrm{n}y2}\,,$$
$$F_{\mathrm{t}x1} = F_{\mathrm{t}x2}, \qquad F_{\mathrm{t}y1} = F_{\mathrm{t}y2}\,, \tag{10.9}$$

which are then composed into the nodal force components by (10.8). In effect, the component of total force on node O normal to the symmetry plane vanishes, consequently the component of velocity normal to the symmetry plane is also zero.

The calculation for a symmetry plane may be simplified still further. The mass of node O in the method described above is twice the contribution of mass from meshes 2 and 3. The tangential force on node O is also twice the tangential force due to the stress contributions from meshes 2 and 3. If the phantom meshes 1 and 4 are discarded altogether, then mass and force contributions arise only from meshes 2 and 3. If the velocity of node O is calculated using (7.7) or (7.16) directly, it is seen that the component of the velocity tangent to the symmetry plane is calculated correctly, but a spurious component normal to the symmetry plane is present, Figure 10.2. If the velocity is now adjusted so that the normal velocity component is set to zero, the result is identical to that obtained by resolving the stresses into normal and tangential components as outlined above, but is obtained with far fewer arithmetic operations.

Note that in cylindrical coordinates the axis $x = 0$ is a symmetry plane, and needs the proper specification of boundary conditions as described above. The

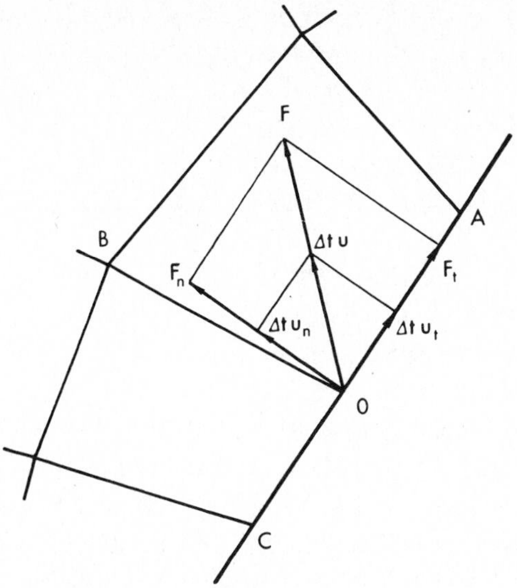

Fig. 10.2. Calculation of nodal velocity at a plane boundary.

boundary condition for a symmetry plane is, except for shear terms, identical with that for a plane rigid surface along which frictionless sliding may occur. This may be generalized in several ways.

If Coulomb friction is present on the surface, then it is necessary to explicitly calculate components of surface traction normal to the surface for each mesh adjacent to the boundary, from which the tangential force due to friction can be calculated. If the magnitude of the normal traction on AO due to the stress in mesh 2 is denoted f_n, given by $(10.6)_1$, then the tangential force on node O due to friction on AO is given by

$$F^{f}_{tx} = \operatorname{sgn}(\Delta\tilde{u})\tfrac{1}{2}\mu f_n(x_A - x_O)\,,$$
$$F^{f}_{ty} = \operatorname{sgn}(\Delta\tilde{u})\tfrac{1}{2}\mu f_n(y_A - y_O)\,, \qquad (10.10)$$

where μ is the Coulomb friction coefficient and $\Delta\tilde{u}$ is the average relative tangential velocity between side AO and the surface. A similar expression holds for side OC. These forces are added to the nodal force contributions arising from stresses in meshes 2 and 3, in order to calculate the velocity of node O. The result is then corrected to set the component of velocity normal to the surface equal to zero. Note that forward time differencing must be used for $\Delta\tilde{u}$ to allow explicit calculation of the velocity.

Viscous friction can be described simply by replacing (10.10) by

$$F^{v}_{tx} = \tfrac{1}{2}\nu\Delta\tilde{u}(x_A - x_O)\,, \qquad F^{v}_{ty} = \tfrac{1}{2}\nu\Delta\tilde{u}(y_A - y_O)\,, \qquad (10.11)$$

where ν is the viscous friction coefficient.

A simple modification allows nodes to separate and reattach to the surface.

The velocity of the node is calculated as above. Before the velocity is corrected so that its component normal to the surface is set to zero, the position of the node is calculated. If the node lies outside the surface, then the node is detached, and friction or viscous sliding forces are henceforth zero. If the node lies on or inside the surface, the node is attached, and its velocity is adjusted so that its final position lies on the surface. Frictional or viscous sliding forces, if present, are henceforth applied.

Another simple modification may be made to allow motion of the boundary surface. After the node velocity is calculated from the nodal forces as above, it is adjusted so that its component normal to the surface equals the specified normal component of surface velocity.

Finally, the calculation may proceed as above even when the bounding surface is curved. Suppose that the boundary is a surface specified in functional form by $\varphi(\boldsymbol{x}, t) = 0$. The velocity of node O is calculated as above, using the nodal forces in the usual way. The result must now be corrected in such a way that the final position of node O once more lies on (or outside) the surface, see Figure 10.3. An error is involved which may be shown to be of $O(\Delta t\, u/R)$ where $\Delta t\, u$ is the magnitude of the final calculated displacement vector and R is the local radius of curvature of the surface. Clearly, large errors may be incurred if the local radius of curvature is not large compared to the node displacement per time step. In particular, discontinuities in slope of the bounding surface generally introduce large errors which usually manifest themselves in numerical oscillations or noise, often of large amplitude.

It should be noted that the calculation of densities, strains and hence of stresses

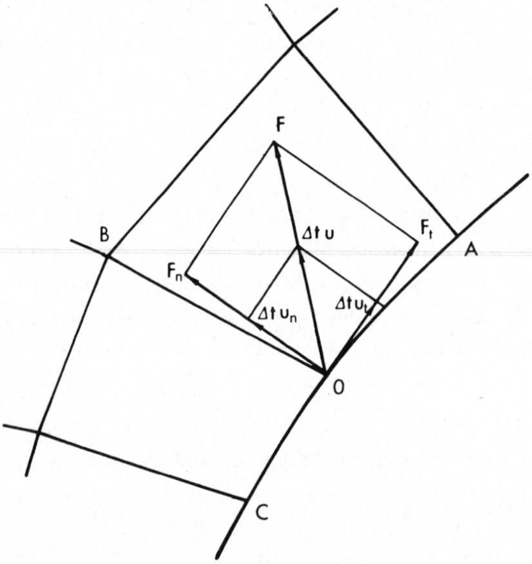

Fig. 10.3. Calculation of nodal velocity at a curved boundary.

in meshes adjacent to a curved boundary require care. For materials with large bulk modulus K, the pressure is sensitive to small errors in density. Consequently it is important that the volume of a mesh adjacent to the boundary be calculated exactly, taking into account the exact shape of the boundary, rather than approximating the boundary by a straight line between nodes, lest errors be introduced into the density through the equation expressing conservation of mass (Section 8). It is compatible with the approximations inherent in the assumption of infinitesimal deviator strain to use only velocities and positions at the four corner nodes of the mesh when calculating strain rates, however.

Summarizing the discussion of external boundary conditions, it is possible to provide phantom meshes around the outside of the boundary, which are endowed with masses, stresses and nodal positions calculated to achieve the prescribed boundary conditions. The calculation of the velocities of boundary nodes is then identical with that for interior nodes. However, the calculation of phantom mesh properties can be cumbersome.

Alternately, boundary node velocities may be calculated from nodal forces including only contributions due to stresses in meshes in the material adjacent to the boundary, plus any specified surface tractions. The result is then corrected to satisfy any boundary velocity constraints. The calculation of the node velocity components via the momentum equation is again identical with that for an interior node. The methods are equivalent, but specification of boundary tractions and subsequent correction of the node velocities to satisfy boundary velocity constraints is usually easier than calculating phantom mesh properties.

An important feature of Lagrangian finite-difference codes is the provision of internal sliding interfaces. Many applications involve the interaction of bodies at an interface where the tangential velocities of the bodies are different. If provision for this relative tangential motion is not made, distortion of the Lagrangian grid adjacent to the interface can become severe enough that the computation cannot continue. More importantly, the forces between the bodies are represented incorrectly, since the materials are prevented from sliding over one another.

Relative motion usually occurs at interfaces between different bodies. However, sharp shear bands may also develop within a body, which may be difficult to resolve with a numerical grid. Occurrence of fracture within a body introduces another situation where relative sliding motion may subsequently occur. Sliding interfaces may be provided which can open or close, providing a means for handling impacts and the opening and subsequent reclosing of fractures. Finally, interfaces upon which opening and sliding are prohibited offer a convenient means of introducing mesh size changes into the numerical grid.

The earliest treatment of sliding interfaces was designed to handle the action of explosive gas products on solid bodies [63]. The boundary of the solid body was designated as a master side. In accelerating boundary nodes of the master side, the pressures in meshes on the slave side are applied as boundary tractions. It is compatible with the approximations of this description to use a surface traction on a master mesh boundary derived from an average of the stresses in meshes on the slave side weighted by the length of their boundary which is coincident with the master mesh boundary [44]. For mesh 2 in Figure 10.4, for example, the force on AO is given simply by

$$F^{b}_{nx} = f_{n8}(y_A - y_G) + f_{n7}(y_G - y_O),$$
$$F^{b}_{ny} = f_{n8}(x_A - x_G) + f_{n7}(x_G - x_O). \tag{10.12}$$

More complicated weightings based on averaging on mesh centers have been used [30] which provide results that do not differ significantly from those obtained as above.

Once the velocities and positions of boundary nodes on the master side have been advanced, boundary nodes on the slave side are advanced, treating the master side as a frictionless prescribed boundary surface.

That is, boundary nodes on the slave side are constrained to lie on master mesh boundaries which have already been determined. Note that it is important to calculate the volume of meshes on the slave side exactly, so that their pressures can be calculated accurately. Mesh 6, for example, must be treated as a pentagon, and not be approximated as a quadrilateral.

The weak form of coupling across the sliding interface, described above, is obviously highly asymmetric with regard to master and slave sides. It is a reasonable approximation when the slave material is of much lower impedance than the master material. Improvements have been made to alleviate this restriction. These involve taking into account the effective mass of the slave material

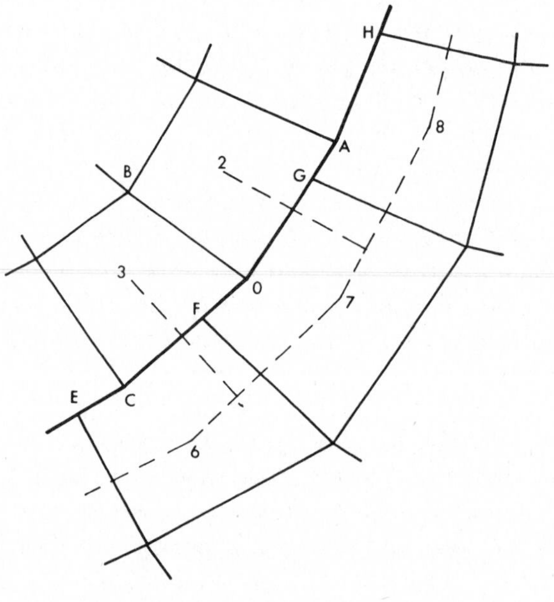

Fig. 10.4. Master–Slave sliding interface.

when master boundary nodes are accelerated, and providing tensor stresses on the slave side to describe solid materials.

For a freely sliding interface, the calculation of master boundary node velocities is divided into two parts, normal and tangential to the interface. Weighted sums of masses as well as normal surface tractions from adjacent slave meshes are included in the calculation of the normal velocity component of master nodes. The tangential velocity component is calculated without mass and force contributions from the slave side. The velocities of slave boundary nodes are calculated as before, treating the master side as a prescribed boundary surface [30, 44].

Coulomb or viscous friction can be accommodated by adding the appropriate nodal forces (10.10) or (10.11) to both master and slave nodes. If the interface does not slide, that is if the materials on each side are considered to be attached to each other, then nodal mass and force distributions from the slave side are used in the calculation of tangential velocity components of master nodes, and slave node velocities are simply found by linear interpolation of master node velocities.

Opening of the interface may be allowed by letting slave nodes separate from the master boundary, as described previously. Subsequent collision can be handled by allowing slave nodes to reattach to the master boundary, treating the latter as a specified boundary surface. Greater accuracy and less numerical noise can be achieved by adjusting normal velocity components of the slave node and its adjacent master nodes in proportion to the impedances of the master and slave materials respectively.

The collision of two bodies may be handled by making their boundaries opposite sides of a sliding interface, which is initially detached. The main modification to the procedures described above involves testing of each slave node against every mesh side lying in the master boundary to determine whether contact has occurred. Most implementations restrict generality in order to reduce the number of arithmetic operations required to do this testing, for example by allowing slave nodes to attach only sequentially along the boundary from a known point of initial contact.

The strong form of coupling across the sliding interface described above still retains some asymmetry with respect to master and slave sides. This could be removed by alternating master and slave sides in successive time steps, in order to achieve a completely symmetric description. In order to do so, it is necessary to smooth the interface after each time step, using, for example, spline interpolation in such a way as to conserve mass, momentum and energy on either side of the interface. Node positions could be weighted in proportion to their nodal masses, and some experimentation with the error assigned to nodal positions and the number of nodes to be included in the fit on either side of the node whose position is being adjusted would be required. A spline fit would have the advantage of providing a good calculation of the surface normal for use in the interface calculation.

11. Rezoning

In previous sections, solution methods for the large deformation dynamic response of nonlinear materials in multidimensions have been developed in considerable detail. In this section the rezoning methods typically employed in solving such problems are developed in a more heuristic fashion where considerable use of one-dimensional examples is employed. Rezoning is used in most Lagrangian large deformation stress wave propagation codes to improve solution accuracy and efficiency. Distortion can be severe in large deformation multidimensional problems and can raise havoc with the solution [45]. Rezoning is required to ameliorate the following difficulties caused by distortion:

(i) Computational grid non-uniformity – which may produce a reduction of the order of accuracy for centered finite differences [46].

(ii) Reduction of adjacent mesh point distances – which may, by virtue of the stability criterion, reduce the time step of the calculation.

(iii) Preposterious conditions such as zero or negative zone volume – which may produce states in the equation of state which are unstable or which are at such high pressures that ensuing calculations are worthless.

Methods to solve the above difficulties in multidimensional codes have progressed from the discarding of selected zones to essentially fully-automatic rezones that can accommodate major changes to the computational grid and can detect when such changes are necessary. Before rezoning techniques were available, it was necessary at times to (i) initiate problems with a distorted mesh to compensate for real distortion and thereby obtain solutions to longer times, or (ii) ignore stability requirements in regions of the problem where distortion is large and where it could be rationalized that the solution in this area was no longer important to the primary result being calculated. These stop-gap measures were clearly unsatisfactory for many large deformation problems and rezoning methods have evolved to provide more satisfactory results.

The evolution of rezoning methods has been slow; spanning at least the last ten years and still requiring refinement. Initial attempts merely discarded meshes that were troublesome. Since most distortion usually occurred near boundaries, highly distorted meshes in these regions were removed and replaced with appropriate boundary meshes. The next iteration [47] in the evolution of rezoning methods was to adjust the position of a single point in the interior of the grid in such a way that the mass, momentum, and possibly the energy of the meshes surrounding this point were conserved and such that distortion was minimized. This procedure could be repeated for several mesh points and at several cycles of the problem to provide a solution. However, this was extremely taxing for the user and required several steps and several days to obtain a solution as it was initially implemented in a batch operating mode on the computer.

A major improvement in rezoning methods [57, 58] involved a procedure for rezoning an entire region of grid. It required user specification of the time when rezoning is to be done and specification of the region needing rezoning; but it provided essentially automatic specification of an optimum new grid for the region and a procedure for specifying all of the state point variables for the new grid such that mass, momentum, and usually energy were conserved. The new grid specification included the possibility of changing the number of rows and/or columns and allowed boundary smoothing. The new grid was obtained by using the boundaries of the old region, smoothing these boundaries if necessary, and solving for new grid point positions by defining potential functions for both spatial coordinates in terms of the logical grid coordinates for the region and by obtaining a solution to the resulting bi-harmonic problem [60]. The new grid was then overlayed on the old grid and the mass, momentum, and energy of every portion of the old zones that intersected each new zone were used to determine the density, velocities, and internal energy of each new zone. Then the constitutive relation was used to complete the specification of the variables for each new grid point. While other methods such as arbitrary Lagrangian-Eulerian coordinates have been used to move the computational grid at each time step [55, 56, 33] the 'one-step' procedure outlined above is attractive because it retains the Lagrangian nature of the problem for as long as possible, minimizes the amount of material property diffusion, and usually minimizes the required computer time to obtain a solution.

The 'one-step' procedure has been implemented in a manner that utilizes the restarting feature of wavecodes [48, 35]. During the running of the wavecode restart files are written. These files are examined after the wavecode is run to determine when rezoning is necessary. If it is decided that rezoning is required for one of these restart files, it is rezoned using the above procedure and a new restart file is generated. The information generated beyond this point in the original wavecode run is discarded and the wavecode is restarted using the rezoned grid and continued while additional restart files are generated which may also be rezoned. Problem solution with this method may involve several job submissions, especially since the determination of appropriate rezoning times is difficult and false continuations may result if rezoning is done too infrequently. Although this can be a trial and error procedure and although solutions can require large demands on user time and several days, many very large deformation problems have been solved using this method [48, 50, 57]. Furthermore, the rezoning procedure was verified by comparing the results to independent solutions in these problems.

Major reductions in the man time required for rezoning problems have recently been obtained by automating the above procedure [52]. This has been accomplished by using a control program along with the wavecode and the rezoning program. The control program automatically runs the wavecode until certain criteria are met at which time the wavecode calculation is interrupted, the rezoning code is run to rezone an appropriate portion of the problem, and the wavecode calculation is restarted and run until these criteria are again met. In essence the difficulty in obtaining a solution is reduced to a specification of rezoning criteria.

The specification of rezoning criteria is still in a development phase. Two different approaches have been used to provide additional rezoning experience and to provide guidance to the development of improved criteria. In one approach called 'semi-automatic' rezoning, the user is required to have a considerable amount of knowledge concerning the solution. For example, the user would specify the region requiring rezoning and the times at which rezoning would be done. In the other approach, called 'fully-automatic' rezoning, very little knowledge of the solution is required a priori. In this case, criteria are used to indicate both the times and regions for rezoning. When a criterion is met at a given zone, rezoning is done at that time in a region in the neighborhood of that zone. The size of this neighborhood is an input quantity which usually extends a distance on the order of 3 or 4 zones from the zone meeting the rezone criterion. In the preliminary version of this automatic rezoning code [52], three criteria are used which examine the distortion of a zone. Reference values for (i) the disparity in the relative lengths of longest and shortest sides of the zone, (ii) the size of the internal angles of the zone and (iii) the area of the zone are computed initially and subsequent values for each of these parameters are calculated and compared to the reference values. If the change in any of these parameters is larger than an input quantity, rezoning is done in the neighborhood of this zone and the values of the above parameters for the newly formed zones are used as new reference values.

Additional criteria or improvements to the above parameters will probably be required before a 'fully-automatic' rezoning procedure will be available. Several possibilities exist including a time step criterion such as a minimum value or a maximum amount of change, adjacent zone area comparisons, row or column line length comparisons, maximum pressure, etc.

In principle, rezoning is straightforward. All that is needed is to redefine a problem in such a way that mass, momentum, and energy are conserved and such that the perturbations on the problem are negligible. In practice, even for one-dimensional problems, it can be very difficult. For example, consider the very simple one-dimensional rezone illustrated in Figure 11.1 where two zones are to be rezoned into one by removing grid line j. A determination of the new velocities at points $j-1$ and $j+1$ and the properties of the new mesh that conserve mass, momentum, and energy are desired. Also, the velocities and material properties of adjacent points should not be influenced by such a rezone.

The density of the new zone can be obtained from mass conservation in a straightforward manner. However, the application of momentum and energy conservation is not so clear. The basic difficulty derives from the fact that there are three unknowns (new velocity at $j-1$, new velocity at $j+1$, and new internal energy of the new mesh) and there are only two equations (momentum and energy conservation). Several different approaches have been used to circumvent

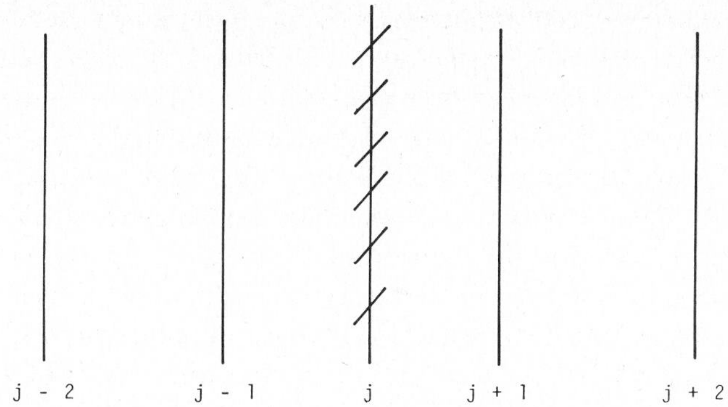

Fig. 11.1. One-dimensional rezoning example in which zone line j is to be removed.

this problem. The way this is done in the WONDY code [61] is essentially to generate another equation by trying to conserve both kinetic and internal energy. Then internal energy conservation is straightforward and the two velocities are determined by a simultaneous solution of the momentum and kinetic energy conservation equations [53]. This requires the solution of a quadratic equation which may have complex roots: if the roots are real, both momentum and kinetic energy are conserved; if the roots are complex, only the real part is used which results in momentum conservation and a minimization of the kinetic energy error. In the PUFF code [62], one of the new velocities is approximated (a mass average is used), the other one is determined from momentum conservation, and the internal energy of the new zone is obtained from total energy conservation.

Another possible approach is to conserve momentum on each side of the zone line removed. This has the effect of generating another momentum equation. An extension of this idea is essentially the basis for the method used in the TOOREZ two-dimensional rezoning code [48]. In a one-dimensional analogy of this code, a momentum zone is defined for each grid point which includes one half the mass of the zones on each side of the grid point and in which the velocity is assumed to be constant and equal to the velocity of the grid point. Then the momentum of new zones is determined by summing the momentum components of all the old momentum zones that intersect the new momentum zone in the same way that mass conservation is done. This results in a procedure that yields the correct number of equations and unknowns for both the interior and boundary points in a two-dimensional rezone.

In one dimension this procedure reduces to that illustrated in Fig. 11.2 where the momentum zone boundaries are indicated by dashed lines. Thus rezoning is accomplished as follows:

1. The new grid, denoted by $i = 1, 2, \ldots$, is determined along with the new momentum zone boundaries which are located at the mid-points between the new grid zone boundaries.

2. The density of all the new zones is determined from the volume of each new

zone and the total amount of mass which this zone intersects when it is overlayed on the old grid. For example, for the new zone defined by zone boundaries i and $i-1$ in Figure 11.2; a portion of the mass defined by old zone boundaries $j-1$ and $j-2$ is included, all of the mass between the j and $j-1$ boundaries is included, plus a portion of that between the $j+1$ and j zone boundaries.

3. The velocity of all new zone boundaries is obtained from momentum conservation by determining the momentum of each new momentum zone. The new momentum equals the sum of the momentum portions from all the old momentum zones which are intersected by the new momentum zone. This is accomplished using momentum zones in the same fashion as outlined above for mass conservation. The velocity of each new zone boundary is obtained by dividing this new momentum by the mass contained in the new momentum zone as obtained from the new densities as determined above.

4. The new specific internal energy for each new zone is usually determined from an expression of total energy conservation. However, if rezoning is not done sufficiently often, grid nonuniformity or other effects may produce errors [46] which upon rezoning result in the single-step conversion of too much kinetic energy to internal energy. This difficulty is described in considerable detail below. Clearly, rezoning should be done with a frequency that allows total energy conservation. However, it is occasionally expedient, especially in multidimensional problems, to take large time intervals before rezoning and use only internal energy conservation.

4a. When internal energy (not total energy) conservation is used, the new specific internal energy for each new zone is obtained in a direct fashion from the

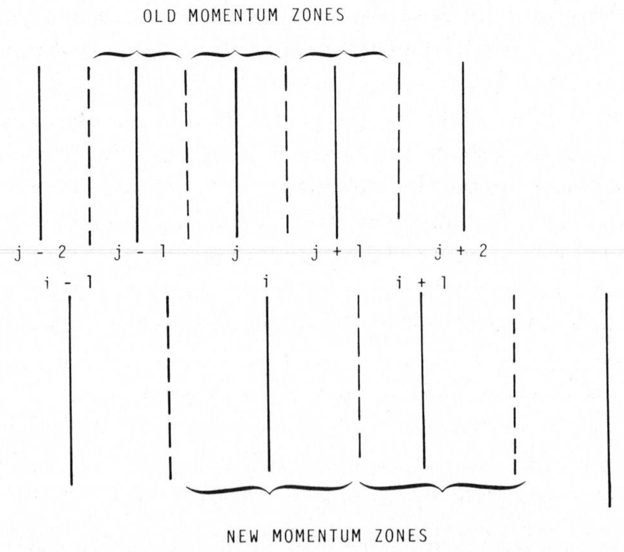

Fig. 11.2. One-dimensional rezoning procedure using momentum zones whose boundaries are indicated by the dashed lines.

new mass (given by the new density and volume) and the sum of the old mass-specific internal energy products for the old zones intersected by the new zone.

4b. When total energy is conserved[9], the constant velocity assumption for each momentum zone is used to calculate the old and new kinetic energies. The new kinetic energy is given by the sum of the product of the new mass and one-half the square of the new velocity for each side of the new zone. The old internal energy and old kinetic energy are obtained by summing over that portion of the old grid that is intersected by the new zone. Then the new specific internal energy is obtained by dividing the new mass into the difference between the old total energy and the new kinetic energy.

5. When the equation of state expresses the spherical portion of the stress as a function of the density and internal energy, the pressure for the new zone is obtained directly using the new density and internal energy. Also the sound speed is obtained directly from a derivative of the equation of state. When a complete equation of state [2, 3] is used, an iteration procedure is required for both the pressure and sound speed. However, this is not a major limitation and such a procedure has been implemented [48].

6. Deviatoric stresses are determined using an approximate scheme which makes it possible to determine the new yield stress even when isotropic work hardening is present[10]. Although force summation provides accurate expressions for each stress component when one-dimensional rezoning is used, such a procedure is too cumbersome when multidimensional rezoning in curvilinear coordinates is considered. As volume averaging of the stress components is accurate for Cartesian coordinates and is a reasonable approximation otherwise (especially when other errors such as those due to centering in the application of energy conservation are considered), the new stress deviator components are obtained by volume averaging.

7. Several other state variables may be present depending upon the constitutive relation being used to describe the material being rezoned. These variables are presently volume averaged. The rationale for this is (i) volume averaging is frequently appropriate, (ii) when mass averaging is appropriate, it can be reduced to volume averaging by using local linearity assumptions [61], and (iii) if sufficient constitutive relation information is available to indicate the need for other averaging methods, the user can easily modify the code to accommodate them.

In summary, the sequence used to implement the above rezoning procedure for one or two dimensions is (i) determine the new grid, (ii) determine the new

[9]Note that in a strict sense it is not possible to exactly conserve total energy because the internal energy and the kinetic energy exist at different values of time because of the way velocity and energy are centered [25].

[10]In the case of isotropic work hardening, the plastic work per unit volume is a function of the yield stress. Thus during rezoning a volume average of the plastic work provides a determination of the new yield stress.

densities for all the new zones in one pass through the new grid, (iii) determine the velocities of all new grid points using momentum zones in another pass through the new grid, and (iv) in a final pass through the new grid, determine the specific internal energy, pressure, stress deviators, sound speed, and any other appropriate variables for all the new zones.

As previously mentioned, rezoning can result in the single-step conversion of kinetic energy into internal energy. This process is illustrated in an extreme example as shown in Figure 11.3 where the velocity distribution for the old grid is such that the velocity is V for all zone boundaries to the left of zone boundary j and is equal to $-V$ for all other zone boundaries. Using the new grid shown in the lower portion of Figure 11.3 and the momentum zones previously defined, it is clear that (i) for all new zone boundaries to the left of i the velocity is V, (ii) for all new boundaries to the right of i the velocity is $-V$, and (iii) the velocity of boundary i is zero. As each old momentum zone had a specific kinetic energy of $\frac{1}{2}V^2$, as all new momentum zones except the one centered about zone boundary i also have a specific kinetic energy of $\frac{1}{2}V^2$, and as the one centered about i has no specific kinetic energy; it is clear that if total energy is conserved, the internal energy for the zones defined by boundaries i and $i-1$, and $i+1$ and i, must increase.

This conversion of kinetic energy to internal energy is real and not a figment of the particular rezoning method being used. For the example shown in Figure 11.3 it is evident that a shock would form which would increase the internal energy of the zones defined by $i-1$ and $i+1$ and would reduce the kinetic energy of the zones to zero as it propagated in both directions from its origin midway between zone boundaries j and $j-1$. As a matter of fact, the conversion of kinetic to internal energy is in principle a desirable trait of rezoning; that is, the conversion

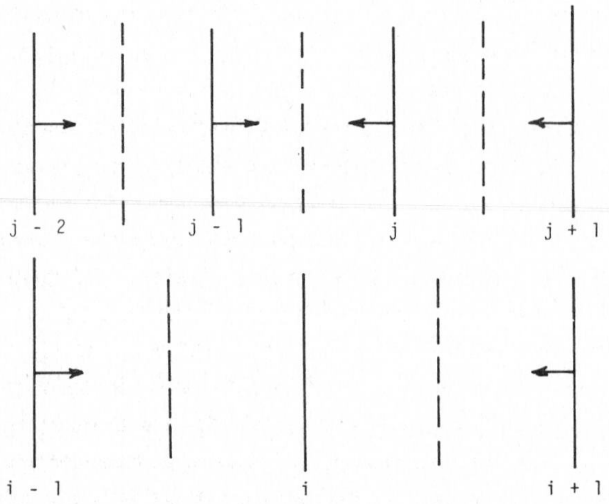

Fig. 11.3. One-dimensional rezone illustrating the conversion of kinetic energy to internal energy.

of random velocities to internal energy is frequently needed. However, if these random velocities are the result of calculational errors, if they become large, and if they converge at a region being rezoned; large stresses may result from this energy conversion which in turn may result in additional errors that ultimately destroy the validity of the calculation. Hence rezoning must be done before such errors become large.

The above energy conversion results from the simple fact that momentum conservation yields average velocities which are less than or equal to those that would have been obtained if a root mean square velocity were used. Hence the final kinetic energy is always less than or equal to the initial kinetic energy.

As previously mentioned only an approximate total energy conservation expression can be written for rezoning because of centering problems. Therefore, errors can be expected in the rezoning process. These errors are fundamental. They have been discussed in terms of entropy errors in problems where a wave is propagated from a region of one zone spacing to another region with different spacing in the absence of rezoning [54].

It is convenient to relate both the energy errors in rezoning and those due a change of zone spacing to entropy errors. The philosophy for doing this is a result of the fact that when the zoning is changed in either case, the information content of the problem changes. For example, the minimum wavelength of any information is limited by the zone spacing. Thus, zone spacing changes in the presence of waves limited by this spacing results in information losses and entropy increases. It is unfortunate that the difficulty of analyzing this energy error forces such a philosophical discussion, however it illustrates one point very clearly, viz., the errors produced during rezoning should be largest in cases when the information is dominated by wavelengths on the order of the mesh spacing. Such is the case when random velocities occur. Thus, we are once again led to the conclusion that rezoning needs to be done frequently. Of course this is balanced by the expense of rezoning, both in terms of computer time and man-time if automatic rezoning is not used.

The procedures and difficulties discussed in the above two sections for one-dimensional rezoning apply directly to two dimensions. As a matter of fact the procedures discussed have not been directly implemented in one dimension – they are one-dimensional analogs of the procedures used in the TOOREZ two-dimensional rezoning code [48]. The primary complications due to the addition of a second dimension result from complications in the grid. That is, new grid definition, old and new momentum zone determination, grid intersection, and boundary treatments are considerably more difficult in two dimensions than they are in one dimension. However, the sequence of steps to determine the new grid, new densities, new velocities, new energy, etc. is the same and the use of another momentum conservation equation to determine the other component of velocity is straightforward.

In the TOOREZ two-dimensional rezoning code [48], rezoning of the entire grid or a region of the grid is done sub-region by sub-region where a sub-region includes only one material and does not contain any sliding interfaces, although the boundary of a sub-region may be coincident with one side of a sliding interface.

The boundaries of an old sub-region are used to obtain the boundaries of the new sub-region. If the number of columns and rows of zones are the same for the old and new grids for this sub-region and if the old boundaries are sufficiently smooth, then the new boundaries are made identical to the old. However it is usually necessary to smooth the boundaries. This can be done in several different ways as long as the resulting new boundary maintains the general curvature of the original boundary and has nearly uniform spacing. Methods have been used where both curvature and spacing iterations are used [57], but the simplified procedure outlined below has been found to be adequate: The total length of the boundary, as determined by summing the distances between each point, is divided by n', the number of new line segments, to obtain the line length D_1 between the first two new points (P'_0 and P'_1) as illustrated in Figure 11.4. The distance D_2 to the next new point is obtained by dividing the distance from P'_1 to P_n along the old line by $n' - 1$ and so on until all the new points are determined for the first iteration. The iteration is continued for a fixed number of iterations (usually 10 or less) or until essentially uniform spacing results ($D_1 \leq 1.1\, D_n$). This procedure rapidly converges, smoothes the boundary, results in nearly uniform spacing along the boundary, and preserves the general shape of the boundary. While some embellishment of the method is required to treat all cases[11], it has been found to be satisfactory for all cases considered including those having extreme distortion [49, 50].

[11]Fixing the position of a few points can be used for boundaries having extreme or multiple curvatures. Also specified boundary curves can be used for special cases.

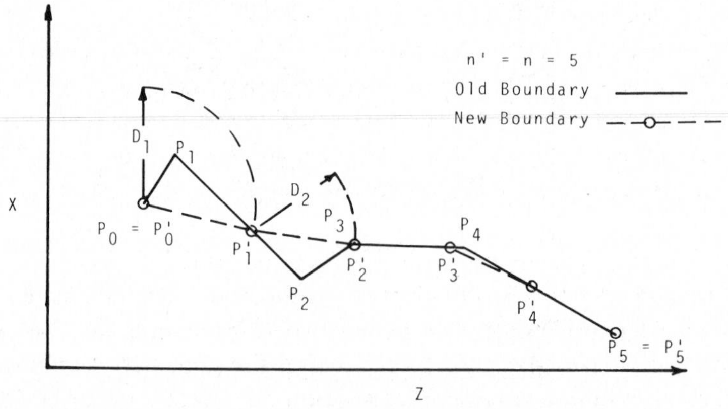

Fig. 11.4. Example of iterative procedure to smooth a two-dimensional boundary where $D_1 \geq D_2 \geq \cdots \geq D_n$ and iteration continues until $D_1 \geq 1.1\, D_n$.

This smoothing procedure is also used when the sub-region boundary is a slide line. In this case, the column of zones adjacent to this sub-region boundary is also rezoned. Whenever boundary smoothing is done, some of the original mass may fall outside the new grid and some of the new grid may cover space originally containing either no mass (free boundary case) or mass from a different sub-region. Original mass falling outside the new grid is discarded and mass is added in places where the new grid covers regions not defined by the original grid. This added mass has the density, velocity, and specific internal energy of the original material closest to the added mass. Thus, when boundary smoothing is used, mass, momentum, and energy conservation are locally violated. These perturbations are usually negligible on the global scale and, if they are found to be too large[12], a fatal error terminates the rezone.

Once the smoothed boundaries are defined, new grid points within these boundaries are determined. Equipotential [60] zoning has been very successful for this purpose. It converges reliably, yields smooth and uniform grids, and provides adequate flexibility for fixing specific points if necessary.

As previously discussed for the one-dimensional case, a momentum zone having a uniform velocity is defined for each point. For interior points, such as P_9 in Figure 11.5, each momentum zone is defined by 4 quadrilateral sub-zones.

[12]Total mass, momentum, and energy before and after rezone are computed and errors are compared to input values.

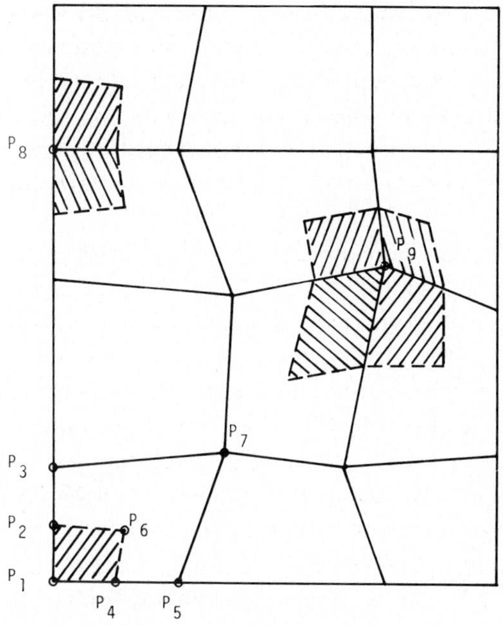

Fig. 11.5. Momentum zones as defined by quadrilateral subzones for two-dimensional rezoning.

Boundary points such as P_8 have 2 sub-zones and corner points (P_1) have one sub-zone. The 4 points defining each sub-zone quadrilateral are the grid point, the mid-point of two associated zone lines and usually the point of intersection of the zone diagonals. For example, consider P_1 in Figure 11.5 where the 4 points are P_1, P_2 which is midway between P_1 and P_3, P_4 which is midway between P_1 and P_5, and P_6 which in this case is the intersection of lines P_1P_7 and P_3P_5. If an angle, as measured interior to a zone, exceeds 180°, one diagonal lies outside the zone. Although this is rare and rezoning should be done before this occurs, it is possible for the old grid. When it occurs, the mid point of the diagonal lying inside the zone is used as the fourth point for the quadrilateral momentum sub-zone.

The only remaining major difficulty introduced by adding the second dimension is determining zone intersections. This can become quite involved especially for the case of momentum sub-zones as there are nearly 4 times as many of these zones in both the old and new grids. While the determination of the region in the old grid that is intersected by each new zone (for mass conservation) or sub-zone (for momentum and energy conservation) requires a search through a large amount of data, it is straightforward [48] and will only be outlined very briefly.

For interior zones in the new grid the search of the old grid begins at the location in the old grid where the last portion of material was found when the search was completed for the new zone adjacent to the one currently being filled. Both the old and new quadrilateral zones or sub-zones are decomposed [48] into four trapezoids to facilitate the search and intersection problem. The area of each quadrilateral is the oriented sum of the areas of the trapezoids, the intersection of the trapezoids forms trapezoids, and the intersection area is also an oriented sum [48]. The intersection volume (using either a unit depth or rotation angle) and mass are computed and, if appropriate, the momentum and energy components are also determined. For the boundary zones of the new grid, the same procedure is used where an additional treatment, due to the fact that material must be added or deleted when boundary smoothing occurs, is incorporated as previously discussed.

The above discussion summarizes a manual two-dimensional rezoning procedure as implemented in the TOOREZ code [48]. As indicated, several approximations are used and there are several sources for error in this procedure. The only feasible way to check the adequacy of the method is to solve large deformation problems and compare the results to independent code solutions and to experiment. This has been done for several problems where rezoning was done a large number of times (order 30) and these comparisons have been favorable [49, 50, 51]. Although such studies cannot prove the accuracy or adequacy of the method, they do provide the user with some confidence that the method can be applied to a wide range of problems.

On the other hand, the use of manual rezoning for problems requiring many rezones is very expensive when measured in terms of the amount of man-time. Typical solutions of this nature require a man-month or so. Hence, steps have been taken to automate the process. The first step was to develop a semi-automatic procedure which required the user to specify the region requiring rezoning and the times at which it was to be done. This was implemented by using a control program [59] which runs the wave propagation code until rezoning is to be done, then runs the rezoning code, and returns to the wave propagation code. The next step was the development of rezoning criteria [52] in an attempt to provide a procedure that would automatically detect where and when rezoning was required with very little user knowledge of the solution. Considerable success has been obtained with the present criteria and great savings in man-time have resulted from the use of this method, but refinements will probably be required before a fully-automatic rezoning code is completed. Such refinements will evolve as more problems are solved with this code.

Glossary

$\boldsymbol{a}$	Displacement vector
A	Helmholtz free energy, mesh area
$\boldsymbol{b}$	Body force vector
C	Specific heat
c	Bulk sound speed
$\boldsymbol{D}$	Stretching tensor
$\boldsymbol{e}$	Infinitesimal strain tensor
$\boldsymbol{E}$	Green's strain tensor
$\mathscr{E}$	Internal energy
f	Evolution function
$\boldsymbol{F}$	Deformation gradient tensor
F	Nodal force
$\boldsymbol{g}$	Temperature gradient vector
$\boldsymbol{G}$	Material temperature gradient vector
$\boldsymbol{h}$	Heat flux vector
$\boldsymbol{H}$	Material heat flux vector
J	Jacobian
$\boldsymbol{K}$	Fourth order modulus tensor
K	Bulk modulus
$\boldsymbol{L}$	Velocity gradient tensor
$\boldsymbol{M}$	Fourth order viscosity tensor
M	Mesh mass
$\boldsymbol{Q}$	Rotation tensor
r	Heat source strength
$\boldsymbol{R}$	Rotation tensor
S	Entropy
t	Time
T	Temperature
$\boldsymbol{u}$	Velocity vector
$\boldsymbol{U}$	Stretch tensor
$\boldsymbol{W}$	Spin tensor
$\boldsymbol{x}$	Spatial coordinate vector
x	Spatial coordinate
$\boldsymbol{X}$	Material coordinate vector
y	Spatial coordinate
α	Internal state variable
β	Artificial viscosity coefficient
$\boldsymbol{\Gamma}$	Grueneisen tensor
Γ	Grueneisen coefficient
ζ	Bulk viscosity coefficient
η	Shear viscosity coefficient
θ	Dilatational strain
$\boldsymbol{\mathscr{K}}$	Heat conduction tensor
$\mathscr{K}$	Heat conduction coefficient
λ	Lamé elasticity coefficient
μ	Lamé elasticity coefficient
ν	Viscous friction coefficient
ξ	Mesh distortion measure
ρ	Density
$\boldsymbol{\sigma}$	Cauchy stress tensor
$\boldsymbol{\Sigma}$	Second Piola–Kirchoff stress tensor

Φ Stress relaxation function

χ Mesh distortion measure

ψ Yield function

Ψ Temperature relaxation function

ω Rotational velocity

Appendix. Two-dimensional differential equations for an isotropic material

$$\frac{\partial u_x}{\partial t} = -u_x \frac{\partial u_x}{\partial x} - u_y \frac{\partial u_x}{\partial y} + \frac{1}{\rho}\frac{\partial \sigma^{e\prime}_{xx}}{\partial x} + \frac{1}{\rho}\frac{\partial \sigma^{e\prime}_{xy}}{\partial y} - \frac{1}{\rho}\frac{\partial p}{\partial x}$$

$$+ \frac{1}{\rho}\frac{\partial \sigma^{v\prime}_{xx}}{\partial x} + \frac{1}{\rho}\frac{\partial \sigma^{v\prime}_{xy}}{\partial y} - \frac{1}{\rho}\frac{\partial q}{\partial x} + b_x + \left[\frac{2\sigma^{e\prime}_{xx} + \sigma^{e\prime}_{yy}}{\rho x} + \frac{2\sigma^{v\prime}_{xx} + \sigma^{v\prime}_{yy}}{\rho x}\right], \tag{A1}$$

$$\frac{\partial u_y}{\partial t} = -u_x \frac{\partial u_y}{\partial x} - u_y \frac{\partial u_y}{\partial y} + \frac{1}{\rho}\frac{\partial \sigma^{e\prime}_{xy}}{\partial x} + \frac{1}{\rho}\frac{\partial \sigma^{e\prime}_{yy}}{\partial y} - \frac{1}{\rho}\frac{\partial p}{\partial y}$$

$$+ \frac{1}{\rho}\frac{\partial \sigma^{v\prime}_{xy}}{\partial x} + \frac{1}{\rho}\frac{\partial \sigma^{v\prime}_{yy}}{\partial y} - \frac{1}{\rho}\frac{\partial q}{\partial y} + b_y + \left[\frac{\sigma^{e\prime}_{xy}}{\rho x} + \frac{\sigma^{v\prime}_{xy}}{\rho x}\right], \tag{A2}$$

$$\frac{\partial x}{\partial t} = u_x, \tag{A3}$$

$$\frac{\partial y}{\partial t} = u_y, \tag{A4}$$

$$\frac{\partial \rho}{\partial t} = -u_x \frac{\partial \rho}{\partial x} - u_y \frac{\partial \rho}{\partial y} - \rho \frac{\partial u_x}{\partial x} - \rho \frac{\partial u_y}{\partial y} - \left[\frac{\rho u_x}{x}\right], \tag{A5}$$

$$\frac{\partial e'_{xx}}{\partial t} = -u_x \frac{\partial e'_{xx}}{\partial x} - u_y \frac{\partial e'_{xx}}{\partial y} + \frac{2}{3}\frac{\partial u_x}{\partial x} - \frac{1}{3}\frac{\partial u_y}{\partial y} - \left[\frac{1}{3}\frac{u_x}{x}\right], \tag{A6}$$

$$\frac{\partial e'_{xy}}{\partial t} = -u_x \frac{\partial e'_{xy}}{\partial x} - u_x \frac{\partial e'_{xy}}{\partial y} + \frac{1}{2}\frac{\partial u_y}{\partial x} + \frac{1}{2}\frac{\partial u_x}{\partial y}, \tag{A7}$$

$$\frac{\partial e'_{yy}}{\partial t} = -u_x \frac{\partial e'_{yy}}{\partial x} - u_y \frac{\partial e'_{yy}}{\partial y} - \frac{1}{3}\frac{\partial u_x}{\partial x} + \frac{2}{3}\frac{\partial u_y}{\partial y} - \left[\frac{1}{3}\frac{u_x}{x}\right], \tag{A8}$$

$$\frac{\partial p}{\partial t} = -u_x \frac{\partial p}{\partial x} - u_y \frac{\partial p}{\partial y} - (K + \Gamma\sigma^{v\prime}_{xx} - \Gamma q)\frac{\partial u_x}{\partial x} - \Gamma\sigma^{v\prime}_{xy}\frac{\partial u_y}{\partial x}$$

$$- \Gamma\sigma^{v\prime}_{xy}\frac{\partial u_x}{\partial y} - (K + \Gamma\sigma^{v\prime}_{yy} - \Gamma q)\frac{\partial u_y}{\partial y} - \Gamma\frac{\partial h_x}{\partial x} - \Gamma\frac{\partial h_y}{\partial y} - \Gamma\rho r - \Phi_{\mathrm{b}}$$

$$- \left[(K - \Gamma\sigma^{v\prime}_{xx} - \Gamma\sigma^{v\prime}_{yy} - \Gamma q)\frac{u_x}{x} + \Gamma\frac{h_x}{x}\right], \tag{A9}$$

$$\frac{\partial \sigma^{e\prime}_{xx}}{\partial t} = -u_x \frac{\partial \sigma^{e\prime}_{xx}}{\partial x} - u_y \frac{\partial \sigma^{e\prime}_{xx}}{\partial y} + \frac{4}{3}\mu\frac{\partial u_x}{\partial x} - \sigma^{e\prime}_{xy}\frac{\partial u_y}{\partial x} + \sigma^{e\prime}_{xy}\frac{\partial u_x}{\partial y} - \frac{2}{3}\mu\frac{\partial u_y}{\partial y} + \Phi_{\mathrm{d}}e'_{xx}$$

$$- \left[\frac{2}{3}\mu\frac{u_x}{x}\right], \tag{A10}$$

$$\frac{\partial \sigma_{xy}^{e\prime}}{\partial t} = -u_x \frac{\partial \sigma_{xy}^{e\prime}}{\partial x} - u_y \frac{\partial \sigma_{xy}^{e\prime}}{\partial y} + \left(\mu + \frac{1}{2}\sigma_{xx}^{e\prime} - \frac{1}{2}\sigma_{yy}^{e\prime}\right)\frac{\partial u_y}{\partial x}$$

$$+ \left(\mu - \frac{1}{2}\sigma_{xx}^{e\prime} + \frac{1}{2}\sigma_{yy}^{e\prime}\right)\frac{\partial u_x}{\partial y} + \Phi_{\mathrm{d}} e'_{xy} - \left[\frac{2}{3}\mu \frac{u_x}{x}\right], \tag{A11}$$

$$\frac{\partial \sigma_{yy}^{e\prime}}{\partial t} = -u_x \frac{\partial \sigma_{yy}^{e\prime}}{\partial x} - u_y \frac{\partial \sigma_{yy}^{e\prime}}{\partial y} - \frac{2}{3}\mu \frac{\partial u_x}{\partial x} + \sigma_{xy}^{e\prime}\frac{\partial u_y}{\partial x} - \sigma_{xy}^{e\prime}\frac{\partial u_x}{\partial y}$$

$$+ \frac{4}{3}\mu \frac{\partial u_y}{\partial y} + \Phi_{\mathrm{d}} e'_{yy} - \left[\frac{2}{3}\mu \frac{u_x}{x}\right], \tag{A12}$$

$$\frac{\partial \alpha_a}{\partial t} = -u_x \frac{\partial \alpha_a}{\partial x} - u_y \frac{\partial \alpha_a}{\partial y} + f_a, \qquad a = 1, 2, \ldots, k, \tag{A13}$$

$$\frac{\partial \mathscr{E}}{\partial t} = -u_x \frac{\partial \mathscr{E}}{\partial x} - u_y \frac{\partial \mathscr{E}}{\partial y}$$

$$+ \frac{1}{\rho}(\sigma_{xx}^{e\prime} - p + \sigma_{xx}^{v\prime} - q)\frac{\partial u_x}{\partial x} + \frac{1}{\rho}(\sigma_{xy}^{e\prime} + \sigma_{xy}^{v\prime})\frac{\partial u_y}{\partial x} + \frac{1}{\rho}(\sigma_{xy}^{e\prime} + \sigma_{xy}^{v\prime})\frac{\partial u_x}{\partial y}$$

$$+ \frac{1}{\rho}(\sigma_{yy}^{e\prime} - p + \sigma_{yy}^{v\prime} - q)\frac{\partial u_y}{\partial y} + \frac{1}{\rho}\frac{\partial h_x}{\partial x} + \frac{1}{\rho}\frac{\partial h_y}{\partial y} + r$$

$$- \left[\frac{1}{\rho}(\sigma_{xx}^{e\prime} + \sigma_{yy}^{e\prime} + p + \sigma_{xx}^{v\prime} + \sigma_{yy}^{v\prime} + q)\frac{u_x}{x} - \frac{1}{\rho}\frac{h_x}{x}\right], \tag{A14a}$$

$$\frac{\partial T}{\partial t} = -u_x \frac{\partial T}{\partial x} - u_y \frac{\partial T}{\partial y} + \left(T\Gamma + \frac{\sigma_{xx}^{v\prime} - q}{\rho C_V}\right)\frac{\partial u_x}{\partial x} + \frac{\sigma_{xy}^{v\prime}}{\rho C_V}\frac{\partial u_y}{\partial x} + \frac{\sigma_{xy}^{v\prime}}{\rho C_V}\frac{\partial u_x}{\partial y}$$

$$+ \left(T\Gamma + \frac{\sigma_{yy}^{v\prime} - q}{\rho C_V}\right)\frac{\partial u_y}{\partial y} + \frac{1}{\rho C_V}\frac{\partial h_x}{\partial x} + \frac{1}{\rho C_V}\frac{\partial h_y}{\partial y} + \frac{r}{C_V} + \Psi$$

$$+ \left[\left(T\Gamma - \frac{\sigma_{xx}^{v\prime} + \sigma_{yy}^{v\prime} + q}{\rho C_V}\right)\frac{u_x}{x} + \frac{1}{\rho C_V}\frac{h_x}{x}\right]. \tag{A14b}$$

Acknowledgments

Appreciation is expressed to J.W. Nunziato, J.M. McGlaun, T.J. Burns, D.L. Hicks and J.W. Swegle all of Sandia National Laboratories for checking the equations, offering advice and counsel, and for proofreading the manuscript.

References

[1] P. Chadwick, *Continuum Mechanics* (Wiley, New York, 1976).
[2] B.D. Coleman and M.E. Gurtin, Thermodynamics with Internal State Variables, *J. Chem Phys.* **47**, 597 (1967).
[3] R.M. Bowen, Thermochemistry of Reacting Materials, *J. Chem. Phys.* **49**, 1625 (1968).

[4] B.D. Coleman and M.E. Gurtin, Growth and Decay of Discontinuities in Fluids with Internal State Variables, *Phys. Fluids* **10**, 1454 (1967).
[5] R.M. Bowen and P.J. Chen, Acceleration Waves in Anisotropic Thermoelastic Materials with Internal State Variables, *Acta Mech.* **15**, 95 (1972).
[6] P.J. Chen and M.E. Gurtin, Growth and Decay of One-Dimensional Shock Waves in Fluids with Internal State Variables, *Phys. Fluids* **14**, 1091 (1971).
[7] W. Herrmann, Dynamic Constitutive Equations, in: *7th U.S. Natl. Cong. Applied Mechanics*, Boulder, CO (June 1974).
[8] D.S. Drumheller and A. Bedford, On the Mechanics and Thermodynamics of Fluid Mixtures, *Arch. Rational Mech. Anal.* **71**, 345 (1979).
[9] J.W. Nunziato and E.K. Walsh, On Ideal Multiphase Mixtures with Chemical Reactions and Diffusion, *Arch. Rational Mech. Anal.* **73**, 285 (1980).
[10] B.D. Coleman and V.J. Mizel, Existence of Caloric Equations of State in Thermodynamics, *J. Chem. Phys.* **40**, 1116 (1964).
[11] B.D. Coleman and W. Noll, Material Symmetry and Thermostatic Inequalities in Finite Deformations, *Arch. Rational Mech. Anal.* **15**, 87 (1964).
[12] J.W. Nunziato and D.S. Drumheller, The Thermodynamics of Maxwellian Materials, *Int. J. Solids Structures* **14**, 545 (1978).
[13] D.B. Hayes, Wave Propagation in a Condensed Medium with *N* Transforming Phases: Application to Solid I–Solid II–Liquid Bismuth, *J. Appl. Phys.* **46**, 3438 (1975).
[14] J.W. Nunziato and M.E. Kipp, Numerical Studies of Initiation, Detonation and Detonation Failure in Nitromethane, to be published (1980).
[15] J.W. Nunziato, K.W. Schuler and D.B. Hayes, Wave Propagation Calculations for Nonlinear Viscoelastic Solids, *Proc. Intl. Conf. on Computational Methods in Nonlinear Mechanics*, Univ. Texas, Austin, TX (Sept. 1974).
[16] L.W. Davison and A.L. Stevens, Thermomechanical Constitution of Spalling Elastic Bodies, *J. Appl. Phys.* **44**, 668 (1973).
[17] L. Seaman, D.R. Curran and D.A. Shockey, Computational Models for Ductile and Brittle Fracture, *J. Appl. Phys.* **47**, 4814 (1976).
[18] D.E. Grady and M.E. Kipp, Continuum Modelling of Explosive Fracture in Oil Shale, *Intl. J. Rock Mech. and Mining Sci.*, to be published (1980).
[19] P. Perzyna, The Constitutive Equations for Rate Sensitive Plastic Materials, *Quart. Appl Math.* **20**, 321 (1963).
[20] W. Herrmann, Development of a High Strain Rate Constitutive equation for 6061-T6 Aluminum, *Intl. Conf. on Mechanical Behavior of Materials*, Kyoto, Japan (Aug. 1971) also Sandia Labs., SLA-73-0897 (1974).
[21] B.M. Butcher, The Description of Strain-Rate Effects in Shocked Porous Materials, in: *Shock Waves*, J.J. Burk and V. Weiss eds. (Syracuse University Press, 1971).
[22] J.L. Anderson, S. Preiser and E.L. Rubin, Conservation Form of the Equations of Hydrodynamics in Curvilinear Coordinate Systems, *J. Comp. Phys.* **2**, 279 (1968).
[23] N.N. Yanenko, *The Method of Fractional Steps* (Springer Berlin–New York, 1971).
[24] G.I. Marchuk, *Methods of Numerical Mathematics* (Springer, Berlin–New York, 1975).
[25] J. von Neumann and R.D. Richtmyer, A Method for the Numerical Calculation of Hydrodynamic Shocks, *J. Appl. Phys.* **21**, 232 (1950).
[26] J.W. Nunziato, K.W. Schuler and D.B. Hayes, Wave Propagation Calculations for Nonlinear Viscoelastic Solids, *Proc. Intl. Conf. on Computational Methods in Nonlinear Mechanics*, Austin, TX (1974).
[27] M.E. Kipp and A.L. Stevens, Numerical Integration of a Spall-Damage Viscoplastic Constitutive Model in a One-Dimensional Wave Propagation Code, Sandia Labs., SAND-76-0061 (1976).
[28] D. Gottlieb, Strang-Type Difference Schemes for Multidimensional Problems, *SIAM J. Numer. Anal.* **9**, 650 (1972).
[29] Y.W. Chang and J. Gvildys, REXCO-HEP: A Two-Dimensional Computer Code for Calculating the Primary System Response in Fast Reactors, Argonne National Lab., ANL/RAS-75-11 (1975).
[30] M.L. Wilkins, Calculation of Elastic-Plastic Flow, Lawrence Radiation Lab., UCRL-7322 (revised) (1969).

[31] W. Herrmann, A Lagrangian Finite-Difference Method for Two-Dimensional Motion Including Material Strength, Air Force Weapons Lab., WL-TR-64-107 (1964).

[32] S.L. Hancock, Finite Difference Equations for PISCES 2DELK A Coupled Euler Lagrange Continuum Mechanics Computer Program, Physics Intl. Co., Tech Memo TCAM 76-2 (1976).

[33] R. Hoffman, Lagrange Explicit Finite-Difference Technology, Science Applications Inc., EPRI RP-307 (1975).

[34] S.L. Thompson, CSQ II – An Eulerian Finite Difference Program for Two-Dimensional Material Response, Sandia Labs., SAND 77-1339 (1979).

[35] J.W. Swegle, TOODY IV – A Computer Program for Two-Dimensional Wave Propagation, Sandia Labs., SAND-78-0552 (1978).

[36] S. Hancock, Equations for Forces in Axisymmetric Lagrange Zones, *Intl. Conf. on Computational Methods in Nonlinear Mechanics*, Austin, TX (1979).

[37] G.L. Goudreau and J.O. Hallquist, Synthesis of Hydrocode and Finite Element Technology for Large Deformation Lagrangian Computation, *Intl. Sem. on Computational Aspects of the Finite Element Method*, Berlin (1979).

[38] T.B. Belytschko, J.M. Kennedy and D.R. Schoeberle, On Finite Element and Difference Formulations of Transient Fluid-Structure Problems, *Proc. Conf. on Computational Methods in Nuclear Eng.*, Charleston, NC (1975).

[39] W. Herrmann, A Comparison of Finite Difference Expressions Used in Lagrangian Fluid Flow Calculations, Air Force Weapons Lab., WL-TR-64-104 (1964).

[40] R. Landshoff, A Numerical Method for Treating Fluid Flow in the Presence of Shocks, Los Alamos Scientific Lab., LA-1930 (1955).

[41] W. Herrmann and D.L. Hicks, Numerical Analysis Methods, in: *Metallurgical Effects at High Strain Rates*, R.W. Rohde et al., eds. (Plenum, New York, 1973).

[42] G. Maenchen and S. Sack, The TENSOR Code, Lawrence Radiation Lab., UCRL-7316 (1963).

[43] E.D. Giroux, HEMP Users Manual, Lawrence Livermore Lab., UCRL-51079 (1971).

[44] L.D. Bertholf and S.E. Benzeley, TOODY II, A Computer Program for Two-Dimensional Wave Propagation, Sandia Labs., SC-RR-68-41 (1968).

[45] W. Herrmann, L.D. Bertholf and S.L. Thompson, Computational Methods for Stress Wave Propagation in Nonlinear Solid Mechanics, in: *Computational Mechanics, Proc. Int. Conf. on Computational Meth. in Nonlinear Mech.* Austin, TX (1974) pp. 91–127.

[46] P.J. Roache, *Computational Fluid Dynamics* (Hermosa Publishers, Albuquerque, NM 1972) p. 289.

[47] P.L. Browne, REZONE. A Proposal for Accomplishing Rezoning in Two-Dimensional Lagrangian Hydrodynamics Problems, Los Alamos Scientific Lab. LA-3455-MS (Sept. 1965).

[48] B.J. Thorne and D.B. Holdridge, The TOOREZ Lagrangian Rezoning Code, Sandia Labs., SLA-73-1057, (April 1974).

[49] L.D. Bertholf, L.D. Buxton, B.J. Thorne, R.K. Byers, A.L. Stevens and S.L. Thompson, Damage in Steel Plates from Hypervelocity Impact. II. Numerical Results and Spall Measurement, *J. Appl. Phys.* **46**, 3776 (1975).

[50] J. Lipkin and M.E. Kipp, Wave Structure Measurement and Analysis in Hypervelocity Impact Experiments, *J. Appl. Phys.* **47**, 1979 (1976).

[51] L.D. Bertholf, M.E. Kipp and W.T. Brown, Two-Dimensional Calculations for the Oblique Impact of Kinetic Energy Projectiles with a Multi-Layered Target, USA Ballistic Research Labs., BRL CR 333 (March 1977).

[52] P. Yarrington and R.K. Byers, AUTOREZ: A Two-Dimensional Lagrangian Automatic Rezoning Wavecode (Preliminary Version), Sandia Labs., SAND-79-1914 (March 1980).

[53] L.D. Bertholf, One-Dimensional Rezoning Methods, Sandia Labs., SC-TM-68-193 (May 1968).

[54] I.G. Cameron, An Analysis of the Errors Caused by Using Artificial Viscosity Terms to Represent Steady-State Shock Waves, *J. Comp. Phys.* **1**; 1 (1966).

[55] J.G. Trulio, Theory and Structure of the AFTON Codes, Air Force Weapons Lab., AFWL-TR-66-19 (June 1966).

[56] R.T. Walsh, Finite Difference Methods, in: *Dynamic Response of Materials to Intense Impulsive Loading*, P.C. Chen and A.K. Hopkins, eds. (U.S. Government Printing Office, 1973) p. 363.

[57] R.L. Elliott, MAGEE REZONE, A Program for Redefining the Mesh in the Two-Dimensional

Time Dependent Lagrangian Hydrodynamic Code MAGEE, Los Alamos Scientific Lab., Private Comnunication (1970).
[58] R.L. Elliott, S.R. Orr and E.C. Pequette, MAGEE Write Up, Appendix F: MAGEE Rezone Equations, Los Alamos Scientific Lab., Unpublished Report, June 17 (1971).
[59] L.D. Buxton and S.L. Thompson, Sandia Labs., Private Communication (1977).
[60] A.M. Winslow, 'Equipotential' Zoning of Two-Dimensional Meshes, Lawrence Radiation Lab., UCRL-7312 (June 1963).
[61] R.J. Lawrence and D.S. Mason, WONDY IV – A Computer Program for One-Dimensional Wave Propagation with Rezoning, Sandia Labs., SC-RR-71-0284 (August 1971).
[62] R.N. Brodie and J.E. Hormuth, The PUFF 66 and PPUFF 66 Computer Programs, AFWL-TR-66-48 (May 1966).
[63] M.L. Wilkins, Calculation of Elastic-Plastic Flow, in: *Methods in Computational Physics*, Vol. 3, B. Alder, S. Fernbach and M. Rotenburg, eds. (Academic Press, New York, 1964).

CHAPTER 9

Implicit Finite Element Methods

M. GERADIN, M. HOGGE

Aerospace Laboratory of the University of Liège, Belgium

S. IDELSOHN*

Universidad Nacional, Rosario, Argentina

*Visiting from CONICET.

Computational Methods for Transient Analysis
Edited by T. Belytschko and T.J.R. Hughes

1. General aspects of implicit time integration

1.1. Explicit versus implicit solutions

Implicit methods for time integration suffer from certain drawbacks when compared to explicit schemes, but have also their own advantages.

From the user's point of view, the following criteria have to be examined:
- which method of time integration is better adapted to the type of problem to be solved?
- what is the computer cost of the solution?
- which method is simpler to use?

The developer of the program has also to examine:
- the complexity and cost of program development;
- the reliability and generality of the algorithms adopted.

In general, the efficiency of explicit programs relies on the fact that they are very specific of one given class of structural problems, and taking advantage of this specificity leads to computer codes which are generally very difficult to challenge if computations are organized in central core memory. The absence of iteration procedure in the solution brings considerable simplification in their use, since the only decision that the user has to take in applying the time marching algorithm is to adapt the step length to his current problem.

Furthermore, practical solutions exist to automatize the choice of the step length [23], in which case an explicit program may be regarded almost as a black box by the user.

For stability reasons, explicit integration is performed with a very small time step size, but the cost of the large number of time steps is compensated by the very low cost of one explicit step. Therefore, explicit programs are best suited to problems where the high frequency content of the structure contributes significantly to the response, as it is the case in transients induced by shocks, blasts or any type of loading with a broad frequency range.

Implicit programs, on the other hand, are often characterized by a much higher degree of generality. They benefit from the fact that the time step length is not so strongly limited by stability considerations and that the cost of one time step is thus a not so critical criterion of performance. This allows for more frequent use of auxiliary memory, and thus a more flexible organization in the library of elements, and the size of the problems to be solved.

The recourse to large time step sizes limits the type of problems that can be solved efficiently with an implicit code to transients with frequency content in the lower range, in which the behavior of the structure is mainly inertial.

When applied to linear problems, implicit codes are neither more difficult to apply nor to implement than explicit codes. If an unconditionally stable algorithm is employed, the time step length is limited only by the physics of the problem (frequency content of the excitation and lower eigenspectrum of the structure), and the only additional cost of implicit integration is the linear solution involved at each step, while explicit integration is straightforward if the mass matrix is limited to a diagonal form.

This comparison is no longer true when dealing with nonlinear problems. Due to the implicit character of the time integration scheme, achieving dynamic equilibrium – which is indispensable to control the accuracy of the solution – involves elaborate methods of solution for large systems of nonlinear equations. The user is then faced with a broad variety of choices such as:

- selection of a given time integration algorithm;
- choice of a tolerance in the verification of dynamic equilibrium;
- choice of a method for nonlinear iteration within a time step or, more generally, of a strategy for nonlinear iterations; this choice results generally from a compromise between the reliability of the strategy adopted and the computer cost of the iteration;
- choice of the time step length, not only according to the physics of the problem, but also to the convergence of the iteration procedure within the time step.

The main drawback in the implicit integration of nonlinear systems is that each iteration involves the solution of a linear system associated to a tangent iteration matrix, and most of the cost of the solution arises from the evaluation of this matrix and from the solution of the associated linear system.

There are thus three ways of reducing the cost of implicit programs applied to nonlinear problems:

1. choosing algorithms which allow for large time step lengths without encountering convergence difficulties;
2. minimizing the cost of the evaluation of the tangent iteration matrix and of the linear solution;
3. definining methods and strategies of iteration which lead to a minimum number of iterations within a time step.

Despite these rather negative remarks, there is a large number of available nonlinear computer codes using implicit time integration [7]. One reason of this success is probably the great flexibility of these codes, mainly by the variety of their finite element library, compared to explicit codes which may undergo a significant deterioration in their performances when their possibilities are extended outside the initial specific range of application.

A supplementary reason is that numerous limitations concerning mass representation, tying of degrees of freedom, starting conditions, etc. are encountered only with explicit codes.

1.2. Essential features of linear and nonlinear implicit problems

Let us consider the problem of transient response of a nonlinear structure, written in the form of a matrix dynamic equilibrium equation

$$\boldsymbol{M}\ddot{\boldsymbol{q}}(t)+\boldsymbol{f}(\boldsymbol{q},\dot{\boldsymbol{q}})=\boldsymbol{g}(\boldsymbol{q},t),\quad \boldsymbol{q}_0,\dot{\boldsymbol{q}}_0 \text{ given}, \tag{1.1}$$

with the following definitions:

$\boldsymbol{q}(t), \dot{\boldsymbol{q}}(t), \ddot{\boldsymbol{q}}(t)$ the n-dimensional time-dependent vectors of local displacements, velocities and accelerations,

$\boldsymbol{M}$ the mass-matrix of the structure, symmetric and positive definite,

$\boldsymbol{f}(\boldsymbol{q},\dot{\boldsymbol{q}})$ the internally resisting forces in the structure, which may depend on displacements and velocities,

$\boldsymbol{g}(\boldsymbol{q},t)$ the external forces, which vary in general with time, but which may also depend on the displacements,

$\boldsymbol{q}_0, \dot{\boldsymbol{q}}_0$ the initial values of displacements and velocities.

The term $\boldsymbol{f}(\boldsymbol{q},\dot{\boldsymbol{q}})$ describing the internal forces includes in fact two contributions: one corresponds to elastic restoring forces, and the second one represents internal dissipation.

Nonlinearities arise in general from inelastic material behavior and/or from adaptation of the geometry to the loading; they may affect both the external forces $\boldsymbol{g}(\boldsymbol{q},t)$ and the internal ones, $\boldsymbol{f}(\boldsymbol{q},\dot{\boldsymbol{q}})$, which result from the volume integration of the stress distributions $\boldsymbol{\sigma}(\boldsymbol{\varepsilon},\dot{\boldsymbol{\varepsilon}})$ implied by the nonlinear state, i.e. at a single finite element level

$$\boldsymbol{f}_{\mathrm{e}}(\boldsymbol{q},\dot{\boldsymbol{q}})=\int_{V_{\mathrm{e}}}\boldsymbol{B}_{\mathrm{e}}^{\mathrm{T}}\boldsymbol{\sigma}(\boldsymbol{\varepsilon},\dot{\boldsymbol{\varepsilon}})\,\mathrm{d}V. \tag{1.2}$$

In (1.2), $\boldsymbol{\varepsilon}$ and $\dot{\boldsymbol{\varepsilon}}$ denote respectively the strains and strain rates in the material, and $\boldsymbol{B}_{\mathrm{e}}$ is the appropriate finite element matrix yielding the strains in terms of nodal displacements.

If linearity is assumed with respect to displacements and velocities, the internal forces take the classical linear form

$$\boldsymbol{f}(\boldsymbol{q},\dot{\boldsymbol{q}})=\boldsymbol{K}\boldsymbol{q}(t)+\boldsymbol{C}\dot{\boldsymbol{q}}(t) \tag{1.3}$$

where $\boldsymbol{K}$ and $\boldsymbol{C}$, the constant stiffness and viscous damping matrices, are both symmetric and positive semi-definite.

Displacements, velocities and accelerations are not independent quantities. Therefore, the spatially discretized system of equations (1.1) to be solved at each time t may be written in the form of an equation for the displacement vector $\boldsymbol{q}(t)$

$$\boldsymbol{r}(\boldsymbol{q})=\boldsymbol{M}\ddot{\boldsymbol{q}}(t)+\boldsymbol{f}(\boldsymbol{q},\dot{\boldsymbol{q}})-\boldsymbol{g}(\boldsymbol{q},t)=\boldsymbol{0} \tag{1.4}$$

where $\boldsymbol{r}$ is the n-dimensional residual vector, which reads in the linear case

$$\boldsymbol{r}(\boldsymbol{q})=\boldsymbol{M}\ddot{\boldsymbol{q}}(t)+\boldsymbol{K}\boldsymbol{q}(t)+\boldsymbol{C}\dot{\boldsymbol{q}}(t)-\boldsymbol{g}(t)=\boldsymbol{0}\,. \tag{1.5}$$

The implicit character of a time integration method is defined as follows: assume that displacements and velocities at time t_{n+1} are approximated using linear difference formulas of the form

$$\dot{\boldsymbol{q}}_{n+1} = \frac{\alpha}{h}\ddot{\boldsymbol{q}}_{n+1} + \boldsymbol{l}(\dot{\boldsymbol{q}}_n, \ddot{\boldsymbol{q}}_n, \ldots), \qquad \boldsymbol{q}_{n+1} = \frac{\beta}{h^2}\ddot{\boldsymbol{q}}_{n+1} + \boldsymbol{m}(\boldsymbol{q}_n, \dot{\boldsymbol{q}}_n, \ddot{\boldsymbol{q}}_n, \ldots), \tag{1.6}$$

where h is the time step size, and where α and β are coefficients specific of the difference formulas.

If both coefficients α and β take non-zero values in (1.6) accelerations and velocities $\ddot{\boldsymbol{q}}_{n+1}$ and $\dot{\boldsymbol{q}}_{n+1}$ becomes functions of the displacements $\boldsymbol{q}_{n+1}$ at time t_{n+1} for which the solution is sought, and the residual vector becomes after substitution an implicit function of $\boldsymbol{q}_{n+1}$ only:

$$\boldsymbol{r}(\boldsymbol{q}_{n+1}) = \boldsymbol{0}. \tag{1.7}$$

In order for the displacements $\boldsymbol{q}_{n+1}$ to satisfy fully the discretized equations (1.7), an equilibrium iteration sequence is usually required if these are nonlinear. Common solution techniques for such systems of simultaneous nonlinear equations imply a certain type of linearization: assuming an approximation $\boldsymbol{q}^k_{n+1}$ $(k = 0, 1, \ldots)$ to $\boldsymbol{q}_{n+1}$, we admit for instance that in its neighbourhood the linear mapping

$$\boldsymbol{r}_{\mathrm{L}}(\boldsymbol{q}^{k+1}_{n+1}) = \boldsymbol{r}(\boldsymbol{q}^k_{n+1}) + \boldsymbol{S}(\boldsymbol{q}^k_{n+1})(\boldsymbol{q}^{k+1}_{n+1} - \boldsymbol{q}^k_{n+1}) \tag{1.8}$$

is a good approximation to (1.7), with the definition of the iteration (Jacobian) matrix

$$\boldsymbol{S}(\boldsymbol{q}^k_{n+1}) = \left.\frac{\partial \boldsymbol{r}}{\partial \boldsymbol{q}}\right|_{\boldsymbol{q}^k_{n+1}} \tag{1.9}$$

which reads in the present case:

$$\boldsymbol{S}(\boldsymbol{q}) = \frac{\partial \boldsymbol{f}}{\partial \boldsymbol{q}} + \frac{\partial \boldsymbol{f}}{\partial \dot{\boldsymbol{q}}}\frac{\partial \dot{\boldsymbol{q}}}{\partial \boldsymbol{q}} + \boldsymbol{M}\frac{\partial \ddot{\boldsymbol{q}}}{\partial \boldsymbol{q}} - \frac{\partial \boldsymbol{g}}{\partial \boldsymbol{q}}. \tag{1.10}$$

The different terms in the jacobian matrix (1.10) may be interpreted as follows:

$\partial \boldsymbol{f}/\partial \boldsymbol{q}$ represents the variation of internal loads with displacements, and is thus the tangent stiffness matrix $\boldsymbol{K}^{\mathrm{t}}$,

$\partial \boldsymbol{f}/\partial \dot{\boldsymbol{q}}$ describes the variation of internal loads with velocities and has thus the meaning of a tangent damping matrix $\boldsymbol{C}^{\mathrm{t}}$,

$\partial \boldsymbol{g}/\partial \boldsymbol{q}$ describes the dependence of external loads with geometry. Unlike the two preceding terms, the last one is non-symmetric and is usually omitted to preserve the symmetrical character of the Jacobian matrix (1.10). With the above definitions, the latter reads

$$S(q) = K^{t}(q) + C^{t}(q)\frac{\partial \dot{q}}{\partial q} + M\frac{\partial \ddot{q}}{\partial q} - \frac{\partial g}{\partial q}. \tag{1.11}$$

When linearity is assumed, the last term in (1.11) vanishes and the tangent matrices K^{t} and C^{t} reduce simply to the constant stiffness and viscous damping matrices K and C, yielding the constant Jacobian

$$S_0 = K + C\frac{\partial \dot{q}}{\partial q} + M\frac{\partial \ddot{q}}{\partial q}. \tag{1.12}$$

2. Different classes of implicit time operators

There are several ways of constructing an implicit time integrator, the most popular one being the one-step formulas. Two important methods pertain to this category: Newmark's family of algorithms and Wilson's method. Both are characterized by an unconditonal stability when applied to linear problems of structural dynamics. However, it has been noted by Park [24] that both methods may degrade when applied to nonlinear problems if nonlinearities are not appropriately discretized. The explanation of this behaviour is obtained when looking at the multistep formulation of both algorithms.

A second way of obtaining implicit time integrators consists to use linear multistep finite difference formulas which may be of first or second order in the time derivatives. Houbolt's operator [14], which was the first attempt to solve by a time marching process the fundamental equations of structural dynamics, pertains to this category. Another worthwhile mentioning method of this type is Park's operator [22], which keeps the desirable stability properties of Houbolt's method while achieving a better accuracy.

The presentation will be limited to these four methods since most of the implicit linear and nonlinear finite element codes currently available are based on one of them.

2.1. One-step formulas

One-step methods are characterized by the fact the state of the system (that is displacements and velocities) at time t_n is calculated as a function of the state of the system at the former time step, t_n only:

$$\begin{aligned} q_{n+1} &= f_1(\ddot{q}_{n+1}, q_n, \dot{q}_n, \ddot{q}_n), \\ \dot{q}_{n+1} &= f_2(\ddot{q}_{n+1}, q_n, \dot{q}_n, \ddot{q}_n). \end{aligned} \tag{2.1}$$

Their implicit character arises from the dependence on the acceleration at the current time step.

2.1.1. Newmark's method [20]

In Newmark's method, a Taylor expansion of displacements and velocities is used to construct linear relationships of type (2.1). They are limited to second-order time derivatives:

$$\begin{aligned} \boldsymbol{q}_{n+1} &= \boldsymbol{q}_n + h\dot{\boldsymbol{q}}_n + \int_{t_n}^{t_{n+1}} (t_{n+1} - \tau)\ddot{\boldsymbol{q}}(\tau)\,\mathrm{d}\tau\,, \\ \dot{\boldsymbol{q}}_{n+1} &= \dot{\boldsymbol{q}}_n + \int_{t_n}^{t_{n+1}} \ddot{\boldsymbol{q}}(\tau)\,\mathrm{d}\tau\,, \end{aligned} \tag{2.2}$$

The approximation is Newmark's scheme consists in evaluating the integrals in (2.2) using a quadrature formula

$$\begin{aligned} \int_{t_n}^{t_{n+1}} \ddot{\boldsymbol{q}}(\tau)\,\mathrm{d}\tau &= (1-\gamma)h\ddot{\boldsymbol{q}}_n + \gamma h\ddot{\boldsymbol{q}}_{n+1} + \boldsymbol{r}_n\,, \\ \int_{t_n}^{t_{n+1}} (t_{n+1} - \tau)\ddot{\boldsymbol{q}}(\tau)\,\mathrm{d}\tau &= (\tfrac{1}{2} - \beta)h^2\ddot{\boldsymbol{q}}_n + \beta h^2\ddot{\boldsymbol{q}}_{n+1} + \boldsymbol{r}'_n\,. \end{aligned} \tag{2.3}$$

β and γ are the free parameters of Newmark's method: the particular choices ($\beta = \frac{1}{6}$, $\gamma = \frac{1}{2}$) correspond to a linear interpolation of the accelerations over the time interval (t_n, t_{n+1}), while ($\beta = \frac{1}{4}$, $\gamma = \frac{1}{2}$) is obtained by assuming an average constant acceleration. The latter choice is known to guarantee unconditional stability for linear systems.

The substitution of the approximations (2.3) into (2.2) yields the difference formulas

$$\begin{aligned} \dot{\boldsymbol{q}}_{n+1} &= \dot{\boldsymbol{q}}_n + (1-\gamma)h\ddot{\boldsymbol{q}}_n + \gamma h\ddot{\boldsymbol{q}}_{n+1}\,, \\ \boldsymbol{q}_{n+1} &= \boldsymbol{q}_n + h\dot{\boldsymbol{q}}_n + (\tfrac{1}{2} - \beta)h^2\ddot{\boldsymbol{q}}_n + \beta h^2\ddot{\boldsymbol{q}}_{n+1}\,, \end{aligned} \tag{2.4}$$

which are combined with the equilibrium equation

$$\boldsymbol{r}(\boldsymbol{q}_{n+1}) = \boldsymbol{0} \tag{2.5}$$

to obtain the solution at time t_{n+1}.

If use is made of the difference approximations (2.4) into the expression (1.11) of the residual vector, the expression of the tangent iteration matrix is obtained in the general form,

$$\boldsymbol{S}^{\mathrm{N}}(\boldsymbol{q}) = \boldsymbol{K}^{\mathrm{t}}(\boldsymbol{q}) + \frac{\gamma}{\beta h}\boldsymbol{C}^{\mathrm{t}}(\boldsymbol{q}) + \frac{1}{\beta h^2}\boldsymbol{M} - \left.\frac{\partial \boldsymbol{g}}{\partial \boldsymbol{q}}\right|_{\boldsymbol{q}} \tag{2.6}$$

In linear situations in particular, external and internal forces depend linearly on displacements and velocities, giving the linear iteration matrix

$$\boldsymbol{S}_0^{\mathrm{N}} = \boldsymbol{K} + \frac{\gamma}{\beta h}\boldsymbol{C} + \frac{1}{\beta h^2}\boldsymbol{M} \tag{2.7}$$

where $\boldsymbol{K}$, $\boldsymbol{M}$ and $\boldsymbol{C}$ are the linear stiffness, mass and viscous damping matrices.

When applied to the linear residual vector (1.5), the Newmark implicit integration scheme is as follows:

1. predict velocities and displacements using the explicit approximations

$$\begin{aligned} \ddot{\tilde{\boldsymbol{q}}}_{n+1} &= 0\,, \\ \dot{\tilde{\boldsymbol{q}}}_{n+1} &= \dot{\boldsymbol{q}}_n + (1-\gamma)h\ddot{\boldsymbol{q}}_n\,, \\ \tilde{\boldsymbol{q}}_{n+1} &= \boldsymbol{q}_n + h\dot{\boldsymbol{q}}_n + (\tfrac{1}{2}-\beta)h^2\ddot{\boldsymbol{q}}_n\,; \end{aligned} \tag{2.8}$$

2. obtain corrections to displacements from the linearized equilibrium equation written at time t_{n+1}

$$\boldsymbol{S}_0^N \Delta\boldsymbol{q}_{n+1} = \boldsymbol{r}(\tilde{\boldsymbol{q}}_{n+1})\,; \tag{2.9}$$

3. increment displacements, velocities and accelerations by the corrector equations resulting from (2.4)

$$\begin{aligned} \boldsymbol{q}_{n+1} &= \tilde{\boldsymbol{q}}_{n+1} + \Delta\boldsymbol{q}_{n+1}\,, \\ \dot{\boldsymbol{q}}_{n+1} &= \dot{\tilde{\boldsymbol{q}}}_{n+1} + \frac{\gamma}{\beta h}\Delta\boldsymbol{q}_{n+1}\,, \\ \ddot{\boldsymbol{q}}_{n+1} &= \frac{1}{\beta h^2}\Delta\boldsymbol{q}_{n+1}\,; \end{aligned} \tag{2.10}$$

4. increment time: $n = n + 1$, $t = t + h$ and go back to step 1.

For a linear system, either factorization or Gauss elimination on the iteration matrix (2.7) can be performed once for all as long as the step size is kept constant. Large systems of equations can easily be handled with using standard linear equation solvers such as Irons' frontal method [35]. A variant of it will be described in section 4 which does not even require storing in central core storage the front of the system.

When nonlinearities are present in the system, the integration procedure described above is no longer exact, and an iterative scheme has to be used in order to restore dynamic equilibrium in the sense of (2.5). Iteration techniques which generalize the implicit scheme (2.9)–(2.10) to nonlinear situations will be described in the next section.

2.1.2. Wilson's method

Wilson's method [27] is a modification of Newmark's algorithm in which a linear variation of acceleration is assumed within a time step:

$$\ddot{\boldsymbol{q}}(\tau) = \ddot{\boldsymbol{q}}_n + \theta(\ddot{\boldsymbol{q}}_{n+1} - \ddot{\boldsymbol{q}}_n), \quad \theta = \frac{\tau - t_n}{h} > 0\,. \tag{2.11}$$

Displacements and velocities are calculated at time t_{n+1} using the difference formulas (2.2) with $\beta = \frac{1}{6}$, $\gamma = \frac{1}{2}$, but equilibrium is expressed at a larger time step $t_{n+\theta}$ ($\theta > 1$) using the extrapolation formula (2.11), i.e.

$$\boldsymbol{r}(\boldsymbol{q}_{n+\theta}) = \boldsymbol{0}\,. \tag{2.12}$$

It is shown in [2] that a value of $\theta > 1.4$ provides unconditional stability while retaining the accuracy properties of the linear acceleration method.

If use is made of (2.11) into (2.2) where $t_{n+\theta}$ replaces t_{n+1}, the extrapolated values of displacements and velocities required by (2.12) are

$$\begin{aligned} \boldsymbol{q}_{n+\theta} &= \boldsymbol{q}_n + \theta h\dot{\boldsymbol{q}}_n + \tfrac{1}{2}\theta^2h^2\ddot{\boldsymbol{q}}_n + \tfrac{1}{6}\theta^3h^2(\ddot{\boldsymbol{q}}_{n+1} - \ddot{\boldsymbol{q}}_n)\,, \\ \dot{\boldsymbol{q}}_{n+\theta} &= \dot{\boldsymbol{q}}_n + \theta h\ddot{\boldsymbol{q}}_n + \tfrac{1}{2}\theta^2h(\ddot{\boldsymbol{q}}_{n+1} - \ddot{\boldsymbol{q}}_n)\,. \end{aligned} \tag{2.13}$$

The expression of the tangent iteration matrix for nonlinear analysis is thus

$$\boldsymbol{S}^{\mathrm{W}}(\boldsymbol{q}_{n+\theta}) = \boldsymbol{K}^{\mathrm{t}}(\boldsymbol{q}_{n+\theta}) + \frac{3}{\theta h}\boldsymbol{C}^{\mathrm{t}}(\boldsymbol{q}_{n+\theta}) + \frac{6}{\theta^2h^2}\boldsymbol{M} - \left.\frac{\partial \boldsymbol{g}}{\partial \boldsymbol{q}}\right|_{\boldsymbol{q}_{n+\theta}} \tag{2.14}$$

and the associate linear iteration matrix reads

$$\boldsymbol{S}_0^{\mathrm{W}} = \boldsymbol{K} + \frac{3}{\theta h}\boldsymbol{C} + \frac{6}{\theta^2h^2}\boldsymbol{M}\,. \tag{2.15}$$

This yields the following implicit integration scheme:

1. predict accelerations, velocities and displacements at time $t_{n+\theta}$ by

$$\begin{aligned} &\ddot{\boldsymbol{q}}_{n+\theta} = (1-\theta)\ddot{\boldsymbol{q}}_n\,, \qquad \dot{\tilde{\boldsymbol{q}}}_{n+\theta} = \dot{\boldsymbol{q}}_n + h(\theta - \tfrac{1}{2}\theta^2)\ddot{\boldsymbol{q}}_n\,, \\ &\tilde{\boldsymbol{q}}_{n+\theta} = \boldsymbol{q}_n + \theta h\dot{\boldsymbol{q}}_n + \tfrac{1}{2}\theta^2h^2(1 - \tfrac{1}{3}\theta)\ddot{\boldsymbol{q}}_n\,; \end{aligned} \tag{2.16}$$

2. obtain corrections to displacements from the linearized equilibrium equation expressed at time $t_{n+\theta}$

$$\boldsymbol{S}_0^{\mathrm{W}}\Delta\boldsymbol{q}_{n+\theta} = \boldsymbol{r}(\tilde{\boldsymbol{q}}_{n+\theta})\,; \tag{2.17}$$

3. calculate new accelerations, velocities and displacements at time t_{n+1} from the linear accelerations formulas

$$\begin{aligned} &\ddot{\boldsymbol{q}}_{n+1} = \frac{6}{\theta^2h^2}\Delta\boldsymbol{q}_{n+\theta}\,, \qquad \dot{\boldsymbol{q}}_{n+1} = \dot{\boldsymbol{q}}_n + \tfrac{1}{2}h(\ddot{\boldsymbol{q}}_n + \ddot{\boldsymbol{q}}_{n+1})\,, \\ &\boldsymbol{q}_{n+1} = \boldsymbol{q}_n + h\dot{\boldsymbol{q}}_n + \tfrac{1}{3}h^2\ddot{\boldsymbol{q}}_n + \tfrac{1}{6}h^2\ddot{\boldsymbol{q}}_{n+1}\,; \end{aligned} \tag{2.18}$$

4. increment time and go back to step 1.

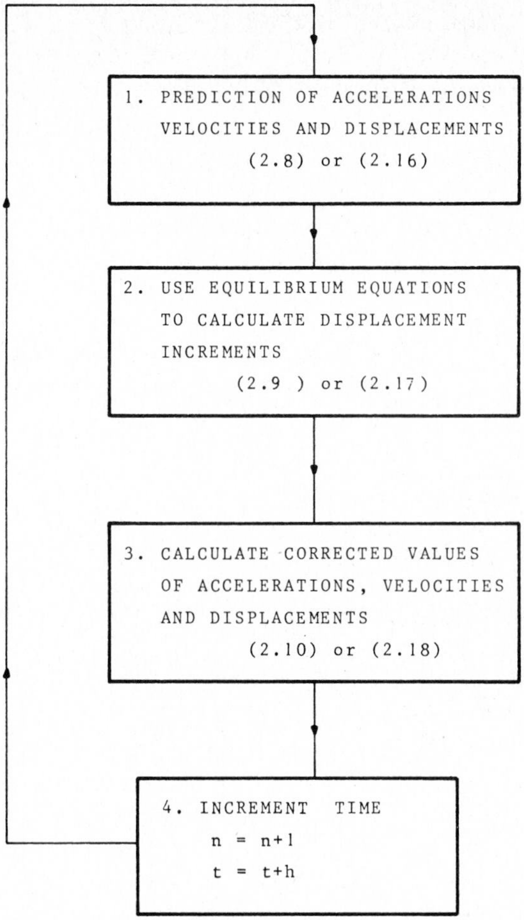

Fig. 2.1. Sequence of operations in each linear time step using one-step formulas.

2.1.3. Summary of operations for one-step methods

In both methods, the integration procedure over one linear time step may thus be summarized as in figure 2.1.

2.2. Multistep formulas [36]

Various multistep integration schemes can be obtained using difference formulas operating on local values of displacements, velocities and accelerations at several successive time steps. For velocities, linear difference formulas may be written in the form

$$\sum_{i=0}^{m} (\alpha_i \boldsymbol{q}_{n+1-i} - h\beta_i \dot{\boldsymbol{q}}_{n+1-i}) = ch^{p+1} \boldsymbol{q}^{(p+1)}(t) \tag{2.19}$$

where p is the consistency order of the formula. The approximation consists in equating the right-hand side of (2.19) to zero. In order to obtain multistep integration formulas applicable to the equilibrium equations (2.5) of structural dynamics, two distinct philosophies may be adopted [9]: specialization of the formulas to second order systems of differential equations or transformation of the equations into equivalent first order systems.

2.2.1. Multistep equations specialized to second order systems

Let us use a difference equation analogous to (2.19) to calculate accelerations

$$\sum_{i=0}^{m} (\gamma_i \boldsymbol{q}_{n+1-i} - h^2 \delta_i \ddot{\boldsymbol{q}}_{n+1-i}) = \boldsymbol{0} \tag{2.20}$$

and construct a linear combination of the residual equations (2.5) in the form

$$\begin{aligned} \boldsymbol{r}^*(\boldsymbol{q}) &= \frac{1}{\delta_0} \sum_{i=0}^{m} \delta_i \boldsymbol{r}(\boldsymbol{q}_{n+1-i}) \\ &= \frac{1}{\delta_0} \sum_{i=0}^{m} \delta_i (\boldsymbol{M}\ddot{\boldsymbol{q}}_{n+1-i} + \boldsymbol{f}_{n+1-i} - \boldsymbol{g}_{n+1-i}) = \boldsymbol{0}\,. \end{aligned} \tag{2.21}$$

According to (2.20) accelerations are easily eliminated to yield an equation involving only displacements and, implicitly, velocities

$$\boldsymbol{r}^*(\boldsymbol{q}) = \frac{1}{\delta_0} \sum_{i=0}^{m} \left[\frac{\gamma_i}{h^2} \boldsymbol{M}\boldsymbol{q}_{n+1-i} + \delta_i (\boldsymbol{f}_{n+1-i} - \boldsymbol{g}_{n+1-i}) \right] = \boldsymbol{0}\,. \tag{2.22}$$

The associate tangent iteration matrix is

$$\begin{aligned} \boldsymbol{S}(\boldsymbol{q}_{n+1}) = \frac{\partial \boldsymbol{r}^*}{\partial \boldsymbol{q}_{n+1}} &= \boldsymbol{K}^{\mathrm{t}}(\boldsymbol{q}_{n+1}) + \boldsymbol{C}^{\mathrm{t}}(\boldsymbol{q}_{n+1}) \left(\frac{\partial \dot{\boldsymbol{q}}}{\partial \boldsymbol{q}} \right)_{n+1} \\ &+ \frac{\gamma_0}{\delta_0 h^2} \boldsymbol{M} - \left(\frac{\partial \boldsymbol{g}}{\partial \boldsymbol{q}} \right)_{n+1}. \end{aligned} \tag{2.23}$$

Use is made next of the difference expression (2.19) to calculate

$$\dot{\boldsymbol{q}}_{n+1} = \frac{\alpha_0}{\beta_0 h} \boldsymbol{q}_{n+1} + \frac{1}{\beta_0 h} \sum_{i=1}^{m} (\alpha_i \boldsymbol{q}_{n+1-i} - \beta_i h \dot{\boldsymbol{q}}_{n+1-i}) \tag{2.24}$$

and thus

$$\left(\frac{\partial \dot{\boldsymbol{q}}}{\partial \boldsymbol{q}} \right)_{n+1} = \frac{\alpha_0}{\beta_0 h}\,. \tag{2.25}$$

The final expression of the tangent iteration matrix is thus

$$\boldsymbol{S}^{\mathrm{MS2}}(\boldsymbol{q}_{n+1}) = \left[\boldsymbol{K}^{\mathrm{t}} + \frac{\alpha_0}{\beta_0 h} \boldsymbol{C}^{\mathrm{t}} + \frac{\gamma_0}{\delta_0 h^2} \boldsymbol{M} - \frac{\partial \boldsymbol{g}}{\partial \boldsymbol{q}} \right]_{\boldsymbol{q}_{n+1}} \tag{2.26}$$

and reduces in the linear case to

$$\boldsymbol{S}_0^{\mathrm{MS2}} = \boldsymbol{K} + \frac{\alpha_0}{\beta_0 h} \boldsymbol{C} + \frac{\gamma_0}{\delta_0 h^2} \boldsymbol{M}\,. \tag{2.27}$$

Assume next a prediction $(\tilde{\boldsymbol{q}}_{n+1}, \dot{\tilde{\boldsymbol{q}}}_{n+1})$ of displacements and velocities. In the linear case, the correction to the displacements is solution of

$$\boldsymbol{S}_0^{\mathrm{MS2}} \Delta \boldsymbol{q}_{n+1} = \boldsymbol{r}^*(\tilde{\boldsymbol{q}}_{n+1}) . \tag{2.28}$$

In practice, the most convenient choice of the predictor formula is

$$\begin{aligned} &\tilde{\boldsymbol{q}}_{n+1} = 0 , \\ &\dot{\tilde{\boldsymbol{q}}}_{n+1} = \frac{1}{\beta_0 h} \sum_{i=1}^{m} (\alpha_i \boldsymbol{q}_{n+1-i} - \beta_i h \dot{\boldsymbol{q}}_{n+1-i}) . \end{aligned} \tag{2.29}$$

Equations (2.22) to (2.29) yield the following integration scheme in linear situations:

1. velocities and displacements are predicted at time t_{n+1} using explicit approximations such as (2.29);
2. the residual vector $\boldsymbol{r}^*(\tilde{\boldsymbol{q}}_{n+1})$ is evaluated according to (2.22);
3. corrections to displacements are calculated by (2.28);
4. displacements and velocities are incremented

$$\begin{aligned} &\boldsymbol{q}_{n+1} = \tilde{\boldsymbol{q}}_{n+1} + \Delta \boldsymbol{q}_{n+1} , \\ &\dot{\boldsymbol{q}}_{n+1} = \dot{\tilde{\boldsymbol{q}}}_{n+1} + \frac{\alpha_0}{\beta_0 h} \Delta \boldsymbol{q}_{n+1} ; \end{aligned} \tag{2.30}$$

5. time is incremented: $n = n + 1$ and the procedure is restarted at step 1.

When nonlinearities are present in the system, the displacement increments are calculated iteratively as described in the next section.

The best known integration scheme pertaining to the category of second-order multistep formulas is Houbolt's method [14]. It uses four-point backwards difference formulas for velocities and accelerations:

$$\begin{aligned} &\dot{\boldsymbol{q}}_{n+1} = (11\boldsymbol{q}_{n+1} - 18\boldsymbol{q}_n + 9\boldsymbol{q}_{n-1} - 2\boldsymbol{q}_{n-2})/6h , \\ &\ddot{\boldsymbol{q}}_{n+1} = (2\boldsymbol{q}_{n+1} - 5\boldsymbol{q}_n + 4\boldsymbol{q}_{n-1} - \boldsymbol{q}_{n-2})/h^2 . \end{aligned} \tag{2.31}$$

Thus, the solution for $\boldsymbol{q}_{n+1}$ requires knowledge of $\boldsymbol{q}_n$, $\boldsymbol{q}_{n-1}$ and $\boldsymbol{q}_{n-2}$ but not of the corresponding derivatives. It corresponds to the general formulas (2.19)–(2.20) with the particular choices:

$$\begin{aligned} &\alpha_0 = 11, \quad \alpha_1 = -18, \quad \alpha_2 = 9, \quad \alpha_3 = -2 , \\ &\beta_0 = 6, \quad \beta_i = 0 \quad (i = 1, 2, 3) , \\ &\delta_0 = 1, \quad \delta_i = 0 \quad (i = 1, 2, 3) , \\ &\gamma_0 = 2, \quad \gamma_i = -5, \quad \gamma_2 = 4, \quad \gamma_3 = -1 . \end{aligned} \tag{2.32}$$

Alike any other multistep algorithm, Houbolt's method is not self-starting: either the time stepping is initiated using a one-step formula, or extra-values $\boldsymbol{q}_{-1}$ and $\boldsymbol{q}_{-2}$ are calculated from the initial values of $\boldsymbol{q}_0$ and $\dot{\boldsymbol{q}}_0$, and expressing equilibrium at $t=0$.

2.2.2. Multistep equations applicable to first-order systems [6, 16]

If a first-order time integrator is used, the initial residual equation is then rewritten in first-order form using velocities as extra-variables

$$\boldsymbol{r}(\boldsymbol{q},\boldsymbol{v}) = \boldsymbol{M}\dot{\boldsymbol{v}} + \boldsymbol{f}(\boldsymbol{q},\boldsymbol{v}) - \boldsymbol{g}(\boldsymbol{q},t) = \boldsymbol{0} \tag{2.33}$$

with

$$\dot{\boldsymbol{q}} - \boldsymbol{v} = \boldsymbol{0}\,. \tag{2.34}$$

Applying the first-order difference formula (2.19) to (2.34) yields

$$\sum_{i=0}^{m} \alpha_i \boldsymbol{q}_{n+1-i} = h \sum_{i=0}^{m} \beta_i \boldsymbol{v}_{n+1-i}\ , \tag{2.35}$$

from which the velocity is expressed in the form

$$\boldsymbol{v}_{n+1} = \frac{\alpha_0}{\beta_0 h}\boldsymbol{q}_{n+1} + \frac{1}{\beta_0 h}\sum_{i=1}^{m}(\alpha_i \boldsymbol{q}_{n+1-i} - \beta_i h \boldsymbol{v}_{n+1-i})\,. \tag{2.36}$$

Similarly, (2.33) allows defining the linear combination of residuals

$$\boldsymbol{r}^*(\boldsymbol{q},\boldsymbol{v}) = \frac{1}{\beta_0}\sum_{i=0}^{m}\beta_i \boldsymbol{r}(\boldsymbol{q}_{n+1-i}, \boldsymbol{v}_{n+1-i}) = \boldsymbol{0}$$

or, making use of (2.35) and (2.33),

$$\boldsymbol{r}^*(\boldsymbol{q},\boldsymbol{v}) = \sum_{i=0}^{m}\frac{\beta_i}{\beta_0}(\boldsymbol{f}_{n+1-i} - \boldsymbol{g}_{n+1-i}) + \sum_{i=0}^{m}\frac{\alpha_i h}{\beta_0}\boldsymbol{M}\boldsymbol{v}_{n+1-i} = \boldsymbol{0}\,. \tag{2.37}$$

The resulting tangent iteration matrix is

$$\begin{aligned}\boldsymbol{S}(\boldsymbol{q}_{n+1}) &= \left(\frac{\partial \boldsymbol{r}^*}{\partial \boldsymbol{q}}\right)_{n+1}\\ &= \frac{\partial \boldsymbol{q}}{\partial \boldsymbol{q}_{n+1}}(\boldsymbol{f}_{n+1} - \boldsymbol{g}_{n+1}) + \left(\frac{\partial \boldsymbol{v}}{\partial \boldsymbol{q}}\right)_{n+1}\left[\frac{\alpha_0 h}{\beta_0}\boldsymbol{M} + \left(\frac{\partial \boldsymbol{f}}{\partial \boldsymbol{v}}\right)_{n+1}\right]\end{aligned}$$

and, with the preceding definitions $\boldsymbol{K}^{\mathrm{t}} = \partial \boldsymbol{f}/\partial \boldsymbol{q}$ and $\boldsymbol{C}^{\mathrm{t}} = \partial \boldsymbol{f}/\partial \boldsymbol{v}$ and the result from (2.36), it becomes

$$\boldsymbol{S}^{\mathrm{MS1}}(\boldsymbol{q}_{n+1}) = \left[\boldsymbol{K}^{\mathrm{t}} + \frac{\alpha_0}{\beta_0 h}\boldsymbol{C}^{\mathrm{t}} + \frac{\alpha_0^2 h^2}{\beta_0^2}\boldsymbol{M} - \frac{\partial \boldsymbol{g}}{\partial \boldsymbol{q}}\right]_{q_{n+1}}. \tag{2.38}$$

In the linear case, it reduces to

$$\boldsymbol{S}_0^{\mathrm{MS1}} = \boldsymbol{K} + \frac{\alpha_0}{\beta_0 h}\boldsymbol{C} + \frac{\alpha_0^2 h^2}{\beta_0^2}\boldsymbol{M}\,. \tag{2.39}$$

The rest of the procedure is the same as in the case of multistep second-order

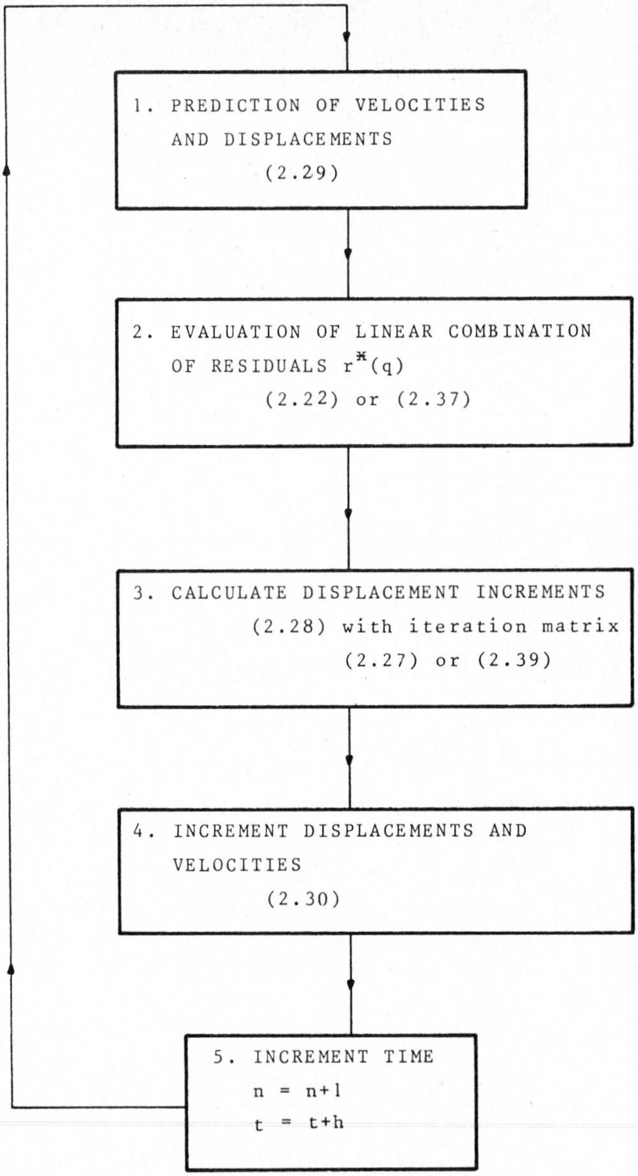

Fig. 2.2. Sequence of operations in each linear time step using multistep formulas.

formulas: the only differences lie in the evaluations of the residual vector (2.37) and of the iteration matrix (2.39).

Park [22] has suggested an integration formula of type (2.36) which exhibits very interesting stability properties when applied to problems of structural dynamics. It results from the composition of Gear's 2- and 3- step formulas [36]:

$$v_{n+1} = \frac{1}{6h}[10q_{n+1} - 15q_n + 6q_{n-1} - q_{n-2}], \tag{2.40}$$

i.e. the particular choices

$$\alpha_0 = 10, \quad \alpha_1 = -15, \quad \alpha_2 = 6, \quad \alpha_3 = -1,$$
$$\beta_0 = 6, \quad \beta_i = 0 \quad (i = 1, 2, 3). \tag{2.41}$$

2.2.3. Summary of operations for multistep methods

The integration procedure is finally very similar to that followed with one-step methods. The main difference lies in the fact that previous values of displacements and residual vectors have to be stored in order to calculate $r^*(q)$ and the new velocities. It may be summarized as in Figure 2.2.

3. Particular aspects of nonlinear implicit integration

3.1. Classes of nonlinear solution techniques

In order to solve the nonlinear system of equilibrium equations

$$r(q_{n+1}) = 0, \tag{3.1}$$

three classes of iterative solution techniques will be considered:

(i) Newton-type iteration (either standard or modified) [3, 13, 21, 25],

(ii) Pseudo-force (constant stiffness) formulation [6, 12, 26],

(iii) Quasi-Newton approximations to the Jacobian matrix [4, 5, 10, 11, 18].

These techniques differ in the way a current residual is expanded around a previous solution, and how the actual Jacobian of the iteration is computed.

3.1.1. Newton-type iteration

In Newton methods, the linearization (1.8) is fully assumed for the behaviour of the residual vector in the neighbourhood of the previous solution q_{n+1}^k. A new solution q_{n+1}^{k+1} is thus seeked by equating $r_L(q_{n+1}^{k+1})$ to zero, i.e. by achieving the path

$$q_{n+1}^{k+1} = q_{n+1}^k - S^{-1}(q_{n+1}^k)r(q_{n+1}^k), \quad k = 0, 1, \ldots. \tag{3.2}$$

Such paths are successively taken unless an appropriate norm of the residual vectors satisfies a given convergence criterion, e.g.

$$\|r(q_{n+1}^{k+1})\| < \varepsilon_R \tag{3.3}$$

where $\|\cdot\|$ stands for the L_2 vector norm $\|x\| = (\Sigma_i x_i^2)^{1/2}$. (See Section 3.5 for a discussion of termination criteria.) Such a 'tangent' (or variable stiffness) iteration procedure is illustrated on Figure 3.1 for a one-dimensional case[1] and is readily

[1]This figure and all forthcoming ones refer strictly to static conditions, i.e. (1.4) with $\ddot{q} = \dot{q} = 0$.

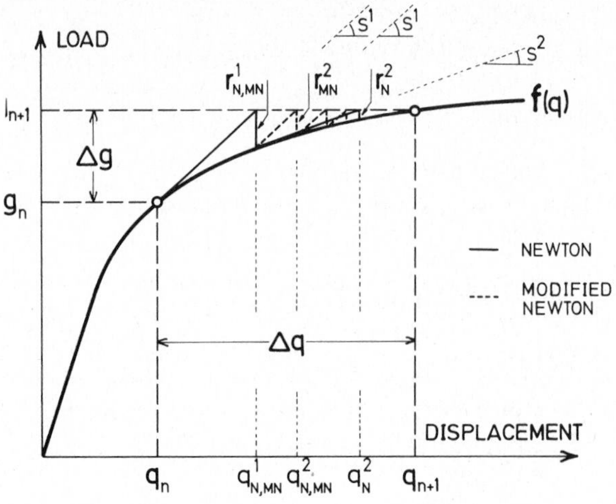

Fig. 3.1. Newton and modified Newton methods in one-dimensional problems.

expensive since the Jacobian matrix is allowed to vary continually and hence each iteration implies factorization or Gauss elimination on a new matrix.

A 'modified' Newton iteration technique is thus very often preferred; it simply assumes that the actual Jacobian matrix may be kept constant for a certain number of paths (see also Figure 3.1 for the one-dimensional case):

$$\boldsymbol{q}^{k+1}_{n+1} = \boldsymbol{q}^{k}_{n+1} - \boldsymbol{S}^{-1}(\boldsymbol{q}^{k_0}_{n+1})\boldsymbol{r}(\boldsymbol{q}^{k}_{n+1}), \quad k \geqslant k_0 . \tag{3.4}$$

This modified iteration procedure can be used from the beginning of each time step (purely modified Newton method) in which case only one evaluation and factorization of the iteration matrix is needed per time step; the reevaluation may either follow the user's own strategy or it can be resorted to only when criterion (3.3) is on its way to be satisfied, i.e.

$$\|\boldsymbol{r}(\boldsymbol{q}^{k_0}_{n+1})\| < \varepsilon_{\mathrm{K}} , \tag{3.5}$$

with a convergence threshold $\varepsilon_{\mathrm{K}} > \varepsilon_{\mathrm{R}}$. The corresponding number of iterations to satisfy dynamic equilibrium in the sense of (3.3) is obviously greater in this latter approach than in the standard tangent iterations. However, iterations of type (3.4) are generally much cheaper since they involve only computations of the current residual vector (corresponding to the previous displacement state):

$$\boldsymbol{r}(\boldsymbol{q}^{k}_{n+1}) = \boldsymbol{M}\ddot{\boldsymbol{q}}^{k}_{n+1} + \boldsymbol{f}(\boldsymbol{q}^{k}_{n+1}, \dot{\boldsymbol{q}}^{k}_{n+1}) - \boldsymbol{g}(\boldsymbol{q}^{k}_{n+1}, t_{n+1}) . \tag{3.6}$$

Iteration techniques (3.2) and (3.4) are usually started with the initial condition

$$\boldsymbol{q}^{0}_{n+1} = \boldsymbol{q}_n \tag{3.7}$$

where $\boldsymbol{q}_n$ is the converged solution of the previous time step.

3.1.2. The pseudo-force approach

Discrete equilibrium equations (3.1) may be written in a different way:

$$M\ddot{q}_{n+1} + C(q_0)\dot{q}_{n+1} + K(q_0)q_{n+1} = g^*_{n+1}(q, t) - f^*_{n+1}(q, \dot{q}) \tag{3.8}$$

where q_0 is a fixed state, i.e. either the initial configuration or any suitable reference state which evenly distributes the residual forces over the expected operational range [6].

In contrast to the previous tangent stiffness iteration approach, we are faced in the present method with a constant stiffness approach corresponding to the reference state by forcing the unknown vector to follow the computational path

$$q^{k+1}_{n+1} = q^k_{n+1} - S_c^{-1}(q_0)r(q^k_{n+1}) \tag{3.9}$$

in which the effective Jacobian matrix is evaluated from the linear stiffness matrix in the reference configuration q_0:

$$S_c(q_0) = K(q_0) + M\left(\frac{\partial \ddot{q}}{\partial q}\right)_{q_0} + C(q_0)\left(\frac{\partial \dot{q}}{\partial q}\right)_{q_0}. \tag{3.10}$$

Figure 3.2 illustrates the procedure for the one-dimensional case. The above iteration sequence should be compared to the modified Newton iteration (3.4) from what it is clear that the two procedures are equivalent: they both operate with a fixed approximate Jacobian. The present approach, however, does not imply the use of tangent stiffness matrices.

3.1.3. Quasi-Newton updates of the Jacobian matrix

The basic idea of quasi-Newton iterative procedures is to substitute to the linearization (1.8), i.e.

$$r(q^{k+1}_{n+1}) = r_L(q^{k+1}_{n+1}) + \text{higher order terms},$$

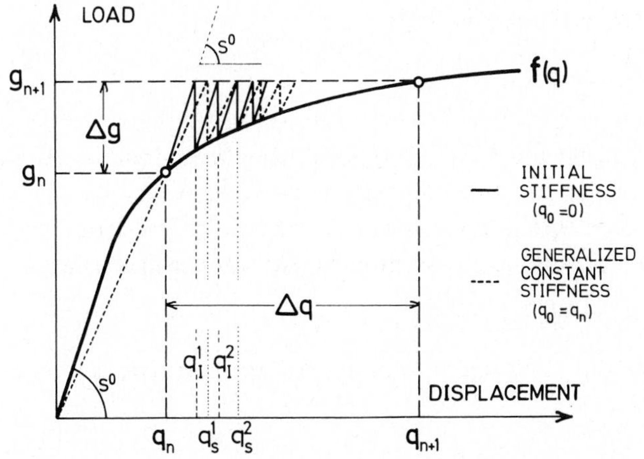

Fig. 3.2. Pseudo-force method in one-dimensional problems.

the quasi-Newton equation

$$\boldsymbol{r}(\boldsymbol{q}_{n+1}^{k+1}) = \boldsymbol{r}(\boldsymbol{q}_{n+1}^{k}) + \boldsymbol{H}(\boldsymbol{q}_{n+1}^{k+1})(\boldsymbol{q}_{n+1}^{k+1} - \boldsymbol{q}_{n+1}^{k}) + \text{higher order terms}\,. \tag{3.11}$$

In that manner, the linearization applies to the Jacobian matrix instead of the residual vector; it provides thus a secant approximation to the iteration matrix from state k to state $k+1$; it can be seen as a procedure midway between full reformation of the matrix (standard Newton method) and use of the matrix in a previous configuration (modified Newton method or pseudo-force approach).

At this stage nothing has been said on the manner in which the approximate Jacobian $\boldsymbol{H}$ is obtained and updated. The quasi-Newton equation (3.11) may be rewritten as

$$\boldsymbol{y}_{n+1}^{k} = \boldsymbol{H}_{n+1}^{k+1}\boldsymbol{d}_{n+1}^{k} \tag{3.12}$$

with the definitions

$$\boldsymbol{y}_{n+1}^{k} = \boldsymbol{r}_{n+1}^{k+1} - \boldsymbol{r}_{n+1}^{k}, \qquad \boldsymbol{d}_{n+1}^{k} = \boldsymbol{q}_{n+1}^{k+1} - \boldsymbol{q}_{n+1}^{k}\,. \tag{3.13}$$

It is desirable that matrices candidates to $\boldsymbol{H}$ satisfy this equation since it is exact if $\boldsymbol{r}$ derives itself from a quadratic functional, and nearly exact if that functional is not quadratic but strictly convex [28, 29].

In other situations it defines a computational path that should be comparable to Newton path. (see Figure 3.3a, b for illustration of the method in one dimensional situations).

For computational effectiveness it is also highly desirable that $\boldsymbol{H}$ be easily obtained recursively from known quantities in a simple manner while satisfying (3.12).

Such a recurrence procedure can be implemented by adding to $\boldsymbol{H}_k$ a single correction matrix of type (from now on we drop the time-step subscript and replace it by the iteration counter)

$$\boldsymbol{H}_{k+1} = \boldsymbol{H}_k + \frac{(\boldsymbol{y}_k - \boldsymbol{H}_k\boldsymbol{d}_k)\boldsymbol{u}_k^{\mathrm{T}}}{\boldsymbol{u}_k^{\mathrm{T}}\boldsymbol{d}_k}\,. \tag{3.14}$$

where $\boldsymbol{u}_k$ is an arbitrary vector not orthogonal to $\boldsymbol{d}_k$, and easily obtained from $\boldsymbol{y}_k$, $\boldsymbol{d}_k$ and $\boldsymbol{H}_k$.

The kth quasi-Newton iteration is thus performed as follows:

(i) compute a direction of search by using the actual Jacobian

$$\boldsymbol{d}_k = -\boldsymbol{H}_k^{-1}\boldsymbol{r}_k\,, \tag{3.15}$$

(ii) evaluate

$$\boldsymbol{y}_k = \boldsymbol{r}(\boldsymbol{q}_k + \boldsymbol{d}_k) - \boldsymbol{r}_k\,,$$

(iii) use a formula of type (3.14) to update $\boldsymbol{H}_k$.

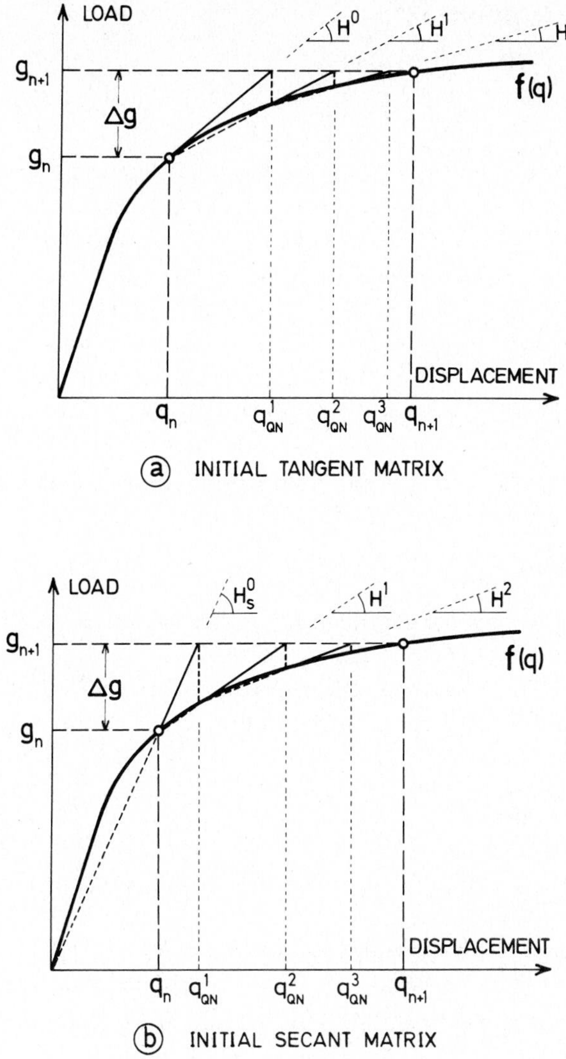

Fig. 3.3. Quasi-Newton method in one-dimensional problems.

The sequence is initialized through a given initial Jacobian computed from the initial tangent or standard stiffness matrix. Obviously the preceding recurrence formula implies the factorization of a new system matrix at each iteration and such a procedure shares the uneconomical characteristics of the standard Newton procedure, unless we obtain directly approximations $\boldsymbol{G}_k$ to the inverse Jacobian $\boldsymbol{G} = \boldsymbol{H}^{-1}$ such that

$$\boldsymbol{d}_k = \boldsymbol{G}_{k+1}\boldsymbol{y}_k \,. \tag{3.16}$$

An inverse updating formula can be obtained from the previous one by using the

Sherman–Morrison formula [28]:

$$(\boldsymbol{A}+\alpha \boldsymbol{a}\boldsymbol{b}^{\mathrm{T}})^{-1}=\boldsymbol{A}^{-1}-\beta \boldsymbol{c}\boldsymbol{d}^{\mathrm{T}} \tag{3.17}$$

with

$$\boldsymbol{c}=\boldsymbol{A}^{-1}\boldsymbol{a},\ \boldsymbol{d}=\boldsymbol{A}^{-\mathrm{T}}\boldsymbol{b}$$

and

$$\beta=\alpha(1+\alpha \boldsymbol{b}^{\mathrm{T}}\boldsymbol{A}^{-1}\boldsymbol{a})^{-1}.$$

The general rank-one inverse update is thus, from (3.14)

$$\boldsymbol{G}_{k+1}=\boldsymbol{G}_k+\frac{(\boldsymbol{d}_k-\boldsymbol{G}_k\boldsymbol{y}_k)\boldsymbol{v}_k^{\mathrm{T}}}{\boldsymbol{v}_k^{\mathrm{T}}\boldsymbol{y}_k} \tag{3.18}$$

where $\boldsymbol{v}_k$ is an arbitrary vector not orthogonal to $\boldsymbol{y}_k$ and easily obtained from $\boldsymbol{d}_k$, $\boldsymbol{y}_k$ and $\boldsymbol{G}_k$.

The recurrence procedure in the inverse formulation is completely similar to the one outlined previously with (3.15) replaced by

$$\boldsymbol{d}_k=-\boldsymbol{G}_k\boldsymbol{r}_k. \tag{3.19}$$

Guidelines for a particular choice of $\boldsymbol{v}_k$ are linked to the behavior of the updating formula with respect to conservation of:
- the symmetrical (or unsymmetrical) character of the matrix to be updated,
- its positive-definite character (in order to prevent illconditioning of the iteration matrix).

Broyden's inverse update [29] corresponds to the choice

$$\boldsymbol{v}_k=\boldsymbol{G}_k^{\mathrm{T}}\boldsymbol{d}_k, \tag{3.20}$$

yielding the formula

$$\boldsymbol{G}_{k+1}^{\mathrm{B}}=\boldsymbol{G}_k+\frac{(\boldsymbol{d}_k-\boldsymbol{G}_k\boldsymbol{y}_k)\boldsymbol{d}_k^{\mathrm{T}}\boldsymbol{G}_k}{\boldsymbol{d}_k^{\mathrm{T}}\boldsymbol{G}_k\boldsymbol{y}_k} \tag{3.21}$$

which obviously does not preserve the eventual symmetry of $\boldsymbol{G}_k$. To keep symmetry, the choice of $\boldsymbol{v}_k$ should be, from (3.18),

$$\boldsymbol{v}_k=\boldsymbol{d}_k-\boldsymbol{G}_k\boldsymbol{y}_k \tag{3.22}$$

which corresponds to Davidon's symmetric update [29] for the inverse Jacobian

$$\boldsymbol{G}_{k+1}^{\mathrm{D}}=\boldsymbol{G}_k+\frac{(\boldsymbol{d}_k-\boldsymbol{G}_k\boldsymbol{y}_k)(\boldsymbol{d}_k-\boldsymbol{G}_k\boldsymbol{y}_k)^{\mathrm{T}}}{(\boldsymbol{d}_k-\boldsymbol{G}_k\boldsymbol{y}_k)^{\mathrm{T}}\boldsymbol{y}_k}. \tag{3.23}$$

None of the preceding rank-one formulas generates positive definite matrices. To this purpose one must resort to rank-two corrections which retain also symmetry such as the Davidon–Fletcher–Powell (DFP) update [29]:

$$\boldsymbol{G}_{k+1}^{\mathrm{DFP}}=\boldsymbol{G}_k+\frac{\boldsymbol{d}_k\boldsymbol{d}_k^{\mathrm{T}}}{\boldsymbol{d}_k^{\mathrm{T}}\boldsymbol{y}_k}-\frac{\boldsymbol{G}_k\boldsymbol{y}_k\boldsymbol{y}_k^{\mathrm{T}}\boldsymbol{G}_k}{\boldsymbol{y}_k^{\mathrm{T}}\boldsymbol{G}_k\boldsymbol{y}_k} \tag{3.24}$$

and the Broyden–Fletcher–Goldfarb–Shanno (BFGS) formula [29]:

$$G_{k+1}^{\mathrm{BFGS}} = \left(I - \frac{d_k y_k^{\mathrm{T}}}{y_k^{\mathrm{T}} d_k}\right) G_k \left(I - \frac{y_k d_k^{\mathrm{T}}}{y_k^{\mathrm{T}} d_k}\right) + \frac{d_k d_k^{\mathrm{T}}}{y_k^{\mathrm{T}} d_k}. \tag{3.25}$$

Only the second one preserves the positive definite character of G_k while the first one does it in its inverse form (for H_k) which corresponds exactly to (3.25) with the substitutions

$$d_k \to y_k, \qquad y_k \to d_k, \qquad G_k \to H_k. \tag{3.26}$$

3.2. *Line search techniques*

Accuracy of iterative procedures is known to be very much dependent on the choice of the optimal step length in the direction of search; hence such a step length σ_k can be evaluated in quasi-Newton (and Newton) methods which cancels the projection of the residual vector in that direction, i.e.

$$d_k^{\mathrm{T}} r(q_k + \sigma_k d_k) = 0\,, \tag{3.27}$$

and the effective solution is then expressed in the form

$$q_{k+1} = q_k + \sigma_k d_k \tag{3.28}$$

where d_k is obtained by one of the techniques (3.2), (3.4), (3.9) or (3.19). However, in the solution of systems of nonlinear algebraic equations, there is no guarantee that Newton or quasi-Newton paths are the best suited paths towards the solution. Besides, line search operations are expensive since one search may involve numerous evaluations of the residual vector to achieve great accuracy and improve the step size. Therefore, this technique should be used with care and does not seem necessary except for rank-two quasi-Newton methods [28].

Numerical experiments lead to the conclusion that statement (3.27) can then be replaced by the test

$$|d_k^{\mathrm{T}} r(q_k + \sigma_k d_k)| / |d_k^{\mathrm{T}} r(q_k)| < \eta \tag{3.29}$$

where η is an appropriate threshold to be adjusted according to the classes of nonlinear problems and the specific updating method (e.g., 0.5 for structural dynamics [4]). In particular, no line search is attempted if (3.29) is satisfied when $\sigma_k = 1$.

3.3. *Algorithm organization for nonlinear solution techniques*

In practice, a new problem is seldom known to be more favourable to one or another of the preceding solution techniques. It is therefore of paramount importance that an implicit computer code contains all possiblities with flexible shifting strategies.

Such an organization is drafted on Figure 3.4 for each implicit time step to be considered, with emphasis of the possible links between the different methods:

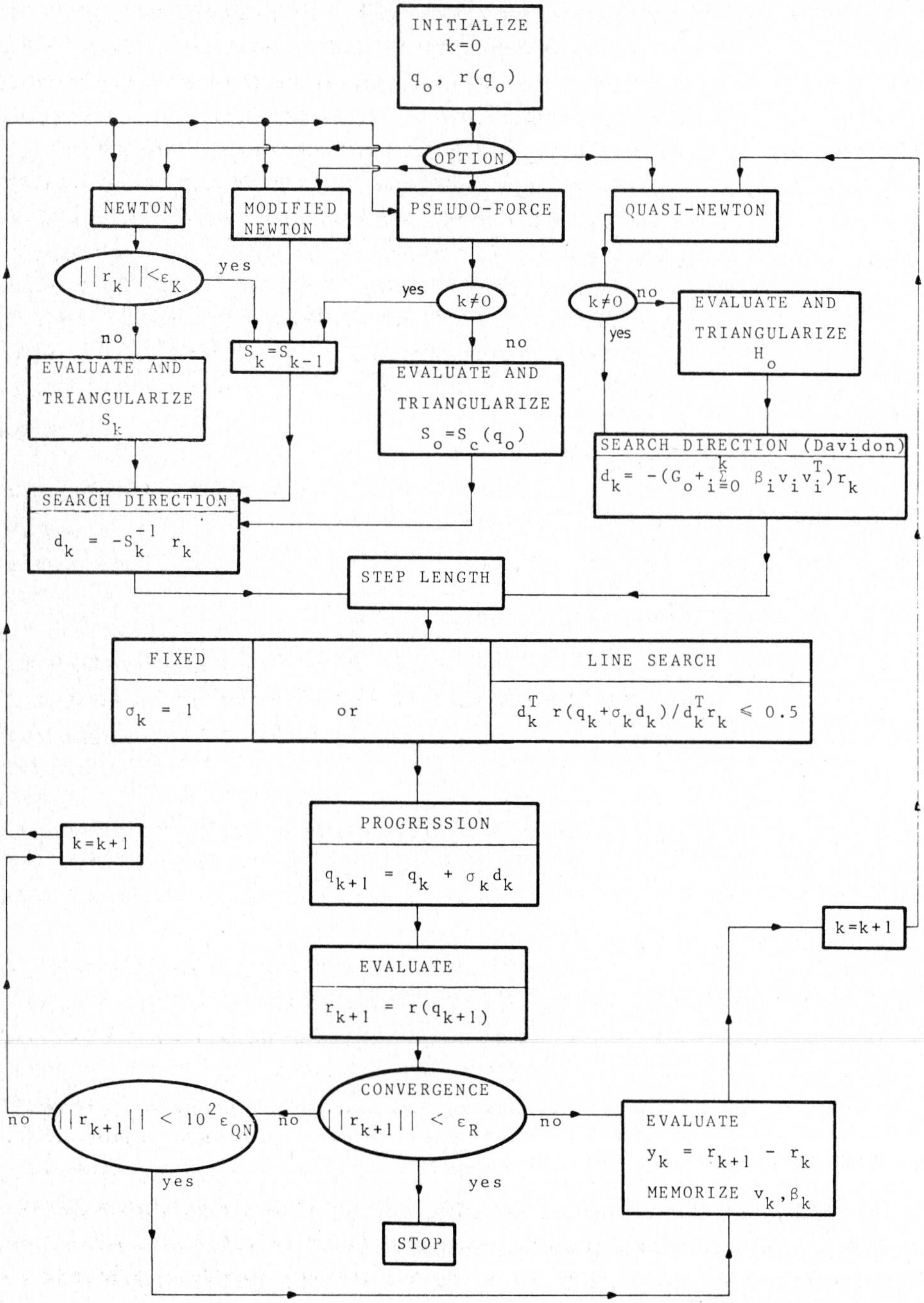

Fig. 3.4. Organization of nonlinear solution techniques at each implicit time station.

– the Newton iteration can be changed into the modified Newton technique when criterion (3.5) is satisfied;
– the pseudo-force method and the modified Newton technique have common characteristics as soon as the constant system matrix has been triangularized;
– quasi-Newton methods can be attempted to form the beginning or, if strong nonlinearities require Newton method, this latter technique should be used for k iterations until the convergence threshold for equilibrium ε_R is reasonably approached, say $\|\boldsymbol{r}_k\| < \varepsilon_{QN}$ with $\varepsilon_{QN} = 10^2\,\varepsilon_R$; the iteration technique is then changed to quasi-Newton updating for the remaining of the convergence procedure.

Implementation of the latter technique requires special attention and is also illustrated on Figure 3.4. The most efficient way of performing the quasi-Newton update consists indeed in applying the correction on the direction of search instead of modifying the approximate Jacobian. In fact, using the inverse update as described by (3.18), at the kth iteration, $\boldsymbol{G}$ can be written (for a single-rank symmetric update) as:

$$\boldsymbol{G}_k = \boldsymbol{G}_0 + \sum_{i=0}^{k} \beta_i \boldsymbol{v}_i \boldsymbol{v}_i^{\mathrm{T}}\,. \tag{3.30}$$

For instance, for Davidon's update (3.23) we have

$$\boldsymbol{v}_i = \sigma_i \boldsymbol{d}_i - \boldsymbol{G}_i \boldsymbol{y}_i \quad \text{and} \quad \beta_i = [(\sigma_i \boldsymbol{d}_i - \boldsymbol{G}_i \boldsymbol{y}_i)^{\mathrm{T}} \boldsymbol{y}_i]^{-1}.$$

If at each iteration the correction vector $\boldsymbol{v}_i$ and coefficient β_i are stored on auxiliary memory, the $(k+1)^{\text{th}}$ direction of search can be obtained from (3.19) as

$$\boldsymbol{d}_k = -\Big(\boldsymbol{G}_0 + \sum_{i=0}^{k} \beta_i \boldsymbol{v}_i \boldsymbol{v}_i^{\mathrm{T}}\Big)\boldsymbol{r}_k\,. \tag{3.31}$$

The new correction vector for Davidon's update is then

$$\boldsymbol{v}_k = \sigma_k \boldsymbol{d}_k - \boldsymbol{G}_0 \boldsymbol{y}_k - \sum_{i=0}^{k} \beta_i \boldsymbol{v}_i \boldsymbol{v}_i^{\mathrm{T}} \boldsymbol{y}_k\,. \tag{3.32}$$

Computational efficiency of this updating technique stems from the fact that, if an initial sparse Jacobian $\boldsymbol{H}_0$ is given, it may be triangularized and stored only once, yielding the successive products $\boldsymbol{G}_0 \boldsymbol{r}_k$ and $\boldsymbol{G}_0 \boldsymbol{r}_{k+1}$ needed in equations (3.31) and (3.32).

In this manner only the nonzero elements of $\boldsymbol{G}_0$ after Gauss elimination, the vectors $\boldsymbol{v}_i$ and the coefficients β_i have to be stored. When the number of correction vectors becomes too large (from our experience, say maximum 10 corrections since convergence will not be reached later on), the algorithm may be restarted with the initial matrix $\boldsymbol{G}_0$.

A last observation is about the theoretical difference one can expect between the cost of Newton iteration and the cost of other iteration techniques. In Newton method, the computation and triangularization of $\boldsymbol{S}_k$ requires $\mathrm{O}(n^3)$ arithmetic operations. In quasi-Newton, for every iteration from the second, this expense is reduced to $O(n^2)$ operations; in modified Newton and pseudo-load iterations, the order is the same but with a lower coefficient since only one back substitution is needed per iteration.

3.4. Common numerical problems encountered in equilibrium iterations

Newton and modified Newton iterations are known to present numerical difficulties (slow convergence or divergence, i.e. out-of-balance loads increase during the solution) in particular situations of nonlinear material or geometrical nonlinearities:

– either divergence or lack of convergence can happen for Newton iteration, respectively in cases of load-deflection curves presenting inflexion points (geometric nonlinearities) and in cases of sudden material stiffenning (unloading of an elasto-plastic material (Figure 3.5a,b).

The first situation is usually resolved by monitoring smaller time steps in that region; in the second one, where the unloading is such that an element suffers a new plastification in the same iteration ($\varepsilon_n - \varepsilon_1, \varepsilon_2 - \varepsilon_3, \ldots$) the remedy is to use the elastic stiffness matrix for the remaining iterations. The associate rate of convergence can be slow but convergence is insured in all cases. (Figure 3.5c);

– modified Newton iteration exhibits slow convergence properties in case of sudden softening of the system during a time step (Figure 3.1) or of slow stiffening (Figure 3.6a).

The obvious remedy in these cases is to choose a smaller time step so that linearization about t_n yields an approximation close enough to the Jacobian matrix at time t_{n+1} and insures convergence.

Plastic unloading with unappropriate choice of the Jacobian matrix can lead to a wrong solution (Figure 3.6b) and the remedy is also in this case to shift to the Jacobian corresponding to the elastic properties.

The pseudo force method shares the numerical difficulties of modified Newton iteration but quasi-Newton techniques, due to their secant approximation of the Jacobian matrix, are able to overcome the cycling behaviour of Newton iteration in case of elastic unloading from a plastic state with new plastification in the iteration, without switching to the initial elastic Jacobian (Figure 3.7) obviously, the recourse to the elastic Jacobian is also an effective technique which can lead to further savings in some cases.

3.5. Convergence criteria in dynamic equilibrium iterations [4, 25]

It should be stressed again that the governing dynamic equilibrium equations must be satisfied in each time step to sufficient accuracy, otherwise solution errors accumulate and can lead to instabilities in a few subsequent steps; therefore solution techniques that would not deal with an iterative correction procedure for equilibrium are not effective in nonlinear analysis, in particular for nonlinear dynamic problems [3].

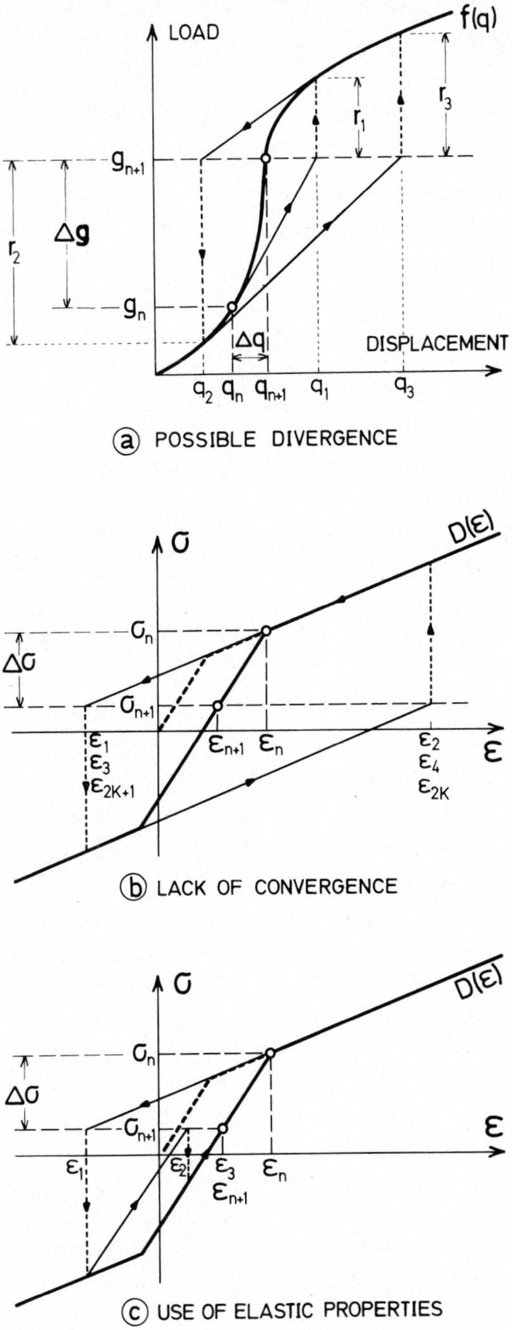

Fig. 3.5. Difficulties with Newton method.

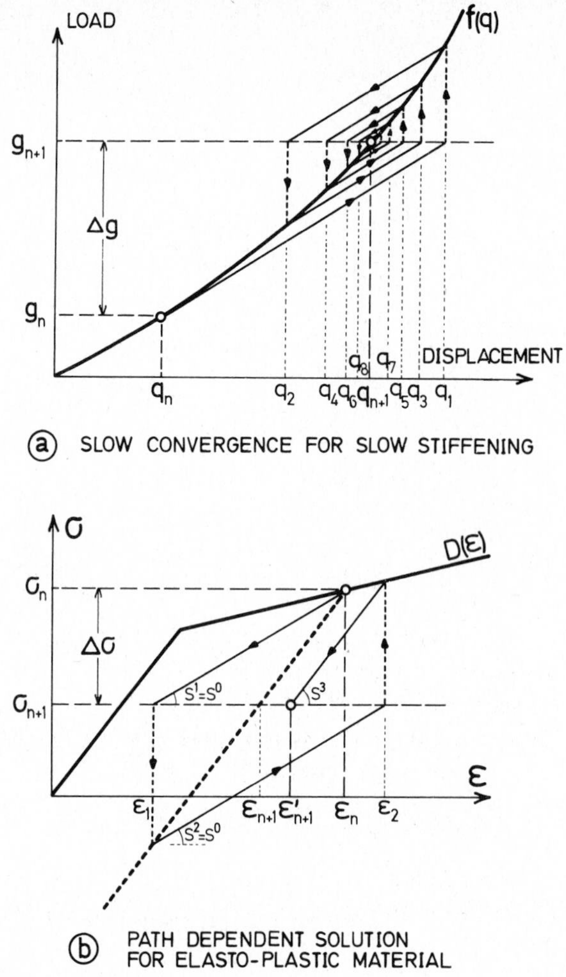

Fig. 3.6. Difficulties with modified Newton method.

Appropriate termination criteria are thus essential for the effectiveness of the iteration procedure: if the convergence tolerance is too loose, it causes divergence of the time marching solution after a few steps; if it is too tight, it leads to unacceptable computational effort.

Convergence criteria can deal with three types of variables: displacements, out-of-balance (residual) forces or step-by-step internal energy. They usually refer some norm (e.g. the Euclidian norm) of the tested variable in the actual configuration to some reference value.

Displacement criteria read

$$\|\Delta \boldsymbol{q}^k\| \leq \varepsilon_D \|\boldsymbol{q}_{REF}\| \tag{3.33}$$

where $\boldsymbol{q}_{REF}$ stands for some previously calculated displacement state, usually $\boldsymbol{q}_n$.

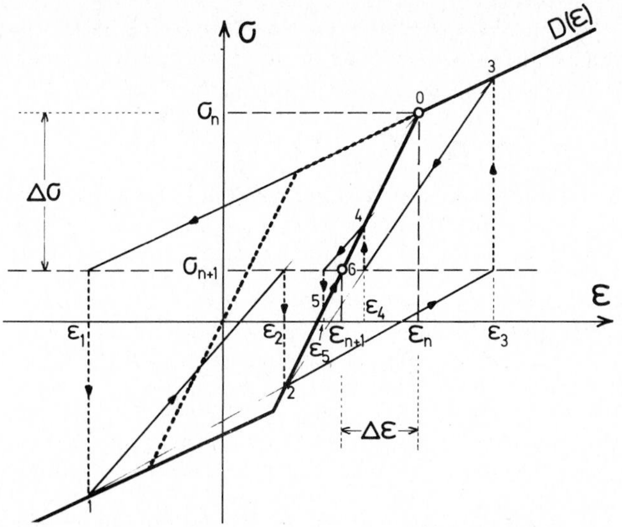

Fig. 3.7. Quasi-Newton method for elastic unloading from a plastic state.

Satisfaction of this criterion does not guarantee that equilibrium is satisfied but only that the displacement solution is stabilized.

Hence convergence criteria based on the out-of-balance forces are more reliable: such a criterion requires that the norm of the residual vector be within a prescribed threshold ε_R of a reference load increment

$$\|\boldsymbol{r}^k\| \leq \varepsilon_R \|\Delta \boldsymbol{g}_{REF}\| . \tag{3.34}$$

The choice of the reference load is manifold:
– it can either be the norm of the load increment encountered during the time step [4]:

$$\Delta \boldsymbol{g}_{REF} = \boldsymbol{g}_{n+1} - \boldsymbol{f}_n - \boldsymbol{M}\ddot{\boldsymbol{q}}_n \tag{3.35}$$

or the corresponding maximum value of this norm encountered during the previous time step (in order to restrain the severity of the criterion for small load increments);
– a convenient choice in our own numerical tests has been found to be

$$\|\Delta \boldsymbol{g}^k_{REF}\| = \|\boldsymbol{g}_{n+1}\| + \|\boldsymbol{f}^k_{n+1}\| \tag{3.36}$$

in which even reactions to prescribed displacements are accounted for, and the presence of internal forces allows for cases where no external forces are applied.

The drawback of force-type convergence checks is the absence of displacement contribution to the termination criterion; problem with small residual loads by nature can suffer from erroneous displacement solutions while the criterion is satisfied (e.g. elasto-plastic material with small strain-hardening modulus when entering the plastic range).

Indications on both displacements and forces near the equilibrium state can be given by looking at the increment of internal energy during each iteration when compared to the corresponding initial increment [4]

$$\boldsymbol{r}_k^{\mathrm{T}}\Delta\boldsymbol{q}_k \leqslant \varepsilon_{\mathrm{E}}\Delta\boldsymbol{q}_1^{\mathrm{T}}(\boldsymbol{g}_{n+1}-\boldsymbol{f}_n-\boldsymbol{M}\ddot{\boldsymbol{q}}_n)\,. \tag{3.37}$$

Clearly, the choice of a particular criterion must be based on the characteristics of the nonlinear problem to be solved, but it appears that a combination of force- and energy-type criteria provides an effective means for both displacement and force convergence. The associate thresholds are also problem-dependent since convergence of 'softening' problems should be carefully checked on displacement and energy variables while stiffening problems rather require tight tolerance on force equilibrium.

3.6. *Numerical integration of nonlinear material constitutive relations* [4, 21, 30]

In any of the residual evaluations (3.6) the vector of internal forces $\boldsymbol{f}_{n+1}^k$ is computed from the stresses that correspond to the current iteration of the $(n+1)^{\text{th}}$ time-step. For a typical finite element and at a given Gauss integration point this stress evaluation can be written as

$$\boldsymbol{\sigma}_{n+1}^k = \boldsymbol{\sigma}_n + \int_{\boldsymbol{\varepsilon}_n}^{\boldsymbol{\varepsilon}_{n+1}^k} \boldsymbol{D}(\boldsymbol{\varepsilon})\,\mathrm{d}\boldsymbol{\varepsilon} \tag{3.38}$$

where $\boldsymbol{\sigma}_n$ are the stresses corresponding to the previous accepted equilibrium configuration at t_n and $\boldsymbol{D}$ is the matrix defining the stress-strain constitutive relation which is not constant during the integration in nonlinear material problems; $\boldsymbol{\varepsilon}_n$ denotes the strains at time t_n and $\boldsymbol{\varepsilon}_{n+1}^k$ the current strains reached at the end of the iteration.

It is important that the final converged result in (3.38) be independent of the particular computational paths followed during the iteration from the last converged strains. This is not the case if one integrates the strain-stress relation over each of the successive increments of strains, i.e.

$$\boldsymbol{\sigma}_{n+1}^k = \boldsymbol{\sigma}_n + \sum_{i=0}^{k-1} \int_{\boldsymbol{\varepsilon}_{n+1}^i}^{\boldsymbol{\varepsilon}_{n+1}^{i+1}} \boldsymbol{D}(\boldsymbol{\varepsilon})\,\mathrm{d}\boldsymbol{\varepsilon} \quad \text{with } \boldsymbol{\varepsilon}_{n+1}^0 = \boldsymbol{\varepsilon}_n\,, \tag{3.39}$$

in place of integrating over the new estimate of the total deformation increment as in (3.38). However, this latter technique requires the choice of an integration path in the strain space, of type

$$\mathrm{d}\boldsymbol{\varepsilon} = (\boldsymbol{\varepsilon}_{n+1}^k - \boldsymbol{\varepsilon}_n)\,\mathrm{d}\gamma$$

where $\gamma \in [0, 1]$ is the normalized strain over the path, which in turn acts on the effective load history. Besides, numerical integration is required in (3.38) which can be rather tedious since numerous paths can be necessary to reach a fixed accuracy to complete the successive total strain increments.

An alternative way is to evaluate stresses by (3.39) in conjunction with a postulate of incremental reversibility [30] (i.e. the constitutive law is reversible within a load increment) which allows to integrate on the additional strain increments encountered at each iteration only.

An explicit Euler integration scheme is then used which enables to predict the material behaviour at every iteration without resorting to an additional iterative integration technique [30].

4. Computer implementation of an implicit finite element code for transient analysis [8]

As already stressed in the introduction, the main drawback of implicit methods, compared to explicit integration algorithms, is the relative complexity of the associate computer programs, specially in the nonlinear case, for the following reasons:

1. implicit methods involve solving a full system of equations of the size of the system, with an iteration matrix which has to be updated regularly for nonlinear problems, or at least at time step changes when solving linear problems;

2. a reevaluation of the tangent iteration matrix requires the computation of the tangent stiffness matrices at the element level. The computational efficiency of the finite element program is of course highly dependent on the cost of this operation and the way it is implemented in combination with the elimination or factorization of the global tangent stiffness matrix;

3. computational strategies for nonlinear analysis are more complex not only to develop, but also simply to use, since the user has freedom in the choice of:
– the type of iteration method (Newton, modified Newton or quasi-Newton) and possibly the combination of them in the same solution,
– the accuracy with which dynamic equilibrium is achieved.

Unfortunately, it is extremely difficult to render these choices fully automated, and in many situations the computational cost or even the convergence of the solution may strongly depend on the choices that have been made.

On the other hand, implicit methods have also specific advantages, e.g.
– in problems of inertial behaviour type, they generally lead to more economical solutions than explicit methods;
– there is no restriction in the type of problems that may be solved; they offer continuous transition between static and dynamic behavior since problems in which the transients in the system establish around a stressed position of equilibrium can be dealt with; there is no limitation with respect to the method of mass discretization; linear and nonlinear constraints may be applied on the variables of the systems, etc.

In what follows, a review is made of the general features of an implicit finite element program for transient analysis.

4.1. Practical organization of an implicit finite element code for transient analysis

The flowchart of Figure 4.1 is a typical example of what is the software organization of a large finite element code for linear and nonlinear transient analyses. The program is intimately connected to a larger finite element system in which operations such as data generation and preprocessing are standardized. The implicit code starts properly with a linear finite element generation in which the linear stiffness, mass and viscous damping matrices are stored on auxiliary memory.

For the two next steps, which correspond to the bulk of the program (calculation of the static equilibrium configurations and of the dynamic response), three types of data are required: the definition of external loads, the computational strategy adopted and the policy for output result recording. Both parts of the program make a constant use of the nonlinear generation modules, and the output results are kept and processed to prepare restart points and time history type of records. An additional facility is provided to plot either on listing or on graphics time history responses. More elaborate graphic representations or manipulations of output results are performed in a separate postprocessing program.

4.2. Linear and nonlinear finite element generation

The finite element generation, that is calculation of internal loads and element matrices, is one of the key problems in the design of nonlinear implicit codes. Three cases have to be considered.

4.2.1. Linear analysis

Under the hypotheses of geometric and material linearity of the elements, the element stiffness, mass and damping matrices can be calculated in advance in a linear generation step and stored once for all on sequential auxiliary memory units. There is no difficulty during the iteration process to recall them whenever the iteration matrix $\boldsymbol{S}_0$, the internal loads $(\boldsymbol{Kq})$, the damping forces $(\boldsymbol{C\dot{q}})$ and the inertia loads $(\boldsymbol{M\ddot{q}})$ have to be reevaluated.

4.2.2. Nonlinear analysis using quasi-Newton and modified Newton iterations

Quasi-Newton and modified Newton iterations do not require a reevaluation of the tangent iteration matrix. Hence a nonlinear finite element generation phase is necessary only to reevaluate the internal loads which may depend on the geometry variations and on the adaptation of material properties. We are thus led back to the same problem as in an explicit code: in general, the evaluation of internal loads can be performed in a core partition of moderate size, even for

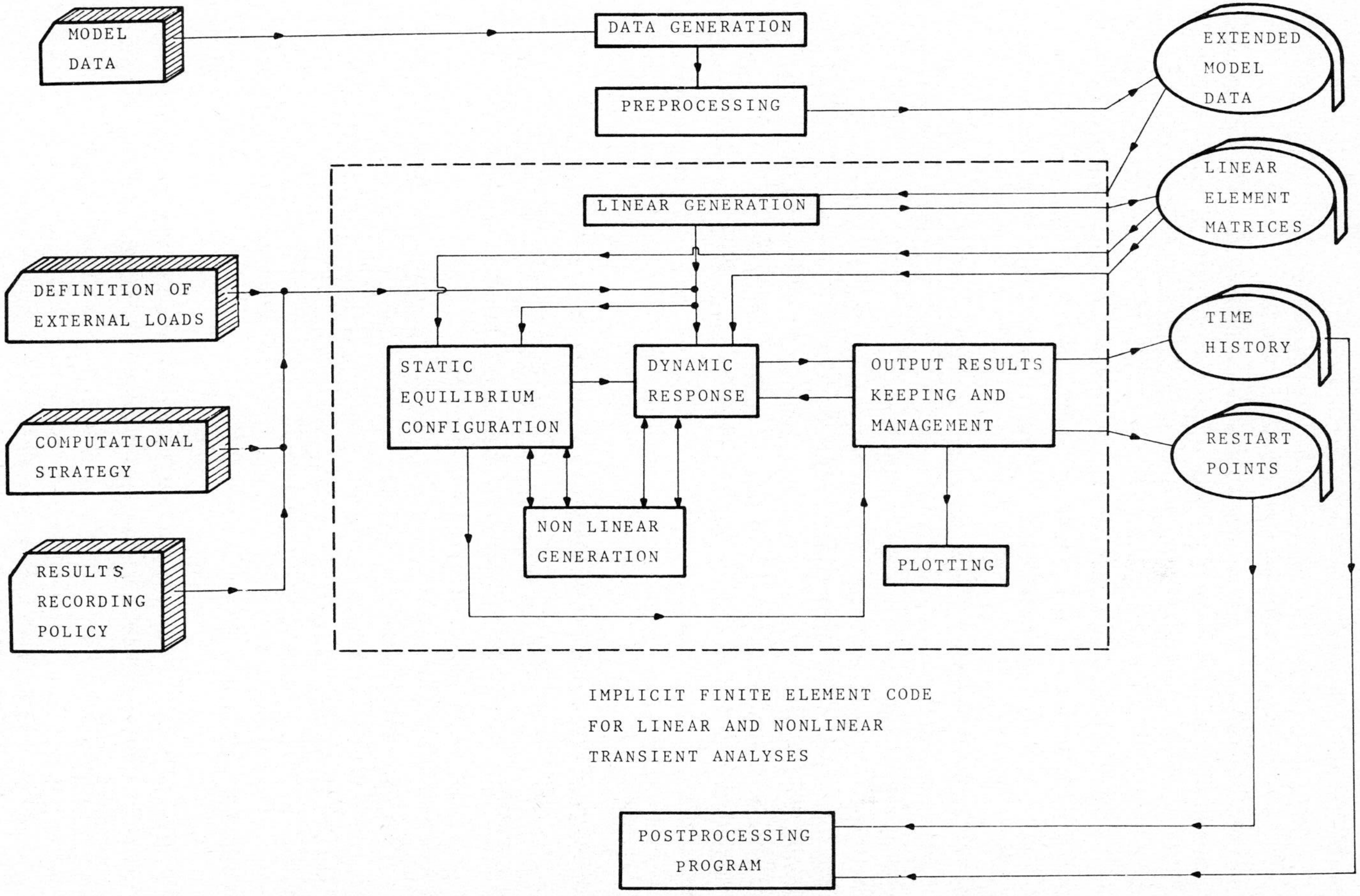

Fig. 4.1. Flow chart of an implicit finite element code for dynamic analysis.

sophisticated elements. Due to the time step size used in implicit integration, however, greater accuracy is required than in explicit codes in the integration of material nonlinear constitutive laws.

4.2.3. Nonlinear analysis using Newton-Raphson iteration

A Newton–Raphson iteration requires also evaluating the element tangent stiffness matrices before assembling them into the tangent iteration matrix on which Gauss elimination is performed. Two techniques may be chosen:

(i) nonlinear finite element generation, assembling the iteration matrix and its Gauss elimination are combined together; the tangent stiffness matrices are then evaluated when required in the evaluation process, avoiding thus computer cost to store them on peripheric devices and to transfer them to and from their storage location;

(ii) the element tangent stiffness matrices are evaluated before assembling the iteration matrix and stored on auxiliary device. They are then recalled from auxiliary memory during the assembling process.

The second procedure minimizes the core space required by the software, and leads to a program which may contain a finite element library of greater varsity. However, these advantages are paid by additional CPU and I/O cost in the assembling and elimination phases of the global iteration matrix.

In practice, the first procedure should rather be adopted if the finite element library is limited to a few models and the linear solution algorithm organized so as to use a minimum core storage in the elimination phase.

4.3. Solution of linear systems of equations

A good choice for the linear equation solver to be incorporated in an implicit integration code is the frontal version of the Gauss elimination method [31]. The reasons for this particular choice are the following:

– its combination with the nonlinear finite element generation (calculation of the element tangent stiffness matrices) yields a very compact solution program;

– it is well adapted to the treatment of special problems such as response to imposed displacements, tying of d.o.f. by nonlinear constraints, etc.

We can write any of the linear systems of equations involved in Section 2 as

$$\boldsymbol{K}\boldsymbol{q} = \boldsymbol{g}\,. \tag{4.1}$$

The frontal algorithm is based on a partitioning of degrees of freedom (d.o.f.) into retained d.o.f., $\boldsymbol{q}_{\mathrm{R}}$, condensed d.o.f., $\boldsymbol{q}_{\mathrm{C}}$, and fixed d.o.f., $\boldsymbol{q}_{\mathrm{F}}$, yielding the partitioned set of equations

$$\begin{bmatrix} \boldsymbol{K}_{\mathrm{RR}} & \boldsymbol{K}_{\mathrm{RC}} & \boldsymbol{K}_{\mathrm{RF}} \\ \boldsymbol{K}_{\mathrm{CR}} & \boldsymbol{K}_{\mathrm{CC}} & \boldsymbol{K}_{\mathrm{CF}} \\ \boldsymbol{K}_{\mathrm{FR}} & \boldsymbol{K}_{\mathrm{FC}} & \boldsymbol{K}_{\mathrm{FF}} \end{bmatrix} \begin{bmatrix} \boldsymbol{q}_{\mathrm{R}} \\ \boldsymbol{q}_{\mathrm{C}} \\ \boldsymbol{q}_{\mathrm{F}} \end{bmatrix} = \begin{bmatrix} \boldsymbol{g}_{\mathrm{R}} \\ \boldsymbol{g}_{\mathrm{C}} \\ \boldsymbol{g}_{\mathrm{F}} \end{bmatrix}. \tag{4.2}$$

In the third equation, g_F is the unknown right-hand side (r.h.s.) vector corresponding to q_F; the latter gives a contribution to the r.h.s. vectors in the two first set of equations as

$$\begin{bmatrix} K_{RR} & K_{RC} \\ K_{CR} & K_{CC} \end{bmatrix} \begin{bmatrix} q_R \\ q_C \end{bmatrix} = \begin{bmatrix} g_R - K_{RF} q_F \\ g_C - K_{CF} q_F \end{bmatrix} = \begin{bmatrix} g_R^* \\ g_C^* \end{bmatrix} . \tag{4.3}$$

The d.o.f. to be condensed are eliminated next by solving the second set of equations

$$q_C = K_{CC}^{-1}[g_C^* - K_{CR} q_R] \tag{4.4}$$

and replaced in the first, yielding finally the condensed equation

$$\begin{aligned}[K_{RR} - K_{RC} K_{CC}^{-1} K_{CR}] q_R &= \bar{K}_{RR} q_R \\ &= g_R^* - K_{RC} K_{CC}^{-1} g_C^* = \bar{g}_R .\end{aligned} \tag{4.5}$$

Equations (4.4) and (4.5) may be used in a repetitive way, on the basis of a sequential organization of the f.e.m. mesh into substructures, to solve large systems of equations in a moderate size partition of core storage. Note that even the front of equations, that is the matrix $\bar{K}_{RR}$, does not need to fit into central core storage except at the very last level of elimination [31]. Saving space in this manner provides in the nonlinear case the working space necessary to perform the generation of tangent stiffness matrices.

When the Gauss elimination technique is to solve the nonlinear equilibrium equations, the iterative procedure (Newton or modified Newton methods) requires defining three black-box subroutines, provided that the external forces are known:

1. Condensation of the tangent iteration matrix:

CALL COND(S, LS, P, NT, . . .)

where S(LS) is a working array, and Q(NT) is the displacement configuration at which the tangent matrix is evaluated.

2. Solution of a linear system:

CALL RESOL(S, LS, G, DQ, NT, . . .)

where G(NT) is the r.h.s. vector, and DQ(NT) the unknown increment. The solution proceeds in three steps: condensation of the r.h.s. G(NT), solution of a condensed system and backwards substitution to obtain the full solution DQ(NT).

3. Calculation of the vector of internal forces:

CALL CHARIN(S, LS, Q, GINT, NT, . . .)

where GINT(NT) is the vector of internal loads evaluated in the displacement configuration Q(NT). In practice, significant savings are made if the vector of internal loads is calculated simultaneously to the tangent stiffness matrix whenever the latter is constructed.

4.4 Nonlinear constraints

In order to solve the nonlinear system of equilibrium equations

$$\boldsymbol{r}(\boldsymbol{q}) = \boldsymbol{0}\,, \tag{4.6}$$

subject to a nonlinear constraint

$$c(\boldsymbol{q}) = 0\,, \tag{4.7}$$

a virtual displacement $\delta\boldsymbol{q}$ is considered which verifies the constraint equation

$$c(\boldsymbol{q}+\delta\boldsymbol{q}) = c(\boldsymbol{q}) + \left(\frac{\partial c}{\partial \boldsymbol{q}}\right)^{\mathrm{T}} \delta\boldsymbol{q} = 0$$

or, making use of (4.7),

$$\left(\frac{\partial c}{\partial \boldsymbol{q}}\right)^{\mathrm{T}} \delta\boldsymbol{q} = 0\,. \tag{4.8}$$

Thus, a Lagrangian multiplier λ may be introduced which transforms the constraint problem (4.6)–(4.7) into the augmented system of $n+1$ equations

$$\boldsymbol{r}(\boldsymbol{q}) + \lambda \frac{\partial c}{\partial \boldsymbol{q}} = \boldsymbol{0}, \qquad c(\boldsymbol{q}) = 0\,. \tag{4.9}$$

This system is then solved incrementally by assuming an approximate solution $(\tilde{\boldsymbol{q}}, \tilde{\lambda})$ from which an improved solution $(\tilde{\boldsymbol{q}} + \Delta\boldsymbol{q}, \tilde{\lambda} + \Delta\lambda)$ is calculated. A first order expansion of (4.9) yields the linear equation

$$\begin{aligned} &\boldsymbol{r}(\tilde{\boldsymbol{q}}) + \left(\frac{\partial \boldsymbol{r}}{\partial \boldsymbol{q}}\right)_{\tilde{\boldsymbol{q}}} \Delta\boldsymbol{q} + \tilde{\lambda}\left(\frac{\partial c}{\partial \boldsymbol{q}}\right)_{\tilde{\boldsymbol{q}}} + \tilde{\lambda}\left(\frac{\partial^2 c}{\partial \boldsymbol{q}^2}\right)_{\tilde{\boldsymbol{q}}} \Delta\boldsymbol{q} + \Delta\lambda\left(\frac{\partial c}{\partial \boldsymbol{q}}\right)_{\tilde{\boldsymbol{q}}} = \boldsymbol{0}\,, \\ &c(\tilde{\boldsymbol{q}}) + \left(\frac{\partial c}{\partial \boldsymbol{q}}\right)^{\mathrm{T}}_{\tilde{\boldsymbol{q}}} \Delta\boldsymbol{q} = 0\,. \end{aligned} \tag{4.10}$$

The correction vector $[\Delta\boldsymbol{q}\ \Delta\lambda]^{\mathrm{T}}$ is thus solution of the augmented tangent equation

$$\boldsymbol{S}(\tilde{\boldsymbol{q}}, \tilde{\lambda}) \begin{bmatrix} \Delta\boldsymbol{q} \\ \Delta\lambda \end{bmatrix} = -\boldsymbol{g}(\tilde{\boldsymbol{q}}, \tilde{\lambda}) \tag{4.11}$$

with the tangent iteration matrix

$$\boldsymbol{S}(\tilde{\boldsymbol{q}}, \tilde{\lambda}) = \begin{bmatrix} \left[\dfrac{\partial \boldsymbol{r}}{\partial \boldsymbol{q}} + \lambda \dfrac{\partial^2 c}{\partial \boldsymbol{q}^2}\right] & \dfrac{\partial c}{\partial \boldsymbol{q}} \\ \left(\dfrac{\partial c}{\partial \boldsymbol{q}}\right)^{\mathrm{T}} & 0 \end{bmatrix}_{(\tilde{\boldsymbol{q}}, \tilde{\lambda})} \tag{4.12}$$

and the extended residual vector

$$g(\tilde{q}, \tilde{\lambda}) = \begin{bmatrix} r(\tilde{q}) + \tilde{\lambda}\left(\dfrac{\partial c}{\partial q}\right)_{\tilde{q}} \\ c(\tilde{q}) \end{bmatrix}. \tag{4.13}$$

The advantage of the implicit solution when such constraints are applied to the system is thus that lagrangian multipliers are treated as ordinary unknowns.

4.5. *Starting procedure for the time integration*

Three situations may arise with respect to the initial conditions of the dynamic response problem to be solved:

(i) The most frequent case corresponds to prescribed initial displacements and velocities. Equilibrium is then achieved at the very first iteration by calculating accelerations such that

$$M\ddot{q}_0 = g_0 - f(q_0, \dot{q}_0). \tag{4.14}$$

It implies thus solving a linear system with the mass matrix as matrix of coefficients, and corresponds to a limiting case where the stiffness and damping terms vanish in the iteration matrix. Solving (4.14) is trivial if the mass matrix is diagonal. Otherwise, the same linear solution scheme is used as for the iteration matrix (see Section 4.3).

(ii) A second important situation corresponds to motion initiated from a static position of equilibrium. Falls into this category the particular problem of the dynamic response of flexible structures such as cable nets under disturbing forces. Velocities and accelerations start then from zero, but the initial displacements are solutions of a nonlinear static equation:

$$f(q_0) = g_0. \tag{4.15}$$

Note that such problems of motion initiated from a static position of equilibrium are only soluble exactly with an implicit code.

(iii) In case of restart of a previous analysis, displacements, velocities and accelerations are normally available and the starting procedure raises no difficulty from the last retrieved time station (see Section 4.7).

4.6. *Impulsive and step loadings*

Impulsive and step loadings both correspond to limiting cases of dynamic loading with steep variation in time: shock on the system and suddenly applied loads respectively.

It is easy to show that an impulse applied to a mechanical system at a given time neither modifies instantaneously its position nor increases its acceleration field, while velocities undergo a jump which is solution of the momentum equation

$$M(\dot{q}_+ - \dot{q}_-) = p \tag{4.16}$$

where $\boldsymbol{p}$ is the momentum of the impulse, defined as

$$\boldsymbol{p} = \int_{t_-}^{t_+} \boldsymbol{g}(\tau)\,\mathrm{d}\tau\,. \tag{4.17}$$

In a similar manner, a step loading does not modify instantaneously the displacements and velocities of the system, but an acceleration jump is observed which is solution of the dynamic equilibrium reduced to the form

$$\boldsymbol{M}(\ddot{\boldsymbol{q}}_+ - \ddot{\boldsymbol{q}}_-) = \Delta\boldsymbol{g} \tag{4.18}$$

where $\Delta\boldsymbol{g}$ is the step load applied at time t, $\Delta\boldsymbol{g} = \boldsymbol{g}_+ - \boldsymbol{g}_-$.

In both impulsive and step loading cases, the singularity in the loading is treated as for (4.14) by solving a particular linear system with the mass matrix as system matrix. The remark regarding its lumped or consistent characteristic holds therefore in the present case also.

An interesting feature of impulsive and step loadings, compared to approximate representations of shocks and suddenly applied loads such as piecewise linear representation, is the fact the load representation does not introduce any restriction on the step size in the time integration scheme around the discontinuity.

4.7. *Saving of output results and restarting facility*

In the calculation of the dynamic structural response, the management of output results and the restarting facility are two difficult problems which are intimately connected, specially for nonlinear applications.

Two simultaneous modes of output results recording are advised:

1. *Time history.* The first mode is user-oriented: it corresponds to the time history of physical quantities previously selected, such as local values of displacements, velocities, accelerations, reactions, element stresses which are memorized with a time frequency that can be specified. This mode of output is normally the most interesting to the user, since his aim is generally to retrace the time evolution of some physical quantities. It is also the most difficult to organize, since it requires recording one given physical quantity for all instants of time, while the normal saving procedure consists to store at the saving frequency all the results that are requested by the user. Postprocessing of the results is thus necessary to present them in the time history form.

2. *Restart points.* The second mode of output result recording is organized in view of a possible restart of the analysis not only at the end of a previous run, but also at some former instants that may be foreseen by the user (changes in excitation forces, refinement of time step, etc.). The necessary information to restart the iteration procedure at a given time is:

– the displacements, velocities and accelerations at all nodes;
– for nonlinear problems, the stress and strain states within all the elements, and all quantities that are necessary to perform the nonlinear finite element generation such as the definition of constitutive laws, the matrices of displacement derivatives, the location of Gauss points, etc.

In practice, this second mode of saving cannot be made with the same frequency as the one corresponding to time history, due to the important amount of information that has to be saved. It is thus organized separately.

5. Sample problems

Sample problems have been split into two classes:

(i) tests of simple structures with simple finite elements that can be run easily and cheaply; in their presentation, attention is focussed on the use of different time integrators and the linear and nonlinear techniques of solution;

(ii) tests of more complex structures or analyses involving more elaborate structural behavior, on which the effects of spatial discretization, solution techniques and choice of time integrators are successively examined.

The present description of test problems has not the pretention of being exhaustive on what is available in the literature to illustrate implicit finite element methods for transient structural response. Additional material can be found in the list of references, e.g. [1, 3, 4, 8, 13, 17, 22, 34].

5.1. Tests of simple structures

5.1.1. Three-dimensional frame under impulse load [32]

The first example deals with a three-dimensional frame submitted to a concentrate step load which is removed after 10^{-2} sec. The finite element discretization is by sixteen beam elements with cubic displacement fields yielding a system with 48 degrees of freedom. Fully linear behaviour is assumed through the analysis.

Table 5.1
Material properties for the frame

	Vertical beams	Horizontal beams
E (N/m^2)	$2.06 \cdot 10^{11}$	
ν	0.3	
ρ (Kg/m^3)	7818.57	7848.00
Cross section (10^{-3} m^2)	5.14	5.68
I_x	$8.495 \cdot 10^{-5}$	$1.2054 \cdot 10^{-4}$
I_y (m^4)	$6.91 \cdot 10^{-6}$	$1.7558 \cdot 10^{-7}$
I_z	$1.7281 \cdot 10^{-7}$	$7.28 \cdot 10^{-6}$

Transient responses have been integrated in time using Houbolt's and Newmark's schemes respectively, the latter with $\beta = \frac{1}{4}$, $\gamma = \frac{1}{2}$. The time step for Newmark's integrator has been chosen twice greater than the one for Houbolt's operator, i.e. respectively $h_N = 10^{-3}$ sec and $h_H = 5 \cdot 10^{-4}$ sec.

Material properties for the constitutive beams are listed in Table 5.1. Figure 5.1 displays the geometric data of the frame, the loading history and two transient

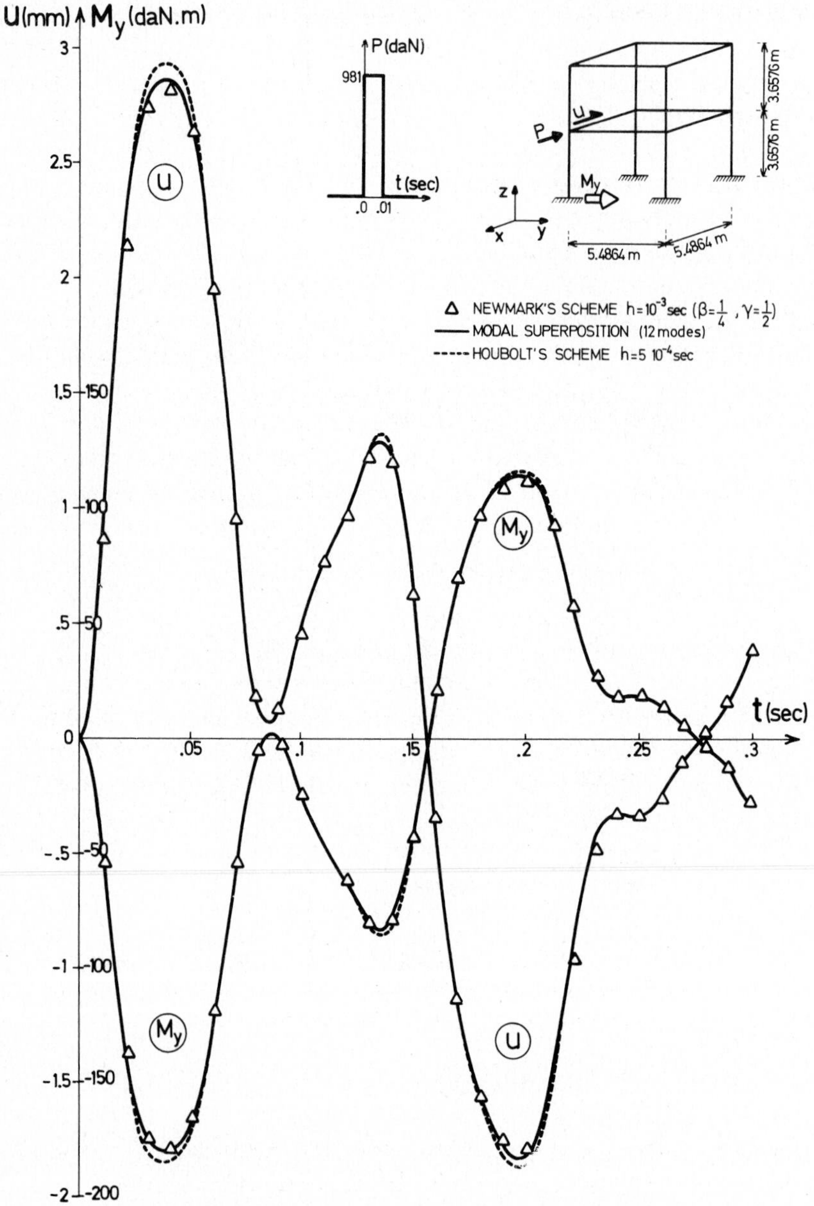

Fig. 5.1. Transient response of a frame under impulse load.

records: the displacement of the point of application of the load in the direction of application and the bending moment at the root of the corresponding vertical beam. Implicit responses are compared to a modal superposition solution [32]. The agreement between all three types of analysis is excellent. Doubling the time step size by Newmark scheme still yields a quite accurate solution.

5.1.2. Stretched cable submitted to transverse loading [17, 25]

In order to appreciate the computational efficiency of quasi-Newton iteration with comparison to Newton and modified Newton methods, a mechanical test problem exhibiting strong geometric nonlinearities has been examined.

It consists of a cable of span L stretched with an initial tension σ_0 between horizontal supports, with no sag and no initial transverse load. The dynamic loading consists of a linearly increasing, uniformly distributed transverse load $p(t) = p_0 t$ while the mechanical data are its extensional rigidity EA_0 and its mass per unit length $\rho_0 A_0$ (Fig. 5.2).

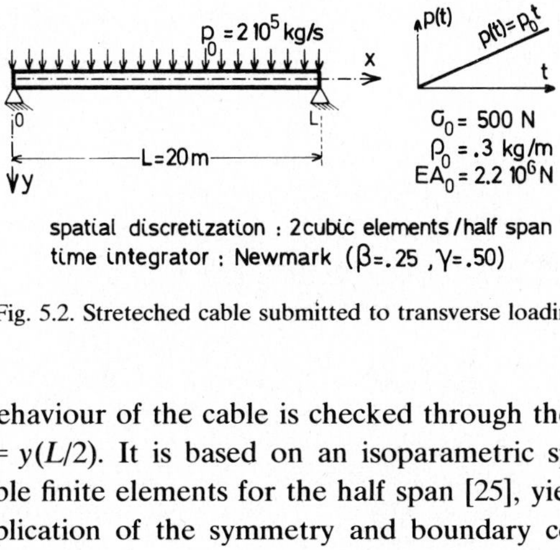

Fig. 5.2. Streteched cable submitted to transverse loading.

The dynamic behaviour of the cable is checked through the vertical motion of midspan node $u = y(L/2)$. It is based on an isoparametric spatial discretization with two cubic cable finite elements for the half span [25], yielding 11 degrees of freedom after application of the symmetry and boundary conditions. The corresponding linear solution is given by an expression deduced from the string theory:

$$y(x, t) = \frac{4p_0 L^2}{\sigma_0} \sum_{n=1}^{\infty} \frac{1}{(n\pi)^3} \left[t - \frac{1}{w_n} \sin w_n t \right] \sin \frac{n\pi x}{L} \tag{5.1}$$

where

$$w_n = \left(\frac{\sigma_0}{\rho_0 A_0} \right)^{1/2} = \frac{n\pi}{L} \quad (n \text{ odd}) .$$

Note that the problem is highly nonlinear: the linear and nonlinear solutions differ rapidly (from $t = 0.032$ sec), and there is a drastic change in the period of the oscillations between the nonlinear and the linear solutions; in the latter case it is constant and equal to

$$T = 2\left(\rho_0 A_0 \frac{L^2}{\sigma_0}\right)^{1/2} = 0.9798\,\text{sec}\,,$$

compared to about 0.04 sec in the nonlinear case.

Comparison of converged solutions (Figure 5.3). Figure 5.3 shows the comparison of the converged solutions (i.e. solutions for which subsequent reductions of the time step have no influence) obtained by three types of schemes for the short-time range: explicit central difference schemes with diagonal and consistent mass representation and Newmark's implicit scheme with $\beta = \frac{1}{4}$, $\gamma = \frac{1}{2}$. It appears that there is no significant difference between the solutions. Implicit ones correspond to a time step $h = 1.10^{-3}$ sec and a modified Newton strategy in each time step with stiffness reevaluation at iteration 1, 2, 5, 8, ... Explicit solutions are very

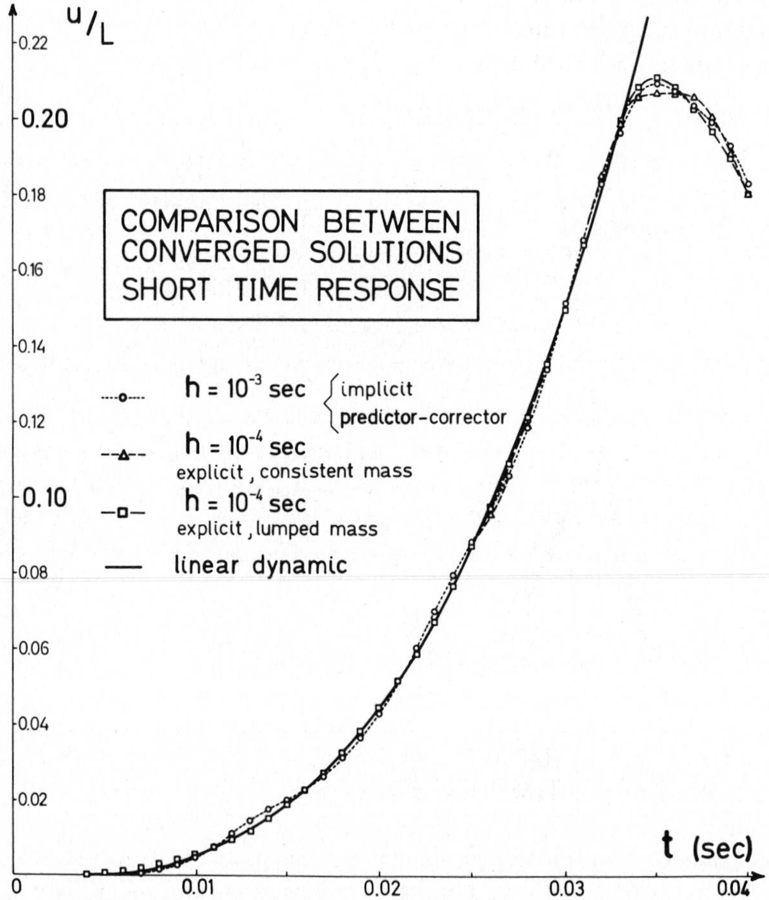

Fig. 5.3. Numerical converged solutions for cable midspan deflection.

similar and correspond to a time step $h = 1.10^{-4}$ sec well below their stability limit. The agreement between all the solutions, though not represented on the figure, remains unchanged throughout the whole time range covered by the analysis, i.e. $t \leq 0.130$ sec.

Solutions by implicit schemes with modified Newton iterations (Figures 5.4 and 5.5). First, various time steps have been tried with the fixed iteration strategy defined above (Figure 5.4.) The following observations are of interest:

(i) As expected, no numerical instability occurs even for rather large time steps ($h = 6 \cdot 10^{-3}$ sec).

(ii) Changes in amplitude and period of oscillation are observed when h is increased. This behaviour is classical even in linear situations.

(iii) The chosen strategy produces convergence within each time step with a minimum of two stiffness reevaluations.

(iv) A critical time step size (here $h = 2 \cdot 10^{-3}$ sec) appears to exist beyond which both accuracy and computational efficiency (number of iterations to converge and thus cost) deteriorate, without however producing instability; computer costs correspond to an IBM 370/178 configuration.

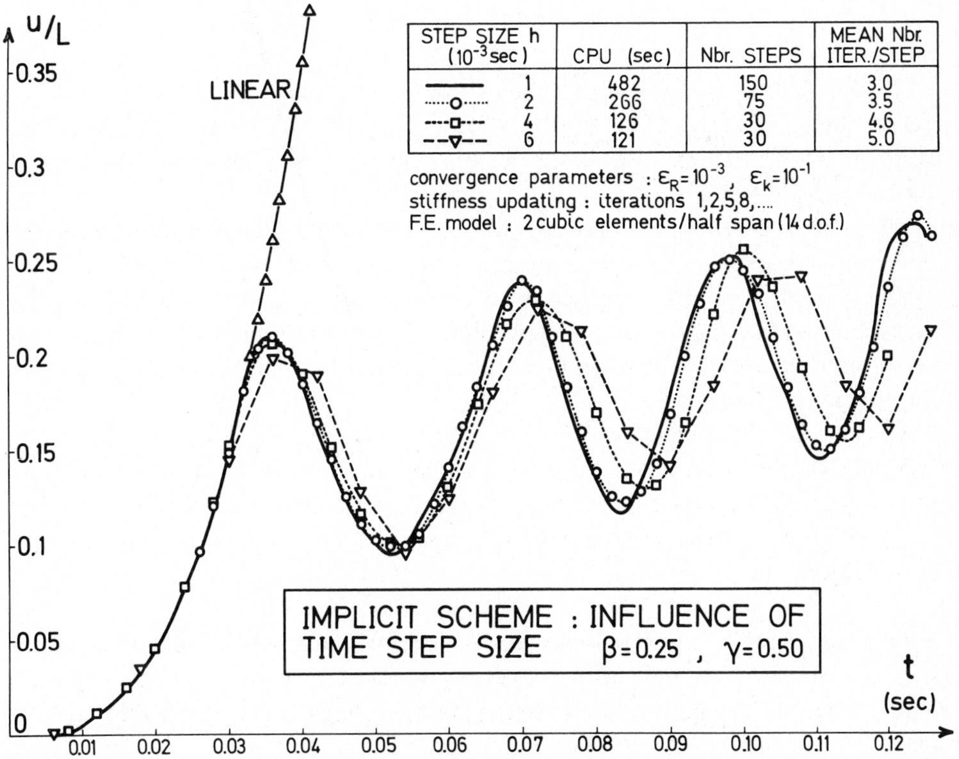

Fig. 5.4. Implicit solutions for cable midspan deflection.

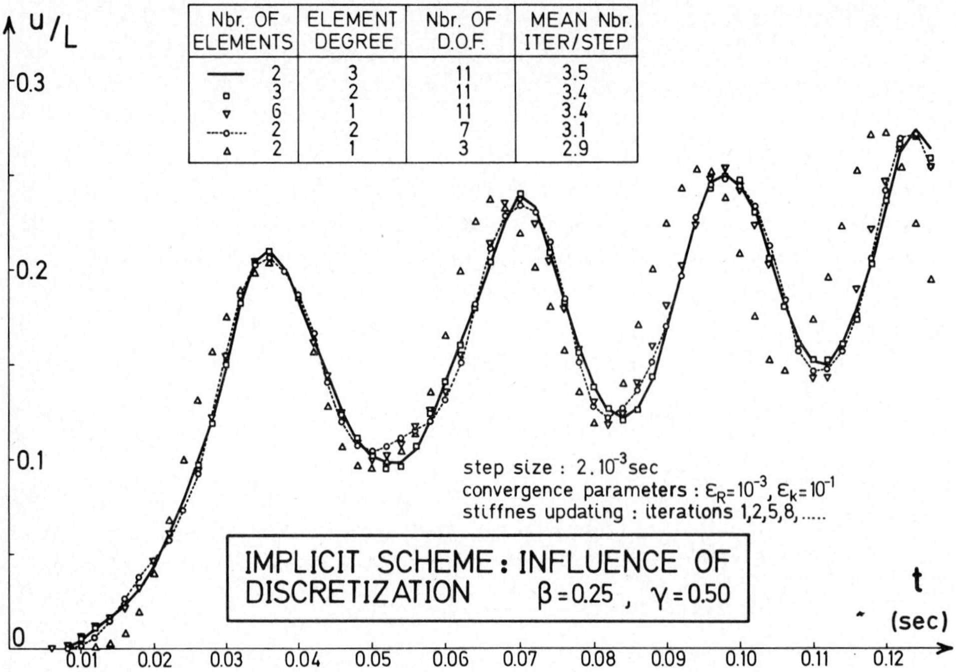

Fig. 5.5. Implicit solutions for cable midspan deflection.

(v) Below that critical value of the time step size, further reduction of cost can be obtained by a reduction of the number of stiffness reevaluations for convergence within a time step while keeping as close as possible to the critical maximum time step size.

The second investigation concerns the influence of the spatial discretization for a fixed time step size and stiffness strategy. It is illustrated in Figure 5.5.

For a fixed number of degrees of freedom the influence of the degree of the elements (i.e. the degree of the interpolating functions) is small on both the cost of the solution and its accuracy; it is only significant for ND = 1, that is when first degree elements are used.

Conversely, the reduction of the number of degrees of freedom has strong influence on the accuracy of the solution without allowing significant savings on the cost of the response, indicating that for such small-size problems the importance of overhead time in implicit programs with an out-of-core organization is quite large.

Solutions by implicit schemes with quasi-Newton iteration. Quasi-Newton iteration with the BFGS update has been applied to the same problem with the step sizes $h = 1, 2$ and $4 \cdot 10^{-3}$ sec. Various strategies have been tried (see Table 5.2), to measure the influence on the convergence rate of:

– the step size,

Table 5.2
Stretched cable. Quasi–Newton iteration (convergence parameter: $\varepsilon_R = 10^{-3}$)

Step size h (10^{-3} sec)	Line search	Stiffness reevaluation (time steps)	Mean number iterations/step
1	yes	5	3.39
2	yes	5	4.31
	yes	10	4.37
	yes	20	4.44
	yes	80	4.50
	no	10	4.93
4	yes	20	5.68

– the use of line search,
– the frequency at which stiffness is reevaluated.

The most instructive cases are those corresponding to $h = 2 \cdot 10^{-3}$ sec. They show that periodic stiffness reevaluation has very limited influence on the convergence of the algorithm. The whole time history of the system can even be computed by performing quasi-Newton iteration without any direct reevaluation of the true Jacobian. They also show that roughly one more iteration per step is required than for Newton iteration. The line search has obviously a beneficial but limited effect on the method since a mean deterioration of 0.5 it/step is observed when skipping it.

5.2. Tests of transient structural response

5.2.1. Spinning cylinder under pressurization [13, 33]

The sudden pressurization of a rotating cylinder is a simple problem which provides check on both the elasto-plastic constitutive modelling and on the ability

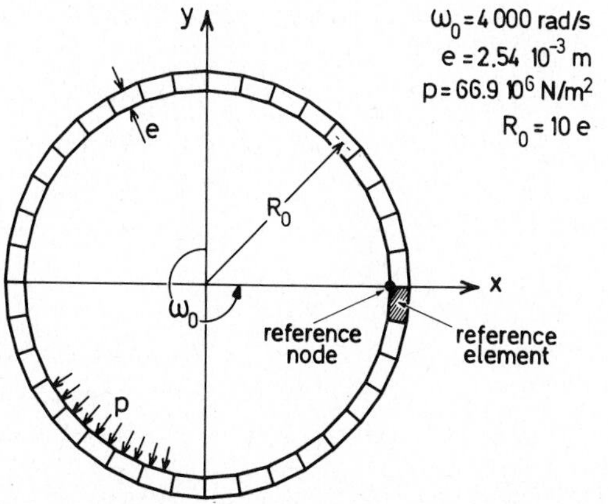

Fig. 5.6. Plane strain finite element model of pressurized spinning cylinder.

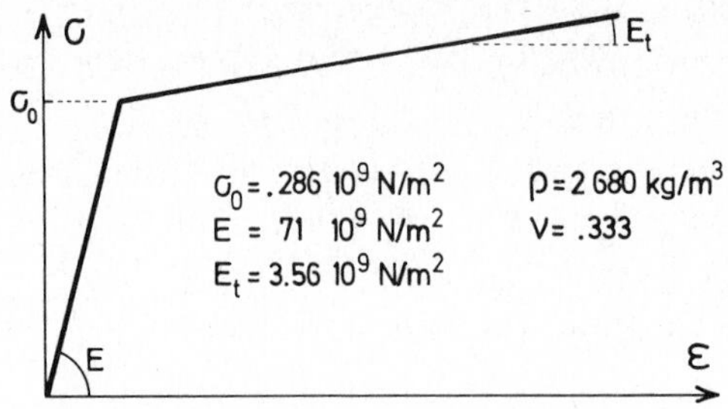

Fig. 5.7. Material properties for cylinder.

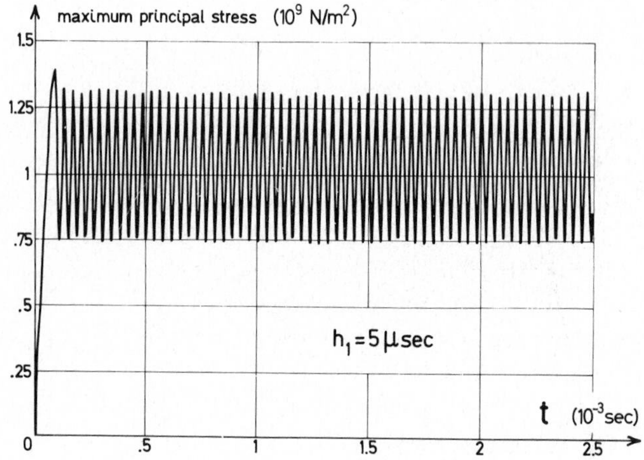

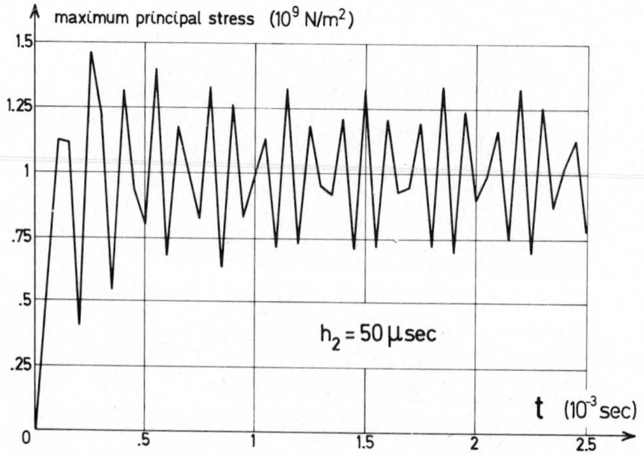

Fig. 5.8. Maximum principal stress in reference element as a function of time step size (Newmark scheme $\beta = \frac{1}{4}$, $\gamma = \frac{1}{2}$) [13].

of the implicit code to deal with large rotations. The test is equivalent to a single degree-of-freedom oscillator with an elastic-plastic spring. The plane strain finite element model is shown in Figure 5.6 and the material properties are given by Figure 5.7. At time zero a spin is established at 4000 rad/sec. and a pressure of $66.9 \cdot 10^6\ \mathrm{N/m^2}$ is applied as a step function in time. As a consequence, the cylinder expands from its initial radius to a deformed one $R_d = 32.51$ mm and then oscillates about this configuration at a frequency of $2.67 \cdot 10^4$ Hz given by $[E/\rho(1-\nu^2)]^{1/2}/2\pi R_d$.

Figures 5.8 and 5.9 plot the results of [13] for a calculation by Newmark's scheme ($\beta = \frac{1}{4}$, $\gamma = \frac{1}{2}$) over a period of 2.5 m sec sufficient for the cylinder to rotate one full revolution ($T = 1.57$ m sec). Two time step sizes of respectively $h_1 = 5\ \mu$ sec and $h_2 = 50\ \mu$ sec have been tried. The overall behaviour of the plotted results for the reference node and the reference element are very similar. However the stress results with the larger time step do not allow for capturing the elastic bracketing of the

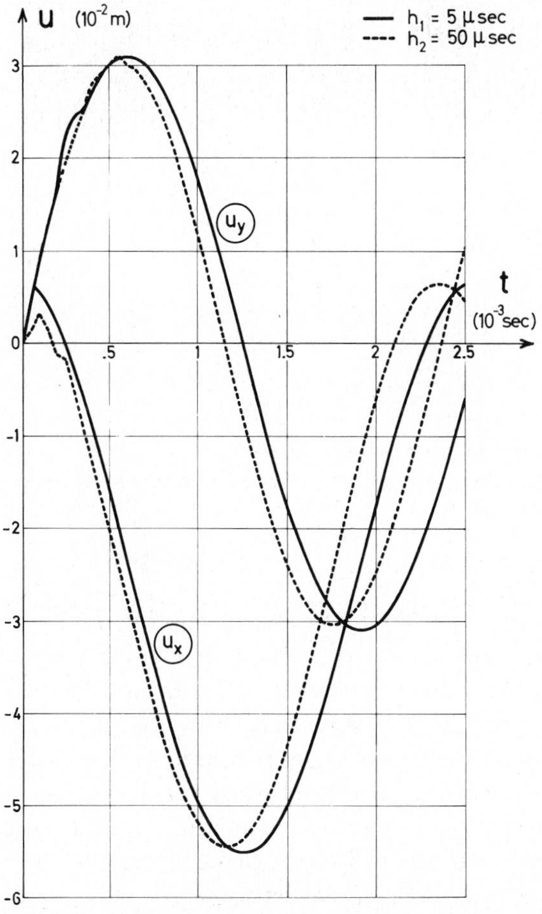

Fig. 5.9. Effect of time step size on displacements of reference node (Newmark scheme $\beta = \frac{1}{4}$, $\gamma = \frac{1}{2}$).

solution and the corresponding displacement results exhibit the well known accuracy deterioration of Newmark scheme (numerical damping and period elongation) while keeping stability of the solution.

5.2.2. Simply-supported beam

The next example is the elastic-plastic transient analysis of a simply-supported beam to which a uniformly distributed pressure is suddenly applied. The beam dimensions and material properties are described on Figure 5.10.

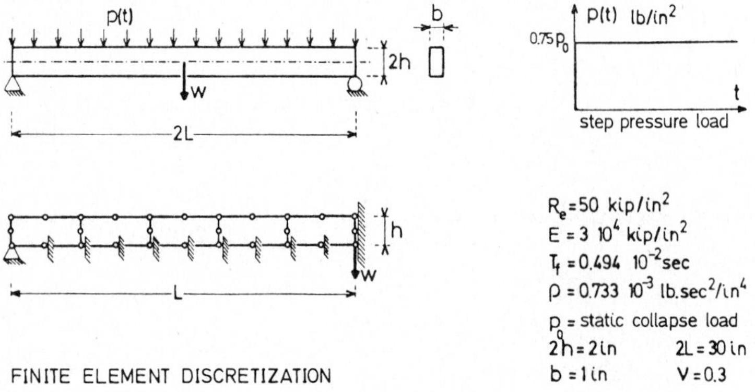

Fig. 5.10. Transient analysis of an elastic-plastic simply supported beam.

The elastic-plastic response is computed for an intensity of step pressure equal to 75% of the static collapse load. One quarter of the beam has been discretized with five quadratic (8 nodes) isoparametric elements, yielding a total number of 42 degrees of freedom. Time integration has been performed with Newmark's scheme ($\beta = \frac{1}{4}$, $\gamma = \frac{1}{2}$) using a time step of $h = 1.5 \cdot 10^{-4}$ sec, which corresponds to 3/100 of the fundamental period for linear undamped vibration. Equilibrium iteration is stopped when

$$\|\boldsymbol{r}_k\|/(\|\boldsymbol{g}^k\| + \|\boldsymbol{f}^k\|) \leq 10^{-3}.$$

Figure 5.11 shows the evolution of the displacement w at midspan for the linear and nonlinear solutions. The characteristic behavior of elastic-plastic material is noteworthy: as soon as plasticity develops in the structure, softening of the material appears producing elongation of the apparent period of vibration and attenuation of its amplitude around a mean value which is much greater than in the linear case (about three times in the present case).

The problem was attacked by several equilibrium iterations techniques resorting to modified Newton and quasi-Newton procedures.

No significant quantitative difference was observed between the corresponding solutions. The interest of the comparison lies thus in the number of iterations and

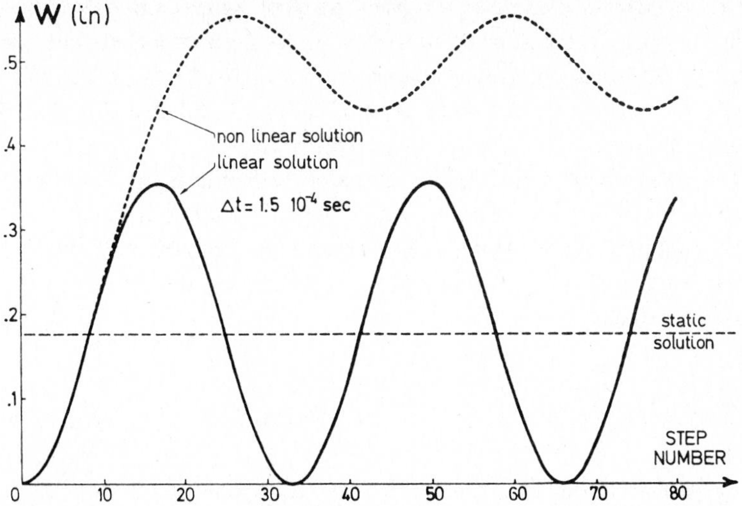

Fig. 5.11. Linear and nonlinear transient responses for the midspan deflection.

CPU times required in each of the procedures. This is given by Table 5.3 for the 30 first steps of the transient response.

In the modified Newton solution, the tangent stiffness is reevaluated at iterations 1, 2, 5 and 8 of each time step. When the material becomes linear, iteration is performed with the linear stiffness matrix.

For quasi-Newton iteration, comparison is given between the following algorithms:

– The rank-one update of Davidon (3.23) has been tested with two strategies: starting the process at each time step either with the linear inverse iteration matrix $\boldsymbol{G}_0$ (in which case only the evaluation of the linear stiffness matrix $\boldsymbol{K}_0$ is

Table 5.3
Simply supported beam – Efficiency of modified and quasi-Newton strategies

		Quasi–Newton		
	Newton	G_D update		G_{BFGS} update
Number of iterations per step	2.9	4.73	2.8	2.67
Total number of stiffness evaluations	43	1	20	20
Total number of residual evaluations	87	172	127	124
Total number of line searches	–	–	13	14
CPU time per iteration (IBM 370/158)	2.58	1.87	2.59	3.12
Total number of iterations (for 30 steps)	87	142	84	80
Total CPU time (30 steps)	224.1	266.8	218.0	249.6

necessary) or with the tangent iteration matrix (which requires one Jacobian evaluation per time step but leads to a reduction of the number of evaluations of residual vectors). Due to the small size of this problem (42 d.o.f.) the difference of cost between a stiffness reevaluation and calculation of the residual vector is not sufficient to improve the total CPU time significantly.

– The BFGS rank-two update (3.25) has also been implemented with recursive substructure correction [11], starting each time step with the effective Jacobian matrix. The rank-two correction does not bring a significant improvement in the convergence rate, and the cost of the double correction makes it less competitive than Davidon's update.

5.2.3. Transient large displacement analysis of a cantilever [3, 34]

The cantilever shown in Figure 5.12a is suddenly subjected to a uniformly distributed load at $t = 0$. The finite element mesh consists of five plane stress elements of isotropic linear elastic material. The problem is however geometrically nonlinear. Time integration is carried out using Newmark's scheme ($\beta = \frac{1}{4}$, $\gamma = \frac{1}{2}$).

To stress the influence of spatial discretizations on transient responses, 4-node and 8-node elements have been used for a linear analysis using a time step size $h = T_f/42$ as in [3], where T_f is the fundamental period of vibration of the cantilever. Figure 5.12b compares the results for the end deflection of the cantilever: it is seen that 8-node elements are indispensable to represent correctly the bending rigidity of the cantilever and, as a corrolary, its fundamental eigenmode and frequency. Note that, according to beam theory, the static deflection is $w/L = 0.3506$.

Figure 5.12c shows the comparison between linear and nonlinear responses for the 8-node element discretizations. The nonlinear analysis was carried out using modified Newton iterations with $\varepsilon_R = 10^{-3}$, and required an average of 3.97 iterations per time step, among which the first two were iterations with stiffness evaluation. The associate computer cost was 737.1 CPU sec (IBM 370/158) for sixty time steps.

The stiffening of the cantilever in the nonlinear case markedly damps out the amplitude and shortens the period of oscillations of the response. Variation of the loading with geometry has been taken into account.

5.2.4. Transient large displacement analysis of an elastic-plastic spherical cap [3, 19, 34]

The last example considered is the structural response of a clamped spherical cap submitted to a sudden pressure loading, and where geometric and material nonlinearities are simultaneously present. The geometric and material properties are summarized on Figure 5.13. Von Mises' plasticity criterion applies with isotropic hardening. The purpose of the analysis is to show the influence on the response of:

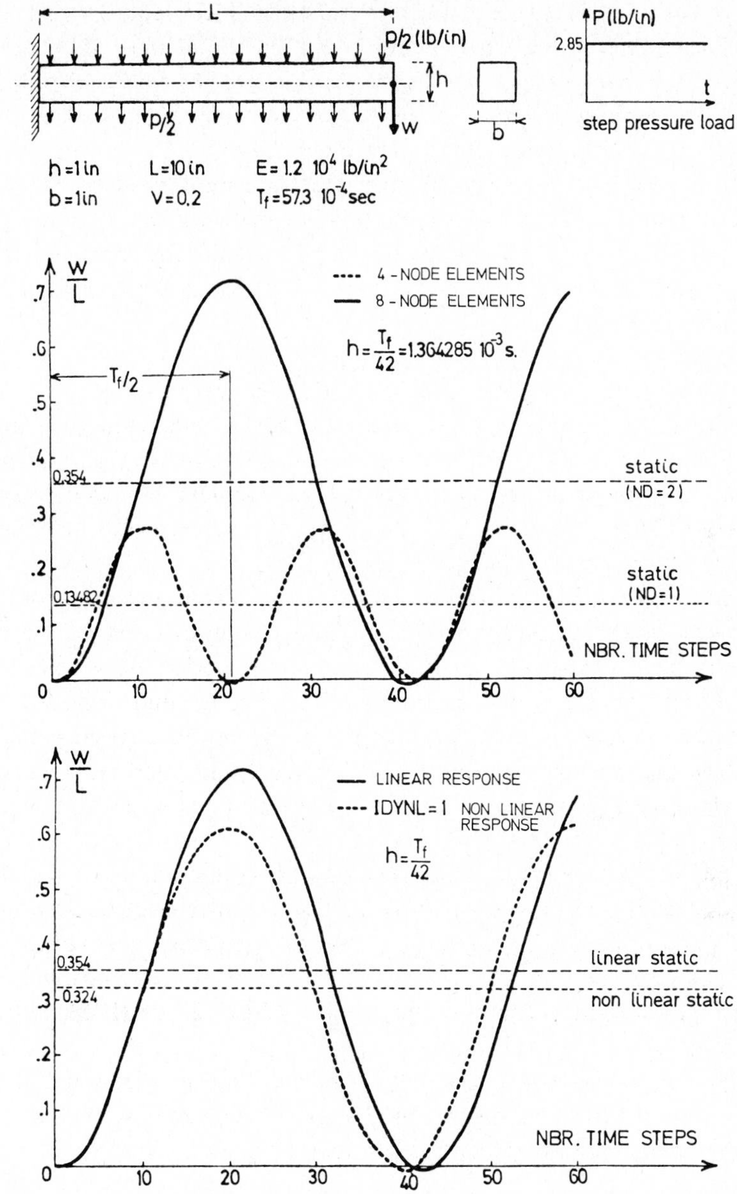

Fig. 5.12. Linear and nonlinear transient responses of a cantilever.

– the spatial discretization by finite element,
– the different nonlinear behaviors,
– the nonlinear iteration techniques.

Discretization by axisymmetrical volume elements. A first spatial discretization uses eight 8-node isoparametric axisymmetrical elements with one single element

$V = .3$
$P_0 = 600 \text{ lb/in}$
$E = 10.5 \ 10^6 \text{ lb/in}^2$
$R_e = 24 \ 10^3 \text{ lb/in}^2$
$\mathcal{E}_t = .21 \ 10^6 \text{lb/in}^2$
$\rho = 2.45 \ 10^{-4} \text{lb.sec}^2/\text{in}^4$

step pressure load

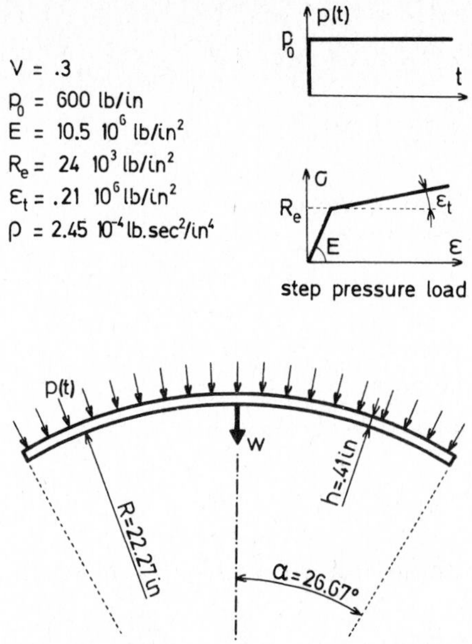

Fig. 5.13. Spherical cap submitted to step pressure loading. Geometry and material properties.

through the thickness. Correct boundary conditions are of prime importance along the clamped edge since two possibilities may arise [19]:

– *conditions A* correspond to the vanishing of the 6 degrees of freedom appearing at the 3 nodes of the edge (Figure 5.14a)

– *conditions B* allow for transverse dilatation of the cap, i.e. the two d.o.f. corresponding to the mean surface of the cap are fixed as well as the in-plane displacement for the two tangent planes at the upper and lower faces of the cap edge (Fig. 5.14b).

These different requirements exercise a great influence on the transverse shear stresses around the clamped edge and lead thus to different developments of

(a) CONDITIONS A   (b) CONDITIONS B

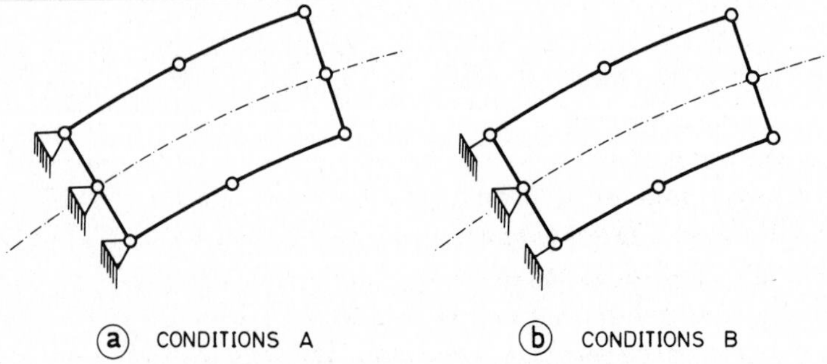

Fig. 5.14. Clamped edge boundary conditions.

plasticity in the nonlinear case through the plasticity criterion. Only 3 Gauss points are used to integrate the constitutive law over the thickness: this relatively crude integration rule is however sufficient to prevent oscillations in the numerical solution when plasticity develops.

Discretization by axisymmetrical shell elements. A second possibility is to model the cap by shell elements: 8 cubic elements are used with 3 or 6 Gauss points through the thickness.

Comparison of the transient responses. Figure 5.15 displays the time history of the axial displacement at the apex of the cap using the two types of spatial modelisation for the following material and geometrical behaviors:
- linear elastic,
- elastic-plastic material, geometrically linear,
- combined material and geometrical nonlinearities.

Time integration is performed with Newmark's scheme ($\beta = \frac{1}{4}$, $\gamma = \frac{1}{2}$) and a relatively large time step of $1.5 \cdot 10^{-5}$ sec has been adopted. (It corresponds approximately to $\frac{1}{36}$ of the fundamental period of vibration for the linear elastic case.) Equilibrium iteration is stopped within each nonlinear time step n when

$$\|\boldsymbol{r}_k^n\| / (\|\boldsymbol{g}_n\| + \|\boldsymbol{f}^n\|) \leq 10^{-3} .$$

Volume elements use boundary conditions of type A to compare with the shell boundary conditions. As expected differences between the two types of discretizations mainly occur when plasticity develops in the structure: the assumption of zero transverse stress in the shell element produces a different perception of the actual state of stress by the von Mises criterion.

Table 5.4 summarizes the associate computer costs of the different analyses. All of them were carried out using modified Newton strategy (if nonlinear) with stiffness evaluation at iterations 1, 2 and then every 3 iterations of a time step. It can be concluded that shell elements are more economical despite the use of 6 Gaussian points for integrating the materially nonlinear behavior.

Comparison of nonlinear behaviors and solution techniques. Figure 5.16 displays the expected influence of nonlinear material and geometrical behaviors for the case of the shell discretization; plasticity alone produces a softening of the structure while the addition of large displacements stiffens the nonlinear response.

As far as nonlinear solution techniques are concerned, very little quantitative difference is observed in the numerical results with the different possible methods of solution. The only interest of such a comparison lies again in computer time and numbers of iterations to obtain the solution (see Table 5.5).

Modified Newton iterations in which the stiffness is reevaluated at iterations 1, 2, 5 and 8 of each time step and the quasi-Newton method using successively the Davidon and BFGS updates have been tried. The performance obtained to

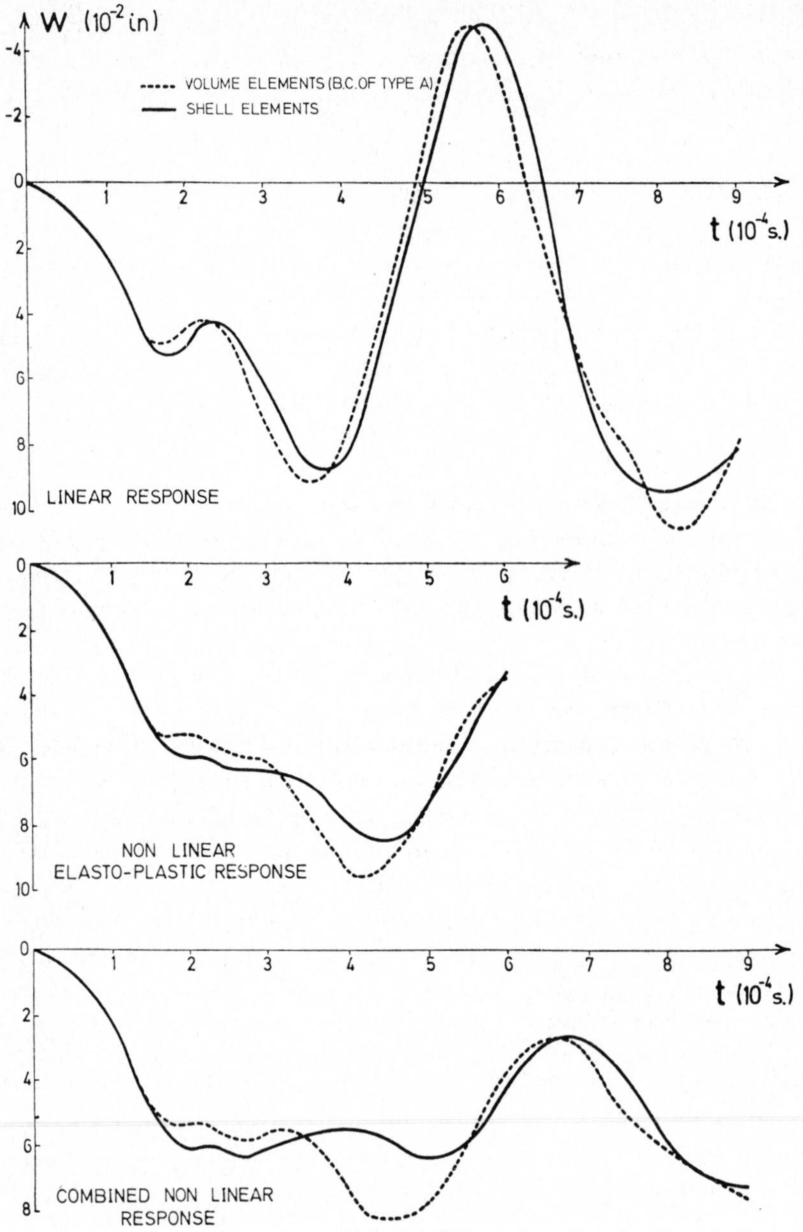

Fig. 5.15. Effect of spatial discretization on transient responses.

integrate the first 17 steps have been summarized in Table 5.5 for the combined nonlinear resonse using shell elements with 3 Gauss points over the thickness.

Quasi-Newton iterations have been performed with and without line search. Davidon's update has been tested using the vectorial correction (starting from K_0 at each time step). The best results were obtained without line search.

Table 5.4
Spherical cap transient analyses

Spatial discretization	Time integration	Material behavior	Number of iterations/step	Computer time
Volume elements	40 steps	L	2.0	73.0
82 D.O.F.	$h = 1.5 \cdot 10^{-5}$ sec	EP	2.85	534.0
3 Gauss points/thickness		C	3.33	741.0
Shell elements	40 steps $h = 1.5 \cdot 10^{-5}$ s	L	2.0	64.0
72 D.O.F.	60 steps $h = 1.0 \cdot 10^{-5}$ s	EP	3.33	891.0
6 Gauss points/thickness	40 steps $h = 1.5 \cdot 10^{-5}$ s	C	3.2	708.0

L = linear, EP = elastic-plastic, C = combined nonlinearities.

The last two columns correspond to the BFGS updates with substructure correction (starting from the tangent stiffness matrix at each time step) [11]. One observes a significant increase in the number of iterations when the process is not restarted at each time step, due to the fact that the number of updates on G_0 becomes excessive.

In spite of the small size of this problem (involving only 72 d.o.f.), the difference of computer costs between the reevaluation of stiffness (with Gauss elimination) and the calculation of the residual vector is yet significant. Quasi-Newton iteration is thus the most efficient procedure.

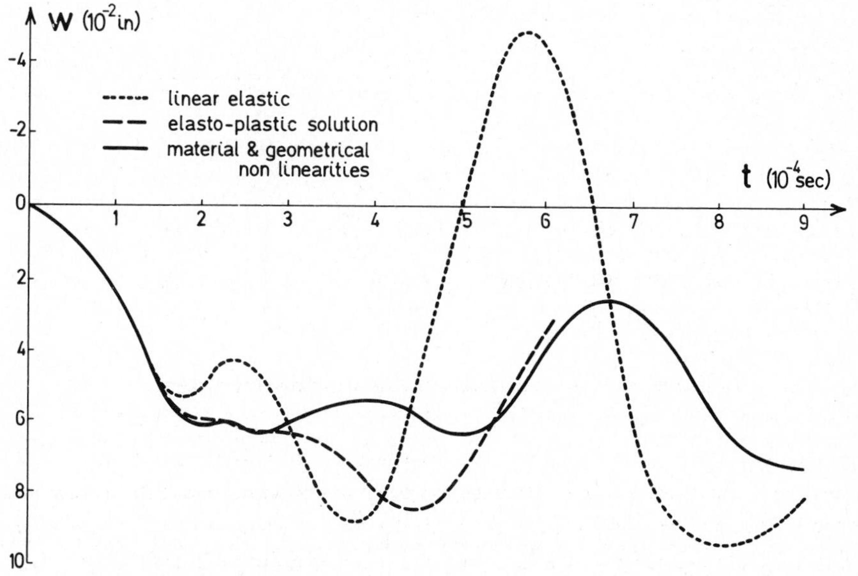

Fig. 5.16. Comparison of material behaviours for shell elements.

Table 5.5
Spherical cap dynamic analysis. Efficiency of Newton and quasi-Newton iterations

		Quasi-Newton			
	Newton	G_D update	G_D update	G_{BFGS} update	G_{BFGS} update
Number of iterations per step	3.35	5.0	4.88	3.29	6.0
Total number of Jacobian evaluations	40	1	1	17	1
Total number of residual evaluations	57	102	111	73	135
Total number of line searches	–	–	11	–	16
CPU. Time per iteration	5.40	2.75	2.97	4.83	3.6
Total number of iterations	57	85	83	56	102
Total CPU time	308.0	233.6	247.0	271.0	368.0

References

[1] J.H. Argyris, P.C. Dunne and T. Angelopoulos, Nonlinear Oscillations using the Finite Element Tehcnique, *Comp. Meth. Appl. Mech. Eng.* **2**, 203–250 (1973).

[2] K.J. Bathe and E.L. Wilson, Stability and Accuracy Analysis of Direct Integration Methods, *Earthquake Eng. Struct. Dyn.* **1**, 283–291 (1973).

[3] K.J. Bathe, E. Ramm and E.L. Wilson, Finite Element Formulations for Large Deformation Dynamic Analysis, *Int. Jnl. Num. Meth. Eng.* **9**, 353–386 (1975).

[4] K.J. Bathe and A. Cimento, Some Practical Procedures for the Solution of Nonlinear Finite Element Applications, *Comp. Meth. in Appl. Mech. Eng.* **22**, 59–85 (1980).

[5] M.A. Crisfield, A Faster Modified Newton–Raphson Iteration, *Comp. Meth. Appl. Mech. Eng.* **20**, 267–278 (1979).

[6] C.A. Felippa and K.C. Park, Direct Integration Methods in Nonlinear Structural Dynamics, *Comp. Meth. Appl. Mech. Eng.* **17/18**, 277–313 (1979).

[7] B. Fredriksson and J. Mackerle, Structural Mechanics Finite Element Computer Programs – Surveys and Availability, Linkoping Inst. of Technology, Sweden, Report LITH–IKP–R-054 (1978).

[8] M. Geradin et al., Module d'Analyse Dynamique Non-linéaire NLDYN, L.T.A.S. Report VF-40, Aerospace Laboratory, Univ. of Liège, Belgium (1979).

[9] M. Geradin, A Classification and Discussion of Integration Operators for Transient Structural Response, AIAA Paper No. 74-105, AIAA 12th Aerospace Sciences Meeting (1974).

[10] M. Geradin, S. Idelsohn and M. Hogge, Computational Strategies for the Solution of Large Nonlinear Problems via Quasi-Newton Methods, *Computers and Structures*, to appear.

[11] M. Geradin, S. Idelsohn and M. Hogge, Nonlinear Structural Dynamics via Newton and Quasi-Newton Methods, *Nucl. Eng. Design* **58**, 339–348 (1980).

[12] W.E. Haisler, J.A. Stricklin and F.J. Stebbins, Development and Evaluation of Solution Procedures for Geometrically Nonlinear Structural Analysis, *AIAA Jnl.* **10**(3), 264–272 (1972).

[13] J.O. Hallquist, NIKE 2D: An Implicit, Finite-Deformation, Finite-Element Code for Analyzing the Static and Dynamic Response of Two-dimensional Solids, Report UCRL-52678, Lawrence Livermore Laboratory (1979).

[14] J.C. Houbolt, A Recurrence Matrix Solution for the Dynamic Response of Elastic Aircraft, *Jnl. Aer. Sci.* **17**, 540–550 (1950).

[15] J.S. Humphreys, On Dynamic Snap Buckling of Shallow Arches, *AIAA Jnl.* **4**(5), 878–886 (1966).

[16] P.S. Jensen, Transient Analysis of Structures by Stiffly Stable Methods, *Comp. Struct.* **4**, 615–626 (1974).

[17] J.R. Lehner and S.C. Batterman, Static and Dynamic Finite Deformations of Cables Using Rate Equations, *Comp. Meth. Appl. Eng.* **2**, 349–366 (1973).

[18] H. Mathies and G. Strang, The Solution of Nonlinear Finite Element Applications, *Int. Jnl. Num. Meth. Eng.* **14**, 1613–1626 (1979).

[19] S. Nagarajan and P. Popov, Elastic-Plastic Dynamic Analysis of Axisymmetric Solids, *Comp. Struct.* **4**, 1117–1134 (1974).

[20] N.M. Newmark, A Method of Computation for Structural Dynamics, *Jnl. Eng. Mch. Div. ASCE* **85** (EM3), Proc. Paper 2094, pp. 67–94 (1959).

[21] D.R.J. Owen, Implicit Finite Element Methods for the Dynamic Transient Analysis of Solids with Particular Reference to Nonlinear Situations, in: *Advanced Structural Dynamics*, J. Donea, ed. (Applied Science Publishers, Barking, U.K., 1980).

[22] K.C. Park, An Improved Stiffly Stable Method for Direct Integration of Nonlinear Structural Dynamic Equations, *Trans. ASME Jnl. Appl. Mech.*, 464–470 (1975).

[23] K.C. Park and P.G. Underwood, A Variable Step Central Difference Method for Structural Dynamics Analysis – Part 1 – Theoretical Aspects, *Comp. Meth. Appl. Eng.* **22**, 241–258 (1980).

[24] K.C. Park, Evaluating Time Integration Methods for Nonlinear Dynamic Analysis, *ASME/AMD* **14**, 35–58 (1975).

[25] G. Sander, M. Geradin, C. Nyssen and M. Hogge, Accuracy Versus Computational Efficiency in Nonlinear Dynamics, *Comp. Meth. Appl. Mech. Eng.* **17/18**, 315–340 (1979).

[26] J.A. Stricklin and W.E. Haisler, Formulations and Solution Procedures for Nonlinear Structural Analysis, *Comp. Struct.* **7**, 125–136 (1977).

[27] E.L. Wilson, I. Farhoomand and K.J. Bathe, Nonlinear Dynamic Analysis of Complex Structures, *Earthquake Eng. Struct. Dyn.* **1**, 241–252 (1973).

[28] M.A. Wolfe, *Numerical Methods for Unconstrained Optimization* (Van Nostrand Reinhold, Wokingham, 1978).

[29] J.E. Dennis and J.J. More, Quasi-Newton Methods, Motivation and Theory, *SIAM Review* **19**(1), 46–89 (1977).

[30] C. Nyssen, Modélisation par éléments Finis du Comportement Non-linéaire des Structures Aérospatiales, Thèse de Doctorat, Univ. of Liège (1978).

[31] M. Geradin, Une Étude Comparative des Méthodes Numériques en Analyse Dynamique des Structures, ATMA, Session 1978, Paris, pp. 167–198.

[32] B.M. Fraeijs de Veubeke, M. Geradin and A. Huck, *Structural Dynamics*, CISM, Udine, 1972, Lecture Series No. 126 (Springer, Wien–New York, 1974).

[33] S.W. Key, J.H. Biffle and R.D. Krieg, A Study of the Computational and Theroretical Differences of Two Finite Strain Elastic-Plastic Constitutive Models, in: *Formulations and Computational Algorithms in Finite Element Analysis*, K.J. Bathe et al., eds. (MIT, 1977).

[34] D.P. Mondkar and G.H. Powell, Evaluation of Solution Schemes for Nonlinear Structures, *Comp. Struct.* **9**, 223–236 (1978).

[35] B.M. Irons, A Frontal Solution Program for Finite Element Analysis, *Int. Jnl. Num. Meth. Eng.* **2**, 5–32 (1970).

[36] C.W. Gear, *Numerical Initial Value Problems in Ordinary Differential Equations* (Prentice-Hall, Englewood Cliffs, NJ, 1971).

CHAPTER 10

Arbitrary Lagrangian–Eulerian Finite Element Methods

Jean DONEA

Commission of the European Communities
Applied Mechanics Division
Joint Research Centre, Ispra Establishment
Ispra, Italy

Computational Methods for Transient Analysis
Edited by T. Belytschko and T.J.R. Hughes

1. Introduction

In determining a method for numerical solution of multidimensional problems in fluid dynamics, a fundamentally important consideration is the relationship between the fluid and the finite grid or mesh of computing zones. Traditionally, there have been two basic viewpoints [1] for both compressible and incompressible flows. The first is Lagrangian, in which the mesh of grid points is embedded in the fluid and moves with it; the second, known as Eulerian, treats the mesh as a fixed reference frame through which the fluid moves. Both the Lagrangian and the Eulerian approaches present advantages and drawbacks. A clear delineation of interfaces and well-resolved details of the flow are afforded by the Lagrangian approach, but it is limited by its inability to cope easily with strong distorsions which often characterize flows of interest. On the contrary, in the Eulerian formulation, strong distorsions can be handled with relative ease, but generally at the expense of precise interface definition and resolution of detail.

Because of the shortcomings of purely Lagrangian and purely Eulerian descriptions, techniques have been developed that succeed to a certain extent in combining the best features of both the Lagrangian and Eulerian approaches. One such technique is the Arbitrary Lagrangian–Eulerian (ALE) method in which the grid points may be moved with the fluid in normal Lagrangian fashion, or be held fixed in Eulerian manner, or be moved in some arbitrarily specified way to give a continuous rezoning capability. Because of this freedom in moving the computational mesh offered by the ALE method, greater distortions in the fluid motion can be handled than would be allowed by a purely Lagrangian method, with more resolution than is afforded by a purely Eulerian method.

Originally, ALE methods for numerical solution of the Navier–Stokes equations have been developed in finite difference formats by Noh [2], Trulio [3] and Hirt, Amsden and Cook [4]. The ALE technique proposed by C.W. Hirt and his colleagues at Los Alamos is particularly noteworthy in that it is applicable to arbitrary finite difference meshes and permits flows at all speeds to be treated.

More recently, ALE finite element methods have been reported by Belytschko and Kennedy [5], Donea et al. [6] and Hughes et al. [7]. These methods were developed in response to the need of very versatile modelling techniques for treating transient fluid-structure systems and have considerable potential for application to a wide variety of problems in reactor safety analysis and other fields. The works reported in [5, 6] deal primarily with ALE finite element methods for inviscid, compressible flow, while reference [7] presents an ALE finite element formulation for viscous, incompressible flow.

The purpose of this chapter is to present a detailed survey of ALE finite element methods in their present stage of development. Particular emphasis has been placed on giving a comprehensive account of the ALE finite element treatment of unsteady compressible flow problems, including the case in which the fluid boundaries involve deforming structures.

In Section 2, we first recall the basic theoretical concepts underlying the Arbitary Lagrangian–Eulerian formulation. This includes the notion of mesh description in mixed co-ordinates and a brief discussion of the kinematics in a moving reference frame. On this basis, we derive ALE differential forms for the basic conservation equations of mass, momentum and energy and develop the associated weak integral forms which provide a basis for the formulation of finite-element models.

Section 3 is devoted to a presentation of spatially discrete models of the ALE conservation equations based upon the Galerkin/finite element method. Specialized forms of the semi-discrete equations are derived for meshes consisting of low-order finite elements in which density, specific internal energy and pressure are assumed elementwise constant.

Numerical time integration of the semi-discrete conservation equations is discussed in Section 4. The emphasis is placed on explicit methods, but a particular scheme is described which allows flows at all speeds to be treated through the inclusion of an optional implicit phase in the explicit calculational cycle.

A major problem with the ALE approach is to find an algorithm which automatically prescribes the displacement of the computational mesh, so as to avoid excessive distorsions and yet allow the fluid boundaries including deforming structures to be tracked satisfactorily. In Section 5, we present such an algorithm and give an example of its application.

The last section is concerned with applications of the ALE finite element method to fluid-structure interaction problems. The finite element method is in itself ideally suited for mixing fluid elements with structural elements, but the freedom in moving the fluid mesh offered by the ALE description is shown to be particularly attractive for the treatment of relative sliding at the fluid–solid interface.

2. The arbitrary Lagrangian–Eulerian formulation

2.1. Mesh description, material, spatial and mixed coordinates

Two basic viewpoints are generally considered in discretizing a fluid region by a finite difference or a finite element method. The first is Lagrangian, in which the mesh of grid points is embedded in the fluid and moves with it; the second, known as Eulerian, treats the mesh as a fixed reference frame through which the fluid moves. The Lagrangian description fixes attention on specific particles of the continuum, whereas the Eulerian description concerns itself with a particular region of space occupied by the continuum. As shown in Figure 1, a representative particle of the continuum occupies a point P_0 in the initial configuration of the continuum at time $t = 0$ and has the position vector

$$\boldsymbol{a} = (a_1, a_2, a_3) . \tag{1}$$

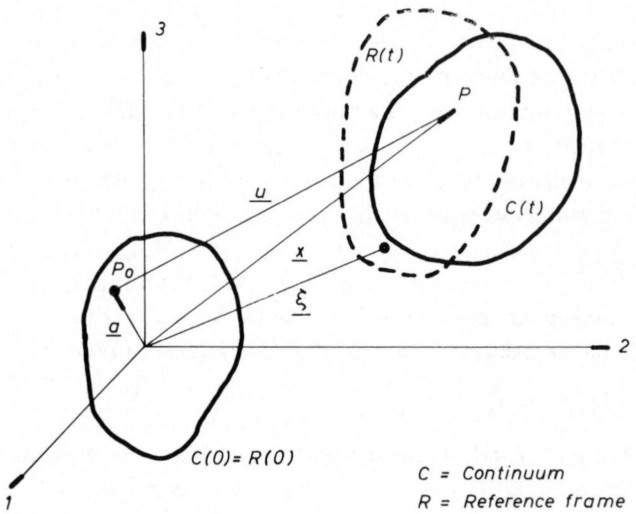

Fig. 1. Kinematics in the ALE description.

The coordinates a_i are called the material coordinates. In the deformed configuration, the particle originally at P_0 is located at the point P and has the position vector

$$\boldsymbol{x} = (x_1, x_2, x_3) . \tag{2}$$

The coordinates x_i which give the current position of the particle, are called spatial coordinates. The vector $\boldsymbol{u}$ joining the points P_0 and P, is the displacement vector. This vector may be expressed as

$$\boldsymbol{u} = \boldsymbol{x} - \boldsymbol{a} . \tag{3}$$

In the Lagrangian description, the motion (or flow) of the continuum is expressed in terms of material coordinates by equations of the form

$$x_i = f_i(a_i, t) . \tag{4}$$

These equations may be interpreted as a mapping of the initial configuration of the continuum into its current configuration. On the other hand, in the Eulerian description, the motion of the continuum is defined by means of the inverse of equations (4) in terms of the spatial coordinates as

$$a_i = F_i(x_i, t)\,. \tag{5}$$

The Eulerian description may thus be viewed as one which provides a tracing to its original position of the particle that now occupies the location $\boldsymbol{x}$. The necessary and sufficient condition for the inverse functions (5) to exist is that the Jacobian determinant

$$J = |\partial x_i/\partial a_j| \tag{6}$$

should not vanish.

Since equations (4) and (5) are the inverses of one another, any physical property of the continuum that is expressed with respect to a specific particle (Lagrangian description) may also be expressed with respect to the particular location in space occupied by the particle (Eulerian description). For example, if the Lagrangian description of a physical property g is given by

$$g = g(a_i, t)\,, \tag{7}$$

the Eulerian description is obtained by replacing a_i in this equation by the function given in expression (5). Thus the Eulerian description of the property g is

$$g = g[F_i(x_i, t), t] = g^*(x_i, t) \tag{8}$$

where the symbol g^* is used to emphasize that the functional form of the Eulerian description is not necessarily the same as the Lagrangian form.

As already mentioned in the introduction, both the Lagrangian and the Eulerian approaches present advantages and drawbacks and techniques have been developed that succeed to a certain extent in combining the best features of both the Lagrangian and the Eulerian descriptions. One such technique is the Arbitrary Lagrangian–Eulerian (ALE) method. The ALE formulation has no basic dependence on particles and treats the computational mesh as a reference frame which may be moving with an arbitrary velocity $\boldsymbol{w}$ in the laboratory system. Depending on the value of the velocity $\boldsymbol{w}$, the following basic viewpoints may be individuated:

(1) $\boldsymbol{w} = \boldsymbol{0}$: The reference frame is fixed in space and this corresponds to the Eulerian viewpoint in which the motion is described in terms of spatial coordinates. A particle is identified by the position $\boldsymbol{x}$ (fixed in space) it occupies at time t.

(2) $\boldsymbol{w} = \boldsymbol{v}$, where $\boldsymbol{v}$ is the particle velocity in the laboratory system. Here, the reference frame moves in space at the same velocity as the particles and this corresponds to the Lagrangian viewpoint. A particle is identified by its initial position vector $\boldsymbol{a}$ at time $t = 0$ in the initial configuration of the continuum.

(3) $\boldsymbol{w} \neq \boldsymbol{v} \neq \boldsymbol{0}$: The reference frame moves in space at a velocity $\boldsymbol{w}$ which is different both from the particle velocity $\boldsymbol{v}$ and from zero. Such a reference frame is called Arbitrary Lagrangian–Eulerian and any point of it is identified by its

instantaneous position vector $\boldsymbol{\xi}$ as indicated in Figure 1. The variables (ξ_i, t) are called the mixed variables. It is emphasized that the position vector $\boldsymbol{\xi}$ is a priori arbitrary and consequently independent from the motion of the particles. In the ALE description, a particle is still identified through its material coordinates a_i in the initial configuration of the continuum. However, this identification process is indirect and takes place through the mixed position vector $\boldsymbol{\xi}$ which is linked to the material variables (a_i, t) by the law of motion of the reference frame. This motion is expressed through equations of the form

$$\xi_i = f_i^*(a_i, t)\,. \tag{9}$$

The ALE description may thus be viewed as a mapping of the initial configuration of the continuum into the current configuration of the reference frame. By analogy with expression (6), one defines the Jacobian

$$\tilde{J} = |\partial \xi_i / \partial a_j| \tag{10}$$

for the transformation between mixed and material coordinates. The Jacobian $\tilde{J}$ provides a mathematical link between the current volume element $\mathrm{d}V$ (function of the mixed variables) in the reference frame and the associated volume element $\mathrm{d}V_0$ (function of the material coordinates) in the initial configuration:

$$\mathrm{d}V = \tilde{J}(\boldsymbol{a}, t)\,\mathrm{d}V_0 \tag{11}$$

with $\tilde{J}(\boldsymbol{a}, 0) = 1$. One may show [8] that the time rate of change of the mixed Jacobian determinant is given by

$$\frac{\partial \tilde{J}}{\partial t} = \tilde{J}\nabla \cdot \boldsymbol{w}\,. \tag{12}$$

2.2. *Kinematics in the ALE description*

Since the ALE description does not refer directly to the particles in motion, it will be useful for describing a flow only if it can be linked to either of the classical Lagrangian or Eulerian descriptions. How this can be accomplished will be briefly illustrated in this paragraph. References [8, 9] should be consulted for a more comprehensive treatment.

Let us consider a physical property $g^{**}(\xi_i, t)$ expressed in mixed representation. In view of the law of motion (9), we may deduce a corresponding functional form in terms of material variables by

$$g^{**}(\xi_i, t) = g^{**}[f_i^*(\boldsymbol{a}, t), t] = \tilde{g}(a_i, t)\,. \tag{13}$$

Now, taking the time derivative of (13) with the material coordinates held constant, we write

$$\left.\frac{\partial \tilde{g}(\boldsymbol{a}, t)}{\partial t}\right|_{\boldsymbol{a}} = \left.\frac{\partial g^{**}(\boldsymbol{\xi}, t)}{\partial t}\right|_{\boldsymbol{\xi}} + \frac{\partial g^{**}(\boldsymbol{\xi}, t)}{\partial \xi_i}\frac{\partial \xi_i}{\partial t}\,. \tag{14}$$

Noting that $\partial\xi_i/\partial t = w_i$ and making use of the identity

$$\nabla\cdot(g^{**}\boldsymbol{w}) = g^{**}\nabla\cdot\boldsymbol{w} + \boldsymbol{w}\cdot\nabla g^{**} \tag{15}$$

which in view of (12) results in

$$\tilde{J}\nabla\cdot(g^{**}\boldsymbol{w}) = \frac{\partial\tilde{J}}{\partial t}g^{**} + \tilde{J}\boldsymbol{w}\cdot\nabla g^{**}, \tag{16}$$

we may rewrite expression (14) in the form

$$\frac{\partial}{\partial t}(\tilde{J}\tilde{g}) = \tilde{J}\left[\frac{\partial g^{**}}{\partial t} + \nabla\cdot(g^{**}\boldsymbol{w})\right]. \tag{17}$$

At this point, we note that $\partial g^{**}/\partial t$ in (17) is a spatial derivative, i.e. taken with the point fixed in position. It follows that its value may be obtained from appropriate differential laws expressed in Eulerian variables. In these conditions, expression (17) appears to be a fundamental relationship which enables us to 'translate' any law expressed in spatial (Eulerian) variables into an equivalent law expressed in mixed variables. This will be illustrated in the next paragraph for the basic conservation laws of mass, momentum and energy.

It is also useful to consider the time rate of change of a property defined as an integral over a finite portion of the moving reference frame. In particular, let a scalar property $G(t)$ be represented by the volume integral

$$G(t) = \int_{V(t)} g^{**}(\boldsymbol{\xi}, t)\,\mathrm{d}V. \tag{18}$$

In view of relations (11) and (13), the time rate of change of G may be expressed in the form

$$\frac{\mathrm{d}G(t)}{\mathrm{d}t} = \frac{\partial}{\partial t}\int_{V(t)} g^{**}(\boldsymbol{\xi}, t)\,\mathrm{d}V = \int_{V_0}\frac{\partial}{\partial t}(\tilde{J}\tilde{g}(\boldsymbol{a}, t))\,\mathrm{d}V_0 \tag{19}$$

which, making use of relations (12) and (14) and applying Gauss' theorem results in

$$\frac{\mathrm{d}G(t)}{\mathrm{d}t} = \int_{V(t)}\frac{\partial g^{**}(\boldsymbol{\xi}, t)}{\partial t}\,\mathrm{d}V + \oint_{S(t)} g^{**}(\boldsymbol{\xi}, t)\boldsymbol{w}\cdot\boldsymbol{n}\,\mathrm{d}S \tag{20}$$

where $\boldsymbol{n}$ denotes the outward normal to the surface S bounding volume V. Equation (20) states that the rate of change of the property $G(t)$ is equal to the sum of the amount created within the instantaneously occupied volume $V(t)$ plus the flux through the bounding surface $S(t)$ induced by the movement of the reference frame. Expression (20) forms the basis for establishing the integral form of the conservation laws in the mixed description. Here again the time derivative $\partial g^{**}/\partial t$ is a spatial derivative and its value may be obtained from appropriate differential laws expressed in Eulerian variables.

2.3. Basic conservation laws in the ALE description

We have shown in the previous paragraph that the ALE form of any differential law may be obtained from the corresponding law expressed in spatial (Eulerian) description. Here, the process will be applied to derive the differential form of the basic conservation laws for mass momentum and energy in the ALE description. We will consider only inviscid, compressible flow.

Using Cartesian tensor notation and neglecting viscous effects, the conservation equations of mass, momentum and energy are expressed in spatial coordinates as

$$\frac{\partial \rho}{\partial t} + \frac{\partial(\rho v_j)}{\partial x_j} = 0\,, \tag{21}$$

$$\frac{\partial(\rho v_i)}{\partial t} + \frac{\partial}{\partial x_j}(\rho v_i v_j) = \rho g_i - \frac{\partial p}{\partial x_i} \quad (i = 1, 2, 3)\,, \tag{22}$$

$$\frac{\partial(\rho e)}{\partial t} + \frac{\partial}{\partial x_j}(\rho v_j e) = \rho v_j g_j - \frac{\partial}{\partial x_j}(p v_j)\,. \tag{23}$$

In these expressions ρ is the fluid density, p the pressure, $\boldsymbol{v}$ the fluid velocity and $\boldsymbol{g}$ a body force. The total specific energy e is defined by

$$e = \tfrac{1}{2}v_i^2 + i \tag{24}$$

where i is the specific internal energy. For compressible fluids, the pressure is defined by an equation of state of the form $p = f(\rho, i)$.

Introducing the Eulerian forms (21)–(23) of the conservation laws into the right-hand side of (17), we obtain the local form of these laws in a reference frame moving at arbitrary velocity $\boldsymbol{w}$:

$$\frac{\partial}{\partial t}(\rho \tilde{J}) = \tilde{J}\frac{\partial}{\partial x_j}(\rho(w_j - v_j))\,, \tag{25}$$

$$\frac{\partial}{\partial t}(\rho v_i \tilde{J}) = \tilde{J}\frac{\partial}{\partial x_j}(\rho v_i(w_j - v_j)) + \tilde{J}\left(\rho g_i - \frac{\partial p}{\partial x_i}\right), \tag{26}$$

$$\frac{\partial}{\partial t}(\rho e \tilde{J}) = \tilde{J}\frac{\partial}{\partial x_j}(\rho e(w_j - v_j)) + \tilde{J}\left(\rho v_j g_j - \frac{\partial}{\partial x_j}(p v_j)\right). \tag{27}$$

One may also write an internal energy equation in the form

$$\frac{\partial}{\partial t}(\rho i \tilde{J}) = \tilde{J}\frac{\partial}{\partial x_j}(\rho i(w_j - v_j)) - \tilde{J}p\frac{\partial v_j}{\partial x_j}\,. \tag{28}$$

In the above relations, the same symbols are used to represent both the Lagrangian (left-hand side) and the mixed (right-hand side) functional forms of the physical quantities. Moreover, the mixed position vector is written $\boldsymbol{x}$ instead of $\boldsymbol{\xi}$. It should also be reminded that the symbol $\partial/\partial t$ indicates a time derivative taken with the material coordinates held fixed. It is easily verified that the particular cases of purely Lagrangian ($\boldsymbol{w} = \boldsymbol{v}$) and purely Eulerian ($\boldsymbol{w} = 0$) descriptions are contained in the above conservation equations. We shall now develop weak variational forms of the ALE conservation laws (25)–(28) which will provide the basis for space discretisation for a finite element model.

2.4. Variational form of momentum equation: the principle of virtual power

The principle of virtual power [10] represents an alternative statement of conservation of momentum and its role corresponds to that of the principle of virtual displacements in solid mechanics. This principle is derived by operating on the ALE differential form (26) of momentum conservation and on the associated stress boundary relation

$$T_i = -n_j \delta_{ji} p = S_i \tag{29}$$

where T_i represents the components of prescribed boundary loads (force per unit area), n_j are the direction cosines of the unit outward normal to the boundary, δ_{ji} is the Kronecker delta and p the fluid pressure.

In order to isolate the acceleration term in the differential form (26) of momentum conservation, we rewrite this equation as

$$\rho \tilde{J} \frac{\partial v_i}{\partial t} + v_i \frac{\partial}{\partial t}(\rho \tilde{J}) = \tilde{J} \frac{\partial}{\partial x_j}(\rho v_i (w_j - v_j)) + \tilde{J}\left(\rho g_i - \frac{\partial p}{\partial x_i}\right) \tag{30}$$

and make use of the mass equation (25) to reduce (30) to the simple form

$$\rho \frac{\partial v_i}{\partial t} = \rho(\boldsymbol{w} - \boldsymbol{v}) \cdot \nabla v_i + \rho g_i - \frac{\partial p}{\partial x_i} \tag{31}$$

which will be used together with the boundary relation (29) to formulate the principle of virtual power. To this aim, let us multiply (31) by an arbitrary admissible variation δv_i of the fluid velocity and integrate over a control volume $V(t)$. Similarly, we multiply (29) by δv_i and integrate along the boundary surface $S(t)$. We then add the two integral equations and the result is

$$\int_{V(t)} \delta v_i \rho \frac{\partial v_i}{\partial t} \, dV = \int_{V(t)} \delta v_i \rho (\boldsymbol{w} - \boldsymbol{v}) \cdot \nabla v_i \, dV + \int_{V(t)} \delta v_i \rho g_i \, dV - \int_{V(t)} \delta v_i \frac{\partial p}{\partial x_i} \, dV + \oint_{S(t)} \delta v_i (T_i - S_i) \, dS . \tag{32}$$

We now integrate by parts the pressure term in (32) and apply the divergence theorem; this yields the principle of virtual power

$$\int_{V(t)} \delta v_i \rho \frac{\partial v_i}{\partial t} \, dV = \int_{V(t)} \delta v_i \rho (\boldsymbol{w} - \boldsymbol{v}) \cdot \nabla v_i \, dV + \int_{V(t)} \delta v_i \rho g_i \, dV + \int_{V(t)} p \frac{\partial}{\partial x_i}(\delta v_i) \, dV + \oint_{S(t)} \delta v_i T_i \, dS \tag{33}$$

which provides the basis for the formulation of spatially discrete models of the momentum equations using finite element approximations.

2.5. Weak forms of mass and energy equations

In view of relations (12) and (15), the ALE differential form (25) of the mass equation may be rewritten as

$$\frac{\partial \rho}{\partial t} = (\boldsymbol{w} - \boldsymbol{v}) \cdot \nabla \rho - \rho \nabla \cdot \boldsymbol{v} . \tag{34}$$

Multiplying this equation by an arbitrary weighting function ψ and integrating over a control volume $V(t)$, we obtain the following weak form:

$$\int_{V(t)} \psi \frac{\partial \rho}{\partial t} \mathrm{d}V = \int_{V(t)} \psi(\boldsymbol{w} - \boldsymbol{v}) \cdot \nabla \rho \, \mathrm{d}V - \int_{V(t)} \psi \rho \nabla \cdot \boldsymbol{v} \, \mathrm{d}V \tag{35}$$

which will be employed to develop spatially discrete models of the mass conservation equation using finite element approximations.

A useful particular form of (35) may be derived by assuming that the density is constant over the control volume $V(t)$. In this case, we have $\nabla \rho = 0$ and the weighting function ψ may be chosen as unity. It follows that (35) reduces to the simple form

$$\int_{V(t)} \frac{\partial \rho}{\partial t} \mathrm{d}V = - \int_{V(t)} \rho \nabla \cdot \boldsymbol{v} \, \mathrm{d}V = - \oint_{S(t)} \rho \boldsymbol{v} \cdot \boldsymbol{n} \, \mathrm{d}S \tag{36}$$

which, in view of expression (20), may be further transformed into

$$\frac{\mathrm{d}}{\mathrm{d}t} \int_{V(t)} \rho \, \mathrm{d}V = \oint_{S(t)} \rho(\boldsymbol{w} - \boldsymbol{v}) \cdot \boldsymbol{n} \, \mathrm{d}S . \tag{37}$$

This integral form of the statement of conservation of mass will be employed in finite element models assuming an elementwise constant density. It also forms the basis for updating the density in ALE finite difference methods applicable to arbitrary meshes (see Hirt et al. [4]).

The weak form of the internal energy equation (28) is derived in complete analogy with what has been done for the mass equation. First, we use relations (12) and (15) to rewrite (28) in the form

$$\frac{\partial}{\partial t}(\rho i) = (\boldsymbol{w} - \boldsymbol{v}) \cdot \nabla(\rho i) - (\rho i + p) \nabla \cdot \boldsymbol{v} . \tag{38}$$

We then multiply this equation by an arbitrary weighting function ψ and integrate over a control volume $V(t)$. This gives the weak integral form

$$\int_{V(t)} \psi \frac{\partial}{\partial t}(\rho i)\,\mathrm{d}V = \int_{V(t)} \psi(\boldsymbol{w}-\boldsymbol{v}) \cdot \nabla(\rho i)\,\mathrm{d}V - \int_{V(t)} \psi(\rho i + p)\nabla \cdot \boldsymbol{v}\,\mathrm{d}V \quad (39)$$

which provides the basis for the finite element model of the internal energy equation.

If the density of internal energy ρi is assumed uniform over the control volume $V(t)$, the following integral statement may be deduced from the weak form (39) of the internal energy equation:

$$\frac{\mathrm{d}}{\mathrm{d}t} \int_{V(t)} \rho i\,\mathrm{d}V = \oint_{S(t)} \rho i(\boldsymbol{w}-\boldsymbol{v}) \cdot \boldsymbol{n}\,\mathrm{d}S - \int_{V(t)} p\nabla \cdot \boldsymbol{v}\,\mathrm{d}V. \quad (40)$$

This integral form will be used in connection with ALE finite element models assuming an elementwise uniform density of internal energy.

3. Spatially discrete models

We now have all the necessary ingredients to derive spatially discrete models of the ALE conservation equations using finite element modelling. The Galerkin formulation of the method of weighted residuals will be employed to generate semi-discrete equations which will then be integrated forward in time to obtain the transient response.

For the purpose of providing a framework for our discussion of spatially discrete models of the ALE conservation equations, we will first define the basic variables in the form of local approximations over a typical finite element. We will use the following notation:

$\{\}$ denotes a column matrix,
$[\,]$ denotes a rectangular or square matrix,
u_i lower case subscripts denote components of a vector,
u_I upper case subscripts denote node numbers,
e denotes a typical finite element,
$\dot{}$ denotes a time derivative,

The usual assumption in discretising partial differential equations in finite element schemes is that the functional dependence of the variables in space and time can be separated, so that any local vector field $u_i(x, t)$ in an element can be described as product of shape functions $N_I(x)$ that are independent of time and nodal values $u_{iI}(t)$ which are independent of x and incorporate the time dependence:

$$u_i(x, t) = N_I(x)u_{iI}(t)\,. \quad (41)$$

Here, the fluid velocity $\boldsymbol{v}$ and the arbitrary grid velocity $\boldsymbol{w}$ will be interpolated over a typical finite element by the same shape functions:

$$v_i(x, t) = N_I(x) v_{iI}(t) , \tag{42}$$

$$w_i(x, t) = N_I(x) w_{iI}(t) . \tag{43}$$

Another set of shape functions will be employed for the local description of the density ρ and for the density of internal energy ρi:

$$\rho(x, t) = \phi_J(x) \rho_J(t); \qquad \rho i(x, t) = \phi_J(x) \rho i_J(t) . \tag{44}$$

While shape functions $[N]$ must ensure inter-element continuity of the velocities $\boldsymbol{v}$ and $\boldsymbol{w}$, this is not necessarily the case for shape functions $[\phi]$. In general, the fluid properties in (44) are interpolated by a polynomial which is at least one order less than that for the velocities.

3.1. Semi-discrete mass equations

Equation (35) is the appropriate weak form of the mass conservation equation and, according to the Galerkin method, the weighting function ψ is chosen as $\psi = \phi_I$ ($I = 1, 2$, etc.), where ϕ_I's are the shape functions defined in (44) for the local description of density. In this way, (35) provides a semi-discrete equation for each of the degrees of freedom in the discrete density field.

The governing equation for a typical degree of freedom ρ_I is obtained from (35) in the form

$$\sum_e \int_{V^e} \phi_I \phi_J \dot{\rho}_J \, \mathrm{d}V = \sum_e \int_{V^e} \phi_I [(\boldsymbol{w} - \boldsymbol{v}) \cdot \rho_J \nabla \phi_J - \phi_J \rho_J \nabla \cdot \boldsymbol{v}] \, \mathrm{d}V \tag{45}$$

where Σ_e indicates a summation over all the elements to which density node I does belong, while the repetition of subscript J means a summation over all density nodes in element e. By formulating one such equation for all the density degrees of freedom, we obtain a global system of semi-discrete equations expressing conservation of mass in the finite element mesh. The global system may be written as

$$[D]\{\dot{\rho}\} = \{F\} \tag{46}$$

and represents a set of ordinary differential equations in time for the nodal values of density.

In these equations,

$[D]$ is a global 'volume' matrix consisting, as shown by (45), of element contributions of the type

$$D^e_{IJ} = \int_{V^e} \phi_I \phi_J \, \mathrm{d}V , \tag{47}$$

$\{\rho^{\cdot}\}$ denotes the total vector of nodal values of the time derivative of the density ρ, and

$\{F\}$ indicates global nodal 'loads' accounting, as shown by the right-hand side of (45), for the transport of mass across the element boundaries.

At the element level, the contribution to a typical component F_I of vector $\{F\}$ is given by

$$F_I^e = \int_{V^e} \phi_I[(\boldsymbol{w} - \boldsymbol{v}) \cdot \rho_J \nabla\phi_J - \phi_J\rho_J\nabla \cdot \boldsymbol{v}]\, \mathrm{d}V \tag{48}$$

where the velocities are evaluated in terms of nodal values as indicated by (42) and (43).

Notice that the semi-discrete mass equations (46) are presented in terms of nodal 'loads' and are thus in a form that is adapted to the use of an explicit method for numerical time integration. However, to make explicit integration schemes viable, the consistent matrix $[D]$ in (47) should be reduced to a diagonal matrix. How this may be accomplished will be discussed later.

3.2. Semi-discrete internal energy equations

Equation (39) is the appropriate weak form of the internal energy equation and the weighting function ψ is again chosen as $\psi = \phi_I$ ($I = 1$, 2, etc.) where the ϕ_I's are the shape functions defined in (44) to describe local spatial variations of the internal energy density ρi.

Proceeding as described in Section 3.1 for the mass equation, a system of ordinary differential equations in time is obtained for the nodal values of the internal energy density in the form

$$[D]\{\rho i^{\cdot}\} = \{G\} \tag{49}$$

where $[D]$ is the global matrix defined in (47), while vector $\{G\}$ indicates global nodal 'loads' consisting of elementary contributions derived from (39) in the form

$$G_I^e = \int_{V^e} \phi_I[(\boldsymbol{w} - \boldsymbol{v}) \cdot \nabla(\rho i) - (\rho i + p)\nabla \cdot \boldsymbol{v}]\, \mathrm{d}V. \tag{50}$$

This expression clearly indicates that the nodal 'loads' in the right-hand side of the semi-discrete energy equations (49) account for the transport of energy across the element boundaries and for the internal work performed by the fluid pressure.

As was the case for the mass equations, the system (49) of semi-discrete internal energy equations is in a form readily adapted to the use of an explicit method for numerical time integration.

3.3. Semi-discrete momentum equations

The principle of virtual power (33) is applied to derive the semi-discrete equations expressing conservation of momentum in the finite element mesh. According to the Galerkin method, the velocity variation in (33) is defined as

$$\delta v_i = N_i(x)\delta v_{iI} \tag{51}$$

where the N_I's are the shape functions defined in (42) to describe the local variations of the fluid velocity. Independent variations of all the velocity degrees of freedom in the mesh are introduced in turn into the principle of virtual power (33) and by invoking the arbitrariness of such variations, the discrete analog to the momentum equations is obtained.

The semi-discrete equation governing a typical velocity node I is found from (33) and (51) in the form

$$\sum_e \int_{V^e} N_I \rho N_J \dot{v}_{iJ}\,\mathrm{d}V = \sum_e \int_{V^e} N_I \rho (w_j - v_j)\cdot \frac{\partial v_i}{\partial x_j}\,\mathrm{d}V + \sum_e \int_{V^e} \left(N_I \rho g_i + p\frac{\partial N_I}{\partial x_i}\right)\mathrm{d}V + \sum_e \oint_{S^e_{\mathrm{ext}}} N_I T_i\,\mathrm{d}S \tag{52}$$

where Σ_e indicates a summation over all the elements to which node I belongs, the repetition of subscript J implies a summation over all velocity nodes in element e, and subscript i denotes components of a vector.

The governing equations for the discrete mesh form a set of ordinary differential equations in time which may be written in condensed form as

$$[M]\{\dot{v}\} = \{F_t\} + \{F_b\} + \{F_p\} + \{\bar{F}\} \tag{53}$$

where:

$[M]$ is the global mass matrix consisting, as shown in (52), of element contributions of the type

$$M^e_{IJ} = \int_{V^e} N_I \rho N_J\,\mathrm{d}V, \tag{54}$$

$\{\dot{v}\}$ denotes the total nodal acceleration vector,

$\{F_t\}$ indicates global nodal loads induced by transport of momentum; as shown by (52), a typical component of vector $\{F_t\}$ is obtained by assembly of element contributions of the form

$$(F^I_{t,i})^e = \int_{V^e} N_I \rho (w_j - v_j)\frac{\partial v_i}{\partial x_j}\,\mathrm{d}V, \tag{55}$$

$\{F_b\}$ denotes global nodal loads due to body forces ρg_i:

$$(F^I_{b,i})^e = \int_{V^e} N_I \rho g_i\,\mathrm{d}V, \tag{56}$$

$\{F_p\}$ represents the nodal loads induced by the fluid pressure p:

$$(F^I_{p,i})^e = \int_{V^e} p \frac{\partial N_I}{\partial x_i} \, dV, \tag{57}$$

$\{\bar{F}\}$ accounts for externally applied loads T_i:

$$(F^I_i)^e = \oint_{S^e_{ext}} N_I T_i \, dS. \tag{58}$$

3.4. *Specialized equations for constant pressure elements*

We will now develop detailed expressions of the semi-discrete conservation equations for two-dimensional quadrilateral and triangular elements in which density, specific internal energy and hence pressure are assumed elementwise constant. In this development we will concentrate on the quadrilateral element and only outline the triangular element.

The treatment of the transport terms poses the major problem in explicit ALE formulations based on low-order elements. This is due to the relatively poor phase characteristics associated with low-order differencing of the convective terms, which cause numerical dispersion and, consequently, spurious oscillations in the results. This problem may be partly rectified by the use of 'upwind' differencing and we will therefore present the specialized semi-discrete conservation equations in a form that permits the introduction of some selective upwinding of the transport terms.

The quadrilateral element is shown in Figure 2 and has four nodes. The velocity field in the element is approximated by the standard bilinear shape functions (see Zienkiewicz for details [11])

$$v_i(x, y) = v_{iI} N_I(\xi, \eta) \tag{59}$$

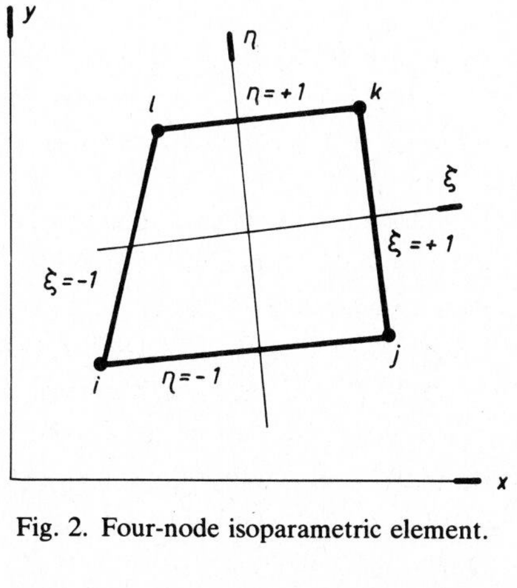

Fig. 2. Four-node isoparametric element.

where the repetition of subscript I implies a summation over the four nodes of the element. This velocity field applies both to the fluid velocity v_i and to the arbitrary grid velocity w_i.

The shape functions for the quadrilateral element are:

$$N_1 = \tfrac{1}{4}(1-\xi)(1-\eta); \qquad N_2 = \tfrac{1}{4}(1+\xi)(1-\eta);$$
$$N_3 = \tfrac{1}{4}(1+\xi)(1+\eta); \qquad N_4 = \tfrac{1}{4}(1-\xi)(1+\eta). \tag{60}$$

where ξ and η are defined by the isoparametric transformation

$$x_i = x_{iI} N_I(\xi, \eta), \tag{61}$$

so that derivatives are given by (see [11])

$$\frac{\partial()}{\partial x} = \frac{1}{j}\left(\frac{\partial y}{\partial \eta}\frac{\partial()}{\partial \xi} - \frac{\partial y}{\partial \xi}\frac{\partial()}{\partial \eta}\right),$$
$$\frac{\partial()}{\partial y} = \frac{1}{j}\left(\frac{\partial x}{\partial \xi}\frac{\partial()}{\partial \eta} - \frac{\partial x}{\partial \eta}\frac{\partial()}{\partial \xi}\right). \tag{62}$$

where j is the Jacobian given by

$$j = \frac{\partial x}{\partial \xi}\frac{\partial y}{\partial \eta} - \frac{\partial x}{\partial \eta}\frac{\partial y}{\partial \xi}. \tag{63}$$

3.4.1. *Mass equation*

In updating the density of the element, the integral form (37) of the conservation of mass is used, which gives

$$\frac{\mathrm{d}}{\mathrm{d}t}\int_{V^e} \rho \,\mathrm{d}V = \oint_{S^e} \rho(w_i - v_i)n_i \,\mathrm{d}S \tag{64}$$

where n_i is a unit normal to the surface S^e. The assumption of constant density in the element and evaluation of the right-hand integral with the shape functions (60) yield the rate of change of the element mass M^e in the form

$$\frac{\mathrm{d}M^e}{\mathrm{d}t} = \frac{1}{2}\sum_{\substack{I=1\\J=I+1}}^{4} \{\rho_{IJ}[y_{JI}(u_{xI}+u_{xJ}) + x_{IJ}(u_{yI}+u_{yJ})]\} \tag{65}$$

where $y_{JI} = y_J - y_I$, $x_{IJ} = x_I - x_J$ and $u_i = w_i - v_i$.

Since the density is assumed constant within an element, discontinuities occur along inter-element boundaries. Therefore, the density ρ_{IJ} along a typical boundary segment IJ is evaluated as an average of the densities in the elements e and e' on either side of the boundary segment. The use of a simple average yields a central-difference approximation to the mass transport term which may generate spurious oscillations in the results or even cause numerical instability if a first-order scheme is employed for explicit time integration. Stable results may be obtained by weighting the average in favour of the value in the element from which the mass is substracted. This is the upstream or donor element convective flux approximation (see Roache [12]). Thus, the density ρ_{IJ} in (65) is evaluated as

$$\rho_{IJ} = \tfrac{1}{2}[(1-\alpha_{IJ})\rho^e + (1+\alpha_{IJ})\rho^{e'}] \tag{66}$$

where α_{IJ} is the donor element weighting factor. As full donor-element weighting ($|\alpha_{IJ}| = 1$) is too diffusive for most circumstances, generally $0 < |\alpha_{ij}| < 1$. The best choice for α_{IJ} is still an open problem and various points of view have been put forth in the recent literature (see Hughes [13] for a survey).

From expression (20), the rate of change of the element volume V^e is given by

$$\frac{\mathrm{d}V^e}{\mathrm{d}t} = \oint_{S^e} w_i n_i \,\mathrm{d}S \tag{67}$$

and may be evaluated in closed form as

$$\frac{\mathrm{d}V^e}{\mathrm{d}t} = \frac{1}{2} \sum_{\substack{I=1\\J=I+1}}^{4} [y_{JI}(w_{xI} + w_{xJ}) + x_{IJ}(w_{yI} + w_{yJ})] \,. \tag{68}$$

Once the element mass and its volume have been updated (see Section 4 for numerical time integration), the new density in the element is readily obtained from the ratio of element mass and volume,

$$\rho^e = \frac{M^e}{V^e} \,. \tag{69}$$

3.4.2. Energy equation

The rate of change of internal energy of the element is given by the integral form (40), which yields

$$\begin{aligned}\frac{\mathrm{d}I^e}{\mathrm{d}t} &= \frac{\mathrm{d}}{\mathrm{d}t} \int_{V^e} \rho i \,\mathrm{d}V \\ &= \oint_{S^e} \rho i (w_i - v_i) n_i \,\mathrm{d}S - \int_{V^e} p \frac{\partial v_i}{\partial x_i} \,\mathrm{d}V\end{aligned} \tag{70}$$

where the pressure p is assumed positive in compression. The assumption of constant ρ, i, and p within the element and the evaluation of the right-hand integrals with the shape functions (60) give

$$\begin{aligned}\frac{\mathrm{d}I^e}{\mathrm{d}t} = \frac{\mathrm{d}}{\mathrm{d}t}(M^e i^e) = \frac{1}{2} \sum_{\substack{I=1\\J=I+1}}^{4} \{&(\rho i)_{IJ}[y_{JI}(u_{xI} + u_{xJ}) + x_{IJ}(u_{yI} + u_{yJ})] \\ &- p[y_{JI}(v_{xI} + v_{xJ}) + x_{IJ}(v_{yI} + v_{yJ})]\}\end{aligned} \tag{71}$$

where the notation defined in (65) has been used. To stabilize the energy equation, a weighted average value of ρi is used, as indicated in formula (66), for the exchanges of energy between neighbouring elements. Once the internal energy of the element has been updated, the new specific internal energy is obtained from the ratio of element internal energy and mass,

$$i^e = \frac{I^e}{M^e} . \tag{72}$$

3.4.3. Momentum equation

The form (53) of the semi-discrete momentum equations is intended for an explicit time integration. However, to make explicit schemes viable, it is imperative to transform the consistent mass matrix (54) into a lumped, diagonal matrix. For the four-node quadrilateral, this may be achieved by adding all terms of each line of the consistent mass matrix and placing the result on the diagonal. This gives

$$M^e_{II} = \int_{V^e} N_I \rho \, \mathrm{d}V , \qquad M^e_{IJ} = 0 \quad (I \neq J) , \tag{73}$$

and a Gauss–Legendre (2×2) quadrature rule may be used to evaluate the integrals. Notice, however, that for plane rectangular elements, exact results are obtained by apportioning one fourth of the mass of the element to each node.

The discretization of the momentum transport terms via the Galerkin approach, (55), yields central-difference type approximations and noisy results may consequently be produced whenever the velocities are large. It is therefore appropriate to introduce some selective 'upwinding' into the transport term to stabilize the momentum equations. Upwind differencing in a finite element context has been developed by Christie et al. [14] and Heinrich et al. [15] who skewed the test functions in the upwind direction. This type of upwinding necessitates higher-order quadrature rules than those used in the Galerkin formulation. Alternate forms of upwinding which are computationally more efficient have been suggested by Hughes [16] and by Belytschko et al. [17].

The technique consists, in essence, in a one-point evaluation of the discrete transport terms in which (55) is altered as follows:

$$(F^I_{t,i})^e = 4N_I(\boldsymbol{\xi}^e)\rho^e(w_j - v_j)_{O^e} \frac{\partial v_i(\boldsymbol{\xi}^e)}{\partial x_j} j(O^e) \tag{74}$$

where, in reference [16], $\boldsymbol{\xi}^e$ is some point in the eth element domain, O^e is the origin of isoparametric coordinates in the eth element and j is the Jacobian determinant (63) of the isoparametric transformation. The location of the point ξ^e determines the degree of upwinding in each element. In reference [17], the evaluation point ξ^e is the origin of isoparametric co-ordinates and the transport term (74) is modified as follows to account for upwinding:

$$F^{I*}_{t,i} = (1 + \gamma_I)F^I_{t,i} \tag{75}$$

where γ_I is an appropriate transport amplification factor.

The nodal loads (56) induced by body forces ρg are evaluated at the element level as

$$(F^I_{b,i})^e = M^e_{II}\rho^e g_i \tag{76}$$

where M^e_{II} is the nodal mass defined by (73).

Finally, the assumption of constant pressure within the element yields the following closed form for the nodal loads (57) induced by the fluid pressure:

$$(F^I_{p,x})^e = \tfrac{1}{2}p(y_J - y_L), \qquad (F^I_{p,y})^e = \tfrac{1}{2}p(x_L - x_J), \tag{77}$$

where it is assumed that J is one node counterclockwise from I and L one node clockwise from I.

The treatment of the three-node triangular element is quite similar, except for the transport terms (55) in the momentum equations which may be evaluated in closed form. For the triangular element, the shape functions are the triangular coordinates

$$\phi_I = \frac{1}{2A}(a_I + b_I x + c_I y) \tag{78}$$

in which

$$a_I = x_J y_K - x_K y_J, \qquad b_I = y_J - y_K, \qquad c_I = x_K - x_J, \tag{79}$$

where nodes I, J, K are numbered in a counterclockwise order. A is the area of the triangle.

In these conditions, the discrete transport term (55) becomes

$$(F^I_{t,i})^e = \int_{V^e} \phi_I \rho \left(u_x \frac{\partial v_i}{\partial x} + u_y \frac{\partial v_i}{\partial y}\right) \mathrm{d}V \tag{80}$$

where

$$u_x = w_x - v_x = \phi_K u_{xK}, \qquad u_y = w_y - v_y = \phi_K u_{yK},$$
$$\frac{\partial v_i}{\partial x} = \frac{1}{2A} b_J v_{iJ}, \qquad \frac{\partial v_i}{\partial y} = \frac{1}{2A} c_J v_{iJ}, \tag{81}$$

and noting the property

$$\int_{V^e} \phi_I \phi_K \, \mathrm{d}V = \frac{A}{12}(1 + \delta_{IK}) \tag{82}$$

where δ_{IK} is the Kronecker delta, we obtain the following closed form expression for the discrete momentum transport term:

$$(F^I_{t,i})^e = \frac{\rho}{24}\{b_J v_{iJ}[u_{xK}(1 + \delta_{IK})] + c_J v_{iJ}[u_{yK}(1 + \delta_{IK})]\} \tag{83}$$

where the summations on J and K extend over the three nodes of the triangular element.

4. Explicit time integration

Since the partial differential equations governing unsteady inviscid, compressible flow are hyperbolic, 'time centered' schemes are in principle required to ensure stability of an explicit time integration procedure. However, if an upwind discretisation of the transport terms is employed, a numerical diffusion is introduced into the semi-discrete conservation equations and stability may be achieved by using simple first-order explicit time integration.

4.1. Central difference method

The most popular method using a centered time derivative is the central difference method, in which the velocities are updated by

$$v(t+\tfrac{1}{2}\Delta t) = v(t-\tfrac{1}{2}\Delta t) + \Delta t\, \dot{v}(t) \tag{84}$$

where Δt is the time step and $\dot{v}(t)$ is the acceleration given by the equations of motion (53) in which the right-hand side (i.e. the nodal forces) must be evaluated at time t. Since the velocities are carried at the half-time points, this necessitates a forward extrapolation of the velocities by

$$v(t) = v(t-\tfrac{1}{2}\Delta t) + \tfrac{1}{2}\Delta t\, \dot{v}(t-\Delta t) \tag{85}$$

so that the momentum transport terms in (53) are for time t like the rest of the terms.

Since pressure is required at integer-time points for use in the momentum equations, density and specific internal energy are updated by a scheme of the form

$$A(t+\Delta t) = A(t) + \Delta t\, \dot{A}(t+\tfrac{1}{2}\Delta t) \tag{86}$$

and a forward extrapolation of density and specific internal energy is needed for the evaluation of

$$\dot{A}(t+\tfrac{1}{2}\Delta t)\,.$$

The central difference method is only conditionally stable and for purposes of numerical stability the time step is limited by

$$\Delta t \leq \frac{l}{(C+|v|)} \tag{87}$$

where l is the minimum element dimension and C the acoustic wave speed in the material.

4.2. A first-order method valid for all flow speeds

An interesting computational scheme based on the first-order explicit time integration and valid for both compressible and incompressible flows has been proposed by Hirt et al. [4] and Stein et al. [18]. In deriving such a scheme, one notes that the transport terms in the semi-discrete conservation equations vanish in the Lagrangian formulation, i.e. when $\boldsymbol{w} = \boldsymbol{v}$. It is this reduction that leads one to first solve the Lagrangian conservation equations in the first step of the calculational cycle, phase I, and then to add the transport contributions as a separate step in the last phase, phase III. The middle step, making up phase II, adds an implicit pressure calculation that permits solutions to be obtained at all flow speeds. In describing the above three phases of the calculational cycle, we closely follow Stein et al. [18] and translate into finite element terms their finite difference formulation.

Phase I: *Explicit Lagrangian calculation.* This step calculates the Lagrangian nodal velocities resulting from the previous cycle's pressure and body forces. From (53) with $\boldsymbol{w} = \boldsymbol{v}$ in (55) and using lumped masses, we obtain

$$\{\tilde{v}\} = \{v^n\} + \frac{\Delta t}{M^n}\,[\{F_b^n\} + \{F_p^n\}] \tag{88}$$

where the tilde superscript indicates an intermediate time for this phase of cycle $n+1$, and the superscript n indicates previous cycle values.

The specific internal energy i is advanced to an intermediate time from the integral equation (70) with $\boldsymbol{w} = \boldsymbol{v}$:

$$\tilde{i} = i^n - \frac{\Delta t}{M_e^n}\int p^n \nabla \cdot \left(\frac{\boldsymbol{v}^n + \tilde{\boldsymbol{v}}}{2}\right) \mathrm{d}V. \tag{89}$$

In this expression, M_e is the element mass and a mean velocity is used, as suggested by Pracht [19], in order to conserve total energy exactly in the finite element formulation.

Phase II: *Implicit Lagrangian calculation.* The explicit Lagrangian phase detailed above yields tilde values for the fluid velocities and specific internal energies, while the corresponding densities and pressures have yet to be calculated before the calculation enters the third (transport) phase to be described later.

If a purely explicit procedure is desired, one may obtain the updated densities and pressures immediately from the equation of conservation of mass and the equation of state of the fluid, viz:

$$\rho^L(1 + \Delta t\, \nabla \cdot \boldsymbol{v}^L) = \rho^n\,, \tag{90}$$

$$p^L = f(\rho^L, i^L)\,, \tag{91}$$

where we have taken $\boldsymbol{v}^L = \tilde{\boldsymbol{v}}$, $i^L = \tilde{i}$ and used a time-advanced discrete form of (64) with $\boldsymbol{w} = \boldsymbol{v}$.

In cases (especially those involving low Mach-numbers) where this procedure is found to be too noisy, the following implicit phase may be employed. The object is to solve (90) and (91) simultaneously with

$$\{v^L\} = \{v^n\} + \frac{\Delta t}{M^n}\{F(p^L)\} \tag{92}$$

and

$$i^L = i^n - \frac{\Delta t}{M_e^n} \int_{V^n} p^L \nabla \cdot \left(\frac{\boldsymbol{v}^n + \boldsymbol{v}^L}{2}\right) \mathrm{d}V. \tag{93}$$

Notice that advanced values (superscript L) of all the fluid variables have been used in the Lagrangian conservation equations (except in (93) which employs an average velocity value), but that the masses and integration volumes are at time level n, a choice made for computational convenience.

Equations (90)–(93) are non-linear relations for the four quantities ρ^L, p^L, $\boldsymbol{v}^L$ and i^L, and we solve them using a pseudo-Newton scheme (see e.g. Pracht and Brackbill [20]). In this scheme, a quantity H is defined which is reduced iteratively to zero for each element:

$$H = p - f(\rho, i). \tag{94}$$

Choosing some arbitrary perturbation $(\Delta p)^e$ from the pressure at level n in element e, one calculates in turn $(\Delta \boldsymbol{v})^e$, $(\Delta i)^e$, $(\Delta f)^e$ and $(\Delta H)^e$ and so numerically evaluates the derivative $(\mathrm{d}H/\mathrm{d}p)^e$ for each element.

The proper iteration is then entered, with starting values p^n, ρ^n, $\tilde{v}$, $\tilde{i}$. H is calculated from (91) and (94) and the new estimated time-advanced pressure determined from

$$p^{\text{new}} = p^{\text{old}} - \omega H \Big/ \left(\frac{\mathrm{d}H}{\mathrm{d}p}\right)^e \tag{95}$$

where ω is a relaxation factor with value close to unity. If the change in p is below a pre-determined tolerance level for every element, then the iteration is complete and the calculation proceeds to phase III. Otherwise, the new pressure values are used to calculate in turn the velocities, densities and internal energies from (92), (90) and (93), respectively and the whole procedure is repeated.

Since the basic objective of the above implicit Lagrangian phase is to obtain new velocities that have been accelerated by pressure forces at the advanced time level, this phase eliminates the usual numerical stability condition that limits sound waves to travel no further than one element per time step.

Phase III: *Rezone or convective-flux calculation.* Phases I and II were assumed Lagrangian, that is, the contributions from the transport terms were not evaluated. In phase III we now add the contributions from these terms to the time-advanced level L values. We assume that the mesh velocities $\boldsymbol{w}$ have in some way been specified; how this may be automatically accomplished is described in Section 5.

The change in volume of the element from time level n to $n+1$ is

$$V^{n+1} - V^n = \Delta t \oint_{S^e} w_i n_i \,\mathrm{d}S. \tag{96}$$

From (64) the updated element mass is

$$M_e^{n+1} = M_e^n + \Delta t \oint_{S^e} \rho^L (w_i - v_i^L) n_i \,\mathrm{d}S \tag{97}$$

and the new density results

$$\rho^{n+1} = M_e^{n+1} / V^{n+1} \tag{98}$$

From (70), the contribution from the transport term to the specific internal energy yields

$$i^{n+1} = \frac{M_e^n}{M_e^{n+1}} i^L + \frac{\Delta t}{M_e^{n+1}} \oint_{S^e} \rho^L i^L (w_i - v_i^L) n_i \,\mathrm{d}S \tag{99}$$

and the transport of momentum contribution to the velocity components at node I is from (55)

$$v_{j,I}^{n+1} = v_{j,I}^L + \frac{\Delta t}{M^n} \int_V N_I \rho^L (w_i - v_i^L) \frac{\partial v_j^L}{\partial x_i} \,\mathrm{d}V. \tag{100}$$

Finally, the new pressure is calculated from the equation of state

$$p^{n+1} = f(\rho^{n+1}, i^{n+1}). \tag{101}$$

This completes the steps contained in a cycle of the time integration procedure.

5. Automatic rezoning

The freedom in moving the fluid mesh offered by the Arbitrary Lagrangian–Eulerian formulation is very attractive. This can however be overshadowed by the burden of specifying grid velocities well suited to the particular problem. As a consequence, the practical implementation of the ALE description, and in particular that of the transport phase of the calculational cycle, requires that an automatic mesh displacement prescription algorithm be supplied. The purpose of this algorithm is to provide the ALE computer code with a capability for automatic and continuous rezoning that conserves, as far as is possible, the regularity of the computational mesh and yet allows the fluid boundaries including deforming structures to be tracked with the accuracy characteristic of Lagrangian methods.

Rezoning techniques are almost entirely based on heuristic developments and their acceptance rests on their performances in practical applications. The particular technique that will be described here has been implemented into EURDYN-1M, an ALE finite-element program for transient dynamic fluid-structure interaction [21], and successful applications have been made to problems in reactor safety analysis.

To achieve the desired continuous and automatic rezone, the components of the grid velocity at a typical ALE node I are computed at each time step by the following relationship:

$$w_{I,i}^{t+\Delta t} = \frac{1}{N}\sum_J w_{J,i}^t + \frac{0.1}{\Delta t}\frac{1}{N}\sum_J L_{IJ}^t \frac{1}{N}\sum_J \frac{\delta_{J,i}^t - \delta_{I,i}^t}{L_{IJ}^t} \tag{102}$$

where N indicates the number of nodes connected to node I via element sides and diagonals, L_{IJ} is the current distance between node I and connected node J, and δ represents total nodal displacements.

The above formula was arrived at by trial and error on the following grounds. The basic idea was to give node I a current grid velocity equal to the mean of the grid velocities of the neighbouring nodes at the previous time step. However, this was occasionally found insufficient to prevent an excessive squeezing of some elements, especially in the vicinity of fluid–fluid interfaces, and led us to introduce a corrective term which enhances the grid velocity when the distance between adjacent nodes tends to become too short.

The components of the grid velocity computed by relation (102) are then limited by the following empirical inequalities which are dictated by numerical stability and accuracy conditions:

$$1 - \gamma^* \le \frac{w_{I,i}^{t+\Delta t}}{v_{I,i}^{t+\Delta t}} \le 1 + \gamma^* . \tag{103}$$

In this expression, $v_{I,i}$ are the components of the fluid velocity, and γ^* is a numerical coefficient satisfying

$$0 \le \gamma^* \le 0.5 . \tag{104}$$

In addition to ALE nodes, a typical finite element mesh generally contains purely Eulerian and purely Lagrangian nodes. By definition, the Eulerian nodes are fixed in space and have a zero grid velocity. In order to preserve one of the advantages of the Lagrangian approach, that of accurately tracking interfaces, Lagrangian nodes are placed along interface lines and they are constrained to move with the fluid by setting $w_{I,i} = v_{I,i}$.

At this point, the assignment of grid velocities is completed and the nodal displacements may be updated by explicit time integration. It is important to note that the presence in the ALE conservation equations of transport terms accounting for arbitrary displacements of the mesh nodes automatically guarantees conservation of mass, momentum and energy during the continuous rezoning procedure. This represents a significant advantage over the rezoning techniques which remap the mesh at fixed times in Lagrangian programs.

Table 1
Material data for overstrong vessel problem

Material	Initial density ρ_0 (g/cm³)	Constitutive law
Water	1.0	$p = p_1\left[1 - \dfrac{0.14(1-V)}{V}\right] + 0.28\, E/V$ $p_1 = A(0.1483 + 2.086\,A - 1.398\,A^2)$ $A = \begin{cases} \dfrac{2.086(1-V) - 1 + \sqrt{[2.086(1-V)-1]^2 + 0.8293(1-V)^2}}{2.796(1-V)}, & V \neq 1.0 \\ 0.0 & \text{when } V = 1.0 \end{cases}$ $V = \rho_0/\rho$; E = energy/unit reference volume (Mbar −cc/cc); $E_0 = 0.0$
Charge	1.67	$p = A\left(1 - \dfrac{W}{R_1 V}\right)e^{-R_1 V} + B\left(1 - \dfrac{W}{R_2 V}\right)e^{-R_2 V} + \rho i \dfrac{W}{V}$ (Mbars) $A = 6.70695$; $B = 0.092646$; $R_1 = 4.6606$; $R_2 = 0.9916$ $W = 0.25$; $V = \rho_0/\rho$; i = specific internal energy; ρ = density $\rho_0 i_0 = 7.796 \cdot 10^{-2}$ Mbar − cc/cc

To illustrate the effectiveness of the above continuous and automatic rezoning procedure, we present a coarse-mesh calculation of an experiment in which a high-explosive charge is detonated in an overstrong cylindrical vessel partially filled with water [22]. The material data for this problem are summarized in Table 1. Figure 3 provides a simplified schematic of the configuration and Figure 4 shows the finite element mesh used to model the experiment. Twenty constant pressure triangular elements are employed to represent the explosive charge and a total 232 quadrilateral elements is used to model the water. The air above the

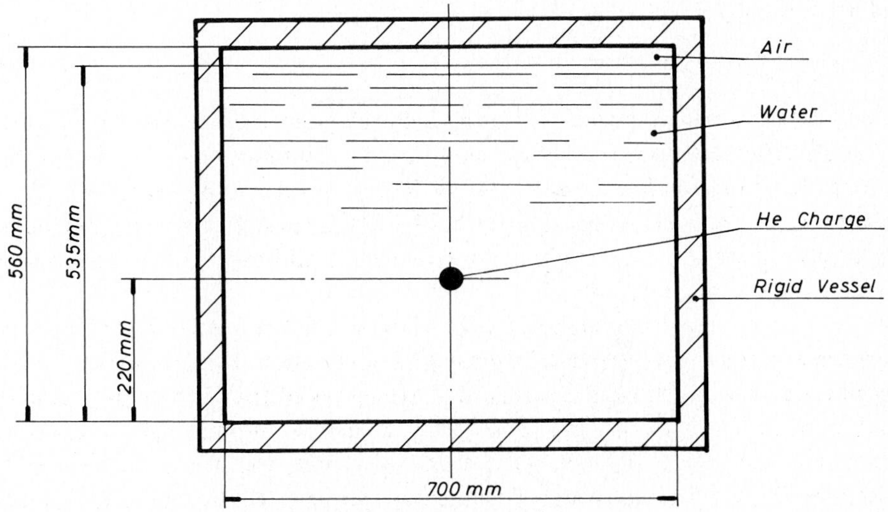

Fig. 3. Schematic of rigid vessel experiment.

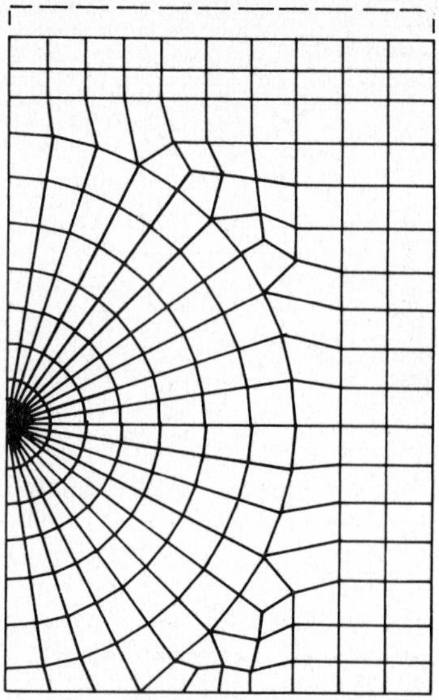

Fig. 4. Rigid vessel experiment: initial finite element mesh.

water surface is modelled as a void. The nodes on the interface between the charge and the water, as well as the nodes defining the free surface, are purely Lagrangian; the remaining nodes are treated in the ALE description and are automatically moved by the program according to formula (102).

The calculation was carried out to 2.5 milliseconds in order to check the ability of the automatic and continuous rezone to accomodate the large distorsions of the water during its interaction with the roof of the rigid vessel. Figures 5 and 6 show the mesh configuration at times $t = 1.0$ ms and $t = 2.5$ ms and it is noted that the

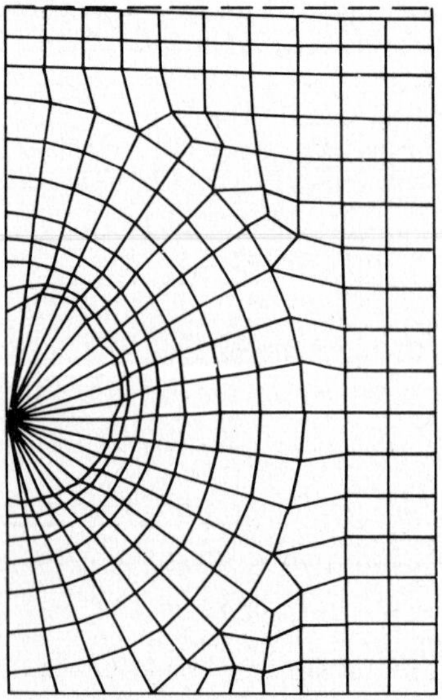

Fig. 5. Rigid vessel experiment: mesh at time $t = 1.0$ ms (ALE description).

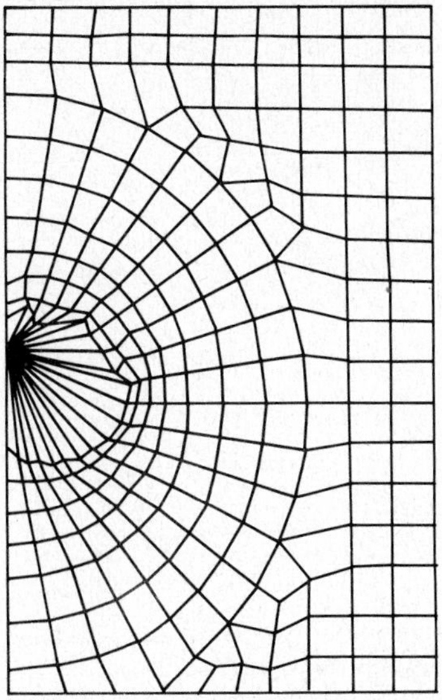

Fig. 6. Rigid vessel experiment: mesh at time $t = 2.5$ ms (ALE description).

automatic mesh displacement prescription algorithm has performed quite well. To further illustrate the ability of the ALE technique to maintain the regularity of the computational mesh while preserving the merits of the Lagrangian approach (clear delineation of interfaces and free surfaces), Figure 7 shows the mesh

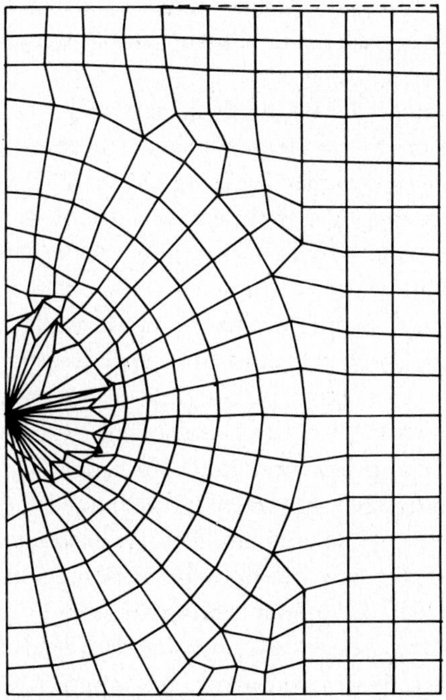

Fig. 7. Rigid vessel experiment: mesh at time $t = 1.0$ ms (purely Lagrangian description).

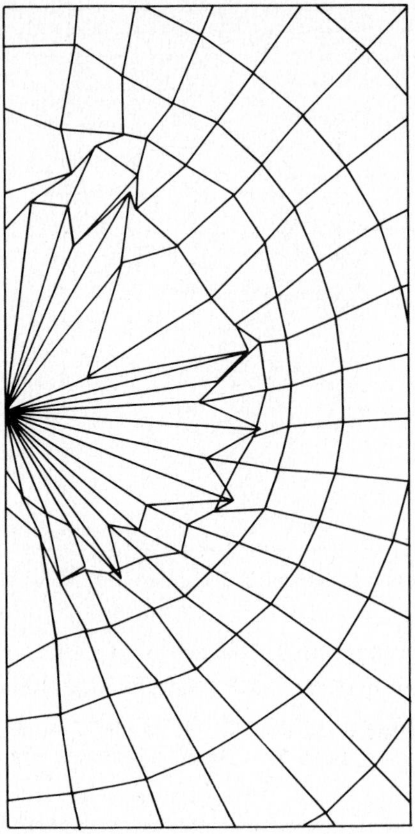

Fig. 7 Bis. Details of charge–water interface (purely Lagrangian description).

configuration obtained at $t = 1.0$ ms by using a purely Lagrangian mesh description. The severe distorsion of the elements close to the explosive charge gives an excellent idea of the benefit obtained by using the ALE description: computations may be performed for much longer times without troublesome degradations of the mesh that lead to inaccurate results and severe restrictions on the time step to assure stability of calculations.

Due to the coarseness of the finite element mesh, detailed comparisons between the ALE code response and the experimental data could not be performed. Nevertheless, we have placed numerical pressure gauges at some selected locations on the floor, wall and roof of the container and compared the predicted impulses to the experimental data. The results are displayed on Figures 8, 9 and 10 and a quite reasonable agreement between the experimental and predicted impulses may be noted. The predicted impulses are systematically below the mean experimental values and this may probably be attributed to the coarseness of the mesh which filters out high frequency signals. Figure 11 gives the wave arrival times on the roof of the vessel, the predicted arrival times are within the experimental records, except near the periphery where slightly late arrival times have been predicted.

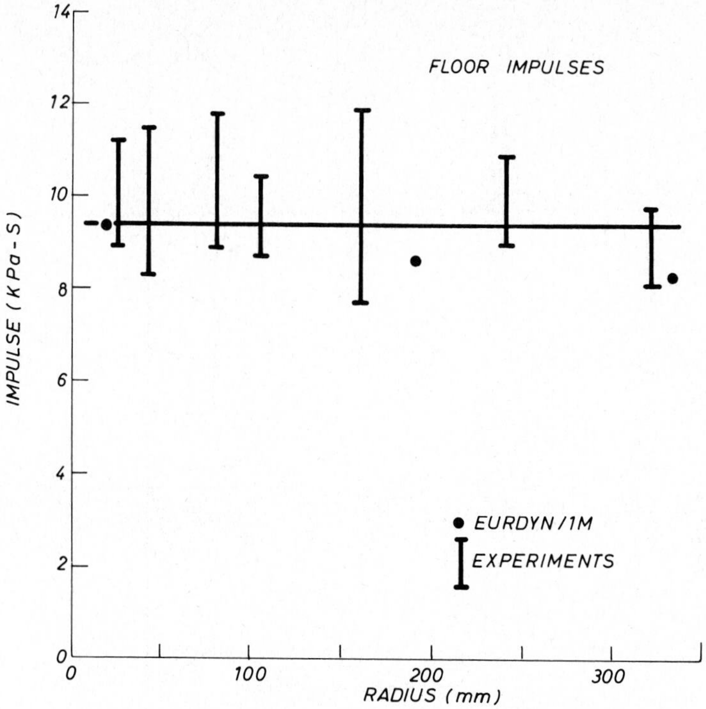

Fig. 8. Rigid vessel experiment: floor impulses at $t = 2.5$ ms.

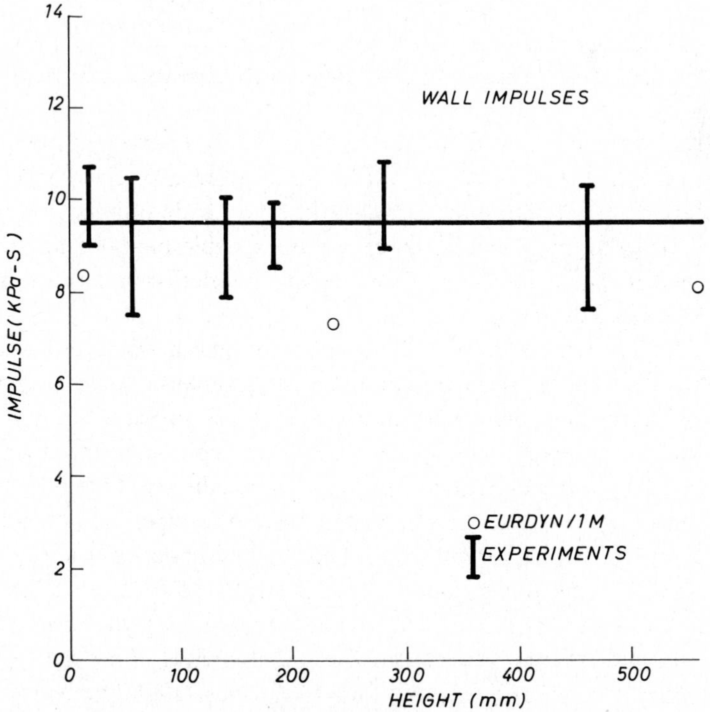

Fig. 9. Rigid vessel experiment: wall impulses at $t = 2.5$ ms.

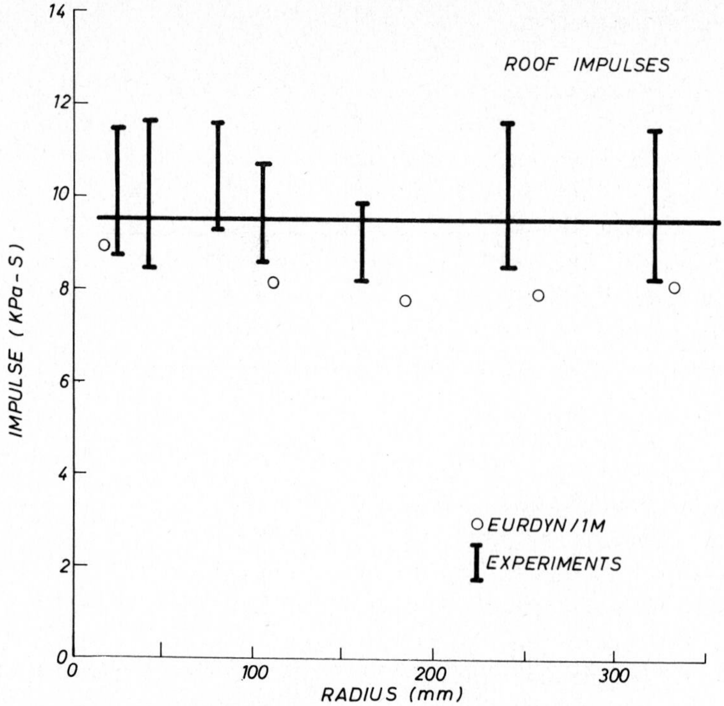

Fig. 10. Rigid vessel experiment: roof impulses at $t = 2.5$ ms.

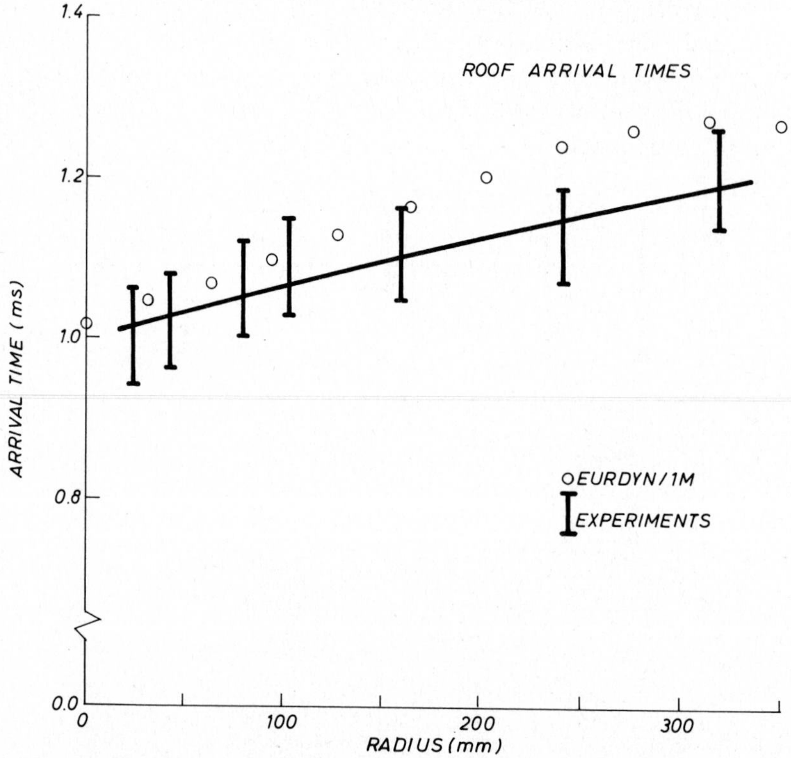

Fig. 11. Rigid vessel experiment: wave arrival times on roof of vessel.

6. Fluid-structure coupling

In recent years, there has been a growing interest in the analysis of the dynamic response of fluid-structure systems and the finite element method is emerging as a very versatile modelling technique for this class of hybrid problems which involve the calculation of both fluid transient and structure dynamics. In fact, if finite element algorithms are used to analyze both the hydrodynamic and the structural parts of a coupled fluid-structure problem, the equations of motion in fluid and solid regions are all expressed in terms of nodal loads and are exactly of the same form. The equations of motion governing the coupled system may thus be solved simultaneously and a strong coupling between fluid and solid responses is achieved.

However, suitable conditions must be prescribed along fluid-solid interfaces to allow relative sliding of the fluid and solid and it is the purpose of this last paragraph to show that fluid-structure coupling may be achieved in a very simple and elegant manner if the fluid is treated in the ALE formulation.

In order to illustrate the coupling procedure along fluid-solid interfaces, let us consider a structural member embedded in a fluid which is allowed to slide along both faces of the structure. As shown in Figure 12, we place two nodes at each point of the interface: one fluid node and one structural node. Since the fluid is treated in the ALE formulation, the movement of the fluid mesh may be chosen completely independent from the movement of the fluid itself. In particular, we may constrain the fluid nodes to remain contiguous to the structural nodes, so that all nodes on the sliding interface remain permanently aligned. It is clear that such a permanent alignment of nodes along the interface greatly facilitates the flow of information between fluid and structural domains and permits fluid-structure

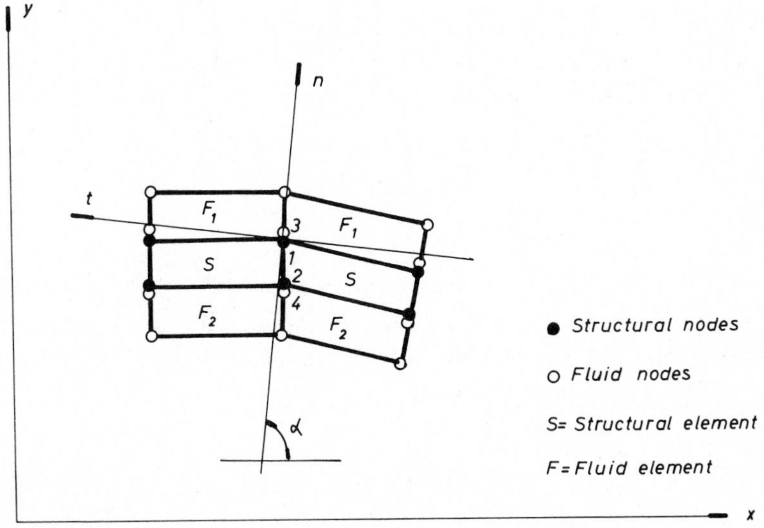

Fig. 12. Sliding interfaces.

coupling to be effected in the simplest and most elegant manner. [Notice, however, that this treatment of the coupling problem only applies to those portions of the structure which are always submerged. Unfortunately, there may exist portions of the structure which only come into contact with the fluid some time after the calculation begins (e.g. structural parts above a fluid free surface) and for such portions some sort of sliding treatment is necessary, as for Lagrangian methods [23].]

The following conditions are prescribed at each point of the interface between an inviscid fluid and a deforming structure:

(a) the grid velocity $\boldsymbol{w}$ of the fluid coincides with the velocity $\boldsymbol{v}_S$ of the solid;

(b) the normal velocity v_F^n of the fluid coincides with the normal velocity v_S^n of the solid;

(c) the tangential velocity v_F^t of the fluid is unconstrained.

In order to specify the above conditions, a local coordinate system (t, n) is set up for each point of the interface (Figure 12), so that t is the tangent to the sliding interface and n is 90° clockwise from t. Whenever a corner occurs in the interface, t is the average of the two tangent directions.

Referring to the node numbering in Figure 12, the condition under point (a) implies that

$$\boldsymbol{w}_3 = \boldsymbol{v}_1 \quad \text{and} \quad \boldsymbol{w}_4 = \boldsymbol{v}_2 \tag{105}$$

where $\boldsymbol{w}$ indicates the grid velocity of the fluid and $\boldsymbol{v}$ the velocity of the solid. These conditions are imposed directly at each time step and serve, together with similar conditions at Lagrangian fluid nodes, as 'boundary conditions' for the automatic fluid mesh displacement algorithm described in Section 5.

The conditions under points (b) and (c) are enforced through the nodal loads. The condition of common normal velocity implies that

$$v_3^n = v_1^n \quad \text{and} \quad v_4^n = v_2^n \tag{106.a}$$

and if the solid element is based on the Kirchoff hypothesis (e.g. a thin shell or a beam element), we have the further equality

$$v_1^n = v_2^n \tag{106.b}$$

so that

$$v_3^n = v_1^n = v_2^n = v_4^n\,. \tag{106.c}$$

Hence the four nodes must have a common normal velocity which is obtained as follows. Be $\boldsymbol{v}_F$ the velocity of a fluid node at the interface and $\boldsymbol{v}_S$ the velocity of the adjacent solid node. Since the two nodes must have a common normal velocity at all times, we prescribe that

$$\frac{\mathrm{d}}{\mathrm{d}t}[(\boldsymbol{v}_F - \boldsymbol{v}_S)\cdot \boldsymbol{n}] = 0 \tag{107.a}$$

or

$$(\dot{\boldsymbol{v}}_F - \dot{\boldsymbol{v}}_S) \cdot \boldsymbol{n} + (\boldsymbol{v}_F - \boldsymbol{v}_S) \cdot \frac{d\boldsymbol{n}}{dt} = 0 . \tag{107.b}$$

Now

$$\frac{d\boldsymbol{n}}{dt} = \frac{\boldsymbol{n}^{t+\Delta t} - \boldsymbol{n}^t}{\Delta t} \tag{107.c}$$

and, since

$$(\boldsymbol{v}_F - \boldsymbol{v}_S) \cdot \boldsymbol{n}^t = 0 , \tag{107.d}$$

equation (107.b) reduces to

$$\dot{v}_F^n - \dot{v}_S^n = \frac{v_S^n - v_F^n}{\Delta t} . \tag{108}$$

Thus, the condition of common normal velocity implies different normal accelerations at the fluid and solid nodes if the normal to the interface changes direction with time. To evaluate these normal accelerations, we pose

$$\dot{v}_F^n = a^n + a_F^n , \qquad \dot{v}_S^n = a^n + a_S^n , \tag{109.a}$$

where a^n is a common normal acceleration and a_F^n, a_S^n are individual complementary accelerations.

The common normal acceleration a^n is obtained from the equation of motion

$$(M_S + M_F)a^n = f^n \tag{109.b}$$

where, referring to Figure 12, M_S is the mass contributed by the structural elements adjacent to nodes 1 and 2, while $M_F = M_{F1} + M_{F2}$ represents the mass contributed by the fluid elements on either side of the interface. The normal load f^n is obtained by transformation into the (t, n) system of the assembled internal nodal loads (f^x, f^y) contributed by the relevant fluid and solid elements:

$$f^n = (f_S^x + f_{F1}^x + f_{F2}^x) \cos \alpha + (f_S^y + f_{F1}^y + f_{F2}^y) \sin \alpha \tag{109.c}$$

where α is the angle between x and n directions, as shown in Figure 12.

In view of relations (109.a) and (108), the complementary accelerations a_F^n and a_S^n are linked by

$$a_F^n - a_S^n = \frac{v_S^n - v_F^n}{\Delta t} = A^n \tag{110.a}$$

and a second relationship between these accelerations may be obtained from the requirement of equilibrium which reads

$$M_F a_F^n + M_S a_S^n = 0 . \tag{110.b}$$

It follows that

$$a_F^n = \frac{M_S}{M_F + M_S} A^n ; \qquad a_S^n = \frac{-M_F}{M_F + M_S} A^n . \tag{110.c}$$

This completes the computation of normal accelerations which ensure a common normal velocity at the fluid and solid nodes on the interface.

The tangential velocities at the fluid and solid nodes are unconstrained and result from the equations of motion

$$\begin{aligned} &M_S \dot{v}_1^{t} = M_S \dot{v}_2^{t} = f_S^{t} , \\ &M_{F1} \dot{v}_3^{t} = f_{F1}^{t} , \qquad M_{F2} \dot{v}_4^{t} = f_{F2}^{t} , \end{aligned} \tag{111}$$

where the tangential loads are again obtained by transformation into the (t, n) system of the assembled internal nodal forces (f^x, f^y).

It may thus be concluded that the problem of fluid-structure coupling becomes very simple, since fluid nodes at the structure can remain attached to it throughout the calculation without imposing any restriction on the fluid tangential velocity. Practical experience with the ALE treatment of sliding at the fluid–solid interface is described by Belytschko, Kennedy and Schoeberle [5, 17] and by Donea et al. [21, 24] for two-dimensional situations. Kulak [25] has reported an ALE finite element formulation for three-dimensional fluid-structure interactions.

Before concluding this section on fluid-structure coupling in the ALE description, we want to draw the attention of the reader to particular problems that arise when the interface between an inviscid fluid and a structure possesses sharp corners.

In ideal fluid flows, the velocity vector can have only one value and it is therefore impossible for a streamline to change direction abruptly. However, sharp corners in inviscid fluid flows are points where these rules are broken. The situation is sketched in Figure 13 which shows a corner in a wall, the wall turning

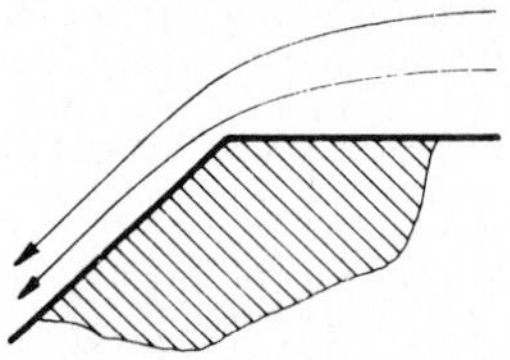

Fig. 13. Illustration of singularity at a corner.

away from the flow. At the corner, the streamline which coincides with the wall changes direction abruptly and the velocity vector has two directions. At this singularity, the velocity of the ideal fluid is infinite. In these conditions, it is easily understood that special precautions must be taken to achieve accuracy in the numerical prediction of the flow near a sharp corner.

As mentioned before, whenever a corner occurs in an interface, the velocity vector at the corner is assumed to be parallel to an average of the two tangent directions. As can be seen from Figure 14, the consequence of this choice is that a spurious transport of fluid takes place across the intersecting parts of the wall: the

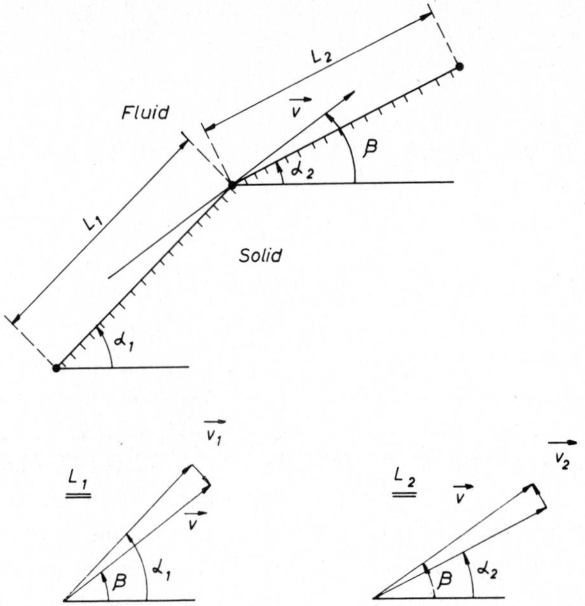

Fig. 14. Treatment of flow around sharp corners.

fluid flows out on portion L_1 of the interface and flows in on portion L_2. These spurious fluxes must indeed be accounted for when expressing the conservation statements of mass, momentum and energy over the fluid finite elements meeting at the corner and this may prevent from achieving conservation in the fluid mesh if the amount of fluid lost on portion L_1 of the interface is not exactly compensated by the amount gained on portion L_2. Fortunately, exact compensation may always be achieved by a suitable choice of the angle β defining the flow direction at the corner. Referring to Figure 14 and recalling that we are using linear fluid elements, the following equality must hold for exact compensation of the spurious fluxes:

$$L_1 v_1 = L_2 v_2 \tag{112}$$

and since

$$v_1 = v \sin(\alpha_1 - \beta), \qquad v_2 = v \sin(\beta - \alpha_2), \tag{113}$$

the optimal value of the angle β is found to be given by

$$\tan \beta = \frac{L_1 \sin \alpha_1 + L_2 \sin \alpha_2}{L_1 \cos \alpha_1 + L_2 \cos \alpha_2}, \tag{114}$$

which reduces to

$$\beta = \tfrac{1}{2}(\alpha_1 + \alpha_2) \tag{115}$$

in the particular case where $L_1 = L_2$. Notice also that angle β in (114) is given by the slope of the line joining the end points of segments L_1 and L_2 of the interface.

As an illustration of fluid-structure coupling in the ALE formulation, we consider the response of a thin cylindrical vessel with hemispherical bottom nearly completely filled with water and impulsively loaded by the detonation of an explosive charge located on the vessel axis. The top of the vessel is clamped on a rigid cover. A schematic of the configuration is depicted in Figure 15.

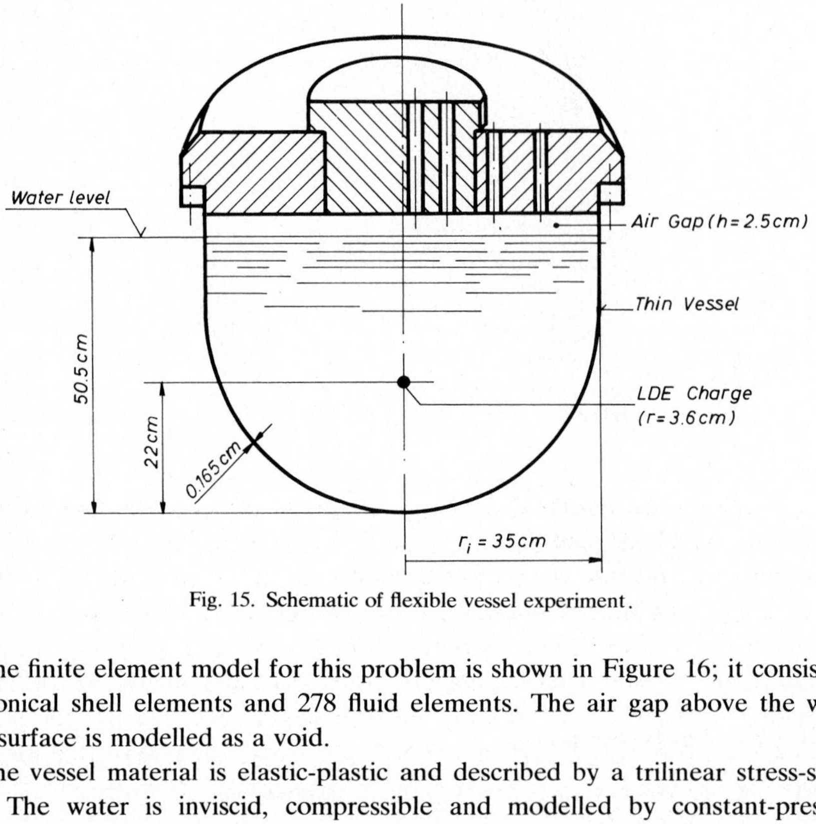

Fig. 15. Schematic of flexible vessel experiment.

The finite element model for this problem is shown in Figure 16; it consists of 23 conical shell elements and 278 fluid elements. The air gap above the water free-surface is modelled as a void.

The vessel material is elastic-plastic and described by a trilinear stress-strain law. The water is inviscid, compressible and modelled by constant-pressure quadrilateral elements. The charge is a low-density explosive of the type described by Cameron, Hoskin and Lancefield [30], and it is modelled by both triangular and quadrilateral elements. The material data for the vessel, the water and the explosive charge are summarized in Table 2.

The interface between the explosive charge and the water as well as the water free surface are taken as purely Lagrangian. The rest of the hydrodynamic domain is treated in the ALE formulation, and the corresponding nodes are automatically moved by the computer program as explained in Section 5. Sliding of water along the vessel is treated as indicated in this section, except near the free surface, where a Lagrangian slide line is introduced to deal with interface nodes which are not aligned.

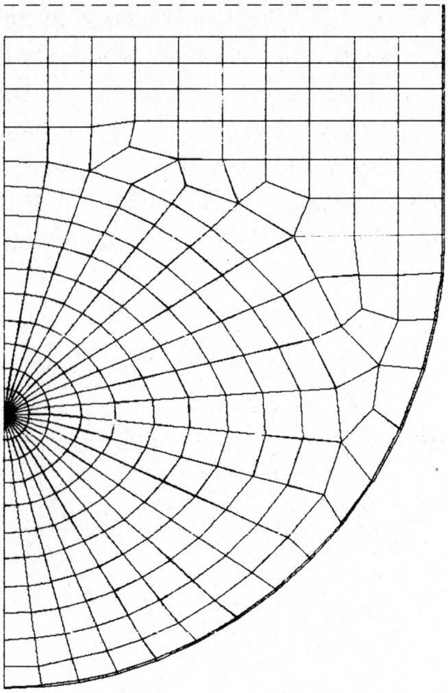

Fig. 16. Flexible vessel, initial finite element mesh.

Figures 17(a)–(d) show the deformed configuration of the computational mesh at times $t = 1.0$, 2.0, 3.0 and 4.0 ms.

An important increase of the charge volume, the progressive impact of water on the vessel cover and the successive deformation of the vessel itself are clearly

Table 2
Material data for flexible vessel problem

Material	Initial density ρ_0 (g/cm^3)	Constitutive law
Vessel	7.9	Trilinear elastic-plastic $\nu = 1/3$; $E^e = 1.93\ 10^5$ MPa $\sigma^1_{\text{yield}} = 2.75\ 10^2$ MPa; $\sigma^2_{\text{yield}} = 3.544\ 10^2$ MPa $E^p_1 = 5.895\ 10^3$ MPa; $E^p_2 = 1.867\ 10^3$ MPa
Water	1.0	See Table 1
Charge	0.27	$p = A\left(1-\dfrac{W}{R_1V}\right)e^{-R_1V} + B\left(1-\dfrac{W}{R_2V}\right)e^{-R_2V} + (\rho_0 i - E_1)\dfrac{W}{V}$ (Mbars) $A = 0.17039$; $B = 0.011595$; $R_1 = 9.0$; $R_2 = 2.4$ $V = \rho_0/\rho$; $W = 0.1$; $E_1 = 0.0102884$; i = specific internal energy $\rho_0 i_0 = 0.01629$ Mbar − cc/cc

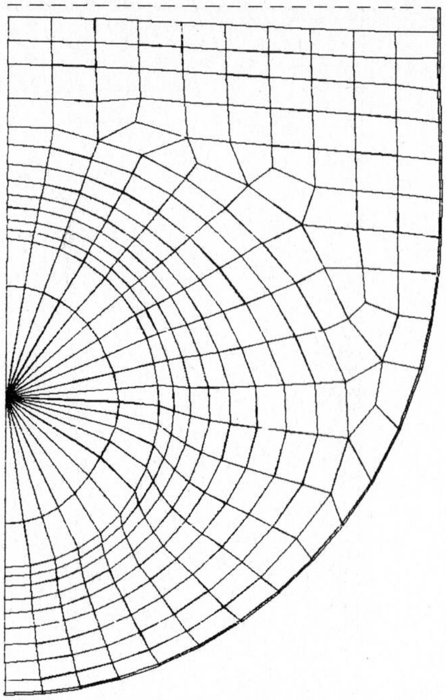

Fig. 17a. Deformed ALE mesh at time $t = 1.0$ ms.

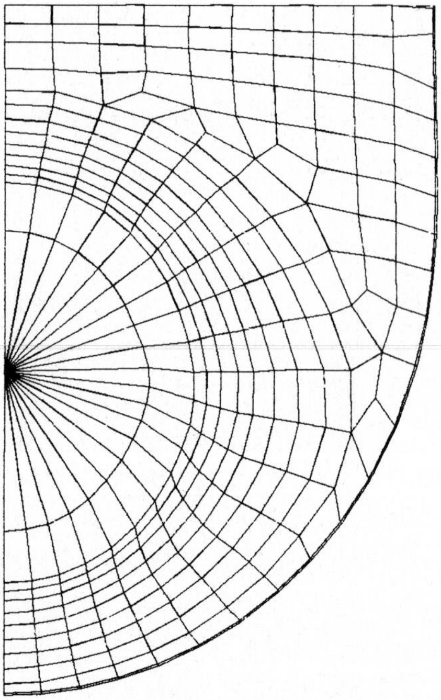

Fig. 17b. Deformed ALE mesh at time $t = 2.0$ ms.

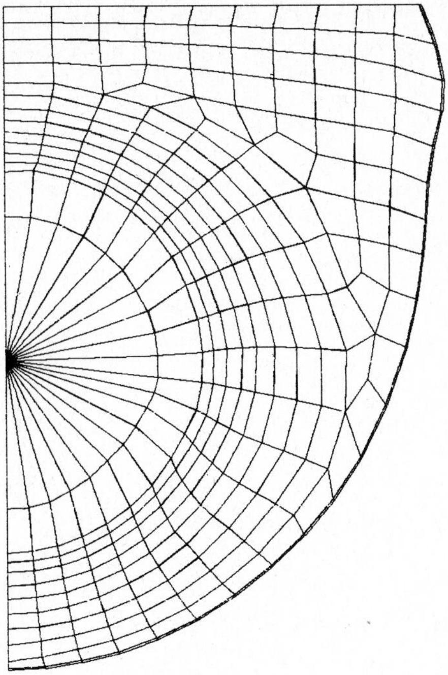

Fig. 17c. Deformed ALE mesh at time $t = 3.0$ ms.

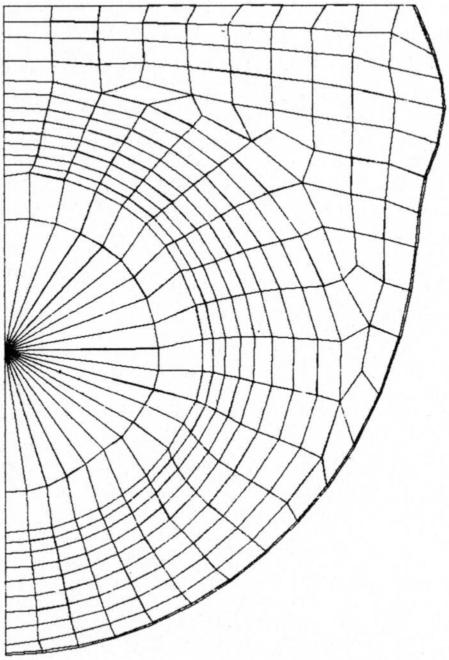

Fig. 17d. Deformed ALE mesh at time $t = 4.0$ ms.

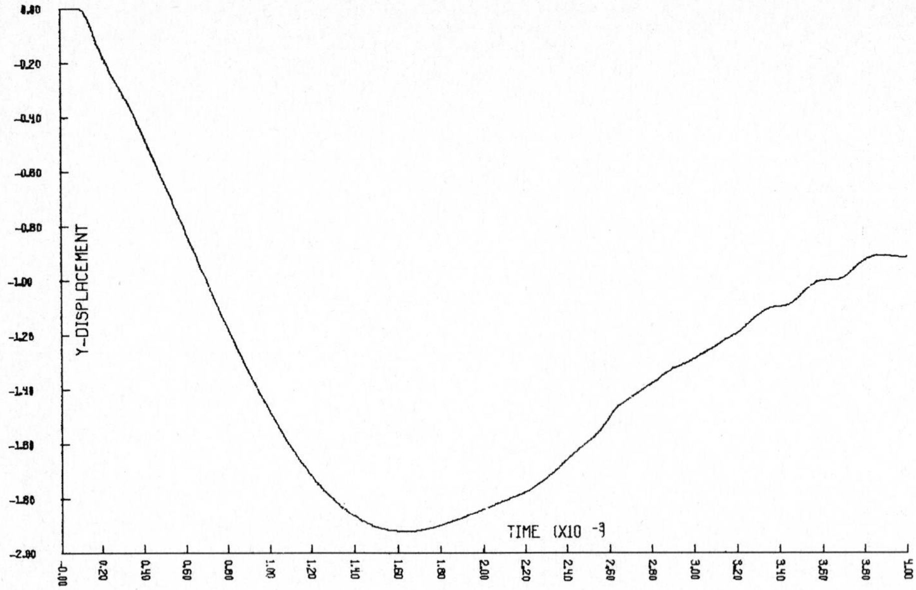

Fig. 18a. Time history of axial displacement of the vessel on the axis of symmetry.

seen in the figures. One also notes that the automatic mesh displacement prescription algorithm performs quite well.

Figures 18(a)–(c) enable us to follow the time history of

(a) the axial displacement of the vessel on the axis of symmetry;

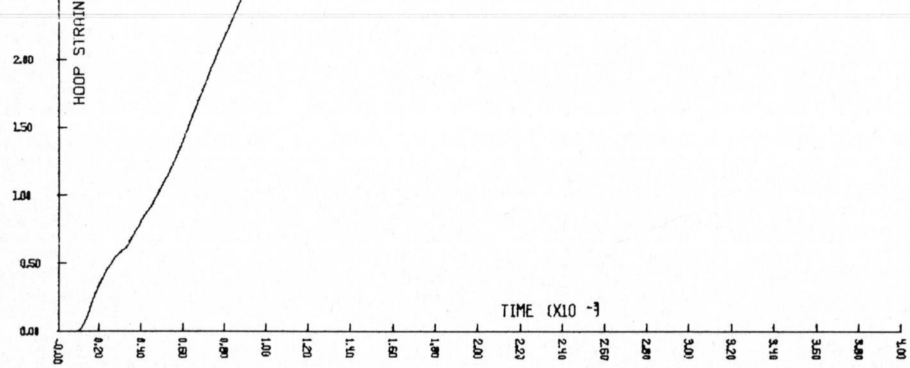

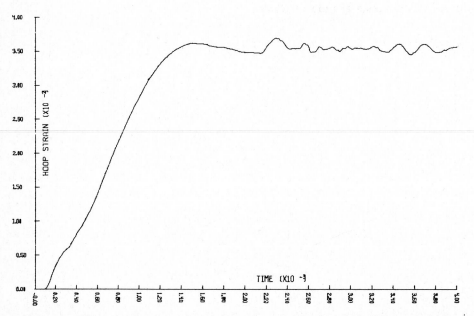

Fig. 18b. Time history of hoop strain at external surface of the vessel near the axis of symmetry.

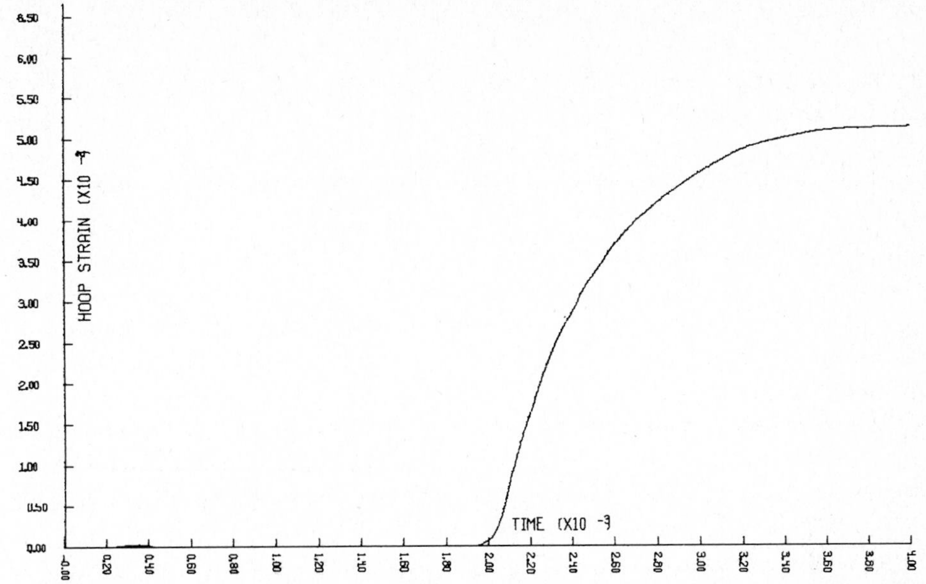

Fig. 18c. Time history of hoop strain at external surface of the vessel in the region of maximum radial displacement.

(b) the hoop strain at the external surface of the vessel near the axis of symmetry;

(c) the hoop strain at the external surface of the vessel in the region of maximum radial deformation ($R_0 = 35.165$ cm; $Z_0 = 49.08$ cm).

Figure 19 shows the meridional distribution of the hoop strain at the external surface of the vessel at time $t = 4$ ms.

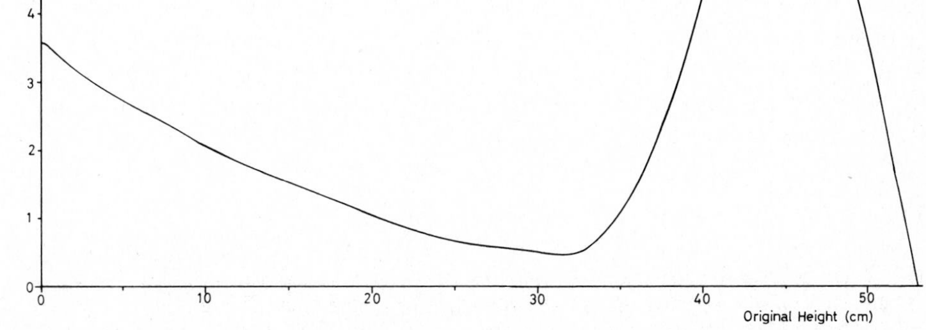

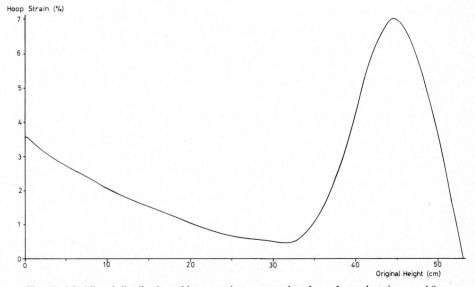

Fig. 19. Meridional distribution of hoop strain at external surface of vessel at time $t = 4.0$ ms.

Conclusion

The discussion presented here has summarized the ALE finite-element method and indicated its considerable potential for application to interacting media problems, such as fluid-structure coupling.

The method combines the basic attributes of the finite element technique – namely, ease in modelling complex geometries and multimaterial problems – and the flexibility in moving the computational mesh offered by the ALE description. The result of this combination is a very versatile modelling technique, with both Lagrangian and Eulerian features, which permits cost-effective computing for long real times, accomodation of large fluid displacements and logically simple, but accurate fluid-structure coupling.

Experience with the use of finite element techniques in fluid dynamics problems is however quite limited and further developmental efforts are required to bring the method to the level of success reached in other disciplines, such as solid and structural mechanics. The development of improved methods for the numerical computation of convection dominated flows is an area which is attracting considerable interest at present and new finite element procedures of upwind-type are being produced [13] to handle problems of this type. These procedures succeed in mitigating the oscillations which occasionally appear in the results produced by the Galerkin method without introducing excessive numerical dissipation.

An area which should receive increased attention is the problem of the development and analysis of methods for automatic and continuous rezoning. It must, in fact, be said that the ALE technique will only be regarded as a viable solution method when rezoning algorithms are developed which are capable of dealing with arbitrary problems and when detailed accuracy and convergence studies of such algorithms are available.

As far as fluid-structure interaction is concerned, the main problem appears to be the enhancement of computational efficiency of time integration procedures. Computations based on a single time integration method and a single time step are in fact often quite uneconomical and the ideal situation would be to have an operator which had the characteristics of an unconditionally stable implicit scheme for the solid elements, where only the low frequency modes are of importance, and which had the characteristics of an explicit scheme for the fluid elements where the high frequencies are important. Several investigators have addressed this problem and mixed time integration methods are being developed with features just enumerated [26]–[29].

References

[1] F.H. Harlow, Numerical Methods for Fluid Dynamics, an Annotated Bibliography, Report No. LA-4281, Los Alamos Scientific Laboratory, Los Alamos, NM, (1969).

[2] W.F. Noh, in: *Methods in Computational Physics*, Vol. 3, B. Alder et al. eds. (Academic Press, New York, 1964) p. 117.

[3] J.G. Trulio, Air Force Weapons Laboratory, AFWL-Tr-66-19 (June 1966).
[4] C.W. Hirt, A.A. Amsden and J.L. Cook, An Arbitrary Lagrangian–Eulerian Computing Method for All Flow Speeds, *J. Comput. Physics*, **14**, 227 (1974).
[5] T. Belytschko and J.M. Kennedy, Computer Models for Subassembly Simulation, *Nucl. Eng. Design*, **49**, 17–38 (1978).
[6] J. Donea, P. Fasoli-Stella and S. Giuliani, Lagrangian and Eulerian Finite Element Techniques for Transient Fluid-Structure Interaction Problems, Paper B1/2, Trans. 4th SMIRT Conf., San Francisco, 15–19 August (1977).
[7] T.J.R. Hughes, W.K. Liu and T.K. Zimmermann, Lagrangian–Eulerian Finite Element Formulation for Incompressible Viscous Flows, *U.S.-Japan Seminar on Interdisciplinary Finite Element Analysis*, Cornell Univ., Ithaca, NY, Aug. 7–11 (1978).
[8] C. Truesdell and R. Toupin, The Classical Field Theories, in: *Encyclopedia of Physics*, Vol III/1 (Springer, Berlin–New York, 1960).
[9] G. Van Goethem, Description Mixte d'Euler–Lagrange et Modèle d'Eléments Finis à Domaine Variable, Doctoral Thesis, University of Louvain, Belgium (1979).
[10] J.J. Connor and C.A. Brebbia, *Finite Element Techniques for Fluid Flow* (Newnes-Butterworths, London, 1977).
[11] O.C. Zienkiewicz, *The Finite Element Method in Engineering Science* (McGraw-Hill, London, 1971).
[12] P.J. Roache, *Computational Fluid Dynamics* (Hermosa Publishers, Albuquerque, NM, 1976).
[13] T.J.R. Hughes (editor), Finite Element Methods for Convection Dominated Flows, *ASME Winter Annual Meeting*, New York, December 2–7, AMD - Vol. 34, (1979).
[14] I. Christie, D.F. Griffiths, A.R. Mitchell and O.C. Zienkiewicz, Finite Element Methods for Second Order Differential Equations with Significant First Derivatives, *Int. J. Num. Meth. Engng.* **10**, 1389–1396 (1976).
[15] J.C. Heinrich, P.S. Huyakorn, O.C. Zienkiewicz and A.R. Mitchell, An Upwind Finite Element Scheme for Two-Dimensional Convective Transport, *Int. J. Num. Meth. Engng.* **11**, 131–145 (1977).
[16] T.J.R. Hughes, A Simple Scheme for Developing Upwind Finite Elements, *Int. J. Num. Meth. Engng.* **12**, 1359–1365 (1978).
[17] T. Belytschko, J.M. Kennedy and D.F. Schoeberle, Quasi-Eulerian Finite Element Formulation for Fluid-Structure Interaction, *Proc. Joint ASME/CSME Pressure Vessels and Piping Conf.*, Montreal, Canada, June 25–30 (1978).
[18] L.R. Stein, R.A. Gentry and C.W. Hirt, Computational Simulation of Transient Blast Loading on Three-dimensional Structures, *Computer Methods in Applied Mechanics and Engineering* **11**, 57–74 (1977).
[19] W.E. Pracht, Calculating Three-Dimensional Fluid Flows at All Speeds with an Eulerian–Lagrangian Computing Mesh, *J. Comput. Physics* **17**, 132–159 (1975).
[20] W.E. Pracht and J.U. Brackbill, BAAL: A Code for Calculating Three-Dimensional Fluid Flows at All Speeds with Eulerian–Lagrangian Computing Mesh, Report LA-6342, Los Alamos Scientific Laboratory (August 1976).
[21] J. Donea, P. Fasoli-Stella, S. Giuliani, J.P. Halleux and A.V. Jones, The Computer Code EURDYN-1M for Transient Dynamic Fluid-Structure Interaction, Part I: Finite Element Modelling, Report EUR 6751, Commission of the European Communities (1980).
[22] N.J.M. Rees, R.B. Tattersall and G. Verzeletti, Results of Repeat Firings of High Explosive Charges in Water-Filled Vessels, Atomic Weapons Research Establishment, AWRE/44/97/1, TRG Report 2909 (R/X), JRC Ispra EE/01/76, AEEW M-1417 (August 1976).
[23] M. Wilkins, Calculation of Elastic-Plastic Flow, in: *Methods in Computational Physics*, B. Alder et al., eds., Vol. 3 (Academic Press, New York, 1964) pp. 211–263.
[24] J. Donea, Finite Element Analysis of Transient Dynamic Fluid-Structure Interaction, in: *Advanced Structural Dynamics*, J. Donea, ed. (Applied Science Publishers, Barking, Essex, England, 1980) pp. 255–290.
[25] R.F. Kulak, A Three-Dimensional Finite-Element Formulation for Fluid-Structure Interaction, Paper B 2/1, *Proc. 5th Int. Conf. Struct. Mech. in Reactor Techn.*, Berlin, August 13–17 (1979).
[26] T. Belytschko, H.J. Yen and R. Mullen, Mixed Methods for Time Integration, *Proc. Int. Conf. on Finite Elements in Non-linear Mechanics*, August 30–September 1, 1978, University of Stuttgart, published in *Comp. Meths. in Appl. Mech. Engng.* **17/18**, 259–275 (1979).

[27] T.J.R. Hughes and W.K. Liu, Implicit–Explicit Finite Elements in Transient Analysis: Stability Theory, *J. Appl. Mechs.* **45**, 371–374 (1978).

[28] T.J.R. Hughes and W.K. Liu, Implicit–Explicit Finite Elements in Transient Analysis. Implementation and Numerical Examples, *J. Appl. Mechs.* **45**, 375–378 (1978).

[29] T.J.R. Hughes, K.S. Pister and R.L. Taylor, Implicit–Explicit Finite Elements in Non-linear Transient Analysis, *Proc. Int. Conf. on Finite Elements in Non-linear Mechanics*, August 30–September 1, 1978, University of Stuttgart, published in *Comp. Meths. in Appl. Mech. Engng.* **17/18**, 159–182 (1979).

[30] I.G. Cameron, N.E. Hoskin, M.J. Lancefield, Paper E2/1, 4th SMIRT Conference, San Francisco, CA (1977).

Author Index

Computational Methods for Transient Analysis
Edited by T. Belytschko and T.J.R. Hughes

Subject Index

Computational Methods for Transient Analysis
Edited by T. Belytschko and T.J.R. Hughes